CHAPTER 10

Degrees of Freedom for an Independent-Samples t Test

$$df_{total} = df_X + df_Y$$

Pooled Variance

$$s^2_{pooled} = \left(\frac{df_X}{df_{total}}\right) s^2_X + \left(\frac{df_Y}{df_{total}}\right) s^2_Y$$

Variance for a Distribution of Means for an Independent-Samples t Test

$$s^2_{M_X} = \frac{s^2_{pooled}}{N_X} \qquad s^2_{M_Y} = \frac{s^2_{pooled}}{N_Y}$$

Variance for a Distribution of Differences Between Means

$$s^2_{difference} = s^2_{M_X} + s^2_{M_Y}$$

Standard Deviation of the Distribution of Differences Between Means

$$s_{difference} = \sqrt{s^2_{difference}}$$

t Statistic for an Independent-Samples t Test

$$t = \frac{(M_X - M_Y) - (\mu_X - \mu_Y)}{s_{difference}}$$

often abbreviated as:

$$t = \frac{(M_X - M_Y)}{s_{difference}}$$

Confidence Interval for an Independent-Samples t Test

$$(M_X - M_Y)_{lower} = -t(s_{difference}) + (M_X - M_Y)_{sample}$$

$$(M_X - M_Y)_{upper} = t(s_{difference}) + (M_X - M_Y)_{sample}$$

Effect Size for an Independent-Samples t Test

$$\text{Cohen's } d = \frac{(M_X - M_Y) - (\mu_X - \mu_Y)}{s_{pooled}},$$

for a t distribution for a difference between means

CHAPTER 11

One-Way Between-Groups ANOVA

$$df_{between} = N_{groups} - 1$$

$$df_{within} = df_1 + df_2 + \ldots + df_{last}$$

(in which df_1 etc. are the degrees of freedom, $N - 1$, for each sample)

$$df_{total} = df_{between} + df_{within}$$

or $df_{total} = N_{total} - 1$

$$GM = \frac{\Sigma(X)}{N_{total}}$$

$SS_{total} = \Sigma(X - GM)^2$ for each score

$SS_{within} = \Sigma(X - M)^2$ for each score

$SS_{between} = \Sigma(M - GM)^2$ for each score

$$SS_{total} = SS_{within} + SS_{between}$$

$$MS_{between} = \frac{SS_{between}}{df_{between}}$$

$$MS_{within} = \frac{SS_{within}}{df_{within}}$$

$$F = \frac{MS_{between}}{MS_{within}}$$

$$R^2 = \frac{SS_{between}}{SS_{total}}$$

Chapter 11 formulas continued on inside back cover

Essentials of Statistics for the Behavioral Sciences

Essentials of Statistics for the Behavioral Sciences

Susan A. Nolan | **Thomas E. Heinzen**

Seton Hall University | William Paterson University

Worth Publishers

A Macmillan Higher Education Company

Senior Vice President, Editorial and Production: Catherine Woods
Publisher: Kevin Feyen
Acquisitions Editor: Daniel DeBonis
Development Editor: Elaine Epstein
Marketing Manager: Lindsay Johnson
Marketing Coordinator: Julie Tompkins
Associate Director of Market Development: Carlise Stembridge
Director of Development for Print and Digital: Tracey Kuehn
Senior Media Editor: Christine Burak
Assistant Media Editor: Anthony Casciano
Associate Managing Editor: Lisa Kinne
Senior Project Editor: Jane O'Neill
Assistant Editor: Nadina Persaud
Photo Editor: Bianca Moscatelli
Interior & Cover Designer: Kevin Kall
Production Manager: Barbara Seixas
Illustrations: Northeastern Graphic, Inc.
Composition: Northeastern Graphic, Inc.
Printing & Binding: RR Donnelley
Cover Painting: Margaret Glew

Library of Congress Control Number: 2013935652

ISBN-13: 978-1-4292-4227-1
ISBN-10: 1-4292-4227-2

Printed in the United States of America

Second Printing

Worth Publishers
41 Madison Avenue
New York, NY 10010
www.worthpublishers.com

To Mrs. Peecha of Greenlodge Elementary School, a creative and inspiring math teacher who taught me that I should "never, never, never, never [literally 10 minutes of 'nevers'] divide by zero." Mrs. Peecha, I remain inspired. Thank you.

—Susan Nolan

To Olga Jenkins, my first algebra teacher, who never imagined that I could or would write a stats book.

—Tom Heinzen

ABOUT THE AUTHORS

Susan Nolan turned to psychology after suffering a career-ending accident on her second workday as a bicycle messenger. A native of Boston, she graduated from the College of Holy Cross and earned her PhD in clinical psychology from Northwestern University. Her research involves experimental investigations of the role of gender in the interpersonal consequences of depression, and studies on gender and mentoring in the fields of science, technology, engineering, and mathematics; her research has been funded by the National Science Foundation. Susan is the Chair of the Department of Psychology as well as Professor of Psychology at Seton Hall University in New Jersey. She has served as a statistical consultant to researchers at universities, medical schools, corporations, and nongovernmental organizations. Susan is a representative from the American Psychological Association to the United Nations in New York City, was the Chair of the 2012 Society for the Teaching of Psychology (STP) Presidential Task force on Statistical Literacy, and was recently named a Master Teacher through the STP Master Teacher Speakers Series. She was elected the 2014–2015 President of the Eastern Psychological Association.

Susan's academic schedule allows her to pursue one travel adventure per year, a tradition that she relishes. Over the years she has ridden her bicycle across the United States (despite her earlier crash), swapped apartments to live in Montréal (her favorite North American city), and explored the Adriatic coast in an intermittently roadworthy 1985 Volkswagen Scirocco. She writes much of the book on her annual trip to Bosnia and Herzegovina, where she and her husband, Ivan Bojanic, own a small house on the Vrbas River in the city of Banja Luka. They currently reside in Jersey City, New Jersey, where Susan roots feverishly, if quietly, for the Boston Red Sox.

Tom Heinzen was a 29-year-old college freshman, began graduate school eight days after the birth of his fourth daughter, and is still amazed that he and his wife somehow managed to stay married. A magna cum laude graduate of Rockford College, he earned his PhD in social psychology at the State University of New York at Albany in just three years.

He published his first book on frustration and creativity in government two years later, was a research associate in public policy until he was fired for arguing over the shape of a graph, consulted for the Johns Hopkins Center for Talented Youth, and then began a teaching career at William Paterson University of New Jersey. He founded the psychology club, established an undergraduate research conference, and has been awarded various teaching honors while continuing to write journal articles, books, plays, and two novels that support the teaching of general psychology and statistics. He is also the editor of *Many Things to Tell You,* a volume of poetry by elderly writers.

He has recently become enamored with the potential of motion graphs and the peculiar personalities who shaped the unfolding story of statistics, such as Stella Cunliffe. He belongs to numerous professional societies, including APA, EPA, APS, and the New York Academy of Science, whose meeting place next to the former Twin Towers offers such a spectacular view of New York City that they have to cover the windows so the speakers don't lose their focus during their talks.

His wife, Donna, is a physician's assistant who has volunteered her time in relief work following Hurricanes Mitch and Katrina, and their daughters work in public health, teaching, and medicine. Tom is an enthusiastic but mediocre tennis player and, as a Yankees and Cubs fan, sympathizes with Susan's tortured New England loyalties.

BRIEF CONTENTS

CONTENTS

PREFACE

When we set out to write the first edition of *Essentials of Statistics for the Behavioral Sciences,* we were excited by the opportunity to both teach and tell the remarkable story of statistics. Happily, we have not been alone. During the last several years, there have been many best-selling books demonstrating the penetrating insights and life-affirming applications of statistics, such as *The Ghost Map, Moneyball, Freakonomics, The Lady Tasting Tea, Outliers,* and *The Signal and the Noise.* People are really "getting it," and our ambition is to help this rising generation of behavioral scientists lead the way to keener insights and more applications.

We understand, of course, that the starting point for many students is anxiety about mathematics. That's why we provide steady, persistent reassurance throughout the text that the core concepts in statistics—the very source of their apprehension—are easily explained with examples from everyday life. By also highlighting engaging examples from the history of statistics, we demonstrate how statistical thinking arose from the desire to answer common everyday questions.

Between us, we have worn many hats in the behavioral sciences, including APA representative to the United Nations, evaluator of government programs, career counselor to our students, and internship coordinator. We both also enjoy a lively artistic side, including editing poetry by elderly writers and creating a photographic documentary of postwar recovery in Bosnia—pursuits that influenced our view of statistics as a highly creative endeavor. Those varied experiences are why we are also eager to show students that statistical skills are highly marketable in a variety of settings—an extra boost of confidence for students anticipating the job market. For all these reasons, we are thrilled that the conversation around statistics has shifted from apology (for being difficult) to opportunity (to live a meaningful, productive life). To have been part of that conversation is a privilege—and the conversation continues with this new edition.

What's New in the Second Edition

In the new edition, we connect students to statistical concepts as efficiently and memorably as possible. We've refocused the book on the core concepts of the course and introduce each topic with a vivid real-world example. Our pedagogy first emphasizes mastering concepts, then gives students multiple step-by-step examples of the process of each statistical method, including the mathematical calculations. The extensive Check Your Learning exercises at the end of each section of the chapter, along with the end-of-chapter problems and the new StatsPortal Web site, give students lots of opportunities to practice. Indeed, there are close to twice as many exercises in the second edition as in the first. We've also clarified our approach by adding the following features throughout the book.

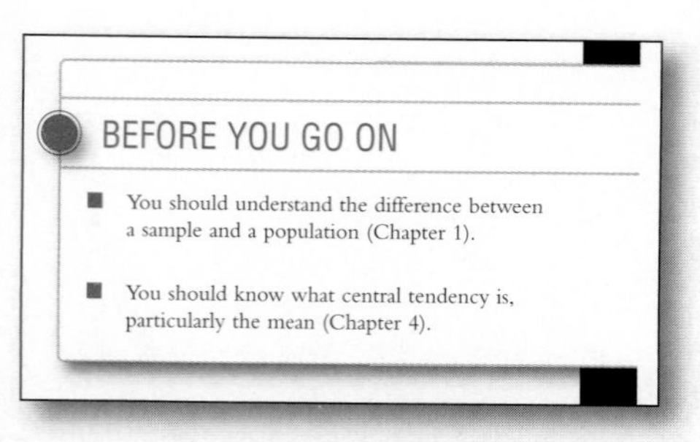

Before You Go On

Each chapter opens with a Before You Go On section that highlights the concepts students need to have mastered before moving on to the next chapter.

Mastering the Formulas and Mastering the Concepts

Some of the most difficult tasks for students new to statistics are identifying the key points and connecting this new knowledge to what they have covered in previous chapters. The unique Mastering the Formula and Mastering the Concept marginal notes provide students with helpful explanations that highlight each formula when it is first introduced and each important concept at its point of relevance. And as a summary of how to apply those concepts, you can't beat Appendix E, Figure E-1: Choosing the Appropriate Hypothesis Test. It's the entire text summarized on a single page; your students will love it.

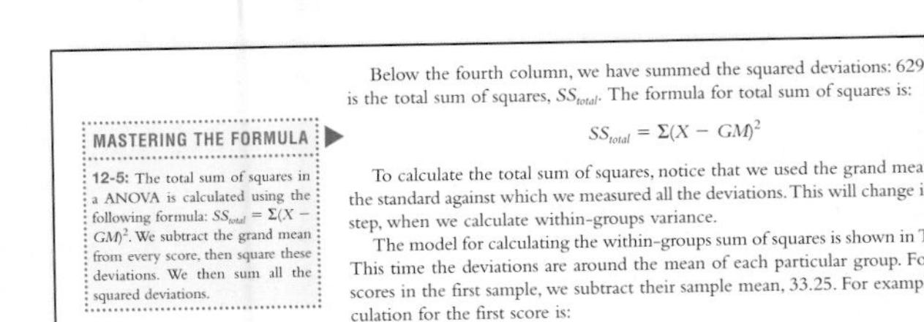

MASTERING THE FORMULA

12-5: The total sum of squares in a ANOVA is calculated using the following formula: $SS_{total} = \Sigma(X - GM)^2$. We subtract the grand mean from every score, then square these deviations. We then sum all the squared deviations.

Below the fourth column, we have summed the squared deviations: 629 is the total sum of squares, SS_{total}. The formula for total sum of squares is:

$$SS_{total} = \Sigma(X - GM)^2$$

To calculate the total sum of squares, notice that we used the grand mea the standard against which we measured all the deviations. This will change i step, when we calculate within-groups variance.

The model for calculating the within-groups sum of squares is shown in T This time the deviations are around the mean of each particular group. Fo scores in the first sample, we subtract their sample mean, 33.25. For exampl culation for the first score is:

$$(28 - 33.25)^2 = 27.563$$

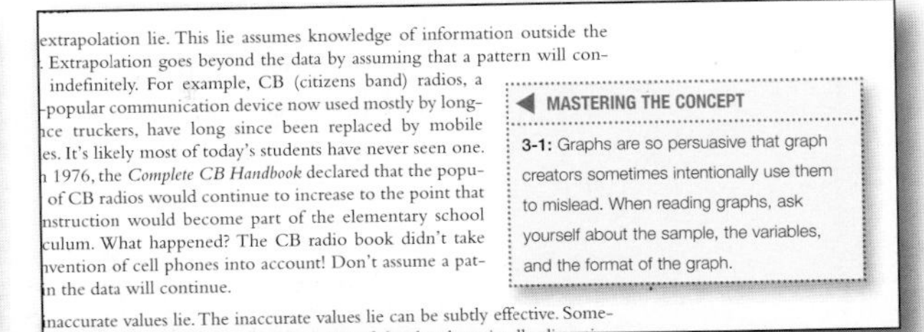

extrapolation lie. This lie assumes knowledge of information outside the Extrapolation goes beyond the data by assuming that a pattern will con- indefinitely. For example, CB (citizens band) radios, a popular communication device now used mostly by long- ce truckers, have long since been replaced by mobile es. It's likely most of today's students have never seen one. 1976, the *Complete CB Handbook* declared that the popu- of CB radios would continue to increase to the point that nstruction would become part of the elementary school culum. What happened? The CB radio book didn't take vention of cell phones into account! Don't assume a pat- n the data will continue.

naccurate values lie. The inaccurate values lie can be subtly effective. Some-

MASTERING THE CONCEPT

3-1: Graphs are so persuasive that graph creators sometimes intentionally use them to mislead. When reading graphs, ask yourself about the sample, the variables, and the format of the graph.

Illustrative, Step-by-Step Examples

The text is filled with real-world examples from a wide variety of sources in the behavioral sciences. We outline statistical techniques in a step-by-step fashion, guiding students through each concept by applying the material creatively and effectively.

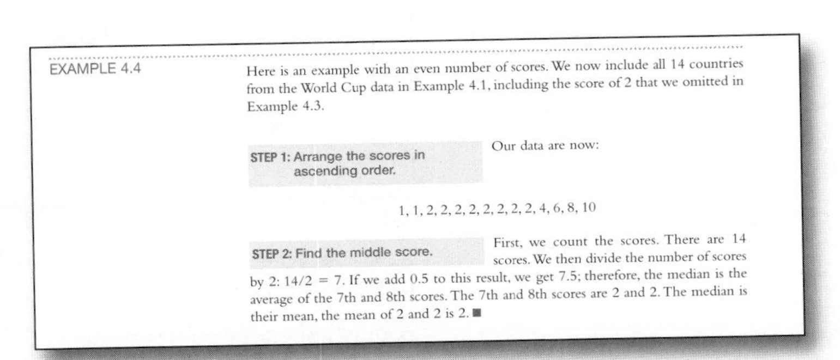

EXAMPLE 4.4

Here is an example with an even number of scores. We now include all 14 countries from the World Cup data in Example 4.1, including the score of 2 that we omitted in Example 4.3.

STEP 1: Arrange the scores in ascending order.

Our data are now:

1, 1, 2, 2, 2, 2, 2, 2, 2, 2, 4, 6, 8, 10

STEP 2: Find the middle score.

First, we count the scores. There are 14 scores. We then divide the number of scores by 2: 14/2 = 7. If we add 0.5 to this result, we get 7.5; therefore, the median is the average of the 7th and 8th scores. The 7th and 8th scores are 2 and 2. The median is their mean, the mean of 2 and 2 is 2. ■

SPSS®

For instructors who integrate SPSS into their course, each chapter includes outlined instructions and screenshots of SPSS output to help students master use of this program with data from the text.

SPSS®

For a paired-samples *t* test, let's use the data from this chapter on performance using a small monitor versus a large monitor. Enter the data in two columns, with each participant having one score in the first column for his or her performance on the small monitor and one score in the second column for his or her performance on the large monitor.

Select **Analyze** → Compare Means → Paired-Samples T Test. Choose the dependent variable under the first condition (small) by clicking it, then clicking the center arrow. Choose the dependent variable under the second condition (large) by clicking it, then clicking the center arrow. Then click "OK." The data and output are shown in the screenshot. Notice that the *t* statistic and confidence interval match ours (5.72 and [−16.34, −5.66]) except that the signs are different. This occurs because of the order in which one score was subtracted from the other score—that is, whether the score on the large monitor was subtracted from the score on the small monitor, or vice versa. The outcome is the same in either case. The *p* value is under "Sig. (2-tailed)" and is .005. We can use this number in Excel to determine the value for p_{rep}, .9657.

	Small	Large
1	122.00	111.00
2	131.00	116.00
3	127.00	113.00
4	123.00	119.00
5	132.00	121.00

Paired Samples Statistics

		Mean	N	Std. Deviation	Std. Error Mean
Pair 1	Small	127.0000	5	4.52769	2.02485
	Large	116.0000	5	4.12311	1.84391

Paired Samples Correlations

		N	Correlation	Sig.
Pair 1	Small & Large	5	.509	.381

Paired Samples Test

		Paired Differences					t	df	Sig. (2-tailed)
					95% Confidence Interval of the Difference				
		Mean	Std. Deviation	Std. Error Mean	Lower	Upper			
Pair 1	Small - Large	11.00000	4.30116	1.92354	5.65940	16.34060	5.719	4	.005

How It Works—Chapter-Specific Worked-Out Exercises

Many students have anxiety as they approach end-of-chapter exercises. To ease that anxiety, the How It Works section provides students with step-by-step worked-out exercises representative of those they will see at the end of the chapter. This section appears just before the end-of-chapter exercises and acts as a model for the more challenging Applying the Concepts and Putting It All Together questions.

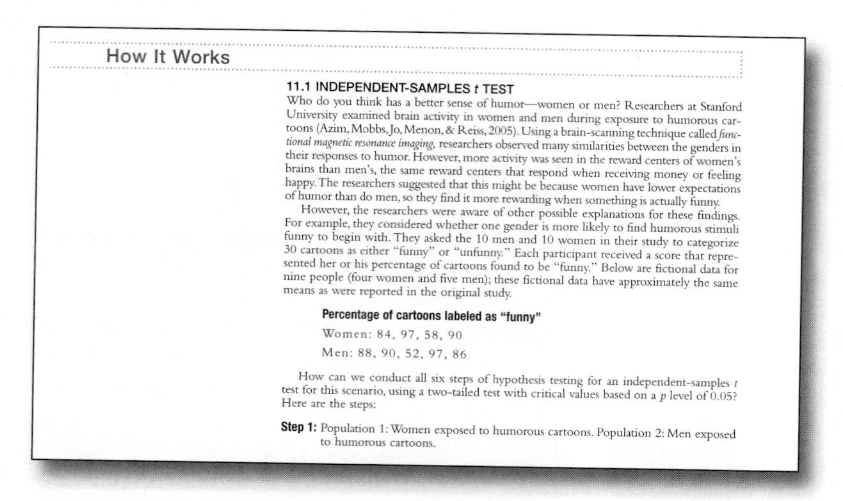

How It Works

11.1 INDEPENDENT-SAMPLES *t* TEST

Who do you think has a better sense of humor—women or men? Researchers at Stanford University examined brain activity in women and men during exposure to humorous cartoons (Azim, Mobbs, Jo, Menon, & Reiss, 2005). Using a brain-scanning technique called *functional magnetic resonance imaging,* researchers observed many similarities between the genders in their responses to humor. However, more activity was seen in the reward centers of women's brains than men's, the same reward centers that respond when receiving money or feeling happy. The researchers suggested that this might be because women have lower expectations of humor than do men, so they find it more rewarding when something is actually funny.

However, the researchers were aware of other possible explanations for these findings. For example, they considered whether one gender is more likely to find humorous stimuli funny to begin with. They asked the 10 men and 10 women in their study to categorize 30 cartoons as either "funny" or "unfunny." Each participant received a score that represented her or his percentage of cartoons found to be "funny." Below are fictional data for nine people (four women and five men); these fictional data have approximately the same means as were reported in the original study.

Percentage of cartoons labeled as "funny"

Women: 84, 97, 58, 90

Men: 88, 90, 52, 97, 86

How can we conduct all six steps of hypothesis testing for an independent-samples t test for this scenario, using a two-tailed test with critical values based on a p level of 0.05? Here are the steps:

Step 1: Population 1: Women exposed to humorous cartoons. Population 2: Men exposed to humorous cartoons.

Building Better Graphs Using Excel

A new appendix guides students through the basics of creating a clear, readable graph with Excel. Using an example from the text, students are guided through the steps of creating a graph and then changing Excel's default choices to meet the criteria for an excellent graph.

Practice

Like a computer game that gradually brings its players to higher levels through repetition, carefully calibrated practice is important to our students' growth. So we constructed levels of challenge that test students' understanding both of concepts and of calculations, and that meet our standards of clarity and effectiveness. Throughout each chapter, we provide multiple opportunities for students to practice with the Check Your Learning exercises, which are placed at the end of every major section. These lead up to full problem sets at the end of each chapter, which feature a total of over 1000 questions. Moreover, the Applying the Concepts and Putting It All Together problems are mostly based on real studies, examples, and data.

Before we were textbook authors, we were teachers who were frustrated that textbooks didn't offer questions that specifically tested conceptual knowledge. We have responded to that frustration by learning to think like game designers. As a result, we created four tiers for all the exercises in the book so students could test themselves on four levels:

- **Clarifying the Concepts** questions help students to master the general concept, the statistical terminology, and the conceptual assumptions of each topic.
- **Calculating the Statistics** exercises provide practice on the basic calculations for each formula and statistic.

- **Applying the Concepts** exercises apply statistical questions to real-world situations across the behavioral sciences and require students to bridge their knowledge of concepts and calculations.
- **Putting It All Together** exercises ask students both to apply the concepts from the chapter to a real-world situation and to connect the chapter's concepts to ideas from previous chapters.

Media and Supplements

StatsPortal

A comprehensive Web resource for teaching and learning statistics

The StatsPortal Web site combines Worth Publishers' high-quality media with an innovative platform for easy navigation. For students, it is the ultimate online study guide, with statistical tools, adaptive quizzing, personalized feedback, and the full text in the e-Book. For instructors, StatsPortal is a full-course space where class documents can be posted, quizzes can be easily assigned and graded, and students' progress can be assessed and recorded. Whether you are looking for the most effective study tools or a robust platform for an online course, StatsPortal is a powerful way to enhance a statistics class for the behavioral sciences.

> StatsPortal to accompany *Essentials of Statistics for the Behavioral Sciences,* Second Edition, can be previewed and purchased at **www.yourstatsportal.com**.
>
> *Essentials of Statistics for the Behavioral Sciences,* Second Edition, and StatsPortal can be ordered together with ISBN-10: 1-4641-6212-3/ISBN-13: 978-1-4641-6212-1. Individual components of StatsPortal may also be available for separate, stand-alone purchase.

StatsPortal for *Essentials of Statistics for the Behavioral Sciences,* Second Edition, includes all the following resources:

- The **LearningCurve** quizzing system was designed based on the latest findings from learning and memory research. It combines adaptive question selection, immediate and valuable feedback, and a game-like interface to engage students in a learning experience that is unique to each student. Each LearningCurve quiz is fully integrated with other resources in StatsPortal through the Personalized Study Plan, so students can review with Worth's extensive library of videos and activities. And question analysis reports allow instructors to track the progress of their entire class.
- An **interactive e-Book** allows students to highlight, bookmark, and make their own notes, just as they would with a printed textbook. Google-style searching and in-text glossary definitions make the text ready for the digital age.
- The **Statistical Video Series** consists of StatClips, StatClips Examples, and Statistically Speaking "Snapshots." The video series can be used to view animated lecture videos, whiteboard lessons, and documentary-style footage that illustrate key statistical concepts and help students visualize statistics in real world scenarios.

- **StatClips lecture videos**, created and presented by Alan Dabney, PhD, Texas A&M University, are innovative visual tutorials that illustrate key statistical concepts. In 3 to 5 minutes, each StatClips video combines dynamic animation, data sets, and interesting scenarios to help students understand the concepts in an introductory statistics course.
- In **StatClips Examples**, Alan Dabney walks students through step-by-step examples related to the StatClips lecture videos to reinforce the concepts through problem solving.
- **SnapShots** videos are abbreviated, student-friendly versions of the Statistically Speaking video series, and they bring the world of statistics into the classroom. In the same vein as the successful PBS series *Against All Odds: Inside Statistics,* Statistically Speaking uses new and updated documentary footage and interviews that show real people using data analysis to make important decisions in their careers and in their daily lives. From business to medicine, from the environment to understanding the census, SnapShots help students see why statistics is important for their careers, and how statistics can be a powerful tool to understand their world.

- **Statistical Applets** allow students to master statistical concepts by manipulating data. The applets can also be used to solve problems.
- **EESEE Case Studies**, taken from the Electronic Encyclopedia of Statistical Exercises and Examples, offer students additional applied exercises and examples.
- A **Data Set from the General Social Survey (GSS)** gives students access to data from one of the most trusted sources of sociological information. Since 1972, the GSS has collected data that reflect changing opinions and trends in the United States. A number of exercises in the text use GSS data, and this data set allows students to explore further.
- **Learning Objectives** give students a framework for self-testing and studying.
- The **Assignment Center** lets instructors easily construct and administer tests and quizzes from the book's Test Bank and course materials. The Test Bank includes a subset of questions from the end-of-chapter exercises with algorithmically generated values, so each student can be assigned a unique version of the question. Assignments can be automatically graded, and the results are recorded in a customizable Gradebook.

Additional Student Supplements

- **Study Guide and SPSS Manual** by Jennifer Coleman and Byron Reischl of Western New Mexico University includes chapter outlines and learning objectives, chapter reviews, and multiple-choice study questions and answers.
- **SPSS: A User-Friendly Approach** by Jeffrey Aspelmeier and Thomas Pierce of Radford University is an accessible introduction to using SPSS. Using a proven teaching method, statistical procedures are made accessible to students by building each section of the text around the storyline from a popular cartoon. Easing anxiety and giving students the necessary support to learn the material, *SPSS: A User-Friendly Approach* provides instructors and students with an informative guide to the basics of SPSS. Available for versions 16, 17, and 18.

- The **iClicker** Classroom Response System is a versatile polling system developed by educators for educators that makes class time more efficient and interactive. iClicker allows you to ask questions and instantly record students' responses, take attendance, and gauge students' understanding and opinions. It can help you gather data on students that you can use to teach statistics, connecting the concepts to students' lives. iClicker is available at a 10% discount when packaged with *Essentials of Statistics for the Behavioral Sciences,* Second Edition.

Take advantage of our most popular supplements!

Worth Publishers is pleased to offer cost-saving packages of *Essentials of Statistics for the Behavioral Sciences,* Second Edition, with our most popular supplements. Below is a list of some of the most popular combinations, available for order through your local bookstore.

***Essentials of Statistics for the Behavioral Sciences,* 2nd ed., & StatsPortal Access Card**
ISBN-10: 1-4641-6212-3 / ISBN-13: 978-1-4641-6212-1

***Essentials of Statistics for the Behavioral Sciences,* 2nd ed., & Study Guide and SPSS Manual**
ISBN-10: 1-4641-4688-8 / ISBN-13: 978-1-4641-4688-6

***Essentials of Statistics for the Behavioral Sciences,* 2nd ed., & *SPSS: A User-Friendly Approach for Versions 17 and 18* by Jeffrey Aspelmeier and Thomas Pierce**
ISBN-10: 1-4641-5214-4 / ISBN-13: 978-1-4641-5214-6

***Essentials of Statistics for the Behavioral Sciences,* 2nd ed., & iClicker**
ISBN-10: 1-4641-5213-6 / ISBN-13: 978-1-4641-5213-9

Instructor Supplements

We understand that one book alone cannot meet the education needs and teaching expectations of the modern classroom. Therefore, we have engaged colleagues to create a comprehensive supplements package that brings statistics to life for students and provides instructors with the resources necessary to effectively supplement their successful strategies in the classroom.

- **Instructor's Resources** by Robin Freyberg, Stern College for Women, Yeshiva University, with contributions by Katherine Makarec of William Paterson University. The contents include Teaching Tips and sample course outlines. Each chapter includes a brief overview, discussion questions, classroom activities, handouts, additional reading suggestions, and online resources.

- **Test Bank** by Jennifer Coleman, Western New Mexico University, with contributions by Kelly M. Goedert, Seton Hall University, and Daniel Cruz, Caldwell College. The Test Bank includes multiple-choice, true/false, fill-in-the-blank, and critical thinking/problem-solving questions for each chapter. It also includes Web Quizzes that are featured on the book's companion Web site.

- **Diploma Computerized Test Bank** (available for Windows or Macintosh on a single CD-ROM). The CD-ROM allows instructors to add an unlimited number of new questions; edit questions; format a test; scramble questions; and include figures, graphs, and pictures. The computerized Test Bank also allows you to export into a variety of formats compatible with many Internet-based testing products.

- Worth Publishers supports multiple **Course Management Systems** with enhanced cartridges that include Test Bank questions and other resources. Cartridges are provided free upon adoption of *Essentials of Statistics for the Behavioral Sciences,* Second Edition, and can be requested through the Instructor Resources tab on the Book Companion Website at **www.worthpublishers.com/nolanessentials2e**.

Acknowledgments

We would like to thank the many people who have contributed directly and indirectly to the writing of this text. We want to thank our students at Seton Hall University and William Paterson University for teaching us how to teach statistics in a way that makes sense.

Tom: The family members who know me on a daily basis and decide to love me anyway deserve more thanks than words can convey: Donna, Rebekah, Nagesh, Debbie, Anthony, Amy, Elizabeth, Mollie, and a mystery guest who will have a name by the time this book is published. The close friends, artists, and colleagues who voiced encouragement and timely support also deserve my deep appreciation: Beth, Army, Culley, and Miran Schultz; Laura Cramer-Berness; Ariana DeSimone; J. Allen Suddeth; Nancy Vail; Gerry Esposito; and Sally Ellyson.

My students, in particular, have always provided a reality check on my teaching methods with the kind of candor that only students engaged in the learning process can bring. And in recent years, we have been building a wonderful department by following the rule to "always hire people who are better than you." Some of those "better than you" colleagues have been a steady source of helpful conversation: Michael Gordon, Amy Learmonth, and Natalie Obrecht. Thank you. And Susan, of course, has been as fine a colleague and friend as I could ever have hoped for.

I also want to thank the people at Worth, all of them. They have a vision for quality textbook publishing that is different from many publishers. I know I speak for Susan as well when I say how deeply we appreciate their level of close cooperation and timely support.

Susan: I am grateful to my Northwestern University professors and classmates for convincing me that statistics can truly be fun. I am also eternally thankful to Beatrix Mellauner for bringing Tom Heinzen and me together as coauthors; it has been a privilege and a pleasure to collaborate with Tom for so many years. I owe thanks, as well, to my Seton Hall colleagues—Kelly Goedert and Marianne Lloyd in particular—who are the sources for an endless stream of engaging examples. Thanks, too, to Marjorie Levinstein, Graduate Assistant extraordinaire, who helps me carve out time to work on this book.

Much of the writing of this book took place during my sabbatical and ensuing summers in Bosnia and Herzegovina; I thank my Bosnian friends for their warmth and hospitality every time I visit. A special thank you to the members of the Bojanic and Nolan clans—especially my parents, Diane and Jim, who have patiently endured the barrage of statistics I often inject into everyday conversation. Finally, I am most grateful to my husband, Ivan Bojanic, for the memorable adventures we've had (and the statistical observations that grew out of many of them); Ivan experienced the evolution of this book through countless road-trip conversations and late-night editorial sessions.

The contributions of the supplements authors are innumerable, and we would like to take a moment to highlight the impressive cast of instructors who have joined our team. Katherine Makarec, Robert Weathersby, and Robin Freyberg are all professionals with a deep interest in creating successful classrooms, and we appreciate the opportunity to work with people of such commitment.

Throughout the writing of the first edition, we relied on the criticism, corrections, encouragement, and thoughtful contributions from reviewers, focus group attendees, survey respondents, and class-testers. We thank them for their expertise and for the time they set aside to help us develop this textbook. We also are grateful to the professors who have used our book and taken the time to provide specific and valuable feedback; in particular, we thank Harvey H. C. Marmurek and Patricia A. Smiley. Special thanks go to Jennifer Coleman at Western New Mexico University and to Kelly Goedert at Seton Hall University for their tireless work in developing the pedagogy with us, providing a responsible accuracy check, and contributing numerous ideas for us to consider as we continue to make this book even better. Special thanks also go to Melanie Maggard at the University of the Rockies and Sherry L. Serdikoff at James Madison University for their invaluable efforts checking the text and exercises.

Tsippa Ackerman
John Jay College

Kenneth Bonanno
Merrimack College

Danuta Bukatko
College of the Holy Cross

Heidi Burross
Pima Community College

Jennifer Coleman
Western New Mexico University

Melanie Conti
College of Saint Elizabeth

Betty Dorr
Fort Lewis College

Nancy Dorr
The College of St. Rose

Kevin Eames
Covenant College

Nancy Gee
SUNY College at Fredonia

Marilyn Gibbons
Texas State University

Elizabeth Haines
William Paterson University

Roberto Heredia
Texas A&M University

Cynthia Ingle
Bluegrass Community and Technical College

E. Jean Johnson
Governors State University

Lauriann Jones-Moore
University of South Florida

Min Ju
SUNY College at New Paltz

Karl Kelley
North Central College

Shelley Kilpatrick
Southwest Baptist University

Megan Knowles
University of Georgia

Paul Koch
Saint Ambrose University

Marika Lamoreaux
Georgia State University

Jennifer Lancaster
St. Francis College

Christine MacDonald
Indiana State University

Suzanne Mannes
Widener University

Walter Marcantoni
Bishop's University

Kelly Marin
Manhattan College

Connie Meinholdt
Ferris State University

William Merriman
Kent State University

Chris Molnar
LaSalle University

Matthew Mulvaney
SUNY College at Brockport

Angela K. Murray
University of Kansas

Aminda O'Hare
University of Kansas

Sue Oliver
Glendale Community College of Arizona

Stephen O'Rourke
The College of New Rochelle

Debra Oswald
Marquette University

Alison Papdakis
Loyola College in Maryland

Laura Rabin
CUNY Brooklyn

Michelle Samuel
Mount St. Mary's College, Chalon

Ken Savitsky
Williams College

Heidi Shaw
Yakima Valley Community College

Ross B. Steinman
Widener University

Colleen Sullivan
Worcester University

Brian Stults
Florida State University

Melanie Tabak
William Penn University

Mark Tengler
University of Houston, Clear Lake

Patricia Tomich
Kent State University

David Wallace
Fayetteville State University

Elizabeth Weiss
The Ohio State University

Charles Woods
Austin Peay State University

Tiffany Yip
Fordham University

Accuracy Reviewers

Verne Bacharach
Appalachian State University

Jeffrey Berman
University of Memphis

Dennis Goff
Randolph College

Linda Henkel
Fairfield University

Melanie Maggard
University of the Rockies

Kathy Oleson
Reed College

Christy Porter
College of William and Mary

Sherry L. Serdikoff
James Madison University

Alexander Wilson
University of New Brunswick

It has truly been a pleasure for us to work with everyone at Worth Publishers. From the moment we signed there, we have been impressed with the passionate commitment of everyone we encountered at Worth at every stage of the publishing process. Senior Vice President Catherine Woods and Publisher Kevin Feyen foster that commitment to quality in the Worth culture.

Our original Development Editor, Michael Kimball, provided an attention to detail that helped us to achieve our vision for this book. Our current Development Editor, Elaine Epstein, along with Director of Development for Print and Digital Tracey Kuehn have been enormously important in shaping this edition; we rely heavily on their expert guidance and eternally good-humored encouragement. We're grateful to Publisher Ruth Baruth for her astute input regarding creative pedagogy in statistics textbooks. Publisher Charles Linsmeier's impressive ability to assess ideas and face problems from multiple angles has contributed to a successful publication. We are also grateful to Acquisitions Editor Daniel DeBonis for his skill and patience in guiding the book to completion.

Senior Project Editor Jane O'Neill, Associate Managing Editor Lisa Kinne, and Production Manager Barbara Seixas managed the production of the text and worked tirelessly to bring the book to fruition. Art Director Babs Reingold's commitment to artistic values in textbook publishing is continually inspiring. Kevin Kall, Designer, united beauty with clarity and content in the interior design. Copyeditor Anna Paganelli and her hawk's eye made our prose more consistent, more accurate, and easier to read. Photo Editor Bianca Moscatelli helped us to select photos that told the stories of statistics. Thanks to each of you for fulfilling Worth's promise to create a book whose aesthetics so beautifully support the specific pedagogical demands of teaching statistics.

Assistant Editor Nadina Persaud, Senior Media Editor Christine Burak, Project Editor Julio Espin, and Production Manager Stacey Alexander guided the development and creation of the supplements package, making life so much better for so many students and instructors. Executive Marketing Manager Katherine Nurre, Marketing Manager Lindsay Johnson, and Associate Director of Market Development Carlise Stembridge quickly understood why we believe so deeply in this book, and each contributed unstinting effort to advocate for this second edition with our colleagues across the country.

We also want to thank the tremendously dedicated Worth team that consistently champions our book while garnering invaluable accolades and critiques from their professor contacts—information that directly leads to a better book. There are far too many of you to thank individually—as a start, we thank our local Worth representative Kathryn Treadway and specialist Tom Kling for their continuing enthusiasm and support.

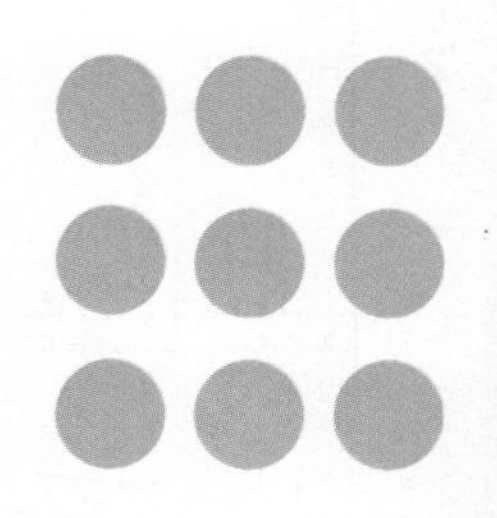

CHAPTER 1

An Introduction to Statistics and Research Design

BEFORE YOU GO ON

- You should be familiar with basic mathematics (see Appendix A).

John Snow's Famous Map Dr. Snow mapped cholera deaths in relation to the Broad Street water well and, in doing so, solved the urgent mystery of how cholera could infect so many people so suddenly. The X's are all neighborhood wells. The X in the red circle is the Broad Street well. Each dot indicates that a person living at this address died of cholera, and a cluster of cases is clearly seen around the Broad Street well (but not around the other wells). Snow was careful to include the other X's to demonstrate that the deaths were closer to one specific source of water.

In just 10 days of the London cholera epidemic of 1854, the number of victims of the disease climbed from 127 to 500, with approximately 37 new deaths each day. Dr. Snow, also of London, had spent years trying to determine how cholera was communicated from one person to another (Vinten-Johansen, Brody, Paneth, Rachman, & Rip, 2003). With the disease in his midst, he tested an idea. On a map, he placed a dot to mark the location of each cholera victim's home. He added an X for each neighborhood water well. The visual presentation of these data revealed that the closer a home was to the well on Broad Street, the X circled in red, the more likely it was that a death from cholera had occurred.

Dr. Snow also examined a sample of Broad Street, the X circled in red, well water under a microscope and discovered white particles floating in it. He proposed a simple solution to the city Board of Guardians: Remove the handle to the water pump to stop the spread of the disease. The odd theory that cholera was communicated in the water supply startled the board. But with Dr. Snow's insistence, the authorities finally removed the pump handle. Cholera deaths declined dramatically.

However, Dr. Snow had a statistical problem. The rate of deaths from cholera started to decline before the handle was removed! How could this happen? The answer is both disturbing and insightful: because many people had died or fled the Broad Street neighborhood, there were fewer people left to be infected.

The Two Branches of Statistics

The statistical genius and research of Snow not only saved lives, it anticipated the two main branches of modern statistics: descriptive statistics and inferential statistics.

- A **descriptive statistic** organizes, summarizes, and communicates a group of numerical observations.
- An **inferential statistic** uses sample data to make general estimates about the larger population.
- A **sample** is a set of observations drawn from the population of interest.
- The **population** includes all possible observations about which we'd like to know something.

Descriptive Statistics

Descriptive statistics *organize, summarize, and communicate a group of numerical observations.* Descriptive statistics describe large amounts of data in a single number or in just a few numbers. Here's an illustration using familiar numbers: body weights. The Centers for Disease Control and Prevention (CDC, 2004) reported that people in the United States weigh more now than they did four decades ago. The average weight for women increased from 140.2 pounds in 1960 to 164.3 in 2002. For men, the average weight went from 166.3 to 191.0 pounds in the same time span. These averages are descriptive statistics because they *describe* the weights of many people in just one number. A single number reporting the average communicates the observations more clearly than would a long list of weights for every person studied by the CDC.

Inferential Statistics

Inferential statistics *use sample data to make general estimates about the larger population.* Inferential statistics infer, or make an intelligent guess about, the population. For example, the CDC made inferences about weight even though it did not actually weigh *everyone* in the United States. Instead, the CDC studied a smaller, representative group of U.S. citizens to make an intelligent guess about the entire population.

MASTERING THE CONCEPT

1.1: Descriptive statistics summarize numerical information about a sample. Inferential statistics draw conclusions about the broader population based on numerical information from a sample.

Distinguishing Between a Sample and a Population

A ***sample*** *is a set of observations drawn from the population of interest.* Researchers usually study a sample, but they are really interested in the ***population***, which *includes all possible observations about which we'd like to know something.* For example, the average weight of the CDC's samples of women and men were used to estimate the average weight for the entire U.S. population, which was the CDC's interest.

Samples are used most often because we are rarely able to study every person (or organization or laboratory rat) in a population. For one thing, it's far too expensive. In addition, it would take too long. Snow did not want to interview every family in the Broad Street neighborhood—people were dying too fast! Fortunately, what he learned from his sample also applied to the larger population.

Descriptive Statistics Summarize Information It is more useful to use a single number to summarize many people's weights than to provide a long, overwhelming list of each person's weight.

CHECK YOUR LEARNING

Reviewing the Concepts

> Descriptive statistics organize, summarize, and communicate large amounts of numerical information.

> Inferential statistics use sample data to draw conclusions about larger populations.

> Samples, or selected observations of a population, are intended to be representative of the larger population.

Clarifying the Concepts

1-1 Are samples or populations used in inferential statistics?

Calculating the Statistics

1-2 a. If your professor calculated the average grade for your statistics class, would that be considered a descriptive statistic or an inferential statistic?

b. If that same class average is used to predict something about how future students might do in statistics, would it be considered a descriptive statistic or an inferential statistic?

Applying the Concepts

1-3 Imagine that the director of the counseling center at your university wants to examine stress levels of students. From the student directory, she randomly chooses 100 of the 12,500 students and assesses their stress levels in a diagnostic interview. She reports that

continued on next page

the average stress level is 18 on a scale of 1–50, a score she knows to be moderately high for college students. She concludes that the students at this institution have a moderately high stress level.

a. What is the sample?

b. What is the population?

c. What is the descriptive statistic?

d. What is the inferential statistic?

Solutions to these Check Your Learning questions can be found in Appendix D.

How to Transform Observations into Variables

Like John Snow, we begin the research process by making observations and transforming them into a useful format. For example, Snow observed the locations of people who had died from cholera and placed these locations on a map that also showed wells in the area. ***Variables*** *are observations of physical, attitudinal, and behavioral characteristics that can take on different values.* Behavioral scientists often study abstract variables such as motivation and self-esteem; they typically begin the research process by transforming their observations into numbers.

- A **variable** is any observation of a physical, attitudinal, or behavioral characteristic that can take on different values.
- A **discrete observation** can take on only specific values (e.g., whole numbers); no other values can exist between these numbers.
- A **continuous observation** can take on a full range of values (e.g., numbers out to several decimal places); an infinite number of potential values exists.
- A **nominal variable** is a variable used for observations that have categories, or names, as their values.
- An **ordinal variable** is a variable used for observations that have rankings (i.e., 1st, 2nd, 3rd . . .) as their values.
- An **interval variable** is a variable used for observations that have numbers as their values; the distance (or interval) between pairs of consecutive numbers is assumed to be equal.
- A **ratio variable** is a variable that meets the criteria for an interval variable but also has a meaningful zero point.

Researchers use both discrete and continuous numerical observations to quantify variables. ***Discrete observations*** *can take on only specific values (e.g., whole numbers); no other values can exist between these numbers.* For example, if we measure the number of times study participants get up early in a particular week, the only possible values would be whole numbers. It is reasonable to assume that each participant could get up early 0 to 7 times in any given week, but not 1.6 or 5.92 times.

Continuous observations *can take on a full range of values (e.g., numbers out to several decimal places); an infinite number of potential values exists.* For example, a person might complete a task in 12.83912 seconds. The possible values are continuous, limited only by the number of decimal places we choose to use.

Discrete Observations

There are two types of observations that are always discrete: nominal variables and ordinal variables. ***Nominal variables*** *are used for observations that have categories, or names, as their values.* For example, when entering data into a statistical computer program, a researcher might code male participants with the number 1 and female participants with the number 2. In this case, the numbers only identify the gender category for each participant. They do not imply any other quality—for instance, that men are better than women because they get the first number, or that women are twice as good as men because they happen to be coded as a 2. Nominal variables are always discrete (whole numbers).

Ordinal variables *are used for observations that have rankings (i.e., 1st, 2nd, 3rd . . .) as their values.* In team sports, for example, a team finishes the season in a particular "place," or rank. Whether the team goes to the playoffs is determined by its rank at the end of the season. It doesn't matter if the team won first place by one game or by many games. Like nominal variables, ordinal variables are always discrete. A team could be 1st or 3rd or 12th, but could not be ranked 1.563.

Continuous Observations

The two types of observations that can be continuous are interval variables and ratio variables. ***Interval variables*** *are used for observations that have numbers as their values; the distance (or interval) between pairs of consecutive numbers is assumed to be equal.* For example,

Scott J. Ferrell/*Congressional Quarterly*/Getty Images

Nominal Variables Just Categorize If you wanted to compare the enthusiasm levels of Republicans (not clapping) and Democrats (clapping), political party would be a nominal variable. Nominal observations merely name categories; the numbers don't have any meaning beyond a name.

temperature is an interval variable because the interval from one degree to the next is always the same. Some interval variables are also discrete variables, such as the number of times one has to get up early each week. This is an interval variable because the distance between numerical observations is assumed to be equal. The difference between 1 and 2 times is the same as the difference between 5 and 6 times. However, this observation is also discrete because, as noted earlier, the number of days in a week cannot be anything but a whole number. Several behavioral science measures are treated as interval measures but also can be discrete, such as some personality and attitude measures.

Sometimes discrete, interval observations, such as counting the number of times one has to get up early each week, are also ***ratio variables****, variables that meet the criteria for interval variables but also have meaningful zero points.* If someone never has to get up early, then zero is a meaningful observation and could represent a variety of life circumstances. Perhaps the person is unemployed, retired, ill, or merely on vacation. The number of times a rat pushes a lever to receive food would also be considered a discrete, ratio variable in that it has a true zero point—the rat might never push the bar (and go hungry). Ratio observations that are not discrete include time running out in a basketball game or crossing the finish line in a race.

Many cognitive studies use the ratio variable of reaction time to measure how quickly we process difficult information. For example, the Stroop test assesses how long it takes to read a list of color words printed in ink of the wrong color (Figure 1-1), such as when the word *red* is printed in blue or the word *blue* is printed in green. If it takes you 1.264 seconds to press a computer key that accurately identifies that the word *red* printed in blue actually reads *red,* then your reaction time is a ratio variable; time always implies a meaningful zero.

You can experience for yourself one way that behavioral scientists transform observations into numbers by taking the Stroop test (Figure 1-1 provides the Web site address). This version of the Stroop test gives response times in whole numbers—for example, 12 seconds—although other versions are more specific and give response times to several decimal places, such as 12.1304 seconds.

FIGURE 1-1
Reaction Time and the Stroop Test

The Stroop test assesses how long it takes to read a list of color words printed in the wrong color, such as the word *red* printed in the color white. Try it and see how tricky (and frustrating) it can be: go to this book's Web site (www.worthpublishers.com/nolanheinzen) and click on "Stroop test."

red white green brown
green red brown white
white brown green red
red white green brown
brown green white red
white brown red green
green white brown red
red brown green white

MASTERING THE CONCEPT

1.2: The three main types of variables are nominal (or categorical), ordinal (or ranked), and scale. The third type (scale) includes both interval variables and ratio variables; the distances between numbers on the measure are meaningful.

Many statistical computer programs refer both to interval numbers and to ratio numbers as *scale observations* because both interval observations and ratio observations are analyzed with the same statistical tests. Specifically, *a* ***scale variable*** *is a variable that meets the criteria for an interval variable or a ratio variable.* Throughout this text, we use the term *scale variable* to refer to variables that are interval or ratio, but it is important to remember the distinction between interval variables and ratio variables. Table 1-1 summarizes the four types of variables.

TABLE 1-1. Quantifying Our Observations

There are four types of variables that we can use to quantify our observations. Two of them, nominal and ordinal, are always discrete. Interval variables can be discrete or continuous; ratio variables are almost always continuous.

	Discrete	Continuous
Nominal	Always	Never
Ordinal	Always	Never
Interval	Sometimes	Sometimes
Ratio	Seldom	Almost always

CHECK YOUR LEARNING

Reviewing the Concepts

> Variables are quantified with discrete or continuous observations.

> Depending on the study, statisticians select nominal, ordinal, or scale (interval or ratio) variables.

Clarifying the Concepts

1-4 What is the difference between discrete observations and continuous observations?

Calculating the Statistics

1-5 Three female students complete a Stroop test. Lorna finishes in 12.67 seconds; Desiree finishes in 14.87 seconds; and Marianne finishes in 9.88 seconds.

a. Are these data discrete or continuous?

b. Is the variable an interval or a ratio observation?

c. On an ordinal scale, what is Lorna's score?

Applying the Concepts

1-6 Eleanor Stampone (1993) randomly distributed pieces of paper to students in a large lecture center. Each paper contained one of three short paragraphs that described the interests and appearance of a female student. The descriptions were identical in every way except for one adjective. The student was described as having either "short," "mid-length," or "very long" hair. At the bottom of each piece of paper, Stampone asked the participants (both female and male) to fill out a measure that indicated the probability that the student described in the scenario would be sexually harassed.

a. What is the nominal variable used in Stampone's hair-length study? Why is this considered a nominal variable?

b. What is the ordinal variable used in the study? Why is this considered an ordinal variable?

c. What is the scale variable used in the study? Why is this considered a scale variable?

Solutions to these Check Your Learning questions can be found in Appendix D.

Variables and Research

A major aim of research is to understand the relations among variables with many different values. It is helpful to first remember that variables vary. For example, when studying a discrete, nominal variable such as gender, we refer to gender as the variable because it can vary—either male or female. The term *level,* along with the terms *value* and *condition,* all refer to the same idea. ***Levels*** *are the discrete values or conditions that variables can take on.* For example, male is a level of the variable gender. Female is another level of the variable gender. In both cases, gender is the variable. Similarly, when studying a continuous, scale variable, such as how fast a runner completes a marathon, we refer to time as the variable. For example, 3 hours, 42 minutes, 27 seconds is one of an infinite number of possible times it would take to complete a marathon. With this in mind, let's explore the three types of variables.

- A **scale variable** is a variable that meets the criteria for an interval variable or a ratio variable.
- A **level** is a discrete value or condition that a variable can take on.
- An **independent variable** has at least two levels that we either manipulate or observe to determine its effects on the dependent variable.
- A **dependent variable** is the outcome variable that we hypothesize to be related to, or caused by, changes in the independent variable.
- A **confounding variable** is any variable that systematically varies with the independent variable so that we cannot logically determine which variable is at work; also called a *confound.*

Independent, Dependent, and Confounding Variables

The three types of variables that we consider in research are independent, dependent, and confounding. Two of these variables are necessary for good research: independent variables and dependent variables. But a confounding variable is the enemy of good research. We usually conduct research to determine if one or more independent variables predict a dependent variable. *An* ***independent variable*** *has at least two levels that we either manipulate or observe to determine its effects on the dependent variable.* For example, if we are studying whether gender predicts one's attitude about politics, then the independent variable is gender.

The ***dependent variable*** *is the outcome variable that we hypothesize to be related to, or caused by, changes in the independent variable.* For example, we hypothesize that the dependent variable (attitudes about politics) depends on the independent variable (gender). If in doubt as to which is the independent variable and which is the dependent variable, then ask yourself which one depends on the other; that one is the dependent variable.

> **MASTERING THE CONCEPT**
>
> **1.3:** We conduct research to see if the independent variable predicts the dependent variable.

By contrast, a ***confounding variable*** *is any variable that systematically varies with the independent variable so that we cannot logically determine which variable is at work.* So how do we decide which is the independent variable and which might be a confounding variable (also called a *confound*)? Well, it all comes down to what *you* decide to study. Let's use an example. Suppose you want to lose weight, so you start using a diet drug *and* begin exercising at the same time; the drug and the exercising are confounded because you cannot logically tell which one is responsible for any weight loss. If we hypothesize that a particular diet drug leads to weight loss, then whether someone uses the diet drug becomes the independent variable and exercise becomes the potentially confounding variable that we would try to control. On the other hand, if we hypothesize that exercise leads to weight loss, then whether someone exercises or not becomes the independent variable and whether that person uses diet drugs along with it becomes the potentially confounding variable that we would try to control. In both of these cases, the dependent variable would be weight loss. But the researcher has to make some decisions about which variables to treat as independent variables, which variables need to be controlled, and which variables to treat as dependent variables. You, the experimenter, are in control of the experiment.

Justin Sullivan/Getty Images

Was the Damage from Wind or Water? During Hurricane Katrina in 2005, high winds were confounded with high water so that often it was not possible to determine whether property damage was due to wind (insured) or to water (not insured).

- **Reliability** refers to the consistency of a measure.
- **Validity** refers to the extent to which a test actually measures what it was intended to measure.
- **Hypothesis testing** is the process of drawing conclusions about whether a particular relation between variables is supported by the evidence.

Reliability and Validity

You probably have a lot of experience in assessing variables—at least on the receiving end. You've taken standardized tests when applying to your university; you've taken short surveys to choose the right product for you, whether jeans or mascara; and you've taken online quizzes—perhaps ones sent through social networking sites like Facebook, such as the Dogster Breed Quiz, which uses a 10-item scale to assess the breed of dog you are most like (2009, http://www.dogster.com/quizzes/what_dog_breed_are_you/).

How good is this quiz? One of the authors took the quiz, choosing on one item a light chicken salad over alternative choices of heavier fare, and was declared to be a bulldog: "You may look like the troublemaker of the pack, but it turns out your tough guy mug is worse than its bite." To determine whether a measure is a good one, we need to know if it is both reliable and valid.

*A **reliable** measure is one that is consistent.* If you were to weigh yourself on your bathroom scale now, and then again in an hour, you would expect your weight to be almost exactly the same. If your weight, as shown on the scale, remains the same when you haven't done anything to change it, then your bathroom scale is reliable. As for the Dogster Breed Quiz, the bulldog author took it twice and was a bulldog the second time as well, one indication of reliability.

But a reliable measure is not necessarily a valid measure. *A **valid** measure is one that measures what it was intended to measure.* Your bathroom scale could be incorrect, but be consistently incorrect—that is, reliable but not valid. A more extreme example is using a ruler when you want to know your weight. You would get a number, and that number might be reliable, but it would not be a valid measure of your weight.

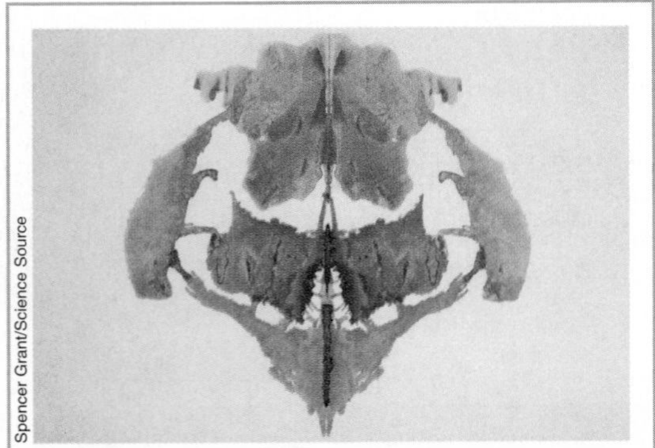
Spencer Grant/Science Source

Reliable and Valid Projective personality tests such as the Rorschach are more reliable than they used to be because of new guidelines, but it is still unclear whether they provide a valid measure. A measure is useful only if it is both reliable (consistent over time) and valid (assesses what it is intended to assess).

And the Dogster Breed Quiz? It's probably not an accurate measure of personality. The quiz, for example, lists an unlikely mix of celebrities with seemingly different personalities as bulldogs—Ellen DeGeneres, Whoopi Goldberg, Jack Black, and George W. Bush! However, we're guessing that no one has done the statistical work to determine whether it is valid or not. When you take such online quizzes, our advice is to view the results as entertaining rather than enlightening.

Any measure with poor reliability cannot have high validity. It is not possible to measure what we intend to measure when the test itself produces varying results. The well-known Rorschach inkblot test is one example of a test whose reliability is questionable, so the validity of the information it produces is difficult to interpret (Wood, Nezworski, Lilienfeld, & Garb, 2003). For instance, two clinicians might analyze the identical set of responses to a Rorschach test and develop quite different interpretations of those responses—meaning it lacks reliability. Reliability can be increased with scoring guidelines, but that doesn't mean validity is increased. Just because two clinicians scoring a Rorschach test designate a person as psychotic, it doesn't necessarily mean that the person *is* psychotic. Reliability is necessary, but not sufficient, to create a valid measure. Nevertheless, the idea that ambiguous images somehow invite revealing information remains attractive to many people; as a result, tests such as the Rorschach are still used frequently, even though there is much controversy about them (Wood et al., 2003).

> **▶ MASTERING THE CONCEPT**
>
> **1.4:** A good variable is both reliable and valid.

CHECK YOUR LEARNING

Reviewing the Concepts

> Independent variables are manipulated or observed by the experimenter.

> Dependent variables are outcomes in response to changes or differences in the independent variable.

> Confounding variables systematically vary with the independent variable, so we cannot logically tell which variable may have influenced the dependent variable.

> Researchers control factors that are not of interest in order to explore the relation between an independent variable and a dependent variable.

> A variable is useful only if it is both reliable (consistent over time) and valid (assesses what it is intended to assess).

Clarifying the Concepts

1-7 The __________ variable predicts the __________ variable.

Calculating the Statistics

1-8 A researcher examines the effects of two variables on memory. One variable is beverage (caffeine or no caffeine) and the other is the subject to be remembered (numbers, word lists, aspects of a story).

a. Identify the independent and dependent variables.

b. How many levels do the variables of "beverage" and "subject to be remembered" have?

Applying the Concepts

1-9 Let's say you wanted to study the impact of declaring a major on school-related anxiety among first-year university students. You recruit 50 students who have not declared a major and 50 students who have declared a major. You have all 100 students complete an anxiety measure.

a. What is the independent variable in this study?

b. What are the levels of the independent variable?

c. What is the dependent variable?

d. What would it mean for the anxiety measure to be reliable?

e. What would it mean for the anxiety measure to be valid?

Solutions to these Check Your Learning questions can be found in Appendix D.

Introduction to Hypothesis Testing

When John Snow suggested that the pump handle be removed from the Broad Street well, he was testing his idea that an independent variable (contaminated well water) led to a dependent variable (deaths from cholera). Behavioral scientists use research to test ideas through a specific statistics-based process called *hypothesis testing.* More formally, ***hypothesis testing*** *is the process of drawing conclusions about whether a particular relation between variables is supported by the evidence.* Typically, we examine data from a sample to draw conclusions about a population. There are many ways to conduct research. In this section, we discuss the process of determining our variables, two different ways to approach research, and two different experimental designs.

Determining what breed of dog you most resemble might seem silly; however, adopting a dog is a very important decision. Can an online quiz such as "Which Dog Is Right for You?" help (2009, http://www.lifescript.com/Quizzes/Pets/Which_Dog_Is_Right_For_You.aspx)? We could conduct a study by having 30 people choose

- An **operational definition** specifies the operations or procedures used to measure or manipulate a variable.
- A **correlation** is an association between two or more variables.
- In **random assignment,** every participant in a study has an equal chance of being assigned to any of the groups, or experimental conditions, in the study.
- An **experiment** is a study in which participants are randomly assigned to a condition or level of one or more independent variables.

a type of dog to adopt, and have another 30 people let the quiz dictate their choice. We would then have to decide how to measure the outcome.

*An **operational definition** specifies the operations or procedures used to measure or manipulate a variable.* We could operationalize a good outcome with a new dog in several ways. Did you keep the dog for more than a year? On a rating scale of satisfaction with your pet, did you get a high score? Does a veterinarian give a high rating to your dog's health?

Do you think a quiz would lead you to make a better choice in dogs? You might hypothesize that the quiz would lead to better choices because it makes you think about important factors in dog ownership, such as outdoor space, leisure time, and your tolerance for dog hair. You already carry many hypotheses like these in your head. You just haven't bothered to test most of them yet, at least not formally. For example, perhaps you believe that North Americans use bank ATMs faster than Europeans do, or that smokers simply lack the willpower to stop. Maybe you are convinced that the parking problem on your campus is part of an uncaring conspiracy by administrators to make your life more difficult. In each of these cases, as shown in the accompanying table, we frame a hypothesis in terms of an independent variable and a dependent variable. The best way to learn about operationalizing a variable is to experience it for yourself. So propose a way to measure each of the variables identified in Table 1-2. We've given you a start with "continent"—North America versus Europe (easy to operationalize)—and how bad the parking problem is (more difficult to operationalize).

Conducting Experiments to Control for Confounding Variables

Once we have decided how to operationalize the variables, we can conduct a study and collect data. There are several different ways to approach research, including experiments and correlational research. A ***correlation*** *is an association between two or more variables.* In Snow's cholera research, it was the idea of a systematic co-relation between two variables (the proximity to the Broad Street well and the number of deaths) that saved so many lives. A correlation is one way to test a hypothesis, but it is not the only way. In fact, when possible, researchers prefer to conduct an experiment rather than a correlational study because it is easier to interpret the results of experiments.

TABLE 1-2. Operationalized Variables

The Independent Variable . . .	Predicts . . .	the Dependent Variable
Continent	____________	who uses ATMs the fastest
Amount of willpower	____________	level of cigarette smoking
Level of caring by administrators	____________	how bad the parking problem is

Conceptual Variable	Operationalized Variable
Continent	North America versus Europe
Who uses ATMs the fastest	____________
Amount of willpower	____________
Level of cigarette smoking	____________
Level of caring by administrators	____________
How bad the parking problem is	Ask students to rate the problem on a scale ranging from 1 (no problem) to 5 (worst problem on campus)

The hallmark of experimental research is random assignment. *With **random assignment**, every participant in the study has an equal chance of being assigned to any of the groups, or experimental conditions, in the study.* And *an **experiment** is a study in which participants are randomly assigned to a condition or level of one or more independent variables.* Random assignment means that neither the participants nor the researchers get to choose the condition. Experiments are the gold standard of hypothesis testing because they are the best way to control confounding variables. Controlling confounding variables allows researchers to infer a cause–effect relation between variables, rather than merely a systematic association between variables. Even when researchers cannot conduct a true experiment, they include as many of the characteristics of an experiment as possible. The critical feature that makes a study worthy of the descriptor *experiment* is random assignment to groups.

Experiments create equality between groups by randomly assigning participants to different levels, or conditions, of the independent variable. Random assignment controls the effects of personality traits, life experiences, personal biases, and other potential confounds by distributing them across each condition of the experiment to an equivalent degree.

EXAMPLE 1.1

It is difficult to control confounding variables, so let's see how random assignment helps to do that. You might wonder whether the hours you spend playing Tetris or Call of Duty are useful. A team of physicians and a psychologist investigated whether video game playing (the independent variable) leads to superior surgical skills (the dependent variable). They reported that surgeons with more video game playing experience were faster and more accurate, on average, when conducting training drills that mimic laparoscopic surgery (a surgical technique that uses a small incision, a small video camera, and a video monitor) than surgeons with no video game playing experience (Rosser et al., 2007).

In the video game and surgery study, the researchers did not randomly assign surgeons to play video games or not. Rather, they asked surgeons to report their video game playing histories and then measured their laparoscopic surgical skills. Can you spot the confounding variable? People may choose to play video games *because* they already have the fine motor skills and eye–hand coordination necessary for surgery, and they enjoy using their skills by playing video games. If that is the case, then, of course, those who play video games will tend to have better surgical skills—they already did before they took up video games!

It would be much more useful to set up an experiment that randomly assigns surgeons to one of the two levels of the independent variable: (1) play video games or (2) do not play video games. Random assignment assures us that our two groups are roughly equal, on average, on all the variables that might contribute to excellent surgical skills, such as fine motor skills, eye–hand coordination, and experience playing other video games. Random assignment attempts to diminish the effects of potential confounds. Specifically, random assignment to groups increases our confidence that the two groups were similar, on average, on aptitude for laparoscopic surgery *prior* to this experiment. (Figure 1-2 visually clarifies the difference between self-selection and random assignment. We explore more specifically how random assignment is implemented in Chapter 5.) If we use random assignment and the "play video games" group has better average laparoscopic surgical skills after the experiment than the "do not play video games group," then we can conclude that playing video games caused the better surgical skills.

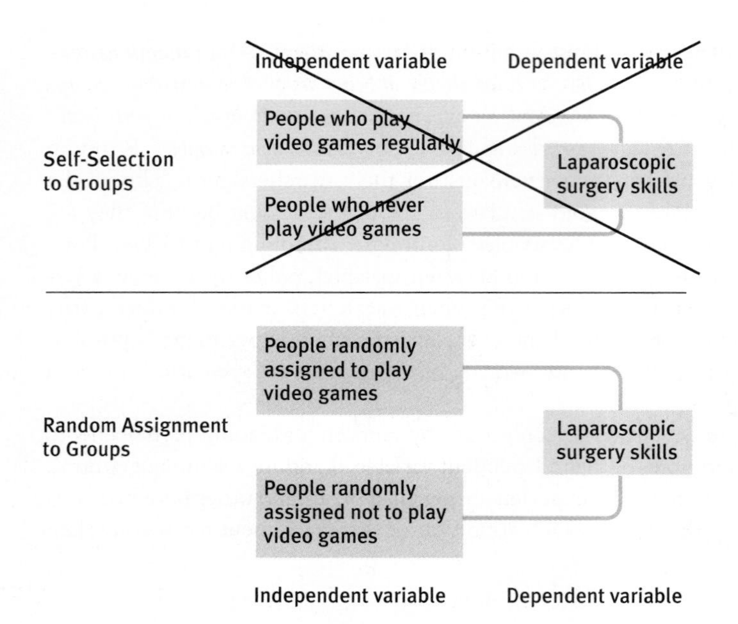

FIGURE 1-2
***Self-Selected* into or *Randomly Assigned* to One of Two Groups: Video Game Players versus Non–Video Game Players.**
This figure visually clarifies the difference between self-selection and random assignment. The design of the first study does not answer the question "Does playing video games improve laparoscopic surgical skills?"

Indeed, many researchers have used experimental designs to explore the causal effects of video game playing. They have found both positive effects, such as improved spatial skills following action games (Feng, Spence, & Pratt, 2007), and negative effects, such as increased hostility after playing violent games with lots of blood (Bartlett, Harris, & Bruey, 2008). ■

Between-Groups Design versus Within-Groups Design

Experiments create meaningful comparison groups in several ways. However, most studies have either a *between-groups* research design or a *within-groups* (also called a *repeated-measures*) research design.

*A **between-groups research design** is an experiment in which participants experience one, and only one, level of the independent variable.* In some between-groups studies, the different levels of the independent variable serve as the only relevant distinction between two (or more) groups that otherwise have been made equivalent through random assignment. An experiment that compares a control group (such as people randomly assigned not to play video games) with an experimental group (such as people randomly assigned to play video games) is an example of a between-groups design.

*A **within-groups research design** is an experiment in which all participants in the study experience the different levels of the independent variable.* An experiment that compares the same group of people before and after they experience a level of an independent variable, such as video game playing, is an example of a within-groups design. The word *within* emphasizes that if you experience one condition of a study, then you remain within the study until you experience all conditions of the study.

- In a **between-groups research design,** participants experience one, and only one, level of the independent variable.
- In a **within-groups research design,** all participants in the study experience the different levels of the independent variable; also called a *repeated-measures design.*

Many applied questions in the behavioral sciences are best studied using a within-groups design. This is particularly true of long-term (often called *longitudinal*) studies that examine how individuals and organizations change over time, or studies involving a naturally occurring event that cannot be duplicated in the laboratory. For example, we obviously cannot randomly assign people to either experience or not experience a hurricane. However, we could use nature's predictability to anticipate hurricane season, collect "before" data, and then collect data once again "after" people experience a hurricane. Such a before/after study is one version of a within-groups design.

Correlational Research

Often, we cannot conduct an experiment because it is unethical or impractical to randomly assign participants to conditions. Snow's cholera research, for example, did not use random assignment; he could not randomly assign some people to drink water from the Broad Street well. His research design was correlational, not experimental.

In correlational studies, we do not manipulate either variable. We merely assess the two variables as they exist. For example, it would be difficult to randomly assign people to either play or not play video games over several years. However, we could observe people over time to see the effects of their actual video game usage. Möller and Krahé (2009) studied German teenagers over a period of 30 months and found that the amount of video game playing when the study started was related to aggression 30 months later. Although these researchers found that video game playing and aggression are related, they do not have evidence that playing video games *causes* aggression. As we will discuss in Chapter 13, there are always alternative explanations in a correlational study; don't be too eager to infer causality just because two variables are correlated.

> **MASTERING THE CONCEPT**
>
> **1.5:** When possible, researchers prefer to use an experiment rather than a correlational study. Experiments use random assignment, which is the only way to determine if one variable causes another.

CHECK YOUR LEARNING

Reviewing the Concepts

- Hypothesis testing is the process of drawing conclusions about whether a particular relation between variables (the hypothesis) is supported by the evidence.
- All variables need to be operationalized—that is, we need to specify how they are to be measured or manipulated.
- Experiments attempt to explain a cause–effect relation between an independent variable and a dependent variable.
- Random assignment to groups, to control for confounding variables, is the hallmark of an experiment.
- Most studies have either a between-groups design or a within-groups design.
- Correlational studies can be used when it is not possible to conduct an experiment.

Clarifying the Concepts

1-10 How do the two types of research discussed in this chapter—experimental and correlational—differ?

1-11 How does random assignment help to address confounding variables?

Calculating the Statistics

1-12 College admissions offices use several methods to operationalize the academic performance of high school students applying to college, including SAT scores. Can you think of other ways to operationalize this variable?

Applying the Concepts

1-13 Expectations matter. Researchers examined how expectations based on stereotypes influence women's math performance (Spencer, Steele, & Quinn, 1999). Some women

continued on next page

were told that a gender difference was found on a certain math test and that women tended to receive lower scores than men did. Other women were told that no gender differences were evident on the test. Women in the first group performed more poorly than men did, on average, whereas women in the second group did not.

a. Briefly outline how researchers could conduct this research as a true experiment using a between-groups design.
b. Why would researchers want to use random assignment?
c. If researchers did not use random assignment but rather chose people who were *already* in those conditions (i.e., who already either believed or did not believe the stereotypes), what might be the possible confounds? Name at least two.
d. How is math performance operationalized here?
e. Briefly outline how researchers could conduct this study using a within-groups design.

Solutions to these Check Your Learning questions can be found in Appendix D.

REVIEW OF CONCEPTS

The Two Branches of Statistics

Statistics is divided into two branches: descriptive statistics and inferential statistics. *Descriptive statistics* organize, summarize, and communicate large amounts of numerical information. *Inferential statistics* draw conclusions about larger populations based on smaller samples of that population. *Samples* are intended to be representative of the larger *population*.

How to Transform Observations into Variables

Observations may be described as either discrete or continuous. *Discrete observations* are those that can take on only certain numbers (e.g., whole numbers, such as 1), and *continuous observations* are those that can take on all possible numbers in a range (e.g., 1.68792). Two types of *variables* can only be discrete: nominal and ordinal. *Nominal variables* use numbers simply to give names to scores. *Ordinal variables* are rank-ordered. Two types of variables can be continuous: interval and ratio. *Interval variables* are those in which the distances between numerical values are assumed to be equal. *Ratio variables* are those that meet the criteria for interval variables but also have a meaningful zero point. *Scale variable* is a term used for both interval and ratio variables, particularly in statistical computer programs.

Variables and Research

Independent variables can be manipulated or observed by the experimenter, and they have at least two *levels,* or conditions. *Dependent variables* are outcomes in response to changes or differences in the independent variable. *Confounding variables* systematically vary with the independent variable, so we cannot logically determine which variable may have influenced the dependent variable. The independent and dependent variables allow researchers to test and explore the relations between variables. A variable is only useful if it is both *reliable* and *valid*. A reliable measure is one that is consistent, and a valid measure is one that assesses what it intends to assess.

Introduction to Hypothesis Testing

Hypothesis testing is the process of drawing conclusions about whether a particular relation between variables is supported by the evidence. *Operational definitions* of the independent and dependent variables are necessary to test a hypothesis. *Experiments* attempt to identify a cause–effect relation between an independent variable and a dependent variable. *Random assignment* to groups, to control for confounding variables, is the hallmark of an experiment. Most studies have either a *between-groups design* or a *within-groups design*. *Correlational studies* can be used when it is not possible to conduct an experiment; they allow us to determine whether there is a predictable relation between two variables.

SPSS®

SPSS is divided into two main screens. The easiest way to move back and forth between these two screens is by using the two tabs located at the lower left labeled "Variable View" and "Data View."

To name the variables, go to "Variable View" and select:

Name. Type in a short version of the variable name—for example, BDI for the Beck Depression Inventory, a common measure of depressive symptoms.

Type. For nominal variables, such as gender, change the type to "String" by clicking the cell in the column labeled "Type" or by clicking the little gray box, choosing "String," and clicking "OK."

To tell SPSS what the variable name means, select:

Label. Type in the full name of the variable, such as Beck Depression Inventory.

To tell SPSS what the numbers assigned to any nominal variable actually mean, select:

Values. In the column labeled "Values," click on the cell next to the appropriate variable, then click on the little gray box on the right of that cell to access the tool that allows you to identify the values (or levels) of the variables. For example, if the nominal variable "Gender" is part of the study, tell SPSS that 1 equals male and 2 equals female. The numbers are the values and the words are the labels. See the accompanying screenshot to see what this looks like.

Now tell SPSS what kind of variables these are by selecting:

Measure. Highlight the type of variable by clicking on the cell in the column labeled "Measure" next to each variable, then clicking on the arrow to access the tool that allows you to identify whether the variable is scale, ordinal, or nominal. Notice that this is not necessary for nominal variables if the type is already listed as "String."

After describing all the variables in the study in "Variable View," switch over to "Data View." The information you entered was automatically transferred to that screen, but now the variables are displayed across the tops of the columns instead of along the left-hand side of the rows. You can now enter the data in "Data View" under the appropriate heading; each participant's data are entered across one row.

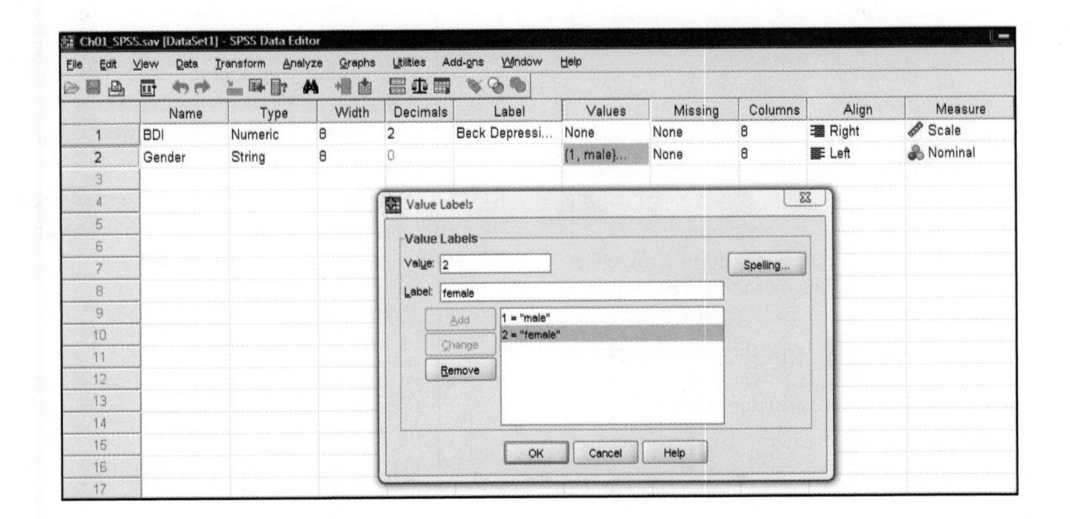

Exercises

Clarifying the Concepts

1.1 What is the difference between descriptive and inferential statistics?

1.2 What is the difference between a sample and a population?

1.3 Identify and define the four types of variables that researchers could use to quantify their observations.

1.4 Describe two ways that statisticians might use the word *scale.*

1.5 Distinguish between discrete and continuous variables.

1.6 What is the relation between an independent variable and a dependent variable?

1.7 What are confounding variables (or simply *confounds*) and how are they controlled using random assignment?

1.8 What is the difference between reliability and validity, and how are the two concepts related?

1.9 To test a hypothesis, we need operational definitions of our independent and dependent variables. What is an operational definition?

1.10 In your own words, define the word *experiment*—first as you would use it in everyday conversation, and then as a researcher would use it.

1.11 What is the difference between experimental research and correlational research?

1.12 What is the difference between a between-groups research design and a within-groups research design?

1.13 In statistics, it is important to pay close attention to language. The following statements are wrong but can be corrected by substituting one word or phrase. For example, the sentence "Only correlational studies can tell us something about causality" could be corrected by changing "correlational studies" to "experiments." Identify the incorrect word or phrase in each of the following statements, and supply the correct word.

a. In a study on exam preparation, every participant had an equal chance of being assigned to study alone or with a group. This was a correlational study.

b. A psychologist was interested in studying the effects of the dependent variable of caffeine on hours of sleep, and she used a scale measure for sleep.

c. A university assessed the reliability of a commonly used scale—a mathematics placement test—to determine if it was truly measuring math ability.

d. In a within-groups experiment on calcium and osteoporosis, participants were assigned to one of two levels of the independent variable: no change in diet or supplementing the diet with calcium.

1.14 The following statements are wrong but can be corrected by substituting one word or phrase. (See the instructions in Exercise 1.13.) Identify the incorrect word or phrase in each of the following statements, and supply the correct word.

a. A researcher examined the effect of the ordinal variable "gender" on the scale variable "hours of reality television watched per week."

b. A psychologist used a between-groups design to study the effects of an independent variable—a workout video—on the dependent variable—the weight—of a group of undergraduate students before and after viewing the video.

c. In a study on the effects of the confounding variable of noise level on the dependent variable of memory, researchers were concerned that the memory measure was not valid.

d. A researcher studied a population of 20 rats to determine whether changes in exposure to light would lead to changes in the dependent variable of amount of sleep.

Calculating the Statistics

1.15 Over the course of 1 week, a grocery store randomly selected 100 customers to complete a survey about their favorite products. Identify the sample and population for this example.

1.16 A researcher studies the average distance that 130 people living in U.S. urban areas walk each week.

a. What is the size of the sample?

b. Identify the population.

c. Is this "average" a descriptive statistic or an inferential statistic if it is used to describe the 130 people studied?

d. How might you operationalize the average distance walked in 1 week as an ordinal measure?

e. How might you operationalize the average distance walked in 1 week as a scale measure?

1.17 Seventy-three people are stopped as they leave a popular grocery store, and the number of fruit and vegetable items they purchased is counted.

a. What is the size of the sample?

b. Identify the population.

c. Is this number of items counted a descriptive statistic or an inferential statistic if it is used to estimate the diets of all shoppers?

d. How might you operationalize the amount of fruit and vegetable items purchased as a nominal measure?

e. How might you operationalize the amount of fruit and vegetable items purchased as an ordinal measure?

f. How might you operationalize the amount of fruit and vegetable items purchased as a scale measure?

1.18 In the fall of 2008, the U.S. stock market plummeted several times, with grave consequences for the world economy. A researcher might assess the economic effect this situation had by seeing how much money people saved in 2009. That amount could be compared to the amount people saved in more economically stable years. How might you operationalize the economic implications at a national level?

1.19 A researcher might be interested in evaluating how the physical and emotional "distance" a person had from Manhattan at the time of the 9/11 terrorist attacks relates to the accuracy of memory about the event.

a. Identify the independent variables and the dependent variable.

b. Imagine that physical distance is assessed as within 100 miles or beyond 100 miles; also, imagine that emotional distance is assessed as knowing no one who was affected, knowing people who were affected but lived, and knowing someone who died in the attacks. How many levels do the independent variables have?

c. How might you operationalize the dependent variable, accuracy of memory about the event?

1.20 A study of the effects of skin tone (light, medium, and dark) on the severity of facial wrinkles in middle age might be of interest to cosmetic surgeons.

a. What would the independent variable be in this study?

b. What would the dependent variable be in this study?

c. How many levels would the independent variable have?

Applying the Concepts

1.21 Average weights in the United States: The CDC reported very large weight increases for U.S. residents of both genders and all age groups over the last four decades. Go to the Web site that reports these data (www.cdc.gov) and search for the article titled "Americans Slightly Taller, Much Heavier Than Four Decades Ago."

a. What were the average weights of 10-year-old girls in 1963 and in 2002?

b. Do you think the CDC weighed every girl in the United States to get these averages? Why would this not be feasible?

c. How does the average weight of 10-year-old girls in 2002 or the average weight of 10-year-old boys in 1963 represent both a descriptive and an inferential statistic?

1.22 Sample versus population in Norway: The Nord-Trøndelag health study surveyed more than 60,000 people in a Norwegian county and reported that "people who have gastrointestinal symptoms, such as nausea, are more likely to have anxiety disorders or depression than people who do not have such symptoms."

a. What is the sample used by these researchers?

b. What is the population to which the researchers would like to extend their findings?

1.23 Types of variables and Olympic swimming: At the 2008 Beijing Summer Olympics, 23-year-old Michael Phelps won eight gold medals, more gold medals than had ever been won before in a single Olympics. One of the events he won was the 200-meter butterfly. For each of the following examples, identify the type of variable—nominal, ordinal, or scale.

a. Phelps of the United States came in first, László Cseh of Hungary came in second, and Takeshi Matsuda of Japan came in third.

b. Phelps finished in 1 minute, 52.03 seconds, a new world record. Cseh finished in 1:52.70, and Matsuda finished in 1:52.97.

c. One might examine whether swimmers were impaired during the race or not. Phelps was blinded when his goggles filled with water. Neither Cseh nor Matsuda suffered any impairment.

1.24 Types of variables and the Kentucky Derby: The Kentucky Derby is perhaps the premier event in U.S. horse racing. For each of the following examples from the Derby, identify the type of variable—nominal, ordinal, or scale.

a. As racing fans, we would be very interested in the variable of finishing position. For example, a stunning upset took place in 2005 when Giacomo, a horse with 50-1 odds, won, followed by Closing Argument and then Afleet Alex.

b. We also might be interested in the variable of finishing time. Giacomo won in 2 minutes, 2.75 seconds.

c. If we were the betting type, we might examine the variable of payoffs. Giacomo was such a long shot that a \$2.00 bet on him to win paid an incredibly high \$102.60.

d. We might be interested in the history of the Derby and the demographic variables of jockeys, such as gender or race. For example, in the first 28 runnings of the Kentucky Derby, 15 of the winning jockeys were African American.

e. In the luxury boxes, high fashion reigns; we might be curious about the variable of hat wearing, observing how many women wear hats and how many do not.

1.25 Discrete versus continuous variables: For each of the following examples, state whether the scale variable is discrete or continuous.

a. The capacity, in terms of songs, of an iPod
b. The playing time of an individual song
c. The cost in cents to download a song legally
d. The number of posted reviews that a CD has on Amazon.com
e. The weight of an MP3 player

1.26 Reliability and validity: Go online and take the personality test found at www.outofservice.com/starwars. This test assesses your personality in terms of the characters from the original *Star Wars* series. (You may have to scroll down to get to the questions.)

a. What does it mean for a test to be reliable? Take the test a second time. Does it seem to be reliable?
b. What does it mean for a test to be valid? Does this test seem to be valid? Explain.
c. The test asks a number of demographic questions at the end, including "In what country did you spend most of your youth?" Can you think of a hypothesis that might have led the developers of this Web site to ask this question?
d. For your hypothesis in part (c), identify the independent and dependent variables.

1.27 Reliability, validity, and wine ratings: You may have been in a wine store and wondered just how useful those posted wine ratings are (they are usually rated on a scale from 50 to 100, with 100 being the top score). After all, aren't ratings subjective? Corsi and Ashenfelter (2001) studied whether wine experts are consistent. Knowing that the weather is the best predictor of price, the researchers wondered how well weather predicted experts' ratings. The variables used for weather included temperature and rainfall, and the variable used for wine experts' ratings was the number they assigned to each wine.

a. Name one independent variable. What type of variable is it? Is it discrete or continuous?
b. Name the dependent variable. What type of variable is it? Is it discrete or continuous?
c. How does this study reflect the concept of reliability?
d. Let's say that you frequently drink wine that has been rated highly by Robert Parker, one of this study's wine experts. His ratings were determined to be reliable, and you find that you usually agree with Parker. How does this observation reflect the concept of validity?

1.28 Operationalizing variables: For each of the following variables—both described at some point in this chapter—state (1) how the researcher operationalized the variable, and (2) one other way in which the researcher could have operationalized the variable.

a. The distance between the well and the homes where people had died (Dr. Snow's study)
b. The length of a woman's hair (Eleanor Stampone's study)

1.29 Identifying and operationalizing variables: For each of the following hypotheses, identify the independent variable and the most likely levels of that independent variable. Then identify the likely dependent variable and a likely way to operationalize that dependent variable. Be specific.

a. Teenagers are better at video games, on average, than are adults in their 30s.
b. Spanking children tends to lead them to be more violent.
c. Weight Watchers leads to more weight loss, on average, for people who go to meetings than for those who only participate online.
d. Students do better in statistics, on average, if they study with other people than if they study alone.
e. Drinking caffeinated beverages with dinner tends to make it harder to get to sleep.

1.30 Between-groups versus within-groups and exercise: Noting marked increases in weight across the population, researchers, nutritionists, and physicians have struggled to find ways to stem the tide of obesity in many Western countries. They have advocated a number of exercise programs, and there has been a flurry of research to determine the effectiveness of these programs. Pretend that you are in charge of a research study to examine the effects of an exercise program on weight loss in comparison with a program that does not involve exercise.

a. Describe how you could study the exercise program using a between-groups research design.
b. Describe how you could study the exercise program using a within-groups design.
c. What is a potential confound of a within-groups design?

1.31 Correlational research and smoking: For decades, researchers, politicians, and tobacco company executives debated the relation between smoking and health problems such as cancer.

a. Why was this research necessarily correlational in nature?
b. What confounding variables might make it difficult to isolate the health effects of smoking tobacco?

c. How might the nature of this research and these confounds "buy time" for the tobacco industry with regard to acknowledging the hazardous effects of smoking?

d. All ethics aside, how could you study the relation between smoking and health problems using a between-groups experiment?

1.32 Experimental versus correlational research and culture: A researcher interested in the cultural values of individualistic and collectivist societies collects data on the rate of relationship conflict experienced by 32 people who test high for individualism and 37 people who test high for collectivism.

a. Is this research experimental or correlational? Explain.

b. What is the sample?

c. Write a possible hypothesis for this researcher.

d. How might we operationalize relationship conflict?

1.33 Experimental versus correlational research and recycling: A researcher wants to know if people's concerns about the environment vary as a function of incentives provided for recycling. Students living on a university campus are recruited to participate in a study. Some students are randomly assigned to a group in which they are rewarded financially for their recycling efforts for one month. The other students are randomly assigned to a group in which they are assessed a fine that is based on the amount of material that they could have, but did not, recycle.

a. Is this research experimental or correlational? Explain.

b. Write a hypothesis for this researcher.

Putting It All Together

1.34 Romantic relationships: In a 2010 report, Goodman and Greaves (2010) reported findings from an analysis of data from participants in a large research project in the United Kingdom, the Millennium Cohort Study. They stated that "... while it is true that cohabiting parents are more likely to split up than married ones, there is very little evidence to suggest that this is due to a causal effect of marriage. Instead, it seems simply that different sorts of people choose to get married and have children, rather than to have children as a cohabiting couple, and that those relationships with the best prospects of lasting are the ones that are most likely to lead to marriage" (p. 1).

a. What is the sample in this study?

b. What is the likely population?

c. Is this a correlational study or an experiment? Explain.

d. What is the independent variable?

e. What is the dependent variable?

f. What is one possible confounding variable? Suggest at least one way in which the confounding variable might be operationalized.

1.35 Experiments, HIV, and cholera: Several studies have documented the susceptibility of people who are HIV-positive to cholera, likely because of weakened immune systems. Researchers in Mozambique (Lucas et al., 2005), a country where an estimated 20% to 30% of the population is HIV-positive, wondered whether an oral vaccine for cholera would work among people who are HIV-positive. Fourteen thousand people in Mozambique who tested positive for HIV were immunized against cholera. Soon thereafter, an epidemic of cholera spread through the region, giving the researchers an opportunity to test their hypothesis.

a. Describe a way in which the researchers could have conducted an experiment to examine the effectiveness of the cholera vaccine among people who are HIV-positive.

b. If the researchers did conduct an experiment, would this have been a between-groups or a within-groups experiment? Explain.

c. The researchers did not randomly assign participants to vaccine or no-vaccine conditions; rather, they conducted a general mass immunization. Why does this limit their ability to draw causal conclusions? Include at least one possible confounding variable.

d. The researchers did not use random assignment when conducting this study. List at least one practical reason and at least one ethical reason that they might not have used random assignment.

e. If we had been conducting the study described here, and were unconcerned with practicality and ethics, describe how we could have used random assignment.

1.36 Ability and wages: Arcidiacono, Bayer, and Hizmo (2008) analyzed data from a National Longitudinal Survey called NLSY79, which includes data from over 12,000 men and women in the United States who were in the 14- to 22-year age range in 1979. The researchers reported that ability is related to wages in early career jobs for university graduates, but not for high school graduates. In line with this, research has found that racial discrimination with respect to wages is more prevalent against high school graduates than college graduates, because when ability is not the primary reason for determining wages, other nonrelevant factors such as race play in. The researchers suggest that their findings might explain why, on average, a black person is more likely to earn a college degree than is a white person of the same ability level.

a. List any independent variables.
b. List any dependent variables.
c. What is the sample in this study?
d. What is the population about which researchers want to draw conclusions?
e. What do the authors mean by "Longitudinal" in this study?
f. The researchers used the Armed Forces Qualification Test (AFQT) as their measure of ability. The AFQT combines scores on word knowledge, paragraph comprehension, arithmetic reasoning, and mathematics knowledge subscales. Can you suggest at least one confounding variable in the relation between ability and wages when comparing college graduates to high school graduates?
g. Suggest at least two other ways in which the researchers might have operationalized ability.

Terms

descriptive statistic (p. 2)
inferential statistic (p. 3)
sample (p. 3)
population (p. 3)
variable (p. 4)
discrete observation (p. 4)
continuous observation (p. 4)
nominal variable (p. 4)
ordinal variable (p. 4)
interval variable (p. 4)
ratio variable (p. 5)
scale variable (p. 6)
level (p. 7)
independent variable (p. 7)
dependent variable (p. 7)
confounding variable (p. 7)
reliability (p. 8)
validity (p. 8)
hypothesis testing (p. 9)
operational definition (p. 10)
correlation (p. 10)
random assignment (p. 11)
experiment (p. 11)
between-groups research design (p. 12)
within-groups research design (p. 12)

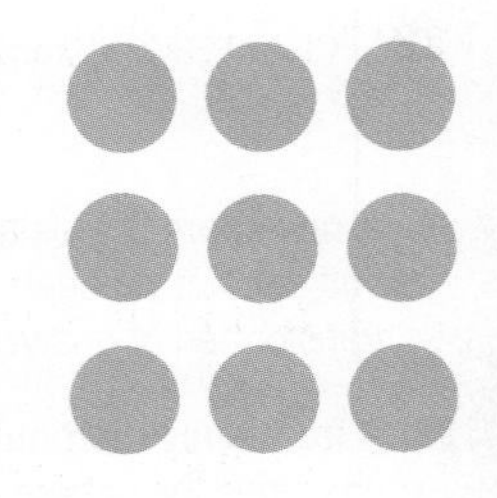

CHAPTER 2

Frequency Distributions

Frequency Distributions
- Frequency Tables
- Grouped Frequency Tables
- Histograms
- Frequency Polygons

Shapes of Distributions
- Normal Distributions
- Skewed Distributions

BEFORE YOU GO ON

- You should understand the different types of variables—nominal, ordinal, and scale (Chapter 1).
- You should understand the difference between a discrete variable and a continuous variable (Chapter 1).

- A **raw score** is a data point that has not yet been transformed or analyzed.
- A **frequency distribution** describes the pattern of a set of numbers by displaying a count or proportion for each possible value of a variable.
- A **frequency table** is a visual depiction of data that shows how often each value occurred, that is, how many scores were at each value. Values are listed in one column, and the numbers of individuals with scores at that value are listed in the second column.

It has been suggested that children who are exposed to fast-paced television programming—quick camera changes, lots of sound effects, multiple plots—have more difficulty with learning and tend to be less imaginative (Healy, 1990). In addition, the popular children's program *Sesame Street* has also been criticized for its fast pacing, which critics believe encourages children to love television but not to love learning (Postman, 1985).

To understand the effects of pacing, researchers created a list that reported the pacing scores for 87 popular children's television programs broadcast in the United States (McCollum & Bryant, 2003). Table 2-1 depicts an excerpt of these pacing scores; you can see that *Mr. Rogers' Neighborhood* was the slowest-paced show (not surprising to those who have seen it), with a pacing score of 14.95. The fastest-paced show was *Bill Nye the Science Guy,* with a pacing score of 56.90.

Do fast-paced programs harm children's ability to learn? The jury is still out. Even though *Sesame Street* is slow-paced relative to other shows, some researchers still regard it as too fast-paced to achieve its educational goals (Schmidt & Vandewater, 2008). However, the researchers who developed the pacing index discovered some other interesting information when they averaged the pacing of children's television shows across different networks. The list in Table 2-2, in order of rank, shows that commercial networks produced the fastest-paced shows and that educational television produced the slowest-paced shows. Perhaps fast pacing helps commercial networks win the competition for viewers' eyes, whereas slow pacing wins the competition for viewers' minds.

In this chapter, we learn how to organize individual data points in a table. Then we go one step further and learn how to use two types of graphs—histograms and frequency polygons—to show the overall pattern of data. Finally, we learn to use these graphs to understand the shape of the distribution of the data points. These tools are important steps for using statistics in the behavioral sciences.

TABLE 2-1. The Pacing of Children's Television Shows

The fast pace of many children's television programs has been criticized for possibly lowering children's ability to concentrate. This table shows a sample of television programs and a pacing index for each one. A higher index indicates a faster-paced program, and a lower index indicates a slower-paced program.

Television Show	Pacing Index
Bill Nye the Science Guy	56.90
Power Rangers	41.90
Tiny Toons	40.70
Charlie Brown	33.10
Scooby Doo	30.60
The Simpsons	30.25
Batman	25.85
Sesame Street	24.80
Blue's Clues	21.85
Mr. Rogers' Neighborhood	14.95

Data from McCollum & Bryant (2003)

TABLE 2-2. The Pacing of Children's Shows by Network

When children's programs were categorized by network and averaged, the data revealed that the fastest-paced programs were offered by commercial networks and the slowest-paced programs were offered by educational networks.

Network	Average Pacing Index
Commercial networks	34.29
Nickelodeon	33.03
Disney	32.55
Public Broadcasting System (PBS)	27.26
The Learning Channel	25.35
Average for all shows	31.86

Data from McCollum & Bryant (2003)

Frequency Distributions

Researchers are usually most interested in the relations between two or more variables, such as the effect of a television show's pacing (independent variable) on children's learning (dependent variable). But to understand the relation between variables, we must first understand each individual variable's data points. The basic ingredients of a data set are called ***raw scores****, data that have not yet been transformed or analyzed.* In statistics, we organize raw scores into a ***frequency distribution****,* which *describes the pattern of a set of numbers by displaying a count or proportion for each possible value of a variable.* For example, a frequency distribution can display the pattern of the scores—the pacing indices—from the excerpted list of television shows in Table 2-1.

There are several different ways to organize the data in terms of a frequency distribution. The first approach, the frequency table, is also the starting point for each of the three other approaches that we will explore. A ***frequency table*** *is a visual depiction of data that shows how often each value occurred; that is, how many scores were at each value.* Once organized into a frequency table, data can be displayed as a grouped frequency table, a histogram, or a frequency polygon.

MASTERING THE CONCEPT

2.1: A frequency table shows the pattern of the data by indicating how many participants had each possible score. The data in a frequency table can be graphed in a frequency histogram or a frequency polygon.

Frequency Tables

EXAMPLE 2.1

The most popular sport in the world is soccer (called football by most people in the world), and a recent book analyzes soccer from the perspectives of several social sciences—statistics, economics, psychology, geography, and sociology. In *Soccernomics,* the authors explore fascinating social science questions (Kuper & Szymanski, 2009). They also present data about where soccer is most popular. Using data on percentages of soccer spectators out of the entire population, they conclude that soccer is most popular in England, followed by Spain, Germany, Italy, and France (in that order). But we wondered: Does popularity coincide with success?

Table 2-3 depicts data from the World Cup Web site (http://www.fifa.com), listing the years in which countries came in first or second in the tournament. The table is in alphabetical order by country. Of the 77 countries that have participated in at least one men's or women's World Cup tournament, only 14 countries have placed first or

World Cup Powerhouses
Enthusiasm for women's soccer in the United States continued to grow after U.S. team's gold-medal victory at the 2012 Olympic Games, especially after losing to an inspired Japanese team in the 2011 World Cup. Like Olympic gold, winning the World Cup is a career-capping ambition. Frequency data from the FIFA website (through 2007) show that some countries have far more top finishes than others. For example, West Germany/Germany had 10 first- or second-place finishes and Brazil had 8. These two teams are shown here playing each other in the 2007 Women's World Cup.

AP Photo/Anja Niedringhaus

second, an indication that some countries dominate. The remaining 63 countries never finished in first or second place. We can use these data to create a frequency table.

At first glance, it is not easy to find a pattern in most lists of numbers. But when we reorder those numbers, a pattern begins to emerge. A frequency table is the best way to create an easy-to-understand distribution of data. In this example, we simply organize the data into a table with two columns, one for the range of responses (the values) and one for the frequencies of each response (the scores).

TABLE 2-3. World Cup Success

This table shows the years in which countries finished in first or second place in the history of the men's and women's World Cup in soccer through 2007. The men's tournament has been held every 4 years since 1930 (except for 1942 and 1946, due to World War II); the women's tournament has been held every 4 years since 1991.

Country	Men First Place	Men Second Place	Women First Place	Women Second Place
Argentina	1978, 1986	1930, 1990		
Brazil	1958, 1962, 1970, 1994, 2002	1950, 1998		2007
China				1999
Czechoslovakia		1934, 1962		
England	1966			
France	1998	2006		
Hungary		1938, 1954		
Italy	1934, 1938, 1982, 2006	1970, 1994		
Norway			1995	1991
Sweden		1958		2003
The Netherlands		1974, 1978		
United States			1991, 1999	
Uruguay	1930, 1950			
West Germany/Germany	1954, 1974, 1990	1966, 1982, 1986, 2002	2003, 2007	1995

Data from www.fifa.com (2012)

There are specific steps to create a frequency table. First, we determine the range of raw scores. For each country, we can count how many first- or second-place finishes these countries have had: 4, 8, 1, 2, 1, 2, 2, 6, 2, 2, 2, 2, 2, and 10. In addition, 63 countries had 0 first- or second-place finishes. We know at a glance that the lowest score is 0. A quick glance also reveals that the highest score is 10; one country finished in first or second place in 10 World Cup tournaments, a most impressive number. Simply noting that the scores range from 0 to 10 brings some clarity to the data set. But we can do even better.

After we identify the lowest and highest scores, we create the two columns that we see in Table 2-4. We examine the raw scores and determine how many countries fall at each value in the range. The appropriate number for each value is recorded in the table. For example, there is one country with 10 first- or second-place finishes, so a 1 is marked there. It is important to note that we include *all* numbers in the range; there are no countries with 9, 7, 5, or 3 top finishes, so we put a 0 next to each one.

TABLE 2-4. Frequency Tables and World Cup Success

This frequency table depicts the numbers of countries that came in first or second in the history of the men's and women's World Cup soccer tournaments. Do there seem to be some stand-out countries?

First- or Second-Place Finishes	Frequency
10	1
9	0
8	1
7	0
6	1
5	0
4	1
3	0
2	8
1	2
0	63

Data from www.fifa.com (2012)

Here is a recap of the steps to create a frequency table:

1. Determine the highest score and the lowest score.
2. Create two columns; label the first with the variable name, and label the second "Frequency."
3. List the full range of values that encompasses all the scores in the data set, from highest to lowest. Include *all* values in the range, even those for which the frequency is 0.
4. Count the number of scores at each value, and write those numbers in the frequency column.

As shown in Table 2-5, we can add a column for percentages. To calculate a percentage, we divide the number of countries at a certain value by the total number of countries, and then multiply by 100. As we observed earlier, 1 out of 77 countries had 10 top finishes.

$$\frac{1}{77}(100) = 1.299$$

So, for the score of 10 top finishes, the percentage for 1 of 77 countries is 1.30%.

Note that when we calculate statistics, we can come up with different answers depending on the number of steps and how we decide to round numbers. In this book, we round off to three decimal places throughout the calculations, but we report the final answers to two decimal places, rounding up or down as

TABLE 2-5. Expansion of a Frequency Table

This frequency table is an expansion of Table 2-4, which depicts numbers of countries that came in first or second in the history of the men's and women's World Cup soccer tournaments. It now includes percentages, which are often more descriptive than actual counts.

First- or Second-Place Finishes	Frequency	Percentage
10	1	1.30
9	0	0.00
8	1	1.30
7	0	0.00
6	1	1.30
5	0	0.00
4	1	1.30
3	0	0.00
2	8	10.39
1	2	2.60
0	63	81.82

Data from www.fifa.com (2012)

▪ A **grouped frequency table** is a visual depiction of data that reports frequencies within a given interval rather than the frequencies for a specific value.

appropriate. Sometimes the numbers don't add up to 100% exactly, due to rounding. If you follow this guideline, then you should get the same answers that we get.

Creating a frequency table for the data gives us more insight into the set of numbers. We can see that two countries, Brazil and West Germany/Germany, are well above the others. Indeed, the subtitle for *Soccernomics* includes the phrase *Why Germany and Brazil Win*. What about England, the country in which soccer is most popular? It's been one of the top two finishers only once, when it won in 1966. It seems clear that the popularity of the sport doesn't necessarily relate to World Cup success. ▪

Grouped Frequency Tables

In the previous example, we used data that counted the numbers of countries, which are whole numbers. In addition, the range was fairly limited—0 to 10. But often data are not so easily understood. Consider these two situations:

1. When data can go to many decimal places, such as in reaction times
2. When data cover a huge range, such as countries' populations

In both of these situations, the frequency table would go on for pages and pages. For example, if someone weighed only 0.0003 pound more than the person at the next weight, that first person would belong to a distinctive, unique category. Using such specific values would lead to two problems: we would be creating an enormous amount of unnecessary work for ourselves, and we wouldn't be able to see trends in the data. Fortunately, we have a technique to deal with these situations: *a* ***grouped frequency table*** *allows us to depict data visually by reporting the frequencies within a given interval rather than the frequencies for a specific value.* The word *interval* is used in more than one way by statisticians. Here, it refers to a range of values (as opposed to an interval variable, the type of variable that has equal distances between values).

EXAMPLE 2.2

The following data exemplify the first of these two situations in which the data aren't easily conveyed in a standard frequency table. These are the pacing indices, to two decimal places, for the 87 television shows, some of which are listed in Table 2-1.

56.90 50.30 46.70 45.95 45.75 44.65 43.25 42.20 41.95 41.90
41.80 40.80 40.70 40.25 40.25 39.10 37.80 37.55 37.00 36.25
36.00 35.90 35.55 35.55 35.50 35.40 34.30 34.00 33.85 33.75
33.55 33.10 32.85 32.75 32.55 32.50 32.40 32.25 31.85 31.60
31.45 31.10 31.00 31.00 30.70 30.65 30.60 30.40 30.30 30.25
30.20 29.85 29.85 29.30 29.30 29.30 29.20 29.20 28.95 28.70
28.55 28.50 28.45 28.20 28.10 27.95 27.55 27.45 27.05 27.05
26.95 26.95 26.75 26.25 25.85 25.35 25.15 25.15 24.80 23.35
23.10 21.85 20.60 19.90 16.50 15.75 14.95

A quick glance at these data does not really tell us the pacing index of the typical television show. A frequency table wouldn't be helpful either. The lowest score is 14.95 and the highest is 56.90. So the table would include 14.95, 14.96, 14.97, and so on, all the way to 56.90! Such a table would be absurdly long and would not convey much more useful information than does the list of the original raw data.

Instead of reporting every single value in the range, we can report intervals, or ranges of values. Here are the five steps to generate a standard grouped frequency table:

STEP 1: Find the highest and lowest scores in the frequency distribution.

In the pacing index example, these scores are 56.90 and 14.95.

STEP 2: Get the full range of data.

If there are decimal places, round both the highest and the lowest scores down to the nearest whole numbers. If they already are whole numbers, use these. Subtract the lowest whole number from the highest whole number and add 1 to get the full range of the data. (Why do we add 1? Try it yourself. If we subtract 14 from 56, we get 42—but count the values from 14 through 56, including the numbers at either end. There are 43 numbers, and we want to know the full range of the data.)

In the pacing index example, 14.95 and 56.90 round down to 14 and 56, respectively; 56 − 14 = 42, and 42 + 1 = 43. The scores fall within a range of 43.

STEP 3: Determine the number of intervals and the best interval size.

There is no consensus about the ideal number of intervals, but most researchers recommend between 5 and 10 intervals, unless the data set is enormous and has a huge range. To find the best interval size, we divide the range by the number of intervals we want, then round to the nearest whole number (as long as the numbers are not too small—that is to many decimal places). For ranges that are wide, the size of intervals could be a multiple of 10 or 100 or 1000; for smaller ranges, it could be as small as 2, 3, or 5, or even less than 1, if the numbers go to many decimal places. Try several interval sizes to determine the best one.

In the pacing index example, we might choose to have about 9 intervals. If we choose 9, the interval size will be 5.

STEP 4: Figure out the number that will be the bottom of the lowest interval.

We want the bottom of that interval to be a multiple of the interval size. For example, if we have 9 intervals of size 5, then we want the bottom interval to start at a multiple of 5. It could start at 0, 10, 55, or 105, depending on the data. We select the multiple of 5 that is below the lowest score.

In the pacing index example, there are 9 intervals of size 5, so the bottom of the lowest interval should be a multiple of 5. The lowest score is 14.95, so the bottom of the lowest interval would be 10. (If the lowest score were 7.22, we would choose 5. Note that this process might lead to one more interval than we planned for; this is perfectly fine. In our case, we have 10, rather than the 9 intervals we had estimated.)

STEP 5: Finish the table by listing the intervals from highest to lowest and then counting the numbers of scores in each.

This step is much like creating a frequency table (without intervals), which we discussed earlier. If we decide on intervals of size 5 and the first one begins at 10, then we count the five numbers that fall in this interval: 10, 11, 12, 13, and 14. The interval in this example runs from 10 to 14. (In reality, it runs from 10 to 14.9999, and the next one begins at 15, five digits higher than the bottom of the preceding interval.) A good rule of thumb is that the *bottom* of the intervals should jump by the chosen interval size, in this case 5.

TABLE 2-6. Grouped Frequency Table

Grouped frequency tables make sense of data sets in which there are many possible values. This grouped frequency table depicts 10 intervals that summarize the pacing indices for 87 children's TV programs.

Interval	Frequency
55.00–59.99	1
50.00–54.99	1
45.00–49.99	3
40.00–44.99	10
35.00–39.99	11
30.00–34.99	25
25.00–29.99	27
20.00–24.99	5
15.00–19.99	3
10.00–14.99	1

In the pacing index example, the lowest interval would be 10 to 14, or 10.00 to 14.99. The next one would be 15.00 to 19.99, and so on.

The grouped frequency table in Table 2-6 gives us a much better sense of the pacing indices of the TV shows in this sample than either the list of raw data or a frequency table without intervals. ■

Histograms

> **MASTERING THE CONCEPT**
>
> **2.2:** The data in a frequency table can be viewed in graph form. In a histogram, bars are used to depict frequencies at each score or interval. In a frequency polygon, a dot is placed above each score or interval to indicate the frequency, and the dots are connected.

Even more than tables, graphs help us to see data at a glance. The two most common methods for graphing scale data for one variable are the histogram and the frequency polygon. Here we learn to construct and interpret the histogram (more common) and the frequency polygon (less common).

Histograms *look like bar graphs but typically depict scale data, with the values of the variable on the x-axis and the frequencies on the y-axis.* Each bar reflects the frequency for a value or an interval. The difference between histograms and bar graphs is that bar graphs typically provide scores for nominal data (e.g., men and women), whereas histograms typically provide frequencies for scale data (e.g., levels of pacing indices). We can construct histograms from frequency tables or from grouped frequency tables. Histograms allow for the many intervals that typically occur with scale data. The bars are stacked one against the next, with the intervals meaningfully arranged from lower numbers (on the left) to higher numbers (on the right). With bar graphs, the categories do not need to be arranged in one particular order and the bars should not touch.

EXAMPLE 2.3

Let's start by constructing a histogram from a frequency table. Table 2-4 depicts data on countries' numbers of World Cup top finishes. We construct a histogram by drawing the x-axis (horizontal) and y-axis (vertical) of a graph. We label the x-axis with the variable of interest—in our case, "first- or second-place finishes"—and we label the

y-axis "frequency." As with most graphs, the lowest numbers start where the axes intersect and the numbers go up—as we go to the right on the x-axis and as we go up on the y-axis. Ideally, the lowest number on each axis is 0, so that the graphs are not misleading. However, if the range of numbers on either axis is far from 0, histograms sometimes use a number other than 0 as the lowest number. Further, if there are negative numbers among the scores (as can be the case in scores for air temperature), the x-axis could have negative numbers.

A **histogram** looks like a bar graph but is typically used to depict scale data with the values of the variable on the x-axis and the frequencies on the y-axis.

Once we've created the graph, we draw a bar for each value. Each bar is *centered on* the value for which it provides the frequency. The height of the bars represents the numbers of scores that fall at each value—the frequencies. If no country had a score at a particular value, no bar is drawn. So, for the value of 2 on the x-axis, a bar centers on 2 with a height of 8 on the y-axis, indicating that eight countries had a first- or second-place finish twice. Figure 2-1 shows the histogram for the World Cup data.

Here is a recap of the steps to construct a histogram from a frequency table:

1. Draw the x-axis and label it with the variable of interest and the full range of values for this variable. (Include 0 unless all of the scores are so far from 0 that this would be impractical.)
2. Draw the y-axis, label it "Frequency," and include the full range of frequencies for this variable. (Include 0 unless it's impractical.)
3. Draw a bar for each value, centering the bar around that value on the x-axis and drawing the bar as high as the frequency for that value, as represented on the y-axis.

Grouped frequency tables can also be depicted as histograms. Instead of listing values on the x-axis, we list the midpoints of intervals. Students commonly make mistakes in determining midpoints. If an interval ranges from 0 to 9, what is the midpoint? If you said 4.5, you're making a *very* common mistake. Remember, this interval really goes from 0.000000 to 9.999999, or as close as you can get to 10, the bottom of the next interval, without actually being 10. Given that there are 10 numbers in this range (0, 1, 2, 3, 4, 5, 6, 7, 8, and 9), the midpoint would be 5 from the bottom. So the midpoint

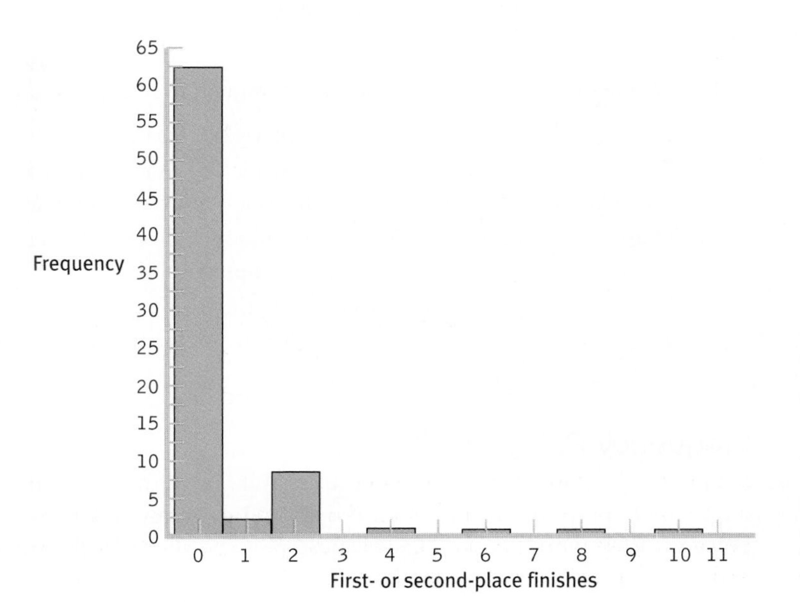

FIGURE 2-1
Histogram for the Frequency Table of World Cup Successes

Histograms are graphic depictions of the information in frequency tables or grouped frequency tables. This histogram shows how many countries had a certain number of first- or second-place finishes in the men's and women's World Cup soccer tournaments through 2007.

for 0 to 9 is 5. A good rule: when determining a midpoint, look at the bottom of the interval that you're interested in and then the bottom of the next interval; then, determine the midpoint of these two numbers. ■

EXAMPLE 2.4

Let's look at the TV pacing index data for which we constructed a grouped frequency histogram. What are the midpoints of the 10 intervals? Let's calculate the midpoint for the lowest interval, 10 to 14.99. We should look at the bottom of this interval, 10.00, and the bottom of the next interval, 15.00. The midpoint of these numbers is 12.50, so that is the midpoint of this interval. The remaining midpoints can be calculated the same way. For the highest interval, 55.00 to 59.99, it helps to imagine that we had one more interval. If we did, it would start at 60.00. The midpoint of 55.00 and 60.00 is 57.50. Using these guidelines, we calculate the midpoints as 12.50, 17.50, 22.50, 27.50, 32.50, 37.50, 42.50, 47.50, 52.50, and 57.50. (A good check is to see if the midpoints should jump by the interval size—in this case, 5.) We now can construct the histogram by placing these midpoints on the *x*-axis and drawing bars that center on them and are as high as the frequency for each interval. The histogram for these data is shown in Figure 2-2.

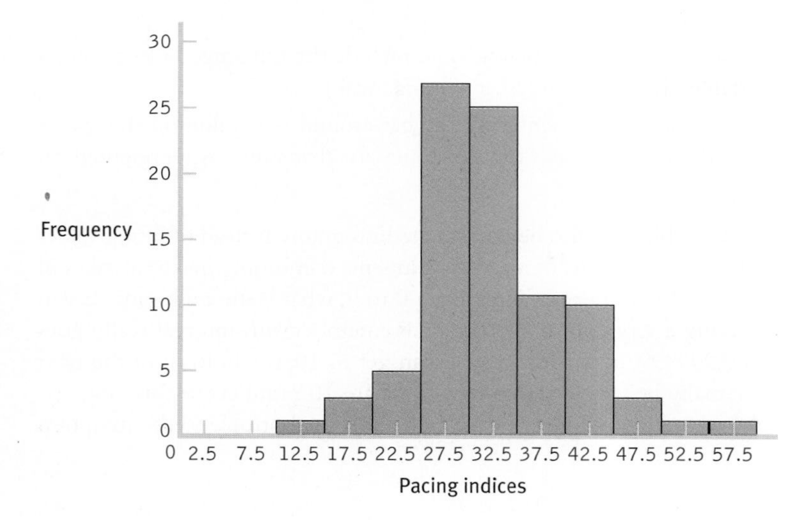

FIGURE 2-2
Histogram for the Grouped Frequency Table of TV Pacing Index Data

Histograms can also depict the data in a grouped frequency table. This histogram depicts the data seen in the grouped frequency table for TV pacing indices.

Here is a recap of the steps to construct a histogram from a grouped frequency table:

1. Determine the midpoint for every interval.
2. Draw the *x*-axis, label it with the variable of interest and with the midpoints for each interval on this variable. (Include 0 unless it's impractical.)
3. Draw the *y*-axis, label it "Frequency," and include the full range of frequencies for this variable. (Include 0 unless it's impractical.)
4. Draw a bar for each midpoint, centering the bar on that midpoint on the *x*-axis and drawing the bar as high as the frequency for that interval, as represented on the *y*-axis. ■

■ A **frequency polygon** is a line graph, with the *x*-axis representing values (or midpoints of intervals) and the *y*-axis representing frequencies; a dot is placed at the frequency for each value (or midpoint), and the dots are connected.

Frequency Polygons

Frequency polygons are constructed in a similar way to histograms. As the name might imply, polygons are many-sided shapes. Histograms look like city skylines, but polygons look more like mountain landscapes. Specifically, a ***frequency polygon*** *is a* *line graph, with the x-axis representing values (or midpoints of intervals) and the y-axis representing frequencies; a dot is placed at the frequency for each value (or midpoint), and the dots are connected.*

EXAMPLE 2.5

For the most part, we make frequency polygons exactly as we make histograms. Instead of constructing bars above each value or midpoint, however, we draw dots and connect them. The other difference is that we need to add an appropriate value (or midpoint) on either end of the graph so that we can draw lines down to 0, grounding the shape. In the case of the TV pacing data, we calculate one more midpoint on each end by subtracting the interval size, 5, from the bottom midpoint ($12.50 - 5 = 7.5$) and adding the interval size, 5, to the top midpoint ($57.5 + 5 = 62.5$). We now can construct the frequency polygon by placing these midpoints on the x-axis, drawing dots at each midpoint that are as high as the frequency for each interval, and connecting the dots. Figure 2-3 shows the frequency polygon for the grouped frequency distribution of TV pacing indices that we constructed previously in Figure 2-2.

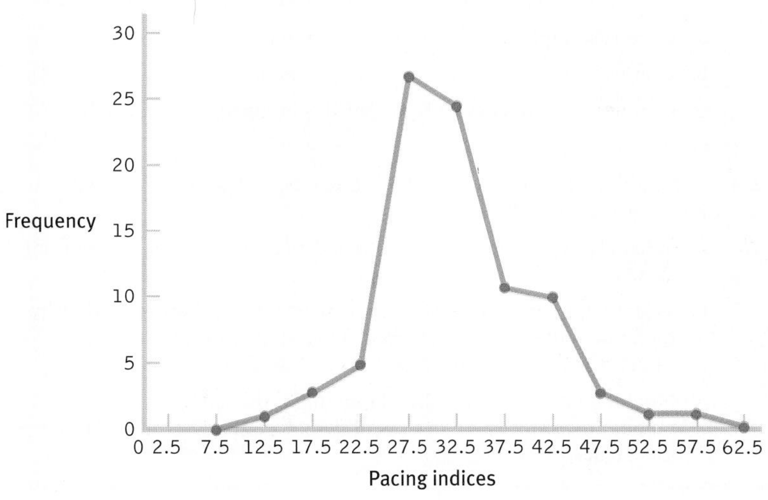

FIGURE 2-3
Frequency Polygon as Another Graphing Option for the TV Pacing Index Data

Frequency polygons are an alternative to histograms. This frequency polygon depicts the same data that were depicted in the histogram in Figure 2-2. In either case, the graph provides an easily interpreted "picture" of the distribution.

Here is a recap of the steps to construct a frequency polygon. When basing a frequency polygon on a frequency table, we place the specific values on the x-axis. When basing it on a grouped frequency table, we place the midpoints of intervals on the x-axis.

1. If based on a grouped frequency table, determine the midpoint for every interval. If based on a frequency table, skip this step.
2. Draw the x-axis and label it with the variable of interest and either the values or the midpoints. (Include 0 unless it's impractical.)
3. Draw the y-axis, label it "frequency," and include the full range of frequencies for this variable. (Include 0 unless it's impractical.)
4. Mark a dot above each value or midpoint depicting the frequency, as represented on the y-axis, for that value or midpoint, and connect the dots.
5. Add an appropriate hypothetical value or midpoint on both ends of the x-axis, and mark a dot for a frequency of 0 for each of these values or midpoints. Connect the existing line to these dots to create a shape rather than a "floating" line. ■

CHECK YOUR LEARNING

Reviewing the Concepts

> The first step in organizing data for a single variable is to list all the values in order of magnitude and count how many times each value occurs.

> There are four techniques for organizing information about a single variable: frequency tables, grouped frequency tables, histograms, and frequency polygons.

Clarifying the Concepts

2-1 Name four different ways to organize raw scores visually.

2-2 What is the difference between frequencies and grouped frequencies?

Calculating the Statistics

2-3 In 2005, *U.S. News & World Report* published a list of the best psychology doctoral programs in the United States. The top 27 programs ranged from Stanford University at number 1 through a tie among the last six universities, which included Johns Hopkins and Northwestern universities. Let's say you're interested in a top program that has ethnic diversity. Seventeen of these schools reported the number of current students who are members of a racial or ethnic minority group. Here are those data:

17 17 8 12 3 59 41 3 32 4 10 59 20 1 9 3 27

a. Construct a grouped frequency table of these data.

b. Construct a histogram for this grouped frequency table.

c. Construct a frequency polygon for this grouped frequency table.

Applying the Concepts

2-4 Consider the data from Check Your Learning 2-3, as well as the table and graphs that you constructed.

a. What can we tell from the graphs and table that we cannot tell from the list of scores?

b. Why might percentages of students who are members of a minority group be more useful than these numbers? (*Hint:* The number of full-time psychology graduate students at these schools ranged from 19 to 258.)

c. Not all schools provided data. How might the schools that provided data on the number of minority students be different from schools that did not provide these data?

Solutions to these Check Your Learning questions can be found in Appendix D.

Shapes of Distributions

- A **normal distribution** is a specific frequency distribution that is a bell-shaped, symmetric, unimodal curve.
- A **skewed distribution** is a distribution in which one of the tails of the distribution is pulled away from the center.
- With **positively skewed** data, the distribution's tail extends to the right, in a positive direction.

We learned how to organize data so that we can better understand the concept of a distribution, a major building block for statistical analysis. We can't get a sense of the overall pattern of data by looking at a list of numbers, but we *can* get a sense of the pattern by looking at a frequency table. We can get an even better sense by creating a graph. Histograms and frequency polygons allow us to see the overall pattern, or shape, of the distribution of data.

The shape of a distribution provides distinctive information. For example, when the U.S.-based General Social Survey (a large data set available to the public via the Internet) asked people about the influence of children's programming—both network television and public television—their responses produced very different patterns for each type of children's programming (Figure 2-4). For example, the most common response for network television shows was that they have a neutral influence, whereas the most common response for public television shows was that they have a positive influence. In this section, we provide you with language that

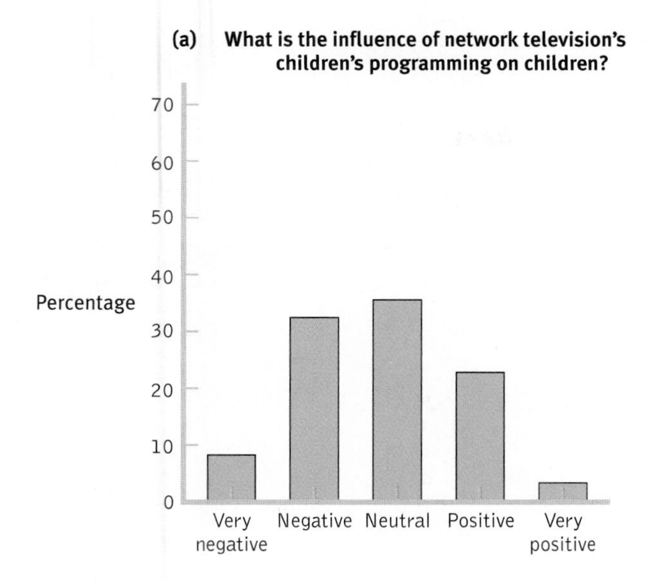

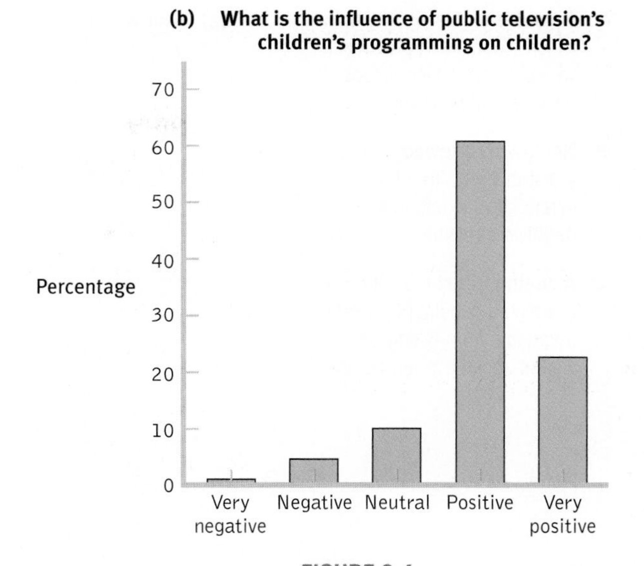

FIGURE 2-4
The Influence of Television Programming on Children

These two histograms tell different stories about the influence of television programming on children. The first histogram (a) describes the influence of network television on children; the second histogram (b) describes the influence of public television on children.

expresses the differences between these patterns. Specifically, you'll learn to describe different shapes of distributions, including normal distributions and skewed distributions.

Normal Distributions

Many, but not all, distributions of variables form a bell-shaped, or *normal,* curve. Statisticians use the word *normal* to describe distributions in a very particular way. *A* ***normal distribution*** *is a specific frequency distribution that is a bell-shaped, symmetric, unimodal curve* (Figure 2-5). People's attitudes toward the network programming of children's shows is an example of a distribution that approaches a normal distribution. There are fewer scores at values that are farther from the center and even fewer scores at the most extreme values (as can be seen in the bar graph in Figure 2-4a). Most scores cluster around the word *neutral* in the middle of the distribution, which would be at the top of the bell.

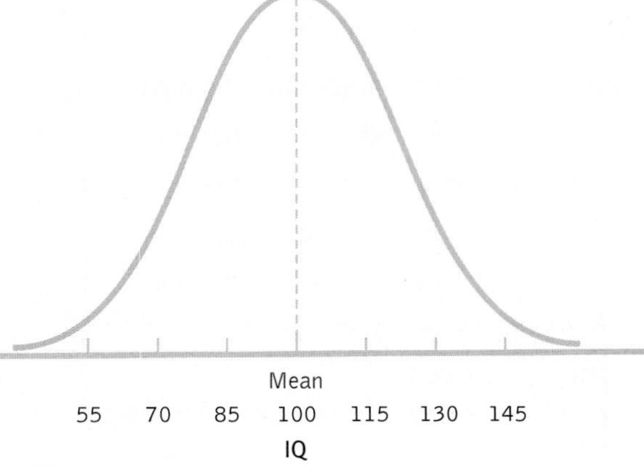

FIGURE 2-5
The Normal Distribution

The normal distribution, shown here for IQ scores, is a frequency distribution that is bell shaped, symmetric, and unimodal. It is central to many calculations in statistics.

Skewed Distributions

Reality is not always normally distributed, which means that the distributions describing those particular observations are not shaped normally. So we need a new term to help us describe such distributions—*skew.* ***Skewed distributions*** *are distributions in which one of the tails of the distribution is pulled away from the center.* Although the technical term for such data is *skewed,* a skewed distribution may also be described as lopsided, off-center, or simply nonsymmetric. Skewed data have an ever-thinning tail in one direction or the other. The distribution of people's attitudes toward the children's programming offered by public television (Figure 2-4b) is an example of a skewed distribution. The scores cluster to the right side of the distribution around the word *positive,* and the tail extends to the left.

When a distribution is ***positively skewed,*** as in Figure 2-6a, *the tail of the distribution extends to the right, in a positive direction.* Positive skew sometimes occurs when there is

- A **floor effect** is a situation in which a constraint prevents a variable from taking values below a certain point.
- **Negatively skewed** data have a distribution with a tail that extends to the left, in a negative direction.
- A **ceiling effect** is a situation in which a constraint prevents a variable from taking on values above a given number.

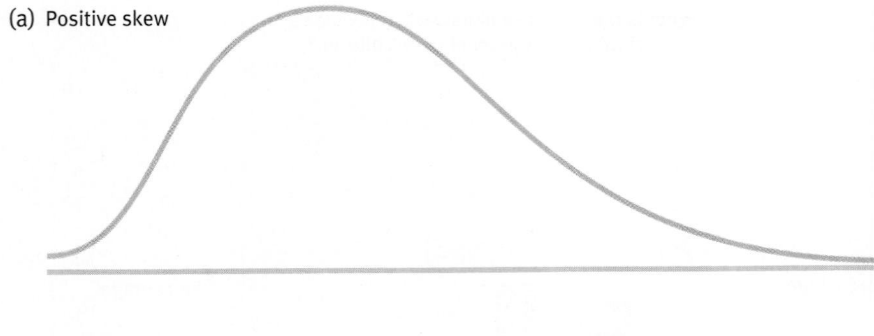

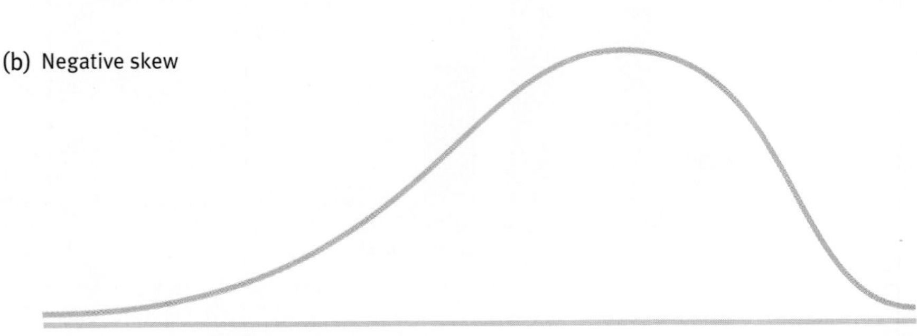

FIGURE 2-6
Two Kinds of Skew

The mnemonic "the tail tells the tale" means that the distribution with the long, thin tail to the right is positively skewed and the distribution with the long, thin tail to the left is negatively skewed.

a ***floor effect****, a situation in which a constraint prevents a variable from taking values below a certain point.* For example, in the "World Cup success" data, scores indicating how many countries came in first or second in the World Cup a certain number of times is an example of a positively skewed distribution with a floor effect. Most countries never came in first or second, which means that the data were constrained at the lower end of the distribution, 0 (that is, they can't go below 0).

> **MASTERING THE CONCEPT**
>
> **2.3:** If a histogram indicates that the data are symmetric and bell shaped, then the data are normally distributed. If the data are not symmetric and the tail extends to the right, the data are positively skewed; if the tail extends to the left, the data are negatively skewed.

The distribution in Figure 2-6b shows ***negatively skewed*** *data, which have a distribution with a tail that extends to the left, in a negative direction.* The distribution of people's attitudes toward public television's programming of children's shows is favorable because it is clustered around the word *positive,* but we describe the shape of that distribution as negatively skewed because the thin tail is to the left side of the distribution. Not surprisingly, negative skew is sometimes the result of a ***ceiling effect****, a situation in which a constraint prevents a variable from taking on values above a given number.* If a professor gives an extremely easy quiz, a ceiling effect might result. A number of students would cluster around 100, the highest possible score, with a few stragglers down in the lower end.

CHECK YOUR LEARNING

Reviewing the Concepts

> A normal distribution is a specific distribution that is unimodal, symmetric, and bell shaped.

> A skewed distribution "leans" either to the left or to the right. A tail to the left indicates negative skew; a tail to the right indicates positive skew.

Clarifying the Concepts

2-5 Distinguish a normal distribution from a skewed distribution.

2-6 When the bulk of data cluster together but the data trail off to the left, you have _________ skew; when that data trail off to the right, you have _________ skew.

Calculating the Statistics

2-7 In Check Your Learning 2-3, you constructed two visual displays of the distribution of racial and ethnic diversity in doctoral psychology programs. What kind of skew is evident in your graphs?

2-8 Alzheimer's disease is typically diagnosed in adults above the age of 70; cases diagnosed sooner are labeled "early onset."

a. Assuming that these early-onset cases represent unique trailing off of data on one side, would this represent positive skew or negative skew?

b. Do these data represent a floor effect or ceiling effect?

Applying the Concepts

2-9 Referring to Check Your Learning 2-8, what implication would identifying such skew have in the screening and treatment process for Alzheimer's disease?

Solutions to these Check Your Learning questions can be found in Appendix D.

REVIEW OF CONCEPTS

Frequency Distributions

There are several ways in which we can depict a *frequency distribution* of a set of *raw scores. Frequency tables* are comprised of two columns, one with all possible values and one with a count of how often each value occurs. *Grouped frequency tables* allow us to work with more complicated data. Instead of containing values, the first column consists of intervals. *Histograms* display bars of different heights indicating the frequency of each value (or interval) that the variable can take on. *Frequency polygons* show frequencies with dots at different heights depicting the frequency of each value (or interval) that the variable can take on. The dots in a frequency polygon are connected to form the shape of the data.

Shapes of Distributions

The *normal distribution* is a specific distribution that is unimodal, symmetric, and bell shaped. Data can also display *skewness.* A distribution that is *positively skewed* has a tail in a positive direction (to the right), indicating more extreme scores above the center. It sometimes results from a *floor effect,* in which scores are constrained and cannot be below a certain number. A distribution that is *negatively skewed* has a tail in a negative direction (to the left), indicating more extreme scores below the center. It sometimes results from a *ceiling effect,* in which scores are constrained and cannot be above a certain number.

SPSS®

The left-hand column in "Data View" is prenumbered, beginning with 1. Each column to the right of that number contains information about a particular variable; each row represents a unique individual. Notice the choices at the top of the "Data View" screen. Enter the pacing index data from p. 26, then, from the menu, select Analyze → Descriptive

Statistics → Frequencies. Select the variables you want SPSS to describe by highlighting them and clicking the arrow in the middle.

We also want to visualize each variable, so after selecting "Frequencies," select Charts → Histograms (click the box next to "with normal curve") → Continue → OK.

For all of the SPSS functions, an output screen automatically appears after "OK" is clicked. The screenshot shown here depicts the part of the SPSS output that includes the histogram. Double-click on the graph to enter the SPSS Chart Editor, and then double-click on each feature in order to make the graph look like we want it to. For example, you might choose to click on the word "Frequency," that by default appears vertically and in sideways lettering along the y-axis. You would then choose "Text Layout" and click on the circle next to "Horizontal" under "Orientation," so that after you click "Apply," you can read the word "Frequency" without having to twist your head. Click and change any feature of the graph that lets your data speak more clearly.

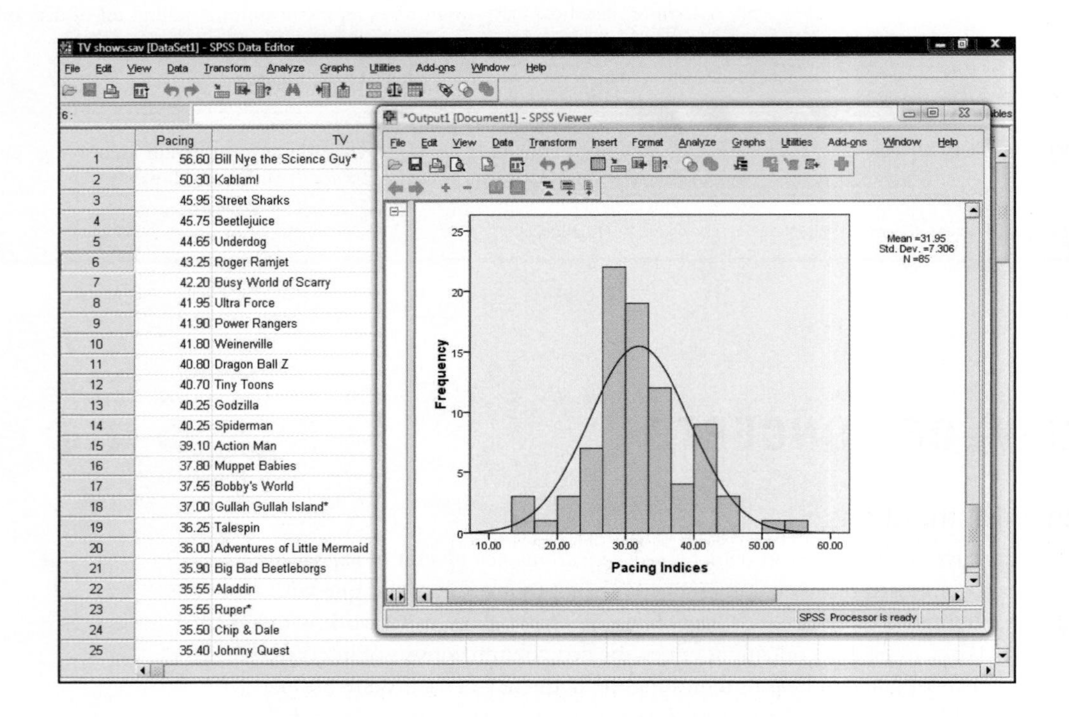

How It Works

2.1 CREATING A FREQUENCY TABLE

Imagine that we ask everyone in a class of 20 first-year college students how many nights they went out to socialize in the previous week. In this case, we might specify that *to socialize* means to leave your place of residence for at least 3 hours after 6:00 P.M. for any purpose unrelated to academic work. This observation allows only a very specific set of possible responses: 0, 1, 2, 3, 4, 5, 6, or 7 nights. If we asked each of the 20 students how many nights a week they typically go out to socialize, we might get a data set that looks like this:

1 2 7 6 1
2 6 5 4 4
0 3 2 2 3
4 3 5 4 4

How can we use these data to create a frequency table? First, we reorder the "nights socializing" data into a table with two columns, one for the range of possible responses (the values) and one for the frequencies of each of the responses (the scores). The frequency table for these data is follows.

Nights	Frequency
7	1
6	2
5	2
4	5
3	3
2	4
1	2
0	1

2.2 CREATING A HISTOGRAM

How can we use these same data to create a histogram? First, we put the number of nights socializing on the *x*-axis and the frequency for each number on the *y*-axis. The bar for each frequency is centered at the appropriate number of nights out. The figure below portrays the histogram for these data.

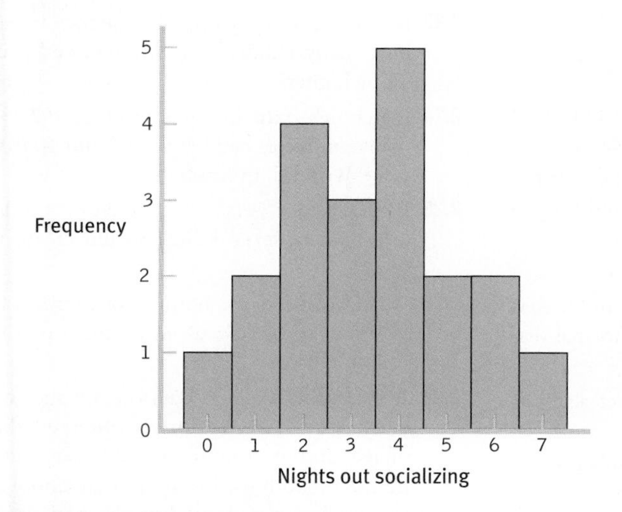

2.3 CREATING A FREQUENCY POLYGON

How can we use these same data to create a frequency polygon? As with the histogram, we put the number of nights socializing on the *x*-axis and the frequency for each number on the *y*-axis. Instead of bars, we now place a dot for each frequency above the appropriate number of nights out. We add additional values at the next whole numbers below and above this range, −1 and 8, and put dots indicating a frequency of 0 at each of these values. The figure below portrays the frequency polygon for these data.

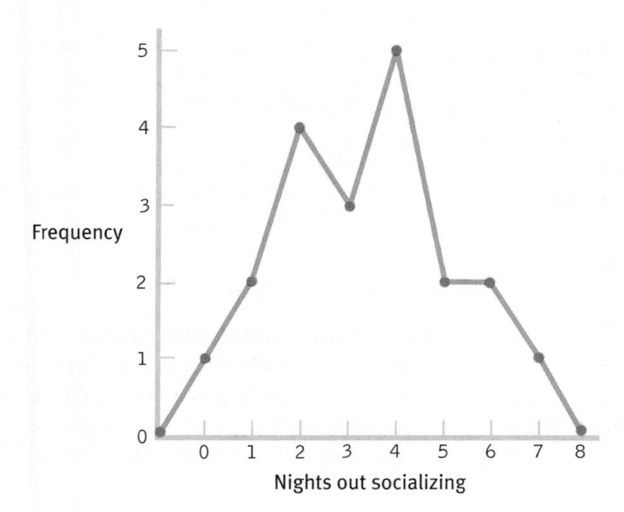

Exercises

Clarifying the Concepts

2.1 What are raw scores?

2.2 What are the steps to create a frequency table?

2.3 What is the difference between a frequency table and a grouped frequency table?

2.4 Describe two ways that statisticians might use the word *interval*.

2.5 What is the difference between a histogram and a bar graph?

2.6 What are the typical labels for the *x*-axis and the *y*-axis in a histogram?

2.7 What is the difference between a histogram and a frequency polygon?

2.8 What is the benefit of creating a visual distribution of data rather than simply looking at a list of the data?

2.9 In your own words, define the word *distribution,* first as you would use it in everyday conversation and then as a statistician would use it.

2.10 What is a normal distribution?

2.11 How do positively skewed distributions and negatively skewed distributions deviate from a normal distribution?

2.12 What is a floor effect and how does it affect a distribution?

2.13 What is a ceiling effect and how does it affect a distribution?

Calculating the Statistics

2.14 Convert the following to percentages: 63 out of 1264; 2 out of 88.

2.15 Convert the following to percentages: 7 out of 39; 122 out of 300.

2.16 Counts are often converted to percentages. Convert 817 out of 22,140 into a percentage. Now convert 4009 out of 22,140 into a percentage. What type of variable (nominal, ordinal, or scale) are these data as counts? What kind of variable are they as percentages?

2.17 Convert 2 out of 2000 into a percentage. Now convert 60 out of 62 into a percentage.

2.18 Throughout this book, final answers are reported to two decimal places. Report the following numbers this way: 1,888.999, 2.6454, and 0.0833.

2.19 Report the following numbers to two decimal places: 0.0391, 198.2219, and 17.886.

2.20 On a test of marital satisfaction, scores could range from 0 to 27.

a. What is the full range of data, according to the calculation procedure described in this chapter?

b. What would the interval size be if we wanted six intervals?

c. List the six intervals.

2.21 If you have data that range from 2 to 68 and you want seven intervals in a grouped frequency table, what would the intervals be?

2.22 A grouped frequency table has the following intervals: 30–44, 45–59, and 60–74. If converted into a histogram, what would the midpoints be?

2.23 Referring to the grouped frequency table in Table 2-6, how many children's shows received pacing scores of 35 or higher?

2.24 Referring to the histogram in Figure 2-1, estimate how many countries had between 2 and 10 first- or second-place World Cup finishes.

2.25 If the average person convicted of murder killed only one person, serial killers would create what kind of skew?

2.26 Would the data for number of murders by those convicted of the crime be an example of a floor effect or ceiling effect?

2.27 A researcher collects data on the ages of college students. As you have probably observed, the distribution of age clusters around 19 to 22 years, but there are extremes on both the low end (high school prodigies) and the high end (nontraditional students returning to school).

a. What type of skew might you expect for such data?

b. Do the skewed data represent a floor effect or a ceiling effect?

2.28 If you have a Facebook account, you are allowed to have up to 5000 friends. At that point, Facebook cuts you off, and you have to "defriend" people to add more. Imagine you collected data from Facebook users at your university about the number of friends each has.

a. What type of skew might you expect for such data?

b. Do the skewed data represent a floor effect or ceiling effect?

Applying the Concepts

2.29 **Frequency tables, histograms, and the National Survey of Student Engagement:** The National Survey of Student Engagement (NSSE) surveys freshmen and seniors about their level of engagement in campus and classroom activities that enhance learning. Hundreds of thousands of students at almost 1000 schools have completed surveys since 1999, when the NSSE was first administered. Among the many questions, students were asked how often they were assigned a paper of 20 pages or more during the academic year. For a sample of 19 institutions classified as national universities that made their data publicly available through the *U.S. News & World Report* Web site, here are the percentages of students who said they were assigned between 5 and 10 twenty-page papers:

0	5	3	3	1	10	2
2	3	1	2	4	2	1
1	1	4	3	5		

a. Create a frequency table for these data. Include a third column for percentages.

b. For what percentage of these schools did exactly 4% of their students report that they wrote between 5 and 10 twenty-page papers that year?

c. Is this a random sample? Explain your answer.

d. Create a histogram of grouped data, using six intervals.

e. In how many schools did 6% or more of the students report that they wrote between 5 and 10 twenty-page papers that year?

f. How are the data distributed?

2.30 Frequency tables, histograms, and the Survey of Earned Doctorates: The Survey of Earned Doctorates regularly assesses the numbers and types of doctorates awarded at U.S. universities. It also provides data on the length of time in years that it takes to complete a doctorate. Below is a modified list of this completion-time data, truncated to whole numbers and shortened to make your analysis easier. These data have been collected every 5 years since 1982.

8	8	8	8	8	7	6	7	7	7	7	7	
6	6	6	6	6	6	7	8	8	8	8	7	
6	6	7	7	7	6	11	13	15	15			
14	12	9	10	10	9	9	9					

a. Create a frequency table for these data.

b. How many schools have an average completion time of 8 years or less?

c. Is a grouped frequency table necessary? Why or why not?

d. Describe how these data are distributed.

e. Create a histogram for these data.

f. At how many universities did students take, on average, 10 or more years to complete their doctorates?

2.31 Frequency tables, histograms, polygons, and alumni donations: Many schools are interested in alumni donations, not only because they want the money but also because it is one of the criteria by which *U.S. News & World Report* ranks U.S. institutions of higher learning. (Higher rates of alumni giving are seen as indicative of the satisfaction of former students with their education.) Despite controversy about the validity of the rankings, an increase in a school's overall ranking by this magazine has been demonstrated to translate into an increase in applications. One set of rankings is for the best national universities: institutions that offer undergraduate, master's, and doctoral degrees and have an emphasis on research. (Harvard tops the list that was published in 2005.) Here are the 2005 percentages of alumni who donated to each of the top 70 national universities in the year prior to publication of these data.

48	61	45	39	46	37	38	34	33	47
29	38	38	34	29	29	36	48	27	25
15	25	14	26	33	16	33	32	25	34
26	32	11	15	25	9	25	40	12	20
32	10	24	9	16	21	12	14	18	20
18	25	18	20	23	9	16	17	19	15
14	18	16	17	20	24	25	11	16	13

a. How was the variable of alumni giving operationalized? What is another way that this variable could be operationalized?

b. Create a grouped frequency table for these data.

c. The data have quite a range, with the lowest scores belonging to Boston University, the University of California at Irvine, and the University of California at San Diego, and the highest belonging to Princeton University. What research hypotheses come to mind when you examine these data? State at least one research question that these data suggest to you.

d. Create a grouped histogram for these data. Be careful when determining the midpoints of your intervals!

e. Create a frequency polygon for these data.

f. Examine these graphs and give a brief description of the distribution. Are there unusual scores? Are the data skewed, and if so, in which direction?

2.32 Frequency tables, histograms, and the NBA: Here are the numbers of wins for the 30 National Basketball Association teams for the 2004–2005 NBA season.

45	43	42	33	33	54	47	44	42	30
59	45	36	18	13	52	49	44	27	26
62	50	37	34	34	59	58	51	45	18

a. Create a grouped frequency table for these data.

b. Create a histogram based on the grouped frequency table.

c. Write a summary describing the distribution of these data with respect to shape and direction of any skew.

d. State one research question that might arise from this data set.

2.33 Types of distributions: Consider these three variables: finishing times in a marathon, number of university dining hall meals eaten in a semester on a three-meal-a-day plan, and scores on a scale of extroversion.

a. Which of these variables is most likely to have a normal distribution? Explain your answer.

b. Which of these variables is most likely to have a positively skewed distribution? Explain your answer, stating the possible contribution of a floor effect.

c. Which of these variables is most likely to have a negatively skewed distribution? Explain your answer, stating the possible contribution of a ceiling effect.

2.34 Type of frequency distribution and type of graph: For each of the types of data described below, first state how you would present individual data values or grouped data when creating a frequency distribution. Then, state which visual display(s) of data would be most appropriate to use. Explain your answers clearly.

a. Eye color observed for 87 people

b. Minutes used on a cell phone by 240 teenagers

c. Time to complete the Boston Marathon for the nearly 22,000 runners who participate

d. Number of siblings for 64 college students

2.35 Grouped frequency table and career services: The director of career services at a large university offers training on résumé construction. In an effort to present up-to-date information, using 23 résumés he just reviewed for a receptionist position in his office, he counts the total number of words used. Here are the data:

226	339	220	295	180	214	257	201
224	237	223	301	267	284	238	251
278	294	266	227	281	312	332	

a. Create a grouped frequency table with four intervals.

b. Is this a random sample? Explain your answer.

c. What does this information tell people who come to his training on résumé construction?

Putting It All Together

2.36 Frequencies, distributions, and numbers of friends: A college student is interested in how many friends the average person has. She decides to count the number of people who appear in photographs on display in dorm rooms and offices across campus. She collects data on 84 students and 33 faculty members. The data are presented below.

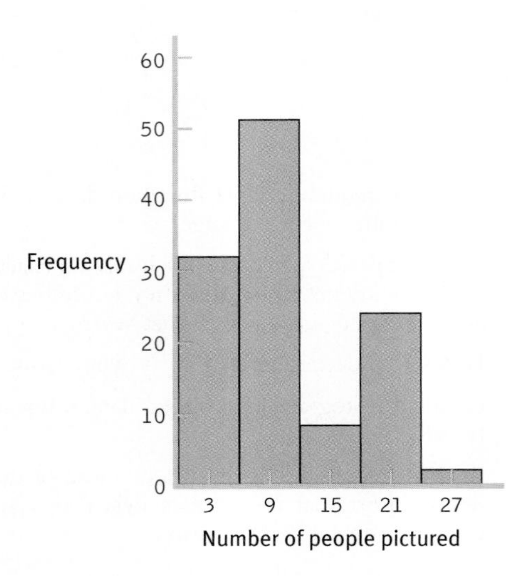

a. What kind of visual display is this?

b. Estimate how many people have fewer than 6 people pictured.

c. Estimate how many people have more than 18 people pictured.

d. Can you think of additional questions you might ask after reviewing the data displayed here?

e. Below is a subset of the data described here. Create a grouped frequency table for these data, using seven groupings.

1	5	3	9	13	0	18	15
3	3	5	7	7	7	11	3
12	20	16	4	17	15	16	10
6	8	8	7	3	17		

f. Create a histogram of the grouped data from (e).

g. Describe how the data depicted in the original graph and the histogram you created in (f) are distributed.

2.37 Frequencies, distributions, and breast-feeding duration: The Centers for Disease Control and other organizations are interested in the health benefits of breast-feeding for infants. The National Immunization Survey includes questions about breast-feeding practices, including: "How long was [your child] breast-fed or fed breast milk?" The data for duration of breast-feeding, in months, for 20 hypothetical mothers are presented below.

0	7	0	12	9	3	2	0	6	10
3	0	2	1	3	0	3	1	1	4

a. Create a frequency table for these data. Include a third column for percentages.

b. Create a histogram of these data.

c. Create a frequency polygon of these data.

d. Create a grouped frequency table for these data with three groups (create groupings around the midpoints of 2.5 months, 7.5 months, and 12.5 months).

e. Create a histogram of the grouped data.

f. Create a frequency polygon of the grouped data.

g. Write a summary describing the distribution of these data with respect to shape and direction of any skew.

h. If you wanted the data to be normally distributed around 12 months, how would the data have to shift to fit that goal? How could you use knowledge about the current distribution to target certain women?

2.38 Developing research ideas from frequency distributions: Below are frequency distributions for two sets of the friends data described in Exercise 2.36, one for the students and one for the faculty members studied.

Interval	Faculty Frequency	Student Frequency
0–3	21	0
4–7	11	26
8–11	1	24
12–15	0	2
16–19	0	27
20–23	0	37
24–27	0	2

a. How would you describe the distribution for faculty members?

b. How would you describe the distribution for students?

c. If you were to conduct a study comparing the numbers of friends that faculty members and students have, what would the independent variable be and what would be the levels of the independent variable?

d. In the study described in (c), what would the dependent variable be?

e. What is a confounding variable that might be present in the study described in (c)?

f. Suggest at least two additional ways to operationalize the dependent variable. Would either of these ways reduce the impact of the confounding variable described in (e)?

2.39 Frequencies, distributions, and graduate advising: In a study of mentoring in chemistry fields, a team of chemists and social scientists identified the most successful U.S. mentors—professors whose students were hired by the top 50 chemistry departments in the United States (Kuck et al., 2007). Fifty-four professors had at least three students go on to such jobs. Here are the data for the 54 professors. Each number indicates the number of students successfully mentored by each different professor.

3	3	3	4	5	9	5	3	3	5	6
3	4	8	6	3	3	3	4	4	4	7
6	3	5	5	7	13	3	3	3	3	3
4	4	4	5	6	7	6	7	8	8	3
3	3	5	3	3	5	3	5	3	3	

a. Construct a frequency table for these data. Include a third column for percentages.

b. Construct a histogram for these data.

c. Construct a frequency polygon for these data.

d. Describe the shape of this distribution.

e. How did the researchers operationalize the variable of mentoring success? Suggest at least two other ways in which they might have operationalized mentoring success.

f. Imagine that researchers hypothesized that an independent variable, the number of publications coauthored by the advisor, predicts the dependent variable of mentoring success. One professor, Dr. Yuan T. Lee, from the University of California at Berkeley, trained 13 future top faculty members. Dr. Lee won a Nobel Prize. Explain how such a prestigious and public accomplishment might present a confounding variable to the hypothesis described above.

g. Dr. Lee had many students who went on to top professorships before he won his Nobel Prize. Several other chemistry Nobel Prize winners in the United States serve as graduate advisors but have not had Dr. Lee's level of success as mentors. What are other possible variables that might predict the dependent variable of attaining a top professor position?

Terms

raw score (p. 23)
frequency distribution (p. 23)
frequency table (p. 23)
grouped frequency table (p. 26)
histogram (p. 28)
frequency polygon (p. 30)
normal distribution (p. 33)
skewed distribution (p. 33)
positively skewed (p. 33)
floor effect (p. 34)
negatively skewed (p. 34)
ceiling effect (p. 34)

CHAPTER 3

Visual Displays of Data

BEFORE YOU GO ON

- You should know how to construct a histogram (Chapter 2).
- You should understand the difference between independent variables and dependent variables (Chapter 1).

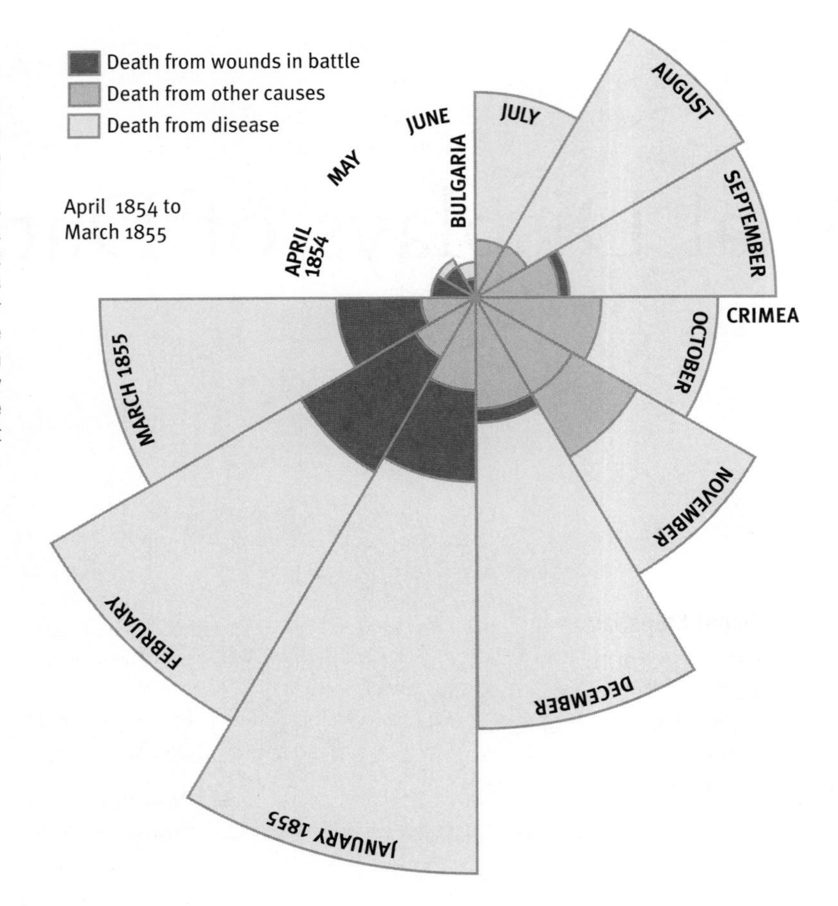

FIGURE 3-1
Graphs That Persuade

Florence Nightingale's coxcomb graphs depict the "Causes of Mortality in the Army in the East" from April 1854 to March 1856. This adaptation of her April 1854 to March 1855 graphic is called a *coxcomb graph* because the data arrangement resembles the shape of a rooster's head. The 12 sections represent the ordinal variable of a year broken into 12 months. The size of the sections representing each month indicates the variable of how many people died in that particular month. The colors correspond to the different causes of death.

Florence Nightingale, a nineteenth-century nurse heroine, was also known as the "passionate statistician" (Diamond & Stone, 1981). The lens of time has softened the image of this sarcastic, sharp-elbowed infighter (Gill, 2005) who created trouble by counting things. She counted supplies in a closet and discovered corruption; she counted physicians' diagnoses and discovered incompetence. She created the visual display in Figure 3-1 after counting the causes of death of British soldiers in the Crimea (now part of Ukraine), and an outraged public demanded change. The army was killing through poor hygiene more of its own soldiers than were dying due to wounds of war. That is one powerful graph!

This chapter shows you how to create graphs, called *figures* in APA-speak, that tell data stories. We recommend *Displaying Your Findings* by Adelheid Nicol and Penny Pexman (2010) for a more detailed look, but a checklist at the end of this chapter will answer most of your questions.

How to Lie with Visual Statistics

Teaching students how to lie may sound like an odd way to teach statistics, but spotting visual tricks can be empowering. We are indebted to Michael Friendly of York University for collecting and managing a Web site (http://www.math.yorku.ca/SCS/Gallery/) that humorously demonstrates the power of graphs both to deceive and to enlighten. He described Figure 3-2 as possibly "the most misleading graph ever published."

"The Most Misleading Graph Ever Published"

The *Ithaca Times* graph in Figure 3-2 appears to answer a simple question: "Why does college have to cost so much?" This graph is chock full of lies.

Lie 1: The rising line represents rising tuition costs over *35 years;* the falling line represents the ranking of Cornell University over only *11 years.*

Lie 2: The *y*-axis compares an ordinal observation (university rank) to a scale observation (tuition). These should be two different graphs.

Lie 3: Cornell's rank begins at a lower point on the *y*-axis than tuition costs, suggesting that an institution already failing to deliver what students are paying for has become dramatically worse.

Lie 4: The graph *reverses* the implied meaning of up and down. A low number in the world of rankings is a good thing. Over this 11-year period, Cornell's ranking *improved* from 15th place to 6th place!

FIGURE 3-2
Graphs That Lie

Michael Friendly describes this graph as a "spectacular example of more graphical sins than I have ever seen in one image" and possibly "the most misleading graph ever published."

Techniques for Misleading with Graphs

When you learn these statistical tricks, you will immediately become a much more critical—and less gullible—consumer of visual statistics:

1. The biased scale lie. *New York* magazine's (see nymag.com) reviewers use five stars to indicate that a restaurant's food, service, and ambience are "ethereal; almost perfect"; three means "generally excellent"; one means "good." So zero stars must mean bad, right? Wrong. Zero "means our critics don't recommend you go out of your way to eat there." Apparently, you can't buy a bad meal in New York City if a *New York* magazine reviewer has eaten there.

2. The sneaky sample lie. You might pick up some useful information from Web sites that rate professors, but be cautious. The students most likely to supply ratings are those who strongly dislike or strongly approve of a particular professor. A self-selected sample means that the information might not apply to you.

3. The interpolation lie. Interpolation occurs when we assume that some value between the data points lies on a straight line between those data points. A 2007 report in *Statistics Canada* reported the lowest rate of break-ins (property crime) since the 1970s (Figure 3-3), but you cannot assume a gradual decline over 30 years. In 1991, there was a dramatic increase in property crime. Make sure that a reasonable number of in-between data points have been reported.

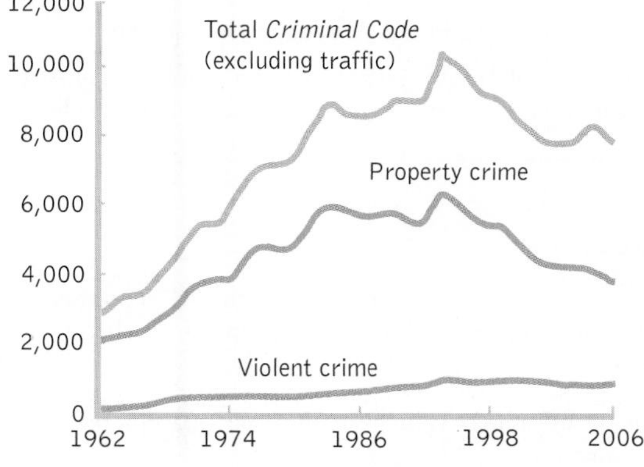

FIGURE 3-3
The Perils of Interpolation

Without seeing all of the data, it is easy to draw false conclusions. Although Canada's property crime rate declined from the late 1970s through 2006, there was a peak in the middle, around 1991. If we saw only the data points for the 1970s and 2006, we might falsely conclude that there was a gradual decline during this time. You can search http://statcan.gc.ca to search for this and many other interesting statistics about Canada.

[Source: http://www.statcan.gc.ca/daily-quotidien/110721/dq11021b-eng.htm]

4. The extrapolation lie. This lie assumes that values beyond the data points will continue indefinitely. In 1976, *The Complete CB Handbook* assumed that elementary schools would have to teach students how to use the once-popular device (now used mostly by long-distance truckers). What happened? Mobile phones. Do not assume that a pattern will continue indefinitely.

5. The inaccurate values lie. This lie tells the truth in one part of the data but visually distorts it in another place. Notice in Figure 3-4 how wide the "highway" is when the accelerating fuel-economy savings is coming at the viewer. The proportional change in distance between the beginning and the end of the highway is many times larger than the proportional change in the size of the data.

MASTERING THE CONCEPT

3.1: Graphs can be misleading until you become a critical thinker. Ask yourself about the sample, the variables, and the format of the graph.

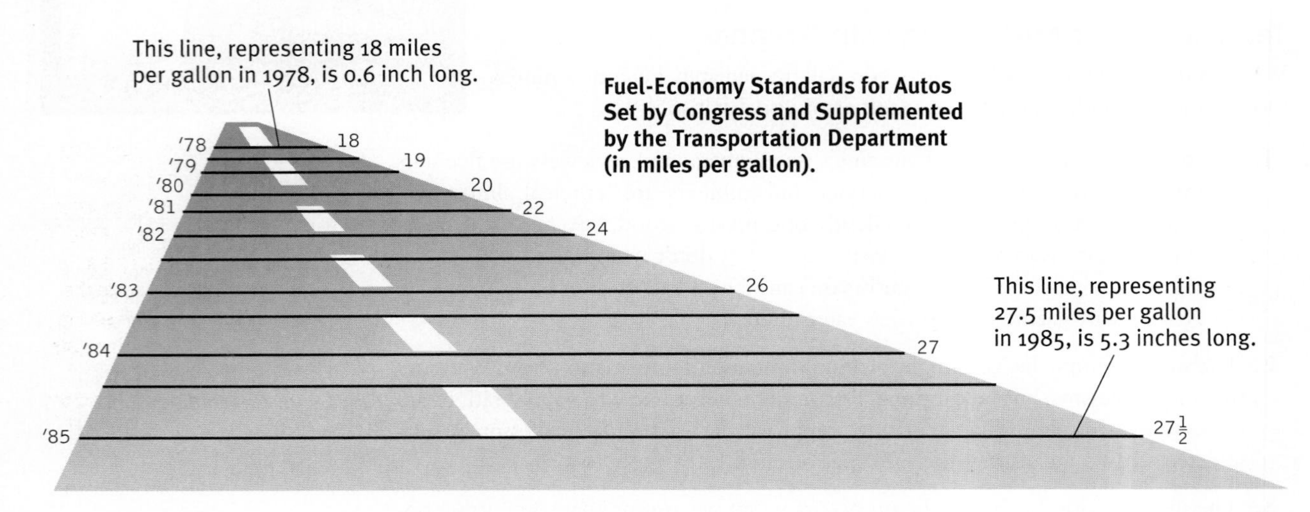

FIGURE 3-4
The Inaccurate Values Lie

The visual lie told here is the result of a "highway" that spreads much farther apart than the data indicate. Michael Friendly (2005) asserts that "this graph, from the *New York Times,* purports to show the mandated fuel economy standards set by the US Department of Transportation. The standard required an increase in mileage from 18 to 27.5, an increase of 53%. The magnitude of increase shown in the graph is 783%, for a whopping lie factor = (783/53) = 14.8!"

CHECK YOUR LEARNING

Reviewing the Concepts

> Creating and understanding graphs is a critical skill in our data-dependent society.

> Graphs can both reveal or obscure information. Don't just glance at a graph; examine it and ask critical questions to be sure the graph creator isn't exaggerating or being misleading.

Clarifying the Concepts

3-1 What is the purpose of a graph?

Calculating the Statistics

3-2 Referring to Figure 3-4, the inaccurate values lie, calculate how much fuel-economy standards changed from 1981 to 1984 in miles per gallon and as a percentage change.

Applying the Concepts

3-3 Which of the two following graphs is misleading? Which seems to be an accurate depiction of the data? Explain your answer.

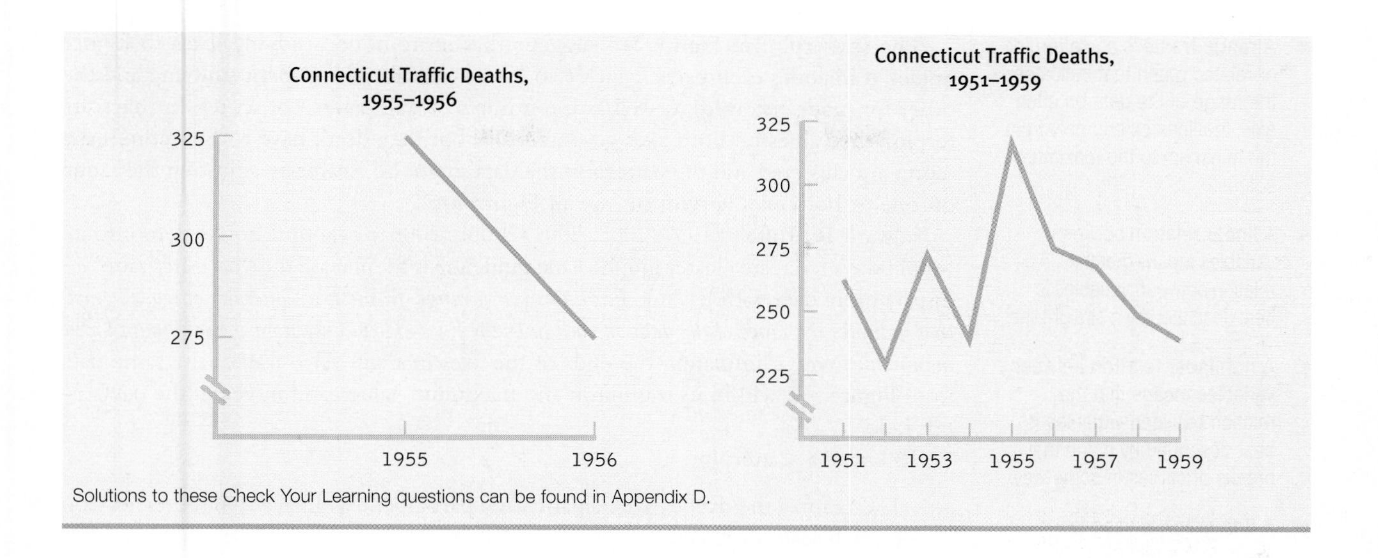

Solutions to these Check Your Learning questions can be found in Appendix D.

Common Types of Graphs

Graphs are powerful because they can display the *relation* between two or more variables in just one image. We first show you how to create scatterplots and line graphs—types of graphs that have two scale variables. Then we learn how to create—and criticize—graphs with just one nominal variable: bar graphs, pictorial graphs, and pie charts.

▪ A **scatterplot** is a graph that depicts the relation between two scale variables. The values of each variable are marked along the two axes, and a mark is made to indicate the intersection of the two scores for each participant. The mark is above the participant's score on the *x*-axis and across from the score on the *y*-axis.

Scatterplots

A ***scatterplot*** *is a graph that depicts the relation between two scale variables. The values of each variable are marked along the two axes. A mark is made to indicate the intersection of the two scores for each participant. The mark is above the participant's score on the x-axis and across from the score on the y-axis.* We suggest that you think through your graph by sketching it by hand before creating it on a computer.

EXAMPLE 3.1

Figure 3-5 describes the relation between the number of hours students spent studying and the students' grades on a statistics exam. In this example, the independent variable (x, on the horizontal axis) is the number of hours spent studying, and the dependent variable (y, on the vertical axis) is the grade on the statistics exam.

MASTERING THE CONCEPT

3.2: Scatterplots and line graphs are used to depict relations between two scale variables.

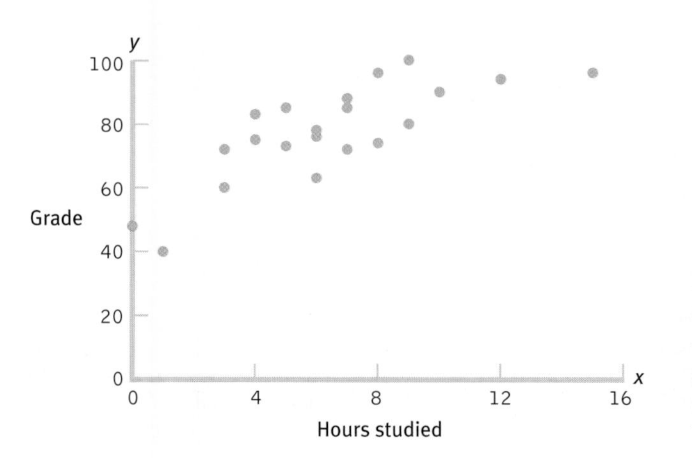

FIGURE 3-5
Scatterplot of Hours Spent Studying and Statistics Grades
This scatterplot depicts the relation between hours spent studying and grades on a statistics exam. Each dot represents one student's score on the independent variable along the *x*-axis and on the dependent variable along the *y*-axis.

- A **range-frame** is a scatterplot or related graph that indicates the range of the data on each axis; the lines extend only from the minimum to the maximum scores.
- A **linear relation** between variables means that the relation between variables is best described by a straight line.
- A **nonlinear relation** between variables means that the relation between variables is best described by a line that breaks or curves in some way.
- A **line graph** is used to illustrate the relation between two scale variables.

The scatterplot in Figure 3-5 suggests that more hours studying leads to higher grades; it includes each participant's two scores (one for hours spent studying and the other for grade received) as well as the group's overall pattern of scores. In this scatterplot, the values on both axes go down to 0 but they don't have to. Sometimes the scores are clustered and the pattern in the data might be clearer by adjusting the range on one or both axes (as you can see in Figure 3-6).

Edward R. Tufte's (1997/2005, 2001/2006b, 2006a) beautiful books demonstrate simple ways to create clearer graphs. One guideline is to increase the "data–ink ratio"—display more data with less ink. For example, *a* ***range-frame*** *is a scatterplot or related graph that indicates the range of the data on each axis; the lines extend only from the minimum to the maximum scores.* Eliminating the ends of the axes in Figure 3-6 frames the same data from Figure 3-5 within its minimum and maximum values, and increases the data-to-ink ratio.

To create a scatterplot:

1. Organize the data by participant; each participant will have two scores, one on each scale variable.
2. Label the horizontal x-axis with the name of the independent variable and its possible values, starting with 0 if practical.
3. Label the vertical y-axis with the name of the dependent variable and its possible values, starting with 0 if practical.
4. Make a mark on the graph above each study participant's score on the x-axis and next to his or her score on the y-axis.
5. To convert to a range-frame, simply erase the axes below the minimum score and above the maximum score.

A scatterplot between two scale variables can tell three possible stories. First, there may be no relation at all; the scatterplot looks like a jumble of random dots. This is an important scientific story if we previously believed that there was a systematic pattern between the two variables.

Second, *a* ***linear relation*** *between variables means that the relation between variables is best described by a straight line.* When the linear relation is positive, the pattern of data points flows upward and to the right. When the linear relation is negative, the pattern of data points flows downward and to the right. The data story about hours studying and statistics grades in Figures 3-5 and 3-6 indicates a positive, linear relation.

A ***nonlinear relation*** *between variables means that the relation between variables is best described by a line that breaks or curves in some way.* *Nonlinear* simply means "not straight,"

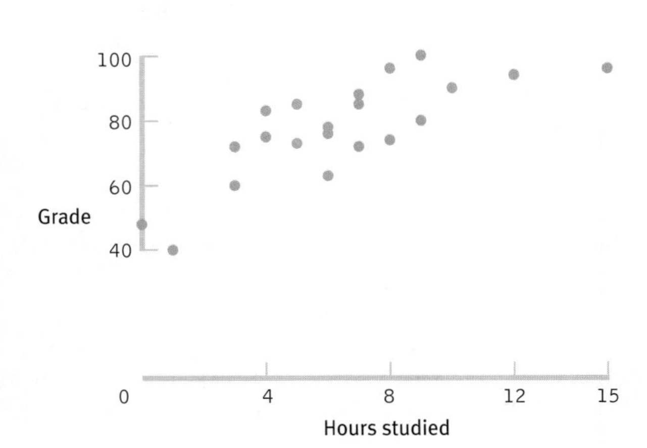

FIGURE 3-6
A Range-Frame Improves on a Scatterplot

A range-frame is a traditional scatterplot that indicates the minimum and maximum observed values on the axes by erasing all ink beyond these points. This simple alteration increases the ratio of ink dedicated to actual data to overall printed ink in this graph.

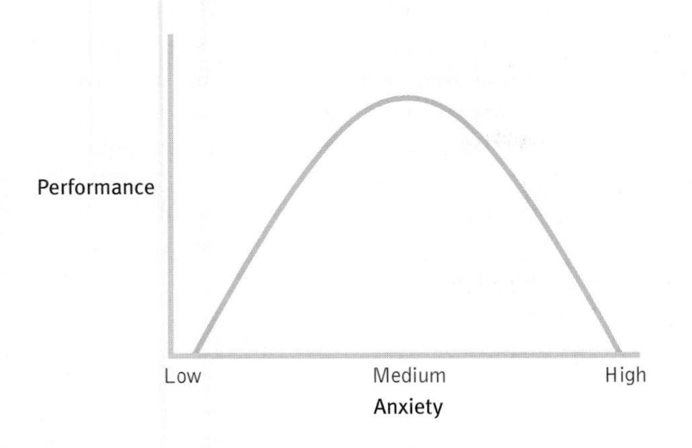

FIGURE 3-7
Nonlinear Relations

The Yerkes–Dodson law predicts that stress/anxiety improves test performance—but only to a point. Too much anxiety leads to an inability to perform at one's best. This inverted U-curve illustrates the concept, but a scatterplot would better clarify the particular relation between these two variables.

so there are many possible nonlinear relations between variables. For example, the Yerkes–Dodson law described in Figure 3-7 predicts the relation between level of arousal and test performance. As professors, we don't want you so relaxed that you don't even show up for the test, but we also don't want you so stressed out that you have a panic attack. You will maximize your performance somewhere in the happy middle described by a nonlinear relation (in this case, an upside-down U-curve). ■

Line Graphs

A ***line graph*** *is used to illustrate the relation between two scale variables.* One type of line graph is based on a scatterplot and allows us to construct a line of best fit that represents the predicted *y* score for each *x* value. A second type of line graph allows us to visualize changes in the values on the *y*-axis over time.

EXAMPLE 3.2

The first type of line graph, based on a scatterplot, is especially useful because the best fit line minimizes the distances between all the data points from that line. That allows us to use the *x* value to predict the *y* value and make predictions based on only one piece of information. For example, we can use the line of best fit in Figure 3-8 to predict that if a student studies for 2 hours, she will earn a test score of about 62; if she studies for 13 hours, she will earn a score of about 100. For now, we can simply eyeball the scatterplot and draw a line of best fit; in Chapter 16, you will learn how to calculate a line of best fit.

Here is a recap of the steps to create a scatterplot with a line of best fit:

1. Label the *x*-axis with the name of the independent variable and its possible values, starting with 0 if practical.
2. Label the *y*-axis with the name of the dependent variable and its possible values, starting with 0 if practical.
3. Make a mark above each study participant's score on the *x*-axis and next to his or her score on the *y*-axis.
4. Visually estimate and sketch the line of best fit through the points on the scatterplot.
5. Maximize the data–ink ratio by converting to a range frame: erase the axes below the minimum score and above the maximum score.

FIGURE 3-8
The Line of Best Fit

The line of best fit allows us to make predictions for a person's value on the *y* variable from his or her value on the *x* variable.

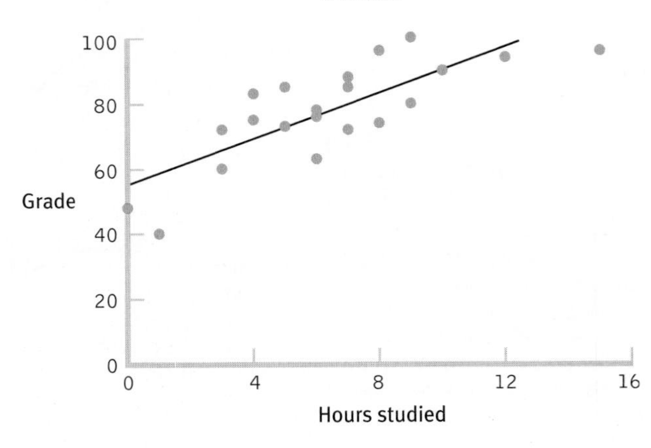

A second situation in which a line graph is more useful than just a scatterplot involves time-related data. A ***time plot,*** or ***time series plot,*** is a graph that plots a scale variable on the *y*-axis as it changes over an increment of time (e.g., second, day, century) labeled on the *x*-axis. As with a scatterplot, marks are placed above each value on the *x*-axis (e.g., at a given minute) at the value for that particular time on the *y*-axis (i.e., the score on the dependent variable). These marks are then connected with a line. With a time plot, it's possible to graph several lines on the same graph, as long as they use the same scale variable on the *y*-axis. With multiple lines, the viewer can compare the trends for different levels of another variable. ■

EXAMPLE 3.3

Figure 3-9, for example, shows positive attitudes and negative attitudes around the world, as expressed on Twitter. The researchers analyzed more than half a billion Tweets over the course of 24 hours (Golder & Macy, 2011) and plotted separate lines for each day of the week. These fascinating data tell many stories. For example, people tend to express more positive and fewer negative attitudes in the morning than later in the day; people express more positive attitudes on the weekends than during the week; and the weekend morning peak in positive attitudes is later than during the week, perhaps an indication that people are sleeping in.

Here is a recap of the steps to create a time plot:

1. Label the *x*-axis with the name of the independent variable and its possible values. The independent variable should be an increment of time (e.g., hour, month, year).
2. Label the *y*-axis with the name of the dependent variable and its possible values, starting with 0 if practical.
3. Make a mark above each value on the *x*-axis at the value for that time on the *y*-axis.
4. Connect the dots.
5. As you did with the scatterplot, maximize the data–ink ratio by converting to a range-frame—erase the *y*-axis below the minimum *y* value and above the maximum *y* value.

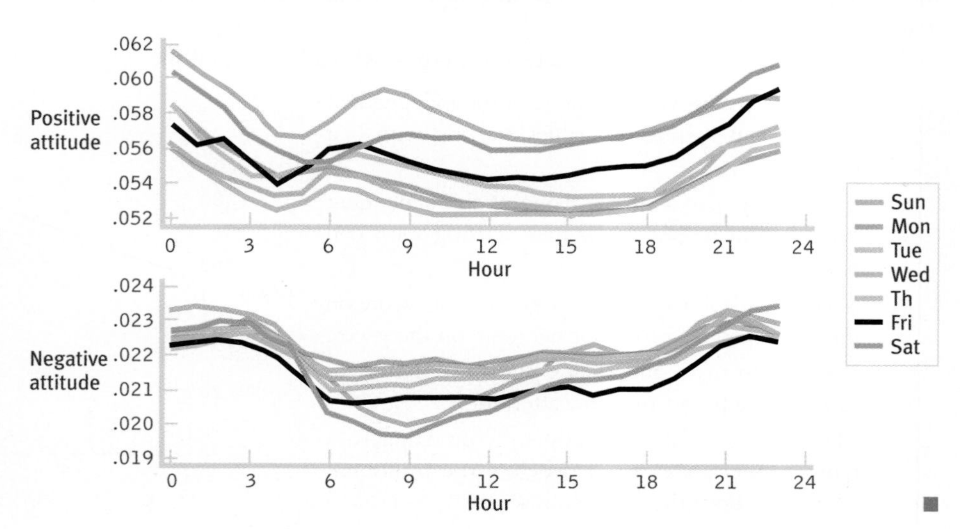

FIGURE 3-9
Hourly Moods as Seen Through Twitter
Researchers tracked positive attitudes and negative attitudes expressed through Twitter over the course of a day and from around the globe (Golder & Macy, 2011). Time plots allow for multiple scale variables on one graph; in this case, there are separate lines for each day of the week.

■

Bar Graphs

*A **bar graph** is a visual depiction of data in which the independent variable is nominal or ordinal and the dependent variable is scale.* The height of each bar typically represents the average value of the dependent variable. The independent variable on the x-axis could be either nominal (such as gender) or ordinal (such as Olympic medal winners who won gold, silver, or bronze medals). We could even combine two independent variables in a single graph by drawing two separate clusters of bars to compare men's and women's running times of the gold, silver, and bronze medalists.

Here is a recap of the variables used to create a bar graph:

1. The x-axis of a bar graph indicates discrete levels of a nominal or an ordinal variable.
2. The y-axis of a bar graph may represent counts or percentages. But the y-axis of a bar graph can also indicate many other scale variables, such as average running speeds, scores on a memory task, or reaction times.

> **MASTERING THE CONCEPT**
>
> **3.3:** Bar graphs depict data for two or more categories. They tell a data story more precisely than either pictorial graphs or pie charts.

Bar graphs are flexible tools for the visual presentation of data. For example, if there are many categories to be displayed along the horizontal x-axis, researchers sometimes create a ***Pareto chart**, a type of bar graph in which the categories along the x-axis are ordered from highest bar on the left to lowest bar on the right.* This ordering allows easier comparisons and easier identification of the most common and least common categories.

EXAMPLE 3.4

Figure 3-10 shows two different ways of depicting the percentage of Internet users in a given country who visited Twitter.com in June 2010. One graph is an alphabetized bar graph; the other is a Pareto chart. How does Canada fare with respect to other countries? Which graph makes it is easier to answer that question?

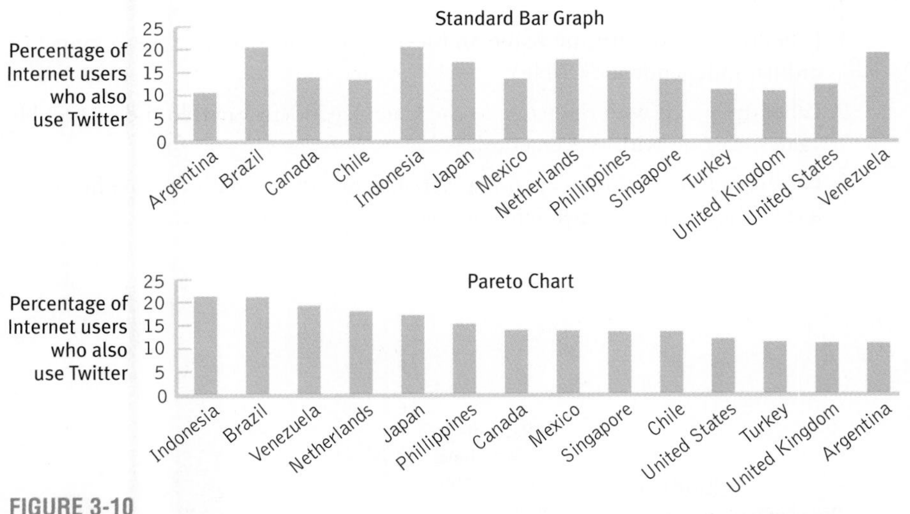

FIGURE 3-10
The Flexibility of the Bar Graph

The standard bar graph provides a comparison of Twitter usage among 14 levels of a nominal dependent variable, country. The Pareto chart, a version of a bar graph, orders the countries from highest to lowest along the horizontal axis, which allows us to more easily pick out the highest and lowest bars. We can more easily know that Canada places in the middle of these countries, and that the United States and the United Kingdom are toward the bottom. We have to do more work to draw these conclusions from the original bar graph. ■

- A **time plot**, or **time series plot**, is a graph that plots a scale variable on the y-axis as it changes over an increment of time (e.g., second, day, century) labeled on the x-axis.
- A **bar graph** is a visual depiction of data in which the independent variable is nominal or ordinal and the dependent variable is scale. Each bar typically represents the average value of the dependent variable for each category.
- A **Pareto chart** is a type of bar graph in which the categories along the x-axis are ordered from highest bar on the left to lowest bar on the right.

EXAMPLE 3.5

Bar graphs can help us understand the answers to interesting questions. For example, researchers wondered whether piercings and tattoos, once viewed as indicators of a "deviant" worldview, had become mainstream (Koch, Roberts, Armstrong, & Owen, 2010). They surveyed 1753 American college students with respect to numbers of piercings and tattoos, as well as about a range of destructive behaviors including academic cheating, illegal drug use, and number of arrests (aside from traffic arrests). The bar graph in Figure 3-11 depicts one finding: the likelihood of having been arrested was fairly similar among all groups, except among those with four or more tattoos, 70.6% of whom reported having been arrested at least once. A magazine article about this research advised parents, "So, that butterfly on your sophomore's ankle is not a sign she is hanging out with the wrong crowd. But if she comes home for spring break covered from head to toe, start worrying" (Jacobs, 2010).

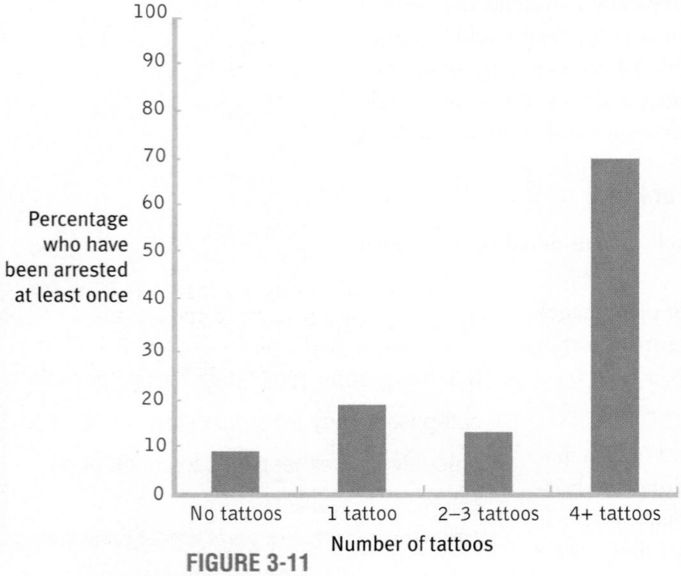

Percentage who have been arrested at least once
100
90
80
70
60
50
40
30
20
10
0
No tattoos
1 tattoo
2–3 tattoos
4+ tattoos
Number of tattoos

FIGURE 3-11
Bar Graphs Highlight Differences Between Averages or Percentages

This bar graph depicts the percentages who have been arrested at least once (other than a traffic arrest) for four groups of U.S. university students: those with no tattoos, one tattoo, two to three tattoos, or four or more tattoos. A bar graph can more vividly depict differences between percentages than just the typed numbers themselves can: 8.5, 18.7, 12.7, and 70.6.

Liars' Alert! The small differences among the students with no tattoos, one tattoo, and two or three tattoos could be exaggerated if a reporter wanted to scare parents. Compare Figure 3-12 to the first three bars of Figure 3-11. Notice what happens when the values on the y-axis do not begin at 0, the intervals change from 10 to 2, and the y-axis ends at 20%. The exact same data leave a very different impression. (*Note:* If the data are very far from 0, and it does not make sense to have the axis go down to 0, indicate this on the graph by including double slashes—called cutmarks—like those shown in Figure 3-12.)

Here is a recap of the steps to create a bar graph. The critical choice for you, the graph creator, is in step 2.

1. Label the x-axis with the name and levels (i.e., categories) of the nominal or ordinal independent variable.
2. Label the y-axis with the name of the scale dependent variable and its possible values, starting with 0 if practical.
3. For every level of the independent variable, draw a bar with the height of that level's value on the dependent variable.

- A **pictorial graph** is a visual depiction of data typically used for an independent variable with very few levels (categories) and a scale dependent variable. Each level uses a picture or symbol to represent its value on the scale dependent variable.
- A **pie chart** is a graph in the shape of a circle, with a slice for every level (category) of the independent variable. The size of each slice represents the proportion (or percentage) of each level.

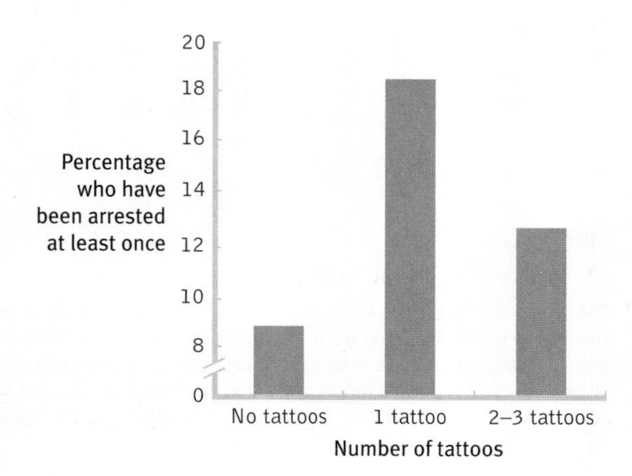

FIGURE 3-12
Deceiving with the Scale

To exaggerate a difference between means, graphmakers sometimes compress the rating scale that they show on their graphs. When possible, label the axis beginning with 0, and when displaying percentages, include all values up to 100%.

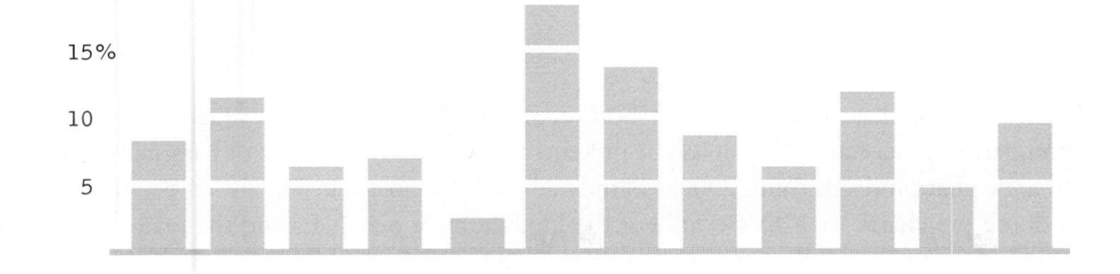

FIGURE 3-13
Redesigning the Bar Graph

Eliminating the frame and the *y*-axis and adding thin white lines through the bars, as suggested by Tufte (2001/2006b), makes this bar graph easier to read and increases the data–ink ratio.

Tufte (2001) has a plan for better bar graphs. In Figure 3-13, Tufte (a) eliminated both the box around the graph and the vertical axis; (b) kept the data labels on the *y*-axis; and (c) replaced the horizontal tick marks with thin white lines through the bars—another increase in the data–ink ratio. ■

Pictorial Graphs

Occasionally, a pictorial graph is acceptable—but be careful. *A* ***pictorial graph*** *is a visual depiction of data typically used for an independent variable with very few levels (categories) and a scale dependent variable. Each level uses a picture or symbol to represent its value on the scale dependent variable.* Eye-catching pictorial graphs are far more common in the popular media than in research journals. They tend to direct attention to the clever artwork rather than to the story that the data tell.

For example, a graphmaker might use stylized drawings of people to indicate population size. Figure 3-14 demonstrates one problem with pictorial graphs. The picture makes the person twice as tall *and* twice as wide (so that the taller person won't look so stretched out). But then the total area of the picture is about four times larger than the shorter one, even though the population is only twice as big—a false impression.

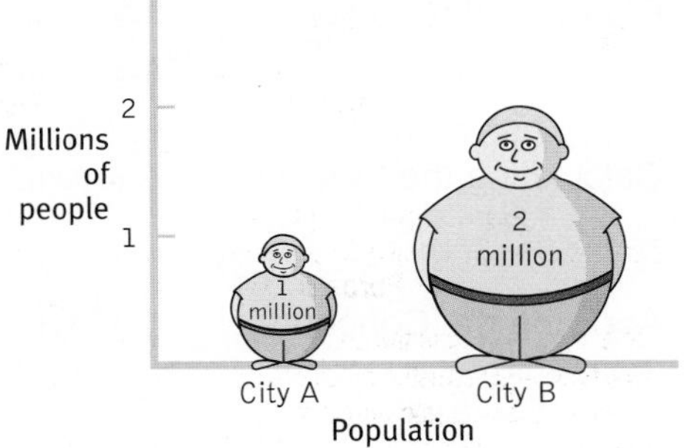

FIGURE 3-14
Distorting the Data with Pictures

In a pictorial graph, doubling the height of a picture is often coupled with doubling the width—which is multiplying by 2 twice. Instead of being twice as big, the picture is *four times* as big!

Pie Charts

A ***pie chart*** *is a graph in the shape of a circle, with a slice for every level (category) of the independent variable. The size of each slice represents the proportion (or percentage) of each category.* A pie chart's slices should *always* add up to 100% (or 1.00, if using proportions). Figure 3-15

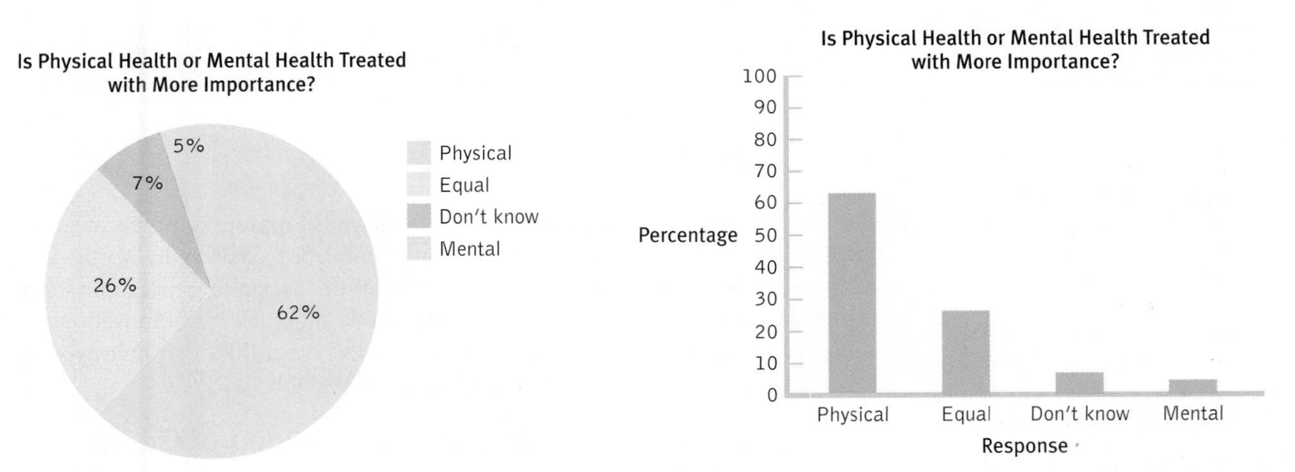

FIGURE 3-15
Pie Chart or Bar Graph?

A research firm hired by the Suicide Prevention Action Network (2004) asked U.S. participants, "Do you think that mental health and physical health are treated with equal importance in our current health care system?" We can see from the pie chart that most people (62%) believe that physical health is treated with more importance than is mental health; however, the bar graph is easier to interpret.

includes a pie chart and a bar graph, both depicting the same data. As suggested by this comparison, data can almost always be presented more clearly in a table or bar graph than in a pie chart. Indeed, Tufte (2006b) bluntly advises: "A table is nearly always better than a dumb pie chart" (p. 178). Because of the limitations of pie charts and the ready alternatives, we do not outline the steps for creating a pie chart here.

CHECK YOUR LEARNING

Reviewing the Concepts

> Scatterplots and line graphs allow us to see relations between two scale variables.
> When examining the relation between variables, it is important to consider linear and nonlinear relations, as well as the possibility that no relation is present.
> Bar graphs, pictorial graphs, and pie charts depict summary values (such as means or percentages) on a scale variable for various levels of a nominal or ordinal variable.
> Bar graphs are preferred; pictorial graphs and pie charts can be misleading.

Clarifying the Concepts

3-4 How are scatterplots and line graphs similar?

3-5 Why should we typically avoid using pictorial graphs or pie charts?

Calculating the Statistics

3-6 What type of visual display of data allows us to calculate or evaluate how a variable is changing over time?

Applying the Concepts

3-7 What is the best type of common graph to depict each of the following data sets and research questions? Explain your answers.

a. Depression severity and amount of stress for 150 university students. Is depression related to stress level?

b. Number of inpatient mental health facilities in Canada as measured every 10 years between 1890 and 2000. Has the number of facilities declined in recent years?

c. Number of siblings reported by 100 people. What size family is most common?

d. Mean years of education for six regions of the United States. Are education levels higher in some regions than in others?

e. Calories consumed in a day and hours slept that night for 85 people. Does the amount of food a person eats predict how long he or she sleeps at night?

Solutions to these Check Your Learning questions can be found in Appendix D.

How to Build a Graph

In this section, you will learn how to choose the most appropriate type of graph and then use a checklist that ensures your graph conforms to APA style. We also discuss innovative graphs that highlight the exciting future of graphing that some social scientists are already using to let their data speak more clearly. Innovations in graphing can help us to deliver a persuasive message, much like that conveyed by Florence Nightingale's coxcomb graph.

> **MASTERING THE CONCEPT**
>
> **3.4:** The best way to determine the type of graph to create is to identify the independent variable and the dependent variable, along with the type of variable that each is—nominal, ordinal, or scale.

Choosing the Appropriate Type of Graph

When deciding what type of graph to use, first examine the variables. Decide which is the independent variable and which is the dependent variable. Also, identify which type of variable—nominal, ordinal, or scale (interval/ratio)—each of them is. Most of the time, the independent variable belongs on the horizontal *x*-axis and the dependent variable goes on the vertical *y*-axis.

After assessing the types of variables that are in the study, use the following guidelines to select the appropriate graph:

1. If there is one scale variable (with frequencies), use a histogram or a frequency polygon (Chapter 2).
2. If there is one scale independent variable and one scale dependent variable, use a scatterplot or a line graph. (Figure 3-9 on page 50 provides an example of how to use more than one line on a time plot.)
3. If there is one nominal or ordinal independent variable and one scale dependent variable, use a bar graph or a Pareto chart.
4. If there are two or more nominal or ordinal independent variables and one scale dependent variable, use a bar graph.

How to Read a Graph

Let's use the graph in Figure 3-16 to confirm your understanding of independent and dependent variables. This study of jealousy (Maner, Miller, Rouby, & Gailliot, 2009) includes two independent variables: level of jealousy (whether or not a participant is low or high in chronic jealousy) and how they were primed (to think about infidelity or a neutral topic). People primed to think about infidelity visualized and wrote about a time when they experienced infidelity-related concerns. People primed to think about a neutral topic wrote a detailed account of four or five things they had done the previous day. The dependent variable is how long participants looked at photographs of an attractive same-sex person; for a jealous heterosexual person, an attractive person of the same sex would be a potential threat to his or her relationship.

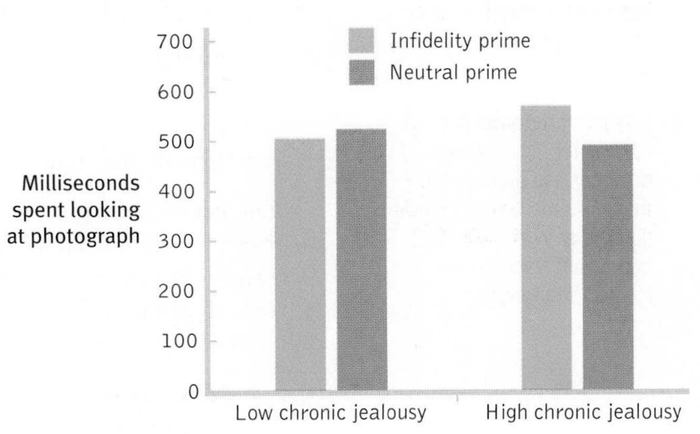

FIGURE 3-16
Two Independent Variables

When we are graphing a data set that has two independent variables, we show one independent variable on the *x*-axis (in this case, chronic jealousy—high or low) and one independent variable in a color-coded key (in this case, type of prime—infidelity or neutral). This graph demonstrates that men high in chronic jealousy looked longer at an attractive man, a potential threat, when primed by thinking about infidelity than when primed by thinking about a neutral topic. The pattern was reversed for men low in chronic jealousy.

1. What variable are the researchers trying to predict? That is, what is the *dependent variable*?
2. Is the dependent variable nominal, ordinal, or scale?
3. What are the units of measurement on the dependent variable? For example, if the dependent variable is IQ as measured by the Wechsler Adult Intelligence Scale, then the possible scores are the IQ scores themselves, ranging from 0 to 145.
4. What variables did the researchers use to predict this dependent variable? That is, what are the *independent variables*?
5. Are these two independent variables nominal, ordinal, or scale?
6. What are the levels for each of these independent variables?

Now check your answers:

1. The dependent variable is time, in milliseconds.
2. Milliseconds is a scale variable.
3. Milliseconds can range from 0 on up. In this case, no average exceeds 600.
4. The first independent variable is level of jealousy; the second independent variable is priming condition.
5. Level of jealousy is an ordinal variable but it can be treated as a nominal variable in the graph; priming condition is a nominal variable.

- **Chartjunk** is any unnecessary information or feature in a graph that detracts from a viewer's ability to understand the data.
- **Moiré vibrations** are any visual patterns that create a distracting impression of vibration and movement.
- **Grids** are chartjunk that take the form of a background pattern, almost like graph paper, on which the data representations, such as bars, are superimposed.
- A **duck** is a form of chartjunk in which a feature of the data has been dressed up to be something other than merely data.
- Computer **defaults** are the options that the software designer has preselected; these are the built-in decisions that the software will implement if you do not instruct it otherwise.

6. The levels for jealousy level are low chronic jealousy and high chronic jealousy. The levels for priming condition are infidelity and neutral.

Because there are two independent variables—both of which are nominal—and one scale dependent variable, we used a bar graph to depict these data.

Guidelines for Creating a Graph

Here is a helpful checklist of questions to ask when you've created a graph or when you encounter a graph. Some we've mentioned previously, and all are wise to follow.

- Does the graph have a clear, specific title?
- Are both axes labeled with the names of the variables? Do all labels read left to right—even the one on the *y*-axis?
- Are all terms on the graph the same terms that are used in the text that the graph is to accompany? Have all abbreviations been eliminated?
- Are the units of measurement (e.g., minutes, percentages) included in the labels?
- Do the values on the axes either go down to 0 or have cut marks (double slashes) to indicate that they do not go down to 0?
- Are colors used in a simple, clear way—ideally, shades of gray instead of other colors?
- Has all chartjunk been eliminated?

The last of these guidelines involves a new term, the graph-corrupting fluff called *chartjunk,* a term coined by Tufte (2001). According to Tufte, ***chartjunk*** *is any unnecessary information or feature in a graph that detracts from a viewer's ability to understand the data.* Chartjunk can take the form of any of three unnecessary features, all demonstrated in the rather frightening graph in Figure 3-17.

1. ***Moiré vibrations*** *are any visual patterns that create a distracting impression of vibration and movement.* They are unfortunately sometimes the default settings for bar graphs in statistical software. Tufte recommends using shades of gray instead of patterns.
2. *A* ***grid*** *is a background pattern, almost like graph paper, on which the data representations, such as bars, are superimposed.* Tufte recommends the use of grids only for hand-drawn drafts of graphs. In final versions of graphs, use only very light grids if necessary.

FIGURE 3-17
Chartjunk Run Amok

Moiré vibrations, such as those seen in the patterns on these bars, might be fun to use, but they detract from the viewer's ability to glean the story of the data. Moreover, the grid pattern behind the bars might appear scientific, but it serves only to distract. Ducks—like the 3-D shadow effect on the bars and the globe clip-art—add nothing to the data, and the colors are absurdly eye straining. Don't laugh; we've had students submit carefully written research papers accompanied by graphs even more garish than this!

3. ***Ducks** are features of the data that have been dressed up to be something other than merely data.* Think of ducks as data in costume. Named for the Big Duck, a store in Flanders, New York, that was built in the form of a very large duck, graphic ducks can be three-dimensional effects, cutesy pictures, fancy fonts, or any other flawed design features. Avoid chartjunk!

MASTERING THE CONCEPT

3.5: Avoid chartjunk—any unnecessary aspect of a graph that detracts from its clarity.

There are several computer-generated graphing programs that have defaults that correspond to many—but not all—of these guidelines. Computer ***defaults** are the options that the software designer has preselected; these are the built-in decisions that the software will implement if you do not instruct it otherwise.* You cannot assume that these defaults represent the APA guidelines for your particular situation. You can usually point the cursor at a part of the graph and click to view the available options.

Franck Fotos/Alamy

Edward Tufte's Big Duck The graphics theorist Edward Tufte took this photograph of the Big Duck, the store in the form of a duck for which he named a type of chartjunk (graphic clutter). In graphs, ducks are any aspects of the graphed data that are "overdressed," obscuring the message of the data. Think of ducks as data in a ridiculous costume.

The Future of Graphs

Thanks to computer technology, we have entered a second golden age of scientific graphing. We mention only three categories here: interactive graphs, clinical applications, and geographic information systems.

Interactive Graphs One informative and haunting graph was published online in the *New York Times* on September 9, 2004, to commemorate the day on which the 1000th U.S. soldier died in Iraq (http://tinyurl.com/55nlu). Titled "A Look at 1000 Who Died," this beautifully designed tribute is formed by photos of each of the dead servicemen and women. One can view these photographs organized by variables such as last name, where they lived, gender, cause of death, and how old they were when they died. Because the photos are the same size, the stacking of the photos also functions as a bar graph. By clicking on two or more months in a row, or on two or more ages in a row, one can visually compare numbers of deaths among levels of a category.

Yet this interactive graph is even more nuanced, because it provides a glimpse into the life stories of these soldiers. By holding the cursor over a photo that catches your eye, you can learn, for example, that Spencer T. Karol, regular duty in the U.S. Army, from Woodruff, Arizona, died on October 6, 2003, at the age of 20 from hostility-inflicted wounds. A thoughtfully designed interactive graph such as this one holds even more power than a traditional flat graph in how it educates, evokes emotion, and provides details that humanize the stories behind the numbers.

Clinical Applications Clinical psychology researchers have developed graphing techniques, illustrated in Figure 3-18, to help therapists identify when the therapy process appears to be leading to a poorer-than-expected outcome (Howard, Moras, Brill, Martinovich, & Lutz, 1996). The dependent variable, rate of actual improvement, is graphed as a line that compares it to the rate of expected improvement (based on previous research). If therapy progresses more slowly than expected, the observed discrepancy points out the need to examine why a particular client is not making better progress.

Geographic Information Systems (GIS) Many companies have published software that enables computer programmers to link Internet-based data to Internet-based maps (Markoff, 2005). These visual tools are all variations on geographic information systems (GIS). The APA sponsors an advanced workshop on how to apply GIS to the social sciences.

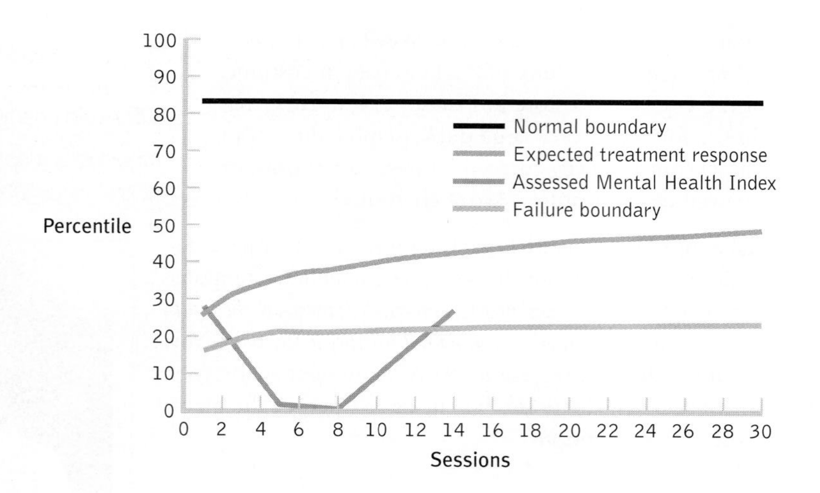

FIGURE 3-18
Graph as Therapy Tool
Some graphs allow therapists to compare the actual rate of a client's improvement with the expected rate given that client's characteristics. This client (assessed Mental Health Index in gray) is doing worse than expected (expected treatment response in red) but has improved enough to be above the failure boundary (in yellow).

Sociologists, geographers, political scientists, consumer psychologists, and epidemiologists (who use statistics to track patterns of disease) have already become familiar with GIS in their respective fields. Organizational psychologists, social psychologists, and environmental psychologists can use GIS to organize workflow, assess group dynamics, and study the design of classrooms. Ironically, this advance in computerized mapping is pretty much what John Snow did without a computer in 1854 when he studied the Broad Street cholera outbreak.

CHECK YOUR LEARNING

Reviewing the Concept

- Graphs should be used when they add information to written text or help to clarify difficult material.
- To decide what kind of graph to use, determine whether the independent variable and the dependent variable are nominal, ordinal, or scale variables.
- A brief checklist will help you create an understandable graph. Label graphs precisely and avoid chartjunk.
- In the near future, online interactive graphs, graphs based on sophisticated prediction models such as those that forecast therapy outcomes, and computerized mapping will become increasingly common.

Clarifying the Concepts

3-8 What is chartjunk?

Calculating the Statistics

3-9 Deciding what kind of graph to use depends largely on how variables are measured. Imagine a researcher is interested in how "quality of sleep" is related to typing performance (measured by the number of errors made). For each of the measures of sleep below, decide what kind of graph to use.

a. Total minutes slept

b. Sleep assessed as sufficient or insufficient

c. Using a scale from 1 (low-quality sleep) to 7 (excellent sleep)

Applying the Concepts

3-10 Imagine that the graph in Figure 3-18 represents data testing the hypothesis that exposure to the sun can impair IQ. Further imagine that the researcher has recruited groups of people and randomly assigned them to different levels of exposure to the sun: 0, 1, 6, and 12 hours per day (enhanced, in all cases, by artificial sunlight when natural light is not available). The mean IQ scores are 142, 125, 88, and 80, respectively. Redesign this chartjunk graph, either by hand or by using software, paying careful attention to the dos and don'ts outlined in this section.

Solutions to these Check Your Learning questions can be found in Appendix D.

REVIEW OF CONCEPTS

How to Lie with Visual Statistics

Learning how visual displays of statistics can mislead or lie will empower you to spot lies for yourself. Because visual displays of data are so easily manipulated, it is important to pay close attention to the details of graphs to be sure the graph creator isn't conveying false information.

Common Types of Graphs

When developing graphing skills, it is important to begin with the basics. Several types of graphs are commonly used by social scientists. *Scatterplots* depict the relation between two scale variables. They are useful when determining whether the relation between the variables is *linear* or *nonlinear.* A *range-frame* is a variant of a scatterplot; it provides more information with less ink by eliminating the axes below the minimum value and above the maximum value. Some *line graphs* expand on scatterplots by including a line of best fit. Others, called *time plots* or *time series plots,* show the change in a scale variable over time.

Bar graphs are used to compare two or more categories of a nominal or ordinal independent variable with respect to a scale dependent variable. A bar graph on which the levels of the independent variable are organized from the highest bar to the lowest bar, called a *Pareto chart,* allows for easy comparison of levels. *Pictorial graphs* are like bar graphs except that pictures are used in place of bars. *Pie charts* are used to depict proportions or percentages on one nominal or ordinal variable with just a few levels. Because both pictorial graphs and pie charts are frequently constructed in a misleading way or are misperceived, bar graphs are almost always preferred to pictorial graphs and pie charts.

How to Build a Graph

We first decide what type of graph to create by examining the independent and dependent variables and by identifying each as nominal, ordinal, or scale. We must then consider a number of guidelines to develop a clear, persuasive graph. It is important that all graphs be labeled thoroughly and appropriately and given a title that allows the graph to tell its story without additional text. For an unambiguous graph, it is imperative that graph creators avoid *chartjunk*: unnecessary information, such as *moiré vibrations, grids,* and *ducks,* that clutters a graph and makes it difficult to interpret. When using software to create graphs, it is important to question the *defaults* built into the software and to override them when necessary to adhere to these guidelines.

Finally, keeping an eye to the future of graphing—including interactive graphs, the use of statistical models to predict therapy outcomes, and computer-generated maps—helps us stay at the forefront of graph-making in the behavioral sciences.

SPSS®

We can request visual displays of data from both the "Data View" screen and the "Variable View" screen. SPSS allows us to create visual displays across several different menus; however, most graphing is done in SPSS using the Chart Builder. This section walks you through the general steps to create a graph, using a scatterplot as an example. But first, enter the data in the screenshot for hours spent studying and exam grades that were used to create the scatterplot in Figure 3-5.

Select Graphs → Chart Builder → Gallery. Under "Choose from:" select the type of graph by clicking on it. For example, to create a scatterplot, click on "Scatter/Dot." Drag a sample graph from the right to the large box above. Usually, you'll want the simplest graph, which tends to be the upper-left sample graph.

Drag the appropriate variables from the "Variables:" box to the appropriate places on the sample graph (e.g., "*x*-axis"). For a scatterplot, drag "hours" to the *x*-axis and "grade" to the *y*-axis. Chart Builder then looks like the screenshot shown here. Click OK and SPSS creates the graph.

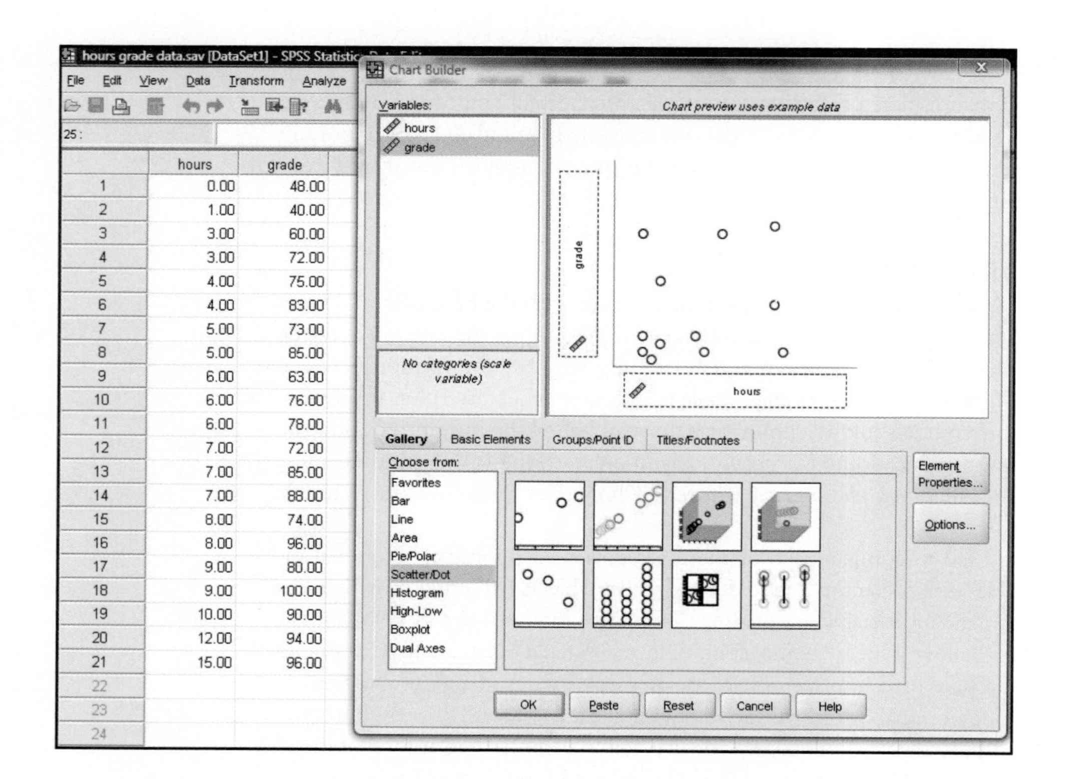

	hours	grade
1	0.00	48.00
2	1.00	40.00
3	3.00	60.00
4	3.00	72.00
5	4.00	75.00
6	4.00	83.00
7	5.00	73.00
8	5.00	85.00
9	6.00	63.00
10	6.00	76.00
11	6.00	78.00
12	7.00	72.00
13	7.00	85.00
14	7.00	88.00
15	8.00	74.00
16	8.00	96.00
17	9.00	80.00
18	9.00	100.00
19	10.00	90.00
20	12.00	94.00
21	15.00	96.00

Remember: You should not rely on the default choices of the software; you are the designer of the graph. Once the graph is created, you can change the graph's appearance by double-clicking on the graph to open the Chart Editor, the tool that allows you to make changes. Then click or double-click on the particular feature of the graph that you want to modify. Clicking once on part of the graph allows you to make some changes. For example, clicking the label of the *y*-axis allows you to retype the label; double-clicking allows you to make other changes, such as making the label horizontal (after double-clicking, select the orientation "Horizontal" under "Text Layout").

How It Works

3.1 CREATING A SCATTERPLOT

Gapminder.org is a wonderful Web site that allows the public to play with a graph and explore the relations between variables over time. We used Gapminder World to find scores for 9 countries plus Hong Kong on two variables.

Country	Children per woman (total fertility)	Life expectancy at birth (years)	Country	Children per woman (total fertility)	Life expectancy at birth (years)
Afghanistan	7.15	43.00	Bolivia	3.59	65.00
India	2.87	64.00	Ethiopia	5.39	53.00
China	1.72	73.00	Iraq	4.38	59.00
Hong Kong	0.96	82.00	Mali	6.55	54.00
France	1.89	80.00	Honduras	3.39	70.00

How can we create a scatterplot to show the relation between these two variables? To create a scatterplot, we put total fertility on the x-axis and life expectancy in years on the y-axis. We then add a dot for each country at the intersection of its fertility rate and life expectancy. The scatterplot is shown in the figure below.

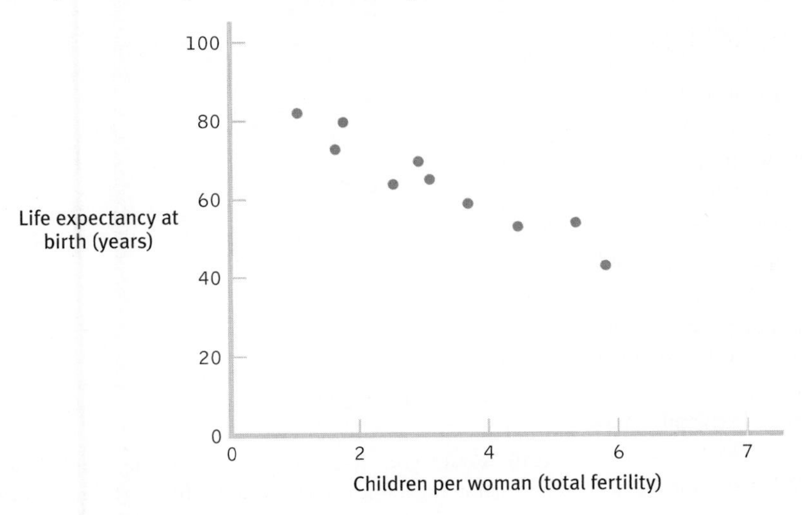

3.2 CREATING A BAR GRAPH

Here is the 2004 gross domestic product (GDP), in trillions of U.S. dollars, for each of the world economic powers that make up what is called the Group of Eight, or G8, nations.

Canada: 0.98
France: 2.00
Germany: 2.71
Italy: 1.67
Japan: 4.62
Russia: 0.58
United Kingdom: 2.14
United States: 11.67

How can we create a bar graph for these data? First, we put the countries on the x-axis. Then, for each country, we draw a bar whose height corresponds to the country's GDP. The following figure shows a bar graph with bars arranged in alphabetical order by country.

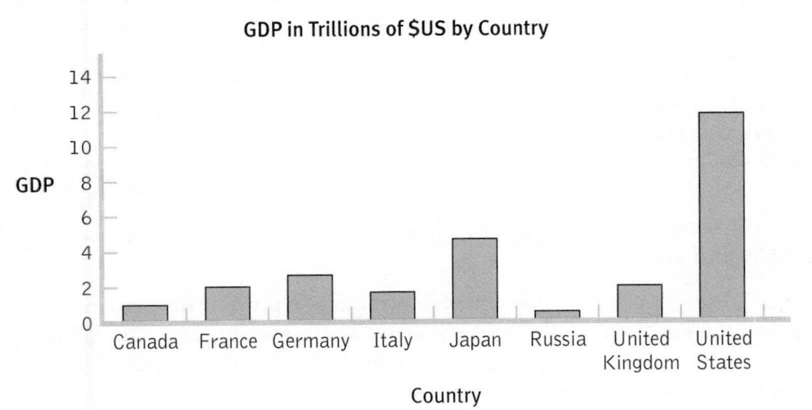

Exercises

Clarifying the Concepts

3.1 What are the five techniques discussed in this chapter for misleading with graphs?

3.2 What are the steps to create a scatterplot?

3.3 How can you convert a scatterplot into a range-frame?

3.4 What does it mean for two variables to be linearly related?

3.5 How can we tell whether two variables are linearly or nonlinearly related?

3.6 What is the difference between a line graph and a time plot?

3.7 What is the difference between a bar graph and a Pareto chart?

3.8 Bar graphs and histograms look very similar. In your own words, what is the difference between the two?

3.9 What are pictorial graphs and pie charts?

3.10 Why are bar graphs preferred over pictorial graphs and pie charts?

3.11 Why is it important to identify the independent variable and the dependent variable before creating a visual display?

3.12 Under what circumstances would the x-axis and y-axis not start at 0?

3.13 Chartjunk comes in many forms. What specifically are moiré vibrations, grids, and ducks?

3.14 Geographic information systems (GIS), such as those provided by computerized graphing technologies, are particularly powerful tools for answering what kinds of research questions?

Calculating the Statistics

3.15 Alumni giving rates, calculated as the total dollars donated per year from 1999 to 2009, represent which kind of variable—nominal, ordinal, or scale? What would be an appropriate graph to depict these data?

3.16 Alumni giving rates for a number of universities, calculated as the number of alumni who donated and the number who did not donate in a given year, represent which kind of variable—nominal, ordinal, or scale? What would be an appropriate graph to depict these data?

3.17 You are exploring the relation between gender and video game performance, as measured by final scores on a game.

a. In this study, what are the independent and dependent variables?

b. Is gender a nominal, ordinal, or scale variable?

c. Is final score a nominal, ordinal, or scale variable?

d. Which graph or graphs would be appropriate to depict the data? Explain why.

3.18 Would you describe the data in the graph below as showing a linear relation, a nonlinear relation, or no relation? Explain.

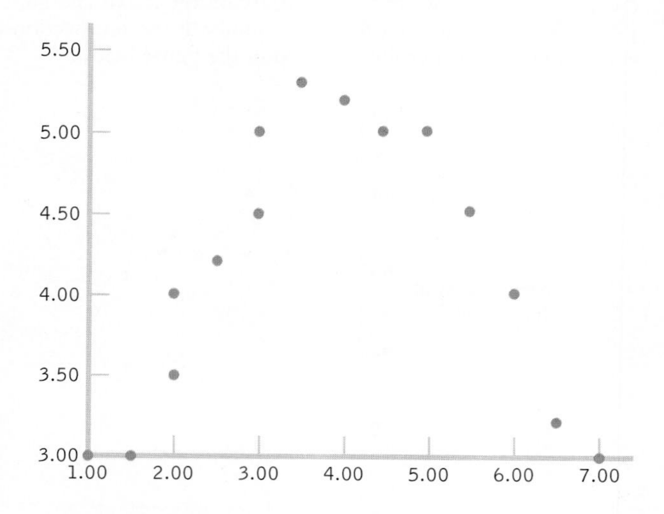

3.19 Would you describe the data in the graph below as showing a linear relation, a nonlinear relation, or no relation? Explain.

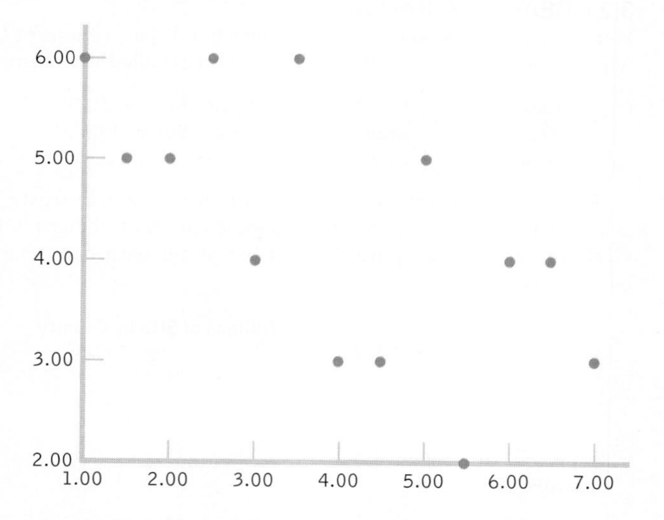

3.20 What elements are missing from the graphs in Exercises 3.18 and 3.19?

3.21 The following figure presents the enrollment of graduate students at a university, across six fall terms, as a percentage of the total student population.

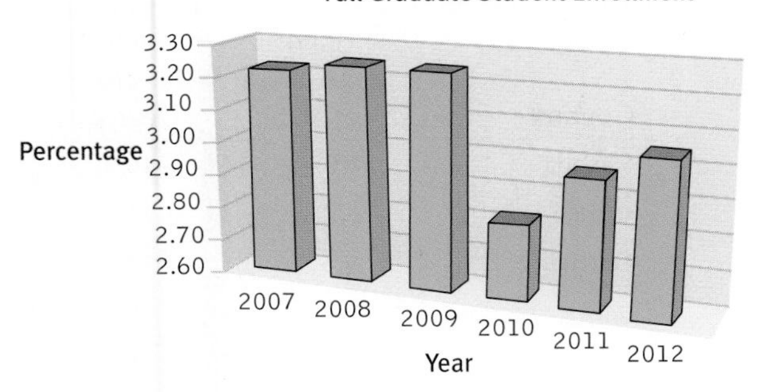

a. What kind of visual display is this?

b. What other type of visual display could have been used?

c. What is missing from the axes?

d. What chartjunk is present?

e. Using this graph, estimate graduate student enrollment, as a percentage of the total student population, in the fall terms of 2007, 2008, and 2009.

f. How would the comparisons between bars change if the y-axis started at 0?

3.22 When creating a graph, we need to make a decision about the numbering of the axes. If you had the following range of data for one variable, how might you label the relevant axis?

337 280 279 311 294 301 342 273

3.23 If you had the following range of data for one variable, how might you label the relevant axis?

0.10 0.31 0.27 0.04 0.09 0.22 0.36 0.18

3.24 The scatterplot in How It Works 3.1 depicts the relation between fertility and life expectancy. Each dot represents a country or, as in the case of Hong Kong, a special administrative region.

a. Approximately, what is the highest life expectancy in years? Approximately, what fertility rate (children per woman) is associated with the highest life expectancy?

b. Does this seem to be a linear relation? Explain why or why not, and explain the relation in plain English.

Applying the Concepts

3.25 Type of graph for the relation between height and attractiveness: A social psychologist studied the effect of height on perceived overall attractiveness. Students were recruited to come to a research laboratory in pairs. The pairs sat together in the waiting room for several minutes and then were brought to separate rooms, where their heights were measured. They also filled out a questionnaire that asked, among other things, that they rate the attractiveness of the person who had been sitting with them in the waiting room on a scale of 1 to 10.

a. In this study, are the independent and dependent variables nominal, ordinal, or scale?

b. Which graph or graphs would be most appropriate to depict the data? Explain why.

c. If height ranged from 58 inches to 71 inches in this study, would your axis start at 0? Explain.

3.26 Type of graph for the effects of cognitive-behavioral therapy on depression: A social worker tracked the depression levels of clients being treated with cognitive-behavioral therapy for depression. For each client, depression was assessed at weeks 1 to 20 of therapy. She calculated a mean for all of her clients at week 1, week 2, and so on, all the way through week 20.

a. What are the independent and dependent variables in this study?

b. Are the variables nominal, ordinal, or scale?

c. Which graph or graphs would be most appropriate to depict the data? Explain why.

3.27 Type of graph for comparative suicide rates: An epidemiologist determined male suicide rates for 20 countries. For example, in 1996, the rate of male suicide in the United States was approximately 19.3 per 100,000 men, while in China that rate was approximately 15.9.

a. What are the variables in this study?

b. Are the variables nominal, ordinal, or scale?

c. Which graph would be most appropriate to depict the data? Explain why.

d. If you wanted to track the suicide rates for three of these countries over 50 years, what type of graph might you use to show these data?

3.28 Scatterplot of daily cycling distances and type of climb: Every summer, the touring company America by Bicycle conducts its Cross-Country Challenge, a 7-week bicycle journey across the United States from San Francisco, California, to Portsmouth, New Hampshire. At some point during the trip, the exhausted cyclists usually start to complain that the organizers are purposely planning for days with lots of hill and mountain climbing to coincide with longer distances. The tour staff counter that no relation exists between climbs and mileage and that the route is organized based on practicalities, such as the location of towns in which riders can stay. The organizers who planned the route (and who also own the company) say that they actually tried to reduce the mileage on the days with the worst climbs. Here are the approximate daily mileages and climbs (in vertical feet), as estimated from one rider's bicycle computer.

Mileage	Climb	Mileage	Climb	Mileage	Climb
83	600	69	2500	102	2600
57	600	63	5100	103	1000
51	2000	66	4200	80	1000
76	8500	96	900	72	900
51	4600	124	600	68	900
91	800	104	600	107	1900
73	1000	52	1300	105	4000
55	2000	85	600	90	1600
72	2500	64	300	87	1100
108	3900	65	300	94	4000
118	300	108	4200	64	1500
65	1800	97	3500	84	1500
76	4100	91	3500	70	1500
66	1200	82	4500	80	5200
97	3200	77	1000	63	5200
92	3900	53	2500		

a. Construct a scatterplot of the cycling data, putting mileage on the x-axis. Be sure to label everything and include a title.

b. We haven't yet learned to calculate inferential statistics on these data, so we can't really know what's going on, but do you think that the amount of vertical climb is related to a day's mileage? If yes, explain the relation in your own words. If no, explain why you think there is no relation.

c. It turns out that inferential statistics do not support the existence of a relation between these variables and that the staff seems to be the most accurate in their appraisal. Why do you think the cyclists and organizers are wrong in opposite directions? What does this say about people's biases and the need for data?

3.29 Scatterplot of gross domestic product and education levels: The Group of Eight (G8) consists of most of the major world economic powers. It meets annually to discuss pressing world problems. In 2005, for example, the agenda included climate change, poverty in Africa, and terrorism. Decisions made by G8 nations can have a global impact; in fact, the eight nations that make up the membership reportedly account for almost two-thirds of the world's economic output. Here are data for seven of the eight G8 nations for gross domestic product (GDP) in 2004 and a measure of education. The measure of education is the percentage of the population between the ages of 25 and 64 that had at least one university degree (Sherman, Honegger, & McGivern, 2003). No data point for education in Russia was available, so Russia is not included.

Country	GDP (in trillions of $US)	Percentage with University Degree
Canada	0.98	19
France	2.00	11
Germany	2.71	13
Italy	1.67	9
Japan	4.62	18
United Kingdom	2.14	17
United States	11.67	27

a. Create a scatterplot of these data, with university degree on the x-axis. Be sure to label everything and to give it a title. Later, we'll use statistical tools to determine the equation for the line of best fit. For now, draw a line of best fit that represents your best guess as to where it would go.

b. In your own words, describe the relation between the variables that you see in the scatterplot.

c. Education is on the x-axis, indicating that education is the independent variable. Explain why it is possible that education predicts GDP. Now reverse your explanation of the direction of prediction, explaining why it is possible that GDP predicts education.

3.30 Time series plot of organ donations: The Canadian Institute for Health Information (CIHI) is a nonprofit organization that compiles data from a range of institutions—from governmental organizations, to hospitals, to universities. Among the many topics that interest public health specialists is the problem of low levels of organ donation. Medical advances have led to ever-increasing rates of transplantation, but organ donation has not kept up with doctors' ability to perform more sophisticated and more complicated surgeries. Data reported by CIHI (2005) provide Canadian transplantation and donation rates for 1994–2004. Here are the donor rates per million deaths.

Year	Donor Rate per Million Deaths	Year	Donor Rate per Million Deaths
1994	14.0	2000	15.3
1995	14.9	2001	13.5
1996	14.2	2002	12.9
1997	14.3	2003	13.5
1998	13.8	2004	13.1
1999	13.8		

a. Construct a time series plot from these data. Be sure to label and title your graph.

b. What story are these data telling?

c. If you worked in public health and were studying the likelihood that families would agree to donate, what research question might you ask about the possible reasons for the trend suggested by these data?

3.31 Bar graph of alumni donation rates at different kinds of universities: U.S. universities are concerned with increasing the percentage of alumni who donate to the school because alumni donation rate is a factor in the *U.S. News & World Report*'s university rankings. What role might type of university play in alumni donation rates? *U.S. News & World Report* lists the top 10 national universities (all of which are private), the top 10 public national universities, and the top 10 liberal arts colleges (also all private) ("America's Best Colleges 2004," 2003). National universities focus on graduate education and research, whereas liberal arts colleges focus on undergraduate education. To give you a sense of the type of institutions in each of these categories, the number one schools for 2004 in the three categories were Harvard University, the University of California at Berkeley, and Williams College, respectively. Here are the 2004 alumni donation rates for the top 10 schools in each of these categories.

Top 10 Private National Schools	Top 10 Public National Schools	Top 10 Liberal Arts Schools
48%	15%	60%
61	14	63
45	26	52
39	16	53
46	25	66
37	26	52
38	15	55
34	9	55
33	12	53
47	32	48

a. What is the independent variable in this example? Is it nominal or scale? If nominal, what are the levels? If scale, what are the units and what are the minimum and maximum values?

b. What is the dependent variable in this example? Is it nominal or scale? If nominal, what are the levels? If scale, what are the units and what are the minimum and maximum values?

c. Construct a bar graph of these data using the default options in your computer software.

d. Construct a bar graph of these data, but change the defaults to satisfy the guidelines for graphs discussed in this chapter. Aim for simplicity and clarity.

e. What does the pattern of the data suggest?

f. Cite at least one research question that you might want to explore next if you worked for one of these universities—your research question should grow out of these data.

g. Explain how these data could be presented as a pictorial graph. (Note that you do not have to construct such a graph.) What kind of picture could you use? What would it look like?

h. What are the potential pitfalls of a pictorial graph? Why is a bar chart usually a better choice?

3.32 Bar graph versus Pareto chart of countries' gross domestic product: In How It Works 3.2, we created a bar graph for the 2004 GDP, in trillions of U.S. dollars, of each of the G8 nations.

a. Explain the difference between a Pareto chart and a standard bar graph in which the countries are in alphabetical order along the x-axis.

b. What is the benefit of the Pareto chart over the standard bar graph?

3.33 Bar graph versus time series plot of graduate school mentoring: Johnson, Koch, Fallow, and Huwe (2000) conducted a study of mentoring in two types of psychology doctoral programs: experimental and clinical. Students who graduated from the two types of programs were asked whether they had a faculty mentor while in graduate school. In response, 48.00% of clinical psychology students who graduated between 1945 and 1950 and 62.31% who graduated between 1996 and 1998 reported having had a mentor; 78.26% of experimental psychology students who graduated between 1945 and 1950 and 78.79% who graduated between 1996 and 1998 reported having had a mentor.

a. What are the two independent variables in this study, and what are their levels?

b. What is the dependent variable?

c. Create a bar graph that depicts the percentages for the two independent variables simultaneously.

d. What story is this graph telling us?

e. Was this a true experiment? Explain your answer.

f. Why would a time series plot be inappropriate for these data? What would a time series plot suggest about the mentoring trend for clinical psychology graduate students and for experimental psychology graduate students?

g. For four time points—1945–1950, 1965, 1985, and 1996–1998—the mentoring rates for clinical psychology graduate students were 48.00, 56.63, 47.50, and 62.31, respectively. For experimental psychology graduate students, the rates were 78.26, 57.14, 57.14, and 78.79, respectively. How does the story we see here conflict with the one that we developed based on just two time points?

3.34 Bar graph versus pie chart of student participation in community activities: The National Survey on Student Engagement (NSSE) has surveyed more than 400,000 students—freshmen and seniors—at 730 U.S. schools since 1999 ("America's Best Colleges 2004," 2003). Among the many questions on the NSSE, students were asked how often they "participated in a community-based project as part of a regular course." For the students at the 19 institutions classified as national universities that made their data publicly available through the *U.S. News &World Report* Web site, here are the data: never, 56%; sometimes, 31%; often, 9%; very often, 5%. (The percentages add up to 101% because of rounding.) Explain why a bar graph would be more suitable for these data than a pie chart.

3.35 Software defaults of graphing programs that portray satisfaction with graduate advisors: The 2000 National Doctoral Program Survey asked 32,000 current and recent Ph.D. students in the United States, across all disciplines, to respond to the statement "I am satisfied with my advisor." The researchers calculated the percentage of students who responded "agree" or "strongly agree": current students, 87%; recent graduates, 86%; former students who left without completing the Ph.D., 48%.

a. Use a software program that produces graphs (e.g., Excel, SPSS, Minitab) to create a bar graph for these data.
b. Play with the options available to you. List aspects of the bar graph that you are able to change to make your graph meet the guidelines listed in this chapter. Be specific, and include the revised graph.

3.36 Types of graph appropriate for behavioral science research: Give an example of a study—real or hypothetical—in the behavioral sciences that might display its data using the following types of graphs. State the independent variable(s) and dependent variable, including levels for any nominal variables.

a. Frequency polygon
b. Line graph (line of best fit)
c. Bar graph (one independent variable)
d. Scatterplot
e. Time series plot
f. Pie chart
g. Bar graph (two independent variables)

3.37 Creating the perfect graph: What advice would you give to the creators of each of the following graphs? Consider the basic guidelines for a clear graph, chartjunk, and the ways to mislead through statistics. Give three pieces of advice for each graph. Be specific—don't just say there's chartjunk; say exactly what you'd change.

a. The shrinking doctor:

b. Workforce participation:

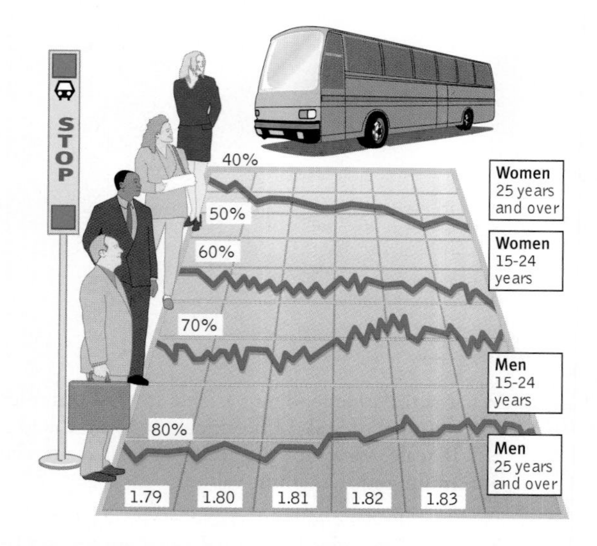

3.38 Graphs in the popular media: Find an article in the popular media (newspaper, magazine, Web site) that includes a graph in addition to the text.

a. Briefly summarize the main point of the article and graph.

b. What are the independent and dependent variables depicted in the graph? What kind of variables are they? If nominal, what are the levels?

c. What descriptive statistics are included in the article or on the graph?

d. In one or two sentences, what story is the graph (rather than the article) trying to tell?

e. How well do the text and graph match up? Are they telling the same story? Are they using the same terms? Explain.

f. Write a paragraph to the graph's creator with advice for improving the graph. Be specific, citing the guidelines from this chapter.

g. Redo the graph, either by hand or by computer, in line with your suggestions.

3.39 Interpreting a graph about two kinds of career regrets: The Yerkes–Dodson graph demonstrates that graphs can be used to describe theoretical relations that can be tested. In a study that could be applied to the career decisions made during college, Gilovich and Medvec (1995) identified two types of regrets—regrets of action and regrets of inaction—and proposed that their intensity changes over time. You can think of these as Type I regrets—things you have done that you wish you had not done (regrets of action)—and Type II regrets—things you have not done that you wish you had done (regrets of inaction). The researchers suggested a theoretical relation between the variables that might look something like the graph below.

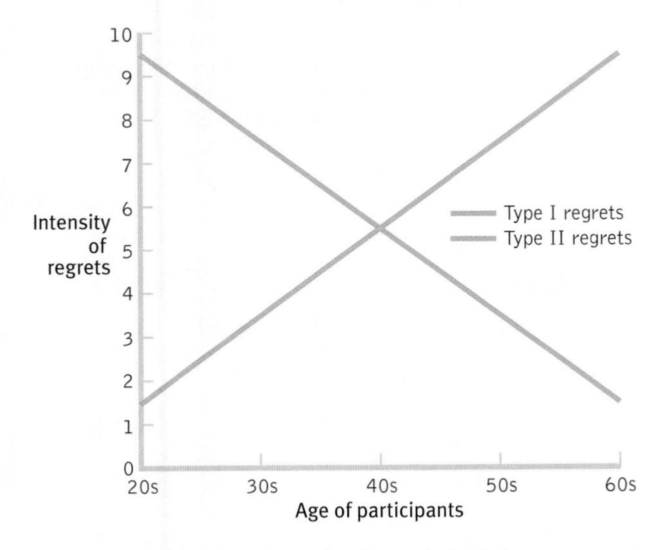

a. Briefly summarize the theoretical relations proposed by the graph.

b. What are the independent and dependent variables depicted in the graph? What kind of variables are they? If nominal or ordinal, what are the levels?

c. What descriptive statistics are included in the text or on the graph?

d. In one or two sentences, what story is the graph trying to tell?

3.40 Critiquing a graph about the frequency of psychology degrees: The American Psychological Association (APA) compiles many statistics about training and careers in the field of psychology. The accompanying graph tracks the numbers of bachelor's, master's, and doctoral degrees between the years 1970 and 2000.

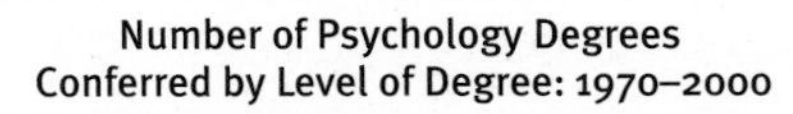

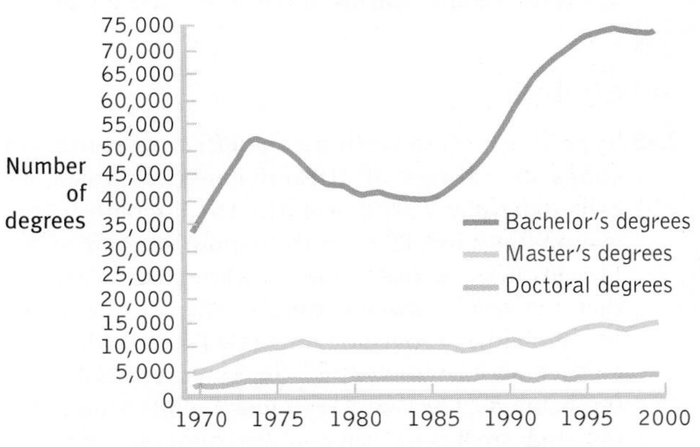

a. What kind of graph is this? Why did the researchers choose this type of graph?

b. Briefly summarize the overall story being told by this graph.

c. What are the independent and dependent variables depicted in the graph? What kind of variables are they? If nominal or ordinal, what are the levels?

d. List at least three things that the graph creators did well (i.e., are in line with the guidelines for graph construction).

e. List at least one thing that the graph creators should have done differently (i.e., is not in line with the guidelines for graph construction).

f. Name at least one variable other than number that might be used to track the prevalence of psychology bachelor's, master's, and doctoral degrees over time.

g. The increase in bachelor's degrees over the years is not matched by an increase in doctoral degrees. List at least one research question that this finding suggests to you.

3.41 Interpreting a graph using a mental health index: The gray line in Figure 3-18 depicts the Mental Health Index (MHI) for a fictional client in relation to several benchmarks. Use the information supplied in Figure 3-18 to answer the following questions:

a. Describe the trajectory of the client's MHI over the course of the therapy sessions.

b. Provide a possible explanation for the trajectory described in part (a).

c. Based on the benchmarks depicted in the graph, would you recommend that the client continue therapy?

3.42 Interpreting a graph about traffic flow: Go to http://maps.google.com/. On a map of your country, click on the traffic button.

a. How is the density and flow of traffic represented in this graph?

b. Describe traffic patterns in different regions of your country.

c. What are the benefits of this interactive graph?

Putting It All Together

3.43 Type of graph describing the effect of romantic songs on ratings of attractiveness: Guéguen, Jacob, and Lamy (2010) wondered if listening to romantic songs would affect the dating behavior of the French heterosexual women who participated in their study. The women were randomly assigned to listen to either a romantic song ("Je l'aime à Mourir," or "I Love Her to Death") or a nonromantic song ("L'heure du Thé," or "Tea Time") while waiting for the study to begin. Later in the study, an attractive male researcher asked each participant for her phone number. Of the women who listened to the romantic song, 52.2% gave their phone number to the researcher, whereas only 27.9% of the women who listened to the nonromantic song gave their phone number to the researcher.

a. What is the independent variable in this study?

b. What is the dependent variable?

c. Is this a between-groups or a within-groups study? Explain your answer.

d. Think back to our discussion of variables in Chapter 1. How did the researcher operationalize "dating behavior" in this study? Do you think this is a valid measure of dating behavior? Explain your answer.

e. What is the best type of graph to depict these results? Explain your answer.

f. Create the graph you described in part (c) using software without changing any of the defaults.

g. Create the graph you described in part (c) a second time, changing the defaults to meet the criteria in the checklist introduced in this chapter.

3.44 Developing research questions from graphs: Graphs not only answer research questions, but can spur new ones. Figure 3-9 on p. 50 depicts the pattern of changing attitudes, as expressed through Twitter.

a. On what day and at what time is the highest average positive attitude expressed?

b. On what day and at what time is the lowest average negative attitude expressed?

c. What research question(s) do these observations suggest to you with respect to weekdays versus weekends? (Remember, 0 on Sunday is midnight—or late Saturday night.) Name the independent and dependent variables.

d. How are the researchers operationalizing mood? Do you think this is a valid measure of mood? Explain your answer.

e. One of the highest average negative attitudes occurs at midnight on Sunday. How does this fit with the research hypothesis you developed in (c)? Does this suggest a new research hypothesis?

Terms

scatterplot (p. 47)
range-frame (p. 48)
linear relation (p. 48)
nonlinear relation (p. 48)
line graph (p. 48)
time plot, *or* time series plot (p. 50)
bar graph (p. 51)
Pareto chart (p. 51)
pictorial graph (p. 53)
pie chart (p. 53)
chartjunk (p. 56)
moiré vibration (p. 56)
grid (p. 56)
duck (p. 57)
defaults (p. 57)

CHAPTER 4

Central Tendency and Variability

BEFORE YOU GO ON

- You should understand what a distribution is (Chapter 2).
- You should be able to interpret histograms and frequency polygons (Chapter 2).

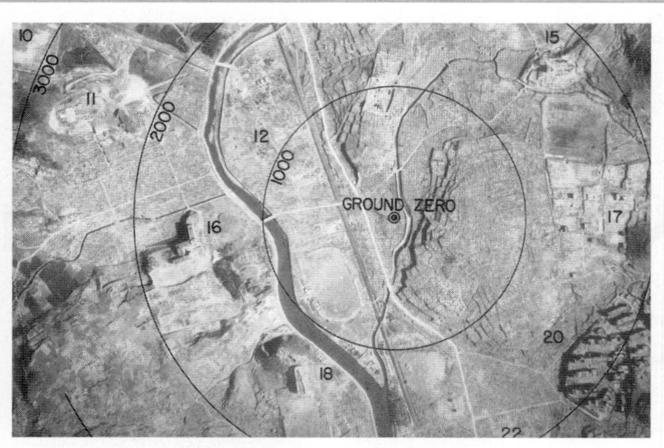

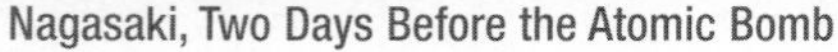

Nagasaki, Two Days Before the Atomic Bomb

Nagasaki, Three Days After the Atomic Bomb

Chance variability matters. On August 9, 1945, during World War II, chance variability in the cloud cover diverted a B-29 bomber from Kokura, Japan, to its secondary target, the city of Nagasaki. When the atomic bomb exploded a few hundred feet above a tennis court, all of the buildings and most of the people who lived in the city of Nagasaki simply disappeared; the people in Kokura survived.

How does any nation recover from such devastation? Five years later, an American statistician named W. Edwards Deming persuaded postwar Japan's leading engineers and businesspeople that a simple statistical idea could recreate their entire industrial-based economy: low variability. Deming's core statistical insight was simple: Low variability means high reliability. Customers were glad to pay for reliable products. Japan's industrial leaders discovered that controlling variability translated into thousands of small manufacturing solutions that, in just a few years, transformed Japan's reputation. By understanding variability, the war-ravaged country once known for turning out cheap junk became known for manufacturing high-quality products.

> **MASTERING THE CONCEPT**
>
> **4.1:** Central tendency refers to three slightly different ways to describe what is happening in the center of a distribution of data: the mean, median, and mode.

Variability is the central idea behind all statistical thinking. In this chapter, we learn about three common measures of variability: range, variance, and standard deviation. But to fully understand variability, we first have to know how to identify the middle, or central, tendency of a distribution.

Central Tendency

Central tendency *refers to the descriptive statistic that best represents the center of a data set, the particular value that all the other data seem to be gathering around*—the "typical" score. Creating a visual representation of the distribution, as we did in Chapter 2, reveals its central tendency. The central tendency is usually at (or near) the highest point in the histogram or the polygon. The way that data cluster around its central tendency can be measured in three different ways: mean, median, and mode. For example, Figure 4-1 shows the histogram for the data on World Cup top finishes (omitting scores for countries with no top finishes). Our guess is that the central tendency is just to the right of the tallest bar.

Mean, the Arithmetic Average

The mean is simple to calculate and is the gateway to understanding statistical formulas. The mean is such an important concept in statistics that we provide you with four distinct ways to think about it: verbally, arithmetically, visually, and symbolically (using statistical notation).

The Mean in Plain English The most commonly reported measure of central tendency is *the **mean**, the arithmetic average of a group of scores.* The mean, often called the average, is used to represent the "typical" score in a distribution. Because this is different from the way we sometimes use the word *average* in everyday conversation, noting that someone is "just" average in athletic ability or that a movie was "only" average, we need to define the mean arithmetically.

The Mean in Plain Arithmetic The mean is calculated by summing all the scores in a data set and then dividing this sum by the total number of scores. You likely have calculated means many times in your life.

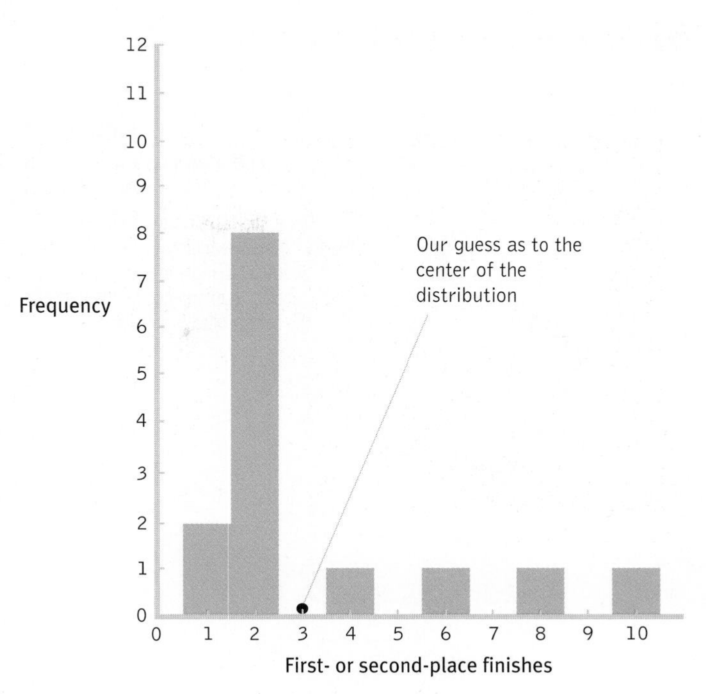

FIGURE 4-1
Estimating Central Tendency with Histograms
Histograms and frequency polygons allow us to see the likely center of a sample's distribution. The arrow points to our guess as to the center of the distribution of World Cup top finishes.

EXAMPLE 4.1

For example, the mean number of top World Cup finishes is calculated by (a) adding the number of top finishes for each country; then (b) dividing by the total number of countries. We'll do this for the 14 countries that had at least 1 top finish, omitting the 63 with 0 top finishes.

STEP 1: Add all of the scores together.

$$4 + 8 + 1 + 2 + 1 + 2 + 2 + 6 + 2 + 2 + 2 + 2 + 2 + 10 = 46$$

STEP 2: Divide the sum of all scores by the total number of scores.

In this case, we divide 46, the sum of all scores, by 14, the number of scores in this sample:

$$46/14 = 3.29 \quad ■$$

Visual Representations of the Mean Think of the mean as the visual point that perfectly balances two sides of a distribution. For example, the mean of 3.29 "top finishes" is represented visually as the point that perfectly balances that distribution, as you can see in the Figure 4-2 histogram.

The Mean Expressed by Symbolic Notation Symbolic notation may sound more difficult than it really is. After all, you just calculated a mean without symbolic notation and without a formula. Fortunately, we need to understand only a handful of symbols to express the ideas necessary to understand statistics.

- **Central tendency** refers to the descriptive statistic that best represents the center of a data set, the particular value that all the other data seem to be gathering around.
- The **mean** is the arithmetic average of a group of scores. It is calculated by summing all the scores in a data set and then dividing this sum by the total number of scores.

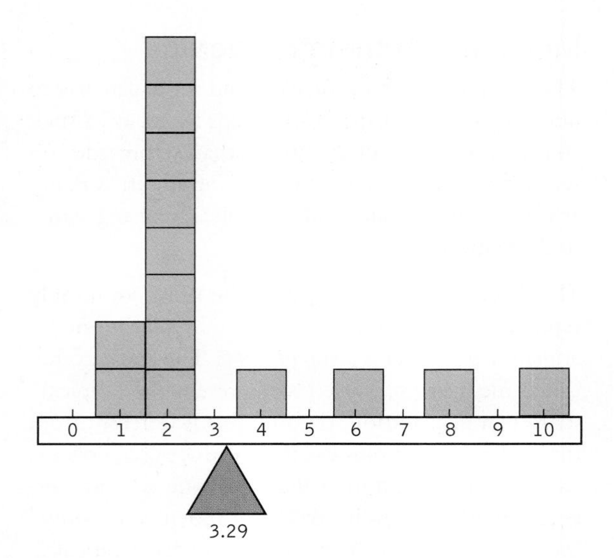

FIGURE 4-2
The Mean as the Fulcrum of the Data

The mean, 3.29, is the balancing point for all the scores for top finishes for the countries that competed in World Cup soccer tournaments. Mathematically, the scores always balance around the mean for any sample.

Here are the several symbols that represent the mean. For the mean of a sample, statisticians typically use M or $\overline{X}$. In this text, we use M; many other texts also use M, but some use $\overline{X}$ (pronounced "X bar"). For a population, statisticians use the Greek letter μ (pronounced "mew") to symbolize the mean. (Latin letters such as M tend to refer to numbers based on samples; Greek letters such as μ tend to refer to numbers based on populations.) *The numbers based on samples taken from a population are called* ***statistics;*** M is a statistic. *The numbers based on whole populations are called* ***parameters;*** μ is a parameter. Table 4-1 summarizes how these terms are used, and Figure 4-3 presents a mnemonic, or memory aid, for these terms.

The formula to calculate the mean of a sample uses the symbol M on the left side of the equation; the right side describes how to perform the calculation. A single score is typically symbolized as X. We know that we're summing all the scores—all the X's—so the first step is to use the summation sign, Σ (pronounced "sigma"), to indicate that we're summing a list of scores. As you might guess, the full expression for summing all the scores would be ΣX. This symbol combination instructs us to add up all of the X's in the sample.

Here are step-by-step instructions for constructing the equations:

Step 1: Add up all of the scores in the sample. In statistical notation, this is ΣX.

Step 2: Divide the total of all of the scores by the total number of scores. The total number of scores in a sample is typically represented by N. (Note that the capital letter N is typically used when we refer to the number of scores in the entire data set; if we

TABLE 4-1. The Mean in Symbols

The mean of a sample is an example of a statistic, whereas the mean of a population is an example of a parameter. The symbols we use depend on whether we are referring to the mean of a sample or of a population.

Number	Used for	Symbol	Pronounced
Statistic	Sample	M or $\overline{X}$	"M" or "X bar"
Parameter	Population	μ	"mew"

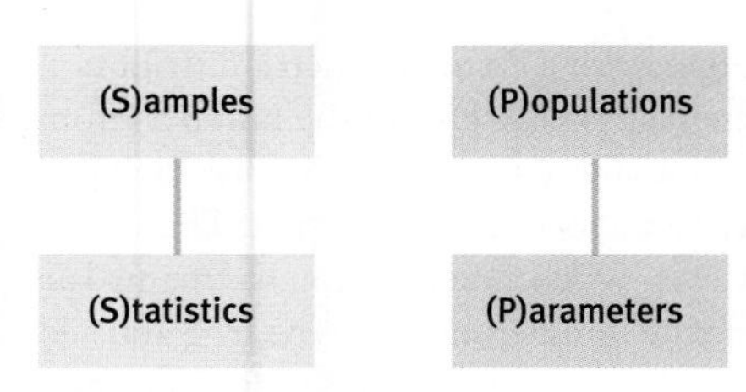

FIGURE 4-3
Samples and Parameters

Try using a mnemonic trick to remember the distinction between samples and parameters. The letter *s* means that numbers based on (*s*)amples are called (*s*)tatistics. The letter *p* means that numbers based on (*p*)opulations are called (*p*)arameters.

break the sample down into smaller parts, as we'll see in later chapters, we typically use the lowercase letter *n*.) The full equation would be:

$$M = \frac{\Sigma X}{N}$$

MASTERING THE FORMULA

4-1: The formula for the mean is: $M = \frac{\Sigma X}{N}$. To calculate the mean, we add up every score, and then divide by the total number of scores.

EXAMPLE 4.2

Let's look at the mean for the World Cup data that we considered earlier in Example 4.1.

STEP 1: We add up every score.

The sum of all scores, as we calculated previously, is 46.

STEP 2: We divide the sum of all scores by the total number of scores.

In this case, we divide the sum of all scores, 46, by the total number of scores, 14. The result is 3.29.

Here's how it would look as a formula:

$$M = \frac{\Sigma X}{N} = \frac{46}{14} = 3.29$$

Language Alert! As demonstrated above, almost all symbols are italicized, but the actual numerical values of the statistics are not italicized. In addition, changing a symbol from uppercase to lowercase often indicates a change in meaning. When you practice calculating means using this formula, be sure to italicize the symbols and use capital letters for *M, X,* and *N.* When writing by hand, indicate that a symbol is italicized by underlining it. ■

- A **statistic** is a number based on a sample taken from a population; statistics are usually symbolized by Latin letters.
- A **parameter** is a number based on the whole population; parameters are usually symbolized by Greek letters.
- The **median** is the middle score of all the scores in a sample when the scores are arranged in ascending order. If there is no single middle score, the median is the mean of the two middle scores.

Median, the Middle Score

The second most common measure of central tendency is the median. *The **median** is the middle score of all the scores when the scores are arranged in ascending order.* We can think of the median as the 50th percentile. The American Psychological Association (APA) suggests abbreviating "median" by writing "*Mdn.*" APA style, by the way, is used across the social and behavioral sciences.

To determine the median, follow these steps:

Step 1: Line up all the scores in ascending order.

Step 2:. Find the middle score. With an odd number of scores, there will be an actual middle score. With an even number of scores, there will be no actual middle score. In this case, calculate the mean of the two middle scores.

David De Lossy/Photodisc Green/Getty Images

Mean versus Median The median is the part of the roadway that divides the directions in which vehicles are permitted to drive. It can be dangerous to confuse the mean and the median, especially when you are calculating the "middle" of the roadway!

Don't bother with a formula when a distribution has only a few data points. Just list the numbers from lowest to highest and note which score has the same number of scores above it and below it. The calculation is easy even when a distribution has many data points. Divide the number of scores (N) by 2 and add ½—that is, 0.5. That number is the ordinal position (rank) of the median, or middle score. As illustrated below, simply count that many places over from the start of your scores and report that number.

EXAMPLE 4.3

Here is an example with an odd number of scores (representing numbers of top finishes for 13 of the 14 countries in the World Cup example; we omit one score, a 2):

4 8 1 2 1 2 2 6 2 2 2 2 10

STEP 1: Arrange the scores in ascending order:

1 1 2 2 2 2 2 2 2 4 6 8 10

STEP 2: Find the middle score. To do this, first we count. There are 13 scores: 13/2 = 6.5. If we add 0.5 to this result, we get 7. Therefore, the median is the 7th score. We now count across to the 7th score. The median is 2. ■

EXAMPLE 4.4

Here is an example with an even number of scores. We now include all 14 countries from the World Cup data in Example 4.1, including the score of 2 that we omitted in Example 4.3.

STEP 1: Arrange the scores in ascending order. The data are now:

1 1 2 2 2 2 2 2 2 2 4 6 8 10

STEP 2: Find the middle score. First, we count the scores. There are 14. We then divide the number of scores by 2: 14/2 = 7. If we add 0.5 to this result, we get 7.5; therefore, the median is the average of the 7th and 8th scores. The 7th and 8th scores are 2 and 2. The median is their mean—the mean of 2 and 2 is 2. ■

- The **mode** is the most common score of all the scores in a sample.
- A **unimodal** distribution has one mode, or most common score.
- A **bimodal** distribution has two modes, or most common scores.
- A **multimodal** distribution has more than two modes, or most common scores.

Mode, the Most Common Score

The *mode* is perhaps the easiest of the three measures of central tendency to calculate. *The* ***mode*** *is the most common score of all the scores in a group of scores.* It is readily picked out on a frequency table, a histogram, or a frequency polygon. Like the median, the APA style for the mode does not have a symbol—in fact, it doesn't even have an abbreviation. When reporting modes, just write, "The mode is. . . ."

EXAMPLE 4.5

Determine the mode for the World Cup data for the 14 countries. Remember that each score represents the number of that country's top finishes in World Cup tournaments. The mode can be found either by searching the list of numbers for the most common score or by constructing a frequency table:

1 1 2 2 2 2 2 2 2 2 4 6 8 10

The mode is: ____

Did you get 2? If you didn't, you might have made a common mistake. The mode is the score that occurs most frequently, not the frequency of that score. So, in the data set above, the score 2 occurs 8 times. The mode is 2, *not* 8.

The mode in this example is particularly easy to determine because it is easy to see that there is one most common score. Sometimes a data set has no specific mode. This is especially true when the scores are reported to several decimal places (and no number occurs twice). When there is no specific mode, we sometimes report the most common interval as the mode; for example, we may say that the mode on a statistics exam is 70–79 (1 interval in the full range of 0–100). When there is more than one mode, we report both, or all, of the most common scores. *When a distribution of scores has one mode, we refer to it as* ***unimodal****. When a distribution has two modes, we call it* ***bimodal****. When a distribution has more than two modes, we call it* ***multimodal***. For example, the salaries of recent U.S. law school graduates are bimodal, as illustrated in Figure 4-4.

As demonstrated in the law salary example, the mode can be used with scale data; however, it is more commonly used with nominal data. For example, Cancer Research UK (2003) reported that lung cancer was the most common cause of cancer death in the United Kingdom (22%). No other type of cancer came close. Colorectal cancer accounted for 10% of cancer deaths, breast cancer for 8%, and each of all the other types for 7% or less. In this data set, the modal type of cancer death is lung cancer. ■

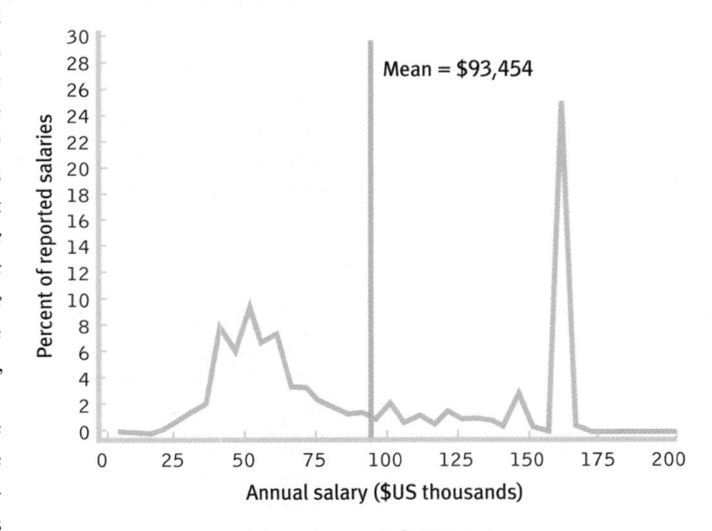

FIGURE 4-4
The Bimodal Salaries of Law School Graduates
With a bimodal or multimodal distribution, neither the mean nor the median is representative of the data. In an example of good statistical journalism, the *New York Times* reported a bimodal distribution for the pay of law school graduates in the United States (Rampell, 2010). The reporter described the two modes as "the $50,000 public service job or the $160,000 'Big Law' job," and noted that very few people actually earn the mean of about $90,000.

How Outliers Affect Measures of Central Tendency

The mean is not the best measure of central tendency when the data are skewed by one or a few statistical outliers—extreme scores that are either very high or very low in comparison with the other scores. To demonstrate the effect of outliers on the mean, as well as the median's resistance to the effect of outliers, let's take a look at the Mundi Index estimates, which compares how many physicians there are per 1000 people in different countries (a statistic called physician density). We'll focus on the top five.

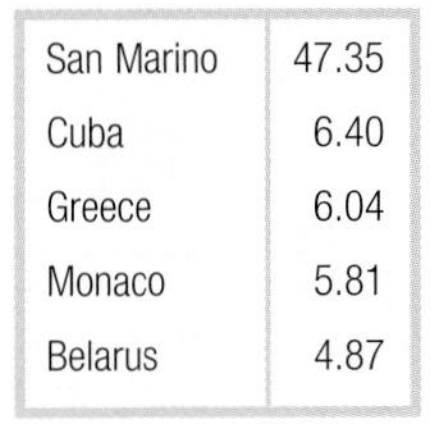

San Marino	47.35
Cuba	6.40
Greece	6.04
Monaco	5.81
Belarus	4.87

EXAMPLE 4.6

To get a sense of overall physician density, we might want to calculate a measure of central tendency for these five countries. We'll use the formula to get a little more practice with the symbols of statistics.

$$M = \frac{\Sigma X}{N} = \frac{\Sigma(47.35 + 6.40 + 6.04 + 5.81 + 4.87)}{5} = \frac{70.83}{5} = 14.17$$

In Figure 4-5, weights are placed to represent each of the scores in the sample, and they demonstrate an important feature of the mean. Like a seesaw, the mean is the point at which all the other scores are perfectly balanced. The physician density scores range from 4.87 to 47.35 and demonstrate the problem with using the mean when there is an outlier: The mean is not typical for any of the five countries in this sample, even though the seesaw is perfectly balanced if we put its fulcrum at the mean of 14.17.

We can't help but notice that the tiny country of San Marino (an enclave within Italy; population 33,000) is very different from the other countries. Four of the scores are between 4.87 and 6.4 physicians per 1000 people, a much smaller range. But San Marino enjoys 47.35 physicians per 1000 people. When there is an outlier like San Marino, it is important to consider what this score does to the mean, especially if we have only a few observations.

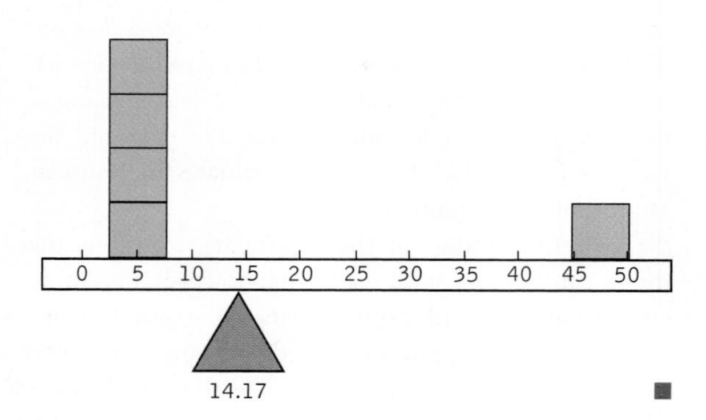

FIGURE 4-5
Outliers and the Mean

With the physician density data, the mean of 14.17 is far above the lowest four scores yet well below the highest. San Marino's score, an outlier, pulls the mean higher, even among the other countries with a high physician density. With such an extreme outlier, the mean does not do a good job of representing the story that these data tell.

■

EXAMPLE 4.7

When we eliminate San Marino's score, the data are now 6.40, 6.04, 5.81, and 4.87, and the new mean is:

$$M = \frac{\Sigma X}{N} = \frac{\Sigma(6.40 + 6.04 + 5.81 + 4.87)}{4} = 5.78$$

The new mean of these scores, 5.78, is a good deal lower than the original mean of the scores that included San Marino's outlier. We see from Figure 4-6 that this new mean, like the original mean, marks the point at which all other scores are perfectly balanced around it. However, this new mean is a little more representative of the scores—5.78 is a more typical score for these four countries.

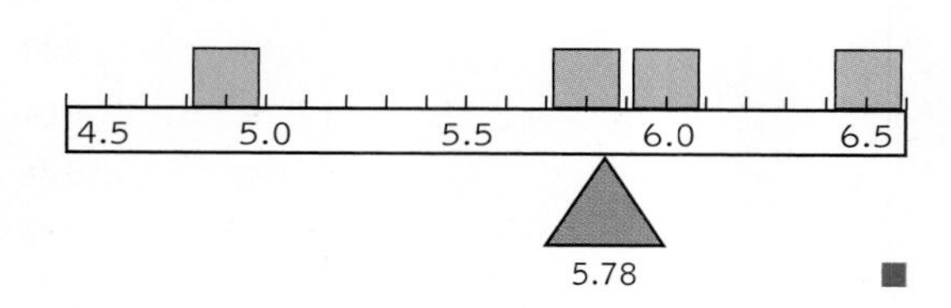

FIGURE 4-6
The Mean Without the Outlier

When the outlier—San Marino—is omitted from the physician density data, the mean becomes more representative of the actual scores in the sample.

■

Which Measure of Central Tendency Is Best?

Different measures of central tendency can lead to different conclusions, but when a decision needs to be made, the choice is usually between the mean and the median. The mean usually wins, especially when there are many observations—but it doesn't always. When distributions are skewed by outliers, for example, or when you have only a handful of observations, the median may provide a better sense of a distribution's central tendency.

The mode is generally used in three situations: (1) when one particular score dominates a distribution; (2) when the distribution is bimodal or multimodal; and (3) when the data are nominal. When you are uncertain as to which measure is the best indicator of central tendency, report all three.

Central tendency communicates an enormous amount of information with a single number, so it is not surprising that measures of central tendency are among the most widely reported of descriptive statistics. Unfortunately, consumers can be tricked by reports that use the mean instead of the median. When you hear a radio report about the central tendency of housing prices, for example, first notice whether it is reporting an average (mean) or a median. You'll see an example of the housing market and a celebrity outlier in the photograph above right.

Mario Magnani/Getty Images

Celebrity Outliers Reports of the cost of a typical Manhattan apartment depend on whether the mean or the median is reported. Film star Gwyneth Paltrow and her husband, Coldplay lead singer Chris Martin, sold their Manhattan apartment in the spring of 2005 for around $7 million. Such a sale would be an outlier and would boost the mean; however, it would not affect the median. Of course, either way, the typical Manhattan apartment is not within the budget of the typical college graduate!

MASTERING THE CONCEPT

4.2: The mean is the most common indicator of central tendency, but it is not always the best. When there is an outlier or few observations, it is usually better to use the median.

CHECK YOUR LEARNING

Reviewing the Concepts

- The central tendency of a distribution is the one number that best describes what is typical in that distribution (often its high point).
- The three measures of central tendency are the mean (arithmetic average), the median (middle score), and the mode (most frequently occurring score).
- The mean is the most commonly used measure of central tendency, but the median is preferred when the distribution is skewed.
- The symbols used in statistics have very specific meanings; changing a symbol even slightly can change its meaning a great deal.

Clarifying the Concepts

4-1 What is the difference between statistics and parameters?

4-2 Does an outlier have the greatest effect on the mean, the median, or the mode?

Calculating the Statistics

4-3 Calculate the mean, median, and mode of the following sets of numbers.

a. 10 8 22 5 6 1 19 8 13 12 8

b. 122.5 123.8 121.2 125.8 120.2 123.8 120.5 119.8 126.3 123.6

c. 0.100 0.866 0.781 0.555 0.222 0.245 0.234

continued on next page

Applying the Concepts

4-4 Let's examine fictional data for 20 seniors in college. Each score represents the number of nights a student spends socializing in one week: 1 0 1 2 5 3 2 3 1 3 1 7 2 3 2 2 2 0 4 6

a. Using the formula, calculate the mean of these scores.

b. If the researcher reported the mean of these scores to the university as an estimate for the whole university population, what symbol would be used for the mean? Why?

c. If the researcher was interested only in the scores of these 20 students, what symbol would be used for the mean? Why?

d. What is the median of these scores?

e. What is the mode of these scores?

f. Are the median and mean similar to or different from each other? What does this tell you about the distribution of scores?

Solutions to these Check Your Learning questions can be found in Appendix D.

Measures of Variability

Immediately after World War II, people often poked fun at Japanese products such as transistor radios—sometimes they worked well; sometimes they worked poorly; sometimes they didn't work at all. They just weren't reliable. After taking statistical reasoning about variability to heart, it took Japan just 3 years to become an industrial powerhouse whose reputation for high-quality, reliable products continues to this day. ***Variability** is a numerical way of describing how much spread there is in a distribution.* One way to numerically describe the variability of a distribution is by computing its *range*. A second and more common way to describe variability is by computing *variance* and its square root, known as *standard deviation*.

▶ MASTERING THE CONCEPT

4.3: Variability is the second most common concept (after central tendency) to help us understand the shape of a distribution. Common indicators of variability are range, variance, and standard deviation.

Range

The range is the easiest measure of variability to calculate. *The **range** is a measure of variability calculated by subtracting the lowest score (the minimum) from the highest score (the maximum). Maximum* and *minimum* are sometimes substituted in this formula to describe the highest and lowest scores, and some statistical computer programs abbreviate these as *max* and *min*. The range is represented in a formula as:

MASTERING THE FORMULA ▶

4-2: The formula for the range is: $range = X_{highest} - X_{lowest}$. We simply subtract the lowest score from the highest score to calculate the range.

$$range = X_{highest} - X_{lowest}$$

Here are the scores for countries' numbers of top finishes in the World Cup that we discussed earlier in the chapter. As before, we'll omit countries with scores of 0 top finishes.

1 1 2 2 2 2 2 2 2 2 4 6 8 10

EXAMPLE 4.8

We can determine the highest and lowest scores either by reading through the data or, more easily, by glancing at the frequency table for these data.

STEP 1: Determine the highest score. In this case, the highest score is 10.

STEP 2: Determine the lowest score.

In this case, the lowest score is 1.

STEP 3: Calculate the range by subtracting the lowest score from the highest score:

$$\text{range} = X_{highest} - X_{lowest} = 10 - 1 = 9.$$

The range can be a useful first indicator of variability, but it is influenced only by the highest and lowest scores. All the other scores in between could be clustered near the highest score, huddled near the center, spread out evenly, or have some other unexpected pattern. We can't know based only on the range. ■

Variance

Variance *is the average of the squared deviations from the mean.* When something varies, it must vary from (or be different from) some standard. That standard is the mean. So when we compute variance, that number describes how far a distribution varies around the mean. A small number indicates a small amount of spread or deviation around the mean, and a larger number indicates a great deal of spread or deviation around the mean.

EXAMPLE 4.9

Students who seek therapy at university counseling centers often do not attend many sessions. For example, in one study, the median number of therapy sessions was 3 and the mean was 4.6 (Hatchett, 2003). Let's examine the spread of fictional scores for a sample of five students: 1, 2, 4, 4, and 10 numbers of therapy sessions, with a mean of 4.2. Next we find out how far each score deviates from the mean by subtracting the mean from every score. First, we label the column that lists the scores with an X. Here, the second column includes the results we get when we subtract the mean from each score, or $X - M$. We call each of these a ***deviation from the mean*** (or just a *deviation*)—*the amount that a score in a sample differs from the mean of the sample.*

X	$X - M$
1	−3.2
2	−2.2
4	−0.2
4	−0.2
10	5.8

But we can't just take the mean of the deviations. If we do (and if you try this, don't forget the signs—negative and positive), we get 0—every time. Are you surprised? Remember, the mean is the point at which all scores are perfectly balanced. Mathematically, the scores *have* to balance out. Yet we know that there *is* variability among these scores. The number representing the amount of variability is certainly not 0!

When we ask students for ways to eliminate the negative signs, two suggestions typically come up: (1) take the absolute value of the deviations, thus making them all positive, or (2) square all the scores, again making them all positive. It turns out that the latter, squaring all the deviations, is how statisticians solve this problem. Once we square the deviations, we can take their average and get a measure of variability. (We will "*un*-square" later on when we calculate the standard deviation.)

- **Variability** is a numerical way of describing how much spread there is in a distribution.
- The **range** is a measure of variability calculated by subtracting the lowest score (the minimum) from the highest score (the maximum).
- **Variance** is the average of the squared deviations from the mean.
- A **deviation from the mean** is the amount that a score in a sample differs from the mean of the sample; also called a *deviation*.

- The **sum of squares**, symbolized as *SS*, is the sum of each score's squared deviation from the mean.
- The **standard deviation** is the square root of the average of the squared deviations from the mean; it is the typical amount that each score varies, or deviates, from the mean.

To recap:

STEP 1: Subtract the mean from every score.

We call these deviations from the mean.

STEP 2: Square every deviation from the mean.

We call these squared deviations.

STEP 3: Sum all of the squared deviations.

This is often called the sum of squared deviations, or the sum of squares, for short.

STEP 4: Divide the sum of squares by the total number in the sample (*N*).

This number represents the mathematical definition of variance—the average of the squared deviations from the mean.

To calculate the variance for the therapy session data, we add a third column to contain the squares of each of the deviations. Then we add all of these numbers up to compute the ***sum of squares*** *(symbolized as SS), the sum of each score's squared deviation from the mean.* In this case, the sum of the squared deviations is 48.80, so the average squared deviation is $48.80/5 = 9.76$. Thus, the variance equals 9.76.

X	$(X - M)$	$(X - M)^2$
1	−3.2	10.24
2	−2.2	4.84
4	−0.2	0.04
4	−0.2	0.04
10	5.8	33.64

Language Alert! We need a few more symbols to use symbolic notation to represent the idea of variance. The symbols that represent the variance of a *sample* include SD^2, s^2, and *MS*. The first two symbols, SD^2 and s^2, both represent the words *standard deviation squared.* The symbolic notation *MS* comes from the words *mean square* (referring to the average of the squared deviations). We'll use SD^2 at this point, but we will alert you when we switch to other symbols for variance later. The variance of the *sample* uses all three symbolic notations; however, the variance of a *population* uses just one symbol: σ^2 (pronounced "sigma squared"). Table 4-2 summarizes the symbols and lan-

TABLE 4-2. Variance and Standard Deviation in Symbols

The variance or standard deviation of a sample is an example of a statistic, whereas the variance or standard deviation of a population is an example of a parameter. The symbols we use depend on whether we are referring to the spread of a sample or of a population.

Number	Used for . . .	Standard Deviation Symbol	Pronounced	Variance Symbol	Pronounced
Statistic	Sample	*SD* or *s*	As written	SD^2, s^2, or *MS*	Letters as written; if superscript ^{2}, then followed by "squared" (e.g., "ess squared")
Parameter	Population	σ	"Sigma"	σ^2	"Sigma squared"

guage used to describe different version of the mean and variance, but we will keep reminding you as we go along.

We already know all the other symbols needed to calculate variance: X to indicate the individual scores, M to indicate the mean, and N to indicate the sample size.

$$SD^2 = \frac{\Sigma(X - M)^2}{N}$$

As you can see, variance is really just a mean—the mean of squared deviations. ■

MASTERING THE FORMULA

4-3: The formula for variance is: $SD^2 = \frac{\Sigma(X - M)^2}{N}$. To calculate variance, subtract the mean (M) from every score (X) to calculate deviations from the mean; then square these deviations, sum them, and divide by the sample size (N). By summing the squared deviations and dividing by sample size, we are taking their mean.

Standard Deviation

EXAMPLE 4.10

Language Alert! Variance and standard deviation refer to the same core idea. The standard deviation is more useful because it *is the typical amount that each score varies from the mean*. Mathematically, the ***standard deviation*** *is the square root of the average of the squared deviations from the mean, or, more simply, the square root of the variance.* The beauty of the standard deviation—compared to the variance—is that we can understand it at a glance.

For example, the numbers of therapy sessions for the five students were 1, 2, 4, 4, and 10, with a mean of 4.2. The typical score does not vary from the mean by 9.76. The variance is based on squared deviations, not deviations, so it is too large. When we ask our students how to solve this problem, they invariably say, "Unsquare it," and that's just what we do. We take the square root of variance to come up with a much more useful number, the standard deviation. The square root of 9.76 is 3.12. Now we have a number that "makes sense" to us. We can now say that the typical number of therapy sessions for students in this sample is 4.2 and the typical amount a student varies from that is 3.12.

As you read journal articles, you often will see the mean and standard deviation reported as: ($M = 4.2$, $SD = 3.12$). A glance at the original data (1, 2, 4, 4, 10) tells us that these numbers make sense: 4.2 does seem to be approximately in the center, and scores do seem to vary from 4.2 by roughly 3.12. The score of 10 is a bit of an outlier—but not so much of one; the mean and the standard deviation are still somewhat representative of the typical score and typical deviation.

We didn't actually need a formula to get the standard deviation. We just took the square root of the variance. Perhaps you guessed the symbols for standard deviation by just taking the square root of those for variance. With a sample, standard deviation is either SD or s. With a population, standard deviation is σ. Table 4-2 presents this information concisely. We can write the formula showing how standard deviation is calculated from variance:

$$SD = \sqrt{SD^2}$$

MASTERING THE FORMULA

4-4: The most basic formula for standard deviation is: $SD = \sqrt{SD^2}$. We simply take the square root of variance.

We can also write the formula showing how standard deviation is calculated from the original X's, M, and N:

$$SD = \sqrt{\frac{\Sigma(X - M)^2}{N}}$$ ■

MASTERING THE FORMULA

4-5: The full formula for standard deviation is: $SD = \sqrt{\frac{\Sigma(X - M)^2}{N}}$. To determine standard deviation, subtract the mean from every score to calculate deviations from the mean. Then, square the deviations from the mean. Sum the squared deviations, then divide by the sample size. Finally, take the square root of the mean of the squared deviations.

CHECK YOUR LEARNING

Reviewing the Concepts

> The simplest way to measure variability is by using the range, which is calculated by subtracting the lowest score from the highest score.

> Variance and standard deviation both measure the degree to which scores in a distribution vary from the mean. The standard deviation is simply the square root of the variance: it represents the typical deviation of a score from the mean.

Clarifying the Concepts

4-5 In your own words, what is variability?

4-6 Distinguish the range from the standard deviation. What does each tell us about the distribution?

Calculating the Statistics

4-7 Calculate the range, variance, and standard deviation for the following data sets (the same ones from the section on central tendency).

a. 10 8 22 5 6 1 19 8 13 12 8

b. 122.5 123.8 121.2 125.8 120.2 123.8 120.5 119.8 126.3 123.6

c. 0.100 0.866 0.781 0.555 0.222 0.245 0.234

Applying the Concepts

4-8 Final exam week is approaching, and students are not eating as well as usual. Four students were asked how many calories of junk food they had consumed between noon and 10:00 P.M. on the day before an exam. The estimated numbers of nutritionless calories, calculated with the help of a nutritional software program, were 450, 670, 1130, and 1460.

a. Using the formula, calculate the range for these scores.

b. What information can't you glean from the range?

c. Using the formula, calculate variance for these scores.

d. Using the formula, calculate standard deviation for these scores.

e. If a researcher was interested only in these four students, what symbols would she use for variance and standard deviation, respectively?

f. If another researcher hoped to generalize from these four students to all students at the university, what symbols would he use for variance and standard deviation?

Solutions to these Check Your Learning questions can be found in Appendix D.

REVIEW OF CONCEPTS

Central Tendency

Three measures of *central tendency* are commonly used in research. (When a numeric description, such as a measure of central tendency, describes a sample, it is a *statistic*; when it describes a population, it is a *parameter*.) The *mean* is the arithmetic average of the data. The *median* is the midpoint of the data set; 50% of scores fall on either side of the median. The *mode* is the most common score in the data set. When there's one mode, the distribution is *unimodal*; when there are two modes, it's *bimodal*; and when there are three or more modes, it's *multimodal*. The mean is highly influenced by outliers, whereas the median and mode are resistant to outliers.

Measures of Variability

The *range* is the simplest measure of *variability* to calculate. It is calculated by subtracting the minimum score in our data set from the maximum score. Variance and standard deviation are much more common measures of variability. They are used when the preferred measure of central tendency is the mean. *Variance* is the average of the squared deviations. It is calculated by subtracting the mean from every score to get *deviations from the mean*, then squaring each of the deviations. (In future chapters, we will use the *sum of squares* of the deviations when using samples to make inferences about a population.) *Standard deviation* is the square root of variance. It is the typical amount that a score deviates from the mean.

SPSS®

The left-hand column in "Data View" is prenumbered, beginning with 1. Each column to the right of that number contains information about a particular variable; each row across from that number represents a unique individual. Notice the choices at the top of the "Data View" screen. Enter the data for countries' top finishes in the World Cup, omitting countries with scores of 0: 1, 1, 2, 2, 2, 2, 2, 2, 2, 2, 4, 6, 8, and 10, as shown on the left of the screenshot below.

To get a numerical description of a variable, select: Analyze → Descriptive Statistics → Frequencies.

Then select the variable of interest, "top finishes," by highlighting it and then clicking the arrow to move it from the left side to the right side. Then select: Statistics → Mean, Median, Mode, Std. deviation, Range → Continue → OK.

Your data and output will look like those in the screenshot shown here.

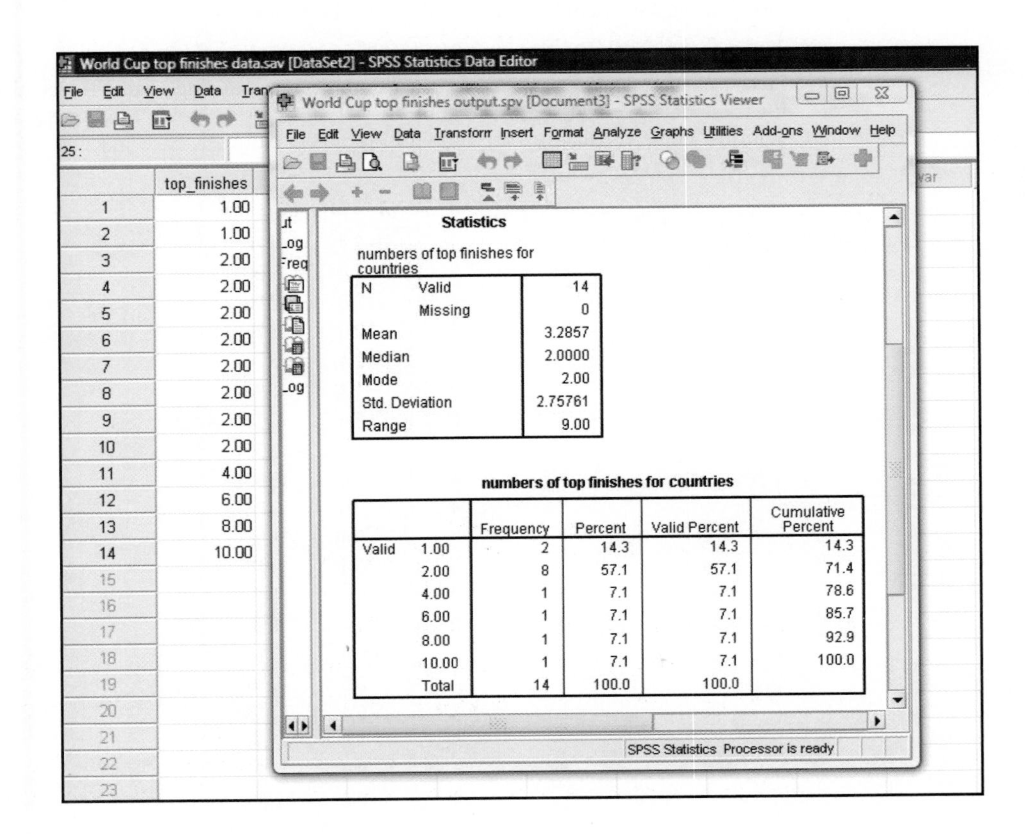

How It Works

4.1 CALCULATING THE MEAN

Here are data for the numbers of nights out socializing in a week for 20 students.

1 2 7 6 1 2 6 5 4 4 0 3 2 2 3 4 3 5 4 4

How can we calculate the mean? First, we add up all of the scores:

$$1 + 2 + 7 + 6 + 1 + 2 + 6 + 5 + 4 + 4 + 0 + 3 + 2 + 2 + 3 + 4 + 3 + 5 + 4 + 4 = 68$$

Then we divide by 20, the number of scores:

$$68/20 = 3.4$$

With the formula $M = \frac{\Sigma X}{N}$, we calculate:

$$M = \frac{(1+2+7+6+1+2+6+5+4+4+0+3+2+2+3+4+3+5+4+4)}{20} = 3.4$$

4.2 CALCULATING THE MEDIAN

Using the data for "nights out socializing," how can we calculate the median? The median is either the middle score or the average of the two middle scores. We first arrange the data from lowest score to highest score:

0 1 1 2 2 2 2 3 3 3 4 4 4 4 4 5 5 6 6 7

With 20 scores (an even number), there are two middle scores, the 10th and 11th scores, which are 3 and 4. We determine the median by taking the average of 3 and 4. The median is 3.5.

4.3 CALCULATING THE MODE

How can we calculate the mode for the "nights out socializing" data? The mode is the most common score. We can determine the mode for these data by looking at the frequency distribution. Five people have a score of 4. The mode is 4.

4.4 CALCULATING VARIANCE

How can we calculate variance for the "nights out socializing" data? To calculate variance for these data, we first subtract the mean, 3.4, from every score. We then square these deviations. These calculations are shown in the table below.

X	$(X - M)$	$(X - M)^2$
1	−2.4	5.76
2	−1.4	1.96
7	3.6	12.96
6	2.6	6.76
1	−2.4	5.76
2	−1.4	1.96
6	2.6	6.76
5	1.6	2.56
4	0.6	0.36
4	0.6	0.36
0	−3.4	11.56
3	−0.4	0.16
2	−1.4	1.96
2	−1.4	1.96
3	−0.4	0.16
4	0.6	0.36
3	−0.4	0.16
5	1.6	2.56
4	0.6	0.36
4	0.6	0.36

We then add all of the scores in the third column to get the sum of squared deviations, or the sum of squares. This sum is 64.8.

We can use the formula to complete our calculations:

$$SD^2 = \frac{\Sigma(X - M)^2}{N} = \frac{64.8}{20} = 3.24$$

The variance is 3.24.

4.5 CALCULATING STANDARD DEVIATION

How can we calculate standard deviation for the "nights out socializing" data? The standard deviation is the square root of the variance. For these data, we calculate standard deviation directly from the variance we calculated above using this formula:

$$SD = \sqrt{SD^2} = \sqrt{3.24} = 1.80$$

The standard deviation is 1.80.

Exercises

Clarifying the Concepts

4.1 Define the three measures of central tendency: mean, median, and mode.

4.2 The mean can be assessed visually and arithmetically. Describe each method.

4.3 Explain how the mean mathematically balances the distribution.

4.4 Explain what is meant by unimodal, bimodal, and multimodal distributions.

4.5 Explain why the mean might not be useful for a bimodal or multimodal distribution.

4.6 What is an outlier?

4.7 How do outliers affect the mean and median?

4.8 In which situations is the mode typically used?

4.9 Explain the concept of standard deviation in your own words.

4.10 Define the symbols used in the equation for variance:

$$SD^2 = \frac{\Sigma(X - M)^2}{N}$$

4.11 Why is the standard deviation typically reported, rather than the variance?

4.12 Find the incorrectly used symbol or symbols in each of the following statements or formulas. For each statement or formula, (1) state which symbol(s) is/are used incorrectly, (2) explain why the symbol(s) in the original statement is/are incorrect, and (3) state which symbol(s) *should* be used.

a. The mean and standard deviation of the sample of reaction times were calculated ($m = 54.2$, $SD^2 = 9.87$).

b. The mean of the sample of high school student GPAs was $\mu = 3.08$.

c. Range $= X_{highest} - X_{lowest}$

Calculating the Statistics

4.13 Use the following data for this exercise:

15 34 32 46 22 36 34 28 52 28

a. Calculate the mean, the median, and the mode.

b. Add another data point, 112. Calculate the mean, median, and mode again. How does this new data point affect the calculations?

c. Calculate the range, variance, and standard deviation for the original data.

4.14 Use the following salary data for this exercise:

$44,751
$52,000
$41,500
$38,862
$51,380
$61,774

a. Calculate the mean, the median, and the mode.

b. Add another salary, $97,582. Calculate the mean, median, and mode again. How does this new salary affect the calculations?

c. Calculate the range, variance, and standard deviation for the original salary data.

d. How does the range change when you include the outlier salary, $97,582?

4.15 The Mount Washington Observatory (MWO) in New Hampshire claims to have the world's worst weather.

Below are some data on the weather extremes recorded at the MWO.

Month	Normal Daily Maximum (°F)	Normal Daily Minimum (°F)	Record Low in °F (Year)	Peak Wind Gust in Miles per Hour (Year)
January	14.0	−3.7	−47 (1934)	173 (1985)
February	14.8	−1.7	−46 (1943)	166 (1972)
March	21.3	5.9	−38 (1950)	180 (1942)
April	29.4	16.4	−20 (1995)	231 (1934)
May	41.6	29.5	−2 (1966)	164 (1945)
June	50.3	38.5	8 (1945)	136 (1949)
July	54.1	43.3	24 (2001)	154 (1996)
August	53.0	42.1	20 (1986)	142 (1954)
September	46.1	34.6	9 (1992)	174 (1979)
October	36.4	24.0	−5 (1939)	161 (1943)
November	27.6	13.6	−20 (1958)	163 (1983)
December	18.5	1.7	−46 (1933)	178 (1980)

a. Calculate the mean and median normal daily minimum temperature across the year.

b. Calculate the mean, median, and mode for the record low temperatures.

c. Calculate the mean, median, and mode for the peak wind-gust data.

d. When no mode appears in the raw data, we can compute a mode by breaking the data into intervals. How might you do this for the peak wind-gust data?

e. Calculate the range, variance, and standard deviation for the normal daily minimum temperature across the year.

f. Calculate the range, variance, and standard deviation for the record low temperatures.

g. Calculate the range, variance, and standard deviation for the peak wind-gust data.

4.16 Here are the *U.S. News & World Report* data again on percentage of alumni giving at the top 70 national universities.

48 61 45 39 46 37 38 34 33 47
29 38 38 34 29 29 36 48 27 25
15 25 14 26 33 16 33 32 25 34
26 32 11 15 25 9 25 40 12 20
32 10 24 9 16 21 12 14 18 20
18 25 18 20 23 9 16 17 19 15
14 18 16 17 20 24 25 11 16 13

a. Calculate the mean of these data, showing that you know how to use the symbols and formula.

b. Determine the median of these data.

c. Describe the variability in these data by computing the range.

Applying the Concepts

4.17 Mean versus median for salary data: Back in Exercises 4.13 and 4.14, we saw how the mean and median changed when an outlier was included in the computations. If you were reporting the "average" salary at a company, how might the mean and median give different impressions to potential applicants?

4.18 Mean versus median for temperature data: For the data presented in Exercise 4.15, the "normal" daily maximum and minimum temperatures recorded at the Mount Washington Observatory are presented for each month. These are likely to be measures of central tendency for each month over time. Explain why these "normal" temperatures might be calculated as means or medians. What would be the reasoning for using one type of statistic over the other?

4.19 Mean versus median for depression scores: A depression research unit recently assessed seven participants chosen at random from the university population. Is the mean or the median a better indicator of the central tendency of these seven participants? Explain your answer.

4.20 Measures of central tendency for weather data: The "normal" weather data from the Mount Washington Observatory are broken down by month. Why might you not want to average across all months in a year? How else could you summarize the year?

4.21 Outliers, central tendency, and data on wind gusts: There appears to be an outlier in the data for peak wind gust recorded on top of Mount Washington (see data in Exercise 4.15). Where do you see an outlier and how does excluding this data point affect the different calculations of central tendency?

4.22 Measures of central tendency for measures of baseball performance: Here are winning percentages for 11 baseball players for their best 4-year pitching performances:

0.755 0.721 0.708 0.773 0.782 0.747
0.477 0.817 0.617 0.650 0.651

a. What is the mean of these scores?

b. What is the median of these scores?

c. Compare the mean and the median. Does the difference between them suggest that the data are skewed very much?

4.23 Mean versus median in "real life": Briefly describe a real-life situation in which the median is preferable to the mean. Give hypothetical numbers for the mean and median in your explanation. Be original! (Don't use home prices or another example from the chapter.)

4.24 Descriptive statistics in the media: Find an advertisement for an anti-aging product either online or in the print media—the more unbelievable the claims, the better!

a. What does the ad promise that this product will do for the consumer?

b. What data does it offer for its promised benefits? Does it offer any descriptive statistics or merely testimonials? If it offers descriptive statistics, what are the limitations of what they report?

c. If you were considering this product, what measures of central tendency would you most like to see? Explain your answer, noting why not all measures of central tendency would be helpful.

d. If a friend with no statistical background was considering this product, what would you tell him or her?

4.25 Descriptive statistics in the media: When there is an ad on TV for a body-shaping product (e.g., an abdominal muscle machine), often a person with a wonderful success story is featured in the ad. The statement "Individual results may vary" hints at what kind of data the advertisement may be presenting.

a. What kind of data is being presented in these ads?

b. What statistics could be presented to help inform the public about how much "individual results might vary"?

4.26 Range for data from the National Survey of Student Engagement: The National Survey of Student Engagement (NSSE) asked U.S. students how often they asked questions in class or participated in classroom discussions. The options were "Never," "Sometimes," "Often," and "Very often." Here are the percentages, reported in 2005, of students who responded "Very often" for the 31 institutions classified as liberal arts colleges that allowed their 2004 data to become public through the *U.S. News & World Report* Web site.

58 45 53 45 65 41 50 46 54
59 52 60 59 62 54 52 53 54
83 60 32 62 50 50 43 32 53
60 52 55 53

a. What is the range of these data?

b. The college whose students participated most is Marlboro College in Vermont, and the two colleges who tied for least participation are Randolph-Macon Woman's College (now called Randolph College) in Virginia and Texas A&M University in Galveston. What research questions do these data suggest to you? State at least one research question generated by these data.

4.27 Descriptive statistics for data from the National Survey of Student Engagement: Here again are the data from the NSSE for a sample of 19 national universities, as reported in 2005. These are the percentages of U.S. students, for each university, who said they were assigned between 5 and 10 twenty-page papers.

0 5 3 3 1 10 2
2 3 1 2 4 2 1
1 1 4 3 5

a. Calculate the mean of these data using the symbols and formula.

b. Calculate the variance of these data using the symbols and formula, but also using columns to show all calculations.

c. Calculate the standard deviation using the symbols and formula.

d. In your own words, describe what the mean and standard deviation of these data tell us about these scores.

4.28 Statistics versus parameters: For each of the following situations, state whether the mean would be a statistic or a parameter. Explain your answer.

a. According to 2006 Canadian census data, the median family income in British Columbia was $62,346, lower than the national average of $63,866.

b. In the 2010–2011 soccer season, the stadiums of teams in the English Premier League had a mean capacity of 38,391 supporters.

c. The General Social Survey (GSS) includes a vocabulary test in which participants are asked to choose the appropriate synonym from a multiple-choice list of five words (e.g., *beast* with the choices *afraid, words, large, animal,* and *separate*). The mean vocabulary test score was 5.98.

d. The National Survey of Student Engagement (NSSE) asked students at participating institutions how often they discussed ideas or readings with professors outside of class. Among the 19 national universities that made their data public, the mean percentage of U.S. students who responded "Very often" was 8%.

4.29 Central tendency and the shapes of distributions: Consider the many possible distributions of grades on a quiz in a statistics class; imagine that the grades could range from 0 to 100. For each of the following situations, give a hypothetical mean and median (that is, make up a mean and a median that might occur with a distribution that has this shape). Explain your answer.

a. Normal distribution
b. Positively skewed distribution
c. Negatively skewed distribution

4.30 Shapes of distributions: For each of the following, state whether the distribution is more likely to be unimodal or bimodal. Explain your answer.

a. Age of patients in a hospital maternity ward
b. University students' depression scores on a Beck Depression Inventory
c. GRE scores of applicants to sociology graduate programs
d. The cost of an AIDS drug that is sold in developed countries in Europe as well as in developing countries in Africa

4.31 Central tendency and outliers from growth chart data: When the average height or average weight of children is plotted to create growth charts, do you think it would be appropriate to use the mean for these data? There are often outliers for height, but why might we not have to be concerned with their effect on these data?

4.32 Measures of central tendency for percentages of advanced degrees: The U.S. Census Bureau collects and analyzes data on numerous aspects of American life by state, including the percentage of people with high school degrees, bachelor's degrees, and advanced degrees. If you wanted to calculate the "average" percentage of people with advanced degrees across all states, would you report a mean, a median, or a mode? Explain your answer clearly.

4.33 Mean versus median for age at first marriage: The mean age at first marriage was 32 for men and 30 for women in Denmark in 2003 (http://www.eurofound.europa.eu/areas/qualityoflife/eurlife/index.php?template=3&radioindic=186&idDomain=5). The median age at first marriage was 27.1 for men and 25.3 for women in the U.S. in 2003 (Simmons & Dye, 2003). Explain why we cannot directly compare these measures of central tendency to make cross-national comparisons.

4.34 Range and world records: Guinness World Records relies on what kind of data for its amazing claims? How does this relate to the calculation of ranges?

Putting It All Together

4.35 Descriptive statistics and basketball wins: Here are the numbers of wins for the 30 National Basketball Association teams in the 2004–2005 season.

45 43 42 33 33 54 47 44 42 30
59 45 36 18 13 52 49 44 27 26
62 50 37 34 34 59 58 51 45 18

a. Create a grouped frequency table for these data.
b. Create a histogram based on the grouped frequency table.
c. Determine the mean, median, and mode of these data. Use symbols and the formula when showing your calculation of the mean.
d. Using software, calculate the range and standard deviation of these data.
e. Write a one- to two-paragraph summary describing the distribution of these data. Mention center, variability, and shape. Be sure to discuss the number of modes (i.e., unimodal, bimodal, multimodal), any possible outliers, and the presence and direction of any skew.
f. State one research question that might arise from this data set.

4.36 Central tendency and outliers for data on global TV viewing habits: Below are approximate, daily average viewing times for 12 countries based on 2007 data from the *Economist* Web site (http://www.economist.com/):

United States	8.2 hours
Turkey	5 hours
Italy	4.05 hours
Japan	3.75 hours
Spain	3.6 hours
Portugal	3.5 hours
Australia	3.2 hours
South Korea	3.16 hours
Canada	3.1 hours
Britain	3 hours
Denmark	3 hours
Finland	2.8 hours

a. Compute the mean and the median across these 12 data points.
b. How are these statistics affected by including or excluding the United States?
c. How do you think these daily "averages" were calculated—using means or medians?
d. Do you think TV viewing habits might vary by other personal or demographic characteristics? Could these represent confounds?
e. How might you collect samples to more specifically describe TV viewing habits as a function of other personal characteristics?

Terms

central tendency (p. 70)
mean (p. 71)
statistic (p. 72)
parameter (p. 72)
median (p. 73)
mode (p. 74)
unimodal (p. 75)
bimodal (p. 75)
multimodal (p. 75)
variability (p. 78)
range (p. 78)
variance (p. 79)
deviation from the mean (p. 79)
sum of squares (p. 80)
standard deviation (p. 81)

Formulas

$M = \frac{\Sigma X}{N}$ (p. 73)

range $= X_{highest} - X_{lowest}$ (p. 78)

$SD^2 = \frac{\Sigma(X - M)^2}{N}$ (p. 81)

$SD = \sqrt{SD^2}$ (p. 81)

$SD = \sqrt{\frac{\Sigma(X - M)^2}{N}}$ (p. 81)

Symbols

M	(p. 72)
$\overline{X}$	(p. 72)
μ	(p. 72)
X	(p. 72)
Σ	(p. 72)
N	(p. 72)
Mdn	(p. 73)
SS	(p. 80)
SD^2	(p. 80)
s^2	(p. 80)
MS	(p. 80)
σ^2	(p. 80)
SD	(p. 81)
s	(p. 81)
σ	(p. 81)

CHAPTER 5

Sampling and Probability

BEFORE YOU GO ON

- You should understand the difference between a sample and a population (Chapter 1).
- You should know how to measure central tendency, especially the mean (Chapter 4).

AP Photo/Michael Conroy

Voter Sampling in the 2000 Presidential Election Sampling errors led to election-night confusion about the winner of the 2000 U.S. presidential election.

No matter how impassioned people get about their politics, it's always going to be tough to beat the statistical drama that unfolded during the 2000 presidential contest between Al Gore and George W. Bush. It all came down to which of them had won the vote in Florida. Based on sampling, Al Gore was declared the winner by ABC, CBS, CNN, Fox News, and NBC—but then the networks retracted their reports. Later, those same networks declared George W. Bush the winner—and retracted those reports, too. Bush was eventually declared the winner, based on a split decision made by the U.S. Supreme Court that overturned a split decision made by the Florida Supreme Court (Konner, Risser, & Wattenberg, 2001).

Sampling errors were at the heart of the confusion. Here's what the networks did wrong:

1. The sample size (both the number of precincts sampled and the size of the samples within the precincts) was too small.
2. The sample was biased. Voters from the Republican-leaning Florida Panhandle, who are in a different time zone, weren't included in early projections.
3. The samples were not independent. All five networks used the same source of information, the Voter News Service.

The result of those errors was that the sample did not mirror, or represent, the larger population. You are less likely to make those mistakes because this chapter teaches you how to (a) sample from populations; (b) assign members of samples to groups; (c) understand why so-called coincidences are fairly common; and (d) form a hypothesis that tests how likely it is that your sample represents the larger population. At the end of the chapter, we add one last caution. Even when you have done everything right, your data— just by chance—still might trick you into errors that are like *false positives* and *false negatives.*

Samples and Their Populations

Almost everything worth evaluating requires a sample, from voting trends to sales patterns at online stores to the effectiveness of flu vaccines. The goal of sampling is simple: collect a sample that represents the population. However, the 2000 presidential election reminds us that sampling is easier in theory than it is in practice.

> **MASTERING THE CONCEPT**
>
> **5.1:** There are two main types of samples in social science research. In the ideal type (a random sample), every member of the population has an equal chance of being selected to participate in a study. In the less ideal but more common type (a convenience sample), researchers use participants who are readily available.

There are two main types of samples: random samples and convenience samples. *A* ***random sample*** *is one in which every member of the population has an equal chance of being selected into the study. A* ***convenience sample*** *is one that uses participants who are readily available.* A random sample remains the ideal and is far more likely to lead to a representative sample, but it is usually expensive and difficult to recruit. A convenience sample—such as college students—is usually less expensive and readily available.

Random Sampling

Imagine that a town has recently experienced a traumatic mass murder and that there are exactly 80 police officers in the local department. You have been hired to determine whether peer counseling or professional counseling is the more effective way to address their concerns in the aftermath of this trauma. Unfortunately, budget constraints dictate that the sample you can recruit must be very small—just 10 people. How do you maximize the probability that 10 officers will accurately represent the larger population of 80 officers?

Here's one way you could do it: Assign each of the 80 officers a number, from 01 to 80. Next, use the random numbers table shown below to choose a sample of 10 police officers. Then, (a) select any point on the table; (b) decide to go across, back, up, or down to read through the numbers.

For example, say you begin with the 6th number of the second row and count across. The first 10 numbers read: 97654 64501. (The spaces between sets of five numbers exist solely to make it easier to read the table.) The first pair of digits is 97, but we would ignore this number because we only have 80 people in our population. The next pair is 65. The 65th police officer in our list would be chosen for our sample. The next two pairs, 46 and 45, would also be in the sample, followed by 01. If we come across a number a second time—45, for example—we ignore it, just as we would ignore 00 and anything above 80. Stick to your decision—you can't rule out Officer Jones because she seems to be adjusting well on her own or rule in Officer McIntyre because you think he needs the help right away.

- A **random sample** is one in which every member of the population has an equal chance of being selected into the study.
- A **convenience sample** is one that uses participants who are readily available.

Excerpt from a Random Numbers Table

This is a small section from a random numbers table used to randomly select participants from a population to be in a sample as well as to randomly assign participants to experimental conditions.

04493	52494	75246	33824	45862	51025	61962
00549	97654	64501	88159	96119	63896	54692
35963	15307	26898	09354	33351	35462	77974
59808	08391	45427	26842	83609	49700	46058

Were you surprised that random sampling selected both the number 46 and the number 45? *Truly random numbers often have strings of numbers that do not seem to be random.* For example, notice the string of three 3's in the third row of the table.

You can also search online for a " random numbers generator," which is how we came up with the following numbers: 10, 23, 27, 34, 36, 67, 70, 74, 77, and 78. You

Matthew Lloyd/Getty Images

Random Dots? True randomness often does not seem random. British artist Damien Hirst farms out the actual painting of many of his works to assistants, including his famous dot paintings. He provides his assistants with instructions, including to arrange the color dots randomly. One assistant painted a series of yellow dots next to each other, which led to a fight with Hirst, who said, "I told him those aren't random. . . .Now I realize he was right, and I was wrong."

might be surprised that 4 of the 10 numbers were in the 70s. Don't be. Random numbers are truly random, even if they don't look random.

Random samples are almost never used in the social sciences because we almost never have access to the whole population. For example, if we were interested in studying the eating behavior of voles, we would never be able to list the whole population of voles from which to then select a random sample. If we were interested in studying the effect of video games on the attention span of teenagers in the United Kingdom, we would never be able to identify all U.K. teenagers from which to choose a random sample.

Convenience Sampling

It is far more convenient (faster, easier, and cheaper) to use voles that we bought from an animal supply company or teenagers from the local school—but there is a significant downside. A convenience sample might not represent the larger population. ***Generalizability*** *refers to researchers' ability to apply findings from one sample or in one context to other samples or contexts.* This principle is also called *external validity,* and it is extremely important—why bother doing the study if it doesn't apply to anyone?

Fortunately, we can increase external validity through ***replication****, the duplication of scientific results in a different context or with a sample that has different characteristics.* In other words, do the study again. And again. Then ask someone else to replicate it, too. That's the slow but trustworthy process by which science creates knowledge that is both reliable and valid. That's also why some of the real "scientific breakthroughs" you hear about are really just the tipping point.

Liars' Alert! We must be even more cautious when we use a ***volunteer sample*** (also called a *self-selected sample*), *a convenience sample in which participants actively choose to participate in a study.* Participants volunteer, or self-select, when they respond to recruitment flyers or choose to complete an online survey, such as polls that recruit people to vote for a favorite reality show contestant or college basketball team. We should be very suspicious of volunteer samples, which may be very different from a randomly selected sample. We should be cautious whenever participants volunteer. The information they provide may not represent the larger population that we are really interested in.

Susan Nolan

Testimonials as Evidence? Does one middle-aged woman's positive experience with Skin's Shangri La—"I'm nearly 60, but no one believes it"—provide evidence that this moisturizer makes for younger-looking skin? Testimonials use a volunteer sample of one person, usually a biased person; moreover, you can bet that the testimonial a company uses in its advertising is the most flattering one.

The Problem with a Biased Sample

Don't be a sucker by not understanding sampling. The colorful cosmetics catalog *Lush Times* includes a description of the amazing skin-rejuvenating powers of the face moisturizer called Skin's Shangri La. It includes a testimonial, which reads in part: "I'm nearly 60, but no one believes it, which proves Skin's Shangri La works!" Let's examine the flaws in this sample of "evidence"—a brief testimonial—for the supposed effectiveness of Skin's Shangri La.

The population of interest is women close to age 60. The sample is the one woman who wrote to Lush. Assuming that it is a real letter, there are two major problems. First, one person is not a trustworthy sample size. Second, this is a volunteer sample. The customer who had this experience chose to write to Lush. Was she likely to write to Lush if she did not feel very strongly about this product? Moreover, would Lush be likely to publish her statement if it wasn't positive?

Let's think more closely about the problem of self-selection. Lush touts its products as meant for people of all ages. But with colorful, cartoonlike drawings and catchy product names such as Candy Fluff and Sonic Death Monkey, it seems likely that teens and 20-somethings are the intended consumers. A 60-year-old woman shopping at Lush may have a more youthful mind-set to begin with.

To create a trustworthy test of the effectiveness of Skin's Shangri La, we could randomly assign a certain number of people to use this product and an equal number of people to use another product (or no product), and then see which group has better skin a certain number of weeks later. Which is more persuasive: a dubious testimonial or a well-designed experiment? If our honest answer is a dubious testimonial, then statistical reasoning once again leads us to ask a better question (nicely answered by social psychologists, by the way) about why anecdotes are sometimes more persuasive than science.

- **Generalizability** refers to researchers' ability to apply findings from one sample or in one context to other samples or contexts; also called *external validity.*
- **Replication** refers to the duplication of scientific results, ideally in a different context or with a sample that has different characteristics.
- A **volunteer sample**, or *self-selected sample,* is a special kind of convenience sample in which participants actively choose to participate in a study.

Random Assignment

Random assignment is the distinctive signature of a scientific study. Why? It levels the playing field when every participant has an equal chance of being assigned to any level of the independent variable. Random assignment is different from random selection. Random selection refers to a method of creating a sample from a population; random assignment refers to a method we can use once we have a sample, whether or not the sample is randomly selected. Random *selection* is almost never used, but random *assignment* is used whenever possible. Random assignment overcomes many of the limitations of a convenience sample.

Random assignment uses procedures similar to those used for random selection. If a study has two levels of the independent variable, as in the study of police officers, then you would need to assign participants to one of two groups. You could decide, arbitrarily, to number the groups 0 and 1 for the "peer-counseling" and "therapist-counseling" groups, respectively. Then, (a) select any point on the table; (b) decide to go across, back, up, or down to read through the numbers—and remember to stick to your decision.

For example, if you began at the first number of the last row and read the numbers across, ignoring any number but 0 or 1, you would find 0010000. Hence, the first two participants would be in group 0, the third would be in group 1, and the next four would be in group 0. (Again, notice the seemingly nonrandom pattern and remember that it *is* random.)

An online random numbers generator lets us tell the computer to give us one set of 10 numbers that range from 0 to 1. We would instruct the program that the numbers should *not* remain unique because we want multiple 0's and multiple 1's. In addition, we would request that the numbers not be sorted because we want to assign participants in the order in which the numbers are generated. When we used an online random numbers generator, the 10 numbers were: 1110100001. In an experiment, we usually want equal numbers in the groups. If the numbers were not exactly half 1's and half 0's, as they are in this case, we could decide in advance to use only the first five 1's or the first five 0's. Just be sure to establish your rule for random assignment ahead of time and then stick to it!

▶ MASTERING THE CONCEPT

5.2: Replication and random assignment to groups help overcome the problems of convenience sampling. With replication, a study is repeated, ideally with different participants or in a different context, to see whether the results are consistent. With random assignment, every participant has an equal chance of being assigned to any level of the independent variable.

CHECK YOUR LEARNING

Reviewing the Concepts

- Data from a sample are used to draw conclusions about the larger population.
- In random sampling, every member of the population has an equal chance of being selected for the sample.
- Convenience samples are far more common than random samples in the behavioral sciences.
- In random assignment, every participant has an equal chance of being assigned to one of the experimental conditions.
- If a study that uses random assignment is replicated in several contexts, we can start to generalize the findings.
- Random numbers may not always appear to be all that random; there may appear to be patterns.

Clarifying the Concepts

5-1 What are the risks of sampling?

Calculating the Statistics

5-2 Use the excerpt from the random numbers table on page 93 to select 6 people out of a sample of 80. Start by assigning each person a number, from 01 to 80. Then select 6 of these people by starting in the fourth row and going across. List the numbers of the 6 people who were selected.

5-3 Use the excerpt from the random numbers table on page 93 to randomly assign these six people to one of two experimental conditions, numbered 0 and 1. This time, start at the top of the first column (with a 0 on top) and go down. When you get to the bottom of that column, start at the top of the second column (with a 4 on top). Using the numbers (0 and 1), list the order in which these people would be assigned to conditions.

Applying the Concepts

5-4 For each of the following scenarios, state whether random selection could have been used from a practical standpoint. Explain your answer, including in it a description of the population to which the researcher likely wants to generalize. Then state whether random assignment could have been used, and explain your answer.

a. A health psychologist examined whether postoperative recovery time was less among patients who received counseling prior to surgery than among those who did not.

b. The head of a school board asked a school psychologist to examine whether children perform better in history classes if they use an online textbook as opposed to a printed textbook.

c. A clinical psychologist studied whether people with diagnosed personality disorders were more likely to miss therapy appointments than were people without diagnosed personality disorders.

Solutions to these Check Your Learning questions can be found in Appendix D.

Probability

You have probably heard phrases such as "the margin of error" or "plus or minus 3 percentage points," especially during an election season. These are another way of saying "We're not 100% sure that we can believe our own results." This could make you cynical about statistics—all this work, and we still don't know for sure what's going on.

You would be more justified, however, in celebrating statistics for being so truthful—statistics allows us to quantify our uncertainty. And let's be honest—most of life is filled with uncertainty. Probability is central to inferential statistics because we are basing a conclusion about a population on data collected from a sample (rather than an anecdote based on just one or two testimonials). When calculating an inferential statistic, we are determining only that it is probable that a given conclusion is true, not that it is certain.

Coincidence and Probability

Probability and statistical reasoning can save us from ourselves by allowing us to distinguish fact from what may otherwise feel like an eerie coincidence, actually fueled by our personal biases. ***Confirmation bias** is our usually unintentional tendency to pay attention to evidence that confirms what we already believe and to ignore evidence that would disconfirm our beliefs.* It is a confirmation bias when an athlete attributes her team's wins to her lucky earrings, ignoring any losses while wearing them or any wins while wearing other earrings. Confirmation biases closely follow illusory correlations. ***Illusory correlation** is the phenomenon of believing one sees an association between variables when no such association exists.* An athlete with a confirmation bias attributing wins to her lucky earrings now believes an illusory correlation. We invite illusory correlations into our lives whenever we ignore the gentle, restraining logic of statistical reasoning.

- **Confirmation bias** is our usually unintentional tendency to pay attention to evidence that confirms what we already believe and to ignore evidence that would disconfirm our beliefs. Confirmation biases closely follow illusory correlations.

- **Illusory correlation** is the phenomenon of believing one sees an association between variables when no such association exists.

Mike Powell/Getty Images

Lucky Charms Many athletes have a lucky article of clothing that they wear on game day because they think it helps them win. Confirmation biases lead us to notice events that match our beliefs (the occasions on which the lucky object is paired with a win) and ignore those that do not (the occasions on which the lucky object is paired with a loss).

For example, the science show *Radiolab* told a remarkable story of coincidence (Abumrad & Krulwich, 2009). A 10-year-old girl named Laura Buxton released a red balloon from her hometown in the north of England. "Almost 10," Laura corrected the host. Laura had written her address on the balloon as well as an entreaty: "Please return to Laura Buxton." The balloon traveled 140 miles to the south of England and was found by a neighbor of another 10-year-old girl, also named Laura Buxton! The second Laura wrote to the first, and they arranged to meet. They both showed up to their meeting wearing jeans and pink sweaters. They were both the same height, had brown hair; and owned a black Labrador retriever, a gray rabbit, and a brown guinea pig with an orange spot. In fact, each brought her guinea pig to the meeting. At the time of the radio broadcast, they were 18 years old and friends. "Maybe we were meant to meet," one of the Laura Buxtons speculated. "If it was just the wind, it was a very, very lucky wind," said the other.

The chances seem unbelievably slim, but confirmation bias and illusory correlations both play a role here—and probability helps us understand why such coincidences occur. The thing is, coincidences are *not* unlikely. We notice and remember strange coincidences but do not notice the uncountable times in which there are not unlikely occurrences—the background, so to speak. We remember the story of the woman who won the lottery twice but forget the many times we bought lottery tickets and lost—and the millions of people like us.

MASTERING THE CONCEPT

5.3: Human biases result from two closely related concepts. When we notice only evidence that confirms what we already believe and ignore evidence that refutes what we already believe, we're succumbing to the confirmation bias. Confirmation biases often follow illusory correlations—when we believe we see an association between two variables, but no association exists.

The *Radiolab* story described this phenomenon as the "blade of grass paradox." Imagine a golfer hitting a ball that flies way down the fairway and lands on a blade of grass. The radio show imagines the blade of grass saying: "Wow. What are the odds that that ball, out of all the billions of blades of grass. . . just landed on me?" Yet, we know that there's almost a 100% chance that some blade of grass was going to be crushed by that ball. It just seems miraculous to the individual blade of grass—or lottery winner.

Let's go back to our story about the Laura Buxtons. A statistician pointed out that the details were "manipulated" to make for a better story. The host had remembered that they were both 10 years old, yet the first Laura reminded him she was still 9 at the time ("almost 10"). The host also admitted there were many discrepancies—one's favorite color was pink and one's was blue, and they had opposite academic interests—biology, chemistry, and geography for one and English, history, and classical civilization for the other. Further, it was not the second Laura Buxton who found the balloon; rather, it was her neighbor. The similarities make a better story. When you add in confirmation bias and illusory correlation, they're the WOW! details that we remember—but it's still just another blade of grass.

Still not convinced? Read on.

Expected Relative-Frequency Probability

When we discuss probability in everyday conversation, we tend to think of what statisticians call ***personal probability****: a person's own judgment about the likelihood that an event will occur; also called subjective probability.* We might say something like "There's a 75% chance I'll finish my paper and go out tonight." We don't mean that the chance we'll go out is precisely 75%. Rather, this is our rating of our confidence that this event will occur. It's really just our best guess.

▶ MASTERING THE CONCEPT

5.4: In everyday life, we use the word *probability* very loosely—saying how likely a given outcome is, in our subjective judgment. Statisticians are referring to something very particular when they refer to probability. For statisticians, probability is the actual likelihood of a given outcome in the long run.

Mathematicians and statisticians, however, use the word *probability* a bit differently. Statisticians are concerned with a different type of probability, one that is more objective. In a general sense, ***probability*** *is the likelihood that a particular outcome—out of all possible outcomes—will occur.* For example, we might talk about the likelihood of getting heads (a particular outcome) if we flip a coin 10 times (all possible outcomes). We use probability because we usually have access only to a sample (10 flips of a coin) when we want to know about an entire population (all possible flips of a coin).

Language Alert! In statistics, we are interested in an even more specific definition of probability—***expected relative-frequency probability****, the likelihood of an event occurring based on the actual outcome of many, many trials.* When flipping a coin, the expected relative-frequency probability of heads, in the long run, is 0.50. Probability refers to the likelihood that something would occur, and frequency refers to how often a given outcome (e.g., heads or tails) occurs out of a certain number of trials (e.g., coin flips). *Relative* indicates that this number is relative to the overall number of trials, and *expected* indicates that it's what we would anticipate, which might be different from what actually occurs.

In reference to probability, the term ***trial*** *refers to each occasion that a given procedure is carried out.* For example, each time we flip a coin, it is a trial. ***Outcome*** *refers to the result of a trial.* For coin-flip trials, the outcome is either heads or tails. ***Success*** *refers to the outcome for which we're trying to determine the probability.* If we are testing for the probability of heads, then success is heads.

Determining Probabilities To determine the probability of heads, we would have to conduct many trials (coin flips), record the outcomes (either heads or tails), and determine the proportion of successes (in this case, heads).

Eyebyte/Alamy

EXAMPLE 5.1

We can think of probability in terms of a formula. We calculate probability by dividing the total number of successes by the total number of trials. So the formula would look like this:

$$\text{probability} = \frac{\textit{successes}}{\textit{trials}}$$

If we flip a coin 2000 times and get 1000 heads, then:

$$\text{probability} = \frac{1000}{2000} = 0.50$$

Here is a recap of the steps to calculate probability:

STEP 1: Determine the total number of trials.

STEP 2: Determine the number of these trials that are considered successful outcomes.

STEP 3: Divide the number of successful outcomes by the number of trials.

People often confuse the terms *probability, proportion,* and *percentage.* Probability, the concept of most interest to us right now, is the proportion that we expect to see in the long run. Proportion is the number of successes divided by the number of trials. In the short run, in just a few trials, the proportion might not reflect the underlying probability. A coin flipped six times might have more or fewer than three heads, leading to a proportion of heads that does not parallel the underlying probability of heads. Both proportions and probabilities are written as decimals.

Percentage is simply probability or proportion multiplied by 100. A flipped coin has a 0.50 probability of coming up heads and a 50% chance of coming up heads. You are probably already familiar with percentages, so simply keep in mind that probabilities are what we would expect in the long run, whereas proportions are what we observe.

One of the central characteristics of expected relative-frequency probability is that it only works in the long run. This is an important aspect of probability, and it is referred to as the *law of large numbers.* Think of the earlier discussion of random assignment in which we used a random numbers generator to create a series of 0's and 1's to assign participants to levels of the independent variable. In the short run, over just a few trials, we can get strings of 0's and 1's and often do not end up with half 0's and half 1's, even though that is the underlying probability. With many trials, however, we're much more likely to get close to 0.50, or 50%, of each, although many strings of 0's or 1's would be generated along the way. In the long run, the results are quite predictable. ■

- **Personal probability** refers to the likelihood of an event occurring based on an individual's opinion or judgment; also called *subjective probability.*
- **Probability** is the likelihood that a particular outcome—out of all possible outcomes—will occur.
- The **expected relative-frequency probability** is the likelihood of an event occurring based on the actual outcome of many, many trials.
- In reference to probability, a **trial** refers to each occasion that a given procedure is carried out.
- In reference to probability, **outcome** refers to the result of a trial.
- In reference to probability, **success** refers to the outcome for which we're trying to determine the probability.

Independence and Probability

A key factor in statistical probability is the fact that the individual trials must be independent. If they are not, then bias might be introduced in the same way that a sample can be biased by choosing too many people who are similar to one another. More

Photofusion Picture Library/Alamy

Gambling and Misperceptions of Probability Many people falsely believe that a slot machine that has not paid off in a long time is "due." A person may continue to feed coins into it in the expectation of an imminent payout. Of course, the slot machine itself, unless rigged, has no memory of its previous outcomes. Each trial is independent of the others.

specifically, if individual trials are not independent, then expected relative-frequency probability will not work out in the long run. This is yet another use of the word *independent,* one of the favorite words of statisticians. Here we use *independent* to mean that the outcome of each trial must not depend in any way on the outcome of previous trials. If we're flipping a coin, then each coin flip is independent of every other coin flip. If we're generating a random numbers list to select participants, each number must be generated without thought to the previous numbers. In fact, this is exactly why humans can't think randomly. We automatically glance at the previous numbers we have generated in order to best make the next one "random." Chance has no memory, and randomness is, therefore, the only way to assure that there is no bias.

CHECK YOUR LEARNING

Reviewing the Concepts

- Probability theory helps us understand that coincidences might not have an underlying meaning; coincidences *are* probable when we think of the vast number of occurrences in the world (billions of interactions between people daily).
- An illusory correlation refers to perceiving a connection where none exists. It is often followed by a confirmation bias, whereby we notice occurrences that fit with our preconceived ideas and fail to notice those that do not.
- Personal probability refers to a person's own judgment about the likelihood that an event will occur; also called called subjective probability.
- Expected relative-frequency probability is the likelihood of an event occurring based on the actual outcome of many, many trials.
- The probability of an event occurring is defined as the expected number of successes (the number of times the event occurred) out of the total number of trials (or attempts) over the long run.
- Short-run proportions might have many different outcomes, whereas long-run proportions are more indicative of the underlying probabilities.

Clarifying the Concepts

5-5 Distinguish the personal probability assessments we perform on a daily basis from the objective probability statisticians use.

Calculating the Statistics

5-6 Calculate the probability for each of the following instances.

a. 100 trials, 5 successes

b. 50 trials, 8 successes

c. 1044 trials, 130 successes

Applying the Concepts

5-7 Consider a scenario in which a student wonders whether men or women are more likely to use the ATM (ABM in Canada) in the student center. He decides to observe those who use the ATM. (Assume that there is no gender difference.)

a. Define success as a woman using the ATM on a given trial. What proportion of successes might this student expect to observe in the short run?

b. What might this student expect to observe in the long run?

Solutions to these Check Your Learning questions can be found in Appendix D.

Inferential Statistics

In Chapter 1, we introduced the two main branches of statistics—descriptive statistics and inferential statistics. The link that connects the two branches is probability. Descriptive statistics allow us to summarize characteristics of the sample, but we must use probability with inferential statistics when we apply what we've learned from the sample, such as in an exit poll, to the larger population. Inferential statistics, also referred to as hypothesis testing, helps us to determine how likely a given outcome is.

- A **control group** is a level of the independent variable that does not receive the treatment of interest in a study. It is designed to match an experimental group in all ways but the experimental manipulation itself.
- An **experimental group** is a level of the independent variable that receives the treatment or intervention of interest in an experiment.
- The **null hypothesis** is a statement that postulates that there is no difference between populations or that the difference is in a direction opposite of that anticipated by the researcher.

Developing Hypotheses

Probability theory can help us find answers for many of our research hypotheses. In a science blog, "TierneyLab," reporter John Tierney and his collaborators asked people to estimate the number of calories in a meal pictured in a photograph (Tierney, 2008a, 2008b). One group was shown a photo of an Applebee's Oriental Chicken Salad and a Pepsi. Another group was shown a photo of the same salad and Pepsi, but it also included a third item—Fortt's crackers, with a label that clearly stated "Trans Fat Free." The researchers wondered if the addition of the "healthy" food item would affect calorie estimates. They tested a sample and used probability to apply their findings from the sample to the population.

Let's put this study in the language of research. The population would include all people living in the United States, a population chosen because of its increasing levels of obesity. The sample was comprised of people living in the Park Slope neighborhood of Brooklyn, an area that Tierney terms "nutritionally correct" for the abundance of organic food in local stores. The independent variable in this case is the presence or absence of the healthy crackers in the photo of the meal. The dependent variable is the number of calories estimated.

Dennis MacDonald/age Fotostock

Using a Sample to Make Probability-Based Judgments About the Population Does the presence of a low-calorie item, such as a diet soda, make a higher-calorie item, such as french fries, seem healthier? Researchers use samples to test hypotheses such as this about a population.

In this case, we might refer to the group that viewed the photo without the healthy crackers as the control group. *A **control group** is a level of the independent variable that does not receive the treatment of interest in a study.* It is designed to match *the **experimental group**—a level of the independent variable that receives the treatment or intervention of interest*—in all ways but the experimental manipulation itself. In this example, the experimental group would be those viewing the photo that included the healthy crackers.

The next step, one that ideally occurs before actually collecting data from the sample, and one that we'll see throughout this book, is the development of the hypotheses to be tested. When we calculate inferential statistics, we're always comparing two hypotheses. One is *the **null hypothesis**—a statement that postulates that there is no difference between populations or that the difference is in a direction opposite of that anticipated by the researcher.* In most circumstances, we can think of the null hypothesis as the boring hypothesis because it proposes that nothing will happen. In many hypothesis tests, the difference being tested is a difference between means. In the healthy food study, the null hypothesis would be that the mean calorie estimate is the same for both populations—

> **MASTERING THE CONCEPT**
>
> **5.5:** Many experiments have an experimental group in which participants receive the treatment or intervention of interest, and a control group in which participants do not receive the treatment or intervention of interest. Aside from the intervention with the experimental group, the two groups are treated identically.

all people in the United States who view the photo without the healthy crackers and all people in the United States who view the photo with the healthy crackers.

> **MASTERING THE CONCEPT**
>
> **5.6:** Hypothesis testing allows us to examine two competing hypotheses. The first, the null hypothesis, posits that there is no difference between populations or that any difference is in the opposite direction from what is predicted. The second, the research hypothesis, posits that there is a difference between populations (or that the difference between populations is in a predicted direction—either higher or lower).

By contrast, the research hypothesis, also called the *alternative hypothesis,* is usually the exciting one. *The* ***research hypothesis*** *is a statement that postulates that there is a difference between populations or sometimes, more specifically, that there is a difference in a certain direction, positive or negative.* This is usually the exciting hypothesis because it proposes a distinctive difference that is worthy of further investigation. In the healthy food study, the research hypothesis would be that, on average, the calorie estimate is different for those viewing the photo with the healthy crackers than for those viewing the photo without the healthy crackers. It also could specify a direction—that the mean calorie estimate is higher (or lower) for those viewing the photo with the healthy crackers than for those viewing the photo with just the salad and Pepsi. Notice that, for all hypotheses, we are very careful to state the comparison group. We do not say merely that the group viewing the photo with the healthy crackers has a higher (or lower) average calorie estimate. We say that it has a higher (or lower) average calorie estimate *than* the group that views the photo without the healthy crackers.

We formulate the null hypothesis and research hypothesis to set them up against each other. We use statistics to determine the probability that there is a large enough difference between the means of the samples that we can conclude there's likely a difference between the means of the underlying populations. So, probability plays into the decision we make about the hypotheses.

Making a Decision About the Hypothesis

When we make a conclusion at the end of a study, the data lead us to conclude one of two things:

1. We decide to *reject* the null hypothesis.
2. We decide to *fail to reject* the null hypothesis.

We always begin our reasoning about the outcome of an experiment by reminding ourselves that we are testing the (boring) null hypothesis. In terms of the healthy food study, the null hypothesis is that there is no mean difference between groups. In hypothesis testing, we determine the probability that we would see a difference between the means of the samples, given that there is no actual difference between the underlying population means.

EXAMPLE 5.2

After we analyze the data, we are able to do one of two things:

1. *Reject the null hypothesis.* "I reject the idea that there is no mean difference between populations." When we reject the null hypothesis that there is *no mean difference,* we can even assert what we believe the difference to be based on the actual findings. We can say that it seems that people who view a photo of a salad, Pepsi, and healthy crackers estimate a lower (or higher, depending on what we found in our study) number of calories, on average, than those who view a photo with only the salad and Pepsi.
2. *Fail to reject the null hypothesis.* "I do not reject the idea that there is no mean difference between populations."

Let's take the first possible conclusion, to reject the null hypothesis. If the group that viewed the photo that included the healthy crackers has a mean calorie estimate that is a good deal higher (or lower) than the control group's mean calorie estimate, then we might be tempted to say that we *accept* the research hypothesis that there is such a mean difference in the populations—that the addition of the healthy crackers makes a difference. Probability plays a central role in determining that the mean difference is large enough that we're willing to say it's real. But rather than *accept* the *research* hypothesis in this case, we *reject* the *null* hypothesis, the one that suggests there is nothing going on. We repeat: when the data suggest that there *is* a mean difference, we *reject* the idea that there is no mean difference.

The second possible conclusion is failing to reject the null hypothesis. There's a very good reason for thinking about this in terms of failing to reject the null hypothesis rather than accepting the null hypothesis. Let's say there's a small mean difference, and we conclude that we cannot reject the null hypothesis (remember, rejecting the null hypothesis is what you want to do!). We determine that it's just not likely enough—or probable enough—that the difference between means is real. It could be that a real difference between means didn't show up in this particular sample just by chance. There are many ways in which a real mean difference in the population might not get picked up by a sample. Again, we repeat: when the data do not suggest a difference, we *fail* to reject the null hypothesis, which is that there is no mean difference.

The way we decide whether to reject the null hypothesis is based directly on probability. We calculate the probability that the data would produce a difference between means this large and in a sample of this size *if* there was nothing going on.

We will be giving you many more opportunities to get comfortable with the logic of formal hypothesis testing before we start applying numbers to it, but here are three easy rules and a table (Table 5-1) that will help keep you on track.

1. Remember: The null hypothesis is that there is no difference between groups, and usually the hypotheses explore the possibility of a *mean* difference.
2. We either *reject* or *fail to reject* the null hypothesis. There are no other options.
3. We never use the word *accept* in reference to formal hypothesis testing.

Hypothesis testing is exciting when you care about the results. You may wonder what happened in Tierney's study. Well, people who saw the photo with just the salad and the Pepsi estimated, on average, that the 934-calorie meal contained 1011 calories. When the 100-calorie crackers were added, the meal actually increased from 934 calories to 1034 calories; however, those who viewed this photo estimated, on average, that the meal contained only 835 calories! Tierney referred to this effect as "a health halo that magically subtracted calories from the rest of the meal." Interestingly, he replicated this study with mostly foreign tourists in New York's Times Square and did not find this effect. He concluded that health-conscious people were more susceptible to bias than other people.

▪ The **research hypothesis** is a statement that postulates that there is a difference between populations or sometimes, more specifically, that there is a difference in a certain direction, positive or negative; also called an *alternative hypothesis.*

TABLE 5-1. Hypothesis Testing: Hypotheses and Decisions

The null hypothesis posits no difference, on average, whereas the research hypothesis posits a difference of some kind. There are only two decisions we can make. We can fail to reject the null hypothesis if the research hypothesis is *not* supported, or we can reject the null hypothesis if the research hypothesis *is* supported.

Hypothesis		Decision
Null hypothesis	No change or difference	Fail to reject the null hypothesis (if research hypothesis is not supported)
Research hypothesis	Change or difference	Reject the null hypothesis (if research hypothesis is supported)

CHECK YOUR LEARNING

Reviewing the Concepts

> In experiments, we typically compare the average of the responses of those who receive the treatment or manipulation (the experimental group) with the average of the responses of similar people who do not receive the manipulation (the control group).

> Researchers develop two hypotheses: a null hypothesis, which theorizes that there is no average difference between levels of an independent variable in the population, and a research hypothesis, which theorizes that there is an average difference of some kind in the population.

> Researchers can draw two conclusions: they can reject the null hypothesis and conclude that they have supported the research hypothesis, or they can fail to reject the null hypothesis and conclude that they have not supported the research hypothesis.

Clarifying the Concepts

5-8 At the end of a study, what does it mean to reject the null hypothesis?

Calculating the Statistics

5-9 State the difference that might be expected, based on the null hypothesis, between the average test grades of students who attend review sessions versus those who do not.

Applying the Concepts

5-10 A university lowers the heat during the winter to save money, and professors wonder whether students will perform more poorly, on average, under cold conditions.

a. Cite the likely null hypothesis for this study.

b. Cite the likely research hypothesis.

c. If the cold temperature appears to decrease academic performance, on average, what will the researchers conclude in terms of formal hypothesis-testing language?

d. If the researchers do not gather sufficient evidence to conclude that the cold temperature leads to decreased academic performance, on average, what will they conclude in terms of formal hypothesis-testing language?

Solutions to these Check Your Learning questions can be found in Appendix D.

Type I and Type II Errors

Exit polling during the 2000 U.S. presidential election taught us that sampling errors can lead us to make a wrong decision. Even when sampling has been properly conducted, however, there are still two ways to make a wrong decision. We can reject the null hypothesis when we should not have rejected it, or we can fail to reject the null hypothesis when we should have rejected it. So let's consider the two types of error using statistical language.

Type I Errors

If we reject the null hypothesis, but it was a mistake to do so, then we have made a Type I error. Specifically, *we commit a **Type I error** when we reject the null hypothesis, but the null hypothesis is correct.* A Type I error is like a false positive in a medical test. For example, if a woman believes she might be pregnant, then she might buy a home pregnancy test. In this case, the null hypothesis would be that she is not pregnant, and the research hypothesis would be that she is pregnant. If the test is positive, the woman rejects the null hypothesis—the one in which she theorizes that she is not pregnant. Based on the test, the woman believes she is pregnant. Pregnancy tests, however, are not perfect. If the woman tests positive and rejects the null hypothesis, it is possible that she is wrong and it is a false positive. Based on the test, the woman believes she is pregnant even though she is not pregnant. A false positive is equivalent to a Type I error.

- A **Type I error** occurs when we reject the null hypothesis, but the null hypothesis is correct.
- A **Type II error** occurs when we fail to reject the null hypothesis, but the null hypothesis is false.

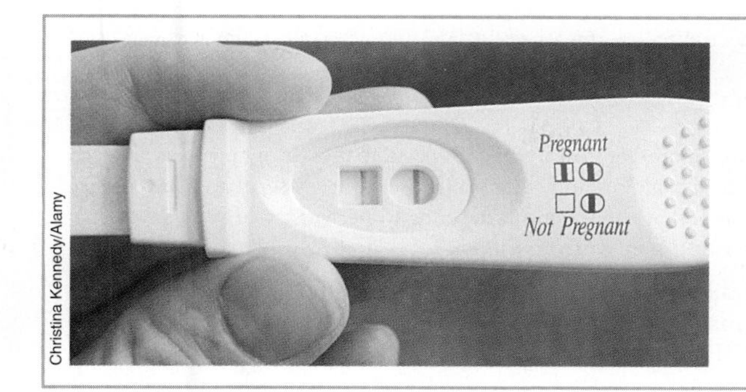

Type I and Type II Errors The results of a home pregnancy test are either positive (indicating pregnancy) or negative (indicating no pregnancy). If the test is positive, but the woman *is not* pregnant, this would be a Type I error in statistics. If the test is negative, but the woman *is* pregnant, this would be a Type II error in statistics. With pregnancy tests, as with hypothesis testing, people are more likely to act on a Type I error than on a Type II error. Although the line on the left is lighter than the line on the right, this particular pregnancy test seems to indicate that this woman is pregnant, but this could be a Type I error.

A Type I error indicates that we rejected the null hypothesis falsely. As you might imagine, the rejection of the null hypothesis typically leads to action, at least until we discover that it is an error. For example, the woman with a false-positive pregnancy test might announce the news to her family and start buying baby clothes. Many researchers consider the consequences of a Type I error to be particularly detrimental because people often take action based on a mistaken finding.

> **MASTERING THE CONCEPT**
>
> **5.7:** In hypothesis testing, there are two types of errors that we risk making. Type I errors, in which we reject the null hypothesis when the null hypothesis is true, are like false positives on a medical test; we think someone *has* a disease, but they really don't. Type II errors, in which we fail to reject the null hypothesis when the null hypothesis is not true, are like false negatives on a medical test; we think someone does *not* have a disease, but they really do.

Type II Errors

If we fail to reject the null hypothesis but it was a mistake to fail to do so, then we have made a Type II error. Specifically, *we commit a* ***Type II error*** *when we fail to reject the null hypothesis, but the null hypothesis is false.* A Type II error is like a false negative in medical testing. In the pregnancy example earlier, the woman might get a negative result on the test and fail to reject the null hypothesis, the one that says she's not pregnant. In this case, she would conclude that she's not pregnant when she really is. A false negative is equivalent to a Type II error.

A Type II error occurs when we incorrectly failed to reject the null hypothesis. A failure to reject the null hypothesis typically results in a failure to take action, because a research intervention or a diagnosis is not received, which is generally less dangerous than incorrectly rejecting the null hypothesis. Yet there are cases in which a Type II error can have serious consequences. For example, the pregnant woman who does not believe she is pregnant because of a Type II error may drink alcohol in a way that unintentionally harms her fetus.

CHECK YOUR LEARNING

Reviewing the Concepts

> When we draw a conclusion from inferential statistics, there is always a chance that we are wrong.

> When we reject the null hypothesis, but the null hypothesis is true, we have committed a Type I error.

> When we fail to reject the null hypothesis, but the null hypothesis is not true, we have committed a Type II error.

Clarifying the Concepts

5-11 Explain how Type I and Type II errors both relate to the null hypothesis.

Calculating the Statistics

5-12 If 7 out of every 280 people in prison are innocent, what is the rate of Type I errors?

continued on next pa

5-13 If the court system fails to convict 11 out of every 35 guilty people, what is the rate of Type II errors?

Applying the Concepts

5-14 Researchers conduct a study on perception by having participants throw a ball at a target first while wearing virtual-reality glasses and then while wearing glasses that allow normal viewing. The null hypothesis is that there is no difference in performance when wearing the virtual-reality glasses versus when wearing the glasses that allow normal viewing.

a. The researchers reject the null hypothesis, concluding that the virtual-reality glasses lead to a worse performance than do the normal glasses. What error might the researchers have made? Explain.

b. The researchers fail to reject the null hypothesis, concluding that it is possible that the virtual-reality glasses have no effect on performance. What error might the researchers have made? Explain.

Solutions to these Check Your Learning questions can be found in Appendix D.

REVIEW OF CONCEPTS

Samples and Their Populations

The gold standard of sample selection is *random sampling*, a procedure in which every member of the population has an equal chance of being chosen for study participation. A random numbers table or a computer-based random numbers generator is used to ensure randomness. For practical reasons, random selection is uncommon in social science research. Many behavioral scientists use a *convenience sample*, a sample that is readily available to them. One kind of convenience sample is the *volunteer sample* (also called a *self-selected sample*), in which participants themselves actively choose to participate in the study. In random assignment, every participant in a study has an equal chance of being assigned to any of the experimental conditions. In conjunction with random assignment, *replication*—the duplication of scientific results—can go a long way toward increasing *generalizability*, our ability to generalize our findings beyond our samples.

Probability

Calculating probabilities is essential because human thinking is dangerously biased. Because of a *confirmation bias*—the tendency to see patterns that we expect to see—we often see meaning in mere coincidence. A confirmation bias often results from an *illusory correlation*, a relation that appears to be present but does not exist. When we think of probability, many of us think of *personal probability*, a person's own judgment about the likelihood that an event will occur. Statisticians, however, are referring to *expected relative-frequency probability*, or the long-run expected outcome if an experiment or trial was repeated many, many times. A *trial* refers to each occasion that a procedure is carried out, and an *outcome* is the result of a trial. A *success* refers to the outcome for which we're trying to determine the probability. *Probability* is a basic building block of inferential statistics. When we draw a conclusion about a population based on a sample, we can only say that it is probable that our conclusion is accurate, not that it is certain.

Inferential Statistics

Inferential statistics, based on probability, start with a hypothesis. The *null hypothesis* is a statement that usually postulates that there is no average difference between populations. The *alternative* or *research hypothesis* is a statement that postulates that there is an average difference between populations. After conducting a hypothesis test, we have only two possible conclusions. We can either reject or fail to reject the null hypothesis. When we conduct inferential statistics, we are often comparing an *experimental group*, the group subjected to an intervention, with a *control group*, the group that is the same as the experimental group in every way except the intervention. We use probability to draw conclusions about a population by estimating the probability that we would find a given difference between sample means if there is no underlying difference between population means.

Type I and Type II Errors

Statisticians must always be aware that their conclusions may be wrong. If a researcher rejects the null hypothesis, but the null hypothesis is correct, the researcher is making a *Type I error*. If a researcher fails to reject the null hypothesis, but the null hypothesis is false, the researcher is making a *Type II error*.

SPSS®

There are many ways to look more closely at the independent and dependent variables. You can request a variety of case summaries by selecting Analyze → Reports → Case Summaries. You can then highlight the variable of interest and click the arrow to move it under "Variables."

If you want to break your first variable down by a second variable, highlight a nominal variable and click the bottom arrow to move it under "Grouping Variable(s)." After making your choices, click on "OK" to see the output screen.

For example, you could use the hours studied and exam grade data from the SPSS section of Chapter 3. You could select "grade" under "Variables" and "hours" under "Grouping Variable(s)." The output, part of which is shown in the screenshot here, tells you all the grades for students who studied a given number of hours. This summary, for instance, tells you that the two students who studied for 3 hours earned grades of 60 and 72 on the exam.

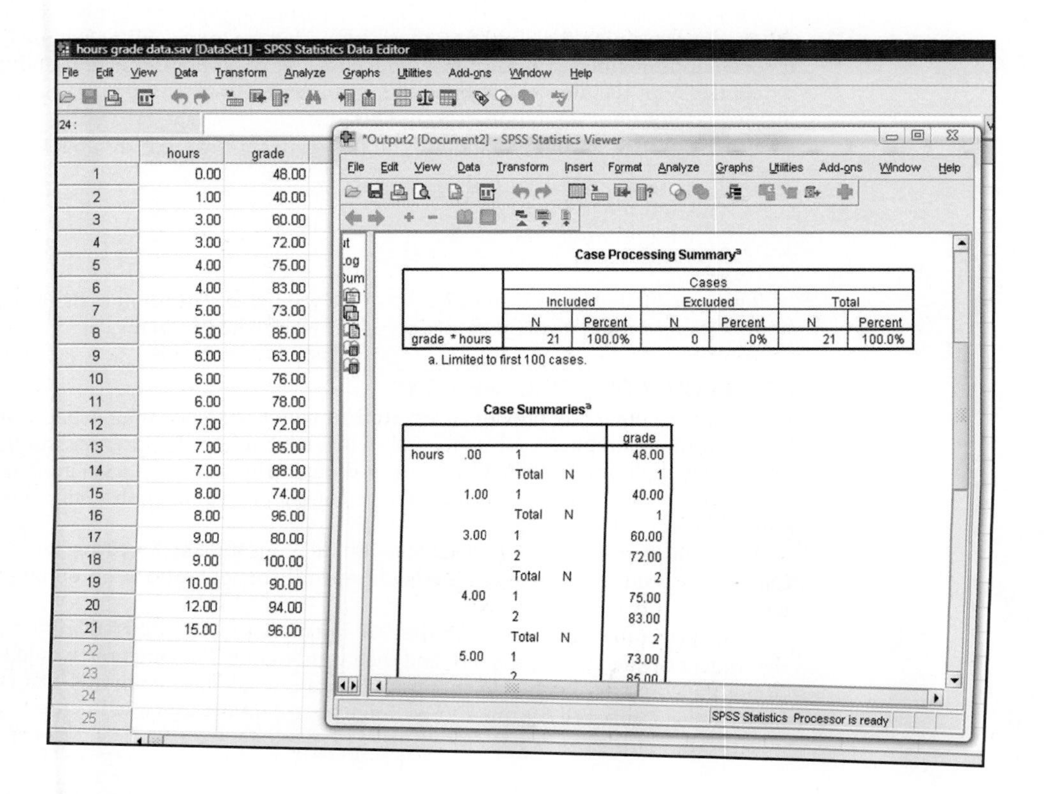

How It Works

5.1 USING RANDOM SELECTION

There are approximately 2000 school psychologists in Australia. A researcher has developed a new diagnostic tool to identify conduct disorder in children and wants to study ways to train school psychologists to administer the tool. How can she recruit a random sample of 30 school psychologists to participate in her study?

She could use an online random numbers generator to randomly select a sample of 30 school psychologists for this study from among the target population of 2000 Australian school psychologists. Let's try it. To do so, she would tell an online random number generator to produce one set of 30 numbers between 0001 and 2000. She would specify that she wants unique numbers, because no school psychologist can be in the study more than once. She can ask the program to sort the numbers, if she wishes, to more easily identify the participants who will comprise her sample.

When we generated a set of 30 random numbers, we got the following:

25 48 84 113 159 165 220 312 319 330
337 452 493 562 613 734 822 860 920 924
931 960 983 1290 1305 1462 1502 1515 1675 1994

Of course, each time we generate a list of random numbers, the list is different. Notice that the typical list of randomly generated numbers does not necessarily appear random. For example, in this list only 7 out of the 30 numbers are over 1000. There are also several cases in which numbers are close in value (e.g., 920 and 924).

5.2 USING RANDOM SELECTION

Imagine that the researcher described in How It Works 5.1 has developed two training modules. One is implemented in a classroom setting and requires that school psychologists travel to a nearby city for training. The other is a more cost-effective Web-based training module. She will administer a test to participants after training to determine how much they learned. How can she randomly assign half of the participants to classroom training and half to Web-based training?

In this case, the independent variable is type of training with two levels: classroom training and Web-based training. The dependent variable is amount of learning as determined by a test. This study is an experiment because participants are randomly assigned to conditions. To determine the condition to which each participant will be assigned, she could use a random numbers generator to produce one set of 30 numbers between 0 and 1. She would not want the numbers to be unique because she wants more than one of each type. She would not want the numbers to be sorted.

When we used an online random numbers generator, we got the following set of 30 numbers:

1 1 0 0 0 0 1 1 1 0
0 1 0 0 0 0 0 0 1 1
0 1 1 1 0 0 1 1 0 0

This set contains 13 ones and 17 zeros. If we wanted exactly 15 in each group, we could stop assigning people to the 0 condition when we reached 15 zeros.

5.3 CALCULATING PROBABILITY

Let's say that a university provides every student with a laptop computer, but students complain that their computers "always" crash when they are on the Internet and have at least three other applications open. One student thought this was an exaggeration and decided to calculate the probability that the campus computers would crash under these circumstances. How could he do this?

He could start by randomly selecting 100 different students to participate in his study. On the 100 students' computers, he could open three applications, go online, and record whether each computer crashed.

In this case, the trials would be the 100 instances (on 100 different laptops) in which the student opened three programs and then went online. The outcome would be whether or not the computer crashed. A success in this case would be a computer that crashed, and let's say that happened 55 times. (You might not consider a crashed computer a success, but in probability theory, a success refers to the outcome for which we want to determine the probability.) He could then take the number of successes (55) and divide by the number of trials:

$$55/100 = 0.55$$

So the probability of a computer crashing when three programs are open and the student goes online is 0.55. Of course, to determine the true expected relative-frequency probability, he'd have to conduct many, many more trials.

Exercises

Clarifying the Concepts

5.1 Why do we study samples rather than populations?

5.2 What is the difference between a random sample and a convenience sample?

5.3 What is generalizability?

5.4 What is a volunteer sample, and what is the risk associated with it?

5.5 What is the difference between random sampling and random assignment?

5.6 What does it mean to replicate research, and how does replication impact our confidence?

5.7 Ideally, an experiment would use random sampling so that the data would accurately reflect the larger population. For practical reasons, this is difficult to do. How does random assignment help make up for a lack of random selection?

5.8 What is the confirmation bias?

5.9 What is an illusory correlation?

5.10 How does the confirmation bias lead to the perpetuation of an illusory correlation?

5.11 In your own words, what is *personal probability*?

5.12 In your own words, what is *expected relative-frequency probability*?

5.13 Statisticians use terms like *trial, outcome,* and *success* in a particular way in reference to probability. What do each of these three terms mean in the context of flipping a coin?

5.14 We distinguish between probabilities and proportions. How does each capture the likelihood of an outcome?

5.15 How is the term *independent* used by statisticians?

5.16 One step in hypothesis testing is to randomly assign some members of the sample to the control group and some to the experimental group. What is the difference between these two groups?

5.17 What is the difference between a null hypothesis and a research hypothesis?

5.18 What are the two decisions or conclusions we can make about our hypotheses, based on the data?

5.19 What is the difference between a Type I error and a Type II error?

Calculating the Statistics

5.20 Forty-three tractor-trailers are parked for the night in a rest stop along a major highway. You assign each truck a number from 1 to 43. Moving from left to right and using the second line in the random numbers table below, select four trucks to weigh as they leave the rest stop in the morning.

00190 27157 83208 79446 92987 61357
23798 55425 32454 34611 39605 39981
85306 57995 68222 39055 43890 36956
99719 36036 74274 53901 34643 06157

5.21 Airport security makes random checks of passenger bags every day. If 1 in every 10 passengers is checked, use the random numbers table in Exercise 5.20 to determine the first 6 people to be checked. Work from top to bottom, starting in the 4th column (the fourth digit from the left in the top line), and allow the number 0 to represent the 10th person.

5.22 Randomly assign eight people to three conditions of a study, numbered 1, 2, and 3, using the random numbers table in Exercise 5.20. Read from right to left, starting in the top row. (*Note:* Assign people to conditions without concern for having an equal number of people in each condition.)

5.23 You are running a study with five conditions, numbered 1 through 5. Assign the first seven participants who arrive at your lab to conditions, not worrying about equal assignment across conditions. Use the random numbers table in Exercise 5.20, and read from left to right, starting in the third row from the top.

5.24 Explain why, given the general tendency people have of exhibiting the confirmation bias, it is important to collect objective data.

5.25 Explain why, given the general tendency people have of perceiving illusory correlations, it is important to collect objective data.

5.26 What is the probability of hitting a target if, in the long run, 71 out of every 489 attempts actually hit the target?

5.27 On a game show, 8 people have won the grand prize and a total of 266 people have competed. Estimate the probability of winning the grand prize.

5.28 Convert the following proportions to percentages:

a. 0.0173

b. 0.8

c. 0.3719

5.29 Convert the following percentages to proportions:

a. 62.7%

b. 0.3%

c. 4.2%

5.30 Using the random numbers table in Exercise 5.20, estimate the probability of the number 6 appearing in a random sequence of numbers. Base your answer on the numbers that appear in the first two rows.

5.31 Indicate whether each of the following statements refers to personal probability or to expected relative-frequency probability. Explain your answers.

a. The chance of a dice showing an even number is 50%.

b. There is a 1 in 4 chance that I'll be late for class tomorrow.

c. The likelihood that I'll break down and eat ice cream while studying is 80%.

d. Planecrashinfo.com reported that the odds of being killed on a single flight on one of the top 25 safest airlines is 1 in 9.2 million.

Applying the Concepts

5.32 Coincidence and the lottery: "Woman wins millions from Texas lottery for 4th time" read the headline about Joan Ginther's amazing luck (Baird, 2010). Two of the tickets were from the same store, whose owner, Bob Solis, said, "This is a very lucky store." Citing concepts from the chapter, what would you tell Ginther and Solis about the roles that probability and coincidence played in their fortunate circumstances?

5.33 Random selection, random assignment, and violence: In France in the fall of 2005, many communities of immigrants from the Middle East and North Africa experienced a great deal of violence, particularly car burnings, committed by young people. Of the cities that shared this demographic, Marseille was one of the few that saw relatively little violence. Consider the following hypothetical research on the French riots: A behavioral science researcher wants to compare Marseille with Lyon, a city that saw a great deal of violence, to determine which characteristics may have moderated violence in Marseille, specifically among high school students. Can she use random selection? Explain. Can she use random assignment? Explain.

5.34 Random selection and a school psychologist career survey: Approximately 21,000 school psychologists are members of the U.S.-based National Association of School Psychologists. Of these, about 5000 have doctoral degrees. A researcher wants to randomly select 100 of the doctoral-level school psychologists for a survey study regarding aspects of their jobs. Use this excerpt from a random numbers table to answer the following questions:

```
04493 52494 75246 33824 45862 51025
00549 97654 64051 88159 96119 63896
35963 15307 26898 09354 33351 35462
59808 08391 45427 26842 83609 49700
```

a. What is the population targeted by this study? How large is it?

b. What is the sample desired by this researcher? How large is it?

c. Describe how the researcher would select the sample. Be sure to explain how the members of the population would be numbered and what sets of digits the researcher should ignore when using the random numbers table.

d. Beginning at the left-hand side of the top line and continuing with each succeeding line, list the first 10 participants that this researcher would select for the study.

5.35 Hypotheses and the school psychologist career survey: Continuing with the study described in Exercise 5.34, once the researcher had randomly selected the sample of 100 school psychologists, she decided to randomly assign 50 of them to receive, as part of their survey materials, a (fictional) newspaper article about the improving job market. She assigned the other 50 to receive a (fictional) newspaper article about the declining job market. The participants then responded to questions about their attitudes toward their careers.

a. What is the independent variable in this experiment, and what are its levels?

b. What is the dependent variable in this experiment?

c. Write a null hypothesis and a research hypothesis for this study.

5.36 Random assignment and school psychologist career survey: Refer to Exercises 5.34 and 5.35 when responding to the following questions:

a. Describe how the researcher would randomly assign the participants to the levels of the independent variable. Be sure to explain how the levels of the independent variable would be numbered and which sets of digits the researcher should ignore when using the random numbers table.

b. Beginning at the left-hand side of the bottom line of the random numbers table in Exercise 5.34 and continuing with the left-hand side of the line above it, list the levels of the independent variable to which the first 10 participants would be assigned. Use 0 and 1 to represent the two conditions.

c. Why do these numbers not appear to be random? Discuss the difference between short-run and long-run proportions.

5.37 Random selection and a survey of psychology majors: Imagine that you have been hired by the psychology department at your school to administer a survey to psychology majors about their experiences in the department. You have been asked to randomly select 60 of these majors from the overall pool of 300. You are working on this project in your dorm room using a random numbers table because the server is down and you cannot use an online random numbers generator. Your roommate offers to write down a list of 60 random numbers between 001 and 300 for you so you can be done quickly. In about three to four sentences, explain to your roommate why she is not likely to create a list of random numbers.

5.38 Random selection and random assignment: For each of the following studies, state (1) whether random selection was likely to have been used, and explain whether it would have been possible. Explain also to what population the researcher wanted to and could generalize and state (2) whether random assignment was likely to have been used, and whether it would have been possible.

a. A researcher recruited 1000 U.S. physicians through the American Medical Association (AMA) to participate in a study of standards of patient confidentiality to compare perceptions of the standard among men versus women.

b. A developmental psychologist wondered whether children born preterm (prematurely) had different social skills at age 5 than children born at full term.

c. A counseling center director wanted to compare the length of therapy in weeks for students who came in for treatment for depression versus students who came in for treatment for anxiety. She wanted to report these data to the university administrators to help develop the next year's budget.

d. An industrial/organizational psychologist wondered whether a new laptop design would affect people's response time when using the computer. He wanted to compare response times when using the new laptop with response times when using two standard versions of laptops, a Mac and a PC.

5.39 Volunteer samples and a college football poll: A volunteer sample is a kind of convenience sample in which participants select themselves to participate. On August 19, 2005, *USA Today* published an online poll on its Web site asking this question about U.S. college football: "Who is your pick to win the ACC conference this year?" Eight options—seven universities, including top vote-getters Virginia Tech and Miami, as well as "other"—were provided.

a. Describe the typical person who might volunteer to be in this sample. Why might this sample be biased, even with respect to the population of U.S. college football fans?

b. What is external validity? Why might external validity be limited in this sample?

c. What other problem can you identify with this poll?

5.40 Samples and *Cosmo* quizzes: *Cosmopolitan* magazine (*Cosmo*, as it's known popularly) publishes many of its well-known quizzes on its Web site. One quiz, aimed at heterosexual women, is titled "Are You Way Too Obsessed with Your Ex?" A question about "your rebound guy" offers these three choices: "Any random guy who will take your mind off the split," "A doppelgänger of your ex," and "The polar opposite of the last guy you dated." Consider whether you want to use the quiz data to determine how obsessed women are with their exes.

a. Describe the typical person who might respond to this quiz. How might data from such a sample be biased, even with respect to the overall *Cosmo* readership?

b. What is the danger of relying on volunteer samples in general?

c. What other problems do you see with this quiz? Comment on the types of questions and responses.

5.41 Samples and political leanings: On its Web site, Advocates for Self-Government offers the "World's Smallest Internet Political Quiz," focusing on the U.S. political spectrum. Using just 10 questions, the quiz identifies a person's political leanings. As of 2012, almost 20 million people had taken the quiz. In 2007, the Web site reported the following breakdown into the five possible categories: centrist, 33.49%; conservative, 8.88%; libertarian, 32.64%; liberal, 17.09%; and statist (big government), 7.89%.

a. Do you think these numbers are representative of the U.S. population? Why or why not?

b. Describe the people most likely to volunteer for this sample. Why might this group be biased in comparison to the overall U.S. population?

c. The Web site says, "Libertarians support maximum liberty in both personal and economic matters." Libertarians are not the predominant political group in the United States. Why, then, might libertarians form one of the largest categories of quiz respondents?

d. This is a huge sample—close to 20 million people. Why is it not enough to have a large sample to conduct a study with high external validity? What would we need to change about this sample to increase external validity?

5.42 Random selection or random assignment: For each of the following hypothetical scenarios, state whether selection or assignment is being described. Is the method of selection or assignment random? Explain your answer.

a. A study of the services offered by counseling centers at Canadian universities studied 20 universities; every Canadian university had an equal chance of being in this study.

b. In a study of phobias, 30 rhesus monkeys were either exposed to fearful stimuli or not exposed to fearful stimuli. Every monkey had an equal chance of being placed in either of the exposure conditions.

c. A study of cell phone usage recruited participants by including an invitation to participate in the study along with their cell phone bills.

d. A study of visual perception recruited 120 Introduction to Psychology students to participate.

5.43 Bias about driver gender: Assume that one of your male friends is complaining about female drivers, and says that men are much better drivers than women. If objective studies of the driving performance of men and women revealed no mean difference between the two groups, what kind of bias has your friend shown?

5.44 Confirmation bias, illusory correlation, and driver gender: Referring to your friend from Exercise 5.43, assume he backs up his claim by recounting two events over the past week in which female drivers have erred (e.g., cutting him off in traffic, not using a turn signal). Explain how the confirmation bias is at work in your friend's statements and how this confirmation bias may be perpetuating an illusory correlation.

5.45 Confirmation bias and negative thought patterns: Explain how the general tendency of a confirmation bias might make it difficult to change negative thought patterns that accompany depression.

5.46 Probability and coin flips: Short-run proportions are often quite different from long-run probabilities.

a. In your own words, explain why we would expect proportions to fluctuate in the short run, but why long-run probabilities are more predictable.

b. What is the expected long-run probability of heads if you flip a coin many, many times? Why?

c. Flip a coin 10 times in a row. What proportion is heads? Do this 5 times. *Note:* You will learn more by actually doing it, so don't just write down numbers!

Proportion for first 10 flips:

Proportion for second 10 flips:

Proportion for third 10 flips:

Proportion for fourth 10 flips:

Proportion for fifth 10 flips:

d. Do the proportions in part (c) match the expected long-run probability in part (b)? Why or why not?

e. Imagine that a friend flipped a coin 10 times, got 9 out of 10 heads, and complained that the coin was biased. How would you explain to your friend the difference between short-term and long-term probability?

5.47 Probability, proportion, percentage, and a deck of cards: A deck of playing cards has four suits and 13 cards in each suit, for a total of 52 cards. Imagine you draw 1 card from the deck, record what the card is, and then put it back in the deck. Let's say you repeat this process 15 times, and 5 of the 15 cards are aces. (A fair deck would only have 4 aces, one for each suit.) Keeping this example in mind, answer the following questions.

a. What does the term *probability* refer to? What is the probability of drawing an ace?

b. What does the term *proportion* refer to? What is the proportion of aces drawn?

c. What does the term *percentage* refer to? What is the percentage of aces drawn?

d. Based on these data (5 out of 15 cards were aces), do you have enough information to determine whether the deck is stacked (i.e., biased)? Why or why not? (*Note:* Four of the 52 cards should be aces.)

5.48 Independent or dependent trials and probability: Gamblers often falsely predict the outcome of a future trial based on the outcome of previous trials. When trials are independent, we cannot predict the outcome of a future trial based on the outcomes of previous trials. For each of the following examples, (1) state whether the trials are independent or dependent and (2) explain why. In addition, (3) state whether it is possible that the quote is accurate or whether it is definitely fallacious, explaining how the independence or dependence of trials influences this.

a. You are playing Monopoly and have rolled a pair of sixes in 4 out of 10 of your last rolls of the dice. You say, "Cool. I'm on a roll and will get sixes again."

b. You are an Ohio State University football fan and are sad because the team has lost two games in a row. You say, "That is really unusual; the Buckeyes are doomed this season. That's what happens with lots of early-season injuries."

c. You have a 20-year-old car that has trouble starting from time to time. It has started every day this week, and now it's Friday. You say, "I'm doomed. It's been reliable all week, and even though I did get a tune-up last week, today is bound to be the day it fails me."

d. It's your first week of your corporate internship and you have to wear nylon stockings to the office if you're wearing a skirt. On the first and second days, you get a run in your stockings almost immediately, an indication of a defect. The third day, you put on yet another new pair of stockings and say, "OK, this pair has to be good. There's no way I'd have three bad pairs in a row. They're even from different stores!"

5.49 Null hypothesis and research hypothesis: For each of the following studies, cite the likely null hypothesis and the likely research hypothesis.

a. A forensic cognitive psychologist wondered whether repetition of false information (versus no repetition) would increase the tendency to develop false memories, on average.

b. A clinical psychologist studied whether ongoing structured assessments of the therapy process (versus no assessment) would lead to better outcomes, on average, among outpatient therapy clients with depression.

c. A corporation recruited an industrial/organizational psychologist to explore the effects of cubicles (versus enclosed offices) on employee morale.

d. A team of developmental cognitive psychologists studied whether teaching a second language to children from birth affects children's ability to speak their native language.

5.50 Decision about the null hypothesis: For each of the following fictional conclusions, state whether the researcher seems to have rejected or failed to reject the null hypothesis (contingent, of course, on inferential statistics having backed up the statement). Explain the rationale for your decision.

a. When false information is repeated several times, people seem to be more likely, on average, to develop false memories than when the information is not repeated.

b. Therapy clients with depression who have ongoing structured assessments of therapy seem to have lower post-therapy depression levels, on average, than do clients who do not have ongoing structured assessments.

c. There is no evidence that employee morale is different, on average, whether employees work in cubicles or in enclosed offices.

d. There is no evidence that a child's native language is different in strength, on average, based on whether the child is raised to be bilingual or not.

5.51 Examine the statements from Exercise 5.50, repeated here. For each, if this conclusion were incorrect, what type of error would the researcher have made? Explain your answer.

a. When false information is repeated several times, people seem to be more likely, on average, to develop false memories than when the information is not repeated.

b. Therapy clients with depression who have ongoing structured assessments of therapy seem to have lower post-therapy depression levels, on average, than do clients who do not have ongoing structured assessments.

c. Employee morale does not seem to be different, on average, whether employees work in cubicles or enclosed offices.

d. A child's native language does not seem to be different in strength, on average, based on whether the child is raised to be bilingual or not.

5.52 Rejecting versus failing to reject an invitation: Imagine you have found a new study partner in statistics class. One day, your study partner asks you to go on a date. This invitation takes you completely by surprise, and you have no idea what to say. You are not attracted to the person in a romantic way, but at the same time you do not want to hurt his or her feelings.

a. Create two possible responses to the person, one in which you *fail to reject the invitation* and another in which you *reject the invitation*.

b. How is your failure to reject the invitation different from rejecting or accepting the invitation?

5.53 Confirmation bias, errors, replication, and horoscopes: A horoscope on astrology.com stated: "A big improvement is in the works, one that you may know nothing about, and today is the day for the big unveiling." A job-seeking recent college graduate might spot some new listings for interesting positions and decide the horoscope was right. If you look for an association, you're likely to find it. Yet, over and over again, careful researchers have failed to find evidence to support the accuracy of astrology (e.g., Dean & Kelly, 2003).

a. Explain to the college graduate how confirmation bias guides his logic in deciding the horoscope was right.

b. If Dean and Kelly and other researchers were wrong, what kind of error would they have made?

c. Explain why replication (i.e., "over and over again") means that this finding is not likely to be an error.

Putting It All Together

5.54 Horoscopes and predictions: People remember when their horoscopes had an uncanny prediction—say, the prediction of a problem in love on the exact day of the breakup of a romantic relationship—and decide that horoscopes are accurate. Yet, Munro & Munro (2000) are among those who have challenged such a conclusion. They reported that 34% of students chose their own horoscope as the best match for them when the horoscopes were labeled with the signs of the zodiac, whereas only 13% chose their own horoscope when the predictions were labeled only with numbers and in a random order. Thirteen percent is not statistically significantly different from 8.3%, which is the percentage we'd expect by chance.

a. What is the population of interest, and what is the sample in this study?

b. Was random selection used? Explain your answer.

c. Was random assignment used? Explain your answer.

d. What is the independent variable and what are its levels? What is the dependent variable? What type of variables are these?

e. What is the null hypothesis and what is the research hypothesis?

f. What decision did the researchers make? (Respond using the language of inferential statistics.)

g. If the researchers were incorrect in their decision, what kind of error did they make? Explain your answer. What are the consequences of this type of error, both in general and in this situation?

5.55 Alcohol abuse interventions: Borsari and Carey (2005) randomly assigned 64 male students who had been ordered, after a violation of university alcohol rules, to meet with a school counselor to one of two conditions. Students were assigned to undergo either (1) a newly developed brief motivational interview (BMI), an intervention in which educational material relates to the students' own experiences, or (2) a standard alcohol education session (AE) in which educational material is presented with no link to students' experiences. Based on inferential statistics, the researchers concluded that those in the BMI group had fewer alcohol-related problems at follow-up, on average, than did those in the AE group.

a. What is the population of interest, and what is the sample in this study?

b. Was random selection used? Explain your answer.

c. Was random assignment used? Explain your answer.

d. What is the independent variable and what are its levels? What is the dependent variable?

e. What is the null hypothesis and what is the research hypothesis?

f. What decision did the researchers make? (Respond using the language of inferential statistics.)

g. If the researchers were incorrect in their decision, what kind of error did they make? Explain your answer. What are the consequences of this type of error, both in general and in this situation?

5.56 Treatment for depression: Researchers conducted a study of 18 patients whose depression had not responded to treatment (Zarate, 2006). Half received one intravenous dose of ketamine, a hypothesized quick fix for depression; half received one intravenous dose of placebo. Far more of the patients who received ketamine improved, as measured by the Hamilton Depression Rating Scale, usually in less than 2 hours, than patients on placebo.

a. What is the population of interest, and what is the sample in this study?

b. Was random selection used? Explain your answer.

c. Was random assignment used? Explain your answer.

d. What is the independent variable and what are its levels? What is the dependent variable?

e. What is the null hypothesis and what is the research hypothesis?

f. What decision did the researchers make? (Respond using the language of inferential statistics.)

g. If the researchers were incorrect in their decision, what kind of error did they make? Explain your answer. What are the consequences of this type of error, both in general and in this situation?

Terms

random sample (p. 92)
convenience sample (p. 92)
generalizability (p. 94)
replication (p. 94)
volunteer sample (p. 94)
confirmation bias (p. 97)
illusory correlation (p. 97)
personal probability (p. 98)
probability (p. 98)
expected relative-frequency probability (p. 98)
trial (p. 98)
outcome (p. 98)
success (p. 98)
control group (p. 101)
experimental group (p. 101)
null hypothesis (p. 101)
research hypothesis (p. 102)
Type I error (p. 104)
Type II error (p. 105)

CHAPTER 6

The Normal Curve, Standardization, and *z* Scores

BEFORE YOU GO ON

- You should be able to create histograms and frequency polygons (Chapter 2).
- You should be able to describe distributions of scores using measures of central tendency, variability, and skewness (Chapters 2 and 4).
- You should understand that we can have distributions of scores based on samples, as well as distributions of scores based on populations (Chapter 5).

A **normal curve** is a specific bell-shaped curve that is unimodal, symmetric, and defined mathematically.

Abraham De Moivre was still a teenager in the 1680s when he was imprisoned in a French monastery for 2 years because of his religious beliefs. After his release, he found his way to Old Slaughter's Coffee House in London. The political squabbles, hustling artists, local gamblers, and insurance brokers didn't seem to bother the young scholar—he probably scratched out his famous formula on coffee-stained paper. The formula was exactly what gamblers needed to predict throws of the dice. It also helped insurance brokers estimate the probability of ships reaching their destinations.

De Moivre's powerful mathematical idea is far easier to understand as a picture, but that visual insight would take another 200 years. For example, Daniel Bernoulli came close in 1769. Augustus De Morgan came even closer when he mailed a sketch to fellow astronomer George Airy in 1849 (Figure 6-1). They were both trying to draw the ***normal curve***, *a specific bell-shaped curve that is unimodal, symmetric, and defined mathematically.*

The astronomers were trying to pinpoint the precise time when a star touched the horizon—but they couldn't get independent observers to agree. Nevertheless, the findings revealed (a) that the pattern of errors was symmetric; and (b) that the midpoint represented a reasonable estimate of reality. Only a few estimates were extremely high or extremely low; most errors clustered tightly around the middle. They were starting to understand what seems obvious to us today: their best estimate of reality was a bell-shaped curve—the foundation of inferential statistics (Stigler, 1999).

In this chapter, we learn about the building blocks of inferential statistics: (a) the characteristics of the normal curve; (b) how to use the normal curve to standardize any variable by using a tool called the z score; and (c) the central limit theorem, which, coupled with a grasp of standardization, allows us to make comparisons between means.

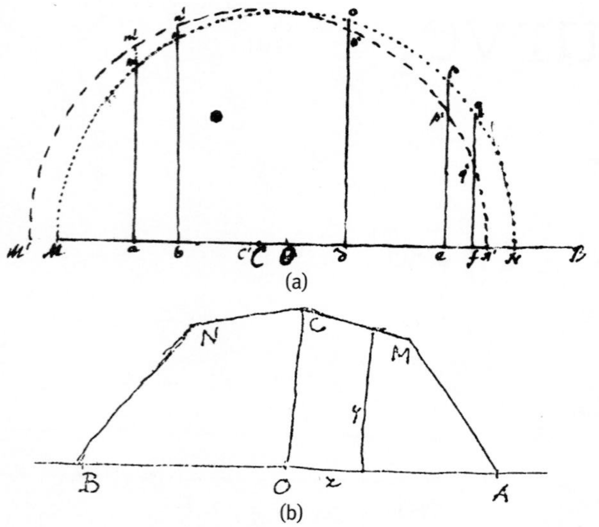

(a)

(b)

FIGURE 6-1
The Bell Curve Is Born

Daniel Bernoulli (a) created an approximation of the bell-shaped curve in this 1769 sketch "describing the frequency of errors." Augustus De Morgan (b) included this sketch in a letter to astronomer George Airy in 1849.

The Normal Curve

EXAMPLE 6.1

In this section, we learn more about the normal curve through a real-life example. Let's examine the heights, in inches, of a sample of 5 students taken from a larger sample of the authors' statistics students:

52 77 63 64 64

Figure 6-2 shows a histogram of those heights, with a normal curve superimposed on the histogram. With so few scores, we can only begin to guess at the emerging shape of a normal distribution. Notice that three of the observations (63 inches, 64 inches, and 64 inches) are represented by the middle bar. This is why it is three times higher than the bars that represent a single observation of 52 inches and another observation of 77 inches.

Now, here are the heights in inches from a sample of 30 students:

52 77 63 64 64 62 63 64 67 52
67 66 66 63 63 64 62 62 64 65
67 68 74 74 69 71 61 61 66 66

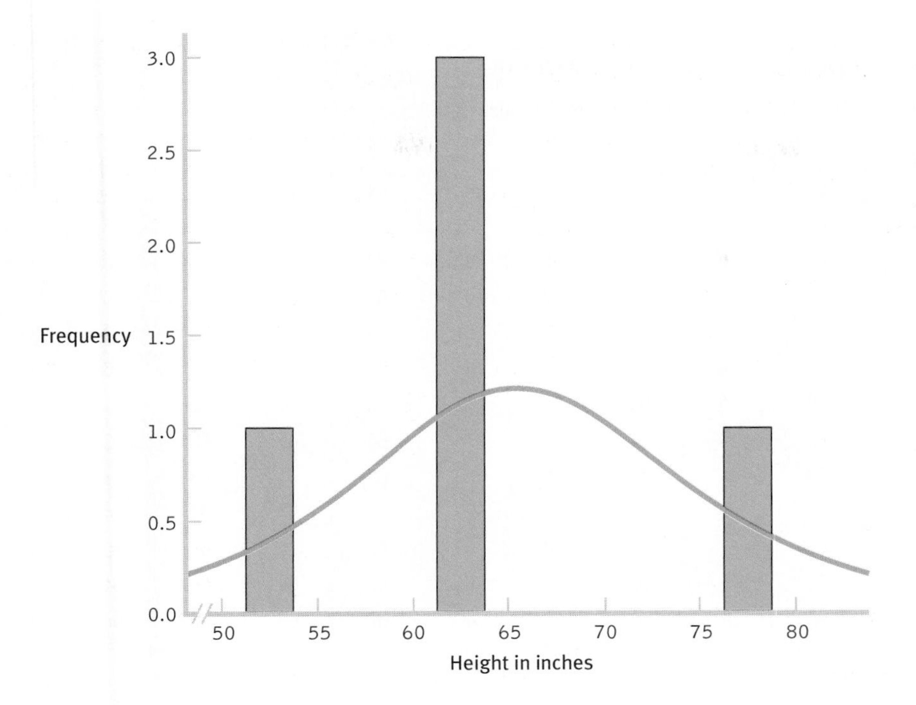

FIGURE 6-2
Sample of 5

Here is a histogram of the heights in inches of 5 students. With so few students, the data are unlikely to closely resemble the normal curve that we would see for an entire population of heights.

Figure 6-3 shows the histogram for these data. Notice that the heights of 30 students resemble a normal curve more so than do the heights of just 5 students, although certainly they don't match it perfectly.

Now, Table 6-1 gives the heights in inches from a random sample of 140 students. Figure 6-4 shows the histogram for these data.

These three images demonstrate why sample size is so important in relation to the normal curve. As the sample size increases, the

> **MASTERING THE CONCEPT**
>
> **6.1:** The distributions of many variables approximate a normal curve, a mathematically defined, bell-shaped curve that is unimodal and symmetric.

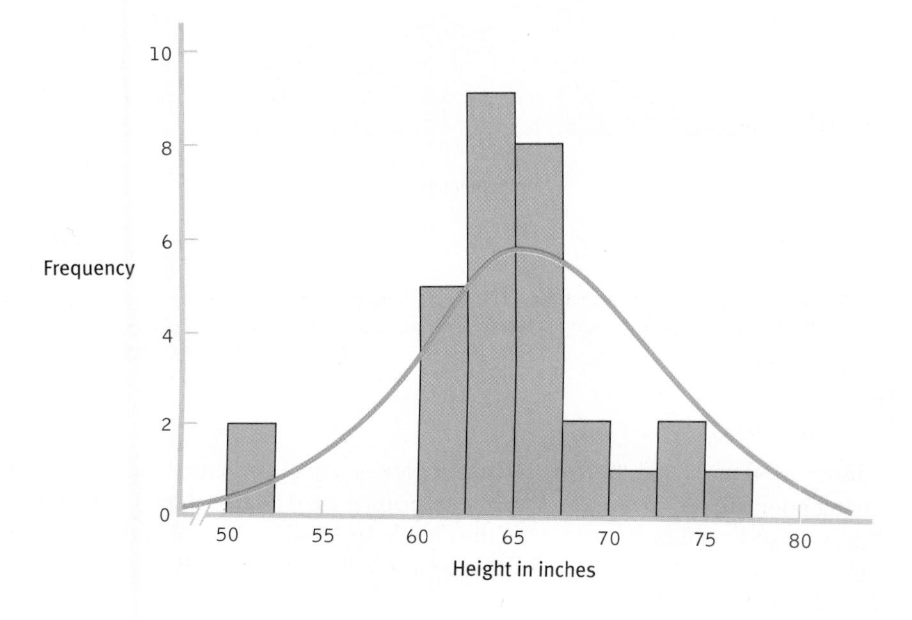

FIGURE 6-3
Sample of 30

Here is a histogram of the heights in inches of 30 students. With a larger sample, the data begin to resemble the normal curve of an entire population of heights.

TABLE 6-1. A Sample of Heights

These are the heights, in inches, of 140 students.

52	77	63	64	64	62	63	64	67	52
67	66	66	63	63	64	62	62	64	65
67	68	74	74	69	71	61	61	66	66
68	63	63	62	62	63	65	67	73	62
63	63	64	60	69	67	67	63	66	61
65	70	67	57	61	62	63	63	63	64
64	68	63	70	64	60	63	64	66	67
68	68	68	72	73	65	61	72	71	65
60	64	64	66	56	62	65	66	72	69
60	66	73	59	60	60	61	63	63	65
66	69	72	65	62	62	62	66	64	63
65	67	58	60	60	67	68	68	69	63
63	73	60	67	64	67	64	66	64	72
65	67	60	70	60	67	65	67	62	66

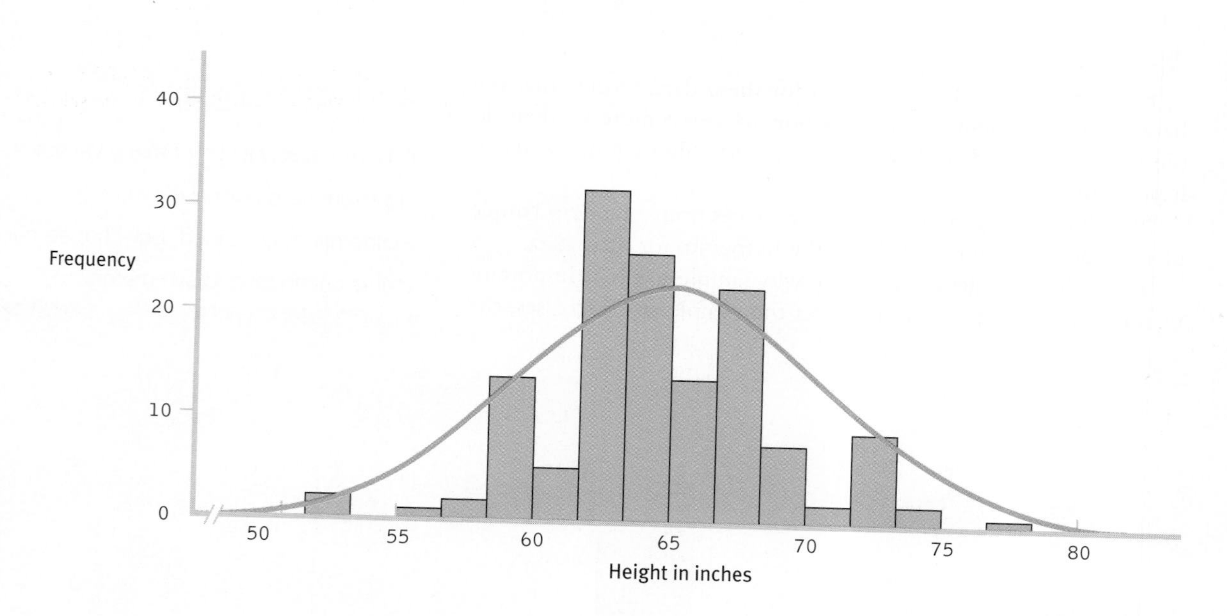

FIGURE 6-4
Sample of 140
Here is a histogram of the heights in inches of 140 students. As the sample increases, the shape of the distribution becomes more and more like the normal curve we would see for an entire population. Imagine the distribution of the data for a sample of 1000 students or of 1 million.

▪ The process of **standardization** converts individual scores from different normal distributions to a shared normal distribution with a known mean, standard deviation, and percentiles.

distribution more and more closely resembles a normal curve (as long as the underlying population distribution is normal). Imagine even larger samples—of 1000 students or of 1 million. As the size of the sample approaches the size of the population, the shape of the distribution tends to be normally distributed. ■

CHECK YOUR LEARNING

Reviewing the Concepts

> The normal curve is a specific, mathematically defined curve that is bell-shaped and symmetric.

> The normal curve describes the distributions of many variables.

> As the size of a sample approaches the size of the population, the distribution resembles a normal curve (as long as the population is normally distributed).

Clarifying the Concepts

6-1 What does it mean to say that the normal curve is unimodal and symmetric?

Calculating the Statistics

6-2 A sample of 225 students completed the Consideration of Future Consequences (CFC) scale. The scores are means of responses to 12 items. Overall CFC scores range from 1 to 5.

a. Here are CFC scores for five students, rounded to the nearest whole or half number: 3.5, 3.5, 3.0, 4.0, and 2.0. Create a histogram for these data, either by hand or by using software.

b. Now create a histogram for these scores of 30 students:

3.5 3.5 3.0 4.0 2.0 4.0 2.0 4.0 3.5 4.5
4.5 4.0 3.5 2.5 3.5 3.5 4.0 3.0 3.0 2.5
3.0 3.5 4.0 3.5 3.5 2.0 3.5 3.0 3.0 2.5

Applying the Concepts

6-3 The histogram below uses the actual (not rounded) CFC scores for all 225 students. What do you notice about the shape of this distribution of scores as the size of the sample increases?

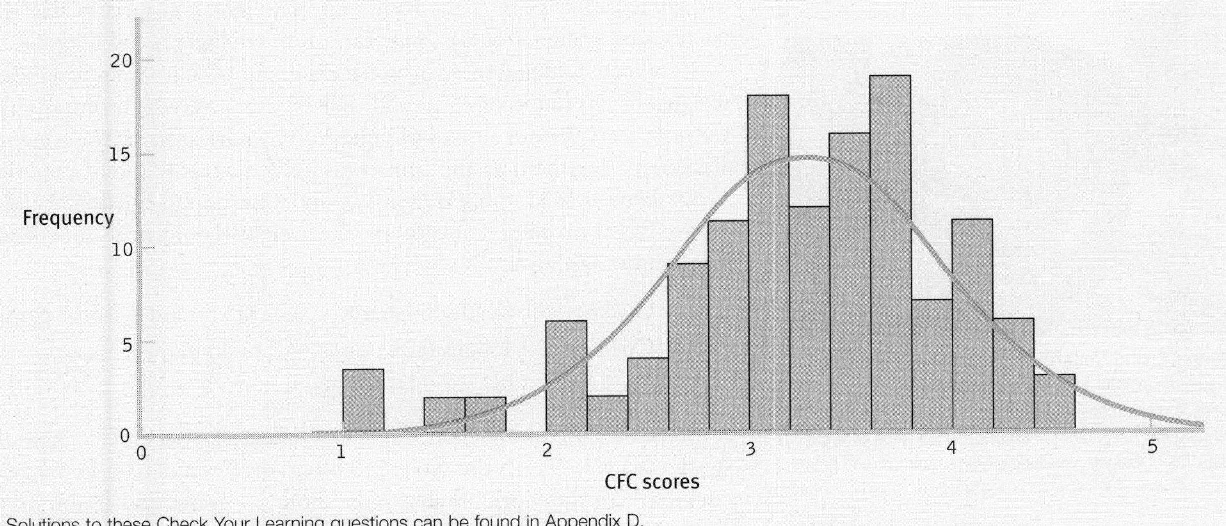

Solutions to these Check Your Learning questions can be found in Appendix D.

Standardization, z Scores, and the Normal Curve

De Moivre's discovery of the normal curve meant that scientists could now make meaningful comparisons. When data are normally distributed, we can compare a score to an entire distribution of scores. To do this, we convert a raw score to a standardized score (for which percentiles are already known). *The process of* ***standardization*** *converts* *individual scores to standard scores for which we know the percentiles (if the data are normally distributed). Standardization does this by converting individual scores from different normal distributions to a shared normal distribution with a known mean, standard deviation, and percentiles.*

In this section, we explain the importance of standardization and introduce the tool that helps us standardize, the z score. We show how we can convert raw scores to z scores, and z scores to raw scores. We demonstrate how the distribution of z scores allows us to know what percentage of the population falls above or below a given z score.

The Need for Standardization

One of the first problems with making meaningful comparisons is that variables are measured on different scales. For example, we measure height in inches but measure weight in pounds. In order to compare heights and weights, we need a way to put different variables on the same standardized scale. Fortunately, we can standardize different variables by using their means and standard deviations to convert any raw score into a z score. *A* ***z score*** *is the number of standard deviations a particular score is from the mean.* A z score is part of its own distribution, the z distribution, just as a raw score, such as a person's height, is part of its own distribution, a distribution of heights. (Note that as with all statistical symbols, the z is italicized.)

MASTERING THE CONCEPT

6.2: z scores give us the ability to convert any variable to a standard distribution, allowing us to make comparisons among variables.

EXAMPLE 6.2

Here is a memorable example of standardization: comparing weights of cockroaches. Different countries use different measures of weight. In the United Kingdom and the United States, the pound is typically used, with variants that are fractions or multiples of the pound, such as the dram, ounce, and ton. In most other countries, the metric system is used, with the gram as the basic unit of weight, and variants that are fractions or multiples of the gram, such as the milligram and kilogram.

If we were told that three imaginary species of cockroaches had mean weights of 8.0 drams, 0.25 pound, and 98.0 grams, which one should we most fear? We can answer this question by standardizing the weights and comparing them on the same measure. A dram is 1/256 of a pound, so 8.0 drams is 1/32 = 0.03125 of a pound. One pound equals 453.5924 grams. Based on these conversions, the weights could be standardized into grams as follows:

Cockroach 1 weighs 8.0 drams = 0.03125 pound = 14.17 grams
Cockroach 2 weighs 0.25 pound = 113.40 grams
Cockroach 3 weighs 98.0 grams

Michael Dick/Animals Animals-Earth Scenes

Standardizing Cockroach Weights Standardization creates meaningful comparisons by converting different scales to a common, or standardized, scale. We can compare the weights of these cockroaches using different measures of weights—including drams, pounds, and grams.

Standardizing allows us to determine that the second cockroach species tends to weigh the most: 113.40 grams. Fortunately, the biggest cockroach in the world weighs only about 35 grams and is about 80 millimeters (3.15 inches) long. Cockroaches 2 and 3 exist only in our imaginations. However, not all conversions are as easy as standardizing weights from different units into grams. That's why statisticians developed the z distribution. ■

Transforming Raw Scores into z Scores

A desire to make meaningful comparisons forces us to convert raw scores into standardized scores. For example, let's say you know that after taking the midterm examination, you are 1 standard deviation above the mean in your statistics class. Is this good news? What if you are 0.5 standard deviation below the mean? Understanding a score's relation to the mean of its distribution gives us important information. For a statistics test, we know that being above the mean is good; for anxiety levels, we know that being above the mean is usually bad. *z* scores create an opportunity to make meaningful comparisons.

▪ A ***z* score** is the number of standard deviations a particular score is from the mean.

The only information we need to convert any raw score to a *z* score is the mean and standard deviation of the population of interest. In the midterm example above, we are probably interested in comparing our grade with the grades of others in this course. In this case, the statistics class is the population of interest. Let's say that your score on the midterm is 2 standard deviations above the mean; your *z* score is 2.0. Imagine that a friend's score is 1.6 standard deviations below the mean; your friend's *z* score is −1.6. What would your *z* score be if you fell exactly at the mean in your statistics class? If you guessed 0, you're correct.

Figure 6-5 illustrates two important features of the *z* distribution. First, the *z* distribution always has a mean of 0. So, if you are exactly at the mean, then you are 0 standard deviations from the mean. Second, the *z* distribution always has a standard deviation of 1. If your raw score is 1 standard deviation above the mean, you have a *z* score of 1.0.

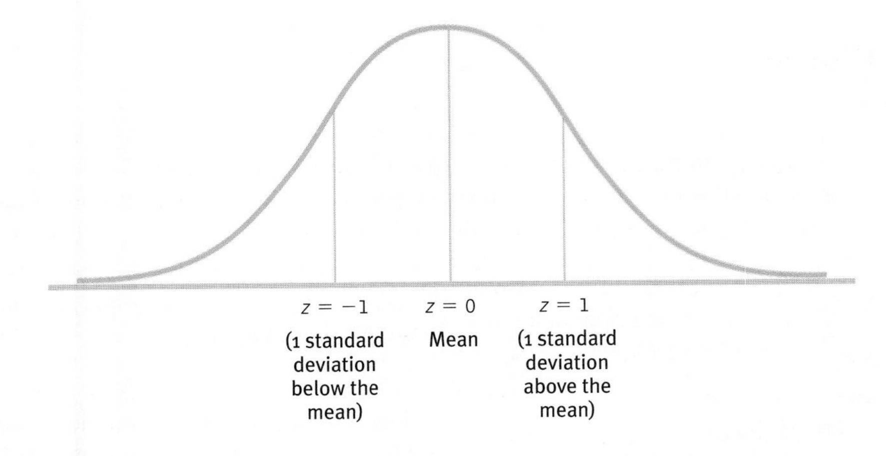

FIGURE 6-5
The *z* Distribution

The *z* distribution always has a mean of 0 and a standard deviation of 1.

Let's calculate *z* scores without a calculator or formula. We'll use the distribution of scores on a statistics exam. (This example is illustrated in Figure 6-6.) If the mean on a statistics exam is 70, the standard deviation is 10, and your score is 80, what is your *z* score? In this case, you are exactly 10 points, or 1 standard deviation, above the mean,

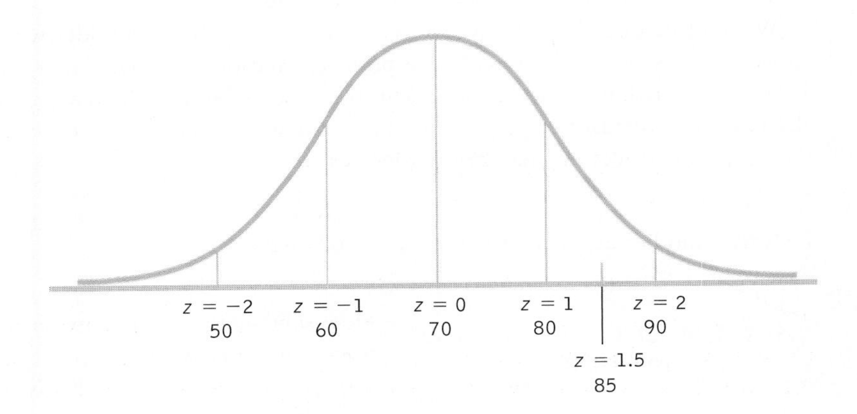

FIGURE 6-6
***z* Scores Intuitively**

With a mean of 70 and a standard deviation of 10, we can calculate many *z* scores without a formula. A raw score of 50 has a *z* score of −2.0. A raw score of 60 has a *z* score of −1.0. A raw score of 70 has a *z* score of 0. A raw score of 80 has a *z* score of 1.0. A raw score of 85 has a *z* score of 1.5.

so your z score is 1.0. Now let's say your score is 50, which is 20 points, or 2 standard deviations, below the mean, so your z score is −2.0. What if your score is 85? Now you're 15 points, or 1.5 standard deviations, above the mean, so your z score is 1.5.

As you can see, we don't need a formula to calculate a z score when we're working with easy numbers. It is important, however, to learn the notation and language of statistics. So let's also convert z scores using a formula for when the numbers are not easy to work with. To calculate a particular z score, there are just two steps.

STEP 1: Determine the distance of a particular person's score (X) from the population mean (μ) as part of the calculation: $X - \mu$.

STEP 2: Express this distance in terms of standard deviations by dividing by the population standard deviation, σ.

MASTERING THE FORMULA

6-1: The formula for a z score is $z = \frac{(X - \mu)}{\sigma}$. We calculate the difference between an individual score and the population mean, then divide by the population standard deviation.

The formula, therefore, is

$$z = \frac{(X - \mu)}{\sigma}$$ ■

EXAMPLE 6.4

Let's take an example that is not so easy to calculate in our heads. The mean height for the population of sophomores at your university is 64.886, with a standard deviation of 4.086. If you are 70 inches tall, what is your z score?

STEP 1: Subtract the population mean from your score.

In this case, subtract the population mean, 64.886, from your score, 70.

STEP 2: Divide by the population standard deviation.

The population standard deviation is 4.086. Here are those steps in the context of the formula:

$$z = \frac{(X - \mu)}{\sigma} = \frac{(70 - 64.886)}{4.086} = 1.25$$

You are 1.25 standard deviations above the mean.

We must be careful not to use a formula mindlessly. Always consider whether the answer makes sense. In this case, 1.25 is positive, indicating that the height is just over 1 standard deviation above the mean. This makes sense because the raw score of 70 is also just over 1 standard deviation above the mean. If you do this quick check regularly, then you can correct mistakes before they cost you. ■

EXAMPLE 6.5

Let's take another example: What if you are 62 inches tall?

STEP 1: Subtract the population mean from your score.

Here, subtract the population mean, 64.886, from your score, 62.

STEP 2: Divide by the population standard deviation.

The population standard deviation is 4.086. Here are those steps in the context of the formula:

$$z = \frac{(X - \mu)}{\sigma} = \frac{(62 - 64.886)}{4.086} = -0.71$$

You are 0.71 standard deviation *below* the mean.

Don't forget the sign of the z score. Changing a z score from negative 0.71 to positive 0.71 makes a big difference!

Masterfile/Radius Images

Estimating z Scores Would you guess that the person on the left has a positive or negative z score for height? What about the person on the right? A person who is very short has a below-average height and thus would have a negative z score. A person who is very tall has an above-average height and thus would have a positive z score. ■

EXAMPLE 6.6

With the height example we've been using, now let's demonstrate that the mean of the z distribution is always 0 and the standard deviation is always 1. The mean is 64.886 and the standard deviation is 4.086. Let's calculate what the z score would be at the mean.

STEP 1: Subtract the population mean from a score right at the mean.

We subtract the population mean, 64.886, from a score right at the mean, 64.886.

STEP 2: Divide by the population standard deviation.

We divide the difference by 4.086. Here are those steps in the context of the formula:

$$z = \frac{(X - \mu)}{\sigma} = \frac{(64.886 - 64.886)}{4.086} = 0$$

If someone is exactly 1 standard deviation above the mean, his or her score would be 64.886 + 4.086 = 68.972. Let's calculate what the z score would be for this person.

STEP 1: Subtract the population mean from a score exactly 1 standard deviation above the mean.

We subtract the population mean, 64.886, from a score exactly 1 standard deviation (4.086) above the mean, 68.972.

STEP 2: Divide by the population standard deviation.

We divide the difference by 4.086. Here are those steps in the context of the formula:

$$z = \frac{(X - \mu)}{\sigma} = \frac{(68.972 - 64.886)}{4.086} = 1$$ ■

Transforming *z* Scores into Raw Scores

If we already know a *z* score, then we can reverse the calculations to determine the raw score. The formula is the same; we just plug in all the numbers instead of the *X*, then solve algebraically. Let's try it with the height example.

EXAMPLE 6.7

The population mean is 64.886, with a standard deviation of 4.086. So, if you have a *z* score of 1.79, what is your height?

$$z = \frac{(X - \mu)}{\sigma} = 1.79 = \frac{(X - 64.886)}{4.086}$$

If we solve for *X*, we get 72.20. For those who prefer to minimize the use of algebra, we can do the algebra on the equation itself to derive a formula that gets the raw score directly. The formula is derived by multiplying both sides of the equation by σ, then adding μ to both sides of the equation. This isolates the *X*, as follows:

MASTERING THE FORMULA

6-2: The formula to calculate the raw score from a *z* score is $X = z(\sigma) + \mu$. We multiply the *z* score by the population standard deviation, then add the population mean.

$$X = z(\sigma) + \mu$$

So, there are two steps to converting a *z* score to a raw score:

STEP 1: Multiply the *z* score by the population standard deviation.

STEP 2: Add the population mean to this product.

Let's try the same problem using this direct formula.

STEP 1: Multiply the *z* score by the population standard deviation.

Multiply the *z* score, 1.79, by the population standard deviation, 4.086.

STEP 2: Add the population mean to this product.

Add the population mean, 64.886, to this product. Here are those steps in the context of the formula:

$$X = 1.79(4.086) + 64.886 = 72.20$$

Regardless of whether we use the original formula or the direct formula, the height is 72.20 inches. As always, think about whether the answer seems accurate. In this case, the answer does make sense because the height is above the mean and the z score is positive. ■

EXAMPLE 6.8

What if your z score is -0.44?

STEP 1: Multiply the *z* score by the population standard deviation.

Multiply the z score, -0.44, by the population standard deviation, 4.086.

STEP 2: Add the population mean to this product.

Add the population mean, 64.886, to this product. Here are those steps in the context of the formula:

$$X = -0.44(4.086) + 64.886 = 63.09$$

Your height is 63.09 inches. Don't forget the negative sign when doing this calculation. ■

As long as we know the mean and standard deviation of the population, we can do two things: (1) calculate the raw score from its z score, and (2) calculate the z score from its raw score.

Now that you understand z scores, let's disprove the saying that "you can't compare apples and oranges." We can take any apple from a normal distribution of apples, find its z score using the mean and standard deviation for the distribution of apples, convert the z score to a percentile, and discover that a particular apple is, say, larger than 85% of all apples. Similarly, we can take any orange from a normal distribution of oranges, find its z score using the mean and standard deviation for the distribution of oranges, convert the z score to a percentile, and discover that this particular orange is, say, larger than 97% of all oranges. The orange (with respect to other oranges) is bigger than the apple (with respect to other apples), and yes, that is an honest comparison. With standardization, we can compare anything, each relative to its own group.

Comstock Images/Jupiter Images

Apples and Oranges Standardization allows us to compare apples with oranges. If we can standardize the raw scores on two different scales, converting both scores to *z* scores, we can then compare the scores directly.

The normal curve also allows us to convert scores to percentiles because 100% of the population is represented under the bell-shaped curve. This means that the midpoint is the 50th percentile. If an individual score on some test is located to the right of the mean, you know that the score lies above the 50th percentile. A score to the left of the mean is below the 50th percentile. To make more specific comparisons, we convert raw scores to z scores and z scores to percentiles using the z distribution. *The* ***z distribution*** *is a normal distribution of standardized scores—* a distribution of z scores. And *the* ***standard normal distribution*** *is a normal distribution of z scores.*

■ The ***z* distribution** is a normal distribution of standardized scores.

■ The **standard normal distribution** is a normal distribution of *z* scores.

Most people are not content merely with knowing whether their own score is above or below the average score. After all, there is likely a big difference between scoring

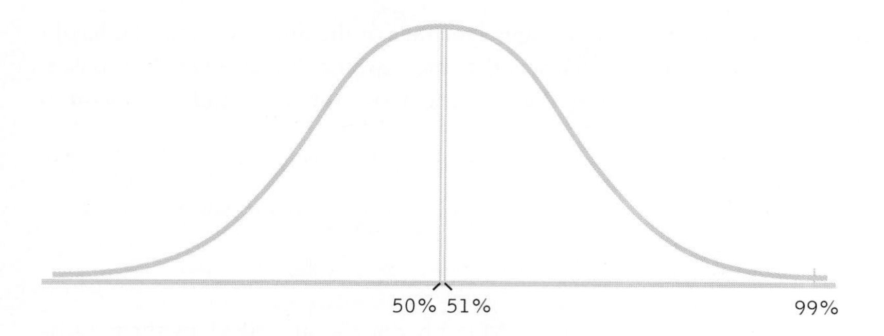

FIGURE 6-7
The All-Encompassing z Distribution

The z distribution theoretically includes all possible scores, so when it's based on a normal distribution, we know that 50% of the scores are above the mean and 50% are below the mean. But the 51st percentile and the 99th percentile are still far from each other, so two people making a comparison usually want more precise information than whether or not they are above average.

at the 51st percentile and scoring at the 99th percentile in height, as shown in Figure 6-7. The standardized z distribution allows us to do the following:

1. Transform raw scores into standardized scores called z scores
2. Transform z scores back into raw scores
3. Compare z scores to each other—even when the underlying raw scores are measured on different scales
4. Transform z scores into percentiles that are more easily understood

Let's begin with an illustration.

Using z Scores to Make Comparisons

Imagine that a friend is taking a course in statistics at the same time that you are, but with a different professor. Each professor has a different grading scheme, so each class produces a different distribution of scores. Thanks to standardization, we can convert each raw score to a z score and compare raw scores from *different* distributions.

EXAMPLE 6.9

For example, let's say that you both took a quiz. You earned 92 out of 100; the distribution of your class had a mean of 78.1 and a standard deviation of 12.2. Your friend earned 8.1 out of 10; the distribution of his class had a mean of 6.8 with a standard deviation of 0.74. Again, we're only interested in the classes that took the test, so these are populations. Who did better?

We standardize the scores in terms of their respective distributions.

$$\text{Your score: } z = \frac{(X - \mu)}{\sigma} = \frac{(92 - 78.1)}{12.2} = 1.14$$

$$\text{Your friend's score: } z = \frac{(X - \mu)}{\sigma} = \frac{(8.1 - 6.8)}{0.74} = 1.76$$

First, let's check our work. Do these answers make sense? Yes—both you and your friend scored above the mean and have positive z scores. Second, we compare the z scores. Although you both scored well above the mean in terms of standard deviations, your friend did better with respect to his class than you did with respect to your class. ■

Making Comparisons z scores create a way to compare students taking different exams from different courses. If each exam score can be converted to a z score with respect to the mean and standard deviation for its particular exam, the two scores can then be compared directly.

Transforming z Scores into Percentiles

So z scores are useful because:

1. z scores give us a sense of where a score falls in relation to the mean of its population (in terms of the standard deviation of its population).
2. z scores allow us to compare scores from different distributions.

Yet we can be even more specific about where a score falls. So an additional and particularly helpful use of z scores is that they also have this property:

3. z scores can be transformed into percentiles.

> **MASTERING THE CONCEPT**
>
> **6.3:** z scores tell us how far a score is from a population mean in terms of the population standard deviation. Because of this characteristic, we can compare z scores to each other, even when the raw scores are from different distributions. We can then go a step further by converting z scores into percentiles and comparing percentiles to each other.

Because the shape of a normal curve is standard, we automatically know something about the percentage of any particular area under the curve. Think of the normal curve and the horizontal line below it as forming a shape. (In fact, it *is* a shape; it's essentially a frequency polygon.) Like any shape, the area below the normal curve can be measured. We can quantify the space below a normal curve in terms of percentages.

Remember that the normal curve is, by definition, symmetric. This means that exactly 50% of scores fall below the mean and 50% fall above the mean. But Figure 6-8 demonstrates that we can be even more specific. Approximately 34% of scores fall between the mean and a z score of 1.0; and because of symmetry, 34% of scores also fall between the mean and a z score of -1.0. We also know that approximately 14% of scores fall between the z scores of 1.0 and 2.0, and 14% of scores fall between the z scores of -1.0 and -2.0. Finally, we know that approximately 2% of scores fall between the z scores of 2.0 and 3.0, and 2% of scores fall between the z scores of -2.0 and -3.0.

By simple addition, we can determine that approximately 68% ($34 + 34 = 68$) of scores fall within 1 standard deviation—or 1 z score—of the mean; that approximately 96% ($14 + 34 + 34 + 14 = 96$) of scores fall within 2 standard deviations of the mean; and that all or nearly all ($2 + 14 + 34 + 34 + 14 + 2 = 100$) scores fall within 3 standard deviations of the mean. So, if you know you are about 1 standard deviation above the mean on your statistics quiz, then you can add the 50% below the mean to the 34% between the mean and the z score of 1.0 that you earned on your quiz, and know that your score corresponds to approximately the 84th percentile.

If you know that you are about 1 standard deviation below the mean, you know that you are in the lower 50% of scores and that 34% of scores fall between your score and the mean. By subtracting, you can calculate that $50 - 34 = 16\%$ of scores fall below yours. Your score corresponds to approximately the 16th percentile. Scores on standardized tests are often expressed as percentiles.

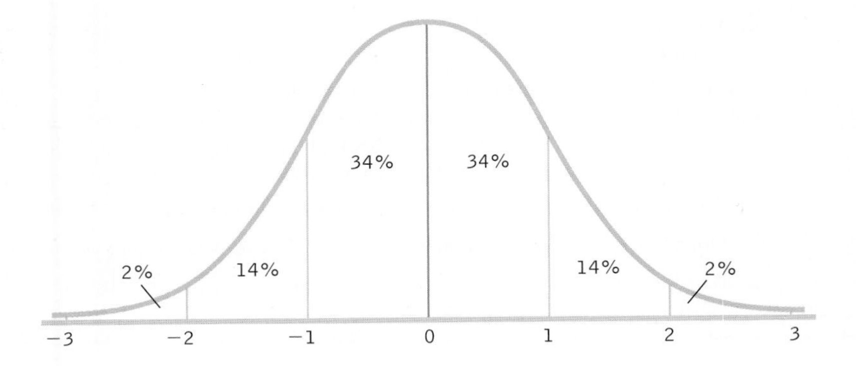

FIGURE 6-8
The Normal Curve and Percentages

The standard shape of the normal curve allows us to know the approximate percentages under different parts of the curve. For example, about 34% of scores fall between the mean and a z score of 1.0.

For now, it's important to understand that the z distribution forms a normal curve with a unimodal, symmetric shape. Because the shape is known and 100% of the population falls beneath the normal curve, we can determine the percentage of any area under the normal curve.

CHECK YOUR LEARNING

Reviewing the Concepts

> Standardization is a way to create meaningful comparisons between observations from different distributions. It can be accomplished by transforming raw scores from different distributions into z scores, also known as standardized scores.

> A z score is the distance that a score is from the mean of its distribution in terms of standard deviations.

> We also can transform z scores to raw scores by reversing the formula for a z score.

> z scores correspond to known percentiles that communicate how an individual score compares with the larger distribution.

Clarifying the Concepts

6-4 Describe the process of standardization.

6-5 What do the numeric value and the sign (negative or positive) of a z score indicate?

Calculating the Statistics

6-6 If the mean of a population is 14 and the standard deviation is 2.5, calculate z scores for the following raw scores:

a. 11.5

b. 18

6-7 Using the same population parameters as in Check Your Learning 6-6, calculate raw scores for the following z scores:

a. 2

b. −1.4

Applying the Concepts

6-8 The Consideration of Future Consequences (CFC) scale assesses how future-oriented students are. Researchers believe that a high CFC score is a positive indicator of a student's career potential. One study found a mean CFC score of 3.51, with a standard deviation of 0.61, for the 664 students in the sample (Petrocelli, 2003).

a. If a student has a CFC score of 2.3, what is her z score? Roughly, to what percentile does this z score correspond?

b. If a student has a CFC score of 4.7, what is his z score? To what percentile does this z score roughly correspond?

c. If a student has a CFC score at the 84th percentile, what is her z score?

d. What is the raw score of the student at the 84th percentile? Use symbolic notation and the formula. Explain why this answer makes sense.

6-9 Samantha has high blood pressure but exercises; she has a wellness score of 84 on a scale with a mean of 93 and a standard deviation of 4.5 (a higher score indicates better health). Nicole is of normal weight but has high cholesterol; she has a wellness score of 332 on a scale with a mean of 312 and a standard deviation of 20.

a. Without using a formula, who would you say is in better health?

b. Using standardization, who is in better health? Provide details using symbolic notation.

c. Based on their z scores, what percentage of people are in better health than Samantha and Nicole, respectively?

Solutions to these Check Your Learning questions can be found in Appendix D.

The Central Limit Theorem

In the early 1900s, W. S. Gossett discovered how the predictability of the normal curve could improve quality control in the Guinness ale factory. One of the practical problems that Gossett faced was related to sampling yeast cultures; too little yeast led to incomplete fermentation, whereas too much yeast led to bitter-tasting beer. To test whether he could sample both accurately and economically, Gossett averaged samples of four observations to see how well they represented a known population of 3000 (Gossett, 1908, 1942; Stigler, 1999).

This small adjustment (taking the *average* of four samples rather than one sample) is possible because of the central limit theorem. *The* ***central limit theorem*** *refers to how a distribution of sample means is a more normal distribution than a distribution of scores, even when the population distribution is not normal.* Indeed, as sample size increases, a distribution of sample means more closely approximates a normal curve. More specifically, the central limit theorem demonstrates two important principles:

1. Repeated sampling approximates a normal curve, *even when the original population is not normally distributed.*
2. A distribution of means is less variable than a distribution of individual scores.

Instead of randomly sampling a single data point, Gossett randomly sampled four data points from the population of 3000 and computed the average. He did this repeatedly and used those many averages to create a distribution of means. *A* ***distribution of means*** *is a distribution composed of many means that are calculated from all possible samples of a given size, all taken from the same population.* Put another way, the numbers that make up the distribution of means are not individual scores; they are *means* of samples of individual scores. Distributions of means are frequently used to understand data across a range of contexts; for example, when a university reports the mean standardized test score of incoming first-year students, that mean would be understood in relation to a distribution of means instead of a distribution of scores.

Gossett experimented with using the average of four data points as his sample, but there is nothing magical about the number four. A mean test score for incoming

- The **central limit theorem** refers to how a distribution of sample means is a more normal distribution than a distribution of scores, even when the population distribution is not normal.
- A **distribution of means** is a distribution composed of many means that are calculated from all possible samples of a given size, all taken from the same population.

MASTERING THE CONCEPT

6.4: The central limit theorem demonstrates that a distribution made up of the means of many samples (rather than individual scores) approximates a normal curve, even if the underlying population is not normally distributed.

Andy Johnstone/*Pacific Coast News*/Newscom.com

Yuri Arcurs/Fotolia.com

A Distribution of Means When we create a distribution of means, we eliminate extreme scores. If we choose just one individual score, there's a chance we'll get an extreme one, such as the length of the fingernails of the woman on the left. But if we select several scores, other, more typical scores—like the fingernails in the sample on the right—will balance out any extreme score. This helps to explain why a distribution of means tends to be less variable than a distribution of scores.

students would have a far larger sample size. The important outcome is that a distribution of means more consistently produces a normal distribution (although with less variance) *even when the population distribution is not normal.*

In this section, we learn how to create a distribution of means, as well as to calculate a z score for a mean (more accurately called a *z statistic* when calculated for means rather than scores). We also learn why the central limit theorem indicates that, when conducting hypothesis testing, a distribution of means is more useful than a distribution of scores.

Creating a Distribution of Means

The central limit theorem underlies many statistical processes that are based on a distribution of means. A distribution of means is more tightly clustered (has a smaller standard deviation) than a distribution of scores.

EXAMPLE 6.10

In an exercise that we conduct in class with our students, we write the numbers in Table 6-1 on 140 individual index cards that can be mixed together in a hat. The numbers represent the heights, in inches, of 140 college students from the authors' classes. As before, we treat these 140 students as the entire population.

1. First, we randomly pull one card at a time and record its score by marking it on a histogram. After recording the score, we return the card to the container representing the population of scores and mix all the cards before pulling the next card. (Not surprisingly, this is known as *sampling with replacement.*) We continue until we have plotted at least 30 scores, drawing a square for each one so that bars emerge above each value. This creates the beginning of a *distribution of scores.* Using this method, we created the histogram in Figure 6-9.
2. Now, we randomly pull three cards at a time, compute the mean of these three scores (rounding to the nearest whole number), and record this mean on a different histogram. As before, we draw a square for each mean, with each stack of squares resembling a bar. Again, we return each set of cards to the population and mix before pulling the next set of three. We continue until we have plotted at least 30 values. This is the beginning of a *distribution of means.* Using this method, we created the histogram in Figure 6-10.

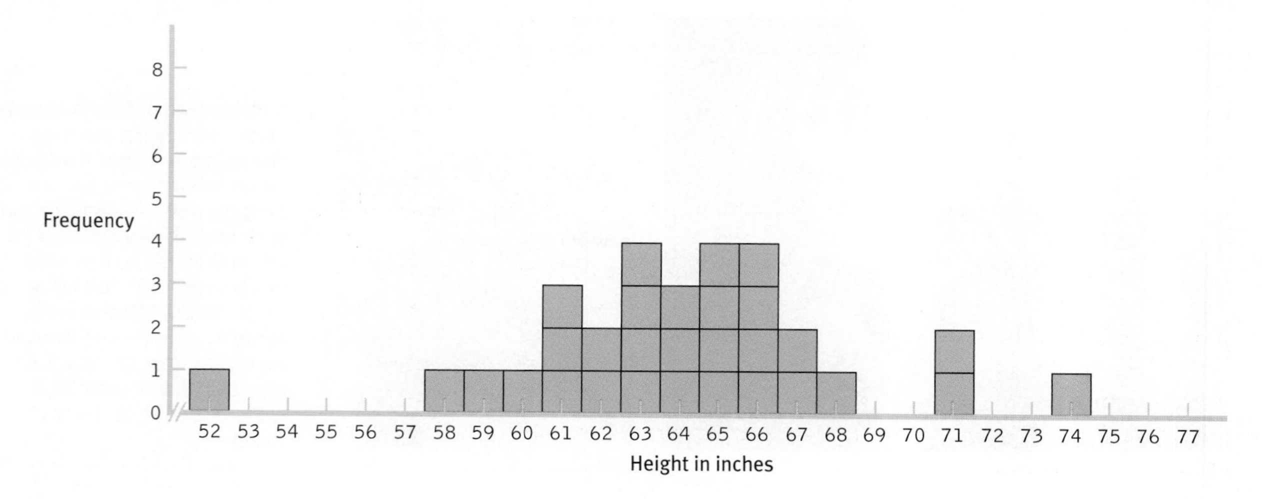

FIGURE 6-9
Creating a Distribution of Scores
This distribution is one of many that could be created by pulling 30 numbers, one at a time, and replacing the numbers between pulls, from the population of 140 heights. If you create a distribution of scores yourself from these data, it should look roughly bell-shaped like this one—that is, unimodal and symmetric.

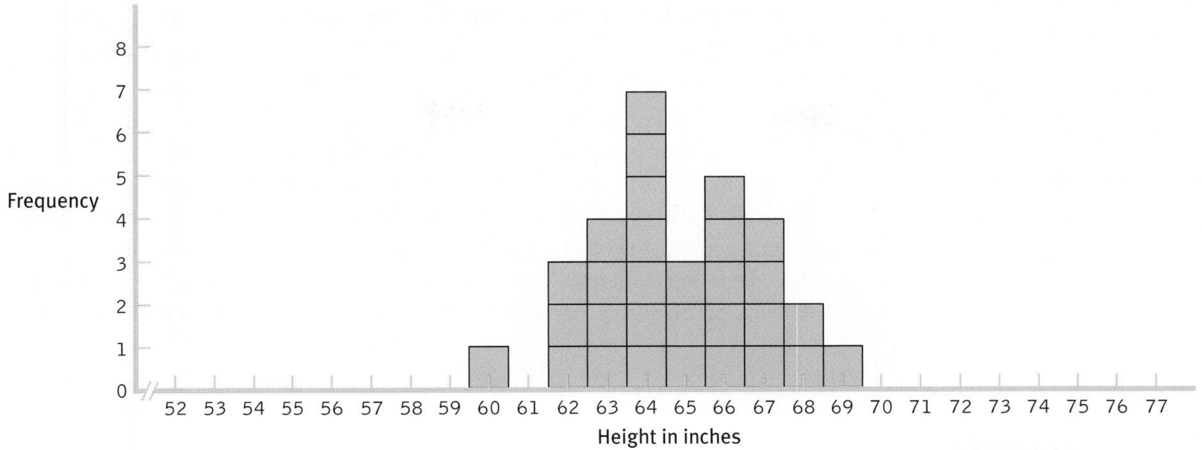

FIGURE 6-10
Creating a Distribution of Means
This distribution is one of many that could be created by pulling 30 means (the average of three numbers at a time) from the population of 140 heights. If you created a distribution of means from these data, it should look roughly bell-shaped. Notice that it is different from the distribution of scores in Figure 6-9: although centered around the same mean, it is narrower; the standard deviation is smaller; and the distribution of means has less spread.

The distribution of scores in Figure 6-9, similar to those we create when we do this exercise in class, ranges from 52 to 74, with a peak in the middle. If we had a larger population, and if we pulled many more numbers, the distribution would become more and more normal. Notice that the distribution is centered roughly around the actual population mean, 64.89. Also notice that all, or nearly all, scores fall within 3 standard deviations of the mean. The population standard deviation of these scores is 4.09. So nearly all scores should fall within this range:

$$64.89 - 3(4.09) = 52.62 \text{ and } 64.89 + 3(4.09) = 77.16$$

In fact, the range of scores in this population of 140 heights is very close to this, 52 through 77.

Is there anything different about the distribution of means in Figure 6-10? Yes, there are not as many means at the far tails of the distribution as in the distribution of scores—we no longer have any values in the 50s or 70s. However, there are no changes in the center of the distribution. The distribution of means is still centered around the actual mean of 64.89. This makes sense. The means of three scores each come from the same set of scores, so the mean of the individual sample means should be the same as the mean of the whole population of scores.

Why does the spread decrease when we create a distribution of means rather than a distribution of scores? When we plotted individual *scores,* an extreme score was plotted on the distribution. But when we plotted *means,* we averaged that extreme score with two other scores. So each time we pulled a score in the 70s, we tended to pull two lower scores as well; when we pulled a score in the 50s, we tended to pull two higher scores as well.

What do you think would happen if we created a distribution of means of 10 scores rather than 3? As you might guess, the distribution would be even narrower, because there would be more scores to balance the occasional extreme score. The mean of each set of 10 scores is likely to be even closer to the actual mean of 64.89. What if we created a distribution of means of 100 scores, or 10,000 scores? The larger the sample size, the smaller the spread of the distribution of means.

All the central limit theorem requires to work its magic is a distribution comprised of many sample means. In fact, distributions of means computed from samples of at least 30 usually produce an approximately normal curve, even when the population distribution is extremely skewed. ■

MASTERING THE CONCEPT

6.5: A distribution of means has the same mean as a distribution of scores from the same population, but a smaller standard deviation.

Characteristics of the Distribution of Means

Because the distribution of means is less variable than the distribution of scores, the distribution of means needs its own standard deviation—a smaller standard deviation than we used for the distribution of individual scores.

The data presented in Figure 6-11 allow us to visually verify that the distribution of means needs a smaller standard deviation. Using the population mean of 64.886 and standard deviation of 4.086, the z scores for the end scores of 60 and 69 are −1.20 and 1.01, respectively—not even close to 3 standard deviations. These z scores are wrong for this distribution. We need to use a standard deviation of sample *means* rather than a standard deviation of individual *scores.*

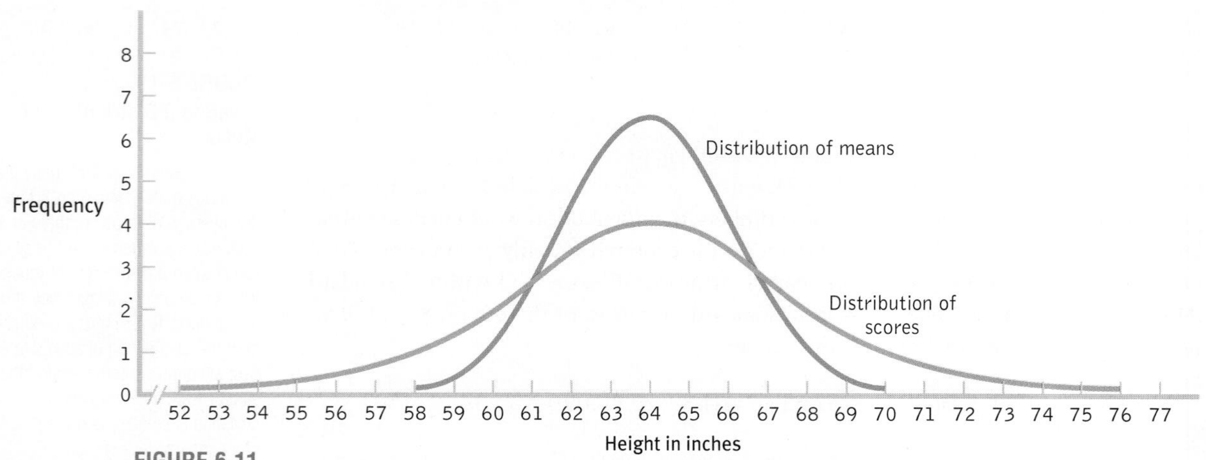

FIGURE 6-11
Using the Appropriate Measure of Spread

Because the distribution of means is narrower than the distribution of scores, it has a smaller standard deviation. This standard deviation has its own name: standard error.

We use slightly modified language and symbols when we describe distributions of means instead of distributions of scores. The mean of a distribution of means is the same as the mean of a population of scores, but it uses the symbol μ_M (pronounced "mew sub em"). The μ indicates that it is the mean of a *population,* and the subscript M indicates that the population is composed of *sample means*—the means of all possible samples of a given size from a particular population of individual scores.

We also need a new symbol and a new name for the standard deviation of the distribution of means—the typical amount that a sample mean varies from the population mean. The symbol is σ_M (pronounced "sigma sub em"). The subscript M again stands for mean; this is the standard deviation of the population of means calculated for *all possible samples* of a given size. The symbol has its own name: ***standard error*** *is the name for the standard deviation of a distribution of means.* Table 6-2 summarizes the alternative names that describe these related ideas.

- **Standard error** is the name for the standard deviation of a distribution of means.

TABLE 6-2. Parameters for Distributions of Scores Versus Means

When we determine the parameters of a distribution, we must consider whether the distribution is composed of scores or means.

Distribution	Symbol for Mean	Symbol for Spread	Name for Spread
Scores	μ	σ	Standard deviation
Means	μ_M	σ_M	Standard error

Fortunately, there is a simple calculation that lets us to know exactly how much smaller the standard error, σ_M, is than the standard deviation, σ. As we've noted, the larger the sample size, the narrower the distribution of means and the smaller the standard deviation of the distribution of means—the standard error. We calculate the standard error by taking into account the sample size used to calculate the means that make up the distribution. The standard error is the standard deviation of the population divided by the square root of the sample size, N. The formula is:

$$\sigma_M = \frac{\sigma}{\sqrt{N}}$$

MASTERING THE FORMULA

6-3: The formula for standard error is: $\sigma_M = \frac{\sigma}{\sqrt{N}}$. We divide the standard deviation for the population by the square root of the sample size.

EXAMPLE 6.11

Imagine that the standard deviation of the distribution of individual scores is 5 and we have a sample of 10 people. The standard error would be:

$$\sigma_M = \frac{\sigma}{\sqrt{N}} = \frac{5}{\sqrt{10}} = 1.58$$

The spread is smaller when we calculate means for samples of 10 people because any extreme scores are balanced by less extreme scores. With a larger sample size of 200, the spread is even smaller because there are many more scores close to the mean to balance out any extreme scores. The standard error would then be:

$$\sigma_M = \frac{\sigma}{\sqrt{N}} = \frac{5}{\sqrt{200}} = 0.35$$

A distribution of means faithfully obeys the central limit theorem. Even if the population of individual scores is *not* normally distributed, the distribution of means will approximate the normal curve if the samples are composed of at least 30 scores. The three graphs in Figure 6-12 depict (a) a distribution of individual scores that is extremely skewed in the positive direction, (b) the less skewed distribution that results

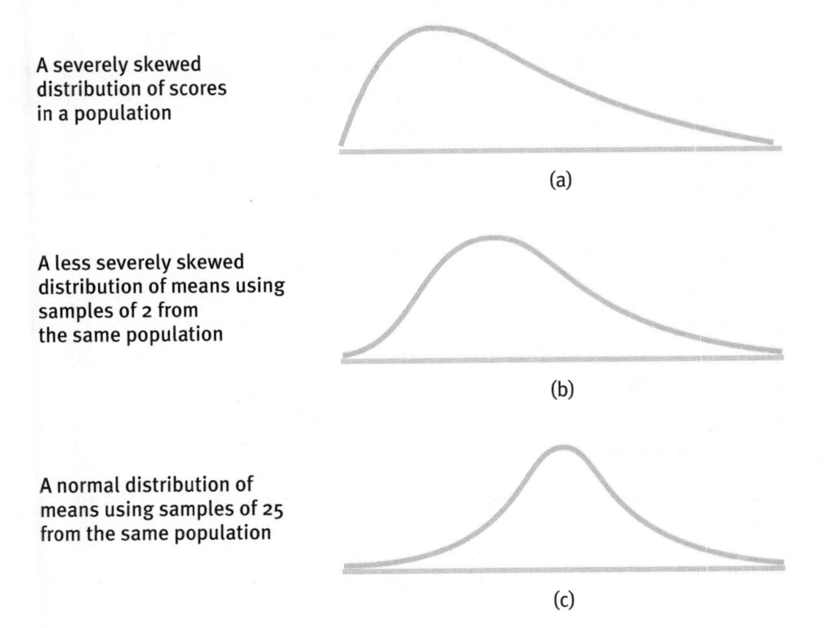

FIGURE 6-12
The Mathematical Magic of Large Samples
Even with a population of individual scores that are not normally distributed, the distribution of means approximates a normal curve as the sample gets larger.

when we create a distribution of means using samples of 2, and (c) the approximately normal curve that results when we create a distribution of means using samples of 25.

We have learned three important characteristics of the distribution of means:

1. As sample size increases, the mean of a distribution of means remains the same.
2. The standard deviation of a distribution of means (called the standard error) is smaller than the standard deviation of a distribution of scores. As sample size increases, the standard error becomes ever smaller.
3. The shape of the distribution of means approximates the normal curve if the distribution of the population of individual scores has a normal shape or if the size of each sample that makes up the distribution is at least 30 (the central limit theorem). ■

Using the Central Limit Theorem to Make Comparisons with *z* Scores

z scores are a standardized version of raw scores based on the population. But we seldom have the entire population to work with, so we typically calculate the mean of a sample and calculate a *z* score based on a distribution of means. When we calculate the *z* score, we simply use a distribution of means instead of a distribution of scores. The *z* formula changes only in the symbols it uses:

MASTERING THE FORMULA

6-4: The formula for *z* based on the mean of a sample is: $z = \frac{(M - \mu_M)}{\sigma_M}$. We subtract the mean of the distribution of means from the mean of the sample, then we divide by the standard error, the standard deviation of the distribution of means.

$$z = \frac{(M - \mu_M)}{\sigma_M}$$

Note that we now use *M* instead of *X* because we are calculating a *z* score for a sample mean rather than for an individual score. Because the *z* score now represents a mean, not an actual score, it is often referred to as a *z statistic*. Specifically, the *z* statistic tells us how many standard errors a sample mean is from the population mean.

EXAMPLE 6.12

Let's consider a distribution for which we know the population mean and standard deviation. Several hundred universities in the United States reported data from their counseling centers (Gallagher, 2009). (For this example, we'll treat this sample as the entire population of interest.) The study found that an average of 8.5 students per institution were hospitalized for mental illness over 1 year. For the purposes of this example, we'll assume a standard deviation of 3.8. Let's say we develop a prevention program to reduce the numbers of hospitalizations and we recruit 30 universities to participate. After 1 year, we calculate a mean of 7.1 hospitalizations at these 30 institutions. Is this an extreme sample mean, given the population?

To find out, let's imagine the distribution of means for samples of 30 hospitalization scores. The means would be collected the same way we collected the means of three heights in the earlier example—just with far more means. It would have the same mean as the population, but the spread would be smaller. Any extreme hospitalization scores would likely be balanced by less extreme scores when each mean is calculated, so the distribution would be less variable. Here are the mean and standard error of the sample of universities, using proper symbolic notation:

$$\mu_M = \mu = 8.5$$

$$\sigma_M = \frac{\sigma}{\sqrt{N}} = \frac{3.8}{\sqrt{30}} = 0.694$$

At this point, we have all the information we need to calculate the z statistic:

$$z = \frac{(M - \mu_M)}{\sigma_M} = \frac{(7.1 - 8.5)}{0.694} = -2.02$$

From this z statistic, we could determine how extreme the mean number of hospitalizations is in terms of a percentage. Then we could draw a conclusion about whether we would be likely to find a mean number of hospitalizations of 7.1 in a sample of 30 universities if the prevention program did *not* work. The useful combination of a distribution of means and a z statistic has led us to a point where we're prepared for inferential statistics and hypothesis testing. ■

CHECK YOUR LEARNING

Reviewing the Concepts

> According to the central limit theorem, a distribution of sample means based on 30 or more scores approximates the normal distribution, even if the original population is not normally distributed.

> A distribution of scores has the same mean as a distribution of means. However, a distribution of scores contains more extreme scores and a larger standard deviation than a distribution of means; this is another principle of the central limit theorem.

> z scores may be calculated from a distribution of scores or from a distribution of means. When we calculate a z score for a mean, we usually call it a z statistic.

> For the measure of spread, the two calculations use different terms: *standard deviation* for a distribution of scores and *standard error* for a distribution of means.

> Just as with z scores, the z statistic tells us about the relative position of a mean within a distribution, and this can be expressed as a percentile.

Clarifying the Concepts

6-10 What are the main ideas behind the central limit theorem?

6-11 Explain what a distribution of means is.

Calculating the Statistics

6-12 The mean of a distribution of scores is 57, with a standard deviation of 11. Calculate the standard error for a distribution of means based on samples of 35 people.

Applying the Concepts

6-13 Let's return to the selection of 30 CFC scores that we considered in Check Your Learning 6-2(b):

3.5 3.5 3.0 4.0 2.0 4.0 2.0 4.0 3.5 4.5

4.5 4.0 3.5 2.5 3.5 3.5 4.0 3.0 3.0 2.5

3.0 3.5 4.0 3.5 3.5 2.0 3.5 3.0 3.0 2.5

a. What is the range of these scores?

b. Take three means of 10 scores each from this sample of scores, one for each row. What is the range of these means?

c. Why is the range smaller for the *means* of samples of 10 scores than for the *individual* scores themselves?

d. The mean of these 30 scores is 3.32. The standard deviation is 0.69. Using symbolic notation and formulas (where appropriate), determine the mean and standard error of the distribution of means computed from samples of 10.

Solutions to these Check Your Learning questions can be found in Appendix D.

REVIEW OF CONCEPTS

The Normal Curve

Three ideas about the normal curve help us to understand inferential statistics. First, the *normal curve* describes the variability of many physical and psychological characteristics. Second, the normal curve may be translated into percentages, allowing us to standardize variables and make direct comparisons of scores on different measures. Third, a distribution of means, rather than scores, produces a more normal curve. The last idea is based on the central limit theorem, by which we know that a distribution of means will be normally distributed as long as the samples from which the means are computed are of a sufficiently large size, usually at least 30.

Standardization, z Scores, and the Normal Curve

The process of *standardization* converts raw scores into z scores. Raw scores from any normal distribution can be converted to the *z distribution*. And a normal distribution of z scores is called the *standard normal distribution*. *z scores* tell us how far a raw score falls from its mean in terms of standard deviation. We can also reverse the formula to convert z scores to raw scores. Standardization using z scores has two important applications. First, standardized scores—that is, z scores—can be converted to percentile ranks (and percentile ranks can be converted to z scores and then raw scores). Second, we can directly compare z scores from different raw-score distributions. z scores work the other way around as well.

The Central Limit Theorem

The z distribution can be used with a *distribution of means* in addition to a distribution of scores. Distributions of means have the same mean as the population of individual scores from which they are calculated, but a smaller spread, which means we must adjust for sample size. The standard deviation of a distribution of means is called the *standard error*. The decreased variability is due to the fact that extreme scores are balanced by less extreme scores when means are calculated. Distributions of means are normally distributed if the underlying population of scores is normal, or the means are computed from sufficiently large samples, usually at least 30 individual scores. This second situation is described by the *central limit theorem*, the principle that a distribution of sample means will be normally distributed even if the underlying distribution of scores is not normally distributed, as long as there are enough scores, usually at least 30, comprising each sample. The characteristics of the normal curve allow us to make inferences from small samples using standardized distributions, such as the z distribution.

SPSS®

SPSS lets us understand each variable, identify its skewness, and explore how well it fits with a normal distribution. Enter the 140 heights from Table 6-1.

We can identify outliers that might skew the normal curve by selecting Analyze → Descriptive Statistics → Explore → Statistics → Outliers. Click "Continue." Choose the variable of interest, "Heights," by clicking it on the left and then using the arrow to move it to the right. Click "OK." The screenshot shown here depicts part of the output.

We encourage you to play with data and explore the many features in SPSS. It is always helpful when we work with our own data. SPSS is easiest to learn when we know the source of every number and why we decided to include it in our study in the first place. It's also much more interesting to test our own ideas!

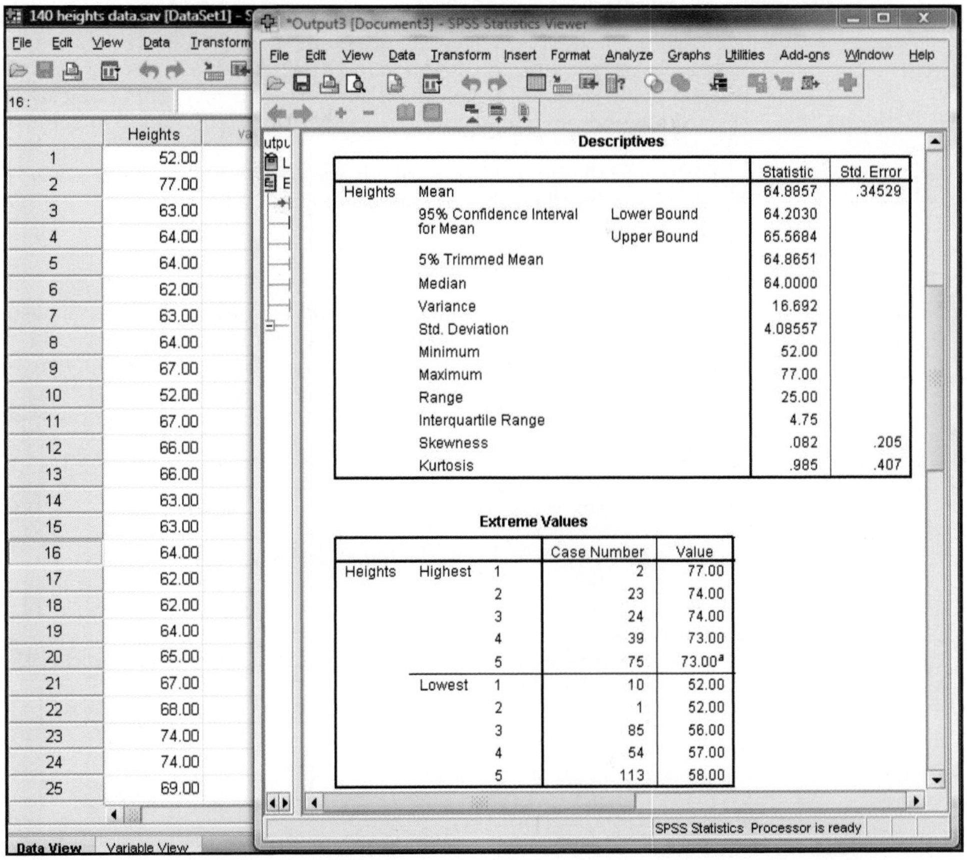

140 heights data.sav [DataSet1]

	Heights
1	52.00
2	77.00
3	63.00
4	64.00
5	64.00
6	62.00
7	63.00
8	64.00
9	67.00
10	52.00
11	67.00
12	66.00
13	66.00
14	63.00
15	63.00
16	64.00
17	62.00
18	62.00
19	64.00
20	65.00
21	67.00
22	68.00
23	74.00
24	74.00
25	69.00

Data View | Variable View

*Output3 [Document3] - SPSS Statistics Viewer

Descriptives

			Statistic	Std. Error
Heights	Mean		64.8857	.34529
	95% Confidence Interval for Mean	Lower Bound	64.2030	
		Upper Bound	65.5684	
	5% Trimmed Mean		64.8651	
	Median		64.0000	
	Variance		16.692	
	Std. Deviation		4.08557	
	Minimum		52.00	
	Maximum		77.00	
	Range		25.00	
	Interquartile Range		4.75	
	Skewness		.082	.205
	Kurtosis		.985	.407

Extreme Values

			Case Number	Value
Heights	Highest	1	2	77.00
		2	23	74.00
		3	24	74.00
		4	39	73.00
		5	75	73.00[a]
	Lowest	1	10	52.00
		2	1	52.00
		3	85	56.00
		4	54	57.00
		5	113	58.00

SPSS Statistics Processor is ready

How It Works

6.1 CONVERTING RAW SCORES TO *z* SCORES

Researchers reported that college students had healthier eating habits, on average, than did those who were neither college students nor college graduates (Georgiou et al., 1997). The researchers found that the 412 college students in the study ate breakfast a mean of 4.1 times per week, with a standard deviation of 2.4. Imagine that this is the entire population of interest.

Using symbolic notation and the formula, how can we calculate the z score for a student who eats breakfast six times per week? We can calculate the z score as follows:

$$z = \frac{(X - \mu)}{\sigma} = \frac{(6 - 4.1)}{2.4} = 0.79$$

Now, how can we calculate the z score for a student who eats breakfast twice a week? We can calculate this z score as follows:

$$z = \frac{(X - \mu)}{\sigma} = \frac{(2 - 4.1)}{2.4} = -0.88$$

6.2 STANDARDIZATION WITH *z* SCORES AND PERCENTILES

Who is doing better financially—Maria Sharapova, with respect to the 10 tennis players with the highest incomes, or Tiger Woods (prior to his scandal-related decline in endorsements), with respect to the 10 golfers with the highest incomes? In 2005, Forbes.com listed the 10 most powerful tennis players in terms of earnings and media exposure, regardless of gender. Sharapova, the first Russian (man or woman) to be ranked number 1 internationally in tennis, ranked second in earnings, with an income of $18.2 million, much of it from endorsements. In 2005, Golfdigest.com listed the top 50 earners in golf, regardless of gender. Woods placed first, with $89.4 million (over one-fourth of it just from Nike endorsements!). But top golfers tend to make more than top tennis players. In comparison to his top 10 peers, did Woods really do better financially than Sharapova did in comparison to her top 10 peers?

For tennis, the mean for the top 10 was $11.58 million, with a standard deviation of $6.58 million. Based on this, Maria Sharapova's z score is:

$$z = \frac{(X - \mu)}{\sigma} = \frac{(18.2 - 11.58)}{6.58} = 1.01$$

We can also estimate her percentile rank. Fifty percent of scores fall below the mean and about 34% fall between the mean and a z score of 1.0, so: 50 + 34 = 84. Sharapova is at approximately the 84th percentile among the top 10 tennis players with the highest incomes.

For golf, the mean for the top 10 was $30.01 million, with a standard deviation of $28.86 million. Based on this, Tiger Woods's z score is:

$$z = \frac{(X - \mu)}{\sigma} = \frac{(89.4 - 30.01)}{28.86} = 2.06$$

We can also estimate his percentile rank. Fifty percent of scores fall below the mean; about 34% fall between the mean and 1 standard deviation above the mean; and about 14% fall between 1 and 2 standard deviations above the mean: 50 + 34 + 14 = 98. Woods is at approximately the 98th percentile among the top 10 golfers with the highest incomes.

In comparison to the top 10 earners in their respective sports, Woods outearned Sharapova.

Exercises

Clarifying the Concepts

6.1 Explain how the word *normal* is used in everyday conversation; then explain how statisticians use it.

6.2 What point on the normal curve represents the most commonly occurring observation?

6.3 How does the size of a sample of scores affect the shape of the distribution of data?

6.4 Explain how the word *standardize* is used in everyday conversation; then explain how statisticians use it.

6.5 What is a z score?

6.6 Give three reasons why z scores are useful.

6.7 What are the mean and standard deviation of the z distribution?

6.8 Why is the central limit theorem such an important idea for dealing with a population that is not normally distributed?

6.9 What does the symbol μ_M stand for?

6.10 What does the symbol σ_M stand for?

6.11 What is the difference between standard deviation and standard error?

6.12 Why does the standard error become smaller simply by increasing the sample size?

6.13 What does a z statistic—a z score based on a distribution of means—tell us about a sample mean?

6.14 Each of the following equations has an error. Identify and fix the error and explain your work.

a. $\sigma_M = \dfrac{\mu}{\sqrt{N}}$

b. $z = \dfrac{(\mu - \mu_M)}{\sigma_M}$ (for a distribution of means)

c. $z = \dfrac{(M - \mu_M)}{\sigma}$ (for a distribution of means)

d. $z = \dfrac{(X - \mu)}{\sigma_M}$ (for a distribution of scores)

Calculating the Statistics

6.15 Create a histogram for these three sets of scores. Each set of scores represents a sample taken from the same population.

a. 6 4 11 7 7

b. 6 4 11 7 7 2 10 7 8 6 6 7 5 8

c. 6 4 11 7 7 2 10 7 8 6 6 7 5 8
7 8 9 7 6 9 3 9 5 6 8 11 8 3
8 4 10 8 5 5 8 9 9 7 8 7 10 7

d. What do you observe happening across these three distributions?

6.16 If a population has a mean of 250 and a standard deviation of 47, calculate *z* scores for each of the following raw scores:

a. 391
b. 273
c. 199
d. 160

6.17 A population has a mean of 1179 and a standard deviation of 164. Calculate *z* scores for each of the following raw scores:

a. 1000
b. 721
c. 1531
d. 1184

6.18 For a population with a mean of 250 and a standard deviation of 47, compute the *z* score for 250. Explain the meaning of the value you obtain.

6.19 For a population with a mean of 250 and a standard deviation of 47, compute the *z* scores for 203 and 297. Explain the meaning of these values.

6.20 For a population with a mean of 250 and a standard deviation of 47, convert each of the following *z* scores to raw scores.

a. 0.54
b. −2.66
c. −1.0
d. 1.79

6.21 For a population with a mean of 1179 and a standard deviation of 164, convert each of the following *z* scores to raw scores.

a. −0.23
b. 1.41
c. 2.06
d. 0.03

6.22 By design, the verbal subtest of the Graduate Record Examination (GRE) has a population mean of 500 and a population standard deviation of 100. Convert the following *z* scores to raw scores *without* using a formula.

a. 1.5
b. −0.5
c. −2.0

6.23 By design, the verbal subtest of the Graduate Record Examination (GRE) has a population mean of 500 and a population standard deviation of 100. Convert the following *z* scores to raw scores using symbolic notation and the formula.

a. 1.5
b. −0.5
c. −2.0

6.24 A study of the Consideration of Future Consequences (CFC) scale found a mean score of 3.51, with a standard deviation of 0.61, for the 664 students in the sample (Petrocelli, 2003). (Treat this sample as the entire population of interest.)

a. If the CFC score is 4.2, what is the *z* score? Use symbolic notation and the formula. Explain why this answer makes sense.
b. If the CFC score is 3.0, what is the *z* score? Use symbolic notation and the formula. Explain why this answer makes sense.
c. If the *z* score is 0, what is the CFC score? Explain.

6.25 As explained in Example 6.8, compare the following "apples and oranges": a score of 45 when the population mean is 51 and the standard deviation is 4, and a score of 732 when the population mean is 765 and the standard deviation is 23.

a. Convert these scores to standardized scores.
b. Using the standardized scores, what can you say about how these two scores compare to each other?

6.26 Compare the following scores:

a. A score of 811 when $\mu = 800$ and $\sigma = 29$ against a score of 4524 when $\mu = 3127$ and $\sigma = 951$
b. A score of 17 when $\mu = 30$ and $\sigma = 12$ against a score of 67 when $\mu = 88$ and $\sigma = 16$

6.27 Assume a normal distribution when answering the following questions.

a. What percentage of scores falls below the mean?
b. What percentage of scores falls between 1 standard deviation below the mean and 2 standard deviations above the mean?
c. What percentage of scores lies beyond 2 standard deviations away from the mean (on both sides)?
d. What percentage of scores is between the mean and 2 standard deviations above the mean?
e. What percentage of scores falls under the normal curve?

6.28 Compute the standard error (σ_M) for each of the following sample sizes, assuming a population mean of 100 and a standard deviation of 20:

a. 45
b. 100
c. 4500

6.29 A population has a mean of 55 and a standard deviation of 8. Compute μ_M and σ_M for each of the following sample sizes:

a. 30
b. 300
c. 3000

6.30 Compute a z statistic for each of the following, assuming the population has a mean of 100 and a standard deviation of 20:

a. A sample of 43 scores has a mean of 101.

b. A sample of 60 scores has a mean of 96.

c. A sample of 29 scores has a mean of 100.

6.31 A sample of 100 people had a mean depression score of 85; the population mean for this depression measure is 80, with a standard deviation of 20. A different sample of 100 people had a mean score of 17 on a different depression measure; the population mean for this measure is 15, with a standard deviation of 5.

a. Convert these means to z statistics.

b. Using the z statistics, what can you say about how these two means compare to each other?

Applying the Concepts

6.32 Normal distributions in real life: Many variables are normally distributed, but not all are. (Fortunately, the central limit theorem saves us when we conduct research on samples from nonnormal populations if the samples are larger than 30!) Which of the following are likely to be normally distributed, and which are likely to be nonnormal? Explain your answers.

a. Scores on the federal or provincial literacy test (required for university admissions) in the population of students admitted to the highly selective University of Toronto.

b. Number of daily calories consumed in the population of secondary school students in New Zealand.

c. Amount of time spent commuting to work in the population of employed adults in San Antonio, Texas.

d. Number of frequent flyer miles earned in a year in the population of North American university students.

6.33 Distributions and getting ready for a date: We asked 150 students (in our statistics classes) how long, in minutes, they typically spent getting ready for a date. The scores ranged from 1 minute to 120 minutes, and the mean was 51.52 minutes. Here are the data for 40 of those students:

30	90	60	60	5	90	30	40	45	60
60	30	90	60	25	10	90	20	15	60
60	75	45	60	30	75	15	30	45	1
20	25	45	60	90	10	105	90	30	60

a. Construct a histogram for the 10 scores in the first row.

b. Construct a histogram for all 40 of these scores.

c. What happened to the shape of the distribution as you increased the number of scores from 10 to 40? What do you think would happen if the data for all 150 students were included? What if we included 10,000 scores? Explain this phenomenon.

d. Are these distributions of scores or distributions of means? Explain.

e. The data here are self-reported. That is, our students wrote down how many minutes they believe that they typically take to get ready for a date. This accounts for the fact that the data include many "pretty" numbers, such as 30, 60, or 90 minutes. What might have been a better way to operationalize this variable?

f. Do these data suggest any hypotheses that you might like to study? List at least one.

6.34 z scores and the GRE: By design, the verbal subtest of the GRE has a population mean of 500 and a population standard deviation of 100 (the quantitative subtest has the same mean and standard deviation).

a. Use symbolic notation to state the mean and the standard deviation of the GRE verbal test.

b. Convert a GRE score of 700 to a z score *without* using a formula.

c. Convert a GRE score of 550 to a z score *without* using a formula.

d. Convert a GRE score of 400 to a z score *without* using a formula.

6.35 The z distribution and hours slept: A sample of 150 statistics students reported the typical number of hours that they sleep on a weeknight. The mean number of hours was 6.65, and the standard deviation was 1.24. (For this exercise, treat this sample as the entire population of interest.)

a. What is *always* the mean of the z distribution?

b. Using the sleep data, demonstrate that your answer to part (a) is the mean of the z distribution. (*Hint:* Calculate the z score for a student who is exactly at the mean.)

c. What is *always* the standard deviation of the z distribution?

d. Using the sleep data, demonstrate that your answer to part (c) is the standard deviation of the z distribution. (*Hint:* Calculate the z score for a student who is exactly 1 standard deviation above or below the mean.)

e. How many hours of sleep do you typically get on a weeknight? What would your z score be compared with that of this population?

6.36 The z distribution applied to admiration ratings: A sample of 148 of our statistics students rated their level of admiration for Hillary Rodham Clinton on a scale of 1 to 7. The mean rating was 4.06, and the standard deviation was 1.70. (For this exercise, treat this sample as the entire population of interest.)

a. Use these data to demonstrate that the mean of the z distribution is always 0.

b. Use these data to demonstrate that the standard deviation of the z distribution is always 1.

c. Calculate the z score for a student who rated his admiration of Hillary Rodham Clinton as 6.1.

d. A student had a z score of -0.55. What rating did she give for her admiration of Hillary Rodham Clinton?

6.37 z statistics and CFC scores: We have already discussed summary parameters for CFC scores for the population of participants in a study by Petrocelli (2003). The mean CFC score was 3.51, with a standard deviation of 0.61. (Remember that we treated the sample of 664 participants as the entire population.) Imagine that you randomly selected 40 people from this population and had them watch a series of videos on financial planning after graduation. The mean CFC score after watching the video was 3.62.

a. Why would it not make sense to compare the mean of this sample with the distribution of scores? Be sure to discuss the spread of distributions in your answer.

b. In your own words, what would the null hypothesis predict? What would the research hypothesis predict?

c. Using symbolic notation and formulas, what are the appropriate measures of central tendency and variability for the distribution from which this sample comes?

d. Using symbolic notation and the formula, what is the z statistic for this sample mean?

e. To what percentile does that z statistic roughly correspond?

6.38 Converting z scores to raw CFC scores: A CFC study found a mean CFC score of 3.51, with a standard deviation of 0.61, for the 664 students in the sample (Petrocelli, 2003).

a. Imagine that your z score on the CFC score is -1.2. What is your raw score? Use symbolic notation and the formula. Explain why this answer makes sense.

b. Imagine that your z score on the CFC score is 0.66. What is your raw score? Use symbolic notation and the formula. Explain why this answer makes sense.

6.39 The normal curve and real-life variables: For each of the following variables, state whether the distribution of scores would likely approximate a normal curve. Explain your answer.

a. Number of movies that a college student watches in a year

b. Number of full-page advertisements in a magazine

c. Human birth weights in Canada

6.40 Percentiles and eating habits: As noted in How It Works 6.1, Georgiou and colleagues (1997) reported that college students had healthier eating habits, on average, than did those who were neither college students nor college graduates. The 412 students in the study ate breakfast a mean of 4.1 times per week, with a standard deviation of 2.4. (For this exercise, again imagine that this is the entire population of interest.)

a. Approximately, what is the percentile for a student who eats breakfast four times per week?

b. Approximately, what is the percentile for a student who eats breakfast six times per week?

c. Approximately, what is the percentile for a student who eats breakfast twice a week?

6.41 z scores and comparisons of sports teams: A common quandary faces sports fans who live in the same city but avidly follow different sports. How does one determine whose team did better with respect to its league division? In 2004, the Boston Red Sox won the World Series; just months later, their local football counterparts, the New England Patriots, won the Super Bowl. In 2005, both teams made the play-offs but lost early on. Which team was better in 2005? The question, then, is: Were the Red Sox better, as compared to other teams in the American League of Major League Baseball (MLB), than the Patriots, as compared to the other teams in the American Football Conference of the National Football League (NFL)? Some of us could debate this for hours, but it's better to examine some statistics. Let's operationalize performance over the season as the number of wins during regular season play.

a. In 2005, the mean number of wins for baseball teams in the American League was 81.71, with a standard deviation of 13.07. Because all teams were included, these are population parameters. The Red Sox won 95 games. What is their z score?

b. In 2005, the mean number of wins for football teams in the American Football Conference was 8.13, with a standard deviation of 3.70. The Patriots won 10 games. What is their z score?

c. Which team did better, according to these data?

d. How many games would the team with the lower z score have had to win to beat the team with the higher z score?

e. List at least one other way we could have operationalized the outcome variable (i.e., team performance).

6.42 z scores and comparisons of admiration ratings: Our statistics students were asked to rate their admiration of Hillary Rodham Clinton on a scale of 1 to 7. They also were asked to rate their admiration of actress, singer, and former *American Idol* judge Jennifer Lopez and their admiration of tennis player Venus Williams on a scale of 1 to 7. As noted earlier, the mean rating of Clinton was 4.06, with a standard deviation of 1.70. The mean rating of Lopez was 3.72, with a standard deviation of 1.90. The mean rating of Williams was 4.58, with a standard deviation of 1.46. One of our students rated her admiration of Clinton and Williams at 5 and her admiration of Lopez at 4.

a. What is the student's z score for her rating of Clinton?

b. What is the student's z score for her rating of Williams?

c. What is the student's z score for her rating of Lopez?

d. Compared to the other statistics students in our sample, which celebrity does this student most admire? (We can tell by her raw scores that she prefers Clinton and Williams to Lopez, but when we take into account the general perception of these celebrities, how does this student feel about them?)

e. How do z scores allow us to make comparisons that we cannot make with raw scores? That is, describe the benefits of standardization.

6.43 Raw scores, z scores, percentiles, and sports teams: Let's look at baseball and football again. We'll look at data for all of the teams in MLB and the NFL, respectively.

a. In 2005, the mean number of wins for MLB teams was 81.00, with a standard deviation of 10.83. The perennial underdogs, the Chicago Cubs, had a z score of -0.18. How many games did they win?

b. In 2005, the mean number of wins for all NFL teams was 8.00, with a standard deviation of 3.39. The New Orleans Saints had a z score of -1.475. How many games did they win?

c. The Pittsburgh Steelers were just below the 84th percentile in terms of NFL wins. How many games did they win? Explain how you obtained your answer.

d. Explain how you can examine your answers in parts (a), (b), and (c) to determine whether the numbers make sense.

6.44 Distributions and life expectancy: Researchers have reported that the projected life expectancy for people diagnosed with human immunodeficiency virus (HIV) and receiving antiretroviral therapy (ART) is 24.2 years (Schackman et al., 2006). Imagine that the researchers determined this by following 250 people with HIV who were receiving ART and calculating the mean.

a. What is the dependent variable of interest?

b. What is the population?

c. What is the sample?

d. For the population, describe what the distribution of *scores* would be.

e. For the population, describe what the distribution of *means* would be.

f. If the distribution of the population were skewed, would the distribution of scores likely be skewed or approximately normal? Explain your answer.

g. Would the distribution of means be skewed or approximately normal? Explain your answer.

6.45 Distributions and personality testing: The revised version of the Minnesota Multiphasic Personality Inventory (MMPI-2) is the most frequently administered self-report personality measure. Test-takers respond to more than 500 true/false statements, and their responses are scored, typically by a computer, on a number of scales (e.g., hypochondriasis, depression, psychopathic deviation). Respondents receive a T score on each scale that can be compared to norms. (You're likely to encounter T scores if you take psychology classes, but it's good to be aware that they're different from the t statistic that you will learn about in a few chapters.) T scores are another way to standardize scores so that percentiles and cutoffs can be determined. The mean T score is always 50, and the standard deviation is always 10. Imagine that you administer the MMPI-2 to 95 respondents who have recently lost a parent; you wonder whether their scores on the depression scale will be, on average, higher than the norms. You find a mean score on the depression scale of 55 in your sample.

a. Using symbolic notation, report the mean and standard deviation of the population.

b. Using symbolic notation and formulas (where appropriate), report the mean and standard error for the distribution of means to which your sample will be compared.

c. In your own words, explain why it makes sense that the standard error is smaller than the standard deviation.

6.46 z statistics and housing costs: Finding an affordable apartment wasn't easy 15–20 years ago, and it hasn't gotten any easier. A Web site for San Mateo County in California published extensive descriptive statistics from its 1998 Quality of Life Survey. The county reported that the mean housing payment (mortgage or rent) was \$1225.15, with a standard deviation of \$777.50. It also reported that the mean cost of an apartment rental, rather than a house rental or a mortgage, was \$868.86. For this exercise, treat the overall mean housing payment as a parameter, and treat the mean apartment rental cost as a statistic based on a sample of 100.

a. Using symbolic notation and formulas (where appropriate), determine the mean and the standard error for the distribution of means for the overall housing payment data.

b. Using symbolic notation and the formula, calculate the z statistic for the cost of an apartment rental.

c. Why is it likely that the z statistic is so large? (*Hint:* Is this distribution likely to be normal? Explain.)

d. Why is it permissible to use the normal curve percentages associated with the z distribution even though the data are not likely normally distributed?

6.47 Distributions and the General Social Survey: The General Social Survey (GSS) is a survey of approximately 2000 adults conducted each year since 1972, for a total of more than 38,000 participants. During several years of the GSS, participants were asked how many close friends they have. The mean for this variable is 7.44 friends, with a standard deviation of 10.98. The median is 5.00 and the mode is 4.00.

a. Are these data for a distribution of scores or a distribution of means? Explain.

b. What do the mean and standard deviation suggest about the shape of the distribution? (*Hint:* Compare the sizes of the mean and the standard deviation.)

c. What do the three measures of central tendency suggest about the shape of the distribution?

d. Let's say that these data represent the entire population. Pretend that you randomly selected a person from this population and asked how many close friends she or he had. Would you compare this person to a distribution of scores or a distribution of means? Explain your answer.

e. Now pretend that you randomly selected a sample of 80 people from this population. Would you compare this sample to a distribution of scores or to a distribution of means? Explain your answer.

f. Using symbolic notation, calculate the mean and standard error of the distribution of means.

g. What is the likely shape of the distribution of means? Explain your answer.

6.48 A distribution of scores and the General Social Survey: Refer to Exercise 6.47. Again, pretend that the GSS sample is the entire population of interest.

a. Imagine that you randomly selected one person from this population who reported that he had 18 close friends. Would you compare his score to a distribution of scores or to a distribution of means? Explain your answer.

b. What is his z score? Based on this z score, what is his approximate percentile?

c. Does it make sense to calculate a percentile for this person? Explain your answer. (*Hint:* Consider the shape of the distribution.)

6.49 A distribution of means and the General Social Survey: Refer to Exercise 6.47. Again, pretend that the GSS sample is the entire population of interest.

a. Imagine that you randomly selected 80 people from this population who had a mean of 8.7. Would you compare this sample mean to a distribution of scores or to a distribution of means? Explain your answer.

b. What is the z statistic for this mean? Based on this z statistic, what is the approximate percentile for this sample?

c. Does it make sense to calculate a percentile for this sample? Explain your answer. (*Hint:* Consider the shape of the distribution.)

6.50 Percentiles, raw scores, and credit card theft: Credit card companies will often call cardholders if the pattern of use indicates that the card might have been stolen. Let's say that you charge an average of $280 a month on your credit card, with a standard deviation of $75. The credit card company will call you anytime your purchases for the month exceed the 98th percentile. What is the dollar amount beyond which you'll get a call from your credit card company?

Putting It All Together

6.51 Probability and medical treatments: The three most common treatments for blocked coronary arteries are medication; bypass surgery; or angioplasty, which is a medical procedure that involves clearing out arteries and that leads to higher profits for doctors than do the other two procedures. The highest rate of angioplasty in the United States is in Elyria, a small city in Ohio. A newspaper article stated that "the statistics are so far off the charts—Medicare patients in Elyria receive angioplasties at a rate nearly four times the national average—that Medicare and at least one commercial insurer are starting to ask questions." The rate, in fact, is three times as high as that of Cleveland, Ohio, which is located just 30 miles from Elyria.

a. What is the population in this example? What is the sample?

b. How did probability play a role in the decision of Medicare and the commercial insurer to begin investigations?

c. How might the z distribution help the investigators to detect possible fraud in this case?

d. If the insurers determine that physicians in Elyria are committing fraud, but the insurers are wrong, what kind of error would they have made? Explain.

e. Does Elyria's extremely high percentile mean that the doctors in town are committing fraud? Cite two other possible reasons for Elyria's status as an outlier.

6.52 Rural friendships and the General Social Survey: Earlier, we considered data from the GSS on numbers of close friends people reported having. The mean for this variable is 7.44, with a standard deviation of 10.98. Let's say that you decide to use the GSS data to test whether people who live in rural areas have a different mean number of friends than does the overall GSS sample. Again, treat the overall GSS sample as the entire population of interest. Let's say that you select 40 people living in rural areas and find that they have an average of 3.9 friends.

a. What is the independent variable in this study? Is this variable nominal, ordinal, or scale?
b. What is the dependent variable in this study? Is this variable nominal, ordinal, or scale?
c. What is the null hypothesis for this study?
d. What is the research hypothesis for this study?
e. Would we compare the sample data to a distribution of scores or to a distribution of means? Explain.
f. Using symbolic notation and formulas, calculate the mean and standard error for the distribution of means.
g. Using symbolic notation and the formula, calculate the z statistic for this sample.
h. What is the approximate percentile for this sample?
i. Let's say that the researchers concluded that people in rural areas have fewer friends than does the general population (thus rejecting the null hypothesis). If they are incorrect, have they made a Type I or a Type II error? Explain.

6.53 Cheating on standardized tests: In their book *Freakonomics*, Levitt and Dubner (2005) describe alleged cheating among teachers in the Chicago Public School system. Certain classrooms had suspiciously strong performances on standardized tests that often mysteriously declined the following year when a new teacher taught the same students. In about 5% of classrooms studied, Levitt and other researchers found blocks of correct answers, among most students, for the last few questions, an indication that the teacher had changed responses to difficult questions for most students. Let's assume cheating in a given classroom if the overall standardized test score for the class showed a surprising change from one year to the next.

a. How are the researchers operationalizing the variable of cheating in this study? Is this a nominal, ordinal, or scale variable?
b. Explain how researchers can use the z distribution to catch cheating teachers.
c. How might a histogram or frequency polygon be useful to researchers aiming to catch cheating teachers?
d. If researchers falsely conclude that teachers are cheating, what kind of error would they be committing? Explain.

Terms

normal curve (p. 116)
standardization (p. 119)
z score (p. 120)
z distribution (p. 125)
standard normal distribution (p. 125)
central limit theorem (p. 129)
distribution of means (p. 129)
standard error (p. 132)

Formulas

$$z = \frac{(X - \mu)}{\sigma} \quad \text{(p. 122)}$$

$$X = z(\sigma) + \mu \quad \text{(p. 124)}$$

$$\sigma_M = \frac{\sigma}{\sqrt{N}} \quad \text{(p. 133)}$$

$$z = \frac{(M - \mu_M)}{\sigma_M} \quad \text{(p. 135)}$$

Symbols

z (p. 120)
μ_M (p. 132)
σ_M (p. 132)

CHAPTER 7

Hypothesis Testing with *z* Tests

BEFORE YOU GO ON

- You should understand how to calculate a *z* statistic for a distribution of scores and for a distribution of means (Chapter 6).
- You should understand that the *z* distribution allows us to determine the percentage of scores (or means) that fall below a particular *z* statistic (Chapter 6).

Experimental Design R. A. Fisher was inspired by the "lady drinking tea" who claimed that she could distinguish the taste of a cup of tea that had been poured tea first versus one that had been poured milk first. His book, *The Design of Experiments,* demonstrated how statistics become meaningful within the context of an experimental design.

When statistician R. A. Fisher offered a cup of tea to Dr. B. Muriel Bristol, the doctor politely declined, but for a strange reason. She preferred the taste of tea when the milk had been poured into the cup first.

"Nonsense. Surely it makes no difference," Fisher replied.

William Roach (who would later marry Dr. Bristol) suggested, "Let's test her." Roach poured cups of tea, some with tea first and others with milk first. But Fisher's mind was awhirl with statistical concerns about how many cups should be used, their order of presentation, and how to control chance variations in temperature or sweetness. The case of the lady drinking tea connected probability to experimental design and became the opening story in Fisher's classic textbook, *The Design of Experiments* (Fisher, 1935/1971).

Fisher's grand insight was that probability could be used to test a hypothesis. Dr. Bristol's chance hit rate for identifying a milk-first cup of tea should have been about 50%; because her rate was somewhere above 50%, we can declare that her tea-tasting ability was significantly different from the chance rate. Roach reportedly said, "Miss Bristol divined correctly more than enough of those cups into which tea had been poured first to prove her case" (Box, 1978, p. 134). But he was in love—so he might have been biased.

We begin our own adventures into the liberating power of hypothesis testing with the simplest hypothesis test, the z test. We learn how the z distribution and the z test make fair comparisons possible through standardization. Specifically, we learn:

1. How to use a z table
2. How to implement the basic steps of hypothesis testing
3. How to conduct a z test to compare a single sample to a known population

The *z* Table

In Chapter 6, we learned that (1) about 68% of scores fall within one z score of the mean, (2) about 96% of scores fall within two z scores of the mean, and (3) nearly all scores fall within three z scores of the mean. These guidelines are useful, but the table of z statistics and percentages is more specific. The z table is printed in its entirety in Appendix B, but an excerpt from it is reproduced in Table 7-1 for your convenience. In this section, we learn how to use the z table to calculate percentages when the z score is not a whole number.

Raw Scores, *z* Scores, and Percentages

Just as the same person might be called "Christina," "Christy," or "Tina," z scores are just one of three different ways to identify the same point beneath the normal curve: raw score, z score, and percentile ranking. The z table is how we transition from one way of naming a score to another. More importantly, the z table gives us a way to state and test hypotheses by standardizing different kinds of observations onto the same scale.

> **MASTERING THE CONCEPT**
>
> **7.1:** We can use the *z* table to look up the percentage of scores between the mean of the distribution and a given *z* statistic.

For example, we can determine the percentage associated with a given z statistic by following two steps.

Step 1. Convert a raw score into a z score.

Step 2. Look up a given z score on the z table to find the percentage of scores *between the mean and that z score.*

TABLE 7-1. Excerpt from the *z* Table

The *z* table provides the percentage of scores between the mean and a given *z* value. The full table includes positive *z* statistics from 0.00 to 4.50. The negative *z* statistics are not included because all we have to do is change the sign from positive to negative. Remember, the normal curve is symmetric: one side always mirrors the other.

z	% Between Mean and *z*
.	.
.	.
.	.
0.97	33.40
0.98	33.65
0.99	33.89
1.00	34.13
1.01	34.38
1.02	34.61
.	.
.	.
.	.

Note that the *z* scores displayed in the *z* table are all positive, but that is just to save space. The normal curve is symmetric, so negative *z* scores (any score below the mean) are the mirror image of positive *z* scores (any score above the mean) (Figure 7-1).

Here is an interesting way to learn how to use the *z* table: A research team (Sandberg, Bukowski, Fung, & Noll, 2004) wanted to know whether very short children tended to have poorer psychological adjustment than taller children and, therefore, should be treated with growth hormone. They categorized 15-year-old boys and girls into one of three groups—short (bottom 5%), average (middle 90%), or tall (top 5%)—based on published norms for a given age and gender (Centers for Disease Control National Center for Health Statistics, 2000; Sandberg et al., 2004). The mean height for 15-year-old boys was approximately 67.00 inches, with a standard deviation of 3.19. For 15-year-old girls, the mean height was approximately 63.80 inches, with a standard deviation of 2.66. We'll consider two 15-year-olds, one taller than average and the other shorter than average.

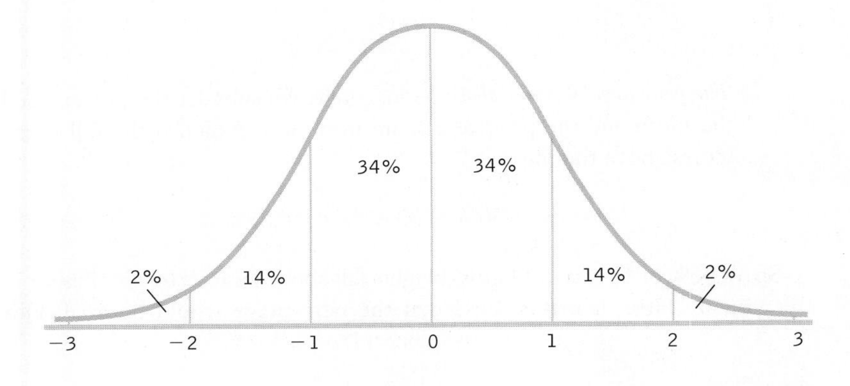

FIGURE 7-1
The Standardized *z* Distribution

We can use a *z* table to determine the percentages below and above a particular *z* score. For example, 34% of scores fall between the mean and a *z* score of 1.

EXAMPLE 7.1

Jessica is 66.41 inches tall (just over 5 feet, 6 inches).

STEP 1: Convert her raw score to a z score, as we learned how to do in Chapter 6.

We use the mean ($\mu = 63.80$) and standard deviation ($\sigma = 2.66$) for the heights of girls:

$$z = \frac{(X - \mu)}{\sigma} = \frac{(66.41 - 63.80)}{2.66} = 0.98$$

STEP 2: Look up 0.98 on the z table to find the associated percentage between the mean and Jessica's z score.

Once we know that the associated percentage is 33.65%, we can determine a number of percentages related to her z score. Here are three.

1. *Jessica's percentile rank, the percentage of scores below her score*: We add the percentage between the mean and the positive z score to 50%, which is the percentage of scores below the mean (50% of scores are on each side of the mean).

Jessica's percentile is 50% + 33.65% = 83.65%

Figure 7-2 shows this visually. As we can do when we are evaluating our calculations of z scores, we can run a quick mental check of the likely accuracy of the answer. We're interested in calculating the percentile of a *positive* z score. Because it is above the mean, we know that the answer must be higher than 50%. And it is.

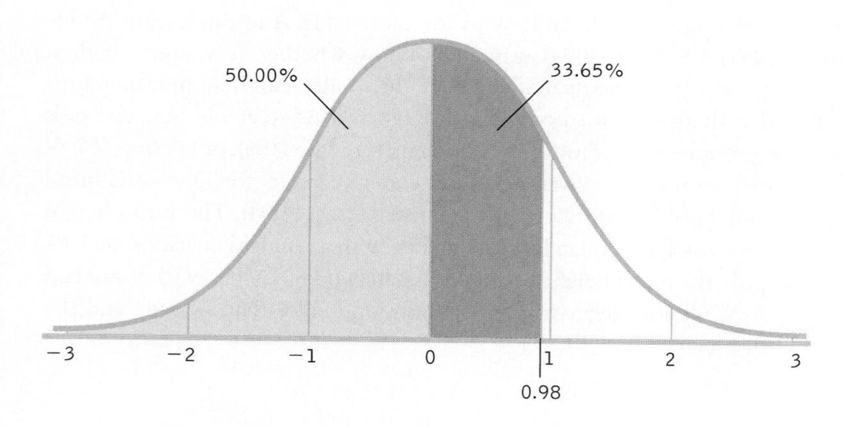

FIGURE 7-2
Calculating the Percentile for a Positive z Score

Drawing curves helps us to determine the appropriate percentage. For a positive z score, we add 50% to the percentage between the mean and that z score to get the total percentage below that z score, the percentile. Here, we add the 50% below the mean to the 33.65% between the mean and the z score of 0.98 to calculate the percentile, which ends up being 83.65%.

2. *The percentage of scores above Jessica's score*: We subtract the percentage between the mean and the positive z score from 50%, which is the full percentage of scores above the mean:

50% − 33.65% = 16.35%

So 16.35% of 15-year-old girls' heights fall above Jessica's height. Figure 7-3 shows this visually. Here, it makes sense that the percentage would be smaller than 50%;

because the z score is positive, we could not have more than 50% above it. As an alternative, a simpler method is to subtract Jessica's percentile rank of 83.35% from 100%. This gives us the same 16.35%. We could also look under the column in the z table labeled "in the tail."

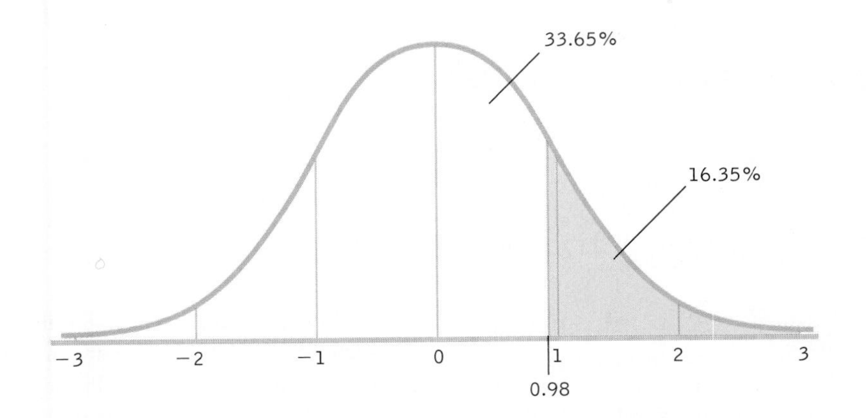

FIGURE 7-3
Calculating the Percentage Above a Positive *z* Score

For a positive *z* score, we subtract the percentage between the mean and that *z* score from 50% (the total percentage above the mean) to get the percentage above that *z* score. Here, we subtract the 33.65% between the mean and the *z* score of 0.98 from 50%, which yields 16.35%.

3. *The scores at least as extreme as Jessica's z score, in both directions*: When we begin hypothesis testing, it will be useful to know the percentage of scores that are at least as extreme as a given z score. In this case, 16.35% of heights are extreme enough to have z scores above Jessica's z score of 0.98. But remember that the curve is symmetric. This means that another 16.35% of the heights are extreme enough to be below a z score of −0.98. So we can double 16.35% to find the total percentage of heights that are as far as or farther from the mean than is Jessica's height:

$$16.35\% + 16.35\% = 32.70\%$$

Thus 32.7% of heights are at least as extreme as Jessica's height in either direction. Figure 7-4 shows this visually.

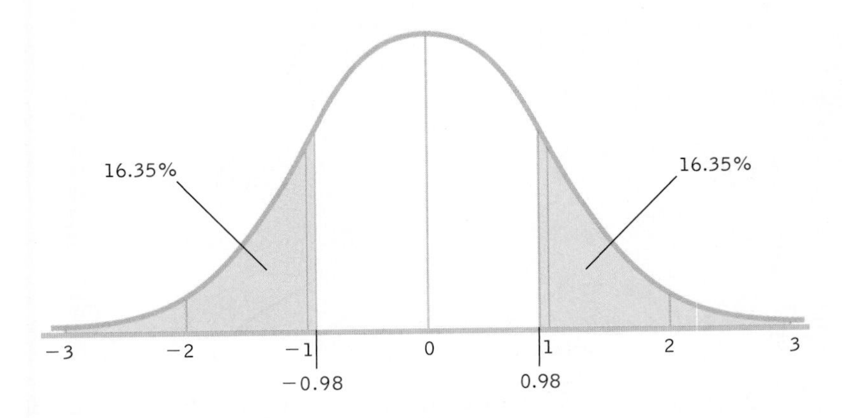

FIGURE 7-4
Calculating the Percentage at Least as Extreme as the *z* Score

For a positive *z* score, we double the percentage above that *z* score to get the percentage of scores that are at least as extreme—that is, at least as far from the mean—as the *z* score is. Here, we double 16.35% to calculate the percentage at least this extreme: 32.70%.

What group would Jessica fall in? Because 16.35% of 15-year-old girls are taller than Jessica, she is not in the top 5%. So she would be classified as average height according to the researchers' definition of *average*. ■

EXAMPLE 7.2

Now let's repeat this process for a score below the mean. Manuel is 61.20 inches tall (about 5 feet, 1 inch) so we want to know if Manuel can be classified as short. Remember, for boys the mean height is 67.00 inches, and the standard deviation for height is 3.19 inches.

STEP 1: Convert his raw score to a z score:

$$z = \frac{(X - \mu)}{\sigma} = \frac{(61.20 - 67.00)}{3.19} = -1.82$$

STEP 2: Calculate the percentile, the percentage above, and the percentage at least as extreme for the negative z score for Manuel's height.

We can again determine a number of percentages related to the z score; however, this time, we need to use the full table in Appendix B. The z table includes only positive z scores, so we look up 1.82 and find that the percentage between the mean and the z score is 46.56%. Of course, percentages are always positive, so don't add a negative sign here!

1. *Manuel's percentile score, the percentage of scores below his score*: For a negative z score, we subtract the percentage between the mean and the z score from 50% (which is the total percentage below the mean):

   Manuel's percentile is 50% − 46.56% = 3.44% (Figure 7-5).

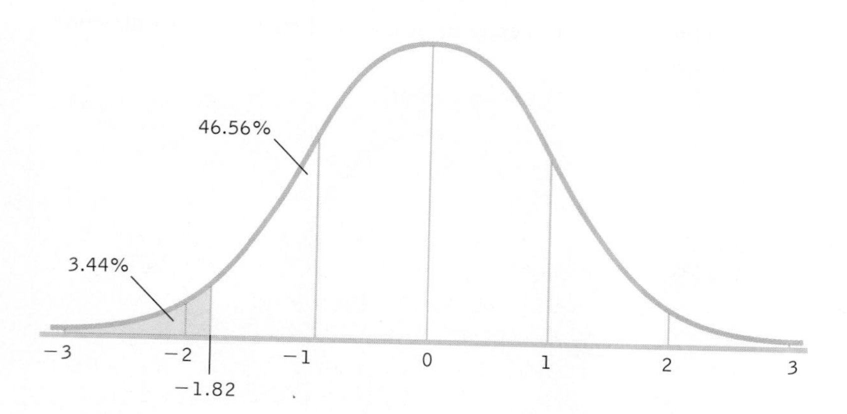

FIGURE 7-5
Calculating the Percentile for a Negative z Score

As with positive z scores, drawing curves helps us to determine the appropriate percentage for negative z scores. For a negative z score, we subtract the percentage between the mean and that z score from 50% (the percentage below the mean) to get the percentage below that negative z score, the percentile. Here we subtract the 46.56% between the mean and the z score of −1.82 from 50%, which yields 3.44%.

2. *The percentage of scores above Manuel's score*: We add the percentage between the mean and the negative z score to 50%, the percentage above the mean:

   50% + 46.56% = 96.56%

So 96.56% of 15-year-old boys' heights fall above Manuel's height (Figure 7-6).

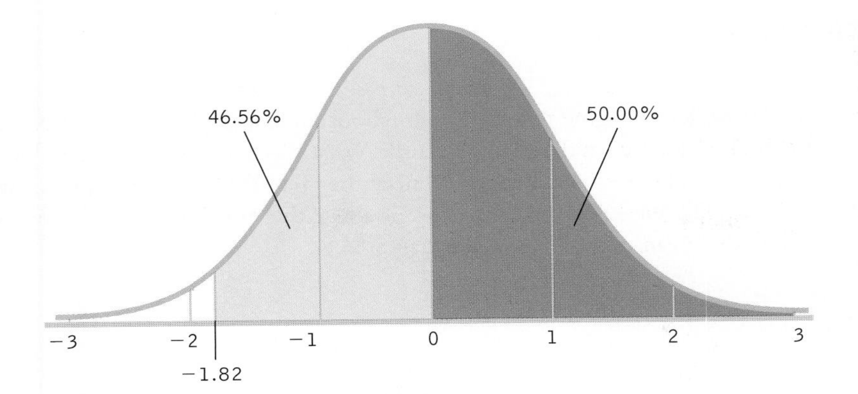

FIGURE 7-6
Calculating the Percentage Above a Negative *z* Score

For a negative *z* score, we add the percentage between the mean and that *z* score to 50% (the percentage above the mean) to get the percentage above that *z* score. Here we add the 46.56% between the mean and the *z* score of −1.82 to the 50% above the mean, which yields 96.56%.

3. *The scores at least as extreme as Manuel's z score, in both directions*: In this case, 3.44% of 15-year-old boys have heights that are extreme enough to have z scores below −1.82. And because the curve is symmetric, another 3.44% of heights are extreme enough to be above a z score of 1.82. So we can double 3.44% to find the total percentage of heights that are as far as or farther from the mean than is Manuel's height:

$$3.44\% + 3.44\% = 6.88\%$$

So 6.88% of heights are at least as extreme as Manuel's in either direction (Figure 7-7).

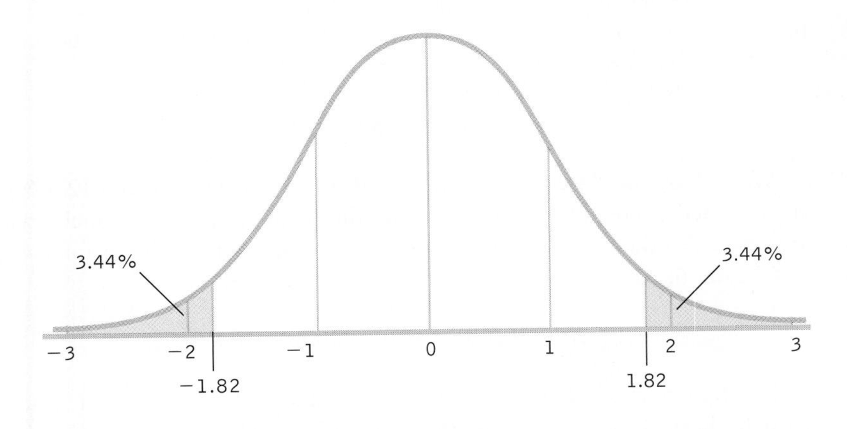

FIGURE 7-7
Calculating the Percentage at Least as Extreme as the *z* Score

With a negative *z* score, we double the percentage below that *z* score to get the percentage of scores that are at least as extreme—that is, at least as far from the mean—as the *z* score is. Here, we double 3.44% to calculate the percentage at least this extreme: 6.88%.

In what group would the researchers classify Manuel? Manuel has a percentile rank of 3.44%. He is in the lowest 5% of heights for boys of his age, so he would be classified as short. Now we can get to the question that drives this research. Does Manuel's short stature doom him to a life of few friends and poor social adjustment? Researchers compared the means of the three groups—short, average, and tall—on several measures of peer relations and social adjustment, but they did not find evidence of differences among the groups (Sandberg et al., 2004). ■

EXAMPLE 7.3

This example demonstrates (a) how to seamlessly shift among raw scores, z scores, and percentile ranks; and (b) why drawing a normal curve makes the calculations much easier to understand.

Many high school students in North America take the Scholastic Aptitude Test (SAT). The parameters for the SAT are set at a mean of 500 and a standard deviation of 100. So let's imagine that Jo, a high school student hoping to attend college, took the SAT and scored at the 63rd percentile. What was her raw score? Begin by drawing a curve, as in Figure 7-8. Then add a line at the point below which approximately 63% of scores fall. We know that this score is above the mean because 50% of scores fall below the mean, and 63% is larger than 50%.

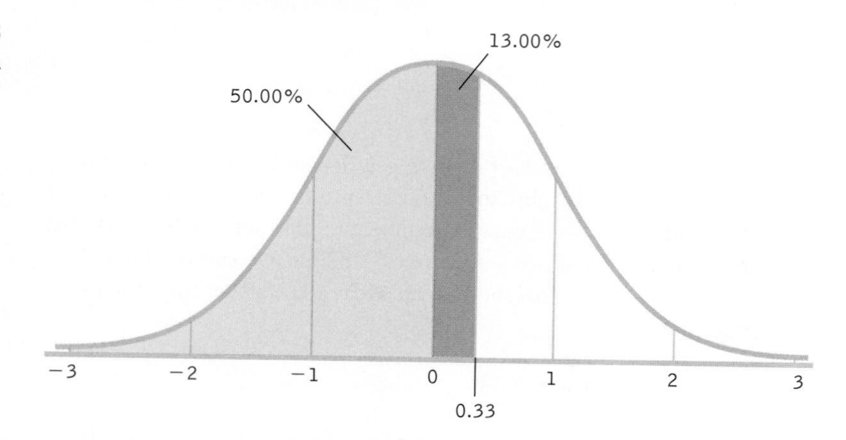

FIGURE 7-8
Calculating a Score from a Percentile

We convert a percentile to a raw score by calculating the percentage between the mean and the *z* score, and looking up that percentage on the *z* table to find the associated *z* score. We then convert the *z* score to a raw score using the formula. Here, we look up 13.00% on the *z* table (12.93% is the closest percentage) and find a *z* score of 0.33, which we can convert to a raw score.

Using the drawing as a guideline, we see that we have to calculate the percentage between the mean and the z score of interest. So, we subtract the 50% below the mean from Jo's score, 63%:

$$63\% - 50\% = 13\%$$

We look up the closest percentage to 13% in the z table (which is 12.93%) and find an associated z score of 0.33. This is above the mean, so we do not label it with a negative sign. We then convert the z score to a raw score using the formula we learned in Chapter 6:

$$X = z(\sigma) + \mu = 0.33(100) + 500 = 533$$

Jo, whose SAT score was at the 63rd percentile, had a raw score of 533. Double check! This score is above the mean of 500, and the percentage is above 50%. ■

The *z* Table and Distributions of Means

Let's shift our focus from the z score of the individual within a group to the group itself. We will use means rather than individual scores because we are now studying a sample of many scores rather than studying one individual score. Fortunately, the z table can also be used to determine percentages and z statistics for distributions of means calculated from many people. The only addition is that we need to calculate the mean and the standard error for the distribution of means before taking that information to the z table.

EXAMPLE 7.4

Many psychology graduate programs require that applicants take the Graduate Record Exam (GRE) subject test in psychology. The test has been used for many years, so we know the actual population mean and standard deviation; for example, the mean was

554 and the standard deviation was 99 for the years 1995 to 1998, numbers that we will treat as population parameters for this example (Matlin & Kalat, 2001). *z* statistics make it possible to compare psychology students at our institution to that known standard. We record the psychology test scores of a representative sample, 90 graduating seniors in our psychology department, and find that the mean score is 568. Before we calculate the *z* statistic, let's use proper symbolic notation to indicate the mean and the standard error of this distribution of means:

$$\mu_M = \mu = 554$$

$$\sigma_M = \frac{\sigma}{\sqrt{N}} = \frac{99}{\sqrt{90}} = 10.436$$

At this point, we have all the information we need to calculate the percentage using the two steps we learned earlier.

STEP 1: We convert to a *z* statistic using the mean and standard error that we just calculated.

$$z = \frac{(M - \mu_M)}{\sigma_M} = \frac{(568 - 554)}{10.436} = 1.34$$

STEP 2: We determine the percentage below this *z* statistic.

Draw it! Draw a curve that includes the mean of the *z* distribution, 0, and this *z* statistic, 1.34 (Figure 7-9). Then shade the area in which we are interested: everything below 1.34. Now we look up the percentage between the mean and the *z* statistic of 1.34. The *z* table indicates that this percentage is 40.99, which we write in the section of the curve between the mean and 1.34. We write 50% in the half of the curve below the mean. We add 40.99% to the 50% below the mean to get the percentile rank, 90.99%. (Subtracting from 100%, only 9.01% of mean scores would be higher than the mean if they come from this population.) Based on this percentage, the mean GRE psychology test score of the sample is quite high. But we still can't arrive at a conclusion about whether our students are performing above the national average of 554 until we conduct a hypothesis test. ■

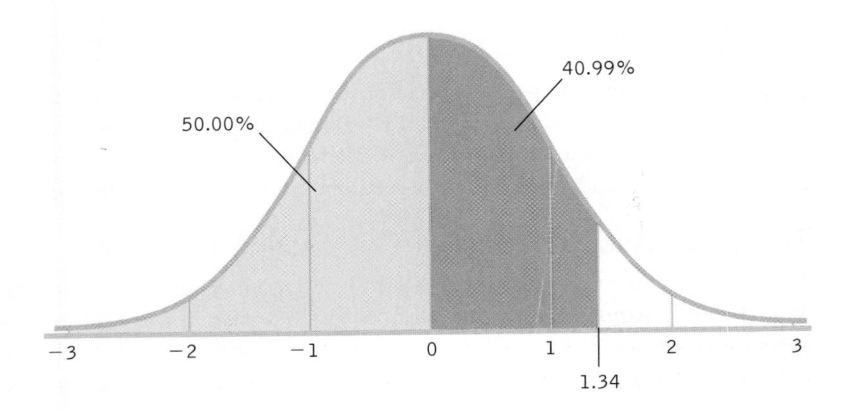

FIGURE 7-9
Percentile for the Mean of a Sample

We can use the *z* table with sample means in addition to sample scores. The only difference is that we use the mean and standard error of the distribution of means rather than the distribution of scores. Here, the *z* score of 1.34 is associated with a percentage of 40.99% between the mean and *z* score. Added to the 50% below the mean, the percentile is 50% + 40.99% = 90.99%.

In the next section, we learn (a) the assumptions for conducting hypothesis testing; (b) the six steps of hypothesis testing using the z distribution; and (c) whether to reject, or fail to reject, the null hypothesis. In the last section, we demonstrate a z test.

CHECK YOUR LEARNING

Reviewing the Concepts

> Raw scores, z scores, and percentile rankings are three ways to describe the same score within a normal distribution.

> If we know the mean and the standard deviation of a population, we can convert a raw score to a z score and then use the z table to determine percentages below, above, or at least as extreme as this z score.

> We can use the z table in reverse as well, taking a percentage and converting it into a z score and then a raw score.

> These same conversions can be conducted on a sample mean instead of on a score. The procedures are identical, but we use the mean and the standard error of the distribution of means, instead of the distribution of scores.

Clarifying the Concepts

7-1 What information do we need to know about a population of interest in order to use the z table?

7-2 How do z scores relate to raw scores and percentile ranks?

Calculating the Statistics

7-3 If the percentage of scores between a z score of 1.37 and the mean is 41.47%, what percentage of scores lies between -1.37 and the mean?

7-4 If 12.93% of scores fall between the mean and a z score of 0.33, what percentage of scores falls below this z score?

Applying the Concepts

7-5 Every year, the Educational Testing Service (ETS) administers the Major Field Test in Psychology (MFTP) to graduating psychology majors. In 2003, Baylor University wondered how its students compared to the national average. On its Web site, Baylor reported that the mean and the standard deviation of the 18,073 U.S. students who took this exam were 156.8 and 14.6, respectively. Thirty-six students in Baylor's Psychology and Neuroscience Department took the exam; these students had a mean score of 164.6.

a. What is the percentile rank for the sample of students at Baylor? Use symbolic notation and write out your calculations.

b. What percentage of samples of this size scored higher than the students at Baylor?

c. What can you say about how Baylor students compared to students across the nation?

Solutions to these Check Your Learning questions can be found in Appendix D.

The Assumptions and the Steps of Hypothesis Testing

The story of the lady tasting tea was an informal experiment; however, the formal process of hypothesis testing is based on particular assumptions about the data. At times it might be safe to violate those assumptions and proceed through the six steps of formal hypothesis testing, but an understanding of the assumptions is essential to making that call.

The Three Assumptions for Conducting Analyses

Think of "statistical assumptions" as the ideal conditions for hypothesis testing. More formally, ***assumptions*** *are the characteristics that we ideally require the population from which we are sampling to have so that we can make accurate inferences.* Why go through all the effort to understand and calculate statistics if you can't believe the story they tell?

The assumptions for the *z* test apply to several other hypothesis tests, especially ***parametric tests***, *inferential statistical analyses based on a set of assumptions about the population*. By contrast, ***nonparametric tests*** *are inferential statistical analyses that are not based on a set of assumptions about the population*. The goal is to match the appropriate statistical test with the characteristics of the data by learning the three main assumptions for parametric tests:

Assumption 1. *The dependent variable is assessed using a scale measure.* If it's clear that the variable is nominal or ordinal, we should not make this assumption and should not use a parametric hypothesis test.

Assumption 2. *The participants are randomly selected.* Every member of the population of interest must have had an equal chance of being selected for the study. This assumption is often violated; it is more likely that participants are a convenience sample. Violating this assumption means that we must be cautious when generalizing from a sample to the population.

Assumption 3. *The distribution of the population of interest must be approximately normal.* Many distributions are approximately normal, but it is important to remember that there are exceptions to this guideline (Micceri, 1989). Because hypothesis tests deal with sample means rather than individual scores, as long as the sample size is at least 30 (in most cases, based on the central limit theorem), it is likely that this assumption is met.

- An **assumption** is a characteristic that we ideally require the population from which we are sampling to have so that we can make accurate inferences.
- A **parametric test** is an inferential statistical analysis based on a set of assumptions about the population.
- A **nonparametric test** is an inferential statistical analysis that is not based on a set of assumptions about the population.
- A **robust** hypothesis test is one that produces fairly accurate results even when the data suggest that the population might not meet some of the assumptions.

Many parametric hypothesis tests can be conducted even if some of the assumptions are not met (Table 7-2), and are robust against violations of some of these assumptions. ***Robust*** *hypothesis tests are those that produce fairly accurate results even when the data suggest that the population might not meet some of the assumptions.*

These three statistical assumptions represent the ideal conditions and are more likely to produce valid research. *Meeting the assumptions improves the quality of research, but not meeting the assumptions doesn't necessarily invalidate research.*

MASTERING THE CONCEPT

7.2: When we calculate a parametric statistic, ideally we have met assumptions regarding the population distribution. For a *z* test, there are three assumptions: the dependent variable should be on a scale measure, the sample should be randomly selected, and the underlying population should have an approximately normal distribution.

The Six Steps of Hypothesis Testing

Hypothesis testing can be broken down into six standard steps.

Step 1: Identify the populations, comparison distribution, and assumptions.

When we first approach hypothesis testing, we consider the characteristics of the data in order to determine the distribution to which we will compare the sample. First, we state the populations represented

TABLE 7-2. The Three Assumptions for Hypothesis Testing

We must be aware of the assumptions for the hypothesis test that we choose, and we must be cautious in choosing to proceed with a hypothesis test even though the data may not meet all of the assumptions. Note that in addition to these three assumptions, for many hypothesis tests, including the *z* test, the independent variable must be nominal.

The Three Assumptions	Breaking the Assumptions
1. Dependent variable is on a scale measure.	Usually OK if the data are not clearly nominal or ordinal.
2. Participants are randomly selected.	OK if we are cautious about generalizing.
3. Population distribution is approximately normal.	OK if the sample includes at least 30 scores.

- A **critical value** is a test statistic value beyond which we reject the null hypothesis; often called a *cutoff.*
- The **critical region** refers to the area in the tails of the comparison distribution in which the null hypothesis can be rejected.
- The probability used to determine the critical values, or cutoffs, in hypothesis testing is a ***p* level**; often called ***alpha***.
- A finding is **statistically significant** if the data differ from what we would expect by chance if there were, in fact, no actual difference.

by the groups to be compared. Then we identify the comparison distribution (e.g., distribution of means). Finally, we identify the hypothesis test and its assumptions (see Appendix Figure E-1 for a quick guide to choosing the appropriate test).

Step 2: State the null and research hypotheses.

Hypotheses are about populations, *not* about samples. The null hypothesis is usually the "boring" one that posits no change or no difference between groups. The research hypothesis is usually the "exciting" one positing that, for example, a given intervention will lead to a change or a difference—for instance, that a particular kind of psychotherapeutic intervention will reduce general anxiety. State the null and research hypotheses in both words and symbolic notation.

Step 3: Determine the characteristics of the comparison distribution.

State the relevant characteristics of the comparison distribution (the distribution based on the null hypothesis). In a later step, we will compare data from the sample (or samples) to the comparison distribution to determine how extreme the sample data are. For z tests, we will determine the mean and standard error of the comparison distribution. These numbers describe the distribution represented by the null hypothesis and will be used when we calculate the test statistic.

Step 4: Determine critical values, or cutoffs.

The critical values, or cutoffs, of the comparison distribution indicate how extreme the data must be, in terms of the z statistic, to reject the null hypothesis. Often called simply *cutoffs,* these numbers are more formally called ***critical values**, the test statistic values beyond which we reject the null hypothesis.* In most cases, we determine two cutoffs, one for extreme samples below the mean and one for extreme samples above the mean.

The critical values, or cutoffs, are based on a somewhat arbitrary standard—the most extreme 5% of the comparison distribution curve: 2.5% on either end. At times, cutoffs are based on a less conservative percentage, such as 10%, or a more conservative percentage, such as 1%. Regardless of the chosen cutoff, the area beyond the cutoff, or critical value, is often referred to as the critical region. Specifically, *the **critical region** refers to the area in the tails of the comparison distribution in which the null hypothesis can be rejected.*

These percentages are typically written as probabilities; that is, 5% would be written as 0.05. *The probabilities used to determine the critical values, or cutoffs, in hypothesis testing are **p levels** (also often called **alphas**).*

Step 5: Calculate the test statistic.

We use the information from step 3 to calculate the test statistic, in this case the z statistic. We can then directly compare the test statistic to the critical values to determine if the sample is extreme enough to warrant a rejection of the null hypothesis.

Step 6: Make a decision.

Based on the statistical evidence, we can now decide whether to reject or fail to reject the null hypothesis. Based on the available evidence, we either reject the null hypothesis if the test statistic is beyond the cutoffs, or we fail to reject the null hypothesis if the test statistic is not beyond the cutoffs.

These six steps of hypothesis testing are summarized in Table 7-3.

When we reject the null hypothesis, we often refer to the results as "statistically significant." A finding is ***statistically significant*** *if the data differ from what we would expect*

TABLE 7-3. The Six Steps of Hypothesis Testing

We use the same six basic steps with each type of hypothesis test.

1. Identify the populations, distribution, and assumptions, and then choose the appropriate hypothesis test.
2. State the null and research hypotheses in both words and symbolic notation.
3. Determine the characteristics of the comparison distribution.
4. Determine the critical values, or cutoffs, that indicate the points beyond which we will reject the null hypothesis.
5. Calculate the test statistic.
6. Decide whether to reject or fail to reject the null hypothesis.

by chance if there were, in fact, no actual difference. The word *significant* is another one of those statistical terms with a very particular meaning. The phrase *statistically significant* does not necessarily mean that the finding is important or meaningful. A small difference between means could be statistically significant but not practically significant or important.

CHECK YOUR LEARNING

Reviewing the Concepts

> When we conduct hypothesis testing, we have to consider the assumptions for that particular test.

> Parametric statistics are those that are based on assumptions about the population distribution; nonparametric statistics have no such assumptions. Parametric statistics are often robust to violations of the assumptions.

> The three assumptions for a z test are that the dependent variable is on a scale measure, the sample is randomly selected, and the underlying population distribution is approximately normal.

> There are six standard steps for hypothesis testing. First, we identify the population, comparison distribution, hypothesis test, and assumptions. Second, we state the null and research hypotheses. Third, we determine the characteristics of the comparison distribution. Fourth, we determine the critical values, or cutoffs, on the comparison distribution. Fifth, we calculate the test statistic. Sixth, we decide whether to reject or fail to reject the null hypothesis.

> The standard practice of statisticians is to consider scores that occur less than 5% of the time based on the null hypothesis as statistically significant, warranting rejection of the null hypothesis; observations that occur more often than 5% of the time do not support this decision, and thus we would fail to reject the null hypothesis in these cases.

Clarifying the Concepts

7-6 Explain the three assumptions made for most parametric hypothesis tests.

7-7 How do critical values help us to make a decision about the hypothesis?

Calculating the Statistics

7-8 If a researcher always sets the critical region as 8% of the distribution, and the null hypothesis is true, how often will he reject the null hypothesis?

7-9 Rewrite each of these percentages as a probability, or p level: a. 15%; b. 3%; c. 5.5%.

Applying the Concepts

7-10 For each of the following scenarios, state whether each of the three basic assumptions for parametric hypothesis tests is met. Explain your answers and label the three assumptions (1) through (3).

a. Researchers compared the ability of experienced, clinical psychologists versus clinical psychology graduate students to diagnose a patient based on a 1-hour interview. For 2 months, either a psychologist or a student interviewed every outpatient at the local community mental health center who had already received diagnoses based on a number of criteria. For each diagnosis, the psychologists and graduate students were given a score of correct or incorrect.

b. Behavioral scientists wondered whether animals raised in captivity would be healthier with diminished human contact. Twenty large cats (e.g., lions, tigers) were randomly selected from all the wild cats living in zoos in North America. Half were assigned to the control group—no change in human interaction. Half were assigned to the experimental group—no humans entered their cages except when the animals were not in them, one-way mirrors were used so that the animals could not see zoo visitors, and so on. The animals received a score for health over 1 year; points were given for various illnesses; very few sickly animals had extremely high scores.

Solutions to these Check Your Learning questions can be found in Appendix D.

An Example of the *z* Test

The story of the lady tasting tea inspired statisticians to use hypothesis testing as a way to understand the many mysteries of human behavior. In this next section, we apply what we've learned about hypothesis testing—including the six steps—to a specific example of a *z* test. (We should note that *z* tests are rarely used in actual behavioral science research because it is very uncommon for us to have one sample and know both the mean and the standard deviation of the population.)

EXAMPLE 7.5

Under Mayor Michael Bloomberg, New York City developed legislation targeted at public health issues, such as a 2003 ban on smoking in restaurants and bars. In 2008, New York became the first U.S. city to require that chain restaurants post calorie counts for all menu items.

The research team of Bollinger, Leslie, and Sorenson (2010) wanted to test the law's effectiveness, so for over a year they gathered data on every transaction at Starbucks coffee shops in several U.S. cities. They determined a population mean of 247 calories in products purchased by customers at stores without calorie postings. Based on the range of 0 to 1208 calories, we estimate a standard deviation of approximately 201 calories, which we'll use as the population standard deviation for this example.

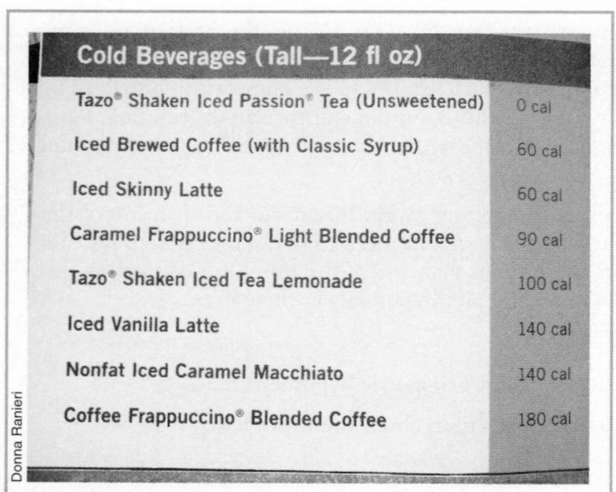

The *z* Test and Starbucks *z* tests are conducted in the rare cases in which we have one sample and we know both the mean and the standard deviation of the population. Do people consume fewer mean calories when they know exactly how many calories are in their favorite latte and muffin? The *z* test allows us to compare average numbers of calories consumed by customers at Starbucks that have calorie counts posted on their menus with average numbers of calories consumed by customers at Starbucks without posted calories.

The researchers also recorded calories for a sample in New York City after calories were posted on Starbucks menus. They reported a mean of 232 calories per purchase, a decrease of 6%. For the purposes of this example, we'll assume a sample size of 1000. Here's how to apply hypothesis testing when comparing a sample of customers at Starbucks with calories posted on their menus to the general population of customers at Starbucks without calories posted on their menus.

We'll use the six steps of hypothesis testing to analyze the calorie data. These six steps will tell us if customers visiting a Starbucks with calories listed on the menu consume fewer calories, on average, than customers visiting a Starbucks without calories listed on the menu. In fact, we use the six-step approach so often in this book that it won't be long before it becomes an automatic way of thinking for you. Each step in the example below is followed by a summary that models how to report hypothesis tests.

STEP 1: Identify the populations, distribution, and assumptions.

First, we identify the populations, comparison distribution, hypothesis test, and assumptions. The *populations* are (1) all customers at those Starbucks with calories posted on the menu (whether or not the customers are in the sample) and (2) all customers at those Starbucks without calories posted on the menu. Because we are studying a sample rather than an individual, the *comparison distribution* is a distribution of means. We compare the mean of the sample of 1000 people visiting Starbucks that have calories posted on the menu (selected from the population of all people visiting those Starbucks with calories posted) to a distribution of all possible means of samples of 1000 people (also selected from the population of all people visiting Starbucks that don't have calories posted on the menu). The *hypothesis test* will be a *z* test because we have only one sample and we know the mean and the standard deviation of the population from the published norms.

Now let's examine the *assumptions* for a z test. (1) The data are on a scale measure, calories. (2) We do not know whether sample participants were selected randomly from among all people visiting those Starbucks with calories posted on the menu. If they were not, this limits the ability to generalize beyond this sample to other Starbucks customers. (3) The comparison distribution should be normal. The individual data points are likely to be positively skewed because the minimum score of 0 is much closer to the mean of 247 than it is to the maximum score of 1208. However, we have a sample size of 1000, which is greater than 30; so based on the central limit theorem, we know that the comparison distribution—the distribution of means—will be approximately normal.

Summary: Population 1: All customers at the Starbucks that have calories posted on the menu. Population 2: All customers at the Starbucks that don't have calories posted on the menu.

The comparison distribution will be a distribution of means. The hypothesis test will be a z test because we have only one sample and we know the population mean and standard deviation. This study meets two of the three assumptions and may meet the third. The dependent variable is scale. In addition, there are more than 30 participants in the sample, indicating that the comparison distribution will be normal. We do not know whether the sample was randomly selected, however, so we must be cautious when generalizing.

STEP 2: State the null and research hypotheses.

Next we state the null and research hypotheses in words and in symbols. Remember, hypotheses are always about populations, not samples. In most forms of hypothesis testing, there are two possible sets of hypotheses: directional (predicting either an increase or a decrease, but not both) or nondirectional (predicting a difference in either direction).

The first possible set of hypotheses is directional. The null hypothesis is that customers at Starbucks that have calories posted on the menu do *not* consume fewer mean calories than customers at Starbucks that don't have calories posted on the menu; in other words, they could have the same or higher mean calories, but not lower. The research hypothesis is that customers at Starbucks that have calories posted on the menu consume fewer mean calories than customers at Starbucks that don't have calories posted on the menu. (Note that the direction of the hypotheses could be reversed.)

The symbol for the null hypothesis is H_0. The symbol for the research hypothesis is H_1. Throughout this text, we use μ for the mean because hypotheses are about populations and their parameters, not about samples and their statistics. So, in symbolic notation, the hypotheses are:

$$H_0: \mu_1 \geq \mu_2$$

$$H_1: \mu_1 < \mu_2$$

For the null hypothesis, the symbolic notation says that the mean calories consumed by those in population 1, customers at Starbucks with calories posted on the menu, is not lower than the mean calories consumed by those in population 2, customers at Starbucks without calories posted on the menu. For the research hypothesis, the symbolic notation says that the mean calories consumed by those in population 1 is lower than the mean calories consumed by those in population 2.

This hypothesis test is considered a one-tailed test. *A* ***one-tailed test*** *is a hypothesis test in which the research hypothesis is directional, positing either a mean decrease or a mean increase in the dependent variable, but not both, as a result of the independent variable.* One-tailed tests are rarely seen in the research literature; they are used only when the researcher is absolutely certain that the effect cannot go in the other direction or the researcher would not be interested in the result if it did.

- A **one-tailed test** is a hypothesis test in which the research hypothesis is directional, positing either a mean decrease or a mean increase in the dependent variable, but not both, as a result of the independent variable.

A **two-tailed test** is a hypothesis test in which the research hypothesis does not indicate a direction of the mean difference or change in the dependent variable, but merely indicates that there will be a mean difference.

The second set of hypotheses is nondirectional. The null hypothesis states that customers at Starbucks with posted calories (whether in the sample or not) consume the same number of calories, on average, as customers at Starbucks without posted calories. The research hypothesis is that customers at Starbucks with posted calories (whether in the sample or not) consume a different average number of calories than do customers at Starbucks without posted calories. The means of the two populations are posited to be different, but neither mean is predicted to be lower or higher.

The hypotheses in symbols would be:

$$H_0: \mu_1 = \mu_2$$
$$H_1: \mu_1 \neq \mu_2$$

For the null hypothesis, the symbolic notation says that the mean number of calories consumed by those in population 1 is the same as the mean number of calories consumed by those in population 2. For the research hypothesis, the symbolic notation says that the mean number of calories consumed by those in population 1 is different from the mean number of calories consumed by those in population 2.

This hypothesis test is considered a two-tailed test. *A **two-tailed test** is a hypothesis test in which the research hypothesis does not indicate a direction of the mean difference or change in the dependent variable, but merely indicates that there will be a mean difference.* Two-tailed tests are much more common than are one-tailed tests. We will use two-tailed tests throughout this book unless we tell you otherwise. If a researcher expects a difference in a certain direction, he or she might have a one-tailed hypothesis; however, if the results are in the opposite direction, the researcher cannot then switch the direction of the hypothesis.

MASTERING THE CONCEPT

7.3: We conduct a one-tailed test if we have a directional hypothesis, such as that the sample will have a higher (or lower) mean than the population. We use a two-tailed test if we have a nondirectional hypothesis, such as that the sample will have a different mean than the population does.

Summary: Null hypothesis: Customers at those Starbucks that have calories posted on the menu consume the same number of calories, on average, as do customers at Starbucks that don't have calories posted on the menu—$H_0: \mu_1 = \mu_2$. Research hypothesis: Customers at Starbucks that have calories posted on the menu consume a different number of calories, on average, than do customers at Starbucks that don't have calories posted on the menu—$H_1: \mu_1 \neq \mu_2$.

STEP 3: Determine the characteristics of the comparison distribution.

Now we determine the characteristics that describe the distribution with which we will compare the sample. For z tests, we must know the mean and the standard error of the population of scores; the standard error for samples of this size is calculated from the standard deviation of the population of scores. Here, the population mean for the number of calories consumed by the general population of Starbucks customers is 247, and the standard deviation is 201. The sample size is 1000. Because we usually use a sample mean in hypothesis testing, rather than a single score, we must use the standard error of the mean instead of the population standard deviation (of the scores). The characteristics of the comparison distribution are determined as follows:

$$\mu_M = \mu = 247$$
$$\sigma_M = \frac{\sigma}{\sqrt{N}} = \frac{201}{\sqrt{1000}} = 6.356$$

Summary: $\mu_M = 247; \sigma_M = 6.356.$

STEP 4: Determine critical values, or cutoffs.

Next we determine critical values, or cutoffs, to which we can compare the test statistic. As stated previously, the research convention is to set the cutoffs to a p level of 0.05. For a two-tailed test, this indicates the most extreme 5%—that is, the 2.5% at the bottom of the comparison distribution and the 2.5% at the top. Because we will calculate a test statistic for the sample—specifically a z statistic—we will report cutoffs in terms of z statistics. We will use the z table to determine the scores for the top and bottom 2.5%.

We know that 50% of the curve falls above the mean, and we know 2.5% falls above the relevant z statistic. By subtracting (50% − 2.5% = 47.5%), we determine that 47.5% of the curve falls between the mean and the relevant z statistic. When we look up this percentage on the z table, we find a z statistic of 1.96. So the critical values are −1.96 and 1.96 (Figure 7-10).

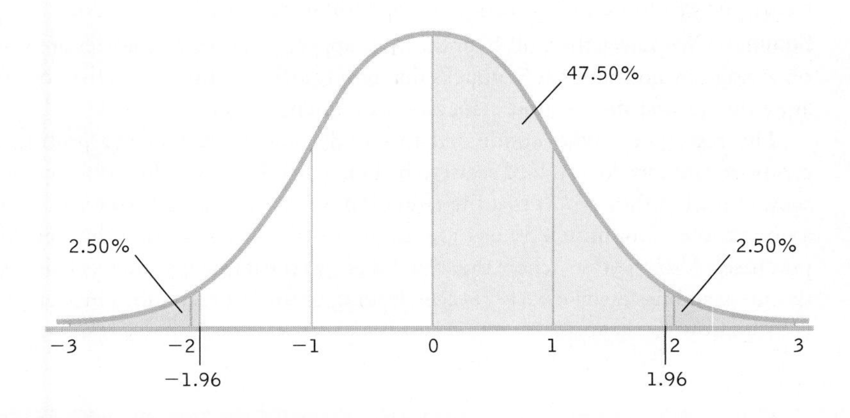

FIGURE 7-10
Determining Critical Values for a *z* Distribution
We typically determine critical values in terms of *z* statistics so that we can easily compare a test statistic to determine whether it is beyond the critical values. Here *z* scores of −1.96 and 1.96 indicate the most extreme 5% of the distribution, 2.5% in each tail.

Summary: The cutoff z statistics are −1.96 and 1.96.

STEP 5: Calculate the test statistic.

In step 5, we calculate the test statistic, in this case a z statistic, to find out what the data really say. We use the mean and standard error calculated in step 3:

$$z = \frac{(M - \mu_M)}{\sigma_M} = \frac{(232 - 247)}{6.356} = -2.36$$

Summary: $z = \frac{(232 - 247)}{6.356} = -2.36$

STEP 6: Make a decision.

Finally, we compare the test statistic to the critical values. We add the test statistic to the drawing of the curve that includes the critical z statistics (Figure 7-11). If the test statistic is in the critical region, we can reject the null hypothesis. In this example, the test statistic, −2.36, is in the critical region, so we reject the null hypothesis. An examination of the means tells us that the mean calories consumed by customers at Starbucks with calories posted on the menu is lower than the mean calories consumed by customers at Starbucks with no calories posted. So, even though we had nondirectional hypotheses, we can report the direction of the finding—that is, it appears that customers consume fewer calories, on average, at Starbucks that post calories on the menu than at Starbucks that do not post calories on the menu.

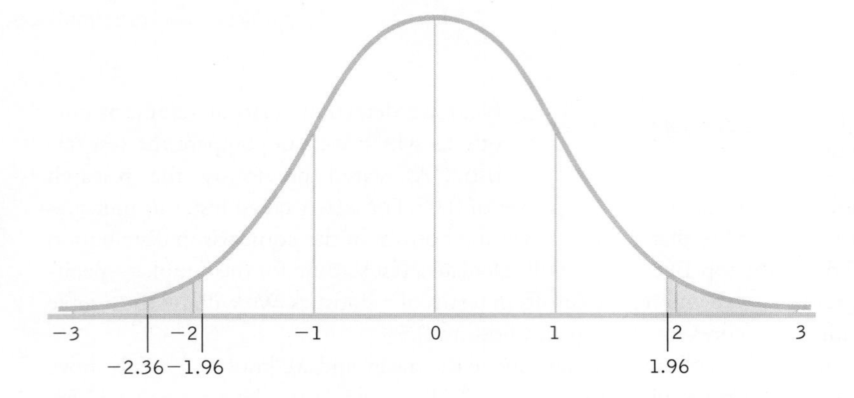

FIGURE 7-11
Making a Decision
To decide whether to reject the null hypothesis, we compare the test statistic to the critical values. In this instance, the *z* score of −2.36 is beyond the critical value of −1.96, so we reject the null hypothesis. Customers at Starbucks with calories posted on the menu consume fewer calories, on average, than do customers at Starbucks without posted calories.

If the test statistic is not beyond the cutoffs, we fail to reject the null hypothesis. This means that we can only conclude that there is no evidence from this study to support the research hypothesis. There might be a real mean difference that is not extreme enough to be picked up by the hypothesis test. We just can't know.

Summary: We reject the null hypothesis. It appears that fewer calories are consumed, on average, by customers at Starbucks that post calories on the menus than by customers at Starbucks that do not post calories on the menu.

The researchers who conducted this study concluded that the posting of calories by restaurants does indeed seem to be beneficial. The 6% reduction may seem small, they admit, but they report that the reduction was larger—a 26% decrease in calories—among those consuming 250 or more calories per visit and among those making food purchases. Also, the researchers theorized that given data such as these, chains might respond by adding lower-calorie choices, leading to further reductions in average calories consumed. ■

CHECK YOUR LEARNING

Reviewing the Concepts

> We conduct a *z* test when we have one sample and we know both the mean and the standard deviation of the population.
> We must decide whether to use a *one-tailed test,* in which the hypothesis is directional, or a *two-tailed test,* in which the hypothesis is nondirectional.
> One-tailed tests are rare in the research literature.

Clarifying the Concepts

7-11 What does it mean to say a test is directional or nondirectional?

Calculating the Statistics

7-12 Calculate the characteristics (μ_M and σ_M) of a comparison distribution for a sample mean based on 53 participants when the population has a mean of 1090 and a standard deviation of 87.

7-13 Calculate the *z* statistic for a sample mean of 1094 based on the sample of 53 people when $\mu = 1090$ and $\sigma = 87$.

Applying the Concepts

7-14 According to the Web site for the Coffee Research Institute, the average coffee drinker in the United States consumes 3.1 cups of coffee daily. Let's assume the population standard deviation is 0.9 cup. Jillian decides to study coffee consumption at her local coffee shop. She wants to know if people sitting and working in a coffee shop will drink a different amount of coffee from what might be expected in the general U.S. population. Throughout the course of 2 weeks, she collects data on 34 people who spend most of the day at the coffee shop. The average number of cups consumed by this sample is 3.17 cups. Use the six steps of hypothesis testing to determine whether Jillian's sample is statistically significantly different from the population mean.

Solutions to these Check Your Learning questions can be found in Appendix D.

REVIEW OF CONCEPTS

The z Table

The z table has several uses when we have normally distributed data. If we know an individual raw score, we can convert it to a z statistic and then determine percentages above, below, or at least as extreme as this score. Alternatively, if we know a percentage, we can look up a z statistic on the table and then convert it to a raw score. The table can be used in the same way with means instead of scores.

The Assumptions and the Steps of Hypothesis Testing

Assumptions are the criteria that are met, ideally, before a hypothesis test is conducted. *Parametric tests* are those that require assumptions about the population, whereas *nonparametric tests* are those that do not. Three basic assumptions apply to many parametric hypothesis tests—the dependent variable should be on a scale measure, the data should be from a randomly selected sample, and the population distribution should be normal (or there should be at least 30 scores in the sample). A *robust* hypothesis test is one that produces valid results even when all assumptions are not met.

There are six steps that apply to every hypothesis test. First, determine the populations, comparison distribution, appropriate hypothesis test, and assumptions. Second, state the null and research hypotheses. Third, determine the characteristics of the comparison distribution to be used to calculate the test statistic. Fourth, determine *critical values*, or *cutoffs*, usually based on a *p level*, or *alpha*, of 0.05, demarcating the most extreme 5% of the comparison distribution, the *critical region*. Fifth, calculate the test statistic. Sixth, use that test statistic to decide to reject or fail to reject the null hypothesis. A finding is deemed *statistically significant* when we reject the null hypothesis.

An Example of the z Test

z tests are conducted in the rare cases in which we have one sample and we know the mean and the standard deviation of the population. We must decide whether to use a *one-tailed test*, in which the hypotheses are directional, or a *two-tailed test*, in which the hypotheses are nondirectional.

SPSS®

SPSS can transform raw data from different scales into standardized data on one scale based on the z distribution, allowing us to look at standardized scores instead of raw scores. We can try this using the numbers of first- or second-place finishes for countries that participated in the World Cup from Chapter 2. The data are: 4, 8, 1, 2, 1, 2, 2, 6, 2, 2, 2, 2, 2, and 10. In addition, 63 countries had 0 first- or second-place finishes. Enter the 77 data points in one column in SPSS. We titled the column "top finishes."

To standardize the variable "top_finishes", select: Analyze → Descriptive Statistics → Descriptives. Select the relevant variable by clicking it on the left side, then clicking the arrow to move it to the right. Check the "Save standardized values as variables" box and click "OK." The new column of standardized variables under the heading "Z top_finishes" is in the sceenshot that follows.

To identify outliers that might skew the normal curve, select: Analyze → Descriptive Statistics → Explore → Statistics → Outliers, and then select the variable of interest. The part of the output that shows the most extreme scores as well as the column of z scores next to the raw scores is in the screenshot here.

SPSS data_World Cup finishes.sav [DataSet2] - SPSS Statistics Data Editor

File Edit View Data Transform Analyze Graphs Utilities Add-ons Window Help

78 : top_finishes

	top_finishes	Ztop_finishes	var	var	var	var	var	var	var
1	4.00	1.98854							
2	8.00	4.32620							
3	1.00	0.23528							
4	2.00	0.81970							
5	1.00	0.23528							
6	2.00	0.81970							
7	2.00	0.81970							
8	6.00	3.15737							
9	2.00	0.81970							
10	2.00	0.81970							
11	2.00	0.81970							
12	2.00	0.81970							
13	2.00	0.81970							
14	10.00	5.49504							
15	0.00	-0.34913							
16	0.00	-0.34913							
17	0.00	-0.34913							
18	0.00	-0.34913							
19	0.00	-0.34913							
20	0.00	-0.34913							
21	0.00	-0.34913							
22	0.00	-0.34913							
23	0.00	-0.34913							

*Output4 [Document4] - SPSS Statistics Viewer

File Edit View Data Transform Insert Format Analyze Graphs Utilities Add-or

Extreme Values

			Case Number	Value
top_finishes	Highest	1	14	10.00
		2	2	8.00
		3	8	6.00
		4	1	4.00
		5	4	2.00[a]
	Lowest	1	77	.00
		2	76	.00
		3	75	.00
		4	74	.00
		5	73	.00[b]

a. Only a partial list of cases with the value 2.00 are shown in the table of upper extremes.

b. Only a partial list of cases with the value .00 are shown in the table of lower extremes.

SPSS Statistics Processor

How It Works

7.1 TRANSITIONING FROM RAW SCORES TO *z* SCORES AND PERCENTILES

Physician assistants (PAs) are increasingly central to the health care system in many countries. Students who graduated from U.S. PA programs in 2004 reported their income (American Academy of Physician Assistants, 2005). The incomes of those working in emergency medicine had a mean of $76,553, a standard deviation of $14,001, and a median of $74,044. The incomes of those working in family/general medicine had a mean of $63,521, a standard deviation of $11,554, and a median of $62,935. How can we compare the income of Gabrielle, who earns $75,500 a year in emergency medicine, with that of Colin, who earns $64,300 a year in family/general medicine?

The *z* distribution should only be used with individual scores if the distribution is approximately normal, as seems to be the case here. For both distributions of incomes, the medians are relatively close to the means of their own distributions, suggesting that the distributions are not skewed. Additionally, the standard deviations are not large compared to the size of the respective means, which suggests that outliers are not inflating the standard deviation, which would indicate skew.

From the information we have, we can calculate Gabrielle's *z* score and her percentile on income—that is, the percentage of PAs working in emergency medicine who make less than she does. Her *z* score is:

$$z = \frac{(X - \mu)}{\sigma} = \frac{(75{,}500 - 76{,}553)}{14{,}001} = -0.08$$

The *z* table tells us that 3.19% of people fall between Gabrielle's income and the mean. Because her score is below the mean, we calculate 50% − 3.19% = 46.81%. Gabrielle's income is in the 46.81th percentile for PAs working in emergency medicine.

Colin's *z* score is:

$$z = \frac{(X - \mu)}{\sigma} = \frac{(64{,}300 - 63{,}521)}{11{,}554} = 0.07$$

The *z* table tells us that 2.79% of people fall between Colin's income and the mean. Because his score is above the mean, we calculate 50% + 2.79% = 52.79%. Colin's income is in the 52.79th percentile for PAs working in general medicine.

Relative to those in their chosen fields, Colin is doing better financially than Gabrielle. Colin's z score of 0.07, which is above the mean for general medicine PAs, is greater than Gabrielle's z score of -0.08, which is below the mean for emergency medicine PAs. Similarly, Colin's income is at about the 53rd percentile, whereas Gabrielle's income is at about the 47th percentile.

7.2 CONDUCTING A *z* TEST

Summary data from the Consideration of Future Consequences (CFC) scale found a mean CFC score of 3.51 with a standard deviation of 0.61 for a large sample (Petrocelli, 2003). (For the sake of this example, let's assume that Petrocelli's sample comprises the entire population of interest.) You wonder whether students who joined a career discussion group might have different CFC scores compared with those of the population. Forty-five students in your psychology department attended these discussion groups and then completed the CFC scale. The mean for this group is 3.7. From this information, how can we conduct all six steps of a two-tailed z test with a p level of 0.05?

Step 1: Population 1: All students who participated in career discussion groups. Population 2: All students who did not participate in career discussion groups.
The comparison distribution will be a distribution of means. The hypothesis test will be a z test because we have only one sample, and we know the population mean and standard deviation. This study meets two of the three assumptions but does not seem to meet the third. The dependent variable is on a scale measure. In addition, there are more than 30 participants in the sample, indicating that the comparison distribution will be normal. The data were not randomly selected, however, so we must be cautious when generalizing.

Step 2: Null hypothesis: Students who participated in career discussion groups had the same mean CFC scores as students who did not participate: $H_0: \mu_1 = \mu_2$. Research hypothesis: Students who participated in career discussion groups had different mean CFC scores from students who did not participate: $H_1: \mu_1 \neq \mu_2$.

Step 3: $\mu_M = \mu = 3.51; \sigma_M = \frac{\sigma}{\sqrt{N}} = \frac{0.61}{\sqrt{45}} = 0.091$

Step 4: The critical z statistics are -1.96 and 1.96.

Step 5: $z = \frac{(M - \mu_M)}{\sigma_M} = \frac{(3.7 - 3.51)}{0.091} = 2.09$

Step 6: Reject the null hypothesis. It appears that students who participate in career discussions have higher mean CFC scores than do students who do not participate.

Exercises

Clarifying the Concepts

7.1 What is a percentile?

7.2 When we look up a z score on the z table, what information can we report?

7.3 How do we calculate the percentage of scores below a particular positive z score?

7.4 How is calculating a percentile for a mean from a distribution of means different from doing so for a score from a distribution of scores?

7.5 In statistics, what do we mean by *assumptions*?

7.6 What sample size is recommended in order to meet the assumption of a normal distribution of means, even when the underlying population of scores is not normal?

7.7 What is the difference between parametric tests and nonparametric tests?

7.8 What are the six steps of hypothesis testing?

7.9 What are critical values and the critical region?

7.10 What is the standard size of the critical region used by most statisticians?

7.11 What does *statistically significant* mean to statisticians?

7.12 What do these symbolic expressions mean: $H_0: \mu_1 = \mu_2$ and $H_1: \mu_1 \neq \mu_2$?

7.13 Using everyday language rather than statistical language, explain why the words *critical region* might have been chosen to define the area in which a z statistic must fall in order to reject the null hypothesis.

7.14 Using everyday language rather than statistical language, explain why the word *cutoff* might have been chosen to define the point beyond which we reject the null hypothesis.

7.15 What is the difference between a one-tailed hypothesis test and a two-tailed hypothesis test in terms of critical regions?

7.16 Why do researchers typically use a two-tailed test rather than a one-tailed test?

Calculating the Statistic

7.17 Calculate the following percentages for a z score of 0.74, with a tail of 22.96%:

a. What percentage of scores falls below this z score?

b. What percentage of scores falls between the mean and this z score?

c. What proportion of scores falls below a z score of −0.74?

7.18 Using the z table in Appendix B, calculate the following percentages for a z score of −0.08:

a. Above this z score

b. Below this z score

c. At least as extreme as this z score

7.19 Using the z table in Appendix B, calculate the following percentages for a z score of 1.71:

a. Above this z score

b. Below this z score

c. At least as extreme as this z score

7.20 Rewrite each of the following percentages as probabilities, or p levels:

a. 5%

b. 83%

c. 51%

7.21 Rewrite each of the following probabilities, or p levels, as percentages:

a. 0.19

b. 0.04

c. 0.92

7.22 If the critical values for a hypothesis test occur where 2.5% of the distribution is in each tail, what are the cutoffs for z?

7.23 For each of the following p levels, what percentage of the data will be in each critical region for a two-tailed test?

a. 0.05

b. 0.10

c. 0.01

7.24 State the percentage of scores in a one-tailed critical region for each of the following p levels:

a. 0.05

b. 0.10

c. 0.01

7.25 If you are performing a z test on a sample of 50 people with an average SAT verbal score of 542 (assume we know the population mean to be 500 and the standard deviation to be 100), calculate the mean and the spread of the comparison distribution (μ_M and σ_M).

7.26 You are conducting a z test on a sample of 132 people for whom you observed a mean SAT verbal score of 490. The population mean is 500, and the standard deviation is 100. Calculate the mean and the spread of the comparison distribution (μ_M and σ_M).

7.27 If the cutoffs for a z test are −1.96 and 1.96, determine whether you would reject or fail to reject the null hypothesis in each of the following cases:

a. $z = 1.06$

b. $z = -2.06$

c. A z score beyond which 7% of the data fall in each tail

7.28 If the cutoffs for a z test are −2.58 and 2.58, determine whether you would reject or fail to reject the null hypothesis in each of the following cases:

a. $z = -0.94$

b. $z = 2.12$

c. A z score for which 49.6% of the data fall between z and the mean

7.29 Use the cutoffs of −1.65 and 1.65 and a p level of approximately 0.10, or 10%. For each of the following values, determine if you would reject or fail to reject the null hypothesis:

a. $z = 0.95$

b. $z = -1.77$

c. A z statistic that 2% of the scores fall above

7.30 You are conducting a z test on a sample for which you observe a mean weight of 150 pounds. The population mean is 160, and the standard deviation is 100.

a. Calculate a z statistic (μ_M and σ_M) for a sample of 30 people.

b. Repeat part (a) for a sample of 300 people.

c. Repeat part (a) for a sample of 3000 people.

Applying the Concepts

7.31 **Percentiles and unemployment rates:** The U.S. Bureau of Labor Statistics' annual report published in 2011 provided adjusted unemployment rates for 10 countries. The mean was 7, and the standard deviation was 1.85. For the following calculations, treat these as the population mean and standard deviation.

a. Australia's unemployment rate was 5.4. Calculate the percentile for Australia—that is, what percentage is less than that of Australia?

b. The United Kingdom's unemployment rate was 8.5. Calculate its percentile—that is, what percentage is less than that of the United Kingdom?

c. The unemployment rate in the United States was 8.9. Calculate its percentile—that is, what percentage is less than that of the United States?

d. The unemployment rate in Canada was 6.5. Calculate its percentile—that is, what percentage is less than that of Canada?

7.32 ***z* distribution and height:** Elena, a 15-year-old girl, is 58 inches tall. Based on what we know, the average height for girls at this age is 63.80 inches, with a standard deviation of 2.66 inches.

a. Calculate her *z* score.

b. What percentage of girls are taller than Elena?

c. What percentage of girls are shorter?

d. How much would she have to grow to be perfectly average?

e. If Sarah is in the 75th percentile for height at age 15, how tall is she? And how does she compare to Elena?

f. How much would Elena have to grow in order to be at the 75th percentile with Sarah?

7.33 **The *z* distribution and height:** Kona, a 15-year-old boy, is 72 inches tall. According to the CDC, the average height for boys at this age is 67.00 inches, with a standard deviation of 3.19 inches.

a. Calculate Kona's *z* score.

b. What is Kona's percentile score for height?

c. What percentage of boys this age are shorter than Kona?

d. What percentage of heights is at least as extreme as Kona's, in either direction?

e. If Ian is in the 30th percentile for height as a 15-year-old boy, how tall is he? How does he compare to Kona?

7.34 **The *z* statistic and height:** Imagine a class of thirty-three 15-year-old girls with an average height of 62.6 inches. Remember, $\mu = 63.8$ inches and $\sigma = 2.66$ inches.

a. Calculate the *z* statistic.

b. How does this sample of girls compare to the distribution of sample means?

c. What is the percentile rank for this sample?

7.35 **The *z* statistic and height:** Imagine a basketball team comprised of thirteen 15-year-old boys. The average height of the team is 69.5 inches. Remember, $\mu = 67$ inches and $\sigma = 3.19$ inches.

a. Calculate the *z* statistic.

b. How does this sample of boys compare to the distribution of sample means?

c. What is the percentile rank for this sample?

7.36 **The *z* distribution and statistics test scores:** Imagine that your statistics professor lost all records of students' raw scores on a recent test. However, she did record students' *z* scores for the test, as well as the class average of 41 out of 50 points and the standard deviation of 3 points (treat these as population parameters). She informs you that your *z* score was 1.10.

a. What was your percentile score on this test?

b. Using what you know about *z* scores and percentiles, how did you do on this test?

c. What was your original test score?

7.37 **The *z* statistic, distributions of means, and height:** Using what we know about the height of 15-year-old girls (again, $\mu = 63.8$ inches and $\sigma = 2.66$ inches), imagine that a teacher finds the average height of 14 female students in one of her classes to be 62.4 inches.

a. Calculate the mean and the standard error of the distribution of mean heights.

b. Calculate the *z* statistic for this group.

c. What percentage of mean heights, based on a sample size of 14 students, would we expect to be shorter than this group?

d. How often do mean heights equal to or more extreme than this size occur in this population?

e. If statisticians define sample means that occur less than 5% of the time as "special" or rare, what would you say about this result?

7.38 **The *z* statistic, distributions of means, and height:** Another teacher decides to average the height of all 15-year-old male students in his classes throughout the day. By the end of the day, he has measured the heights of 57 boys and calculated an average of 68.1 inches (remember, for this population $\mu = 67$ inches and $\sigma = 3.19$ inches).

a. Calculate the mean and the standard error of the distribution of mean heights.

b. Calculate the *z* statistic for this group.

c. What percentage of groups of people would we expect to have mean heights, based on samples of this size (57), taller than this group?

d. How often do mean heights equal to or more extreme than 68.1 occur in this population?

e. How does this result compare to the statistical significance cutoff of 5%?

7.39 **Directional versus nondirectional hypotheses:** For each of the following examples, identify whether the research has expressed a directional or a nondirectional hypothesis:

a. A researcher is interested in studying the relation between the use of antibacterial products and the dryness of people's skin. He thinks these products might alter the moisture in skin compared to other products that are not antibacterial.

b. A student wonders if grades in a class are in any way related to where a student sits in the classroom. In particular, do students who sit in the front row get better grades, on average, than the general population of students?

c. Cell phones are everywhere, and we are now available by phone almost all of the time. Does this translate into a change in the closeness of our long-distance relationships?

7.40 Null hypotheses and research hypotheses: For each of the following examples (the same as those in Exercise 7.39), state the null hypothesis and the research hypothesis, in both words and symbolic notation:

a. A researcher is interested in studying the relation between the use of antibacterial products and the dryness of people's skin. He thinks these products might alter the moisture in skin compared to other products that are not antibacterial.

b. A student wonders if grades in a class are in any way related to where a student sits in the classroom. In particular, do students who sit in the front row get better grades, on average, than the general population of students?

c. Cell phones are everywhere, and we are now available by phone almost all of the time. Does this translate into a change in the nature or closeness of our long-distance relationships?

7.41 The *z* distribution and Hurricane Katrina: Hurricane Katrina hit New Orleans on August 29, 2005. The National Weather Service Forecast Office maintains online archives of climate data for all U.S. cities and areas. These archives allow us to find out, for example, how the rainfall in New Orleans that August compared to the other months of 2005. The table below shows the National Weather Service data (rainfall in inches) for New Orleans in 2005.

Month	Rainfall
January	4.41
February	8.24
March	4.69
April	3.31
May	4.07
June	2.52
July	10.65
August	3.77
September	4.07
October	0.04
November	0.75
December	3.32

a. Calculate the *z* score for August. (*Note:* These are raw data for the population, rather than summaries, so you have to calculate the mean and the standard deviation first.)

b. What is the percentile for the rainfall in August? Does this surprise you? Explain.

c. When results surprise us, it is worthwhile to examine individual data points more closely or even to go beyond the data. The daily climate data as listed by this source for August 2005 shows the code "M" next to August 29, 30, and 31 for all climate statistics. The code indicates that "[REMARKS] ALL DATA MISSING AUGUST 29, 30, AND 31 DUE TO HURRICANE KATRINA." Pretend you were hired as a consultant to determine the percentile for that August. Write a brief paragraph for your report, explaining why the data you generated are likely to be inaccurate.

d. What raw scores mark the cutoff for the top and bottom 10% for these data? Based on these scores, what months had extreme data for 2005? Why should we not trust these data?

7.42 Percentiles and IQ scores: IQ scores are designed to have a mean of 100 and a standard deviation of 15. IQ testing is one way in which people are categorized as having different levels of mental disability; there are four levels of mental retardation between the IQ scores of 0 and 70.

a. People with IQ scores of 20–35 are said to have severe mental retardation and can learn only basic skills (e.g., how to talk, basic self-care). What percentage of people fall in this range?

b. People with IQ scores of 50–70 have scores in the topmost category of IQ scores that indicate an impairment. They are said to have mild mental retardation. They can attain as high as a sixth-grade education and are often self-sufficient. What percentage of people fall in this range?

c. A person has an IQ score of 66. What is her percentile?

d. A person falls at the 3rd percentile. What is his IQ score? Would he be classified as having a mental disability?

7.43 Step 1 of hypothesis testing for a study of the Wechsler Adult Intelligence Scale: Boone (1992) examined scores on the Wechsler Adult Intelligence Scale–Revised (WAIS–R) for 150 adult psychiatric inpatients. He determined the "intrasubtest scatter" score for each inpatient. Intrasubtest scatter refers to patterns of responses in which respondents are almost as likely to get easy questions wrong as hard ones. In the WAIS–R, we expect more wrong answers near the end, as the questions become more difficult, so high levels of intrasubtest scatter would be an unusual pattern of responses. Boone wondered if psychiatric patients have different response patterns than nonpatients have. He compared the intrasubtest scatter for 150 patients to population data from the WAIS-R standardization group. Assume that he had access to both means and standard deviations for this population. Boone reported that "the standardization group's intrasubtest scatter was significantly greater than those reported for the psychiatric inpatients" and concluded that such scatter is normal.

a. What are the two populations?

b. What would the comparison distribution be? Explain.

c. What hypothesis test would you use? Explain.

d. Check the assumptions for this hypothesis test. Label your answers (1) through (3).

e. What does Boone mean when he says *significantly*?

7.44 Step 2 of hypothesis testing for a study of the Wechsler Adult Intelligence Scale: Refer to the scenario described in Exercise 7.43.

a. State the null and research hypotheses for a two-tailed test in both words and symbols.

b. Imagine that you wanted to replicate this study. Based on the findings described in Exercise 7.43, state the null and research hypotheses for a one-tailed test in both words and symbols.

7.45 Step 1 of hypothesis testing for a study of college football: Let's consider whether U.S. college football teams are more likely or less likely to be mismatched in the upper National Collegiate Athletic Association (NCAA) divisions. Overall, the 53 Division I-A (now called the Football Bowl Subdivision) games (the highest division) had a mean spread (winning score minus losing score) of 16.189 in a particular week, with a standard deviation of 12.128. We took a sample of 4 games that were played that week in the next-highest league, Division I-AA (now called the Football Championship Subdivision), to see if the mean spread was different; one of the many leagues within Division I-AA, the Patriot League, played 4 games that weekend.

a. List the independent variable and the dependent variable in this example.

b. Did we use random selection? Explain.

c. Identify the populations of interest in this example.

d. State the comparison distribution.

e. Check the assumptions for this test.

7.46 Step 2 of hypothesis testing and college football: Refer to Exercise 7.45.

a. State the null hypothesis and the research hypothesis for a two-tailed test in both words and symbols.

b. One of our students hypothesized that the spread would be bigger among the Division I-AA teams because "some of them are really bad and would get crushed." State the one-tailed null hypothesis and research hypothesis, based on our student's prediction, in both words and symbols.

7.47 Steps 3 through 6 of hypothesis testing and college football: Refer to Exercise 7.45. Remember, the population mean is 16.189, with a standard deviation of 12.128. The results for the four Division I-AA Patriot League games are as follows:

Holy Cross, 27/Bucknell, 10
Lehigh, 23/Colgate, 15
Lafayette, 31/Fordham, 24
Georgetown, 24/Marist, 21

a. Conduct steps 3 through 6 of hypothesis testing. (You already conducted steps 1 and 2 in Exercises 7.45(e) and 7.46(a), respectively.)

b. Would you be willing to generalize these findings beyond the sample? Explain.

Putting It All Together

7.48 The Graded Naming Test and sociocultural differences: *z* tests are often used when a researcher wants to compare his or her sample to known population norms. The Graded Naming Test (GNT) asks respondents to name objects in a set of 30 black-and-white drawings. The test, often used to detect brain damage, starts with easy words like *kangaroo* and gets progressively more difficult, ending with words like *sextant*. The GNT population norm for adults in England is 20.4. Roberts (2003) wondered whether a sample of Canadian adults had different scores than adults in England. If they were different, the English norms would not be valid for use in Canada. The mean for 30 Canadian adults was 17.5. For the purposes of this exercise, assume that the standard deviation of the adults in England is 3.2.

a. Conduct all six steps of a *z* test. Be sure to label all six steps.

b. Some words on the GNT are more commonly used in England. For example, a *mitre*, the headpiece worn by bishops, is worn by the Archbishop of Canterbury in public ceremonies in England. No Canadian participant correctly responded to this item, whereas 55% of English adults correctly responded. Explain why we should be cautious about applying norms to people different from those on whom the test was normed.

c. When we conduct a one-tailed test instead of a two-tailed test, there are small changes in steps 2 and 4 of hypothesis testing. (*Note:* For this example, assume that those from populations other than the one on which it was normed will score lower, on average. That is, hypothesize that the Canadians will have a lower mean.) Conduct steps 2, 4, and 6 of hypothesis testing for a one-tailed test.

d. Under which circumstance—a one-tailed or a two-tailed test—is it easier to reject the null hypothesis? Explain.

e. If it becomes easier to reject the null hypothesis under one type of test (one-tailed versus two-tailed), does this mean that there is a bigger difference between the groups with a one-tailed test than with a two-tailed test? Explain.

f. When we change the p level that we use as a cutoff, there is a small change in step 4 of hypothesis testing. Although 0.05 is the most commonly used p level, other values, such as 0.01, are often used. For this example, conduct steps 4 and 6 of hypothesis testing for a two-tailed test and p level of 0.01, determining the cutoff and drawing the curve.

g. With which p level—0.05 or 0.01—is it easiest to reject the null hypothesis? Explain.

h. If it is easier to reject the null hypothesis with certain p levels, does this mean that there is a bigger difference between the samples with one p level versus the other p level? Explain.

7.49 Patient adherence and orthodontics: A research report (Behenam & Pooya, 2007) begins, "There is probably no other area of health care that requires a cooperation to the extent that orthodontics does," and explores factors that affected the number of hours per day that Iranian patients wore their orthodontic appliances. The patients in the study reported that they used their appliances, on average, 14.78 hours per day, with a standard deviation of 5.31. We'll treat this group as the population for the purposes of this example. Let's say a researcher wanted to study whether a DVD with information about orthodontics led to an increase in the amount of time patients wore their appliances, but decided to use a two-tailed test to be conservative. Let's say he studied the next 15 patients at his clinic, asked them to watch the DVD, and then found that they wore their appliances, on average, 17 hours per day.

a. What is the independent variable? What is the dependent variable?

b. Did the researcher use random selection to choose his sample? Explain your answer.

c. Conduct all six steps of hypothesis testing. Be sure to label all six steps.

d. If the researcher's decision in step 6 were wrong, what type of error would he have made? Explain your answer.

7.50 Radiation levels on Japanese farms: Fackler (2012) reported that Japanese farmers have become skeptical of the Japanese government's assurances that radiation levels were within legal limits in the wake of the 2011 tsunami and radiation disaster at Fukushima. After reports of safe levels in Onami, more than 12 concerned farmers tested their crops and found dangerously high levels of cesium.

a. If the farmers wanted to conduct a z test comparing their results to the cesium levels found in areas that had not experienced radiation, what would their sample be? Be specific.

b. Conduct step 1 of hypothesis testing.

c. Conduct step 2 of hypothesis testing.

d. Conduct step 4 of hypothesis testing for a two-tailed test and a p level of 0.05.

e. Imagine that the farmers calculated a z statistic of 3.2 for their sample. Conduct step 6 of hypothesis testing.

f. If the farmers' conclusions were incorrect, what type of error would they have made? Explain your answer.

Terms

assumption (p. 154)
parametric test (p. 155)
nonparametric test (p. 155)
robust (p. 155)
critical value (p. 156)
critical region (p. 156)
p level (p. 156)
statistically significant (p. 156)
one-tailed test (p. 159)
two-tailed test (p. 160)

Symbols

H_0 (p. 159)
H_1 (p. 159)

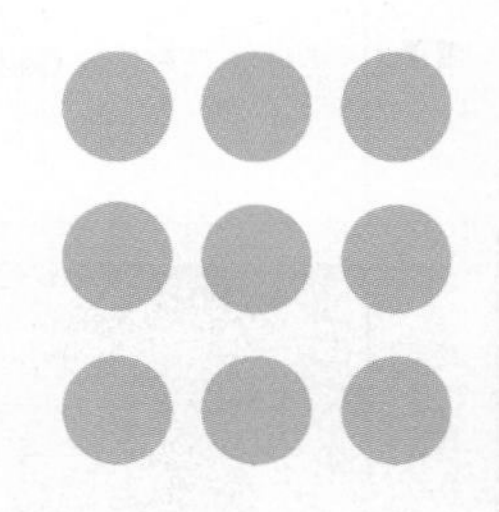

CHAPTER 8

Confidence Intervals, Effect Size, and Statistical Power

BEFORE YOU GO ON

- You should know how to conduct a *z* test (Chapter 7).
- You should understand the concept of statistical significance (Chapter 7).

"Math Class Is Tough" Teen Talk Barbie, with her negative proclamation about math class, was a lightning rod for discussions about gender stereotypes. Some of Barbie's negative press related to the fact that Barbie's message might doom girls to even poorer performance in mathematics. The media tend to play up gender differences instead of the less interesting (and more frequent) realities of gender similarities.

"Want to go shopping? OK, meet me at the mall."
"Math class is tough."

With these and 268 other phrases, Teen Talk Barbie was introduced to the market in July 1992. By September, it was being publicly criticized for its negative message about girls and math. At first the Mattel toy company refused to pull it from store shelves, citing more positive phrases in Barbie's repertoire, such as "I'm studying to be a doctor." But the bad press escalated, and by October Mattel had backed down. The controversy endured, however, and even showed up in a 1994 *Simpsons* episode when Lisa Simpson boycotted the fictional Malibu Stacy doll that had been programmed to say, "Thinking too much gives you wrinkles."

The controversy over gender differences in mathematical reasoning ability began after a study on the topic was published in the prestigious journal *Science*. Participants included about 10,000 male and female students in grades 7 through 10 who were in the top 2% to 3% on standardized tests of mathematics (Benbow & Stanley, 1980). In this sample, the boys' average score on the mathematics portion of the SAT test was 32 points higher than the girls' average score, a statistically significant difference.

Based on this gender difference, the study gained vast media attention (Jacob & Eccles, 1982). But the danger of reporting such a mean difference is the implication that all or most of the members of one group are different from all or most of the members of the other group. As we see in Figure 8-1, such an assertion is far from the truth. Part of this misunderstanding is caused by the language of statistics: "statistically significant" does *not* mean "very important."

The misunderstanding that derived from Benbow and Stanley's study (1980) spread from researchers to the media to the general public. It was exacerbated by the release of Teen Talk Barbie, particularly when members of the Barbie Liberation Organization, a guerrilla art group, switched the computer chips in talking Barbies and GI Joes and returned them to store shelves. Suddenly, GI Joe was telling us, in a voice uncannily like Barbie's, that "math class is tough."

It took a meta-analysis—a study of all the studies about a particular topic—to clarify the research. Janet Hyde and colleagues conducted the meta-analysis by compiling the results from 259 mean differences in mathematical reasoning ability (Hyde, Fennema, & Lamon, 1990). These data represented 1,968,846 male participants and 2,016,836 female participants. Here's what the researchers discovered:

- Mean gender differences in overall mathematical reasoning ability were small.
- When the extreme tails of the distribution were eliminated (such as scores from participants in remedial or gifted programs), the size of the gender difference was even smaller *and* reversed direction, favoring women and girls rather than men and boys.

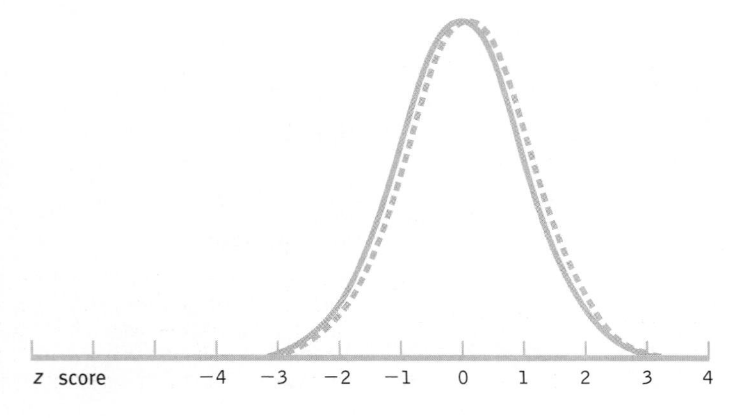

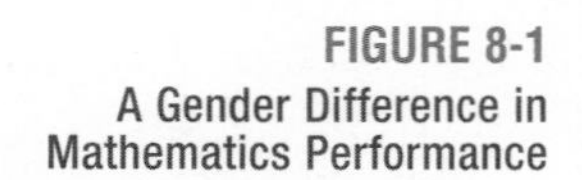

FIGURE 8-1
A Gender Difference in Mathematics Performance

This graph represents the overlap that would be expected if distributions for males and females differed, on average, by the amount that Hyde and colleagues (1990) reported in their meta-analysis of gender differences in mathematics performance.

The authors of this meta-analysis included a graph of two distributions—one for male participants and one for female participants (see Figure 8-1). Are you surprised that a small (but statistically significant) gender difference is almost completely overlapping? This is a case in which hypothesis testing alone inadvertently encouraged a profound misunderstanding (Jacob & Eccles, 1986).

Fortunately, statisticians have developed ways to move beyond the flaws in hypothesis testing. We explore three ways in this chapter. First, we compute confidence intervals, which provide a range of plausible mean differences. Second, we calculate effect sizes, which indicate the size of differences. Finally, we estimate the statistical power of our study to be sure that we have a sufficient sample size to detect a real difference. If we discover that we don't have enough statistical power, then it may not be worth continuing the experiment.

- A **point estimate** is a summary statistic from a sample that is just one number used as an estimate of the population parameter.
- An **interval estimate** is based on a sample statistic and provides a range of plausible values for the population parameter.

Confidence Intervals

In studies on gender differences in mathematics performance, researchers calculate a mean difference by subtracting a mean score for girls from a mean score for boys. All three summary statistics—the mean for boys, the mean for girls, and the difference between them—are point estimates. *A **point estimate** is a summary statistic from a sample that is just one number used as an estimate of the population parameter.* A point estimate, however, is rarely exactly accurate. We can increase accuracy by using an interval estimate when possible.

MASTERING THE CONCEPT

8.1: We can use a sample to calculate a point estimate—one plausible number, such as a mean—for the population. More realistically, we also can use a sample to calculate an interval estimate—a range of plausible numbers, such as a range of means—for the population.

Interval Estimates

*An **interval estimate** is based on a sample statistic and provides a range of plausible values for the population parameter.* Interval estimates are frequently used by the media, often when reporting political polls, and are usually constructed by adding and subtracting a margin of error from a point estimate.

EXAMPLE 8.1

For example, a 2009 Marist poll asked 938 adult respondents in the United States to select from five choices the word or phrase that they found "most annoying in conversation" (http://maristpoll.marist.edu/-107-whatever-takes-top-honors-as-most-annoying/). "Whatever" was chosen by 47% of respondents, ahead of "you know" (25%), "it is what it is" (11%), "anyway" (7%), and "at the end of the day" (2%). The margin of error was reported to be ±3.2% (plus or minus 3.2%).

Because 47 − 3.2 = 43.8 and 47 + 3.2 = 50.2, the interval estimate for "whatever" is 43.8% to 50.2% (see Figure 8-2). Interval estimates provide a range of plausible values, not just one statistic.

Pay attention to whether the interval estimates overlap. "You know" came in second, with 25%, giving an interval estimate of 21.8% to 28.2%. There's no overlap with the first-place word, a strong indication that "whatever" really was most annoying in the population as well as in the sample. However, if "you know" had received 42% of the vote, it would

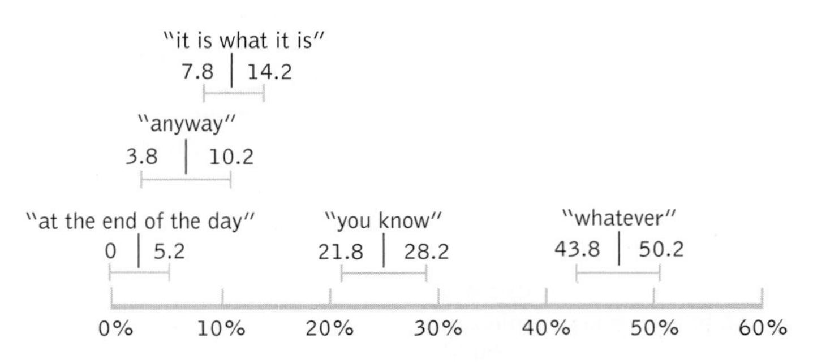

FIGURE 8-2
Intervals and Overlap

When two intervals, like those for "whatever" and "you know," do not overlap, we conclude that the population means are likely different. In the population, it seems that "whatever" really is more annoying than "you know." However, when two intervals do overlap, like those for "it is what it is" and "anyway," then it is plausible that the two phrases are deemed equally annoying in the population.

▪ A **confidence interval** is an interval estimate based on the sample statistic; it includes the population mean a certain percentage of the time if we sample from the same population repeatedly.

have placed only 5% behind "whatever," and it would have had an interval estimate of 38.8% to 45.2%. This range would have overlapped with the one for "whatever," an indication that both expressions could plausibly have been equally annoying in the population.

Language Alert! The terms "margin of error," "interval estimates," and "confidence intervals" all represent the same core idea. Specifically, *a* ***confidence interval*** *is an interval estimate based on the sample statistic; it includes the population mean a certain percentage of the time if we sample from the same population repeatedly.* (*Note:* We are not saying that we are confident that the population mean falls in the interval; we are merely saying that we expect to find the population mean within a certain interval a certain percentage of the time—usually 95%—when we conduct this same study with the same sample size.)

The confidence interval is centered around the mean of the sample. A 95% confidence level is most commonly used, indicating the 95% that falls *between* the two tails (i.e., 100% − 5% = 95%). Note the terms used here: the confidence *level* is 95%, but the confidence *interval* is the range between the two values that surround the sample mean. ▪

Calculating Confidence Intervals with *z* Distributions

The symmetry of the *z* distribution makes it easy to calculate confidence intervals.

EXAMPLE 8.2

We already conducted hypothesis testing in Example 7.5 in Chapter 7 for a study on calories consumed by patrons of different Starbucks stores that either did or did not post the calories on their menus (Bollinger, Leslie, & Sorensen, 2010). Here is how confidence intervals can help us listen more closely to the story these data tell. The population mean was 247 calories, and we considered 201 to be the population standard deviation. The 1000 people in the sample consumed a mean of 232 calories. When we conducted hypothesis testing, we centered the curve around the mean according to the null hypothesis, the population mean of 247. We determined critical values based on this mean and compared the sample mean to these cutoffs. The test statistic (−2.36) was beyond the critical *z* statistic, so we rejected the null hypothesis. The data led us to conclude that people going to those Starbucks that posted calories consumed fewer calories, on average, than people going to those Starbucks that did not post calories.

There are several steps to calculating a confidence interval.

STEP 1: Draw a picture of a distribution that will include the confidence interval.

We draw a normal curve (Figure 8-3) that has the *sample* mean, 232, at its center, instead of the *population* mean, 247.

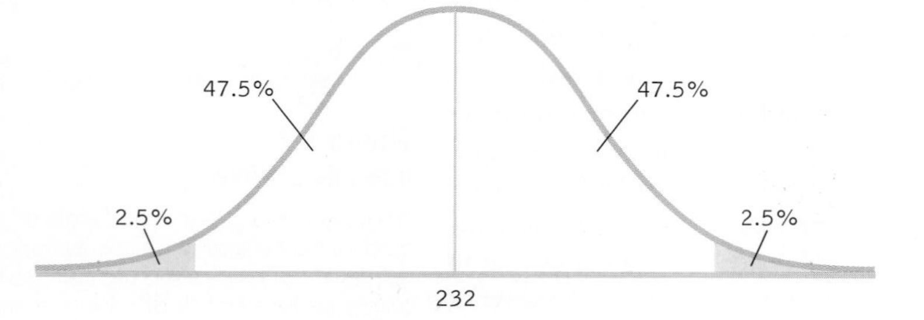

FIGURE 8-3
A 95% Confidence Interval, Part I

To begin calculating a confidence interval for a *z* distribution, we draw a normal curve, place the sample mean at its center, and indicate the percentages within and beyond the confidence interval.

STEP 2: Indicate the bounds of the confidence interval on the drawing.

We draw a vertical line from the mean to the top of the curve. For a 95% confidence interval, we also draw two small vertical lines to indicate the middle 95% of the normal curve (2.5% in each tail, for a total of 5%).

The curve is symmetric, so half of the 95% falls above and half falls below the mean. Half of 95 is 47.5, so we write 47.5% in the segments on either side of the mean. In the tails beyond the two lines that indicate the end of the middle 95%, we also write the appropriate percentages.

STEP 3: Determine the *z* statistics that fall at each line marking the middle 95%.

To do this, we turn back to the versatile *z* table in Appendix B. The percentage between the mean and each of the *z* scores is 47.5%. When we look up this percentage in the *z* table, we find a *z* statistic of 1.96. (Note that this is identical to the cutoffs for the *z* test; this will always be the case because the *p* level of 0.05 corresponds to a confidence level of 95%.) We can now add the *z* statistics of −1.96 and 1.96 to the curve, as seen in Figure 8-4.

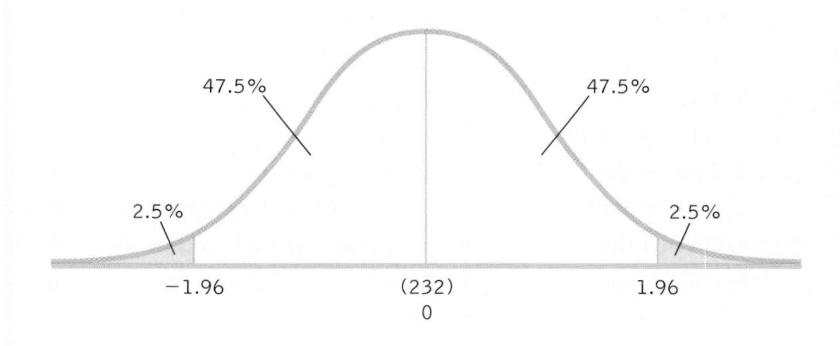

FIGURE 8-4
A 95% Confidence Interval, Part II

The next step in calculating a confidence interval is identifying the *z* statistics that indicate each end of the interval. Because the curve is symmetric, the *z* statistics will have the same magnitude—one will be negative and one will be positive (−1.96 and 1.96).

STEP 4: Turn the *z* statistics back into raw means.

We use the formula for this conversion, but first we identify the appropriate mean and standard deviation. There are two important points to remember. First, we center the interval around the *sample* mean (not the *population* mean). So we use the sample mean of 232 in the calculations. Second, because we have a sample *mean* (rather than an individual *score*), we use a distribution of means. So we calculate standard error as the measure of spread:

$$\sigma_M = \frac{\sigma}{\sqrt{N}} = \frac{201}{\sqrt{1000}} = 6.356$$

Notice that this is the same standard error that we calculated in Example 7.5 in Chapter 7 when we conducted a hypothesis test.

Using this mean and standard error, we calculate the raw mean at each end of the confidence interval, and add them to the curve, as in Figure 8-5:

$$M_{lower} = -z(\sigma_M) + M_{sample} = -1.96(6.356) + 232 = 219.54$$

$$M_{upper} = z(\sigma_M) + M_{sample} = 1.96(6.356) + 232 = 244.46$$

The 95% confidence interval, reported in brackets as is typical, is [219.54, 244.46].

Mastering the Formula

8-1: The formula for the lower bound of a confidence interval using a *z* distribution is $M_{lower} = -z(\sigma_M) + M_{sample}$, and the formula for the upper bound is $M_{upper} = z(\sigma_M) + M_{sample}$. The first symbol in each formula refers to the mean at that end of the confidence interval. To calculate each bound, we multiply the *z* statistic by the standard error, then add the sample mean. The *z* statistic for the lower bound is negative, and the *z* statistic for the upper bound is positive.

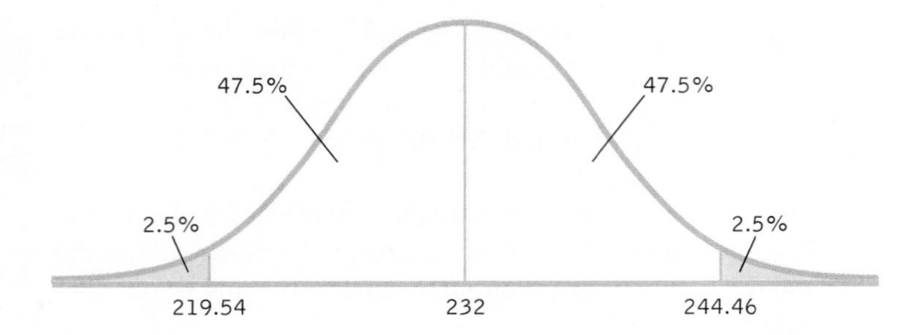

FIGURE 8-5
A 95% Confidence Interval, Part III

The final step in calculating a confidence interval is converting the z statistics that indicate each end of the interval into raw means.

STEP 5: Check that the confidence interval makes sense.

The sample mean should fall exactly in the middle of the two ends of the interval.

$$219.54 - 232 = -12.46 \text{ and } 244.46 - 232 = 12.46$$

We have a match. The confidence interval ranges from 12.46 below the sample mean to 12.46 above the sample mean. We can think of this number, 12.46, as the margin of error.

To recap the steps for the creation of a confidence interval for a z statistic:

1. *Draw* a normal curve with the *sample* mean in the center.
2. *Indicate* the bounds of the confidence interval on either end, and write the percentages under each segment of the curve.
3. *Look up* the z statistics for the lower and upper ends of the confidence interval in the z table. These are always -1.96 and 1.96 for a 95% confidence interval.
4. *Convert* the z statistics to raw means for each end of the confidence interval.
5. *Check* your answer; each end of the confidence interval should be exactly the same distance from the sample mean.

If we were to sample 1000 customers at the Starbucks that post calories on their menus from the same population over and over, the 95% confidence interval would include the population mean 95% of the time. Note that the population mean for customers at the Starbucks that do not post calories, 247, falls outside of this interval. So, it is not plausible that the sample of customers at Starbucks that post calories comes from the population according to the null hypothesis—customers at Starbucks that do not post calories. The conclusions from both the z test and the confidence interval are the same, but the confidence interval gives us more information—an interval estimate, not just a point estimate. Moreover, there is some evidence that reporting confidence intervals instead of the results of hypothesis testing can lead to more accurate interpretations of the findings (Coulson, Healey, Fidler, & Cumming, 2010). ■

CHECK YOUR LEARNING

Reviewing the Concepts

> A point estimate is just a single number, such as a mean, that provides a plausible value for the population parameter. An interval estimate provides a range of plausible values for the population parameter.

> A confidence interval is one kind of interval estimate and can be created around a sample mean using a z distribution.

> The confidence interval confirms the results of the hypothesis test while adding more detail.

Clarifying the Concepts

8-1 Why are interval estimates better than point estimates?

Calculating the Statistics

8-2 If 21% of voters want to raise taxes, with a margin of error of 4%, what is the interval estimate? What is the point estimate?

Applying the Concepts

8-3 In How It Works 7.2, we conducted a z test based on the following information adapted from a study by Petrocelli (2003) that used the Consideration of Future Consequences (CFC) scale as the dependent variable. The population mean CFC score was 3.51, with a standard deviation of 0.61. The sample was 45 students who joined a career discussion group, and the study examined whether this might have changed CFC scores. The mean for this group was 3.7.

a. Calculate the 95% confidence interval.

b. Explain what this confidence interval tells us.

c. Why is this confidence interval superior to the hypothesis test that we conducted in Chapter 7?

Solutions to these Check Your Learning questions can be found in Appendix D.

Effect Size

As we learned when we looked at the research on gender differences in mathematical reasoning ability, "statistically significant" does *not* mean that the findings from a study represent a meaningful difference. "Statistically significant" only means that those findings are unlikely to occur if in fact the null hypothesis is true. Calculating an effect size moves us a little closer to what we are most interested in: Is the pattern in a data set meaningful or important?

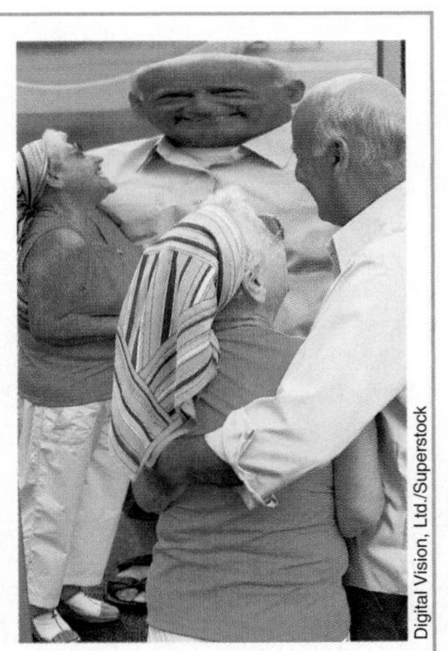

Misinterpreting Statistical Significance Statistical significance that is achieved by merely collecting a large sample can make a research finding appear to be far more important than it really is, just as a curved mirror can exaggerate a person's size.

Digital Vision, Ltd./Superstock

The Effect of Sample Size on Statistical Significance

The almost completely overlapping curves in Figure 8-1 were "statistically significant" because the sample size was so big. Increasing sample size always increases the test statistic if all else stays the same. For example, researchers reported data for psychology test scores on the Graduate Record Examination (GRE) over several years: $\mu = 554$, $\sigma = 99$ (Matlin & Kalat, 2001). In a fictional study, Example 7.4 in Chapter 7, we reported that 90 graduating seniors had a mean of 568. Based on the sample size of 90, we reported the mean and standard error for the distribution of means as:

$$\mu_M = 554; \sigma_M = \frac{\sigma}{\sqrt{N}} = \frac{99}{\sqrt{90}} = 10.436$$

The test statistic calculated from these numbers was:

$$z = \frac{(M - \mu_M)}{\sigma_M} = \frac{(568 - 554)}{10.436} = 1.34$$

What would happen if we increased the sample size to 200? We'd have to recalculate the standard error to reflect the larger sample, and then recalculate the test statistic to reflect the smaller standard error.

$$\mu_M = 554; \sigma_M = \frac{\sigma}{\sqrt{N}} = \frac{99}{\sqrt{200}} = 7.000$$

$$z = \frac{(M - \mu_M)}{\sigma_M} = \frac{(568 - 554)}{7.000} = 2.00$$

What if we increased the sample size to 1000?

$$\mu_M = 554; \sigma_M = \frac{\sigma}{\sqrt{N}} = \frac{99}{\sqrt{1000}} = 3.131$$

$$z = \frac{(M - \mu_M)}{\sigma_M} = \frac{(568 - 554)}{3.131} = 4.47$$

What if we increased it to 100,000?

$$\mu_M = 554; \sigma_M = \frac{\sigma}{\sqrt{N}} = \frac{99}{\sqrt{100,000}} = 0.313$$

$$z = \frac{(M - \mu_M)}{\sigma_M} = \frac{(568 - 554)}{0.313} = 44.73$$

Rob Walls/Alamy

Larger Samples Give Us More Confidence in Our Conclusions Stephen, a British student studying in the United States, is told that he won't be able to find his favorite candy bar, Yorkie. He tests this hypothesis in 3 stores and finds no Yorkie bars. Another British student, Victoria, also warned by her friends, looks for her favorite, Curly Wurly. She tests her hypothesis in 25 U.S. stores and finds no Curly Wurly bars. Both conclude that their friends were right. Do you feel more confident that Stephen or Victoria really won't be able to find the favorite candy bar?

Notice that each time we increased the sample size, the standard error decreased and the test statistic increased. The original test statistic, 1.34, was not beyond the critical values of 1.96 and −1.96. However, the remaining test statistics (2.00, 4.47, and 44.73) were increasingly more extreme than the positive critical value. In their study of gender differences in mathematics performance, researchers studied 10,000 participants, a very large sample (Benbow & Stanley, 1980). It is not surprising, then, that a small difference would be a statistically significant difference.

Let's consider, logically, why it makes sense that a large sample should allow us to reject the null hypothesis more readily than a small sample. If we randomly selected 5 people among all those who had taken the GRE and they had scores well above the national average, we might say, "It could be chance." But if we selected 1000 people with GRE scores well above the national average, it is very unlikely that we just happened to choose 1000 people with high scores.

> **▶ MASTERING THE CONCEPT**
>
> **8.2:** As sample size increases, so does the test statistic (if all else stays the same). Because of this, a small difference might not be statistically significant with a small sample but might be statistically significant with a large sample.

But just because a real difference exists does not mean it is a large, or meaningful, difference. The difference we found with 5 people might be the same as the difference we found with 1000 people. As we demonstrated with multiple z tests with different sample sizes, we might fail to reject the null hypothesis with a small sample but then reject the null hypothesis for the same-size difference between means with a large sample.

Cohen (1990) used the small but statistically significant correlation between height and IQ to explain the difference between statistical significance and practical importance. The sample size was big: 14,000 children. Imagining that height and IQ were causally related, Cohen calculated that a person would have to grow by 3.5 feet to increase IQ by 30 points (2 standard deviations). Or, to increase height by 4 inches, a person would have to increase IQ by 233 points! Height may have been statistically significantly related to IQ, but there was no practical real-world application. A larger sample size should influence the level of confidence that the story is true, but it shouldn't increase our confidence that the story is important. *Statistical significance does not indicate practical importance.*

What Effect Size Is

Effect size can tell us whether a statistically significant difference might also be an important difference. ***Effect size*** *indicates the size of a difference and is unaffected by sample size.* Effect size tells us how much two populations *do not* overlap. Simply put, the less overlap, the bigger the effect size.

Effect size indicates the size of a difference and is unaffected by sample size.

The amount of overlap between two distributions can be decreased in two ways. First, Figure 8-6 shows that overlap decreases and effect size increases when means are further apart. Then Figure 8-7 shows that overlap decreases and effect size increases when variability within each distribution of scores is smaller.

When we discussed gender differences in mathematical reasoning ability, you may have noticed that we described the size of the findings as "small" (Hyde, 2005). Because

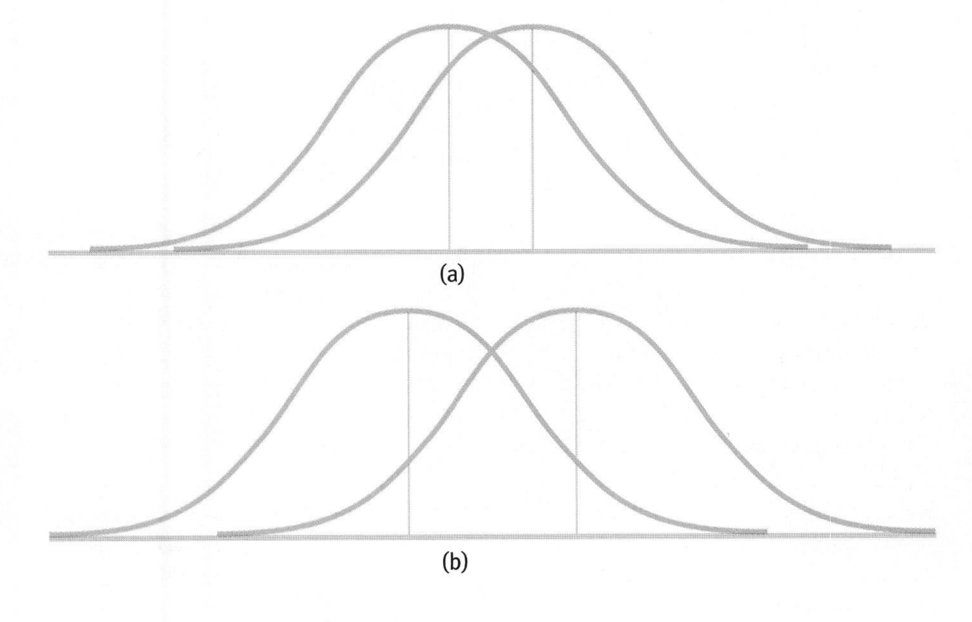

FIGURE 8-6
Effect Size and Mean Differences

When two population means are farther apart, as in (b), the overlap of the distributions is less and the effect size is bigger.

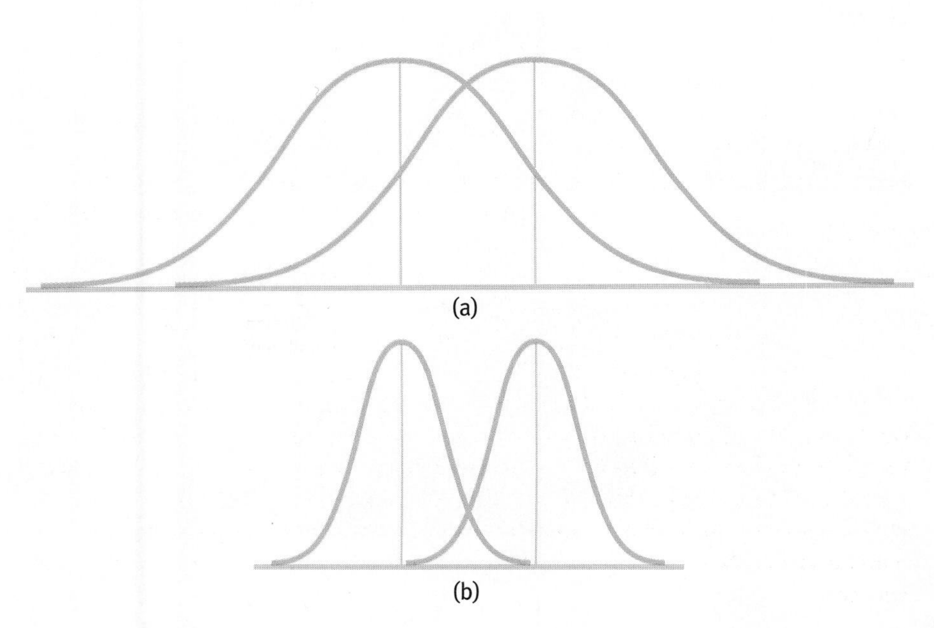

FIGURE 8-7
Effect Size and Standard Deviation

When two population distributions decrease their spread, as in (b), the overlap of the distributions is less and the effect size is bigger.

effect size is a standardized measure based on scores rather than means, we can compare the effect sizes of different studies with one another, even when the studies have different sample sizes.

EXAMPLE 8.3

Figure 8-8 demonstrates why we use scores instead of means to calculate effect size. First, assume that each of these distributions is based on the same underlying population. Second, notice that all means represented by the vertical lines are identical. The

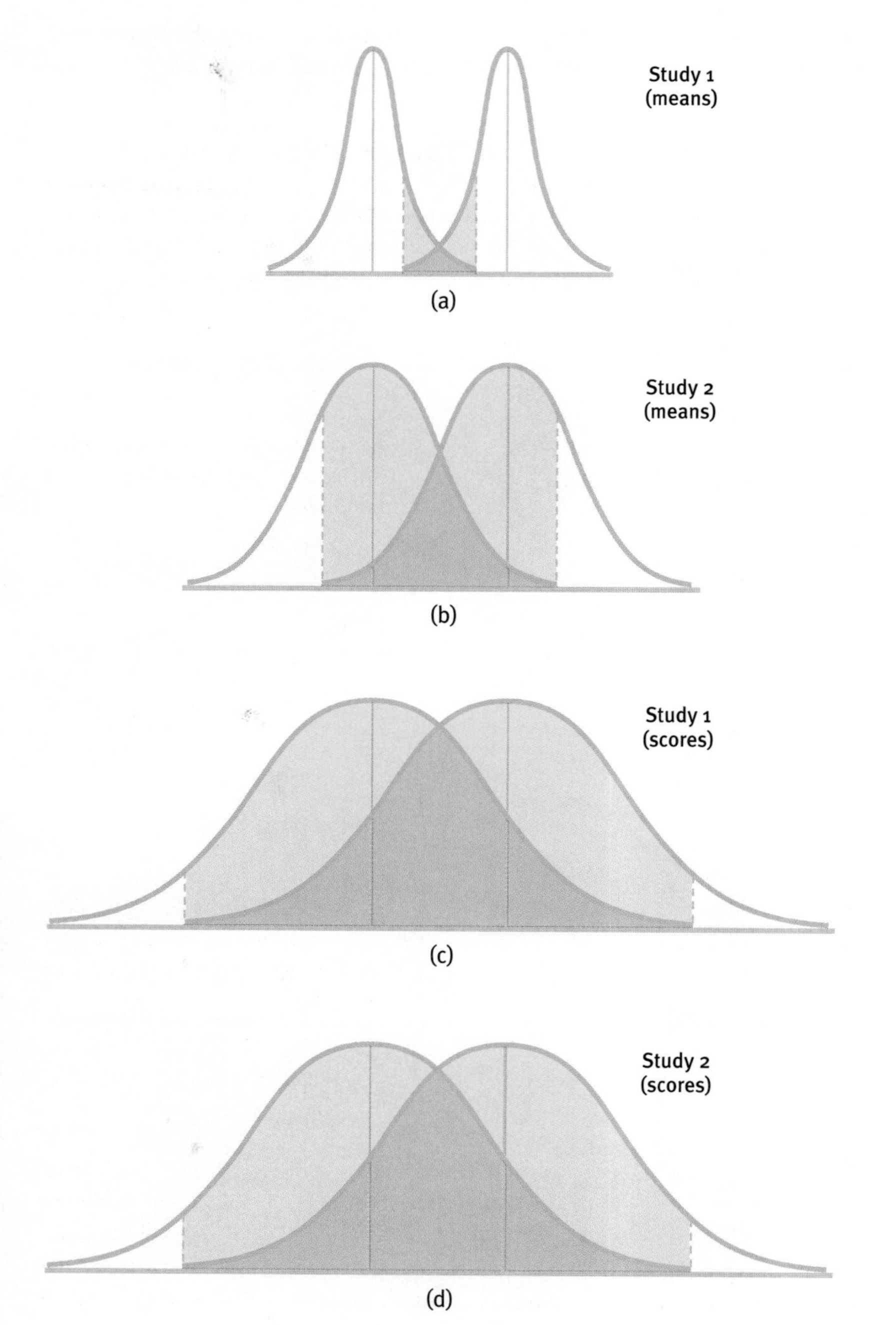

FIGURE 8-8
Making Fair Comparisons

The top two pairs of curves (a and b) depict two studies, study 1 and study 2. The first study (a) compared two samples with very large sample sizes, so each curve is very narrow. The second study (b) compared two samples with much smaller sample sizes, so each curve is wider. The first study has less overlap, but that doesn't mean it has a bigger effect than study 2; we just can't compare the effects. The bottom two pairs (c and d) depict the same two studies, but used standard deviation for individual scores. Now they are comparable and we see that they have the same amount of overlap—the same effect sizes.

differences are due only to the spread of the distributions. The small degree of overlap in the tall, skinny distributions of means in Figure 8-8a is the result of a large sample size. The greater degree of overlap in the somewhat wider distributions of means in Figure 8-8b is the result of a smaller sample size. By contrast, the distributions of scores in Figures 8-8c and 8-8d represent scores rather than means for these two studies. Because these flatter, wider distributions include actual scores, sample size is not an issue in making comparisons.

In this case, the amounts of real overlap in Figures 8-7c and 8-7d are identical. We can directly compare the amount of overlap and see that they have the same effect size. ■

Cohen's *d* is a measure of effect size that assesses the difference between two means in terms of standard deviation, not standard error.

Cohen's *d*

There are many different effect-size statistics, but they all neutralize the influence of sample size. When we conduct a z test, the effect-size statistic is typically Cohen's d, developed by Jacob Cohen (Cohen, 1988). ***Cohen's d*** *is a measure of effect size that assesses the difference between two means in terms of standard deviation, not standard error.* In other words, Cohen's d allows us to measure the difference between means using standard deviations, much like a z statistic. We accomplish this by using standard deviation in the denominator (rather than using standard error).

EXAMPLE 8.4

Let's calculate Cohen's d for the situation for which we constructed a confidence interval. We simply substitute standard deviation for standard error. When we calculated the test statistic for the 1000 customers at Starbucks with posted calories, we first calculated standard error:

$$\sigma_M = \frac{\sigma}{\sqrt{N}} = \frac{201}{\sqrt{1000}} = 6.356$$

We calculated the z statistic using the population mean of 247 and the sample mean of 232:

$$z = \frac{(M - \mu_M)}{\sigma_M} = \frac{(232 - 247)}{6.356} = -2.36$$

To calculate Cohen's d, we simply use the formula for the z statistic, substituting σ for σ_M (and μ for μ_M, even though these means are always the same). So we use 201 instead of 6.356 in the denominator. The Cohen's d is now based on the spread of the distribution of scores, rather than the distribution of means.

$$d = \frac{(M - \mu)}{\sigma} = \frac{(232 - 247)}{201} = -0.07$$

MASTERING THE FORMULA

8-2: The formula for Cohen's d for a z statistic is: Cohen's $d = \frac{(M - \mu)}{\sigma}$. It is the same formula as for the z statistic, except we divide by the population standard deviation rather than by standard error.

Now that we have the effect size, often written in shorthand as $d = -0.07$, what does it mean? First, we know that the two sample means are 0.07 standard deviation apart, which doesn't sound like a big difference—and it isn't. Cohen developed guidelines for what constitutes a small effect (0.2), a medium effect (0.5), or a large effect (0.8). Table 8-1 displays these guidelines, along with the amount of overlap between

TABLE 8-1. Cohen's Conventions for Effect Sizes: *d*

Jacob Cohen published guidelines (or conventions), based on the overlap between two distributions, to help researchers determine whether an effect is small, medium, or large. These numbers are not cutoffs, merely rough guidelines to aid researchers in their interpretation of results.

Effect Size	Convention	Overlap
Small	0.2	85%
Medium	0.5	67%
Large	0.8	53%

> **MASTERING THE CONCEPT**
>
> **8.3:** Because a statistically significant effect might not be an important one, we should calculate effect size in addition to conducting a hypothesis test. We can then report whether a statistically significant effect is small, medium, or large.

two curves that is indicated by an effect of that size. No sign is provided because it is the magnitude of an effect size that matters; an effect size of −0.5 is the same size as one of 0.5.

Based on these numbers, the effect size for the study of Starbucks customers (−0.07) is not even at the level of a small effect. As we pointed out in Chapter 7, however, the researchers hypothesized that even a small effect might spur eateries to provide more low-calorie choices. Sometimes a small effect can be meaningful. ■

Meta-Analysis

Many researchers consider meta-analysis to be the most important recent advancement in social science research (e.g., Newton & Rudestam, 1999). *A* ***meta-analysis*** *is a study that involves the calculation of a mean effect size from the individual effect sizes of many studies.* Meta-analysis provides added statistical power by considering many studies simultaneously and helps to resolve debates fueled by contradictory research findings (Lam & Kennedy, 2005).

The logic of meta-analysis process is surprisingly simple. There are just four steps:

Step 1: Select the topic of interest, and decide exactly how to proceed *before* beginning to track down studies.

Step 2: Locate every study that has been conducted and meets the criteria.

Step 3: Calculate an effect size, often Cohen's *d*, for every study.

Step 4: Calculate statistics—ideally, summary statistics, a hypothesis test, a confidence interval, and a visual display of the effect sizes (Rosenthal, 1995).

In Step 4, researchers calculate a mean effect size for all studies, the central goal of a meta-analysis. They also apply many of the other statistical insights we've learned: medians, standard deviations, confidence intervals and hypothesis testing, and graphs. The goal of hypothesis testing with meta-analysis is to reject the null hypothesis that the mean effect size is 0.

CHECK YOUR LEARNING

Reviewing the Concepts

- As sample size increases, the test statistic becomes more extreme and it becomes easier to reject the null hypothesis.
- A statistically significant result is not necessarily one with practical importance.
- Effect sizes are calculated with respect to scores, rather than means, so are not contingent on sample size.

> The size of an effect is based on the difference between two group means and the amount of variability within each group.
> Effect size for a z test is measured with Cohen's d, which is calculated much like a z statistic.
> A meta-analysis is a study of studies that provides a more objective measure of an effect size than an individual study does.

Clarifying the Concepts

8-4 Distinguish statistical significance and practical importance.

8-5 What is effect size?

Calculating the Statistics

8-6 Using IQ as a variable, where we know the mean is 100 and the standard deviation is 15, calculate Cohen's d for an observed mean of 105.

Applying the Concepts

8-7 In Check Your Learning 8-3, we calculated a confidence interval based on CFC data. The population mean CFC score was 3.51, with a standard deviation of 0.61. The mean for the sample of 45 students who joined a career discussion group is 3.7.

a. Calculate the appropriate effect size for this study.

b. Citing Cohen's conventions, explain what this effect size tells us.

c. Based on the effect size, does this finding have any consequences or implications for anyone's life?

Solutions to these Check Your Learning questions can be found in Appendix D.

Statistical Power

The effect size statistic tells us that the public controversy over gender differences in mathematical ability was justified: The observed gender differences had no practical importance. Calculating statistical power is another way to limit such controversies from developing in the first place.

Power is a word that statisticians use in a very specific way. ***Statistical power*** *is a measure of the likelihood that we will reject the null hypothesis, given that the null hypothesis is false.* In other words, statistical power is the probability that we will reject the null hypothesis

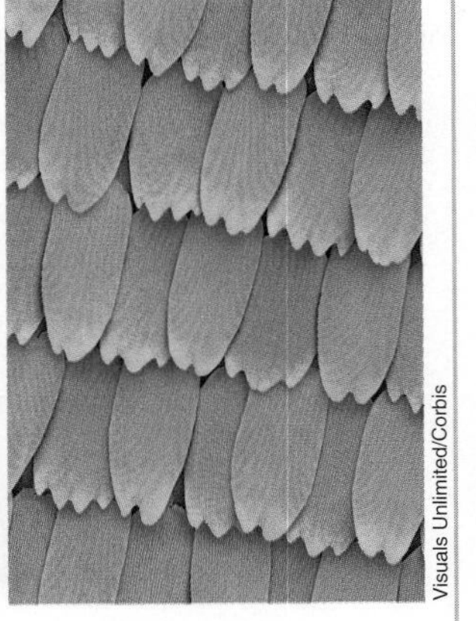

Reuters/Corbis

Visuals Unlimited/Corbis

Statistical Power Statistical power, like the progressive powers of a microscope used to show the fine details of a butterfly's wing, refers to the likelihood that we will detect differences that really exist.

- A **meta-analysis** is a study that involves the calculation of a mean effect size from the individual effect sizes of many studies.
- **Statistical power** is a measure of the likelihood that we will reject the null hypothesis given that the null hypothesis is false.

when we *should* reject the null hypothesis—the probability that we will not make a Type II error.

The calculation of statistical power ranges from a probability of 0.00 to a probability of 1.00 (or from 0% to 100%). Statisticians have historically used a probability of 0.80 as the minimum for conducting a study. If we have an 80% chance of correctly rejecting the null hypothesis, then it is appropriate to conduct the study. Let's look at statistical power for a one-tailed z test. (We use a one-tailed test to simplify calculations.)

MASTERING THE CONCEPT

8.4: Statistical power is the likelihood of rejecting the null hypothesis when we should reject the null hypothesis. Researchers consider a probability of 0.80—an 80% chance of rejecting the null hypothesis if we should reject it—to be the minimum for conducting a study.

The Importance of Statistical Power

To understand statistical power, we need to consider several characteristics of the two populations of interest: the population we believe the sample represents (population 1) and the population to which we're comparing the sample (population 2). We represent these two populations visually as two overlapping curves. Let's consider a variation on an example from Chapter 4—a study aimed at determining whether an intervention changes the mean number of sessions attended at university counseling centers (Hatchett, 2003).

EXAMPLE 8.5

STEP 1: Determine the information needed to calculate statistical power—the hypothesized mean for the sample, the sample size, the population mean, the population standard deviation, and standard error based on this sample size.

In this example, let's say that we hypothesize that a sample of 9 counseling center students will have a mean number of 6.2 sessions after the intervention. The population mean number of sessions attended is 4.6, with a population standard deviation of 3.12. The sample mean of 6.2 is an increase of 1.6 over the population mean, the equivalent of a Cohen's d of about 0.5, a medium effect. Because we have a sample of 9, we need to convert the standard deviation to standard error; to do this, we divide the standard deviation by the square root of the sample size and find that standard error is 1.04. The numbers needed to calculate statistical power are summarized in Table 8-2.

You might wonder how we came up with the hypothesized sample mean, 6.2. We never know, particularly prior to a study, what the actual effect will be. Researchers typically estimate the mean of population 2 by examining the existing research literature or by deciding how large an effect size would make the study worthwhile (Murphy

TABLE 8-2. The Ingredients for the Calculation of Statistical Power

To calculate statistical power for a z test, we must know several "ingredients" before beginning.

Ingredients for Calculating Power	Counseling Center Study
Mean of population 1 (expected sample mean)	$M = 6.2$
Planned sample size	$N = 9$
Mean of population 2	$\mu_{M_1} = \mu = 4.6$
Standard deviation of the population	$\sigma = 3.12$
Standard error (using the planned sample size)	$\sigma_M = \frac{\sigma}{\sqrt{N}} = \frac{3.12}{\sqrt{9}} = 1.04$

& Myors, 2004). In this case, we hypothesized a medium effect—a Cohen's d of 0.5—which translated to an increase in the mean of 1.6, from 4.6 to 6.2.

STEP 2: Determine a critical value in terms of the z distribution and in terms of the raw mean so that statistical power can be calculated.

For this example, the distribution of means for population 1, centered around 6.2, and the distribution of means for population 2, centered around 4.6, are shown in Figure 8-9. This figure also shows the critical value for a one-tailed test with a p level of 0.05. The critical value in terms of the z statistic is 1.65, which is converted to a raw mean of 6.316.

$$M = 1.65(1.04) + 4.6 = 6.316$$

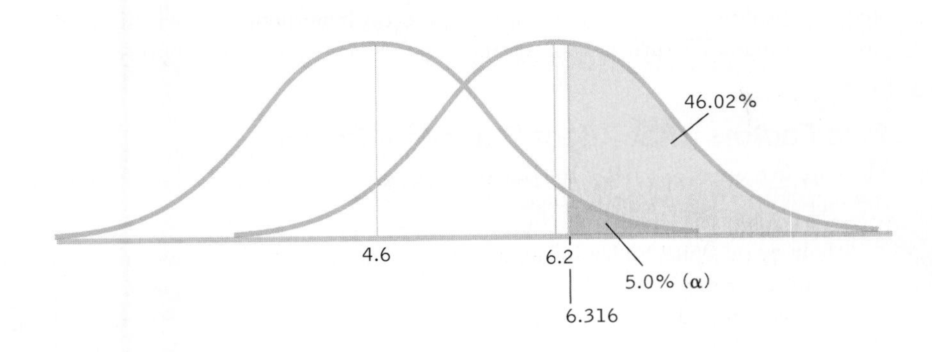

FIGURE 8-9
Statistical Power: The Whole Picture

Now we can visualize statistical power in the context of two populations. Statistical power is the percentage of the distribution of means for population 1 that is above the cutoff. Alpha is the percentage of the distribution of means for population 2 that is above the cutoff; alpha is set by the researcher and is usually 0.05, or 5%.

The p level is shaded in dark purple and marked as 5.0% (α), the percentage version of a proportion of 0.05. The critical value of 6.316 marks off the upper 5% of the distribution based on the null hypothesis, for population 2.

This critical value, 6.316, has the same meaning as it did in hypothesis testing. If the test statistic for a sample falls above this cutoff, then we reject the null hypothesis. Notice that the mean we estimated for population 1 does *not* fall above the cutoff. If the actual difference between the two populations is what we expect, then we can already see that there is a good chance we will not reject the null hypothesis with a sample size of 9. This indicates that we might not have enough statistical power.

STEP 3: Calculate the statistical power—the percentage of the distribution of means for population 1 (the distribution centered around the hypothesized sample mean) that falls above the critical value.

The proportion of the curve above the critical value, shaded in light purple in Figure 8-9, is statistical power, which can be calculated with the use of the z table. Remember, statistical power is the chance that we will reject the null hypothesis if we *should* reject the null hypothesis.

Statistical power in this case is the percentage of the distribution of means for population 1 (the distribution centered around 6.2) that falls above the critical value of 6.316. We convert this critical value to a z statistic based on the hypothesized mean of 6.2.

$$z = \frac{(6.316 - 6.2)}{1.04} = 0.112$$

We look up this z statistic on the z table to determine the percentage above a z statistic of 0.112. That percentage, the area shaded in light purple in Figure 8-9, is 45.62%.

From Figure 8-9, we see the critical value as determined in reference to population 2; in raw score form, it is 6.316. We see that the percentage of the distribution of means for population 2 that falls above 6.316 is 0.05, or 5%, the usual p level that was introduced in Chapter 7. The p level is the chance of making a Type I error. When we turn to the distribution of means for population 1, the percentage above that same cutoff is the statistical power. Given that population 1 exists, 45.62% of the time that we select a sample of size 9 from this population, we will be able to reject the null hypothesis. This is far below the 80% considered adequate, and we would be wise to increase the sample size.

On a practical level, statistical power calculations tell researchers how many participants are needed to conduct a study whose findings we can trust. Remember, however, that statistical power is based, to some degree, on hypothetical information, and it is just an estimate. We turn next to several factors that affect statistical power. ■

Five Factors That Affect Statistical Power

Here are five ways to increase the power of a statistical test, from the easiest to the most difficult:

1. Increase alpha. Increasing alpha is like changing the rules by widening the goal posts in football or the goal in soccer. In Figure 8-10, we see how statistical power increases when we increase a p level of 0.05 (Figure 8-10a) to 0.10 (Figure 8-10b). This has the side effect of increasing the probability of a Type I error from 5% to 10%, however, so researchers rarely choose to increase statistical power in this manner.

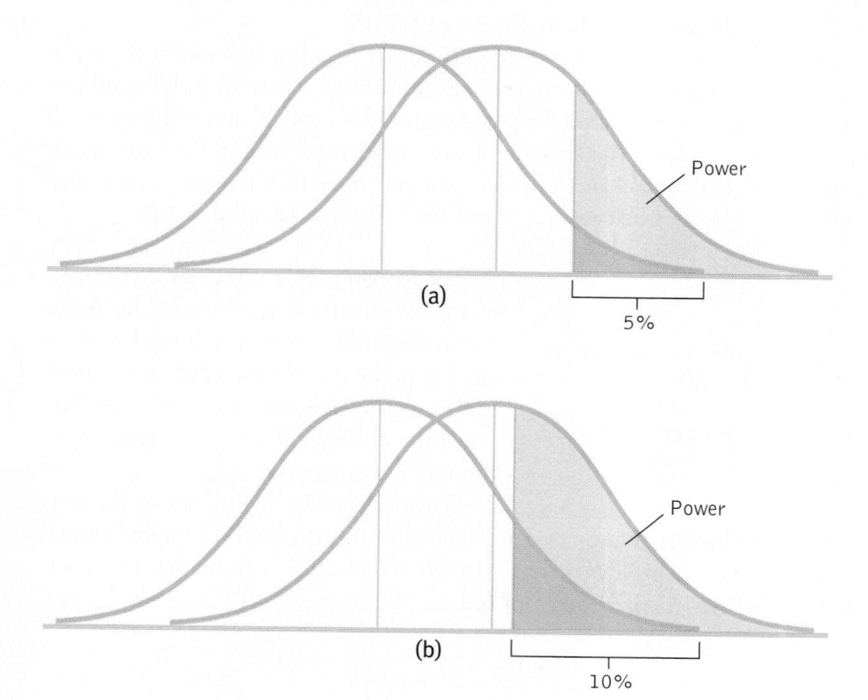

FIGURE 8-10
Increasing Alpha

As we increase alpha from the standard of 0.05 to a larger level, such as 0.10, statistical power increases. Because this also increases the probability of a Type I error, this is not usually a good method for increasing statistical power.

2. Turn a two-tailed hypothesis into a one-tailed hypothesis. We have been using a simpler one-tailed test, which provides more statistical power. However, researchers usually begin with the more conservative two-tailed test. In Figure 8-11, we see the difference between the less powerful two-tailed test (Figure 8-11a) and the more powerful one-tailed test (Figure 8-11b). The curves in part (a), with a two-tailed test, show less statistical power than do the curves in part (b). However, it is usually best to be conservative and use a two-tailed test.

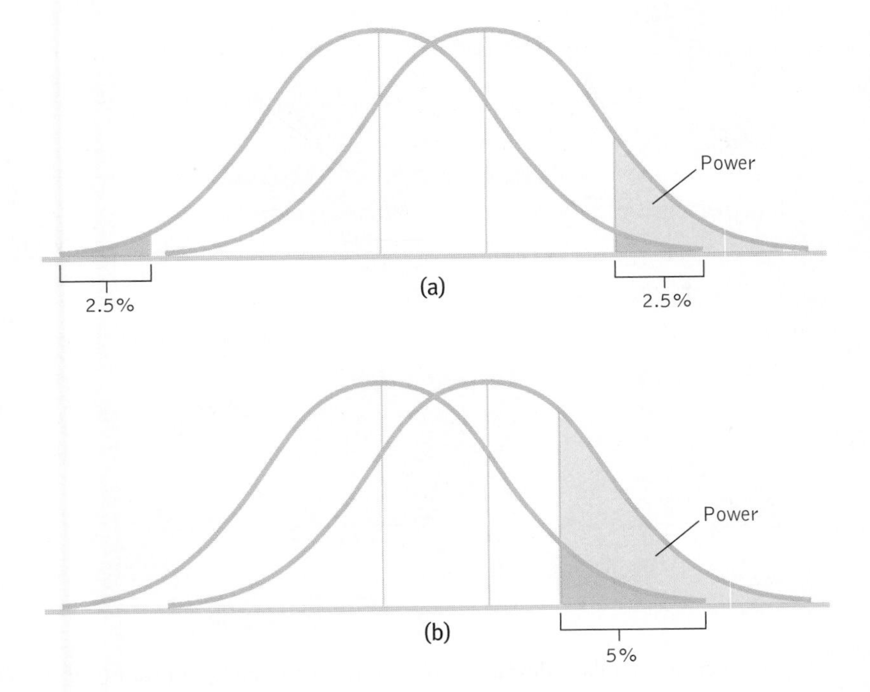

FIGURE 8-11
Two-Tailed Versus One-Tailed Tests

A two-tailed test divides alpha into two tails. When we use a one-tailed test, putting the entire alpha into just one tail, we increase the chances of rejecting the null hypothesis, which translates into an increase in statistical power.

3. Increase *N*. As we demonstrated earlier in this chapter, increasing sample size leads to an increase in the test statistic, making it easier to reject the null hypothesis. The curves in Figure 8-12a represent a small sample size; those in Figure 8-12b represent a larger sample size. The curves are narrower in part (b) than in part (a) because a larger sample size means smaller standard error. We have direct control over sample size, so simply increasing *N* is often an easy way to increase statistical power.

FIGURE 8-12
Increasing Sample Size or Decreasing Standard Deviation

As sample size increases, from part (a) to part (b), the distributions of means become more narrow and there is less overlap. Less overlap means more statistical power. The same effect occurs when we decrease standard deviation. As standard deviation decreases, also reflected from part (a) to part (b), the curves are narrower and there is less overlap—and more statistical power.

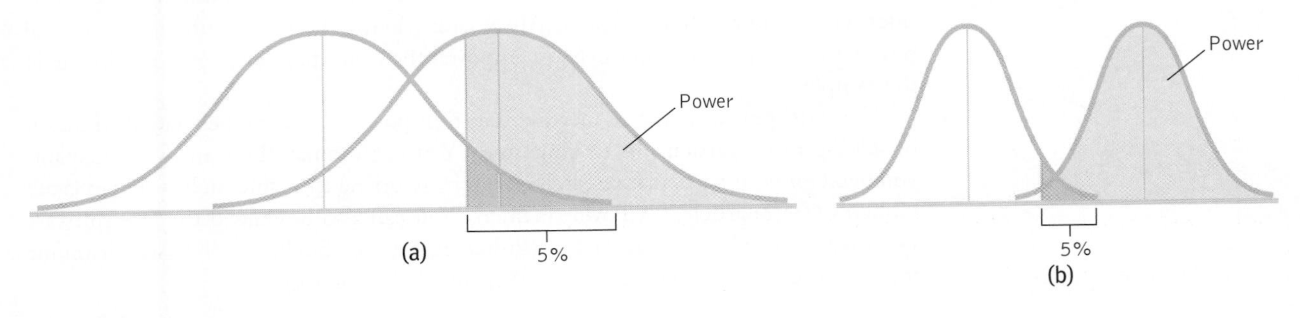

4. Exaggerate the mean difference between levels of the independent variable. As seen in Figure 8-13, the mean of population 2 is farther from the mean of population 1 in part (b) than it is in part (a). The difference between means is not easily changed, but it can be done. For instance, if we were studying the effectiveness of group therapy for social phobia, we could increase the length of therapy from 12 weeks to 6 months. It is possible that a longer program might lead to a larger change in means than would the shorter program.

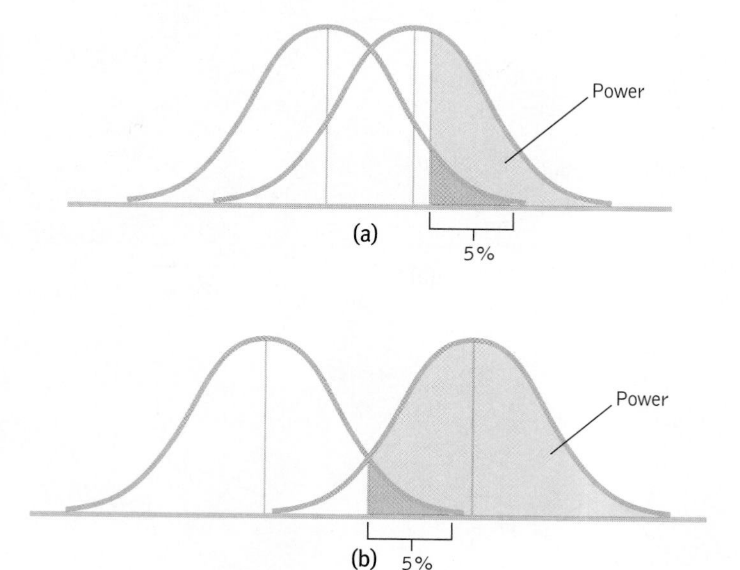

FIGURE 8-13
Increasing the Difference between the Means

As the difference between means becomes larger, there is less overlap between curves. Here, the lower pair of curves has less overlap than the upper pair. Less overlap means more statistical power.

5. Decrease standard deviation. We see the same effect on statistical power if we find a way to decrease the standard deviation as when we increase sample size. Look again at Figure 8-12, which reflects an increase in sample size. The curves can become narrower not just because the denominator of the standard error calculation is larger, but also because the numerator is smaller. When standard deviation is smaller, standard error is smaller, and the curves are narrower. We can reduce standard deviation in two ways: (1) using reliable measures from the beginning of the study, thus reducing error, or (2) sampling from a more homogeneous group in which participants' responses are more likely to be similar to begin with.

Because statistical power is affected by so many variables, it is important to consider when reading about research. Always ask whether there was sufficient statistical power to detect a real finding. Most importantly, were there enough participants in the sample?

The most practical way to increase statistical power for many behavioral studies is by adding more participants to your study. You can estimate how many participants you need for your particular research design by referring to a table such as that in Jacob Cohen's (1992) article, "A Power Primer." You can also download the free software G*Power, available for Mac or PC (Erdfelder, Faul, & Buchner, 1996; search online for G*Power or find the link on the Web site for this book).

Statistical power calculators like G*Power are versatile tools that are typically used in one of two ways. First, we calculate power *after* conducting a study from several pieces of information. For most electronic power calculators, including G*Power, we determine power by inputting the effect size and sample size along with some of the information that we outlined earlier in Table 8-2. We calculate power after determining effect size and other characteristics, so G*Power refers to these calculations as "post hoc," which means "after the fact." Second, we use them in reverse, *before* conducting a study, by calculating the sample size necessary to achieve a given power. In this case, we use a power calculator to determine the sample size necessary to achieve the statistical power that we want *before* we conduct the study. G*Power refers to such calculations as "a priori," which means "prior to."

The controversy over the 1980 Benbow and Stanley study of gender differences in mathematical ability demonstrates why it is important to go beyond hypothesis testing. The study used 10,000 participants, so it had plenty of statistical power. That means we can trust that the statistically significant difference they found was real. But the effect size informed us that this statistically significant difference was trivial. Combining all four ways of analyzing data (hypothesis testing, confidence intervals, effect size, and power analysis) helps us listen to the data story with all of its wonderful nuances and suggestions for future research.

CHECK YOUR LEARNING

Reviewing the Concepts

- Statistical power is the probability that we will reject the null hypothesis if we should reject it.
- Ideally, a study is not conducted unless the researcher has 80% statistical power; that is, at least 80% of the time we will correctly reject the null hypothesis.
- Statistical power is affected by several factors, but most directly by sample size.
- Before conducting a study, researchers often determine the number of participants they need to ensure statistical power of 0.80.
- To get the most complete story about the data, it is best to combine the results of hypothesis testing with information gained from confidence intervals, effect size, and power.

Clarifying the Concepts

8-8 What are three ways to increase statistical power?

Calculating the Statistics

8-9 Check Your Learning 8-3 and 8-7 discussed a study aimed at changing CFC scores through a career discussion group. Imagine that those in the discussion group of 45 students have a mean CFC score of 3.7. Let's say that you know that the population mean CFC score is 3.51, with a standard deviation of 0.61. Calculate statistical power for this as a one-tailed test.

Applying the Concepts

8-10 Refer to Check Your Learning 8-9.

a. Explain what the number obtained in your statistical power calculation means.

b. Describe how the researchers might increase statistical power.

Solutions to these Check Your Learning questions can be found in Appendix D.

REVIEW OF CONCEPTS

Confidence Intervals

A summary statistic, such as a mean, is a *point estimate* of the population mean. A more useful estimate is an *interval estimate,* a range of plausible numbers for the population mean. The most commonly used interval estimate is the *confidence interval,* which can be created around a mean using a z distribution. The confidence interval provides the same information as a hypothesis test but also gives us a range of values.

Effect Size

Knowing that a difference is statistically significant does not provide information about the size of the effect. A study with a large sample might find a small effect to be statistically significant, whereas a study with a small sample might fail to detect a large effect. To understand the importance of a finding, we must calculate an *effect size.* Effect sizes are independent of sample size because they are based on distributions of scores rather than distributions of means. One common effect-size measure is *Cohen's d,* which can be used when a z test has been conducted. A *meta-analysis* is a study of studies in which the researcher chooses a topic, decides on guidelines for a study's inclusion, tracks down every study on a given topic, and calculates an effect size for each. A mean effect size is calculated and reported, often along with a standard deviation, median, hypothesis testing, confidence interval, and appropriate graphs.

Statistical Power

Statistical power is a measure of the likelihood that we will correctly reject the null hypothesis; that is, the chance that we will not commit a Type II error when the research hypothesis is true. Statistical power is affected most directly by sample size, but it is also affected by other factors. Researchers often use a computerized statistical power calculator to determine the appropriate sample size to achieve 0.80 statistical power.

How It Works

8.1 CALCULATING CONFIDENCE INTERVALS

The Graded Naming Test (GNT) asks respondents to name objects in a set of 30 black-and-white drawings in order to detect brain damage. The GNT population norm for adults in England is 20.4. Researchers wondered whether a sample of Canadian adults had different scores from adults in England (Roberts, 2003). If the scores were different, the English norms would not be valid for use in Canada. The mean for 30 Canadian adults was 17.5. Assume that the standard deviation of the adults in England is 3.2. How can we calculate a 95% confidence interval for these data?

Given $\mu = 20.4$ and $\sigma = 3.2$, we can start by calculating standard error:

$$\sigma_M = \frac{\sigma}{\sqrt{N}} = \frac{3.2}{\sqrt{30}} = 0.584$$

We then find the z values that mark off the most extreme 0.025 in each tail, which are -1.96 and 1.96. We calculate the lower end of the interval as:

$$M_{lower} = -z(\sigma_M) + M_{sample} = -1.96(0.584) + 17.5 = 16.36$$

We calculate the upper end of the interval as:

$$M_{upper} = z(\sigma_M) + M_{sample} = 1.96(0.584) + 17.5 = 18.64$$

The 95% confidence interval around the mean of 17.5 is [16.36, 18.64].

8.2 CALCULATING EFFECT SIZE

The Graded Naming Test (GNT) study has a population norm for adults in England of 20.4. Researchers found a mean for 30 Canadian adults of 17.5, and we assumed a standard deviation of adults in England of 3.2 (Roberts, 2003). How can we calculate effect size for these data?

The appropriate measure of effect size for a z statistic is Cohen's d, which is calculated as:

$$d = \frac{M - \mu}{\sigma} = \frac{17.5 - 20.4}{3.2} = -0.91$$

Based on Cohen's conventions, this is a large effect size.

Exercises

Clarifying the Concepts

8.1 What specific danger exists when reporting a statistically significant difference between two means?

8.2 In your own words, define the word *confidence*—first as you would use it in everyday conversation and then as a statistician would use it in the context of a confidence interval.

8.3 Why do we calculate confidence intervals?

8.4 What are the five steps to create a confidence interval for the mean of a z distribution?

8.5 In your own words, define the word *effect*—first as you would use it in everyday conversation and then as a statistician would use it.

8.6 What effect does increasing the sample size have on standard error and the test statistic?

8.7 Relate effect size to the concept of overlap between comparison distributions.

8.8 What does it mean to say an effect-size statistic neutralizes the influence of sample size?

8.9 What are Cohen's guidelines for small, medium, and large effects?

8.10 How does statistical power relate to Type II errors?

8.11 In your own words, define the word *power*—first as you would use it in everyday conversation and then as a statistician would use it.

8.12 How are statistical power and effect size different but related?

8.13 Traditionally, what minimum percentage chance of correctly rejecting the null hypothesis is suggested in order to proceed with an experiment?

8.14 Explain how increasing alpha increases statistical power.

8.15 List five factors that affect statistical power. For each, indicate how a researcher can leverage that factor to increase power.

8.16 What are the four basic steps of a meta-analysis?

8.17 What is the goal of a meta-analysis?

8.18 In statistics, concepts are often expressed in symbols and equations. For $M_{lower} = -z(\sigma) + M_{sample}$, (i) identify the incorrect symbol, (ii) state what the correct symbol is, and (iii) explain why the initial symbol was incorrect.

8.19 In statistics, concepts are often expressed in symbols and equations. For $d = \frac{(M - \mu)}{\sigma_M}$, (i) identify the incorrect symbol, (ii) state what the correct symbol is, and (iii) explain why the initial symbol was incorrect.

Calculating the Statistics

8.20 In 2008, the Gallup poll asked people whether or not they were suspicious of steroid use among Olympic athletes. Thirty-five percent of respondents indicated suspicion when they saw an athlete break a track-and-field record, with a 4% margin of error. Calculate an interval estimate.

8.21 In 2008, twenty-two percent of Gallup respondents indicated that they were suspicious of steroid use by athletes who broke world records in swimming. Calculate an interval estimate using a margin of error at 3.5%.

8.22 In 2006, approximately 47% of Americans, when surveyed by a Gallup poll, felt that having a gun in the home made them safer than having no gun. The margin of error reported was 3%. Construct an interval estimate.

8.23 For each of the following confidence levels, indicate how much of the distribution would be placed in the cutoff region for a one-tailed test.

a. 80%
b. 85%
c. 99%

8.24 For each of the following confidence levels, indicate how much of the distribution would be placed in the cutoff region for a two-tailed test.

a. 80%
b. 85%
c. 99%

8.25 For each of the following confidence levels, look up the critical z value for a one-tailed test.

a. 80%

b. 85%

c. 99%

8.26 For each of the following confidence levels, look up the critical z values for a two-tailed test.

a. 80%

b. 85%

c. 99%

8.27 Calculate the 95% confidence interval for the following fictional data regarding daily TV viewing habits: $\mu = 4.7$ hours; $\sigma = 1.3$ hours; sample of 78 people with a mean of 4.1 hours.

8.28 Calculate the 80% confidence interval for the same fictional data regarding daily TV viewing habits: $\mu = 4.7$ hours; $\sigma = 1.3$ hours; sample of 78 people with mean of 4.1 hours.

8.29 Calculate the 99% confidence interval for the same fictional data regarding daily TV viewing habits: $\mu = 4.7$ hours; $\sigma = 1.3$ hours; sample of 78 people with mean of 4.1 hours.

8.30 Calculate standard error for each of the following sample sizes when $\mu = 1014$ and $\sigma = 136$:

a. 12

b. 39

c. 188

8.31 For a given variable, imagine we know that the population mean is 1014 and the standard deviation is 136. A sample mean of 1057 is obtained. Calculate the z statistic for this mean, using each of the following sample sizes:

a. 12

b. 39

c. 188

8.32 Calculate the effect size for the mean of 1057 observed in Exercise 8.31 where $\mu = 1014$ and $\sigma = 136$.

8.33 Calculate the effect size for each of the following average SAT math scores. Remember, SAT math is standardized such that $\mu = 500$ and $\sigma = 100$.

a. 61 people sampled have a mean of 480.

b. 82 people sampled have a mean of 520.

c. 6 people sampled have a mean of 610.

8.34 For each of the effect-size calculations in Exercise 8.33, identify the size of the effect using Cohen's guidelines. Remember, for SAT math, $\mu = 500$ and $\sigma = 100$.

a. 61 people sampled have a mean of 480.

b. 82 people sampled have a mean of 520.

c. 6 people sampled have a mean of 610.

8.35 For each of the following d values, identify the size of the effect, using Cohen's guidelines.

a. $d = 0.79$

b. $d = -0.43$

c. $d = 0.22$

d. $d = -0.04$

8.36 For each of the following d values, identify the size of the effect, using Cohen's guidelines.

a. $d = 1.22$

b. $d = -1.22$

c. $d = 0.13$

d. $d = -0.13$

8.37 For each of the following z statistics, calculate the p value for a two-tailed test.

a. 2.23

b. −1.82

c. 0.33

8.38 A meta-analysis reports an average effect size of $d = 0.11$, with a confidence interval of $d = 0.08$ to $d = 0.14$.

a. Would a hypothesis test (assessing the null hypothesis that the average effect size is 0) lead us to reject the null hypothesis? Explain.

b. Use Cohen's conventions to describe the average effect size of $d = 0.11$.

8.39 A meta-analysis reports an average effect size of $d = 0.11$, with a confidence interval of $d = -0.06$ to $d = 0.28$. Would a hypothesis test (assessing the null hypothesis that the average effect size is 0) lead us to reject the null hypothesis? Explain.

8.40 Assume you are conducting a meta-analysis over a set of five studies. The effect sizes for each study follows: $d = 0.67$; $d = 0.03$; $d = 0.32$; $d = 0.59$; $d = 0.22$.

a. Calculate the mean effect size for these studies.

b. Use Cohen's conventions to describe the mean effect size you calculated in part (a).

Applying the Concepts

8.41 Distributions and the Burakumin: A friend reads in her *Introduction to Psychology* textbook about a minority group in Japan, the Burakumin, who are racially the same as other Japanese people but are viewed as outcasts because their ancestors were employed in positions that involved the handling of dead animals (e.g., butchers). In Japan, the text reported, mean IQ scores of Burakumin were 10 to 15 points below mean IQ scores of other Japanese. In the United States, where Burakumin experienced no discrimination, there was no mean difference (from Ogbu, 1986, as reported in Hockenbury & Hockenbury, 2003). Your friend says to you: "Wow—when I taught English in Japan last summer, I had a Burakumin student. He seemed smart; perhaps I was fooled." What should your friend consider about the two distributions, the one for Burakumin people and the one for other Japanese people?

8.42 Sample size, z statistics, and the Consideration of Future Consequences scale: Here are summary data from a z test regarding scores on the Consideration of Future Consequences scale (Petrocelli, 2003): the population mean (μ) is 3.51 and the population standard deviation (σ) is 0.61. Imagine that a sample of students had a mean of 3.7.

a. Calculate the test statistic for a sample of 5 students.

b. Calculate the test statistic for a sample of 1000 students.

c. Calculate the test statistic for a sample of 1,000,000 students.

d. Explain why the test statistic varies so much even though the population mean, population standard deviation, and sample mean do not change.

e. Why might sample size pose a problem for hypothesis testing and the conclusions we are able to draw?

8.43 Sample size, z statistics, and the Graded Naming Test: In an exercise in Chapter 7, we asked you to conduct a z test to ascertain whether the Graded Naming Test (GNT) scores for Canadian participants differed from the GNT norms based on adults in England. We also used these data in the How It Works section of this chapter. The mean for a sample of 30 adults in Canada was 17.5. The normative mean for adults in England is 20.4, and we assumed a population standard deviation of 3.2. With 30 participants, the z statistic was -4.97, and we were able to reject the null hypothesis.

a. Calculate the test statistic for 3 participants. How does the test statistic change compared to when N of 30 was used? Conduct step 6 of hypothesis testing. Does your conclusion change? If so, does this mean that the actual difference between groups changed? Explain.

b. Conduct steps 3, 5, and 6 for 100 participants. How does the test statistic change?

c. Conduct steps 3, 5, and 6 for 20,000 participants. How does the test statistic change?

d. What is the effect of sample size on the test statistic?

e. As the test statistic changes, has the underlying difference between groups changed? Why might this present a problem for hypothesis testing?

8.44 Cheating with hypothesis testing: Unsavory researchers know that one can cheat with hypothesis testing. That is, they know that a researcher can stack the deck in her or his favor, making it easier to reject the null hypothesis.

a. If you wanted to make it easier to reject the null hypothesis, what are three specific things you could do?

b. Would it change the actual difference between the samples? Why is this a potential problem with hypothesis testing?

8.45 Overlapping distributions and the LSATs: A Midwestern U.S. university reported that its behavioral science majors tended to outperform its humanities majors on the LSAT standardized test for law school admissions. Sadie, an English major, and Kofi, a sociology major, both just took the LSAT.

a. Can we tell which student will do better on the LSAT? Explain your answer.

b. Draw a picture that represents what the two distributions, that for social science majors and that for humanities majors at this institution, might look like with respect to one another.

8.46 Confidence intervals, effect sizes, and tennis serves: Let's assume the average speed of a serve in men's tennis is around 135 mph, with a standard deviation of 6.5 mph. Because these statistics are calculated over many years and many players, we will treat them as population parameters. We develop a new training method that will increase arm strength, the force of the tennis swing, and the speed of the serve, we hope. We recruit 9 professional tennis players to use our method. After 6 months, we test the speed of their serves and compute an average of 138 mph.

a. Using a 95% confidence interval, test the hypothesis that our method makes a difference.

b. Compute the effect size and describe its strength.

c. Calculate statistical power using an alpha of 0.05, or 5%, and a one-tailed test.

d. Calculate statistical power using an alpha of 0.10, or 10%, and a one-tailed test.

e. Explain how power is affected by alpha in the calculations in (c) and (d).

8.47 Confidence intervals and football wins: In an exercise in Chapter 7, we asked whether college football teams tend to be more likely or less likely to be mismatched in the upper National Collegiate Athletic Association (NCAA) divisions. During week 11 of the fall 2006 college football season, the population of 53 Division I-A games had a mean spread (winning score minus losing score) of 16.189, with a standard deviation of 12.128. We took a sample of four games that were played that week in the next-highest league, Division I-AA, to see if the spread were different; one of the many leagues within Division I-AA, the Patriot League, played four games that weekend. Their mean was 8.75.

a. Calculate the 95% confidence interval for this sample.

b. State in your own words what we learn from this confidence interval.

c. What information does the confidence interval give us that we also get from a hypothesis test?

d. What additional information does the confidence interval give us that we do not get from a hypothesis test?

8.48 Confidence intervals and football wins (continued): Using the football data presented in Exercise 8.47, practice evaluating data, using confidence intervals.

a. Compute the 80% confidence interval.

b. How do your conclusion and the confidence interval change as you move from 95% confidence to 80% confidence?

c. Why don't we talk about having 100% confidence?

8.49 Effect size and football wins: In Exercises 8.47 and 8.48, we considered the study of week 11 of the fall 2006 college football season, during which the population of 53 Division I-A games had a mean spread (winning score minus losing score) of 16.189, with a standard deviation of 12.128. The sample of four games that were played that week in the next highest league, Division I-AA, had a mean of 8.75.

a. Calculate the appropriate measure of effect size for this sample.

b. Based on Cohen's conventions, is this a small, medium, or large effect?

c. Why is it useful to have this information in addition to the results of a hypothesis test?

8.50 Effect size and football wins (continued): In Exercise 8.49, you calculated an effect size for data from week 11 of the fall 2006 college football season with 4 games. Imagine that you had a sample of 20 games. How would the effect size change? Explain why it does or does not change.

8.51 Confidence intervals, effect sizes, and Valentine's Day spending: According to the Nielsen Company, Americans spend $345 million on chocolate during the week of Valentine's Day. Let's assume that we know the average married person spends $45, with a population standard deviation of $16. In February 2009, the U.S. economy was in the throes of a recession. Comparing data for Valentine's Day spending in 2009 with what is generally expected might give us some indication of the attitudes of American citizens.

a. Compute the 95% confidence interval for a sample of 18 married people who spent an average of $38.

b. How does the 95% confidence interval change if the sample mean is based on 180 people?

c. If you were testing a hypothesis that things had changed under the financial circumstances of 2009 as compared to in previous years, what conclusion would you draw in part (a) versus part (b)?

d. Compute the effect size based on these data and describe the size of the effect.

8.52 More about confidence intervals, effect sizes, and tennis serves: Let's assume the average speed of a serve in women's tennis is around 118 mph, with a standard deviation of 12 mph. We recruit 100 amateur tennis players to use our method this time, and after 6 months we calculate a group mean of 123 mph.

a. Using a 95% confidence interval, test the hypothesis that our method makes a difference.

b. Compute the effect size and describe its strength.

8.53 Confidence intervals, effect sizes, and tennis serves (continued): As in the previous exercise, assume the average speed of a serve in women's tennis is around 118 mph, with a standard deviation of 12 mph. But now we recruit only 26 amateur tennis players to use our method. Again, after 6 months we calculate a group mean of 123 mph.

a. Using a 95% confidence interval, test the hypothesis that our method makes a difference.

b. Compute the effect size and describe its strength.

c. How did changing the sample size from 100 (in Exercise 8.52) to 26 affect the confidence interval and effect size? Explain your answer.

8.54 Statistical power and football wins: In several exercises in this chapter, we considered the study of week 11 of the fall 2006 college football season, during which the population of 53 Division I-A games had a mean spread (winning score minus losing score) of 16.189, with a standard deviation of 12.128. The sample of four games that were played that week in the next-highest league, Division I-AA, had a mean of 8.75.

a. Calculate statistical power for this study using a one-tailed test and a p level of 0.05.

b. What does the statistical power suggest about how we should view the findings of this study?

c. Using G*Power or an online power calculator, calculate statistical power for this study for a one-tailed test with a p level of 0.05.

8.55 Meta-analysis, mental health treatments, and cultural contexts: A meta-analysis examined studies that compared two types of mental health treatments for ethnic and racial minorities—the standard available treatments and treatments that were adapted to the clients' cultures (Griner & Smith, 2006). An excerpt from the abstract follows:

> *Many previous authors have advocated traditional mental health treatments be modified to better match clients' cultural*

contexts. Numerous studies evaluating culturally adapted interventions have appeared, and the present study used meta-analytic methodology to summarize these data. Across 76 studies the resulting random effects weighted average effect size was d = .45, indicating a . . . benefit of culturally adapted interventions (p. 531).

a. What is the topic chosen by the researchers conducting the meta-analysis?

b. What type of effect size statistic did the researchers' calculate for each study in the meta-analysis?

c. What was the mean effect size? According to Cohen's conventions, how large is this effect?

d. If a study chosen for the meta-analysis did not include an effect size, what summary statistics could the researchers use to calculate an effect size?

8.56 Meta-analysis, mental health treatments, and cultural contexts (continued): The research paper on culturally targeted therapy describe in Exercise 8.55 reported the following:

Across all 76 studies, the random effects weighted average effect size was d = .45 (SE = .04, p < .0001), with a 95% confidence interval of d = .36 to d = .53. The data consisted of 72 nonzero effect sizes, of which 68 (94%) were positive and 4 (6%) were negative. Effect sizes ranged from d = −48 to d = 2.7 (Griner & Smith, 2006, p. 535).

a. What is the confidence interval for the effect size?

b. Based on the confidence interval, would a hypothesis test lead us to reject the null hypothesis that the effect size is zero? Explain.

c. Why would a graph, such as a histogram, be useful when conducting a meta-analysis like this one? (*Hint:* Consider the problems when using a mean as the measure of central tendency.)

Putting It All Together

8.57 Fantasy baseball: Your roommate is reading *Fantasyland: A Season on Baseball's Lunatic Fringe* (Walker, 2006) and is intrigued by the statistical methods used by competitors in fantasy baseball leagues (in which competitors select a team of baseball players from across all major league teams, winning in the fantasy league if their eclectic roster of players outperforms the chosen mixes of other fantasy competitors). Among the many statistics reported in the book is a finding that Major League Baseball (MLB) players who have a third child show more of a decline in performance than players who have a first child or a second child. Your friend remembers that Johnny Damon had a third child and drops him from consideration for his fantasy team.

a. Explain to your friend why a difference between means doesn't provide information about any specific individual player. Include a drawing of overlapping curves as part of your answer. On the drawing, mark places on the x-axis that might represent a player from the distribution of those who recently had a third child (mark with an X) scoring *above* a player from the distribution of those who recently had a first or second child (mark with a Y).

b. Explain to your friend that a statistically significant difference doesn't necessarily indicate a large effect size. How might a measure of effect size, such as Cohen's d, help us understand the importance of these findings and compare them to other predictors of performance that might have larger effects?

c. Given that the reported association is true, can we conclude that having a third child *causes* a decline in performance? Explain your answer. What confounding variables might lead to the difference observed in this study?

d. Given the relatively limited numbers of MLB players (and the relatively limited numbers of those who recently had a child—whether first, second, or third), what general guess would you make about the likely statistical power of this analysis?

8.58 Hours of Sleep: The table below provides information about hours of sleep.

Mean of population 1 (from which the sample comes)	14.9 hours of sleep
Sample size	37 infants
Mean of population 2	16 hours of sleep
Standard deviation of the population	1.7 hours of sleep
Standard error	$\sigma_M = \frac{\sigma}{\sqrt{N}} = \frac{1.7}{\sqrt{37}} = 0.279$

a. Calculate statistical power for a one-tailed test (α = 0.05, or 5%) aimed at determining if those in the sample sleep fewer hours, on average, than those in the population.

b. Recalculate statistical power with alpha of 0.01, or 1%. Explain why changing alpha affects power. Explain why we should not use a larger alpha to increase power.

c. Without performing any computations, describe how statistical power is affected by performing a two-tailed test for this example. Why are two-tailed tests recommended over one-tailed tests?

d. The easiest way to affect the outcome of a hypothesis test is to increase sample size. Similarly, true results may sometimes be missed because a sufficient sample was not used in the research. Perform the hypothesis test on these data with a sample of 37. Then, perform the same hypothesis test but assume that the mean was based on only 4 infants.

e. The easiest way to increase statistical power is to increase sample size. Similarly, statistical power decreases with a smaller sample size. For these data, compute the statistical power of the one-tailed statistical test with alpha of 0.05 when N is 4. How does that value compare to when N was 37?

Terms

point estimate (p. 173)
interval estimate (p. 173)
confidence interval (p. 174)
effect size (p. 179)
Cohen's d (p. 181)
meta-analysis (p. 182)
statistical power (p. 183)

Formulas

$M_{lower} = -z(\sigma_M) + M_{sample}$ (p. 175)

$M_{upper} = z(\sigma_M) + M_{sample}$ (p. 175)

Cohen's $d = \frac{(M - \mu)}{\sigma}$ for a z distribution (p. 181)

Symbols

Cohen's d (or just d) (p. 181)
α (p. 185)

CHAPTER 9

The Single-Sample *t* Test and the Paired-Samples *t* Test

BEFORE YOU GO ON

- You should know the six steps of hypothesis testing (Chapter 7).
- You should know how to determine a confidence interval for a *z* statistic (Chapter 8).
- You should understand the concept of effect size and know how to calculate Cohen's *d* for a *z* test (Chapter 8).

Holiday Weight Gain and Two-Group Studies Two-group studies indicate that the average holiday weight gain by college students is less than many people believe—only about 1 pound.

Radius Images/Alamy

In many parts of the world, the winter holiday season is a time when family food traditions take center stage. Popular wisdom suggests that during this season many Americans put on 5 to 7 pounds. But before-and-after studies suggest a weight gain of just over 1 pound (Hull, Radley, Dinger, & Fields, 2006; Roberts & Mayer, 2000; Yanovski et al., 2000). A 1-pound weight gain over the holidays might not seem so bad, but weight gained over the holidays tends to stay (Yanovski et al., 2000).

> **MASTERING THE CONCEPT**
>
> **9.1:** There are three types of *t* tests. We use a single-sample *t* test when we compare a sample mean to a population mean but do not know the population standard deviation. We use a paired-samples *t* test when we compare two samples and every participant is in both samples—a within-groups design. We use an independent-samples *t* test, discussed in Chapter 10, when we compare two samples and every participant is in only one sample—a between-groups design.

The fact that researchers used two groups in their study—students before the holidays and students after the holidays—is important for this chapter. We will learn about an expansion of the *z* distribution—the more versatile *t* distributions. With a *t* distribution we can compare one sample to a population when we don't know all the details about the parameters, and we can compare two samples to each other. There are two ways to compare two samples: we can use a within-groups design (as when the same people are weighed before and after the holidays) or a between-groups design (as when different people are in the pre-holiday sample and the post-holiday sample). For a within-groups design, we use a paired-samples *t* test. The steps for a paired-samples *t* test are similar to those for a single-sample *t* test, which is why we learn about both of these hypothesis tests in this chapter. (For a between-groups design, we use an independent-samples *t* test, which we will learn about in Chapter 10.)

The *t* Distributions

The *t* distributions (note the plural) help us specify how confident we can be in our research findings. We want to know if we can generalize what we have learned about one sample to a larger population. The *t* test, based on the *t* distributions, tells us how confident we can be that the sample differs from the larger population.

The *t* distributions are more versatile than the *z* distribution because we can use them when (a) we don't know the population standard deviation, and (b) we are com-

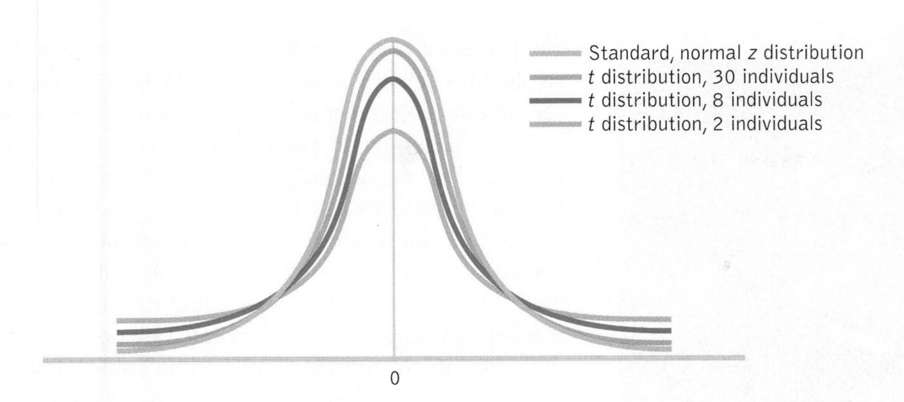

FIGURE 9-1
The Wider and Flatter *t* Distributions

For smaller samples, the *t* distributions are wider and flatter than the *z* distribution. As the sample size increases, however, the *t* distributions approach the shape of the *z* distribution. For instance, in this figure, the *t* distribution most similar to the *z* distribution is that for a sample of 30 individuals. This makes sense because a distribution derived from a larger sample size would be more likely to be similar to that of the entire population than one derived from a smaller sample size.

paring two samples. Figure 9-1 demonstrates that there are many *t* distributions—one for each possible sample size. Just as you are less likely to believe gossip from only one or two people, we become less certain about what the population distribution really looks like when we have a small sample size. The uncertainty of a small sample size means that the *t* distributions become flatter and more spread out. However, as the sample size gets larger, the *t* distributions begin to merge with the *z* distribution because we gain confidence as more participants are added to a study.

Estimating Population Standard Deviation from a Sample

Before we conduct a single-sample *t* test, we must estimate the population standard deviation. To do this, we use the sample standard deviation to estimate the population standard deviation. Estimating the standard deviation is the only practical difference between conducting a *z* test with the *z* distribution and conducting a *t* test with a *t* distribution. Here is the sample standard deviation formula that we have used up until now:

$$SD = \sqrt{\frac{\Sigma(X - M)^2}{N}}$$

We need to make a correction to this formula to account for the fact that there is likely to be some level of error when we estimate the population standard deviation from a sample. Specifically, any given sample is likely to have somewhat less spread than does the entire population. One tiny alteration of this formula leads to a slightly larger and more accurate standard deviation. Instead of dividing by *N*, we divide by $(N - 1)$ to get the mean of the squared deviations. Subtraction is the key. For example, if the numerator was 90 and the denominator (N) was 10, the answer would be 9; if we divide by $(N - 1) = (10 - 1) = 9$, the answer would be 10, a slightly larger value. So the formula is:

$$s = \sqrt{\frac{\Sigma(X - M)^2}{(N - 1)}}$$

MASTERING THE FORMULA

9-1: The formula for standard deviation when estimating from a sample is: $s = \sqrt{\frac{\Sigma(X - M)^2}{(N - 1)}}$. We subtract 1 from the sample size in the denominator to correct for the probability that the sample standard deviation slightly underestimates the actual standard deviation in the population.

Notice that we call this standard deviation *s* instead of *SD*. We still use Latin rather than Greek letters because it is a statistic (from a sample) rather than a parameter (from a population). From now on, we will calculate the standard deviation in this way because we will be estimating the population standard deviation.

Let's apply the new formula for standard deviation to a situation that involves a familiar activity: multitasking. Employees were observed at one of two high-tech

Ocean/Corbis

Multitasking If multitasking reduces productivity in a sample, we can statistically determine the probability that multitasking reduces productivity among a much larger population.

companies for over 1000 hours (Mark, Gonzalez, & Harris, 2005). The employees spent just 11 minutes, on average, on one project before an interruption. Moreover, after each interruption, they needed an average of 25 minutes to get back to the original project! So maybe the reality is that multitasking actually reduces overall productivity.

Suppose you were a manager at one of these firms and decided to reserve a period from 1:00 to 3:00 each afternoon during which employees could not interrupt one another, but might still be interrupted by people outside the company. To test the intervention, you observe five employees and develop a score for each—time spent on a selected task before being interrupted. Here are the fictional data: 8, 12, 16, 12, and 14 minutes. In this case, we treat 11 minutes as the population mean, but we do not know the population standard deviation.

EXAMPLE 9.1

To calculate the estimated standard deviation for the population, there are two steps.

STEP 1: Calculate the sample mean. Even though we know the population mean (i.e., 11), we use the sample mean to calculate the corrected sample standard deviation. The mean for these scores is:

$$M = \frac{(8 + 12 + 16 + 12 + 14)}{5} = 12.4$$

STEP 2: Use the sample mean in the corrected formula for the standard deviation.

$$s = \sqrt{\frac{\Sigma(X - M)^2}{(N - 1)}}$$

Remember, the easiest way to calculate the numerator under the square root sign is by first organizing the data into columns, as shown here:

X	$X - M$	$(X - M)^2$
8	−4.4	19.36
12	−0.4	0.16
16	3.6	12.96
12	−0.4	0.16
14	1.6	2.56

Thus, the numerator is:

$$\Sigma(X - M)^2 = \Sigma(19.36 + 0.16 + 12.96 + 0.16 + 2.56) = 35.2$$

A Simple Correction: $N - 1$ When estimating variability, subtracting one person from a sample of four makes a big difference. Subtracting one person from a sample of thousands makes only a small difference.

And given a sample size of 5, the corrected standard deviation is:

$$s = \sqrt{\frac{\Sigma(X - M)^2}{(N - 1)}} = \sqrt{\frac{35.2}{(5 - 1)}} = \sqrt{8.8} = 2.97 \quad ■$$

Calculating Standard Error for the *t* Statistic

We now have an estimate of the standard deviation of the distribution of scores, but not an estimate of the spread of a distribution of means, the standard error. As we did with the *z* distribution, we make the spread smaller to reflect the fact that a distribution of means is less variable than a distribution of scores. We do this in exactly the same way that we adjusted for the *z* distribution. We divide s by $\sqrt{N}$. The formula for the standard error as estimated from a sample, therefore, is:

$$s_M = \frac{s}{\sqrt{N}}$$

Notice that we have replaced σ with s because we are using the corrected sample standard deviation rather than the population standard deviation.

MASTERING THE FORMULA

9-2: The formula for standard error when we estimate from a sample is: $s_M = \frac{s}{\sqrt{N}}$. It only differs from the formula for standard error we learned previously in that we use s instead of σ because we're working from a sample instead of a population.

EXAMPLE 9.2

Here's how we convert the corrected standard deviation of 2.97 to a standard error. The sample size was 5, so we divide by the square root of 5:

$$s_M = \frac{s}{\sqrt{N}} = \frac{2.97}{\sqrt{5}} = 1.33$$

So the standard error is 1.33. Just as the central limit theorem predicts, the standard error for the distribution of sample means is smaller than the standard deviation of sample scores. (*Note:* This step can lead to a common mistake. Because we implemented a correction when calculating *s*, students often want to implement an extra correction here by dividing by $\sqrt{N-1}$. Do not do this! We still divide by $\sqrt{N}$ in this step. There is no need for a further correction to the standard error.) ■

Using Standard Error to Calculate the *t* Statistic

We now have the tools necessary to conduct the simplest type of *t* test, the single-sample *t* test. When conducting a single-sample *t* test, we use the ***t* statistic**, *the distance of a sample mean from a population mean in terms of the estimated standard error.* We introduce the formula for that *t* statistic here, and in the next section we go through all six steps for a single-sample *t* test. The formula is identical to that for the *z* statistic, except that it uses estimated standard error. Here is the formula for the *t* statistic for a distribution of means:

MASTERING THE FORMULA

9-3: The formula for the *t* statistic is: $t = \frac{(M - \mu_M)}{s_M}$. It only differs from the formula for the *z* statistic in that we use s_M instead of σ_M because we use the sample to estimate standard error rather than the actual population standard error.

$$t = \frac{(M - \mu_M)}{s_M}$$

Note that the denominator is the only difference between this formula for the *t* statistic and the formula used to compute the *z* statistic for a sample mean. The corrected denominator makes the *t* statistic smaller and thereby reduces the probability of observing an extreme *t* statistic. That is, a *t* statistic is not as extreme as a *z* statistic; in scientific terms, it's more conservative.

EXAMPLE 9.3

The *t* statistic for the sample of 5 scores representing minutes until interruptions is:

$$t = \frac{(M - \mu_M)}{s_M} = \frac{(12.4 - 11)}{1.33} = 1.05$$

As part of the six steps of hypothesis testing, the *t* statistic can help us make an inference about whether the ban on internal interruptions affected the average number of minutes until an interruption. ■

CHECK YOUR LEARNING

Reviewing the Concepts

> *t* distributions are used when we do not know the population standard deviation and are comparing only two groups.

> The two groups may be a sample and a population, or two samples as part of a within-groups design or a between-groups design.

> The formula for the *t* statistic for a single-sample *t* test is the same as the formula for the *z* statistic for a distribution of means, except that we use estimated standard error in the denominator rather than the actual standard error for the population.

> We calculate estimated standard error by dividing by $N - 1$, rather than *N*, when calculating standard error.

Clarifying the Concepts

9-1 What is the *t* statistic?

Calculating the Statistics

9-2 Calculate the standard deviation for a sample (*SD*) and as an estimate of the population (*s*) using the following data: 6, 3, 7, 6, 4, 5.

9-3 Calculate standard error for *t* for the data given in Check Your Learning 9-2.

Applying the Statistics

9-4 In the discussion of a study on multitasking (Mark et al., 2005), we imagined a follow-up study in which we measured time until a task was interrupted. For each of the five employees, let's now examine time until work on the initial task was resumed at 20, 19, 27, 24, and 18 minutes. Remember that the original research showed it took 25 minutes on average for an employee to return to a task after being interrupted.

a. What distribution will be used in this situation? Explain your answer.

b. Determine the appropriate mean and standard deviation (or standard error) for this distribution. Show all your work; use symbolic notation and formulas where appropriate.

c. Calculate the *t* statistic.

Solutions to these Check Your Learning questions can be found in Appendix D.

The Single-Sample *t* Test

A ***single-sample t test*** *is a hypothesis test in which we compare data from one sample to a population for which we know the mean but not the standard deviation.* We begin with the single-sample *t* test because understanding it will help us when using the more sophisticated *t* tests that let us compare two samples.

- The ***t* statistic** indicates the distance of a sample mean from a population mean in terms of the estimated standard error.
- A **single-sample *t* test** is a hypothesis test in which we compare data from one sample to a population for which we know the mean but not the standard deviation.
- **Degrees of freedom** is the number of scores that are free to vary when estimating a population parameter from a sample.

The *t* Table and Degrees of Freedom

When we use the *t* distributions, we use the *t* table. There are different *t* distributions for every sample size and the *t* table takes sample size into account. However, we do not look up the actual sample size on the table. Rather, we look up ***degrees of freedom***, *the number of scores that are free to vary when estimating a population parameter from a sample.*

Language Alert! The phrase "free to vary" refers to the number of scores that can take on different values if we know a given parameter.

EXAMPLE 9.4

For example, the manager of a baseball team needs to assign nine players to particular spots in the batting order but only has to make eight decisions ($N - 1$). Why? Because only one option remains after making the first eight decisions. So before the manager makes any decisions, there are $N - 1$, or $9 - 1 = 8$, degrees of freedom. After the second decision, there are $N - 1$, or $8 - 1 = 7$, degrees of freedom, and so on.

As in the baseball example, there is always one score that cannot vary once all of the others have been determined. For example, if we know that the mean of four scores is 6 and we know that three of the scores are 2, 4, and 8, then the last score must be 10. So the degrees of freedom is the number of scores in the sample minus 1. Degrees of freedom is written in symbolic notation as *df*, which is always italicized. The formula for degrees of freedom for a single-sample *t* test, therefore, is:

MASTERING THE CONCEPT

9.2: Degrees of freedom refers to the number of scores that can take on different values if we know a given parameter. For example, if you know that the mean of three scores is 10, only two scores are free to vary. Once you know the values of two scores, you know the value of the third. If you know that two of the scores are 9 and 10, then you know that the third must be 11.

$$df = N - 1$$ ▪

MASTERING THE FORMULA

9-4: The formula for degrees of freedom for a single-sample *t* test is: $df = N - 1$. To calculate degrees of freedom, we subtract 1 from the sample size.

Table 9-1 is an excerpt from a *t* table; the full table is in Appendix B. Notice the relation between degrees of freedom and the critical value needed to declare statistical significance. In the column corresponding to a one-tailed test at a *p* level of 0.05 with only 1 degree of freedom (two observations), the critical *t* value is 6.314. With only

TABLE 9-1. Excerpt from the *t* Table

When conducting hypothesis testing, we use the *t* table to determine critical values for a given *p* level, based on the degrees of freedom and whether the test is one- or two-tailed.

	One-Tailed Tests			Two-Tailed Tests		
df	0.10	0.05	0.01	0.10	0.05	0.01
1	3.078	6.314	31.821	6.314	12.706	63.657
2	1.886	2.920	6.965	2.920	4.303	9.925
3	1.638	2.353	4.541	2.353	3.182	5.841
4	1.533	2.132	3.747	2.132	2.776	4.604
5	1.476	2.015	3.365	2.015	2.571	4.032

1 degree of freedom, the two means have to be extremely far apart and/or the standard deviation has to be very small to declare a statistically significant difference. But with 2 degrees of freedom (three observations), the critical *t* value drops to 2.920. It is easier to reach the critical *t* value because we're more confident with three observations than with just two.

The pattern continues when we have four observations (with *df* of 3). The critical *t* value needed to declare statistical significance decreases from 2.920 to 2.353. The level of confidence in the observations increases and the critical value decreases.

The *t* distributions become closer to the *z* distribution as sample size increases—after all, if we kept enlarging the sample, we would eventually study the entire population and wouldn't need a pesky *t* test in the first place. But in the real world of research, the corrected standard deviation of a large enough sample will be so similar to the actual standard deviation of the population that the *t* distribution is the same as the *z* distribution.

Check it out for yourself by comparing the *z* and *t* tables in Appendix B. For example, the *z* statistic for the 95th percentile—a percentage between the mean and the *z* statistic of 45%—is between 1.64 and 1.65. At a sample size of infinity, the *t* statistic for the 95th percentile is 1.645. Infinity (∞) indicates a very large sample size; a sample size of infinity itself is, of course, impossible.

Let's remind ourselves why the *t* statistic merges with the *z* statistic as sample size increases. More participants in a study—if they are a representative sample—correspond to increased confidence that we are making an accurate observation. So don't think of the *t* distributions as completely separate from the *z* distribution. Rather, think of the *z* statistic as a single-blade Swiss Army knife and the *t* statistic as a multi-blade Swiss Army knife that still includes the single blade that is the *z* statistic.

Let's determine the cutoffs, or critical *t* value(s), for a research study using the full *t* table in Appendix B.

MASTERING THE CONCEPT

9.3: As sample size increases, the *t* distributions more and more closely approximate the *z* distribution. You can think of the *z* statistic as a single-blade Swiss Army knife and the *t* statistic as a multi-blade Swiss Army knife that includes the single blade that is the *z* statistic.

EXAMPLE 9.5

The study: A researcher knows the mean number of calories a rat will consume in half an hour if unlimited food is available. He wonders whether a new food will lead rats to consume a different number of calories—either more or fewer. He studies 38 rats and uses a *p* level of 0.05.

The cutoff(s): This is a two-tailed test because the research hypothesis allows for change in either direction. There are 38 rats, so the degrees of freedom is:

$$df = N - 1 = 38 - 1 = 37$$

We want to look in the *t* table under two-tailed tests, in the column for 0.05 and in the row for a *df* of 37; however, there is no *df* of 37. In this case, we err on the side of being more conservative and choose the more extreme (i.e., larger) of the two possible critical *t* values, which is always the smaller *df*. Here, we look next to 35, where we see a value of 2.030. Because this is a two-tailed test, we will have critical values of −2.030 and 2.030. ■

Zigy Kaluzny/Getty Images

Nonparticipation in Therapy
Clients missing appointments can be a problem for their therapists. A *t* test can compare the consequences between those who do and those who do not commit themselves to participating in therapy for a set period.

The Six Steps of the Single-Sample *t* Test

Now we have all the tools necessary to conduct a single-sample *t* test. So let's consider a hypothetical study and conduct all six steps of hypothesis testing.

EXAMPLE 9.6

Chapter 4 presented data that included the mean number of sessions attended by clients at a university counseling center. We noted that one study reported a mean of 4.6 sessions (Hatchett, 2003). Let's imagine that the counseling center hoped to increase participation rates by having students sign a contract to attend at least 10 sessions. Five students sign the contract and attend 6, 6, 12, 7, and 8 sessions, respectively. The researchers are interested only in their university, so treat the mean of 4.6 sessions as a population mean.

STEP 1: Identify the populations, distribution, and assumptions.

Population 1: All clients at this counseling center who sign a contract to attend at least 10 sessions. Population 2: All clients at this counseling center who do not sign a contract to attend at least 10 sessions.

The comparison distribution will be a distribution of means. The hypothesis test will be a single-sample *t* test because we have only one sample and we know the population mean but not the population standard deviation.

This study meets one of the three assumptions and may meet the other two: (1) The dependent variable is scale. (2) We do not know whether the data were randomly selected, however, so we must be cautious with respect to generalizing to other clients at this university who might sign the contract. (3) We do not know whether the population is normally distributed, and there are not at least 30 participants. However, the data from the sample do not suggest a skewed distribution.

STEP 2: State the null and research hypotheses.

Null hypothesis: Clients at this university who sign a contract to attend at least 10 sessions attend the same number of sessions, on average, as clients who do not sign such a contract—$H_0: \mu_1 = \mu_2$.

Research hypothesis: Clients at this university who sign a contract to attend at least 10 sessions attend a different number of sessions, on average, than do clients who do not sign such a contract—$H_1: \mu_1 \neq \mu_2$.

STEP 3: Determine the characteristics of the comparison distribution.

$$\mu_M = 4.6; s_M = 1.114$$

Calculations:

$$\mu_M = \mu = 4.6$$

$$M = \frac{\Sigma X}{N} = \frac{(6 + 6 + 12 + 7 + 8)}{5} = 7.8$$

X	*X* − *M*	(*X* − *M*)²
6	−1.8	3.24
6	−1.8	3.24
12	4.2	17.64
7	−0.8	0.64
8	0.2	0.04

The numerator of the standard deviation formula is the sum of squares:

$$\Sigma(X - M)^2 = \Sigma\ (3.24 + 3.24 + 17.64 + 0.64 + 0.04) = 24.8$$

$$s = \sqrt{\frac{\Sigma(X - M)^2}{(N - 1)}} = \sqrt{\frac{24.8}{(5 - 1)}} = \sqrt{6.2} = 2.490$$

$$s_M = \frac{s}{\sqrt{N}} = \frac{2.490}{\sqrt{5}} = 1.114$$

STEP 4: Determine the critical values, or cutoffs.

$$df = N - 1 = 5 - 1 = 4$$

For a two-tailed test with a p level of 0.05 and df of 4, the critical values are −2.776 and 2.776 (as seen in the curve in Figure 9-2).

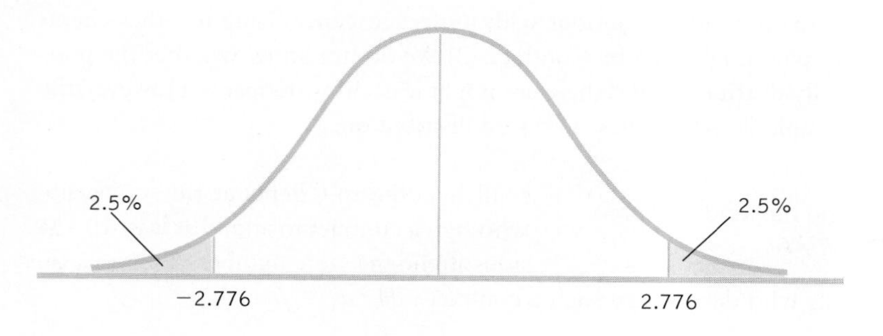

FIGURE 9-2
Determining Cutoffs for a *t* Distribution

As with the *z* distribution, we typically determine critical values in terms of *t* statistics rather than means of raw scores so that we can easily determine whether the test statistic is beyond one of the cutoffs. Here, the cutoffs are −2.776 and 2.776, and they mark off the most extreme 5%, with 2.5% in each tail.

STEP 5: Calculate the test statistic.

$$t = \frac{(M - \mu_M)}{s_M} = \frac{(7.8 - 4.6)}{1.114} = 2.873$$

STEP 6: Make a decision. Reject the null hypothesis. It appears that counseling center clients who sign a contract to attend at least 10 sessions do attend more sessions, on average, than do clients who do not sign such a contract (Figure 9-3).

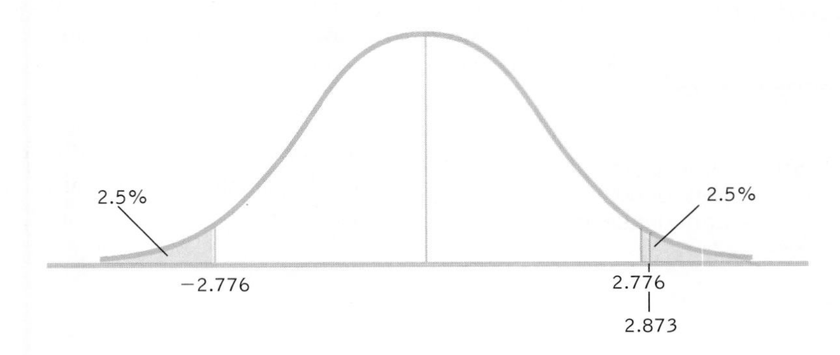

FIGURE 9-3
Making a Decision
To decide whether to reject the null hypothesis, we compare the test statistic to the critical *t* values. In this case, the test statistic, 2.873, is beyond the cutoff of 2.776, so we can reject the null hypothesis.

After completing the hypothesis test, we want to present the primary statistical information in a report. There is a standard American Psychological Association (APA) format for the presentation of statistics across the behavioral sciences so that the results are easily understood by the reader.

1. Write the symbol for the test statistic (e.g., t).
2. Write the degrees of freedom, in parentheses.
3. Write an equal sign and then the value of the test statistic, typically to two decimal places.
4. Write a comma and then indicate the p value by writing "$p =$" and then the actual value. (Unless we use software to conduct the hypothesis test, we will not know the actual p value associated with the test statistic. In this case, we simply state whether the p value is beyond the critical value by saying $p < 0.05$ when we reject the null hypothesis or $p > 0.05$ when we fail to reject the null hypothesis.)

In the counseling center example, the statistics would read:

$$t(4) = 2.87, p < 0.05$$

The statistics typically follow a statement about the finding: for example, "It appears that counseling center clients who sign a contract to attend at least 10 sessions do attend more sessions, on average, than do clients who do not sign such a contract, $t(4) = 2.87, p < 0.05$." The report would also include the sample mean and standard deviation (not standard error) to two decimal points. Here, the descriptive statistics would read: ($M = 7.80$, $SD = 2.49$). By convention, we use SD instead of s to symbolize the standard deviation. ■

Calculating a Confidence Interval for a Single-Sample *t* Test

As with a *z* test, the APA recommends that researchers report confidence intervals and effect sizes, in addition to the results of hypothesis tests, whenever possible.

EXAMPLE 9.7

> **MASTERING THE CONCEPT**
>
> **9.4:** Whenever researchers conduct a hypothesis test, the APA encourages that, if possible, they also calculate a confidence interval and an effect size.

We can calculate a confidence interval with the single-sample *t* test data. The population mean was 4.6. We used the sample to estimate the population standard deviation to be 2.490 and the population standard error to be 1.114. The five students in the sample attended a mean of 7.8 sessions.

When we conducted hypothesis testing, we centered the curve around the mean according to the null hypothesis—the population mean of 4.6. Now we can use the same information to calculate the 95% confidence interval around the sample mean of 7.8.

STEP 1: Draw a picture of a *t* distribution that includes the confidence interval.

We draw a normal curve (Figure 9-4) that has the sample mean, 7.8, at its center (instead of the population mean, 4.6).

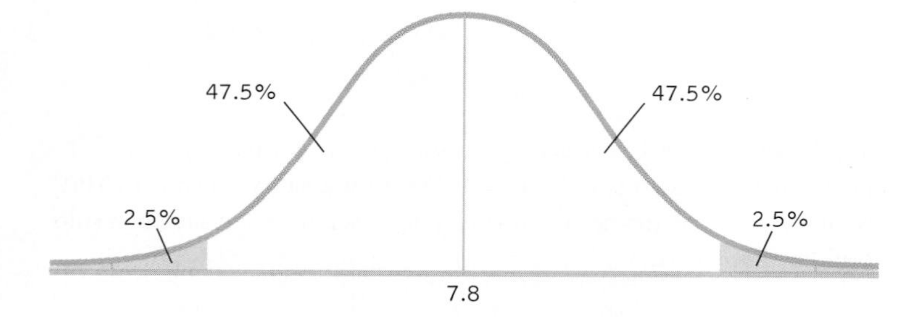

FIGURE 9-4
A 95% Confidence Interval for a Single-Sample *t* Test, Part I
To begin calculating a confidence interval for a single-sample *t* test, we place the sample mean, 7.8, at the center of a curve and indicate the percentages within and beyond the confidence interval.

STEP 2: Indicate the bounds of the confidence interval on the drawing.

We draw a vertical line from the mean to the top of the curve. For a 95% confidence interval, we also draw two much smaller vertical lines indicating the middle 95% of the *t* distribution (2.5% in each tail for a total of 5%). We then write the appropriate percentages under the segments of the curve.

STEP 3: Look up the *t* statistics that fall at each line marking the middle 95%.

For a two-tailed test with a *p* level of 0.05 and a *df* of 4, the critical values are −2.776 and 2.776. We can now add these *t* statistics to the curve, as seen in Figure 9-5.

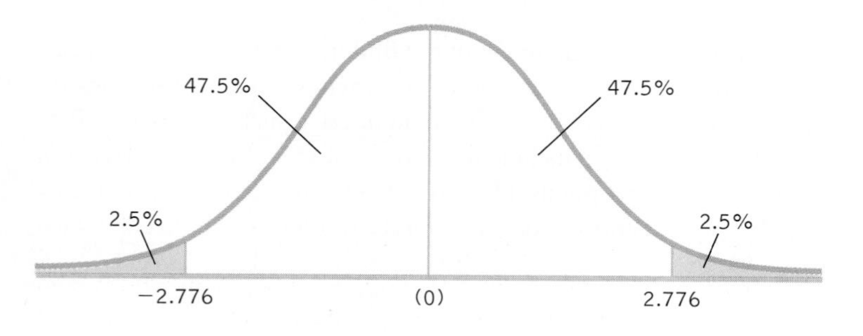

FIGURE 9-5
A 95% Confidence Interval for a Single-Sample *t* Test, Part II
The next step in calculating a confidence interval for a single-sample *t* test is to identify the *t* statistics that indicate each end of the interval. Because the curve is symmetric, the *t* statistics have the same magnitude—one is negative, −2.776, and one is positive, 2.776.

STEP 4: Convert the *t* statistics back into raw means.

As we did with the z test, we can use formulas for this conversion, but first we identify the appropriate mean and standard deviation. There are two important points to remember. First, we center the interval around the sample mean, so we use the sample mean of 7.8 in the calculations. Second, because we have a sample mean (rather than an individual score), we use a distribution of means. So we use the standard error of 1.114 as the measure of spread.

Using this mean and standard error, we calculate the raw mean at each end of the confidence interval, and add them to the curve, as in Figure 9-6. The formulas are exactly the same as for the z test except that z is replaced by t, and σ_M is replaced by s_M.

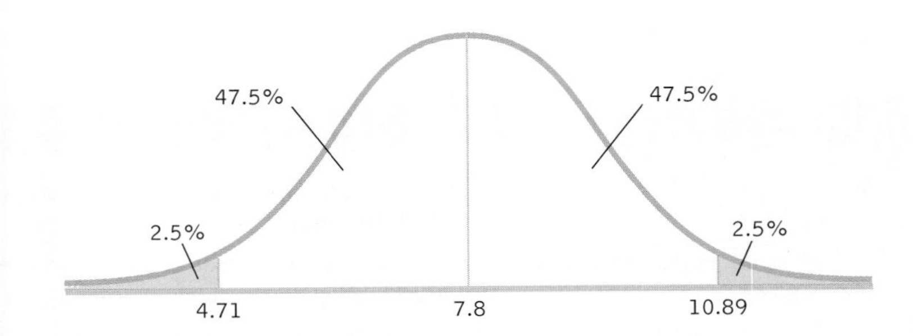

FIGURE 9-6
A 95% Confidence Interval for a Single-Sample *t* Test, Part III
The final step in calculating a confidence interval for a single-sample t test is to convert the t statistics that indicate each end of the interval into raw means, 4.71 and 10.89.

$$M_{lower} = -t(s_M) + M_{sample} = -2.776(1.114) + 7.8 = 4.71$$

$$M_{upper} = t(s_M) + M_{sample} = 2.776(1.114) + 7.8 = 10.89$$

The 95% confidence interval, reported in brackets as is typical, is [4.71, 10.89].

MASTERING THE FORMULA

9-5: The formula for the lower bound of a confidence interval for a single-sample t test is $M_{lower} = -t(s_M) + M_{sample}$. The formula for the upper bound of a confidence interval for a single-sample t test is $M_{upper} = t(s_M) + M_{sample}$. The only differences from those for a z test are that in each formula z is replaced by t, and σ_M is replaced by s_M.

STEP 5: Check that the confidence interval makes sense.

The sample mean should fall exactly in the middle of the two ends of the interval.

$$4.71 - 7.8 = -3.09 \text{ and } 10.89 - 7.8 = 3.09$$

We have a match. The confidence interval ranges from 3.09 below the sample mean to 3.09 above the sample mean. If we were to sample five students from the same population over and over, the 95% confidence interval would include the population mean 95% of the time. Note that the population mean, 4.6, does not fall within this interval. This means it is not plausible that this sample of students who signed contracts came from the population according to the null hypothesis—students seeking treatment at the counseling center who did not sign a contract. We conclude that the sample comes from a different population, one in which students attend more mean sessions than does the general population. As with the z test, the conclusions from the hypothesis test and the confidence interval are the same, but the confidence interval gives us more information—an interval estimate, not just a point estimate. ■

Calculating Effect Size for a Single-Sample *t* Test

As with a z test, we can calculate the effect size (Cohen's d) for a single-sample t test.

EXAMPLE 9.8

Let's calculate the effect size for the counseling center study. Similar to what we did with the z test, we simply use the formula for the t statistic, substituting s for s_M (and μ for μ_M, even though these means are always the same). This means we use 2.490

MASTERING THE FORMULA

9-6: The formula for Cohen's d for a t statistic is: Cohen's $d = \frac{(M - \mu)}{s}$. It is the same formula as for the z statistic, except that we divide by the population standard deviation (s) rather than the population standard error (s_M).

instead of 1.114 in the denominator. Cohen's d is based on the spread of the distribution of individual scores, rather than the distribution of means.

$$\text{Cohen's } d = \frac{(M - \mu)}{s} = \frac{(7.8 - 4.6)}{2.490} = 1.29$$

The effect size, $d = 1.29$, tells us that the sample mean and the population mean are 1.29 standard deviations apart. According to the conventions we learned in Chapter 8 (0.2 is a small effect; 0.5 is a medium effect; 0.8 is a large effect), this is a large effect. We can add the effect size when we report the statistics as follows: $t(4) = 2.87, p < 0.05, d = 1.29$. ■

CHECK YOUR LEARNING

Reviewing the Concepts

- A single-sample t test is used to compare data from one sample to a population for which we know the mean but not the standard deviation.
- We consider degrees of freedom, or the number of scores that are free to vary, instead of N when we assess estimated test statistics against distributions.
- As sample size increases, our confidence in the estimates improves, degrees of freedom increase, and the critical value for t drops, making it easier to reach statistical significance. In fact, as sample size grows, the t distributions approach the z distribution.
- The single-sample t test follows the same six steps of hypothesis testing as the z test, except that we estimate the standard deviation from the sample before we calculate standard error.
- We can calculate a confidence interval and an effect size, Cohen's d, for a single-sample t test.

Clarifying the Concepts

9-5 Explain the term *degrees of freedom*.

9-6 Why is a single-sample t test more useful than a z test?

Calculating the Statistics

9-7 Compute degrees of freedom for each of the following:

a. An experimenter times how long it takes 35 rats to run through a maze with 8 pathways.

b. Test scores for 14 students are collected and averaged over 4 semesters.

9-8 Identify the critical t value for each of the following tests:

a. A two-tailed test with alpha of 0.05 and 11 degrees of freedom

b. A one-tailed test with alpha of 0.01 and N of 17

Applying the Concepts

Solutions to these Check Your Learning questions can be found in Appendix D.

9-9 Let's assume that according to university summary statistics, the average student misses 3.7 classes during a semester. Imagine the data you have been working with (6, 3, 7, 6, 4, 5) are the number of classes missed by a group of students. Conduct all six steps of hypothesis testing, assuming a two-tailed test with a p level of 0.05. (*Note:* You already completed the work for step 3 in Check Your Learning 9-2 and 9-3.)

The Paired-Samples *t* Test

As we learned in the chapter opening, researchers found that weight gain over the holidays is far less than had once been thought. Guess what? The dreaded "freshman 15" also appears to be a myth. One study found that male university students gained an

Jeffrey Blackler/Alamy

Before and After Many companies use before-and-after photos to encourage consumers to purchase their products. Statistics can help us overcome the persuasive powers of anecdotal evidence. We can use a paired-samples *t* test to compare weight before and after participation in an advertised program to determine if the mean difference is statistically significant.

average of 3.5 pounds between the beginning of the fall semester and November, and female students gained an average of 4.0 pounds (Holm-Denoma, Joiner, Vohs, & Heatherton, 2008). These types of before-and-after comparisons can be tested by using the paired-samples *t* test.

The ***paired-samples t test*** (also called *dependent-samples t test) is used to compare two means for a within-groups design, a situation in which every participant is in both samples.* The paired-samples *t* test can be used in many situations. For example, if a participant is in both conditions (such as a memory task before ingesting a caffeinated beverage and again after ingesting a non-caffeinated beverage), then her score in one depends on her score in the other.

The steps for the paired-samples *t* test are almost the same as the steps for the single-sample *t* test. The major difference in the paired-samples *t* test is that we must create difference scores for every participant. Because we'll be working with difference scores, we need to learn about a new distribution—a distribution of the means of these difference scores, or a distribution of mean differences.

Distributions of Mean Differences

We already learned about a distribution of scores and a distribution of means. Now we need to develop a distribution of mean *differences* so that we can establish a distribution that specifies the null hypothesis. Let's use pre- and post-holiday weight data to demonstrate how to create a distribution of mean differences, the distribution that accompanies a within-groups design.

Imagine that many college students' weights were measured before and after the winter holidays to determine if they gained or lost weight. You plan to gather data on a sample of three people from among this population of many college students, and there are two cards for each person in the population on which weights are listed—

▪ The **paired-samples *t* test** is used to compare two means for a within-groups design, a situation in which every participant is in both samples; also called a *dependent-samples t test.*

one before the holidays and one after the holidays. So you have many pairs of cards, one pair for each student in the population. Let's walk through the steps to create a distribution of mean differences from the data on these cards. It is this distribution of mean differences to which we will compare our sample of three people.

Step 1. Randomly choose three pairs of cards, replacing each pair of cards before randomly selecting the next.

Step 2. For each pair, subtract the first weight from the second weight to calculate a difference score.

Step 3. Calculate the mean of the differences in weights for these three people.

Then complete these three steps again. Randomly choose another three people from the population of many college students, calculate their difference scores, and calculate the mean of the three difference scores. And then complete these three steps again, and again, and again.

Let's use the three steps in an example.

Step 1. We randomly select one pair of cards and find that the first student weighed 140 pounds before the holidays and 144 pounds after the holidays. We replace those cards and randomly select another pair; the second student had before and after scores of 126 and 124, respectively. We replace those cards and randomly select another pair; the third student had before and after scores of 168 and 168, respectively.

Step 2. For the first student, the difference between weights, subtracting the before score from the after score, is $144 - 140 = 4$. For the second student, the difference between weights is $124 - 126 = -2$. For the third student, the difference between weights is $168 - 168 = 0$.

Step 3. The mean of these three difference scores—4, −2, and 0—is 0.667.

We would then choose three more students and calculate the mean of their difference scores. Eventually, we would have many mean differences to plot on a curve of mean differences—some positive, some negative, and some right at 0.

But this would only be the beginning of what this distribution of mean differences would look like. If we were to calculate the whole distribution of mean differences, then we would do this an uncountable number of times. When the authors of this book calculated 30 mean differences for pairs of weights, we got the distribution in Figure 9-7. If no mean difference is found when comparing weights from before and after the holidays, as with the data we used to create Figure 9-7, the dis-

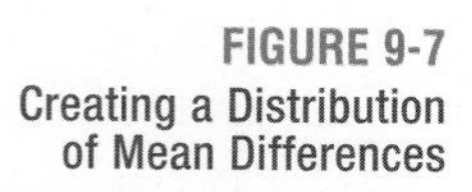

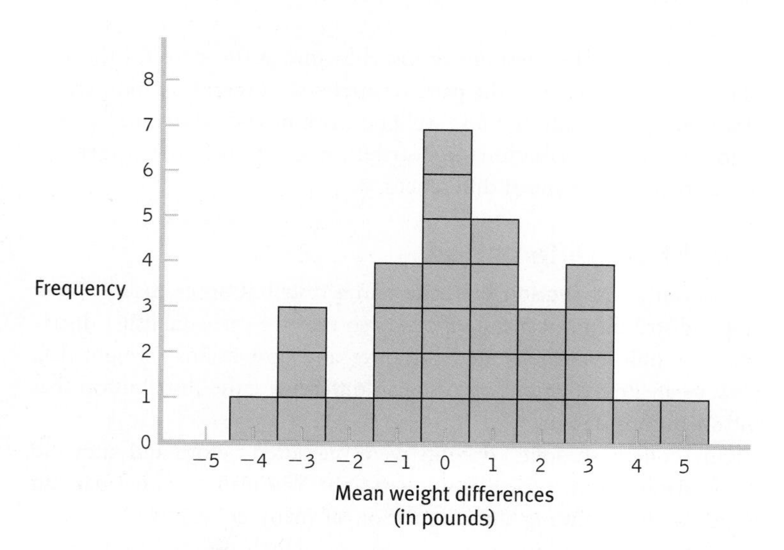

FIGURE 9-7
Creating a Distribution of Mean Differences

This distribution is one of many that could be created by pulling 30 mean differences, the average of three differences between pairs of weights, pulled one at a time from a population of pairs of weights—one pre-holiday and one post-holiday. The population used here is one based on the null hypothesis—that there is no average difference in weight from before the holidays to after the holidays.

tribution would center around 0. According to the null hypothesis, we would expect no mean difference in weight—or a mean difference of 0—from before the holidays to after the holidays.

Large Monitors and Productivity Does a large monitor increase your productivity? Microsoft researchers and cognitive psychologists (Czerwinski et al., 2003) reported a 9% increase in productivity when research volunteers used a large 42-inch display versus a 15-inch display. Every participant used both displays and thus was in both samples. A paired-samples *t* test is the appropriate hypothesis test for this two-group design.

Blend Images/Alamay

The Six Steps of the Paired-Samples *t* Test

In a paired-samples *t* test, each participant has two scores—one in each condition. When we conduct a paired-samples *t* test, we write the pairs of scores in two columns, side by side next to the same participant. We then subtract each score in one column from its paired score in the other column to create difference scores. Ideally, a positive difference score indicates an increase, and a negative difference score indicates a decrease. Typically, we subtract the first score from the second so that the difference scores match this logic. Next we walk through the six steps of the paired-samples *t* test.

EXAMPLE 9.9

For example, behavioral scientists at Microsoft studied how 15 volunteers performed on a set of tasks under two conditions—while using a 15-inch computer monitor and while using a 42-inch monitor (Czerwinski et al., 2003), the latter of which allows the user to have multiple programs in view at the same time.

Here are five participants' fictional data, which reflect the actual means reported by researchers. Note that a smaller number is good—it indicates a faster time. The first person completed the tasks on the small monitor in 122 seconds and on the large monitor in 111 seconds; the second person in 131 and 116; the third in 127 and 113; the fourth in 123 and 119; and the fifth in 132 and 121.

STEP 1: Identify the populations, distribution, and assumptions.

The paired-samples *t* test is like the single-sample *t* test in that we analyze a single sample of scores. For the paired-samples *t* test, however, we analyze difference scores. For the paired-samples *t* test, one population is reflected by each condition, but the comparison distribution is a *distribution of mean difference scores* (rather than a distribution of means). The comparison distribution is based on the null hypothesis that posits no mean difference. So the mean of the comparison distribution is 0. For the paired-samples *t* test, the three assumptions are the same as for the single-sample *t* test.

MASTERING THE CONCEPT

9.5: The steps for the paired-samples *t* test are very similar to those for the single-sample *t* test. The main difference is that we are comparing the sample *mean difference between scores* to the mean difference for the population according to the null hypothesis, rather than comparing the sample *mean of individual scores* to the population mean according to the null hypothesis.

Summary: Population 1: People performing tasks using a 15-inch monitor. Population 2: People performing tasks using a 42-inch monitor.

The comparison distribution is a distribution of mean difference scores based on the null hypothesis. The hypothesis test is a paired-samples *t* test because we have two samples of scores and a within-groups design.

This study meets one of the three assumptions and may meet the other two: (1) The dependent variable is time, which is scale. (2) The participants were not randomly selected, however, so we must be cautious with respect to generalizing our findings. (3) We do not know whether the population is normally distributed, and there are not at least 30 participants. However, the data from this sample do not suggest a skewed distribution.

STEP 2: State the null and research hypotheses.

This step is identical to that for the single-sample *t* test.

Summary: Null hypothesis: People who use a 15-inch screen will complete a set of tasks in the same amount of time, on average, as people who use a 42-inch screen—$H_0: \mu_1 = \mu_2$. Research hypothesis: People who use a 15-inch screen will complete a set of tasks in a different amount of time, on average, than people who use a 42-inch screen—$H_1: \mu_1 \neq \mu_2$.

STEP 3: Determine the characteristics of the comparison distribution.

This step is similar to that for the single-sample *t* test. We determine the appropriate mean and standard error of the comparison distribution—the distribution based on the null hypothesis. With the paired-samples *t* test, however, we have a sample of difference scores and a comparison distribution of mean differences (instead of a sample of individual scores and a comparison distribution of means). According to the null hypothesis, there is no difference. So the mean of the comparison distribution is always 0, as long as the null hypothesis posits no difference.

For the paired-samples *t* test, standard error is calculated exactly as it is calculated for the single-sample *t* test, only we use the difference scores rather than the scores in each condition. To get the difference scores in the current example, we want to know what happens when we go from the control condition (small screen) to the experimental condition (large screen), so we subtract the first score from the second score. This means that a negative difference indicates a decrease in time when the screen goes from small to large. (The test statistic will be the same if we reverse the order in which we subtract, but the sign will change.)

Summary: $\mu_M = 0; s_M = 1.923$

Calculations: (Notice that we crossed out the original scores once we created the column of difference scores. We did this to remind ourselves that all remaining calculations involve the differences scores, not the original scores.)

X	*Y*	Difference	Difference − mean difference	Squared deviation
122	111	−11	0	0
131	116	−15	−4	16
127	113	−14	−3	9
123	119	−4	7	49
132	121	−11	0	0

The mean of the difference scores is:

$$M_{difference} = -11$$

The numerator is the sum of square, *SS* (that we learned about in Chapter 4):

$$0 + 16 + 9 + 49 + 0 = 74$$

The standard deviation, *s*, is:

$$s = \sqrt{\frac{74}{(5-1)}} = \sqrt{18.5} = 4.301$$

The standard error, s_M, is:

$$s_M = \frac{4.301}{\sqrt{5}} = 1.924$$

STEP 4: Determine the critical values, or cutoffs.

This step is the same as that for the single-sample *t* test, except that the degrees of freedom is the number of *participants* (not the number of scores) minus 1.

Summary: $df = N - 1 = 5 - 1 = 4$

The critical values, based on a two-tailed test and a *p* level of 0.05, are −2.776 and 2.776, as seen in the curve in Figure 9-8.

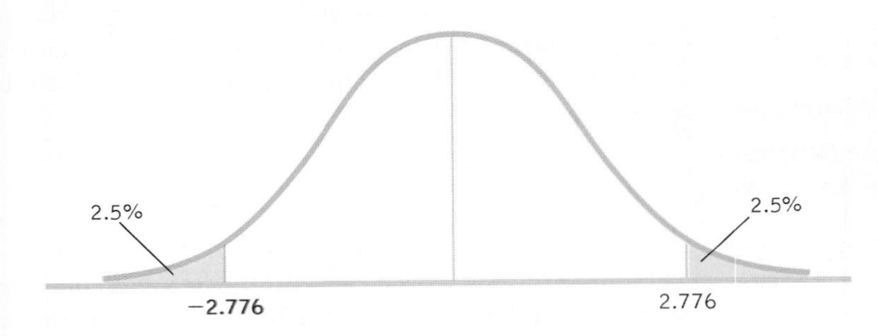

FIGURE 9-8
Determining Cutoffs for a Paired-Samples *t* Test
We typically determine critical values in terms of *t* statistics rather than means of raw scores so that we can easily determine whether the test statistic is beyond one of the cutoffs.

STEP 5: Calculate the test statistic.

This step is identical to that for the single-sample *t* test, except that we use means of difference scores instead of means of individual scores. We subtract the mean difference score according to the null hypothesis, 0, from the mean difference score calculated for the sample. We then divide by standard error.

Summary: $t = \frac{(-11 - 0)}{1.923} = -5.72$

STEP 6: Make a decision.

This step is identical to that for the single-sample *t* test.

Summary: Reject the null hypothesis. When we examine the means (M_X = 127; M_Y = 116), it appears that, on average, people perform faster when using a 42-inch monitor than when using a 15-inch monitor (as shown by the curve in Figure 9-9).

The statistics, as reported in a journal article, follow the same APA format as for a single-sample *t* test. (*Note:* Unless we use software, we can only indicate whether the *p* value is less than or greater than the cutoff *p* level of 0.05.) In the current example, the statistics would read:

$$t(4) = -5.72, p < 0.05$$

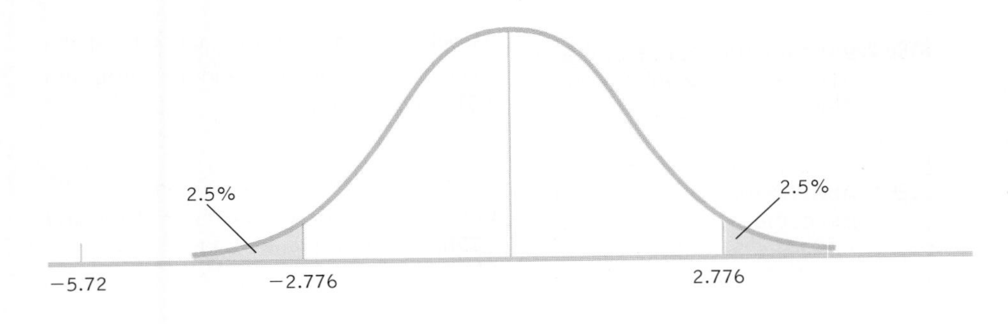

FIGURE 9-9
Making a Decision
To decide whether to reject the null hypothesis, we compare the test statistic to the critical values. In this figure, the test statistic, −5.72, is beyond the cutoff of −2.776, so we can reject the null hypothesis.

We also include the means and the standard deviations for the two samples. We calculated the means in step 6 of hypothesis testing, but we would also have to calculate the standard deviations for the two samples to report them.

The researchers note that the faster time with the large display might not *seem* much faster but that, in their research, they have had great difficulty identifying *any* factors that lead to faster times (Czerwinski et al., 2003). Based on their previous research, therefore, this is an impressive difference. ■

Calculating a Confidence Interval for a Paired-Samples *t* Test

MASTERING THE CONCEPT

9.6: As with a *z* test and a single-sample *t* test, we can calculate a confidence interval and an effect size for a paired-samples *t* test.

When we conduct a paired-samples *t* test, the APA encourages the use of confidence intervals and effect sizes (as with the *z* test and the single-sample *t* test). We calculate both the confidence interval and the effect size for the example of productivity with small versus large computer monitors.

EXAMPLE 9.10

Let's start by determining the confidence interval. First, let's recap the information we need. The population mean difference according to the null hypothesis was 0, and we used the sample to estimate the population standard deviation to be 4.301 and the standard error to be 1.924. The five participants in the study sample had a mean difference of −11. We will calculate the 95% confidence interval around the sample mean difference of −11.

STEP 1: Draw a picture of a *t* distribution that includes the confidence interval.

We draw a normal curve (Figure 9-10) that has the *sample* mean difference, −11, at its center instead of the *population* mean difference, 0.

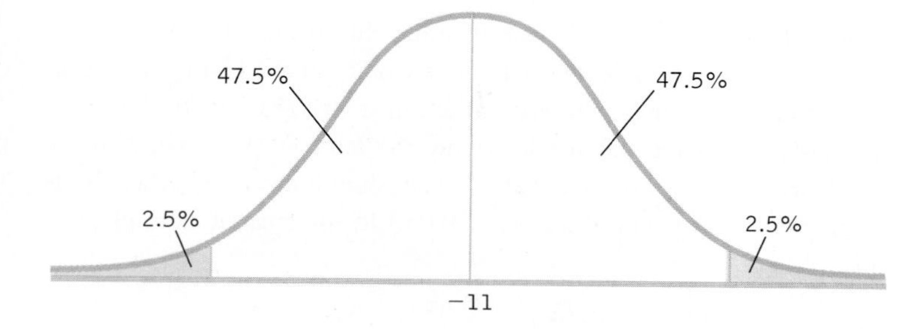

FIGURE 9-10
A 95% Confidence Interval for a Paired-Samples *t* Test, Part I
We start the confidence interval for a distribution of mean differences by drawing a curve with the sample mean difference, −11, in the center.

STEP 2: Indicate the bounds of the confidence interval on the drawing.

As before, 47.5% fall on each side of the mean between the mean and the cutoff, and 2.5% fall in each tail.

STEP 3: Add the critical *t* statistics to the curve.

For a two-tailed test with a *p* level of 0.05 and 4 *df,* the critical values are −2.776 and 2.776, as seen in Figure 9-11.

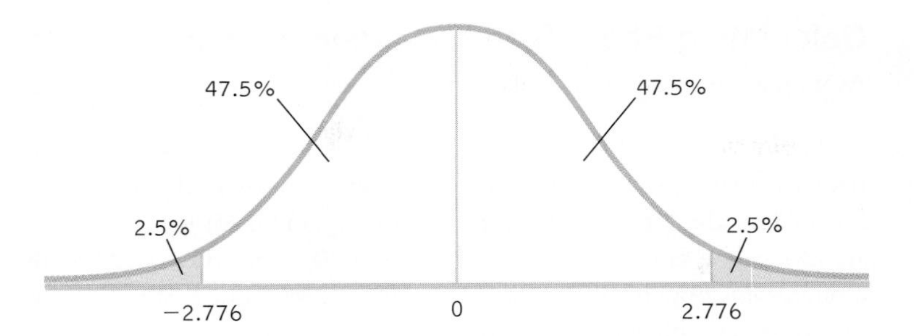

FIGURE 9-11
A 95% Confidence Interval for a Paired-Samples *t* Test, Part II

The next step in calculating a confidence interval for mean differences is identifying the *t* statistics that indicate each end of the interval. Because the curve is symmetric, the *t* statistics have the same magnitude—one is negative, −2.776, and one is positive, 2.776.

STEP 4: Convert the critical *t* statistics back into raw mean differences.

As we do with other confidence intervals, we use the sample mean difference (−11) in the calculations and the standard error (1.924) as the measure of spread. We use the same formulas as for the single-sample *t* test, recalling that these means and standard errors are calculated from differences between two scores for each participant. We added these raw mean differences to the curve in Figure 9-12.

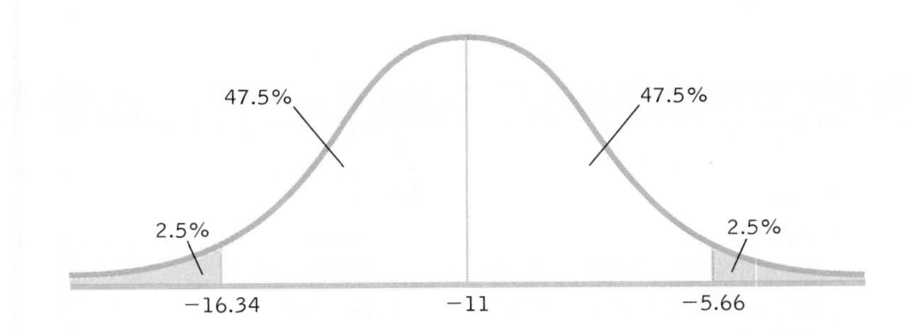

FIGURE 9-12
A 95% Confidence Interval for a Paired-Samples *t* Test, Part III

The final step in calculating a confidence interval for mean differences is converting the *t* statistics that indicate each end of the interval to raw mean differences, −16.34 and −5.66.

$$M_{lower} = -t(s_M) + M_{sample} = -2.776(1.924) + (-11) = -16.34$$

$$M_{upper} = t(s_M) + M_{sample} = 2.776(1.924) + (-11) = -5.66$$

The 95% confidence interval, reported in brackets as is typical, is [−16.34, −5.66].

MASTERING THE FORMULA

9-7: The formula for the lower bound of a confidence interval for a paired-samples t test is $M_{lower} = -t(s_M) + M_{sample}$. The formula for the upper bound of a confidence interval for a paired-samples t test is $M_{upper} = t(s_M) + M_{sample}$. These are the same as for a single-sample t test, but remember that the means and standard errors are calculated from differences between pairs of scores, not individual scores.

STEP 5: Verify that the confidence interval makes sense.

The sample mean difference should fall exactly in the middle of the two ends of the interval.

$$-11 - (-16.34) = 5.34 \text{ and } -11 - (-5.66) = -5.34$$

We have a match. The confidence interval ranges from 5.34 below the sample mean difference to 5.34 above the sample mean difference. If we were to sample five people from the same population over and over, the 95% confidence interval would include the population mean 95% of the time. Note that the population mean difference according to the null hypothesis, 0, does not fall within this interval. This means it is not plausible that the difference between those using the 15-inch monitor and those using the 42-inch monitor is 0.

As with other hypothesis tests, the conclusions from both the paired-samples t test and the confidence interval are the same, but the confidence interval gives us more information—an interval estimate, not just a point estimate. ■

Calculating Effect Size for a Paired-Samples *t* Test

As with a z test, we can calculate the effect size (Cohen's d) for a paired-samples t test.

EXAMPLE 9.11

Let's calculate effect size for the computer monitor study. Again, we simply use the formula for the t statistic, substituting s for s_M (and μ for μ_M, even though these means are always the same). This means we use 4.301 instead of 1.924 in the denominator. Cohen's d is now based on the spread of the distribution of individual differences between scores, rather than the distribution of mean differences.

MASTERING THE FORMULA

9-8: The formula for Cohen's d for a paired-samples t statistic is: Cohen's $d = \frac{(M - \mu)}{s}$. It is the same formula as for the single-sample t statistic, except that the mean and standard deviation are for difference scores rather than individual scores.

$$\text{Cohen's } d = \frac{(M - \mu)}{s} = \frac{(-11 - 0)}{4.301} = -2.56$$

The effect size, $d = -2.56$, tells us that the sample mean difference and the population mean difference are 2.56 standard deviations apart. This is a large effect. Recall that the sign has no effect on the size of an effect: −2.56 and 2.56 are equivalent effect sizes. We can add the effect size when we report the statistics as follows: $t(4) = -5.72$, $p < 0.05$, $d = -2.56$. ■

CHECK YOUR LEARNING

Reviewing the Concepts

> The paired-samples t test is used when we have data for all participants under two conditions—a within-groups design.

> In the paired-samples t test, we calculate a difference score for every individual in the study. The statistic is calculated based on those difference scores.

> We use the same six steps of hypothesis testing that we used with the z test and with the single-sample t test.

> We can calculate a confidence interval and an effect size for a paired-samples t test.

Clarifying the Concepts

9-10 How do we conduct a paired-samples t test?

9-11 Explain what an individual difference score is, as it is used in a paired-samples t test.

9-12 How does creating a confidence interval for a paired-samples t test give us the same information as hypothesis testing with a paired-samples t test?

9-13 How do we calculate Cohen's d for a paired-samples t test?

Calculating the Statistics

9-14 Below are energy-level data (on a scale of 1 to 7, where 1 = feeling of no energy and 7 = feeling of high energy) for five students before and after lunch. Calculate the mean difference for these people so that loss of energy is a negative value.

Before lunch	After lunch
6	3
5	2
4	6
5	4
7	5

9-15 Assume that researchers asked five participants to rate their mood on a scale from 1 to 7 (1 being lowest, 7 being highest) before and after watching a funny video clip. The researchers reported that the average difference between the "before" mood score and the "after" mood score was $M = 1.0$, $s = 1.225$. They calculated a paired-samples t test, $t(4) = 1.13$, $p > 0.05$ and failed to reject the null hypothesis using a two-tailed test with a p level of 0.05.

a. Calculate the 95% confidence interval for this t test and describe how it results in the same conclusion as the hypothesis test.

b. Calculate and interpret Cohen's d.

Applying the Concepts

9-16 Using the energy-level data presented in Check Your Learning 9-14, test the hypothesis that students have different energy levels before and after lunch. Perform the six steps of hypothesis testing for a two-tailed paired-samples t test.

9-17 Using the energy-level data, let's go beyond hypothesis testing.

a. Calculate the 95% confidence interval and describe how it results in the same conclusion as the hypothesis test.

b. Calculate and interpret Cohen's d.

Solutions to these Check Your Learning questions can be found in Appendix D.

REVIEW OF CONCEPTS

The *t* Distributions

The t distributions are similar to the z distribution, except that we must estimate the standard deviation from the sample. When estimating the standard deviation, we make a mathematical correction to adjust for the increased likelihood of error. After estimating the standard deviation, the *t statistic* is calculated like the z statistic for a distribution of means. The t distributions can be used to compare the mean of a sample to a population mean when we don't know the population standard deviation (single-sample t test), to compare two samples with a within-groups design (paired-samples t test), and to compare two samples with a between-groups design (independent-samples t test). (We will learn about the independent-samples t test in Chapter 10.)

The Single-Sample *t* Test

Like z tests, *single-sample t tests* are conducted in the rare cases in which we have one sample that we're comparing to a known population. The difference is that we only have to know the mean of the population to conduct a single-sample t test. There are many t distributions, one for every possible sample size. We look up the appropriate critical values on the t table based on *degrees of freedom,* a number calculated from the sample size. We can calculate a confidence interval and an effect size (Cohen's d), for a single-sample t test.

The Paired-Samples *t* Test

A *paired-samples t test* is used when we have two samples, and the same participants are in both samples; to conduct the test, we calculate a difference score for every individual in the study. The comparison distribution is a distribution of mean difference scores instead of the distribution of means that we used with a single-sample *t* test. Aside from the comparison distribution, the steps of hypothesis testing are similar to those for a single-sample *t* test. As with a *z* test and a single-sample *t* test, we can calculate a confidence interval and an effect size (Cohen's *d*) for a paired-samples *t* test.

SPSS®

Let's conduct a single-sample *t* test using the data on number of counseling sessions attended that we tested earlier in this chapter. The five scores were: 6, 6, 12, 7, and 8.

Select Analyze → Compare Means → One-Sample T Test. Highlight the dependent variable (sessions) and click the arrow in the center to choose it. Type the population mean to which we're comparing the sample, 4.6, next to "Test Value" and click "OK." If you conduct this test using SPSS, you'll notice that the *t* statistic, 2.874, is almost identical to the one we calculated, 2.873. The difference is due solely to the rounding decisions. You'll also notice that the confidence interval is different from the one we calculated. This is an interval around the difference between the two means, rather than around the mean of the sample. The *p* value is under "Sig (2-tailed)." The *p* value of .045 is less than the chosen *p* level of .05, an indication that this is a statistically significant finding.

For a paired-samples *t* test, let's use the data from this chapter on performance using a small monitor versus a large monitor. Enter the data in two columns, with each participant having one score in the first column for his or her performance on the small monitor and one score in the second column for his or her performance on the large monitor.

Select Analyze → Compare Means → Paired-Samples T Test. Choose the dependent variable under the first condition (small) by clicking it, then clicking the center arrow. Choose the dependent variable under the second condition (large) by clicking it, then clicking the center arrow. Click "OK." The screenshot here shows the data and output. You'll notice that the *t* statistic and confidence interval match ours (−5.72 and [−16.34, −5.66]) except that the signs are different. This occurs because of the order in which one score was subtracted from the other score—that is, whether the score on the large monitor was subtracted from the score on the small monitor, or vice versa. The outcome is the same in either case. The *p* value is under "Sig. (2-tailed)" and is .005.

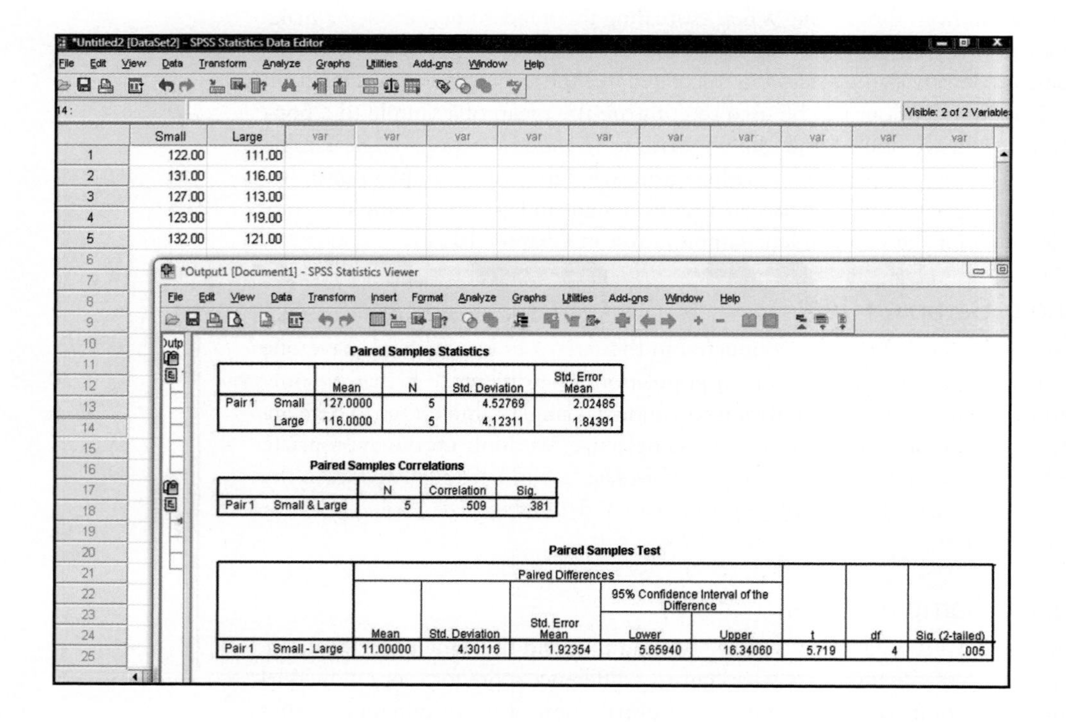

	Small	Large
1	122.00	111.00
2	131.00	116.00
3	127.00	113.00
4	123.00	119.00
5	132.00	121.00

Paired Samples Statistics

		Mean	N	Std. Deviation	Std. Error Mean
Pair 1	Small	127.0000	5	4.52769	2.02485
	Large	116.0000	5	4.12311	1.84391

Paired Samples Correlations

		N	Correlation	Sig.
Pair 1	Small & Large	5	.509	.381

Paired Samples Test

		Paired Differences					t	df	Sig. (2-tailed)
					95% Confidence Interval of the Difference				
		Mean	Std. Deviation	Std. Error Mean	Lower	Upper			
Pair 1	Small - Large	11.00000	4.30116	1.92354	5.65940	16.34060	5.719	4	.005

HOW IT WORKS

9.1 CONDUCTING A SINGLE-SAMPLE *t* TEST

In How It Works 7.2, we conducted a z test for data from the Consideration of Future Consequences (CFC) scale (Petrocelli, 2003). How can we conduct all six steps for a single-sample t test for the same data using a p level of 0.05 and a two-tailed test? To start, we use the population mean CFC score of 3.51, but pretend that we no longer know the population standard deviation. As before, we wonder whether behavioral sciences students who joined a career discussion group might have improved CFC scores, on average, compared with the population. Forty-five students attended these discussion groups, and had a mean CFC score of 3.7 with a standard deviation of 0.52.

Step 1: Population 1: All students in career discussion groups. Population 2: All students who did not participate in career discussion groups.
The comparison distribution will be a distribution of means. The hypothesis test will be a single-sample t test because we have only one sample and we know the population mean, but we do not know the population standard deviation. This study meets two of the three assumptions and may meet the third. The dependent variable is scale. In addition, there are more than 30 participants in the sample, indicating that the comparison distribution will be normal. The data were not randomly selected, however, so we must be cautious when generalizing.

Step 2: Null hypothesis: Students who participated in career discussion groups had the same CFC scores, on average, as students who did not participate—$H_0: \mu_1 = \mu_2$. Research hypothesis: Students who participated in career discussion groups had different CFC scores, on average, than students who did not participate—$H_1: \mu_1 \neq \mu_2$.

Step 3: $\mu_M = \mu = 3.51;\ s_M = \frac{s}{\sqrt{N}} = \frac{0.52}{\sqrt{45}} = 0.078$

Step 4: $df = N - 1 = 45 - 1 = 44$

The critical values, based on 44 degrees of freedom (because 44 is not in the table, we look up the more conservative degrees of freedom of 40), a p level of 0.05, and a two-tailed test, are -2.021 and 2.021.

Step 5: $t = \frac{(M - \mu_M)}{s_M} = \frac{(3.7 - 3.51)}{0.078} = 2.44$

Step 6: Reject the null hypothesis. It appears that students who participate in career discussion groups have higher CFC scores, on average, than do students who do not participate.

The statistics, as presented in a journal article, would read:

$t(44) = 2.44, p < 0.05$

(*Note:* If we had used software, we would report the actual p value instead of just whether the p value is larger or smaller than the critical p value.)

9.2 CONDUCTING A PAIRED-SAMPLES *t* TEST

Salary Wizard is an online tool that allows you to look up incomes for specific jobs for cities in the United States. We looked up the 25th percentile for income for six jobs in two cities: Boise, Idaho, and Los Angeles, California. The data are below.

	Boise	Los Angeles
Executive chef	$53,047.00	$62,490.00
Genetics counselor	$49,958.00	$58,850.00
Grants/proposal writer	$41,974.00	$49,445.00
Librarian	$44,366.00	$52,263.00
Public schoolteacher	$40,470.00	$47,674.00
Social worker (with bachelor's degree)	$36,963.00	$43,542.00

How can we conduct a paired-samples *t* test to determine whether income in one of these cities differs, on average, from income in the other? We'll use a two-tailed test and a *p* level of 0.05.

Step 1: Population 1: Job types in Boise, Idaho. Population 2: Job types in Los Angeles, California.
The comparison distribution will be a distribution of mean differences. The hypothesis test will be a paired-samples *t* test because we have two samples, and all participants are in both samples.
This study meets the first of the three assumptions and may meet the third. The dependent variable, income, is scale. We do not know whether the population is normally distributed, there are not at least 30 participants, and there is not much variability in the data in the samples, so we should proceed with caution. The data were not randomly selected, so we should be cautious when generalizing beyond this sample of job types.

Step 2: Null hypothesis: Jobs in Boise pay the same, on average, as jobs in Los Angeles—H_0: $\mu_1 = \mu_2$. Research hypothesis: Jobs in Boise pay different incomes, on average, than do jobs in Los Angeles—H_1: $\mu_1 \neq \mu_2$.

Step 3: $\mu_M = \mu = 0; s_M = 438.919$

Boise	Los Angeles	Difference (D)	$(D - M_{difference})$	$(D - M_{difference})^2$
$53,047.00	$62,490.00	9443	1528.667	2,336,822.797
$49,958.00	$58,850.00	8892	977.667	955,832.763
$41,974.00	$49,445.00	7471	−443.333	196,544.149
$44,366.00	$52,263.00	7897	−17.333	300.433
$40,470.00	$47,674.00	7204	−710.333	504,572.971
$36,963.00	$43,542.00	6579	−1335.333	1,783,114.221

$M_{difference} = 7914.333$

$SS = \Sigma(D - M_{difference})^2 = 5,777,187.334$

$$s = \sqrt{\frac{\Sigma(D - M_{difference})^2}{(N-1)}} = \sqrt{\frac{5,777,185.203}{(6-1)}} = 1074.913$$

$$s_M = \frac{s}{\sqrt{N}} = \frac{1074.913}{\sqrt{6}} = \frac{1074.913}{2.449} = 438.919$$

Step 4: $df = N - 1 = 6 - 1 = 5$

The critical values, based on 5 degrees of freedom, a *p* level of 0.05, and a two-tailed test, are −2.571 and 2.571.

Step 5: $t = \frac{(M_{difference} - \mu_{difference})}{s_M} = \frac{(7914.333 - 0)}{438.919} = 18.03$

Step 6: Reject the null hypothesis. It appears that jobs in Los Angeles pay more, on average, than do jobs in Boise.

The statistics, as they would be presented in a journal article, are:

$t(5) = 18.03, p < 0.05$

Exercises

Clarifying the Concepts

9.1 When should we use a *t* distribution?

9.2 Why do we modify the formula for calculating standard deviation when using *t* tests (and divide by $N - 1$)?

9.3 How is the calculation of standard error different for a *t* test than for a *z* test?

9.4 Explain why the standard error for the distribution of sample means is smaller than the standard deviation of sample scores.

9.5 Define the symbols in the formula for the t statistic: $t = \frac{(M - \mu_M)}{s_M}$

9.6 When is it appropriate to use a single-sample t test?

9.7 What does the phrase "free to vary," referring to a number of scores in a given sample, mean for statisticians?

9.8 How is the critical t value affected by sample size and degrees of freedom?

9.9 Why do the t distributions merge with the z distribution as sample size increases?

9.10 Explain what each part of the following statistical phrase means, as it would be reported in APA format: $t(4) = 2.87, p = 0.032$.

9.11 What do we mean when we say we have a distribution of mean differences?

9.12 When do we use a paired-samples t test?

9.13 Explain the distinction between the terms *independent samples* and *paired samples* as they relate to t tests.

9.14 How is a paired-samples t test similar to a single-sample t test?

9.15 How is a paired-samples t test different from a single-sample t test?

9.16 Why is the population mean almost always equal to 0 for the null hypothesis in the two-tailed, paired-samples t test?

9.17 If we calculate the confidence interval around the sample mean difference used for a paired-samples t test, and it includes the value of 0, what can we conclude?

9.18 If we calculate the confidence interval around the sample mean difference used for a paired-samples t test, and it does not include the value of 0, what can we conclude?

9.19 Why is a confidence interval more useful than a single-sample t test or a paired-samples t test?

Calculating the Statistics

9.20 We use formulas to describe calculations. Find the error in symbolic notation in each of the following formulas. Explain why it is incorrect and provide the correct symbolic notation.

a. $z = \frac{(X - M)}{\sigma}$

b. $X = z(\sigma) - \mu_M$

c. $\sigma_M = \frac{\sigma}{\sqrt{N-1}}$

d. $t = \frac{(M - \mu_M)}{\sigma_M}$

9.21 For the data 93, 97, 91, 88, 103, 94, 97, calculate the standard deviation under both of these conditions:

a. For the sample

b. As an estimate of the population

c. Calculate the standard error for t using symbolic notation.

d. Calculate the t statistic, assuming $\mu = 96$.

9.22 For the data 1.01, 0.99, 1.12, 1.27, 0.82, 1.04, calculate the standard deviation under both of these conditions. (*Note:* You will have to carry some calculations out to the third decimal place to see the difference in calculations.)

a. For the sample

b. As an estimate of the population

c. Calculate the standard error for t using symbolic notation.

d. Calculate the t statistic, assuming $\mu = 0.96$.

9.23 Identify the critical t value in each of the following circumstances:

a. One-tailed test, $df = 73$, p level of 0.10

b. Two-tailed test, $df = 108$, p level of 0.05

c. One-tailed test, $df = 38$, p level of 0.01

9.24 Calculate degrees of freedom and identify the critical t value for a single-sample t test in each of the following circumstances:

a. Two-tailed test, $N = 8$, p level of 0.10

b. One-tailed test, $N = 42$, p level of 0.05

c. Two-tailed test, $N = 89$, p level of 0.01

9.25 Identify critical t values for each of the following tests:

a. A single-sample t test examining scores for 26 participants to see if there is any difference compared to the population, using a p level of 0.05

b. A one-tailed, single-sample t test performed on scores on the Marital Satisfaction Inventory for 18 people who went through marriage counseling, as compared to the population of people who had not been through marital counseling, using a p level of 0.01

c. A one-tailed, paired-samples t test performed on before-and-after scores on the Marital Satisfaction Inventory for 18 people who went through marriage counseling, using a p level of 0.01

d. A two-tailed, paired-samples t test performed on before-and-after scores on the Marital Satisfaction Inventory for 64 people who went through marriage counseling, using a p level of 0.05

e. A two-tailed, single-sample t test, using a p level of 0.05, with 34 degrees of freedom

9.26 Assume we know the following for a two-tailed, single-sample t test, at a p level of 0.05: $\mu = 44.3$, $N = 114$, $M = 43$, $s = 5.9$.

a. Calculate the t statistic.

b. Calculate a 95% confidence interval.

c. Calculate effect size using Cohen's d.

9.27 Assume we know the following for a two-tailed, single-sample t test: $\mu = 7$, $N = 41$, $M = 8.5$, $s = 2.1$.

a. Calculate the t statistic.

b. Calculate a 99% confidence interval.

c. Calculate effect size using Cohen's d.

9.28 Assume 8 participants completed a mood scale before and after watching a funny video clip.

a. Identify the critical t value for a one-tailed, paired-samples t test with a p level of 0.01.

b. Identify the critical t values for a two-tailed, paired-samples t test with a p level of 0.01.

9.29 The following are scores for 8 students on two different exams.

Exam I	Exam II
92	84
67	75
95	97
82	87
73	68
59	63
90	88
72	78

a. Calculate the paired-samples t statistic for these exam scores.

b. Using a two-tailed test and a p level of 0.05, identify the critical t values and make a decision regarding the null hypothesis.

c. Assume you instead collected exam scores from 1000 students whose mean difference score and standard deviation were exactly the same as for these 8 students. Using a two-tailed test and a p level of 0.05, identify the critical t values and make a decision regarding the null hypothesis.

d. How did changing the sample size affect the decision regarding the null hypothesis?

9.30 The following are mood scores for 12 participants before and after watching a funny video clip (higher values indicate better mood).

Before	After	Before	After
7	2	4	2
5	4	7	3
5	3	4	1
7	5	4	1
6	5	5	3
7	4	4	3

a. Calculate the paired-samples t statistic for these mood scores.

b. Using a one-tailed hypothesis test that the video clip improves mood and a p level of 0.05, identify the critical t values and make a decision regarding the null hypothesis.

c. Using a two-tailed hypothesis test with a p level of 0.05, identify the critical t values and make a decision regarding the null hypothesis.

9.31 Consider the following data:

Score 1	Score 2	Score 1	Score 2
45	62	15	26
34	56	51	56
22	40	28	33
45	48		

a. Calculate the paired-samples t statistic, assuming a two-tailed test.

b. Calculate the 95% confidence interval, assuming a two-tailed test.

c. Calculate the effect size for the mean difference.

9.32 Consider the following data.

Score 1	Score 2
23	16
30	12
28	25
30	27
14	6

a. Calculate the paired-samples t statistic, assuming a two-tailed test.

b. Calculate the 95% confidence interval.

c. Calculate the effect size.

9.33 Assume we know the following for a paired-samples t test: $N = 13$, $M_{difference} = -0.77$, $s = 1.42$.

a. Calculate the t statistic.

b. Calculate a 95% confidence interval for a two-tailed test.

c. Calculate effect size using Cohen's d.

9.34 Assume we know the following for a paired-samples t test: $N = 32$, $M_{difference} = 1.75$, $s = 4.0$.

a. Calculate the t statistic.

b. Calculate a 95% confidence interval for a two-tailed test.

c. Calculate effect size using Cohen's d.

Applying the Concepts

9.35 The relation between the z distribution and the t distribution: For each of the problems described below, which are the same as some of those described in Exercise 9.25, identify what the critical z value would have been if there had been just one sample and we knew the mean and standard deviation of the population:

a. A single-sample t test examining scores for 26 participants to see if there is any difference compared to the population, using a p level of 0.05

b. A one-tailed, single-sample t test performed on scores on the Marital Satisfaction Inventory for 18 people who went through marriage counseling, using a p level of 0.01

c. A two-tailed, single-sample t test, using a p level of 0.05, with 34 degrees of freedom

d. Comparing the critical t values with the critical z values, explain how and why these are different.

9.36 t statistics and standardized tests: On its Web site, the Princeton Review claims that students who have taken its course improve their Graduate Record Examination (GRE) scores, on average, by 210 points. (No other information is provided about this statistic.) Treating this average gain as a population mean, a researcher wonders whether the far cheaper technique of practicing for the GRE on one's own would lead to a different average gain. She randomly selects five students from the pool of students at her university who plan to take the GRE. The students take a practice test before and after 2 months of self-study. They reported (fictional) gains of 160, 240, 340, 70, and 250 points. (Note that many experts suggest that the results from self-study are similar to those from a structured course if you have the self-discipline to go solo. Regardless of the format, preparation has been convincingly demonstrated to lead to increased scores, on average.)

a. Using symbolic notation and formulas (where appropriate), determine the appropriate mean and standard error for the distribution to which we will compare this sample. Show all steps of your calculations.

b. Using symbolic notation and the formula, calculate the t statistic for this sample.

c. As an interested consumer, what critical questions would you want to ask about the statistic reported by the Princeton Review? List at least three questions.

9.37 Single-sample t test, military training, and anger: Bardwell, Ensign, and Mills (2005) assessed the moods of 60 male U.S. Marines following a month-long training exercise conducted in cold temperatures and at high altitudes. Negative moods, including fatigue and anger, increased substantially during the training and lasted up to 3 months after the training ended. Mean mood scores were compared to population norms for three groups: college men, adult men, and male psychiatric outpatients. Let's examine anger scores for six Marines at the end of training; these scores are fictional, but their mean and standard deviation are very close to the actual descriptive statistics for the sample: 14, 12, 13, 12, 14, 15.

a. The population mean anger score for college men is 8.90. Conduct all six steps of a single-sample t test. Report the statistics as you would in a journal article.

b. Now calculate the test statistic to compare this sample mean to the population mean anger score for adult men ($M = 9.20$). You do not have to repeat all the steps from part (a), but conduct step 6 of hypothesis testing and report the statistics as you would in a journal article.

c. Now calculate the test statistic to compare this sample mean to the population mean anger score for male psychiatric outpatients ($M = 13.5$). Do not repeat all the steps from part (a), but conduct step 6 of hypothesis testing and report the statistics as you would in a journal article.

d. What can we conclude overall about Marines' moods following high-altitude, cold-weather training?

9.38 t Tests and retail: Many communities worldwide are lamenting the effects of so-called big box retailers (e.g., Walmart) on their local economies, particularly on small, independently owned shops. Do these large stores affect the bottom lines of locally owned retailers? Imagine that you decide to test this premise. You assess earnings at 20 local stores for the month of October, a few months before a big box store opens. You then assess earnings the following October, correcting for inflation.

a. What are the two populations?

b. What is the comparison distribution? Explain.

c. What hypothesis test would you use? Explain.

d. Check the assumptions for this hypothesis test.

e. What is one flaw in drawing conclusions from this comparison over time?

f. State the null and research hypotheses in both words and symbols.

9.39 Paired-samples t tests, confidence intervals, and hockey goals: Below are the numbers of goals scored by the lead scorers of the New Jersey Devils hockey team in the 2007–2008 and 2008–2009 seasons. On average, did the Devils play any differently in 2008–2009 than they did in 2007–2008?

Player	2007–2008	2008–2009
Elias	20	31
Zajac	14	20
Pandolfo	12	5
Langenbrunner	13	29
Gionta	22	20
Parise	32	45

a. Conduct the six steps of hypothesis testing using a two-tailed test and a p level of 0.05.

b. Report the test statistic in APA format.

c. Calculate the confidence interval for the paired-samples t test you conducted in part (a). Compare the confidence interval to the results of the hypothesis test.

d. Calculate the effect size for the mean difference between the 2007–2008 and 2008–2009 seasons.

9.40 Paired-samples *t* test and graduate admissions: Is it harder to get into graduate programs in psychology or history? We randomly selected five institutions from among all U.S. institutions with graduate programs. The first number for each is the minimum grade point average (GPA) for applicants to the psychology doctoral program, and the second is for applicants to the history doctoral program. These GPAs were posted on the Web site of the well-known college guide company Peterson's.

Wayne State University:	3.0, 2.75
University of Iowa:	3.0, 3.0
University of Nevada–Reno:	3.0, 2.75
George Washington University:	3.0, 3.0
University of Wyoming:	3.0, 3.0

a. The participants are not people; explain why it is appropriate to use a paired-samples t test for this situation.

b. Conduct all six steps of a paired-samples t test. Be sure to label all six steps.

c. Calculate the effect size and explain what this adds to your analysis.

d. Report the statistics as you would in a journal article.

9.41 Attitudes toward statistics and the paired-samples *t* test: A professor wanted to know if her students' attitudes toward statistics changed by the end of the course, so she asked them to fill out an "Attitudes Toward Statistics" scale at the beginning of the term and at the end of the term.

a. What kind of t test should she use to analyze the data?

b. If the average (mean) at the end of the class was higher than it was at the beginning, is that necessarily a statistically significant improvement?

c. Which situation makes it easier to declare that a finding that a certain mean difference is statistically significant: a class with 7 students or a class with 700 students? Explain your answer.

9.42 Paired-samples *t* tests, confidence intervals, and wedding-day weight loss: It seems that 14% of engaged women buy a wedding dress at least one size smaller than their current size. Why? Cornell researchers reported an alarming tendency for women who are engaged to sometimes attempt to lose an unhealthy amount of weight prior to their wedding (Neighbors & Sobal, 2008). The researchers found that engaged women weighed, on average, 152.1 pounds. The average ideal wedding weight reported by 227 women was 136.0 pounds. The data below represent the fictional weights of 8 women on the day they bought their wedding dress and on the day they got married. Did women lose weight for their wedding day?

Dress Purchase	Wedding Day
163	158
144	139
151	150
120	118
136	132
158	152
155	150
145	146

a. Conduct the six steps of hypothesis testing using a one-tailed test and a p level of 0.05.

b. Report the test statistic in APA format.

c. Calculate the confidence interval for the paired-samples t test that you conducted in part (a). Compare the confidence interval to the results of the hypothesis test.

Putting It All Together

9.43 Death row: The Florida Department of Corrections publishes an online death row fact sheet. It reports the average time on death row prior to execution as 11.72 years but provides no standard deviation. This mean is a parameter because it is calculated from the entire population of executed prisoners in Florida. Has the time spent on death row changed in recent years? According to the execution list linked to the same Web site, the six prisoners executed in Florida during the years 2003, 2004, and 2005 spent 25.62, 13.09, 8.74, 17.63, 2.80, and 4.42 years on death row, respectively. (All were men, although Aileen Wuornos, the serial killer portrayed by Charlize Theron in the 2003 film *Monster*, was among the three prisoners executed by the state of Florida in 2002; Wuornos spent 10.69 years on death row.)

a. Using symbolic notation and formulas (where appropriate), determine the appropriate mean and standard error for the distribution of means. Show all steps of your calculations.

b. Using symbolic notation and the formula, calculate the *t* statistic for time spent on death row for the sample of recently executed prisoners.

c. The execution list provides data on all prisoners executed since the death penalty was reinstated in Florida in 1976. Included for each prisoner are the name, race, gender, date of birth, date of offense, date sentenced, date arrived on death row, data of execution, number of warrants, and years on death row. State at least one hypothesis, other than year of execution, which could be examined using a *t* distribution and the comparison mean of 11.72 years on death row. Be specific about your hypothesis (and if you are interested, you can search for the data online).

d. What additional information would you need to calculate a *z* score for the length of time Aileen Wuornos spent on death row?

e. Write hypotheses to address the question "Has the time spent on death row changed in recent years?"

f. Using these data as "recent years" and the mean of 11.72 years as the comparison, answer the question based on the *t* statistic calculated in part (b), using alpha of 0.05.

g. Calculate the confidence interval for this statistic based on the data presented.

h. What conclusion would you make about your hypotheses based on this confidence interval? What can you say about the size of this confidence interval?

i. Calculate the effect size using Cohen's *d*.

j. Evaluate the size of this effect.

9.44 Single-sample *t* test and paid days off: The number of paid days off (e.g., vacation, sick leave) taken by eight employees at a small local business is compared to the national average. You are hired by the new business owner to help her determine what to expect for paid days off. In general, she wants to set some standard for her employees and for herself. Let's assume your search on the Internet for data on paid days off leaves you with the impression that the national average is 15 days. The data for the eight local employees during the last fiscal year are: 10, 11, 8, 14, 13, 12, 12, and 27 days.

a. Write hypotheses for your research.

b. Which type of test would be appropriate to analyze these data in order to answer your question?

c. Before doing any computations, do you have any concerns about this research? Are there any questions you might like to ask about the data you have been given?

d. Calculate the appropriate *t* statistic. Show all of your work in detail.

e. Draw a statistical conclusion for this business owner.

f. Calculate the confidence interval.

g. Calculate and interpret the effect size.

h. Consider all the results you have calculated. How would you summarize the situation for this business owner? Identify the limitations of your analyses, and discuss the difficulties of making comparisons between populations and samples. Make reference to the assumptions of the statistical test in your answer.

i. After further investigation, you discover that one of the data points, 27 days, was actually the owner's number of paid days off. Calculate the *t* statistic and draw a statistical conclusion, adapting for this new information by deleting that value. What changed in the re-analysis of the data?

j. Calculate and interpret the effect size, adapting for this new information by deleting the outlier of 27 days. What changed in the re-analyses of the data?

9.45 Hypnosis and the Stroop effect: In Chapter 1, you were given an opportunity to complete the Stroop test, in which color words are printed in the wrong color; for example, the word *red* might be printed in the color blue. The conflict that arises when we try to name the color of ink the words are printed in but are distracted when the color word does not match the ink color increases reaction time and decreases accuracy. Several researchers have suggested that the Stroop effect can be decreased by hypnosis. Raz, Fan, and Posner (2005) used brain-imaging techniques to demonstrate that posthypnotic suggestion led highly hypnotizable people to see Stroop words as nonsense words. Imagine that you are working with Raz and colleagues and your assignment is to determine whether reaction times decrease (remember, a decrease is a good thing; it indicates that participants are faster) when highly hypnotizable people receive a posthypnotic suggestion to view the words as nonsensical. You conduct the experiment on six participants, once in each condition, and receive the following data; the first number is reaction time in seconds without the posthypnotic suggestion, and the second number is reaction time with the posthypnotic suggestion:

Participant 1:	12.6, 8.5
Participant 2:	13.8, 9.6
Participant 3:	11.6, 10.0
Participant 4:	12.2, 9.2
Participant 5:	12.1, 8.9
Participant 6:	13.0, 10.8

a. What is the independent variable and what are its levels? What is the dependent variable?

b. Conduct all six steps of a paired-samples *t* test as a two-tailed test. Be sure to label all six steps.

c. Report the statistics as you would in a journal article.

d. Now let's look at the effect of switching to a one-tailed test. Conduct steps 2, 4, and 6 of hypothesis testing for a one-tailed paired-samples *t* test. Under which circumstance—a one-tailed or a two-tailed test—is it easier to reject the null hypothesis? If it becomes easier to reject the null hypothesis under one type of test (one-tailed versus two-tailed), does this mean that there is a bigger mean difference between the samples? Explain.

e. Now let's look at the effect of *p* level. Conduct steps 4 and 6 of hypothesis testing for a *p* level of 0.01 and a two-tailed test. With which *p* level—0.05 or 0.01—is it easiest to reject the null hypothesis with a two-tailed test? If it is easier to reject the null hypothesis with certain *p* levels, does this mean that there is a bigger mean difference between the samples? Explain.

f. Now let's look at the effect of sample size. Calculate the test statistic using only participants 1–3 and determine the new critical values. Is this test statistic closer to or farther from the cutoff? Does reducing the sample size make it easier or more difficult to reject the null hypothesis? Explain.

Terms

t statistic (p. 202)
single-sample *t* test (p. 203)
degrees of freedom (p. 203)
paired-samples *t* test (p. 211)

Formulas

$$s = \sqrt{\frac{\Sigma(X - M)^2}{(N - 1)}}$$ (p. 199)

$$s_M = \frac{s}{\sqrt{N}}$$ (p. 201)

$$t = \frac{(M - \mu_M)}{s_M}$$ (p. 202)

$$df = N - 1$$ (p. 203)

$$M_{lower} = -t(s_M) + M_{sample}$$ (pp. 209, 217)

$$M_{upper} = t(s_M) + M_{sample}$$ (pp. 209, 217)

$$\text{Cohen's } d = \frac{(M - \mu)}{s}$$ (pp. 210, 218)

Symbols

t (p. 198)
s_M (p. 201)
df (p. 203)

CHAPTER 10

The Independent-Samples *t* Test

BEFORE YOU GO ON

- You should know the six steps of hypothesis testing (Chapter 7).
- You should understand the differences between a distribution of scores (Chapter 2), a distribution of means (Chapter 6), and a distribution of mean differences (Chapter 9).
- You should know how to conduct a single-sample *t* test and a paired-samples *t* test, including the calculations for the corrected versions of standard deviation and variance (Chapter 9).
- You should understand the basics of determining confidence intervals (Chapter 8).
- You should understand the concept of effect size and know the basics of calculating Cohen's *d* (Chapter 8).

Stella Cunliffe Stella Cunliffe created a remarkable career through her statistical reasoning and became the first female president of the Royal Statistical Society. As a statistician, she used hypothesis testing to improve quality control at the Guinness Brewing Company and to shape public policy in the criminology division at the British Home Office.

Stella Cunliffe built a path of success through two male-dominated industries: beer making and statistics. However, after living what she described as "a free and in many ways exciting life," she wondered what she had gotten herself into after World War II when she accepted a job at the Guinness Brewing Company.

Cunliffe was a practical statistician whose insights helped her bypass ancient social norms for gender and become the first woman elected president of the Royal Statistical Society. In her presidential speech, she reminded her audience that applied "statistics are concerned much more with people than with vague ideas" and that it was "impossible to find human beings without prejudices and without the delightful idiosyncrasies which make them so fascinating" (1976, p. 4).

For example, the quality of handmade beer barrels came down to a simple decision by the quality control worker: accept or reject. But that decision was biased because accepting meant kicking a barrel downhill and rejecting required pushing a barrel uphill, clearly the more arduous task. Cunliffe "de-biased" these judgments by moving the quality-control workstation so that it required equal effort either to accept or to reject a barrel—and saved Guinness a great deal of money. We can "de-bias" many experiments in the same way: by equalizing initial conditions through random assignment to independent groups.

In Chapter 9, we learned how to conduct two types of *t* tests: (1) a single-sample *t* test (comparing one sample to a population for which we know the mean but not the standard deviation) and (2) a paired-samples *t* test (such as a before/after design in which the same participants are in both groups). Here is a third situation calling for a *t* test: a two-group study in which each participant is in only one group. The scores for each group are independent of what happens in the other group. We also demonstrate how to determine a confidence interval and calculate an effect size for situations in which we have two independent groups.

Language Alert! The independent-samples *t* test is also called a between-groups *t* test.

Conducting an Independent-Samples *t* Test

The ***independent-samples t test*** *is used to compare two means for a between-groups design, a situation in which each participant is assigned to only one condition.* This test uses a distribution of differences between means. This affects the *t* test in a few minor ways. As we will see, the biggest difference is that it takes more work to estimate the appropriate standard error. It's not difficult—just a bit time consuming.

> **MASTERING THE CONCEPT**
>
> **10.1:** An independent-samples *t* test is used when we have two groups and a between-groups research design—that is, every participant is in only one of the two groups.

A Distribution of Differences Between Means

Because we have different people in each condition of the study, we cannot create a difference score for each person. We're looking at overall differences between two independent groups, so we need to develop a new type of distribution, a distribution of *differences between means.*

Let's use the Chapter 6 data about heights to demonstrate how to create a distribution of differences between means. Let's say that we were planning to collect data on two groups of three people each and wanted to determine the comparison distribution for this research scenario. Remember that in Chapter 6, we used the example of a population of 140 college students from the authors' classes. We described writing the height of each student on a card and putting the 140 cards in a bowl.

EXAMPLE 10.1

Let's use that example to create a distribution of differences between means. We'll walk through the steps for this process.

STEP 1: We randomly select three cards, replacing each after selecting it, and calculate the mean of the heights listed on them. This is the first group.

STEP 2: We randomly select three other cards, replacing each after selecting it, and calculate their mean. This is the second group.

STEP 3: We subtract the second mean from the first.

That's really all there is to it—except we repeat these three steps many more times. So there are two samples and two sample means, but we're building just *one* curve of differences between means.

Here's an example using the three steps.

STEP 1: We randomly select three cards, replacing each after selecting it, and find that the heights are 61, 65, and 72. We calculate a mean of 66 inches. This is the first group.

STEP 2: We randomly select three other cards, replacing each after selecting it, and find that the heights are 62, 65, and 65. We calculate a mean of 64 inches. This is the second group.

STEP 3: We subtract the second mean from the first: $66 - 64 = 2$. (Note that it's fine to subtract the first from the second, as long as we're consistent in the arithmetic.)

We repeat the three-step process. Let's say that, this time, we calculate means of 65 and 68 for the two samples. Now the difference between means would be $65 - 68 = -3$. We might repeat the three steps a third time and find means of 63 and 63, for a difference of 0. Eventually, we would have many differences between means—some positive, some negative, and some right at 0—and could plot them on a curve. But this would only be the beginning of what this distribution would look like. If we were to calculate the whole distribution, then we would do this many,

▪ An **independent-samples *t* test** is used to compare two means for a between-groups design, a situation in which each participant is assigned to only one condition.

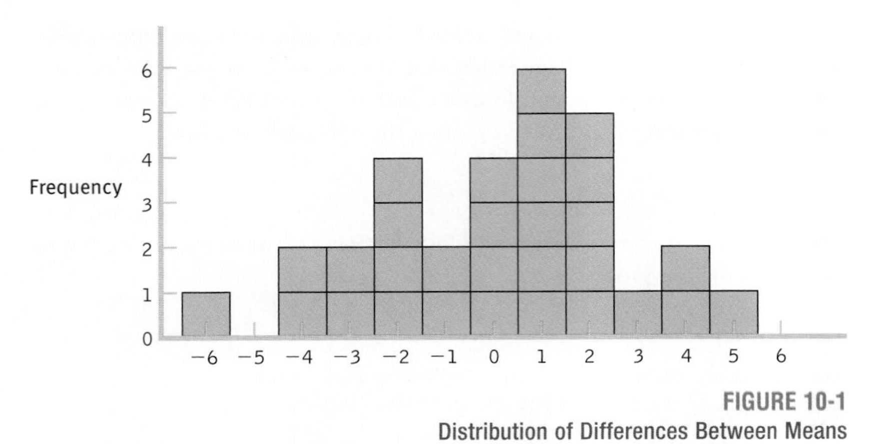

FIGURE 10-1
Distribution of Differences Between Means

This curve represents the beginning of the development of a distribution of differences between means. It includes only 30 differences, whereas the actual distribution would include all possible differences.

many more times. When creating the beginning of a distribution of differences between means, the authors calculated 30 differences between means, as shown in Figure 10-1. ■

The Six Steps of an Independent-Samples *t* Test

EXAMPLE 10.2

Does the price of a product influence how much you like it? If you're told that your sister's new flat-screen television cost \$3000, do you perceive the picture quality to be sharper than if you're told it cost \$1200? If you think your friend's new shirt is from a high-end designer like Dolce & Gabbana, do you covet it more than if he tells you it's from a trendy but low-priced mass-retailer like Target?

Economics researchers from Northern California, not far from prime wine country, wondered whether this would be true for wine (Plassmann, O'Doherty, Shiv, & Rangel, 2008). In part of their study, they randomly assigned some wine drinkers to taste wine that was said to cost \$10 per bottle and others to taste *the same wine* at a supposed price of \$90 per bottle. (Note that we're altering some aspects of the design and statistical analysis of this study for teaching purposes.) The researchers asked participants to rate how much they liked the wine; they also used functional magnetic resonance imaging (fMRI), a brain-scanning technique, to determine whether differences were evident in areas of the brain that are typically activated when people experience a stimulus as pleasant (e.g., the medial orbitofrontal cortex). Which wine do you think participants preferred, the \$10 wine or the \$90 one?

We will conduct an independent-samples *t* test on fictional data, using the ratings of how much nine people like the wine they were randomly assigned to taste (four tasting wine from the "\$10" bottle and five tasting wine from the "\$90" bottle). Remember, everyone is actually tasting wine from the *same* bottle! These fictional data have approximately the same means as were reported in the original study. Notice that we do not need to have the same number of participants in each sample, although it is best if the sample sizes are fairly close.

Mean "liking ratings" of the wine:

"\$10" wine: 1.5 2.3 2.8 3.4

"\$90" wine: 2.9 3.5 3.5 4.9 5.2

STEP 1: Identify the populations, distribution, and assumptions. In terms of determining the populations, this step is similar to that for the paired-samples *t* test: There are two populations—those told they are drinking wine from a $10 bottle and those told they are drinking wine from a $90 bottle. The comparison distribution for an independent-samples *t* test, however, will be a distribution of differences between means (rather than a distribution of mean difference scores). Table 10-1 summarizes the distributions we have encountered with the hypothesis tests we have learned so far.

TABLE 10-1. Hypothesis Tests and Their Distributions

We must consider the appropriate comparison distribution when we choose which hypothesis test to use.

Hypothesis Test	Number of Samples	Comparison Distribution
z test	One	Distribution of means
Single-sample *t* test	One	Distribution of means
Paired-samples *t* test	Two (same participants)	Distribution of mean difference scores
Independent-samples *t* test	Two (different participants)	Distribution of differences between means

As usual, the comparison distribution is based on the null hypothesis. As with the paired-samples *t* test, the null hypothesis for the independent-samples *t* test posits no mean difference. So the mean of the comparison distribution would be 0; this reflects a mean difference between means of 0. We compare the difference between the sample means to a difference of 0, which is what would occur if there was no difference between groups. The assumptions for an independent-samples *t* test are the same as for the single-sample *t* test and the paired-samples *t* test.

Summary: Population 1: People told they are drinking wine from a $10 bottle. Population 2: People told they are drinking wine from a $90 bottle.

The comparison distribution will be a distribution of differences between means based on the null hypothesis. The hypothesis test will be an independent-samples *t* test because we have two samples composed of different groups of participants. This study meets one of the three assumptions. (1) The dependent variable is a rating on a liking measure, which can be considered a scale variable. (2) We do not know whether the population is normally distributed, and there are not at least 30 participants. However, the sample data do not suggest that the underlying population distribution is skewed. (3) The wine drinkers in this study were not randomly selected from among all wine drinkers, so we must be cautious with respect to generalizing these findings.

STEP 2: State the null and research hypotheses. This step for an independent-samples *t* test is identical to that for the previous *t* tests.

Summary: Null hypothesis: On average, people drinking wine they were told was from a $10 bottle give it the same rating as people drinking wine they were told was from a $90 bottle—$H_0: \mu_1 = \mu_2$. Research hypothesis: On average, people drinking wine they were told was from a $10 bottle give it a different rating than people drinking wine they were told was from a $90 bottle—$H_1: \mu_1 \neq \mu_2$.

STEP 3: Determine the characteristics of the comparison distribution.

This step for an independent-samples *t* test is similar to that for previous *t* tests: We determine the appropriate mean and the appropriate standard error of the comparison distribution—the distribution based on the null hypothesis. According to the null hypothesis, no mean difference exists between the populations; that is, the difference between means is 0. So the mean of the comparison distribution is always 0, as long as the null hypothesis posits no mean difference.

Because we have two samples for an independent-samples *t* test, however, it is more complicated to calculate the appropriate measure of spread. There are five stages to this process. First, let's consider them in words; then we'll learn the calculations. These instructions are basic, and you'll understand them better when you do the calculations, but they'll help you to keep the overall framework in mind. (These verbal descriptions are keyed by letter to the calculation stages below.)

a. Calculate the corrected variance for each sample. (Notice that we're working with variance, not standard deviation.)
b. Pool the variances. Pooling involves taking an average of the two sample variances while accounting for any differences in the sizes of the two samples. Pooled variance is an estimate of the common population variance.
c. Convert the pooled variance from squared standard deviation (i.e., variance) to squared standard error (another version of variance) by dividing the pooled variance by the sample size, first for one sample and then again for the second sample. These are the estimated variances for each sample's distribution of means.
d. Add the two variances (*squared* standard errors), one for each distribution of sample means, to calculate the estimated variance of the distribution of differences between means.
e. Calculate the square root of this form of variance (*squared* standard error) to get the estimated standard error of the distribution of differences between means.

Notice that stages (a) and (b) are an expanded version of the usual first calculation for a *t* test. Instead of calculating one corrected estimate of standard deviation, we're calculating two for an independent-samples *t* test—one for each sample. Also, for an independent-samples *t* test, we use variances instead of standard deviations. Because there are two calculations of variance, we combine them (i.e., the pooled variance). Stages (c) and (d) are an expanded version of the usual second calculation for a *t* test. Once again, we convert to standard error (only this time it is squared because we are working with variances) and combine the variances from each sample. In stage (e), we take the square root so that we have standard error. Let's examine the calculations.

(a) We calculate corrected variance for each sample (corrected variance is the one we learned in Chapter 9 that uses $N - 1$ in the denominator). First, we calculate variance for X, the sample of people told they are drinking wine from a \$10 bottle. Be sure to use the mean of the ratings of the \$10 wine drinkers only, which is 2.5. Notice that the symbol for this variance uses s^2, instead of SD^2 (just as the standard deviation uses s instead of SD in the previous *t* tests). Also, we included the subscript X to indicate that this is variance for the first sample, whose scores are arbitrarily called X. (Remember, don't take the square root. We want variance, not standard deviation.)

Pooled variance is a weighted average of the two estimates of variance—one from each sample—that are calculated when conducting an independent-samples *t* test.

X	$X - M$	$(X - M)^2$
1.5	−1.0	1.00
2.3	−0.2	0.04
2.8	0.3	0.09
3.4	0.9	0.81

$$s_X^2 = \frac{\Sigma(X-M)^2}{N-1} = \frac{(1.00+0.04+0.09+0.81)}{4-1} = \frac{1.94}{3} = 0.647$$

Now we do the same for *Y*, the people told they are drinking wine from a $90 bottle. Remember to use the mean for *Y*; it's easy to forget and use the mean we calculated earlier for *X*. The mean for *Y* is 4.0. The subscript *Y* indicates that this is the variance for the second sample, whose scores are arbitrarily called *Y*. (We could call these scores by any letter, but statisticians tend to call the scores in the first two samples *X* and *Y*.)

Y	$Y - M$	$(Y - M)^2$
2.9	−1.1	1.21
3.5	−0.5	0.25
3.5	−0.5	0.25
4.9	0.9	0.81
5.2	1.2	1.44

$$s_Y^2 = \frac{\Sigma(Y-M)^2}{N-1} = \frac{(1.21+0.25+0.25+0.81+1.44)}{5-1} = \frac{3.96}{4} = 0.990$$

(b) We pool the two estimates of variance. Because there are often different numbers of people in each sample, we cannot simply take their mean. We mentioned earlier in this book that estimates of spread taken from smaller samples tend to be less accurate. So we weight the estimate from the smaller sample a bit less and weight the estimate from the larger sample a bit more. We do this by calculating the proportion of degrees of freedom represented by each sample. Each sample has degrees of freedom of $N - 1$. We also calculate a total degrees of freedom that sums the degrees of freedom for the two samples. Here are the calculations:

$$df_X = N - 1 = 4 - 1 = 3$$

$$df_Y = N - 1 = 5 - 1 = 4$$

$$df_{total} = df_X + df_Y = 3 + 4 = 7$$

MASTERING THE FORMULA

10-1: There are three degrees of freedom calculations for an independent-samples *t* test. We calculate the degrees of freedom for each sample by subtracting 1 from the number of participants in that sample: $df_X = N - 1$ and $df_Y = N - 1$. Finally, we sum the degrees of freedom from the two samples to calculate the total degrees of freedom: $df_{total} = df_X + df_Y$.

Using these degrees of freedom, we calculate a sort of average variance. ***Pooled variance*** *is a weighted average of the two estimates of variance—one from each sample—that are calculated when conducting an independent-samples t test.* The estimate of variance from the larger sample counts for more in the pooled variance than does the estimate from the smaller sample because larger samples tend to lead to somewhat

MASTERING THE FORMULA

10-2: We use all three degrees of freedom calculations, along with the variance estimates for each sample, to calculate pooled variance: $s^2_{pooled} = \left(\frac{df_X}{df_{total}}\right)s^2_X + \left(\frac{df_Y}{df_{total}}\right)s^2_Y$. This formula takes into account the size of each sample. A larger sample has a larger degrees of freedom in the numerator, and that variance therefore has more weight in the pooled variance calculations.

MASTERING THE FORMULA

10-3: The next step in calculating the *t* statistic for a two-sample, between-groups design is to calculate the variance version of standard error for each sample by dividing variance by sample size. We use the pooled version of variance for both calculations because it's more likely to be accurate than the individual variance estimate for each sample. For the first sample, the formula is: $s^2_{M_X} = \frac{s^2_{pooled}}{N_X}$. For the second sample, the formula is: $s^2_{M_Y} = \frac{s^2_{pooled}}{N_Y}$. Note that because we're dealing with variance, the square of standard deviation, we divide by *N*, the square of $\sqrt{N}$—the denominator for standard error.

MASTERING THE FORMULA

10-4: To calculate the variance of the distribution of differences between means, we sum the variance versions of standard error that we calculated in the previous step: $s^2_{difference} = s^2_{M_X} + s^2_{M_Y}$.

MASTERING THE FORMULA

10-5: To calculate the standard deviation of the distribution of differences between means, we take the square root of the previous calculation, the variance of the distribution of differences between means. The formula is: $s_{difference} = \sqrt{s^2_{difference}}$.

more accurate estimates than do smaller samples. Here's the formula for pooled variance, and the calculations for this example:

$$s^2_{pooled} = \left(\frac{df_X}{df_{total}}\right)s^2_X + \left(\frac{df_Y}{df_{total}}\right)s^2_Y = \left(\frac{3}{7}\right)0.647 + \left(\frac{4}{7}\right)0.990 = 0.277 + 0.566 = 0.843$$

(*Note:* If we had exactly the same number of participants in each sample, this would be an unweighted average—that is, we could compute the average in the usual way by summing the two sample variances and dividing by 2.)

(c) Now that we have pooled the variances, we have an estimate of spread. This is similar to the estimate of the standard deviation in the previous *t* tests, but now it's based on two samples (and it's an estimate of variance rather than standard deviation). The next calculation in the previous *t* tests was dividing standard deviation by $\sqrt{N}$ to get standard error. In this case, we divide by *N* instead of $\sqrt{N}$. Why? Because we are dealing with variances, not standard deviations. Variance is the square of standard deviation, so we divide by the square of $\sqrt{N}$, which is simply *N*. We do this once for each sample, using pooled variance as the estimate of spread. We use pooled variance because an estimate based on two samples is better than an estimate based on one. The key here is to divide by the appropriate *N:* 4 for the first sample and 5 for the second sample.

$$s^2_{M_X} = \frac{s^2_{pooled}}{N_X} = \frac{0.843}{4} = 0.211$$

$$s^2_{M_Y} = \frac{s^2_{pooled}}{N_Y} = \frac{0.843}{5} = 0.169$$

(d) In stage (c), we calculated the variance versions of standard error for each sample, but we want only one such measure of spread when we calculate the test statistic. So, we combine the two variances, similar to the way in which we combined the two estimates of variance in stage (b). This stage is even simpler, however. We merely add the two variances together. When we sum them, we get the variance of the distribution of differences between means, symbolized as $s^2_{difference}$. Here are the formula and the calculations for this example:

$$s^2_{difference} = s^2_{M_X} + s^2_{M_Y} = 0.211 + 0.169 = 0.380$$

(e) We now have paralleled the two calculations of the previous *t* tests by doing two things: (1) We calculated an estimate of spread (we made two calculations using a formula we learned in Chapter 9, one for each sample, then combined them), and (2) We then adjusted the estimate for the sample size (again, we made two calculations, one for each sample, then combined them). The main difference is that we have kept all calculations as variances rather than standard deviations. At this final stage, we convert from variance form to standard deviation form. Because standard deviation is the square root of variance, we do this by simply taking the square root:

$$s_{difference} = \sqrt{s^2_{difference}} = \sqrt{0.380} = 0.616$$

Summary: The mean of the distribution of differences between means is: $\mu_X - \mu_Y = 0$. The standard deviation of the distribution of differences between means is: $s_{difference} = 0.616$.

STEP 4: Determine critical values, or cutoffs.

This step for the independent-samples *t* test is similar to those for previous *t* tests, but we use the total degrees of freedom, df_{total}.

Summary: The critical values, based on a two-tailed test, a *p* level of 0.05, and a df_{total} of 7, are −2.365 and 2.365 (as seen in the curve in Figure 10-2).

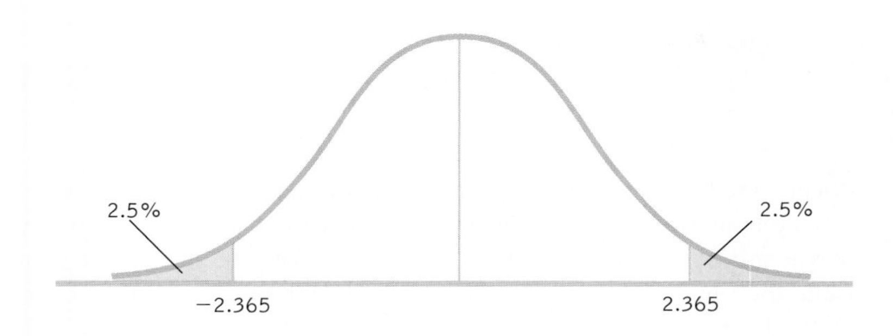

FIGURE 10-2
Determining Cutoffs for an Independent-Samples *t* Test

To determine the critical values for an independent-samples *t* test, we use the total degrees of freedom, df_{total}. This is the sum of the degrees of freedom for each sample, which is $N - 1$ for each sample.

STEP 5: Calculate the test statistic.

This step for the independent-samples *t* test is similar to the fifth step in previous *t* tests. Here we subtract the population difference between means based on the null hypothesis from the difference between means for the samples. The formula is:

$$t = \frac{(M_X - M_Y) - (\mu_X - \mu_Y)}{s_{difference}}$$

MASTERING THE FORMULA

10-6: We calculate the test statistic for an independent-samples *t* test using the following formula: $t = \frac{(M_X - M_Y) - (\mu_X - \mu_Y)}{s_{difference}}$. We subtract the difference between means according to the null hypothesis, usually 0, from the difference between means in the sample. We then divide this by the standard deviation of the differences between means. Because the difference between means according to the null hypothesis is usually 0, the formula for the test statistic is often abbreviated as: $t = \frac{M_X - M_Y}{s_{difference}}$.

As in previous *t* tests, the test statistic is calculated by subtracting a number based on the populations from a number based on the samples, then dividing by a version of standard error. Because the population difference between means (according to the null hypothesis) is almost always 0, many statisticians choose to eliminate the latter part of the formula. So the formula for the test statistic for an independent-samples *t* test is often abbreviated as:

$$t = \frac{M_X - M_Y}{s_{difference}}$$

You might find it easier to use the first formula, however, as it reminds us that we are subtracting the population difference between means according to the null hypothesis (0) from the actual difference between the sample means. This format more closely parallels the formulas of the test statistics we calculated in Chapter 9.

Summary: $t = \frac{(2.5 - 4.0) - 0}{0.616} = -2.44$

STEP 6: Make a decision.

This step for the independent-samples *t* test is identical to that for the previous *t* tests. If we reject the null hypothesis, we need to examine the means of the two conditions so that we know the direction of the effect.

Summary: Reject the null hypothesis. It appears that those told they are drinking wine from a $10 bottle give it lower ratings, on average, than those told they are drinking from a $90 bottle (as shown by the curve in Figure 10-3).

This finding documents the fact that people report liking a more expensive wine better than a less expensive one—even when it's the same wine! The researchers

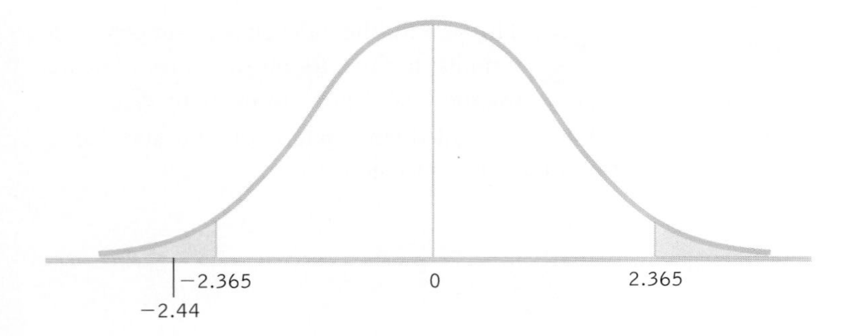

FIGURE 10-3
Making a Decision

As in previous *t* tests, in order to decide whether or not to reject the null hypothesis, we compare the test statistic to the critical values. In this figure, the test statistic, −2.44, is beyond the lower cutoff, −2.365. We reject the null hypothesis. It appears that those told they are drinking wine from a $10 bottle give it lower ratings, on average, than those told they are drinking wine from a $90 bottle.

documented a similar finding with a narrower gap between prices—$5 and $45. Naysayers are likely to point out, however, that participants drinking an expensive wine may report liking it better than participants drinking an inexpensive wine simply because they are expected to say they like it better because of its price. However, the fMRI that was conducted, which is a more objective measure, yielded a similar finding. Those drinking the more expensive wines showed increased activation in brain areas such as the medial orbitofrontal cortex, essentially an indication in the brain that people are enjoying an experience. Expectations really do seem to influence us. ■

Reporting the Statistics

To report the statistics as they would appear in a journal article, follow standard APA format, including the degrees of freedom, the value of the test statistic, and the *p* value associated with the test statistic. (Note that because the *t* table in Appendix B only includes the *p* values of 0.10, 0.05, and 0.01, we cannot use it to determine the actual *p* value for the test statistic. Unless we use software, we can only report whether or not the *p* value is less than the critical *p* level.) In the current example, the statistics would read:

$$t(7) = -2.44, p < 0.05$$

In addition to the results of hypothesis testing, we would also include the means and standard deviations for the two samples. We calculated the means in step 3 of hypothesis testing, and we also calculated the variances (0.647 for those told they were drinking from a $10 bottle and 0.990 for those told they were drinking from a $90 bottle). We can calculate the standard deviations by taking the square roots of the variances. The descriptive statistics can be reported in parentheses as:

($10 bottle: $M = 2.5$, $SD = 0.80$; $90 bottle: $M = 4.0$, $SD = 0.99$)

CHECK YOUR LEARNING

Reviewing the Concepts

- When we conduct an independent-samples *t* test, we cannot calculate individual difference scores. That is why we compare the mean of one sample with the mean of the other sample.
- The comparison distribution is a distribution of differences between means.
- We use the same six steps of hypothesis testing that we used with the *z* test and with the single-sample and paired-samples *t* tests.
- Conceptually, the *t* test for independent samples makes the same comparisons as the other *t* tests. However, the calculations are different, and critical values are based on degrees of freedom from two samples.

Clarifying the Concepts

10-1 In what situation do we conduct a paired-samples *t* test? In what situation do we conduct an independent-samples *t* test?

10-2 What is pooled variance?

Calculating the Statistics

10-3 Imagine you have the following data from two independent groups:

Group 1: 3, 2, 4, 6, 1, 2
Group 2: 5, 4, 6, 2, 6

Compute each of the following calculations needed to complete your final calculation of the independent-samples *t* test.

a. Calculate the corrected variance for each group.

b. Calculate the degrees of freedom and pooled variance, s^2_{pooled}.

c. Calculate the variance version of standard error for each group.

d. Calculate the variance of the distribution of differences between means, then convert this number to standard deviation.

e. Calculate the test statistic.

Applying the Concepts

10-4 In Check Your Learning 10-3, you calculated several statistics; now let's consider a context for those numbers. Steele and Pinto (2006) examined whether people's level of trust in their direct supervisor was related to their level of agreement with a policy supported by that leader. They found that the extent to which subordinates agreed with their supervisor was statistically significantly related to trust and showed no relation to gender, age, time on the job, or length of time working with the supervisor. We have presented fictional data to re-create these findings, where group 1 represents employees with low trust in their supervisor and group 2 represents the high-trust employees. The scores presented are the level of agreement with a decision made by a leader, from 1 (strongly disagree) to 7 (strongly agree).

Group 1 (low trust in leader): 3, 2, 4, 6, 1, 2
Group 2 (high trust in leader): 5, 4, 6, 2, 6

a. State the null and research hypotheses.

b. Identify the critical values and make a decision.

c. Write your conclusion in a formal sentence that includes presentation of the statistic in APA format.

d. Explain why your results are different from those in the original research, despite having a similar mean difference.

Solutions to these Check Your Learning questions can be found in Appendix D.

Beyond Hypothesis Testing

After working at Guinness, Stella Cunliffe was hired by the British government's criminology department. She noticed that adult male prisoners who had short prison sentences returned to prison at a very high rate—an apparent justification for longer prison sentences. But Cunliffe noticed that the returning prisoners were almost all older people with mental health problems who had been sent to prison because the mental hospitals would not take them. Because of observations like this, good researchers know that they need ways to evaluate and interpret data that go beyond hypothesis testing.

Two ways that researchers can evaluate the findings of a hypothesis test are by calculating a confidence interval and an effect size.

MASTERING THE CONCEPT

10.2: As we can with the *z* test, the single-sample *t* test, and the paired-samples *t* test, we can determine a confidence interval and calculate a measure of effect size—Cohen's *d*—when we conduct an independent-samples *t* test.

Calculating a Confidence Interval for an Independent-Samples *t* Test

Confidence intervals for the independent-samples *t* test are centered around the *difference between means* (rather than the means themselves). So we use the difference between means for the samples and the standard error for the difference between means, $s_{difference}$, which we calculate in an identical manner to the one used in hypothesis testing.

We use the formula for the independent-samples *t* statistic when calculating the raw differences between means. To do this, we use algebra on the original formula for an independent-samples *t* test to isolate the upper and lower mean differences. Here is the original *t* statistic formula:

$$t = \frac{(M_X - M_Y) - (\mu_X - \mu_Y)}{s_{difference}}$$

We replace the population mean difference, $(\mu_X - \mu_Y)$, with the sample mean difference, $(M_X - M_Y)_{sample}$, because this is what the confidence interval is centered around. We also indicate that the first mean difference in the numerator refers to the bounds of the confidence intervals, the upper bound in this case:

MASTERING THE FORMULA

10-7: The formulas for the upper and lower bounds of the confidence interval are $(M_X - M_Y)_{upper} = t(s_{difference}) + (M_X - M_Y)_{sample}$ and $(M_X - M_Y)_{lower} = -t(s_{difference}) + (M_X - M_Y)_{sample}$, respectively. In each case, we multiply the *t* statistic that marks off the tail by the standard error for the differences between means and add it to the difference between means in the sample. Remember that one of the *t* statistics is negative.

$$t_{upper} = \frac{(M_X - M_Y)_{upper} - (M_X - M_Y)_{sample}}{s_{difference}}$$

With algebra, we isolate the upper bound of the confidence interval to create the following formula:

$$(M_X - M_Y)_{upper} = t(s_{difference}) + (M_X - M_Y)_{sample}$$

We create the formula for the lower bound of the confidence interval in exactly the same way, using the negative version of the *t* statistic:

$$(M_X - M_Y)_{lower} = -t(s_{difference}) + (M_X - M_Y)_{sample}$$

EXAMPLE 10.3

Let's calculate the confidence interval that parallels the hypothesis test we conducted earlier, comparing ratings of those who are told they are drinking wine from a $10 bottle and ratings of those told they are drinking wine from a $90 bottle (Plassmann et al., 2008). Previously, we calculated the difference between the means of these samples to be $2.5 - 4.0 = -1.5$; the standard error for the differences between means, $s_{difference}$, to be 0.616; and the degrees of freedom to be 7. (Note that the order of subtraction in calculating the difference between means is irrelevant; we could just as easily have subtracted 2.5 from 4.0 and gotten a positive result, 1.5.) Here are the five steps for determining a confidence interval for a difference between means:

STEP 1: Draw a normal curve with the sample difference between means in the center (as shown in Figure 10-4).

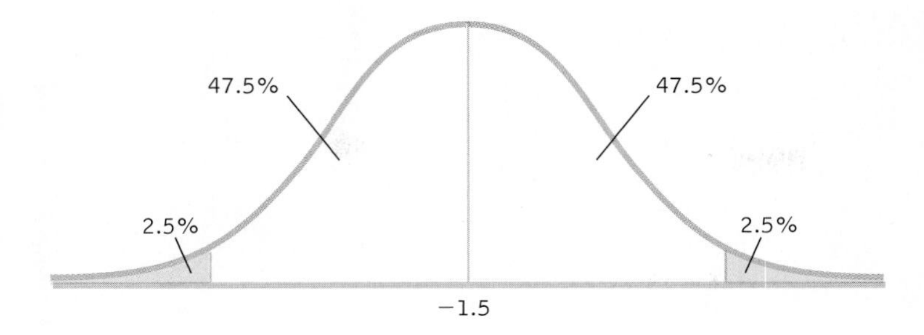

FIGURE 10-4
A 95% Confidence Interval for Differences Between Means, Part I

As with a confidence interval for a single-sample *t* test, we start the confidence interval for a difference between means by drawing a curve with the sample difference between means in the center.

STEP 2: Indicate the bounds of the confidence interval on either end, and write the percentages under each segment of the curve.

(See Figure 10-4.)

STEP 3: Look up the *t* statistics for the lower and upper ends of the confidence interval in the *t* table.

Use a two-tailed test and a p level of 0.05 (which corresponds to a 95% confidence interval). Use the degrees of freedom—7—that we calculated earlier. The table indicates a t statistic of 2.365. Because the normal curve is symmetric, the bounds of the confidence interval fall at t statistics of -2.365 and 2.365. (Note that these cutoffs are identical to those used for the independent-samples t test because the p level of 0.05 corresponds to a confidence level of 95%.) We add those t statistics to the normal curve, as in Figure 10-5.

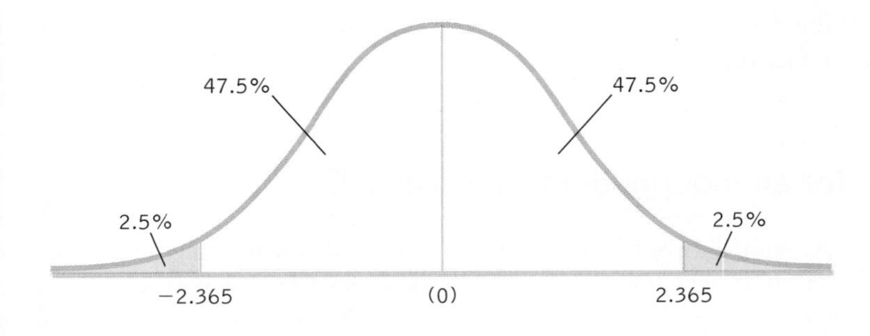

FIGURE 10-5
A 95% Confidence Interval for Differences Between Means, Part II

The next step in calculating a confidence interval is identifying the *t* statistics that indicate each end of the interval. Because the curve is symmetric, the *t* statistics have the same magnitude—one is negative and one is positive.

STEP 4: Convert the *t* statistics to raw differences between means for the lower and upper ends of the confidence interval.

For the lower end, the formula is:

$$(M_X - M_Y)_{lower} = -t(s_{difference}) + (M_X - M_Y)_{sample}$$
$$= -2.365(0.616) + (-1.5) = -2.96$$

For the upper end, the formula is:

$$(M_X - M_Y)_{upper} = t(s_{difference}) + (M_X - M_Y)_{sample}$$
$$= 2.365(0.616) + (-1.5) = -0.04$$

The confidence interval is [−2.96, −0.04], as shown in Figure 10-6.

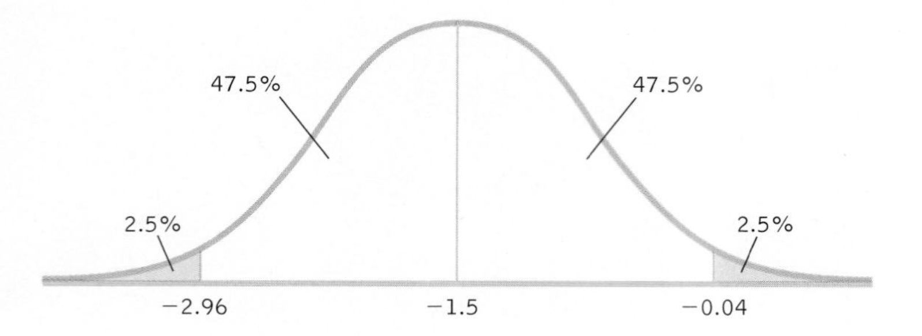

FIGURE 10-6
A 95% Confidence Interval for Differences Between Means, Part III

The final step in calculating a confidence interval is converting the *t* statistics that indicate each end of the interval into raw differences between means.

STEP 5: Check your answer. Each end of the confidence interval should be exactly the same distance from the sample mean.

$$-2.96 - (-1.5) = -1.46$$

$$-0.04 - (-1.5) = 1.46$$

The interval checks out. The bounds of the confidence interval are calculated as the difference between sample means, plus or minus 1.46.

Also, the confidence interval does not include 0. Thus, it is not plausible that there is no difference between means. We can conclude that people told they are drinking wine from a $10 bottle give different ratings, on average, than those told they are drinking wine from a $90 bottle.

When we conducted the independent-samples *t* test earlier, we rejected the null hypothesis and drew the same conclusion as we did with the confidence interval. But the confidence interval provided more information because it is an interval estimate rather than a point estimate. ■

Calculating Effect Size for an Independent-Samples *t* Test

EXAMPLE 10.4

As with all hypothesis tests, it is recommended that the results be supplemented with an effect size. For an independent-samples *t* test, as with other *t* tests, we can use Cohen's *d* as the measure of effect size. The fictional data provided means of 2.5 for those told they were drinking wine from a $10 bottle and 4.0 for those told they were drinking wine from a $90 bottle (Plassmann et al., 2008). Previously, we calculated a standard error for the difference between means, $s_{difference}$, of 0.616. Here are the calculations we performed:

Stage a (variance for each sample):

$$s_X^2 = \frac{\Sigma(X - M)^2}{N - 1} = 0.647; \; s_Y^2 = \frac{\Sigma(Y - M)^2}{N - 1} = 0.990$$

Stage b (combining variances):

$$s_{pooled}^2 = \left(\frac{df_X}{df_{total}}\right) s_X^2 + \left(\frac{df_Y}{df_{total}}\right) s_Y^2 = 0.843$$

Stage c (variance form of standard error for each sample):

$$s_{M_X}^2 = \frac{s_{pooled}^2}{N_X} = 0.211; \; s_{M_Y}^2 = \frac{s_{pooled}^2}{N_Y} = 0.169$$

Stage d (combining variance forms of standard error):

$$s_{difference}^2 = s_{M_X}^2 + s_{M_Y}^2 = 0.380$$

Stage e (converting the variance form of standard error to the standard deviation form of standard error):

$$s_{difference} = \sqrt{s_{difference}^2} = 0.616$$

Because the goal is to disregard the influence of sample size in order to calculate Cohen's *d,* we want to use the standard deviation in the denominator, not the standard error. So we can ignore the last three stages, all of which contribute to the calculation of standard error. That leaves stages a and b. It makes more sense to use the one that includes information from both samples, so we focus our attention on stage b. Here is where many students make a mistake. What we have calculated in stage b is pooled *variance,* not pooled *standard deviation.* We must take the square root of the pooled variance to get the pooled standard deviation, the appropriate value for the denominator of Cohen's *d.*

$$s_{pooled} = \sqrt{s_{pooled}^2} = \sqrt{0.843} = 0.918$$

The test statistic that we calculated for this study was:

$$t = \frac{(M_X - M_Y) - (\mu_X - \mu_Y)}{s_{difference}} = \frac{(2.5 - 4.0) - 0}{0.616} = -2.44$$

For Cohen's *d,* we simply replace the denominator with standard deviation, s_{pooled}, instead of standard error, $s_{difference}$.

$$\text{Cohen's } d = \frac{(M_X - M_Y) - (\mu_X - \mu_Y)}{s_{pooled}} = \frac{(2.5 - 4.0) - 0}{0.918} = -1.63$$

For this study, the effect size is reported as: $d = -1.63$. The two sample means are 1.63 standard deviations apart. According to the conventions we learned in Chapter 8 (0.2 is a small effect; 0.5 is a medium effect; 0.8 is a large effect), this is a large effect. ■

MASTERING THE FORMULA

10-8: We use pooled standard deviation to calculate Cohen's *d* for a two-sample, between-groups design. We calculate pooled standard deviation by taking the square root of the pooled variance that we calculated as part of the independent-samples *t* test: $s_{pooled} = \sqrt{s_{pooled}^2}$.

MASTERING THE FORMULA

10-9: For a two-sample, between-groups design, we calculate Cohen's *d* using the following formula: Cohen's $d = \frac{(M_X - M_Y) - (\mu_X - \mu_Y)}{s_{pooled}}$. The formula is similar to that for the test statistic in an independent-samples *t* test, except that we divide by pooled standard deviation, rather than standard error, because we want a measure of variability not altered by sample size.

CHECK YOUR LEARNING

Reviewing the Concepts

> A confidence interval can be created with a *t* distribution around a difference between means.
> We can calculate an effect size, Cohen's *d,* for an independent-samples *t* test.

Clarifying the Concepts

10-5 Why do we calculate confidence intervals?

continued on next page

10-6 How does considering the conclusions in terms of effect size help to prevent incorrect interpretations of the findings?

Calculating the Statistics

10-7 Use the hypothetical data on level of agreement with a supervisor, as listed here, to calculate the following:

Group 1 (low trust in leader): 3, 2, 4, 6, 1, 2

Group 2 (high trust in leader): 5, 4, 6, 2, 6

a. Calculate the 95% confidence interval.

b. Calculate effect size using Cohen's *d*.

Applying the Concepts

Solutions to these Check Your Learning questions can be found in Appendix D.

10-8 Explain what the confidence interval calculated in Check Your Learning 10-7 tells us. Why is this confidence interval superior to the hypothesis test that we conducted?

10-9 Interpret the meaning of the effect size calculated in Check Your Learning 10-7. What does this add to the confidence interval and hypothesis test?

REVIEW OF CONCEPTS

Conducting an Independent-Samples *t* Test

We use *independent-samples t tests* when we have two samples and different participants are in each sample. Because the samples are comprised of different people, we cannot calculate difference scores, so the comparison distribution is a distribution of differences between means. Because we are working with two separate samples of scores (rather than one set of difference scores) when we conduct an independent-samples *t* test, we need additional steps to calculate an estimate of spread. As part of these steps, we calculate estimates of variance from each sample, and then combine them to create a *pooled variance*. We can present the statistics in APA style as we did with other hypothesis tests.

Beyond Hypothesis Testing

As with other forms of hypothesis testing, it is useful to replace or supplement the independent-samples *t* test with a confidence interval. A confidence interval can be created around a difference between means using a *t* distribution. To understand the importance of a finding, we must also calculate an effect size. With an independent-samples *t* test, as with other *t* tests, a common effect-size measure is Cohen's *d*.

SPSS®

We can conduct an independent-samples *t* test using SPSS for the wine-tasting data we presented earlier in this chapter. We start by creating two columns of data, one for stated cost of the bottle ($10 versus $90) and one for the liking rating. For wine cost, we could, for example, give a "1" to each person told that the wine is from a $10 bottle, and give a "2" to each person told that the wine is from a $90 bottle. We can use the "Values" function in the Variable View to tell SPSS that 1 = $10 and 2 = $90. We place each participant's data in one row—a score for the stated cost of the wine in the first column and a liking rating in the second column. We can now conduct the hypothesis test.

Select Analyze → Compare Means → Independent-Samples T Test. Choose the dependent variable, "rating," by clicking it, then clicking the arrow in the upper center. Choose the independent variable, "cost," by clicking it, then clicking the arrow in the lower center. Click the "Define Groups" button, then provide the values for each level of the independent variable. For example, enter "1" for group 1 and "2" for group 2. Then click "OK."

Part of the output is shown in the screenshot below. Toward the top, we see means and standard deviations for participants told they were drinking wine from a \$10 bottle and participants told they were drinking wine from a \$90 bottle. For example, the output tells us that the mean for those told they were drinking wine from a \$10 bottle is 2.5, with a standard deviation of 0.80416. We can see that the t statistic is -2.436, with a p value (under "Sig. (2-tailed)") of 0.045. The t statistic is the same as the one we calculated earlier, -2.44.

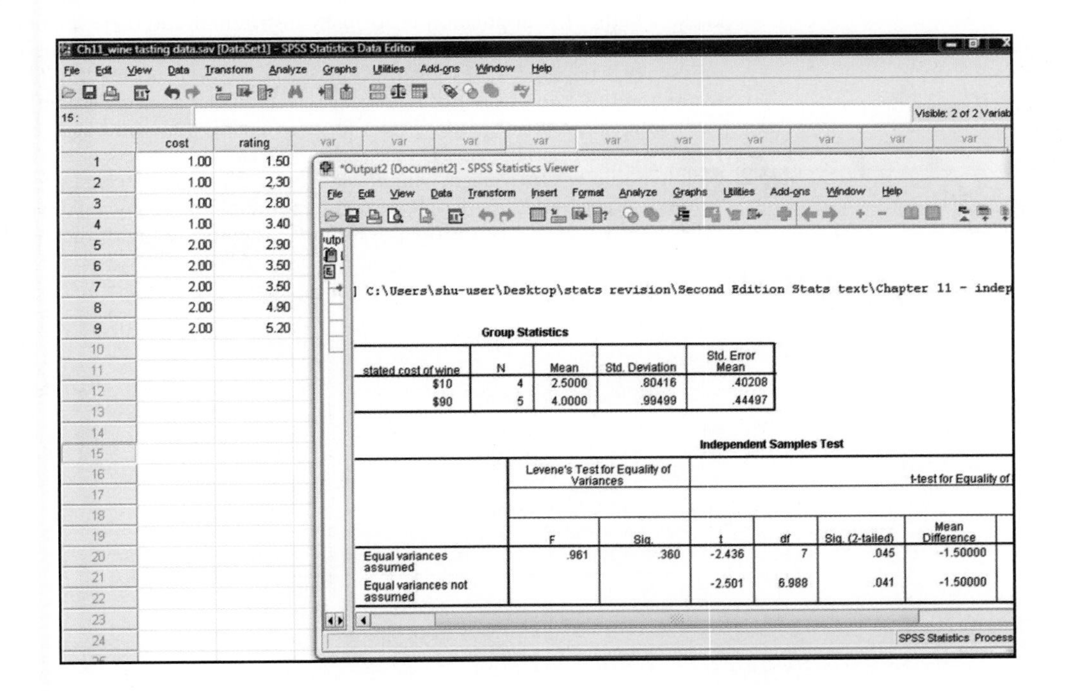

How It Works

10.1 INDEPENDENT-SAMPLES *t* TEST

Who do you think has a better sense of humor—women or men? Researchers at Stanford University examined brain activity in women and men during exposure to humorous cartoons (Azim, Mobbs, Jo, Menon, & Reiss, 2005). Using a brain-scanning technique called *functional magnetic resonance imaging* (fMRI), researchers observed more activity in the reward centers of women's brains than men's, the same reward centers that respond when receiving money or feeling happy. The researchers suggested that this might be because women have lower expectations of humor than do men, so they find it more rewarding when something is actually funny.

However, the researchers were aware of other possible explanations for these findings. For example, they considered whether one gender is more likely to find humorous stimuli funny to begin with. In this study, men and women indicated the percentage of 30 cartoons that they perceived to be either "funny" or "unfunny." Below are fictional data for nine people (four women and five men); these fictional data have approximately the same means as in the original study.

Percentage of cartoons labeled as "funny"

Women: 84, 97, 58, 90

Men: 88, 90, 52, 97, 86

How can we conduct all six steps of hypothesis testing for an independent-samples t test for this scenario, using a two-tailed test with critical values based on a p level of 0.05? Here are the steps:

Step 1: Population 1: Women exposed to humorous cartoons. Population 2: Men exposed to humorous cartoons.

The comparison distribution will be a distribution of differences between means based on the null hypothesis. The hypothesis test will be an independent-samples t test because we have two samples composed of different groups of participants. This study meets one of the three assumptions. (1) The dependent variable is a percentage of cartoons categorized as "funny," which is a scale variable. (2) We do not know whether the population is normally distributed, and there are not at least 30 participants. Moreover, the data suggest some negative skew; although this test is robust with respect to this assumption, we must be cautious. (3) The men and women in this study were not randomly selected from among all men and women, so we must be cautious with respect to generalizing these findings.

Step 2: Null hypothesis: On average, women categorize the same percentage of cartoons as "funny" as men—H_0: $\mu_1 = \mu_2$. Research hypothesis: On average, women categorize a different percentage of cartoons as "funny" as compared with men—H_1: $\mu_1 \neq \mu_2$.

Step 3: $(\mu_1 - \mu_2) = 0$; $s_{difference} = 11.641$

Calculations:

a. $M_x = 82.25$

X	$X - M$	$(X - M)^2$
84	1.75	3.063
97	14.75	217.563
58	−24.25	588.063
90	7.75	60.063

$$s_X^2 = \frac{\Sigma(X - M)^2}{N - 1} = \frac{(3.063 + 217.563 + 588.063 + 60.063)}{4 - 1} = 289.584$$

$M_Y = 82.6$

Y	$Y - M$	$(Y - M)^2$
88	5.4	29.16
90	7.4	54.76
52	−30.6	936.36
97	14.4	207.36
86	3.4	11.56

$$s_Y^2 = \frac{\Sigma(Y - M)^2}{N - 1} = \frac{(29.16 + 54.76 + 936.36 + 207.36 + 11.56)}{5 - 1} = 309.800$$

b. $df_X = N - 1 = 4 - 1 = 3$

$df_Y = N - 1 = 5 - 1 = 4$

$df_{total} = df_X + df_Y = 3 + 4 = 7$

$$s_{pooled}^2 = \left(\frac{df_X}{df_{total}}\right)s_X^2 + \left(\frac{df_Y}{df_{total}}\right)s_Y^2 = \left(\frac{3}{7}\right)289.584 + \left(\frac{4}{7}\right)309.800 = 124.107 + 177.029 = 301.136$$

$$s_{M_X}^2 = \frac{s_{pooled}^2}{N_X} = \frac{301.136}{4} = 75.284$$

c. $$s_{M_Y}^2 = \frac{s_{pooled}^2}{N_Y} = \frac{301.136}{5} = 60.227$$

d. $s_{difference}^2 = s_{M_X}^2 + s_{M_Y}^2 = 75.284 + 60.227 = 135.511$

e. $s_{difference} = \sqrt{s_{difference}^2} = \sqrt{135.511} = 11.641$

Step 4: The critical values, based on a two-tailed test, a p level of 0.05, and a df_{total} of 7, are −2.365 and 2.365 (as seen in the curve in Figure 10-2 on page 237).

Step 5: $t = \frac{(82.25 - 82.6) - (0)}{11.641} = -0.03$

Step 6: Fail to reject the null hypothesis. We conclude that there is no evidence from this study to support the research hypothesis that either men or women are more likely than the opposite gender, on average, to find cartoons funny.

10.2 CONFIDENCE INTERVALS FOR AN INDEPENDENT-SAMPLES *t* TEST

How would we calculate a 95% confidence interval for the independent-samples t test we conducted in How It Works 10.1?

Previously, we calculated the difference between the means of these samples to be 82.25 − 82.6 = −0.35; the standard error for the differences between means, $s_{difference}$, to be 11.641; and the degrees of freedom to be 7. (Note that the order of subtraction in calculating the difference between means is irrelevant; we could just as easily have subtracted 82.25 from 82.6 and gotten a positive result, 0.35.)

1. We draw a normal curve with the sample difference between means in the center.
2. We indicate the bounds of the 95% confidence interval on either end and write the percentages under each segment of the curve—2.5% in each tail.
3. We look up the t statistics for the lower and upper ends of the confidence interval in the t table, based on a two-tailed test, a p level of 0.05 (which corresponds to a 95% confidence interval), and the degrees of freedom—7—that we calculated earlier. Because the normal curve is symmetric, the bounds of the confidence interval fall at t statistics of −2.365 and 2.365. We add those t statistics to the normal curve.
4. We convert the t statistics to raw differences between means for the lower and upper ends of the confidence interval.

   $(M_X - M_Y)_{lower} = -t(s_{difference}) + (M_X - M_Y)_{sample} = -2.365(11.641) + (-0.35) = -27.88$

   $(M_X - M_Y)_{upper} = t(s_{difference}) + (M_X - M_Y)_{sample} = 2.365(11.641) + (-0.35) = 27.18$

   The confidence interval is [−27.88, 27.18].

10.3 EFFECT SIZE FOR AN INDEPENDENT-SAMPLES *t* TEST

How can we calculate an effect size for the independent-samples t-test we conducted in How It Works 10.1? In How It Works 10.1, we calculated means of 82.25 for women and 82.6 for men. Previously, we calculated a standard error for the difference between means, $s_{difference}$, of 11.641. This time, we'll take the square root of the pooled variance to get the pooled standard deviation, the appropriate value for the denominator of Cohen's d.

$$s_{pooled} = \sqrt{s^2_{pooled}} = \sqrt{301.136} = 17.353$$

For Cohen's d, we simply replace the denominator of the formula for the test statistic with the standard deviation, s_{pooled}, instead of the standard error, $s_{difference}$.

$$\text{Cohen's } d = \frac{(M_X - M_Y) - (\mu_X - \mu_Y)}{s_{pooled}} = \frac{(82.25 - 82.6) - (0)}{17.353} = -0.02$$

According to Cohen's conventions, this is not even near the level of a small effect.

Exercises

Clarifying the Concepts

10.1 When is it appropriate to use the independent-samples t test?

10.2 Explain random assignment and what it controls.

10.3 What are independent events?

10.4 Explain how the paired-samples t test evaluates individual differences and the independent-samples t test evaluates group differences.

10.5 As they relate to comparison distributions, what is the difference between *mean differences* and *differences between means*?

10.6 As measures of variability, what is the difference between standard deviation and variance?

10.7 What is the difference between s_X^2 and s_Y^2?

10.8 What is pooled variance?

10.9 Why would we want the variability estimate based on a larger sample to count more (to be more heavily weighted) than one based on a smaller sample?

10.10 Define the symbols in the following formula: $s_{difference}^2 = s_{M_X}^2 + s_{M_Y}^2$.

10.11 How do confidence intervals relate to margin of error?

10.12 What is the difference between pooled variance and pooled standard deviation?

10.13 How does the size of the confidence interval relate to the precision of the prediction?

10.14 Why does the effect-size calculation use standard deviation rather than standard error?

10.15 Explain how we determine standard deviation (needed to calculate Cohen's d) from the several steps of calculations we made to determine standard error.

10.16 For an independent-samples t test, what is the difference between the formula for the t statistic and the formula for Cohen's d?

10.17 How do we interpret effect size using Cohen's d?

Calculating the Statistic

10.18 Below are several sample means. For each class, calculate the differences between the means for students who sit in the front versus the back of a classroom.

Mean test grades	Students in the front	Students in the back
Class 1	82	78
Class 2	79.5	77.41
Class 3	71.5	76
Class 4	72	71.3

10.19 Consider the following data from two independent groups:

Group 1: 97, 83, 105, 102, 92

Group 2: 111, 103, 96, 106

a. Calculate s^2 for group 1 and for group 2.

b. Calculate df_X, df_Y, and df_{total}.

c. Determine the critical values for t, assuming a two-tailed test with a p level of 0.05.

d. Calculate pooled variance, s_{pooled}^2.

e. Calculate the variance version of standard error for each group.

f. Calculate the variance and the standard deviation of the distribution of differences between means.

g. Calculate the t statistic.

h. Calculate the 95% confidence interval.

i. Calculate Cohen's d.

10.20 Consider the following data from two independent groups:

Liberals: 2, 1, 3, 2

Conservatives: 4, 3, 3, 5, 2, 4

a. Calculate s^2 for each group.

b. Calculate df_X, df_Y, and df_{total}.

c. Determine the critical values for t, assuming a two-tailed test with a p level of 0.05.

d. Calculate pooled variance, s_{pooled}^2.

e. Calculate the variance version of standard error for each group.

f. Calculate the variance and the standard deviation of the distribution of differences between means.

g. Calculate the t statistic.

h. Calculate the 95% confidence interval.

i. Calculate Cohen's d.

10.21 Find the critical t values for the following data sets:

a. Group 1 has 21 participants and group 2 has 16 participants. You are performing a two-tailed test with a p level of 0.05.

b. You studied 3-year-old children and 6-year-old children, with samples of 12 and 16, respectively. You are performing a two-tailed test with a p level of 0.01.

c. You have a total of 17 degrees of freedom for a two-tailed test and a p level of 0.10.

Applying the Concepts

10.22 **Making a decision:** Numeric results for several independent-samples t tests are presented here. Decide whether each test is statistically significant, and report each result in the standard APA format.

a. A total of 73 people were studied, 40 in one group and 33 in the other group. The test statistic was calculated as 2.13 for a two-tailed test with a p level of 0.05.

b. One group of 23 people was compared to another group of 18 people. The t statistic obtained for their data was 1.77. Assume you were performing a two-tailed test with a p level of 0.05.

c. One group of nine mice was compared to another group of six mice, using a two-tailed test at a p level of 0.01. The test statistic was calculated as 3.02.

10.23 The independent-samples *t* test, hypnosis, and the Stroop effect: Using data from Exercise 9.45 on the effects of posthypnotic suggestion on the Stroop effect (Raz, Fan, & Posner, 2005), let's conduct an independent-samples *t* test. For this test, we will pretend that two sets of people participated in the study, a between-groups design, whereas previously we considered data from a within-groups design. The first score for each original participant will be in the first sample—those not receiving a posthypnotic suggestion. The second score for each original participant will be in the second sample—those receiving a posthypnotic suggestion.

Sample 1: 12.6, 13.8, 11.6, 12.2, 12.1, 13.0

Sample 2: 8.5, 9.6, 10.0, 9.2, 8.9, 10.8

a. Conduct all six steps of an independent-samples *t* test. Be sure to label all six steps.

b. Report the statistics as you would in a journal article.

c. What happens to the test statistic when you switch from having all participants in both samples to having two separate samples? Given the same numbers, is it easier to reject the null hypothesis with a within-groups design or with a between-groups design?

d. In your own words, why do you think it is easier to reject the null hypothesis in one of these situations than in the other?

e. Calculate the 95% confidence interval.

f. State in your own words what we learn from this confidence interval.

g. What information does the confidence interval give us that we also get from the hypothesis test?

h. What additional information does the confidence interval give us that we do not get from the hypothesis test?

i. Calculate the appropriate measure of effect size.

j. Based on Cohen's conventions, is this a small, medium, or large effect size?

k. Why is it useful to have this information in addition to the results of a hypothesis test?

10.24 An independent-samples *t* test and getting ready for a date: In an example we sometimes use in our statistics classes, several semesters' worth of male and female students were asked how long, in minutes, they spend getting ready for a date. The data reported below reflect the actual means and the approximate standard deviations for the actual data from 142 students.

Men: 28, 35, 52, 14

Women: 30, 82, 53, 61

a. Conduct all six steps of an independent-samples *t* test. Be sure to label all six steps.

b. Report the statistics as you would in a journal article.

c. Calculate the 95% confidence interval.

d. Calculate the 90% confidence interval.

e. How are the confidence intervals different from each other? Explain why they are different.

f. Calculate the appropriate measure of effect size.

g. Based on Cohen's conventions, is this a small, medium, or large effect size?

h. Why is it useful to have this information in addition to the results of a hypothesis test?

10.25 An independent-samples *t* test, gender, and talkativeness: "Are Women Really More Talkative Than Men?" is the title of a 2007 article that appeared in the journal *Science.* In the article, Mehl and colleagues report the results of a study of 396 men and women. Each participant wore a microphone that recorded every word he or she uttered. The researchers counted the number of words uttered by men and women and compared them. The data below are fictional but they re-create the pattern that Mehl and colleagues observed:

Men: 16,345 17,222 15,646 14,889 16,701

Women: 17,345 15,593 16,624 16,696 14,200

a. Conduct all six steps of an independent-samples *t* test. Be sure to label all six steps.

b. Report the statistics as you would in a journal article.

c. Calculate the 95% confidence interval.

d. Express the confidence interval in writing, according to the format discussed in the chapter.

e. State in your own words what we learn from this confidence interval.

f. Calculate the appropriate measure of effect size.

g. Based on Cohen's conventions, is this a small, medium, or large effect size?

h. Why is it useful to have this information in addition to the results of a hypothesis test?

10.26 An independent-samples *t* test, "all-inclusive" resorts, and alcohol consumption: At some vacation destinations, "all-inclusive" resorts allow you to pay a flat rate and then eat and drink as much as you want. There has been concern about whether these deals might lead to excessive consumption of alcohol by young adults on spring break trips. You decide to spend your spring break collecting data on this issue. Of course, you need to take all of your friends on this funded research trip, because you need a lot of research assistants! You collect data on the number of drinks consumed in a day by people staying at all-inclusive resorts and by those staying at noninclusive resorts. Your data include the following:

All-inclusive resort guests: 10, 8, 13

Noninclusive resort guests: 3, 15, 7

a. Conduct all six steps of an independent-samples *t* test. Be sure to label all six steps.

b. Report the statistics as you would in a journal article.

c. Is there a shortcut you could or did use to compute your hypothesis test? (*Hint:* There are equal numbers of participants in the two groups.)

d. Calculate the 95% confidence interval.

e. State in your own words what we learn from this confidence interval.

f. Express the confidence interval, in a sentence, as a margin of error.

g. Calculate the appropriate measure of effect size.

h. Based on Cohen's conventions, is this a small, medium, or large effect size?

i. Why is it useful to have this information in addition to the results of a hypothesis test?

10.27 An independent-samples *t* test and "mother hearing": Some people claim that women can experience "mother hearing," an increased sensitivity to and awareness of noises, in particular those of children. This special ability is often associated with being a mother, rather than simply being female. Using hypothetical data, let's put this idea to the test. Imagine we recruit women to come to a sleep experiment where they think they are evaluating the comfort of different mattresses. While they are asleep, we introduce noises to test the minimum volume needed for the women to be awakened by the noise. Here are the data in decibels (dB):

Mothers: 33, 55, 39, 41, 67

Nonmothers: 56, 48, 71

a. Conduct all six steps of an independent-samples *t* test. Be sure to label all six steps.

b. Report the statistics as you would in a journal article.

c. Calculate the 95% confidence interval.

d. State in your own words what we learn from this confidence interval.

e. Explain why interval estimates are better than point estimates.

f. Calculate the appropriate measure of effect size.

g. Based on Cohen's conventions, is this a small, medium, or large effect size?

h. Why is it useful to have this information in addition to the results of a hypothesis test?

10.28 Choosing a hypothesis test: For each of the following three scenarios, state which hypothesis test you would use from among the four introduced so far: the *z* test, the single-sample *t* test, the paired-samples *t* test, and the independent-samples *t* test. (*Note:* In the actual studies described, the researchers did not always use one of these tests, often because the actual experiment had additional variables.) Explain your answer.

a. A study of 40 children who had survived a brain tumor revealed that they were more likely to have behavioral and emotional difficulties than were children who had not experienced such a trauma (Upton & Eiser, 2006). Parents rated children's difficulties, and the ratings data were compared with known means from published population norms.

b. Talarico and Rubin (2003) recorded the memories of 54 students just after the terrorist attacks in the United States on September 11, 2001—some memories related to the terrorist attacks on that day (called *flashbulb memories* for their vividness and emotional content) and some everyday memories. They found that flashbulb memories were no more consistent over time than everyday memories, even though they were perceived to be more accurate.

c. The HOPE VI Panel Study (Popkin & Woodley, 2002) was initiated to test a U.S. program aimed at improving troubled public housing developments. Residents of five HOPE VI developments were examined at the beginning of the study so researchers could later ascertain whether their quality of life had improved. Means at the beginning of the study were compared to known national data sources (e.g., the U.S. Census, the American Housing Survey) that had summary statistics, including means and standard deviations.

10.29 Choosing a hypothesis test: For each of the following three scenarios, state which hypothesis test you would use from among the four introduced so far: the *z* test, the single-sample *t* test, the paired-samples *t* test, and the independent-samples *t* test. (*Note:* In the actual studies described, the researchers did not always use one of these tests, often because the actual experiment had additional variables.) Explain your answer.

a. Taylor and Ste-Marie (2001) studied eating disorders in 41 Canadian female figure skaters. They compared the figure skaters' data on the Eating Disorder Inventory to the means of known populations, including women with eating disorders. On average, the figure skaters were more similar to the population of women with eating disorders than to those without eating disorders.

b. In an article titled "A Fair and Balanced Look at the News: What Affects Memory for Controversial

Arguments," Wiley (2005) found that people with a high level of previous knowledge about a given controversial topic (e.g., abortion, military intervention) had better average recall for arguments on both sides of that issue than did those with lower levels of knowledge.

c. Engle-Friedman and colleagues (2003) studied the effects of sleep deprivation. Fifty students were assigned to one night of sleep loss (students were required to call the laboratory every half-hour all night) and then one night of no sleep loss (normal sleep). The next day, students were offered a choice of math problems with differing levels of difficulty. Following sleep loss, students tended to choose less challenging problems.

10.30 Null and research hypotheses: Using the research studies described here (from Exercise 10.29), create null hypotheses and research hypotheses appropriate for the chosen statistical test:

a. Taylor and Ste-Marie (2001) studied eating disorders in 41 Canadian female figure skaters. They compared the figure skaters' data on the Eating Disorder Inventory to the means of known populations, including women with eating disorders. On average, the figure skaters were more similar to the population of women with eating disorders than to those without eating disorders.

b. In article titled "A Fair and Balanced Look at the News: What Affects Memory for Controversial Arguments," Wiley (2005) found that people with a high level of previous knowledge about a given controversial topic (e.g., abortion, military intervention) had better average recall for arguments on both sides of that issue than did those with lower levels of knowledge.

c. Engle-Friedman and colleagues (2003) studied the effects of sleep deprivation. Fifty students were assigned to one night of sleep loss (students were required to call the laboratory every half-hour all night) and then one night of no sleep loss (normal sleep). The next day, students were offered a choice of math problems with differing levels of difficulty. Following sleep loss, students tended to choose less challenging problems.

10.31 Independent-samples *t* test and walking speed: The New York City Department of City Planning (2006) studied pedestrian walking speeds. The report stated that pedestrians who were en route to work walked a median of 4.41 feet per second, whereas tourist pedestrians walked a median of 3.79 feet per second. They did not report results of any hypothesis tests.

a. Why would an independent-samples *t* test be appropriate in this situation?

b. What would the null hypothesis and research hypothesis be in this situation?

Putting It All Together

10.32 Gender and number words: Chang, Sandhofer, and Brown (2011) wondered whether mothers used number words more with their preschool sons than with their preschool daughters. Each participating family included one mother and one child—either female or male. They speculated that early exposure to more number words might predispose children to like mathematics. They reported the following: "An independent-samples *t* test revealed statistically significant differences in the percentages of overall numeric speech used when interacting with boys compared with girls, $t(30) = 2.40$, $p < .05$, $d = .88$. That is, mothers used number terms with boys an average of 9.49% of utterances ($SD = 6.78\%$) compared with 4.64% of utterances with girls ($SD = 4.43\%$)" (pp. 444–445).

a. Is this a between-groups or within-groups design? Explain your answer.

b. What is the independent variable? What is the dependent variable?

c. How many children were in the total sample? Explain how you determined this.

d. Is the sample likely randomly selected? Is it likely that the researchers used random assignment?

e. Were the researchers able to reject the null hypothesis? Explain.

f. What can you say about the size of the effect?

g. Describe how you could design an experiment to test whether exposure to more number words in preschool leads children to like mathematics more when they enter school.

10.33 School lunches: Alice Waters, owner of the Berkeley, California, restaurant Chez Panisse, has long been an advocate for the use of simple, fresh, organic ingredients in home and restaurant cooking. She has also turned her considerable expertise to school cafeterias. Waters (2006) praised changes in school lunch menus that have expanded nutritious offerings, but she hypothesizes that students are likely to circumvent healthy lunches by avoiding vegetables and smuggling in banned junk food unless they receive accompanying nutrition education and hands-on involvement in their meals. She has spearheaded an Edible Schoolyard program in Berkeley, which involves public school students in the cultivation and preparation of fresh foods, and states that such interactive education is necessary to combat growing levels of childhood obesity. "Nothing less," Waters writes, "will change their behavior."

a. In your own words, what is Waters predicting? Citing the confirmation bias, explain why Waters' program, although intuitively appealing, should not be instituted nationwide without further study.

b. Describe a simple between-groups experiment with a nominal independent variable with two levels and a scale dependent variable to test Waters' hypothesis. Specifically identify the independent variable, its levels, and the dependent variable. State how you will operationalize the dependent variable.

c. Which hypothesis test would be used to analyze this experiment? Explain your answer.

d. Conduct step 1 of hypothesis testing.

e. Conduct step 2 of hypothesis testing.

f. State at least one other way you could operationalize the dependent variable.

g. Let's say, hypothetically, that Waters discounted the need for the research you propose by citing her own data that the Berkeley school in which she instituted the program has lower rates of obesity than other California schools. Describe the flaw in this argument by discussing the importance of random selection and random assignment.

10.34 Perception and portion sizes: Researchers at the Cornell University Food and Brand Lab conducted an experiment at a fitness camp for adolescents (Wansink & van Ittersum, 2003). Campers were given either a 22-ounce glass that was tall and thin or a 22-ounce glass that was short and wide. Campers with the short glasses tended to pour more soda, milk, or juice than campers with the tall glasses.

a. Is it likely that the researchers used random selection? Explain.

b. Is it likely that the researchers used random assignment? Explain.

c. What is the independent variable, and what are its levels?

d. What is the dependent variable?

e. What hypothesis test would the researchers use? Explain.

f. Conduct step 1 of hypothesis testing.

g. Conduct step 2 of hypothesis testing.

h. How could the researchers redesign this study so that they could use a paired-samples t test?

Terms

independent-samples t test (p. 230)
pooled variance (p. 235)

Formulas

$df_{total} = df_X + df_Y$ (p. 235)

$s^2_{pooled} = \left(\frac{df_X}{df_{total}}\right)s^2_X + \left(\frac{df_Y}{df_{total}}\right)s^2_Y$ (p. 236)

$s^2_{M_X} = \frac{s^2_{pooled}}{N_X}$ (p. 236)

$s^2_{M_Y} = \frac{s^2_{pooled}}{N_Y}$ (p. 236)

$s^2_{difference} = s^2_{M_X} + s^2_{M_Y}$ (p. 236)

$s_{difference} = \sqrt{s^2_{difference}}$ (p. 236)

$t = \frac{(M_X - M_Y) - (\mu_X - \mu_Y)}{s_{difference}}$ often abbreviated as: $t = \frac{M_X - M_Y}{s_{difference}}$ (p. 237)

$(M_X - M_Y)_{upper} = t(s_{difference}) + (M_X - M_Y)_{sample}$ (p. 240)

$(M_X - M_Y)_{lower} = -t(s_{difference}) + (M_X - M_Y)_{sample}$ (p. 240)

$s_{pooled} = \sqrt{s^2_{pooled}}$ (p. 243)

Cohen's $d = \frac{(M_X - M_Y) - (\mu_X - \mu_Y)}{s_{pooled}}$ for a t distribution for a difference between means (p. 243)

Symbols

s^2_{pooled} (p. 236)
$s^2_{difference}$ (p. 236)
$s_{difference}$ (p. 236)

CHAPTER 11

One-Way ANOVA

BEFORE YOU GO ON

- You should understand the z distribution and the t distributions. You should also be able to differentiate among distributions of scores (Chapter 6), means (Chapter 6), mean differences (Chapter 9), and differences between means (Chapter 10).
- You should know the six steps of hypothesis testing (Chapter 7).
- You should understand what variance is (Chapter 4).
- You should be able to differentiate between between-groups designs and within-groups designs (Chapter 1).
- You should understand the concept of effect size (Chapter 8).

In 1986, California created a task force to promote self-esteem in schoolchildren, hoping to solve social problems such as drug abuse and teenage pregnancy. It seemed to make sense because people with high self-esteem are more satisfied with their lives, experience more positive feelings, and are less likely to be anxious or depressed (Myers & Diener, 1995; Twenge & Campbell, 2001). Experiments, however, revealed a dark side to building self-esteem.

In one experiment, researchers (Forsyth, Lawrence, Burnette, & Baumeister, 2007) randomly assigned university students who had earned D's and F's on their midterm exam to one of three groups. They received regular emails that provided them with (1) review questions (the control group); (2) review questions plus self-esteem-bolstering messages; or (3) review questions plus encouragement to take responsibility for their learning (see Noel, Forsyth, & Kelley, 1987). Three separate experiments could have been conducted, but a one-way analysis of variance (ANOVA) was a more efficient approach, focusing attention on the most important finding (Figure 11-1): Building self-esteem lowered test scores. In this study, building self-esteem was an intervention that backfired!

FIGURE 11-1
Comparing Three Groups

Researchers compared three groups in this one study, which allowed them to discover that a self-esteem intervention can backfire. The main reason we use ANOVA is to compare three or more groups in a single study.

For researchers, a three-group ANOVA is a bargain: three experiments for the price of one! In this chapter, we will learn about (a) the distributions used with ANOVA (the *F* distributions); (b) how to conduct an ANOVA when we have a between-groups design; (c) the effect-size statistic used with between-groups ANOVA; (d) how to conduct a post hoc (or follow-up) test to determine exactly which groups are different from one another; and (e) how to apply those same skills to a within-groups ANOVA.

Using the *F* Distributions with Three or More Samples

The self-esteem experiment revealed the possible dangers of building self-esteem because the researchers compared two groups with two different ways of motivating students to a third group, a control group. A three-group comparison is slightly more complicated than a two-group comparison, so it requires comparison distributions that can accommodate that complexity: the *F* distributions.

Type I Errors When Making Three or More Comparisons

When comparing three or more groups, it is tempting to conduct an easy-to-understand *t* test on each of the possible comparisons. Unfortunately, there's a big downside: You may not be able to believe your own results. Why? Conducting numerous *t* tests greatly increases the probability of a Type I error (a false positive: rejecting the null hypothesis when the null hypothesis is true). The three possible comparisons in the self-esteem experiment demonstrate why conducting many *t* tests inflates the possibility of making a Type I error:

Group 1 with group 2
Group 1 with group 3
Group 2 with group 3

That's 3 comparisons, and if there were four groups, there would be 6 comparisons. With five groups, there would be 10 comparisons, and so on. With only 1 comparison, there is a 0.05 chance of a Type I error in any given analysis if the null hypothesis is true, and a 0.95 chance of not having a Type I error when the null hypothesis is true. Those are pretty good odds, and we would tend to believe the conclusions in that study. However, Table 11-1 shows what happens when we conduct more studies on the same sample. The chances of not having a Type I error on the first analysis *and* not having a Type I error on the second analysis are $(0.95)(0.95) = (0.95)^2 = 0.903$, or about 90%. This means that the chance of having a Type I error is almost 10%. With three analyses, the chance of not having a Type I error is $(0.95)(0.95)(0.95) = (0.95)^3 = 0.857$, or about 86%. This means that there is about a 14% chance of having at least one Type I error. And so on, as we see in Table 11-1. ANOVA is a more powerful approach because it lets us test differences among three or more groups in just one test.

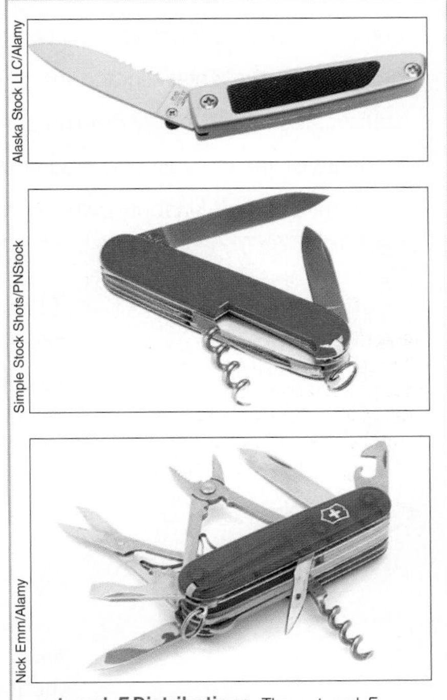

***z*, *t*, and *F* Distributions** The *z*, *t*, and *F* distributions are three increasingly complex variations on one great idea: the normal curve.

The *F* Statistic as an Expansion of the *z* and *t* Statistics

We use *F* distributions because they allow us to conduct a single hypothesis test with multiple groups. *F* distributions are more conservative versions of the *z* distribution and the *t* distributions. Just as the *z* distribution is still part of the *t* distributions, the *t* distributions are also part of the *F* distributions—and they all rely on the characteristics of the normal bell-shaped curve. The distributions are like progressively more complex versions of the Swiss Army knife: The *z* distribution has just one blade; the *t* distributions add a second

TABLE 11-1. The Probability of a Type I Error Increases as the Number of Statistical Comparisons Increases

As the number of samples increases, the number of *t* tests necessary to compare every possible pair of means increases at an even greater rate. And with that, the probability of a Type I error quickly becomes far larger than 0.05.

Number of Means	Number of Comparisons	Probability of a Type I Error
2	1	0.05
3	3	0.143
4	6	0.265
5	10	0.401
6	15	0.537
7	21	0.659

blade; and the versatile F distributions can do everything the z and the t can do—and many more complex statistical tasks.

The hypothesis tests that we have learned so far—the z test and the three types of t tests—are calculated in similar ways. The numerator describes how far apart comparison groups are from each other (between-groups variability); the denominator describes other sources of variability, such as individual differences and chance (within-groups variability). For example, the average height of men is greater than the average height of women: between-groups variability. Yet not all men are the same height and not all women are the same height: within-groups variability. However, many women are taller than many men, so there is considerable overlap between the two distributions. The F statistic calculates between-group and within-group variance to conduct the hypothesis test called ***analysis of variance*** (***ANOVA***; pronounced "ah-**noe**-vah"), *a hypothesis test typically used with one or more nominal (and sometimes ordinal) independent variables (with at least three groups overall) and a scale dependent variable.*

▶ MASTERING THE CONCEPT

11.1: The F statistic is used when we're comparing means for more than two groups. Like the z statistic and the t statistic, it's calculated by dividing some measure of variability among means by some measure of variability within groups.

The *F* Distributions for Analyzing Variability to Compare Means

Comparing the height between men and women demonstrates that the ***F statistic*** *is a ratio of two measures of variance: (1) between-groups variance, which indicates differences among sample means, and (2) within-groups variance, the average of the sample variances.*

$$F = \frac{\text{between-groups variance}}{\text{within-groups variance}}$$

Let's begin with the numerator, called ***between-groups variance*** because it *is an estimate of the population variance based on the differences among the means.* A big number in the numerator indicates a great deal of distance (or spread) between the means, suggesting that they come from different populations. A small number in the numerator indicates very little distance (or spread) between the means, suggesting that they come from the same population. With more than two means, we can't use simple subtraction to find a number that indicates how spread apart they are, so we calculate the variance among the sample means. For example, if we wanted to compare how fast people talk in Philadelphia, Memphis, Chicago, and Toronto, then the number representing between-groups variance (in this case, the between-cities variance) is an estimate of the variability among the average number of words per minute spoken by the people representing each of those four cities.

The denominator of the F statistic is called *the* ***within-groups variance****, an estimate of the population variance based on the differences within each of the three (or more) sample distributions.* For example, not everyone living in Philadelphia, Memphis, Chicago, or Toronto speaks at the same pace. There are within-city differences in talking speeds, so within-groups variance refers to the average of the four variances.

To calculate the F statistic, we simply divide the between-groups variance by the within-groups variance. If the F statistic is a large number (when the between-groups variance is much larger than the within-groups variance), then we can infer that the sample means are different from one another. But we cannot make that inference when the F statistic is close to the number 1 (the between-groups variance is about the same as the within-groups variance).

To summarize, we can think of within-groups variance as reflecting the difference between means that we'd expect just by chance. Variability exists within any population, so we would expect some difference among means just by chance. Between-groups variance reflects the difference between means that we found in the data. If this difference is much larger than the within-groups variance—what we'd expect by chance—then we can reject the null hypothesis and conclude that there is some difference between means.

- **Analysis of variance (ANOVA)** is a hypothesis test typically used with one or more nominal (and sometimes ordinal) independent variables (with at least three groups overall) and a scale dependent variable.

The *F* Table

The *F* table is an expansion of the *t* table. Just as there are many *t* distributions represented in the *t* table—one for each possible sample size—there are many *F* distributions. Both tables include a wide range of sample sizes (represented by degrees of freedom) but the *F* table adds a third factor: the number of samples. (The *t* statistic is limited to two samples.) There is an *F* distribution for every possible combination of sample size (represented by one type of degrees of freedom) and number of samples (represented by another type of degrees of freedom).

The *F* table for two samples can even be used as a *t* test; the numbers are the same except that the *F* is based on variance and the *t* on the square root of the variance, the standard deviation. For example, if we look in the *F* table under two samples for a sample size of infinity for the equivalent of the 95th percentile, we see 2.71. If we take the square root of this, we get 1.646. We can find 1.645 on the *z* table for the 95th percentile and on the *t* table for the 95th percentile with a sample size of infinity. (The slight differences are due only to rounding decisions.) The connections between the *z, t,* and *F* distributions are summarized in Table 11-2.

- The ***F* statistic** is a ratio of two measures of variance: (1) between-groups variance, which indicates differences among sample means, and (2) within-groups variance, which is essentially an average of the sample variances.
- **Between-groups variance** is an estimate of the population variance based on the differences among the means.
- **Within-groups variance** is an estimate of the population variance based on the differences within each of the three (or more) sample distributions.

The Language and Assumptions for ANOVA

Here is a simple guide to the language that statisticians use to describe different kinds of ANOVAs (Landrum, 2005). The word *ANOVA* is almost always preceded by two adjectives that indicate: (1) the number of independent variables; and (2) the research design (between-groups or within-groups).

Study 1. For example, what would you call an ANOVA with year in school as the only independent variable and Consideration of Future Consequences (CFC) scores as the dependent variable? Answer: A one-way between-groups ANOVA. *A **one-way ANOVA** is a hypothesis test that includes both one nominal independent variable with more than two levels and a scale dependent variable. A **between-groups ANOVA** is a hypothesis test in which there are more than two samples, and each sample is composed of different participants.*

- A **one-way ANOVA** is a hypothesis test that includes both one nominal independent variable with more than two levels and a scale dependent variable.
- A **between-groups ANOVA** is a hypothesis test in which there are more than two samples, and each sample is composed of different participants.

TABLE 11-2. Connections Among Distributions

The *z* distribution is subsumed under the *t* distributions in certain specific circumstances, and both the *z* and *t* distributions are subsumed under the *F* distributions in certain specific circumstances.

	When Used	Links Among the Distributions
z	One sample; μ and σ are known	Subsumed under the *t* and *F* distributions
t	(1) One sample; only μ is known (2) Two samples	Same as *z* distribution if there is a sample size of ∞ (or a very large sample size)
F	Three or more samples (but can be used with two samples)	Square of *z* distribution if there are only two samples and a sample size of ∞ (or a very large sample size); square of *t* distribution if there are only two samples

- A **within-groups ANOVA** is a hypothesis test in which there are more than two samples, and each sample is composed of the same participants; also called a *repeated-measures ANOVA*.
- **Homoscedastic** populations are those that have the same variance; homoscedasticity is also called *homogeneity of variance*.
- **Heteroscedastic** populations are those that have different variances.

Study 2. What if you wanted to test the same group of students every year? Answer: You would use a one-way within-groups ANOVA. *A* ***within-groups ANOVA*** *is a hypothesis test in which there are more than two samples, and each sample is composed of the same participants.* (This test is also called a *repeated-measures ANOVA*.)

Study 3. And what if you wanted to add gender to the first study, something we explore in the next chapter? Now you have two independent variables: year in school *and* gender. Answer: You would use a two-way, between-groups ANOVA.

All ANOVAs, regardless of type, share the same three assumptions that represent the optimal conditions for valid data analysis.

Assumption 1. *Random selection* is necessary if we want to generalize beyond a sample. Because of the difficulty of random sampling, researchers often substitute convenience sampling and then replicate their experiment with a new sample.

Assumption 2. A *normally distributed population* allows us to examine the distributions of the samples to get a sense of what the underlying population distribution might look like. This assumption becomes less important as the sample size increases.

Assumption 3. *Homoscedasticity* (also called homogeneity of variance) assumes that the samples all come from populations with the same variances. (*Heteroscedasticity* means that the populations do not all have the same variance.) ***Homoscedastic*** *populations are those that have the same variance.* ***Heteroscedastic*** *populations are those that have different variances.* (Note that homoscedasticity is also often called *homogeneity of variance*.)

What if your study doesn't match these ideal conditions? You may have to throw away your data—but that is usually not necessary. You also can (a) report and justify your decision to violate those assumptions; or (b) conduct a more conservative nonparametric test (see Chapter 15).

CHECK YOUR LEARNING

Reviewing the Concepts

> The F statistic, used in an analysis of variance (ANOVA), is essentially an expansion of the z statistic and the t statistic that can be used to compare more than two samples.

> Like the z statistic and the t statistic, the F statistic is a ratio of a difference between group means (in this case, using a measure of variability) to a measure of variability within samples.

> One-way between-groups ANOVA is an analysis in which there is one independent variable with at least three levels and in which different participants are in each level of the independent variable. A within-groups ANOVA differs in that all participants experience all levels of the independent variable.

> The assumptions for ANOVA are that participants are randomly selected, the populations from which the samples are drawn are normally distributed, and those populations have the same variance (an assumption known as homoscedasticity).

Clarifying the Concepts

11-1 The F statistic is a ratio of what two kinds of variance?

11-2 What are the two types of research designs for a one-way ANOVA?

Calculating the Statistics

11-3 Calculate the F statistic, writing the ratio accurately, for each of the following cases:

a. Between-groups variance is 8.6 and within-groups variance is 3.7.

b. Within-groups variance is 123.77 and between-groups variance is 102.4.

c. Between-groups variance is 45.2 and within-groups variance is 32.1.

Applying the Concepts

11-4 Consider the research on multitasking that we explored in Chapter 9 (Mark, Gonzalez, & Harris, 2005). Let's say we compared three conditions to see which one would lead to the quickest resumption of a task following an interruption. In one condition, the control group, no changes were made to the working environment. In the second condition, a communication ban was instituted from 1:00 to 3:00 P.M. In the third condition, a communication ban was instituted from 11:00 A.M. to 3:00 P.M. We recorded the time, in minutes, until work on an interrupted task was resumed.

a. What type of distribution would be used in this situation? Explain your answer.

b. In your own words, explain how we would calculate between-groups variance. Focus on the logic rather than on the calculations.

c. In your own words, explain how we would calculate within-groups variance. Focus on the logic rather than on the calculations.

Solutions to these Check Your Learning questions can be found in Appendix D.

One-Way Between-Groups ANOVA

The self-esteem study (Forsyth et al., 2007) led to a startling conclusion—that a self-esteem intervention can backfire—because researchers were able to use ANOVA to compare three groups in a single study. In this section, we use a new example to apply the principles of ANOVA to hypothesis testing.

Everything About ANOVA but the Calculations

To introduce the steps of hypothesis testing for a one-way between-groups ANOVA, we use an international study about whether the economic makeup of a society affects the degree to which people behave in a fair manner toward others (Henrich et al., 2010).

EXAMPLE 11.1

The researchers studied people in 15 societies from around the world. For teaching purposes, we'll look at data from four types of societies—foraging, farming, natural resources, and industrial.

1. *Foraging.* Several societies, including ones in Bolivia and Papua New Guinea, were categorized as foraging in nature. Most food was acquired through hunting and gathering.
2. *Farming.* Some societies, including ones in Kenya and Tanzania, primarily practiced farming and tended to grow their own food.
3. *Natural resources.* Other societies, such as in Colombia, built their economies by extracting natural resources, such as trees and fish. Most food was purchased.
4. *Industrial.* In industrial societies, which include the major city of Accra in Ghana as well as rural Missouri in the United States, most food was purchased.

The researchers wondered which groups would behave more or less fairly toward others—the first and second groups, which grew their own food, or the third and fourth groups, which depended on others for food. The researchers measured fairness through several games. In the Dictator Game, for example, two players were given a sum of money equal to approximately the daily minimum wage for that society. The first player (the dictator) could keep all of the money or give any portion of it to the other person. The proportion of money given to the second player constituted the measure of fairness. For example, it would be considered fairer to give the second player 40% of the money than to give him or her only 10% of the money.

The Dictator Game Here a researcher introduces a fairness game to a woman from Papua New Guinea, one of the foraging societies. Using games, researchers were able to compare fairness behaviors among different types of societies—those that depend on foraging, farming, natural resources, or industry. Because there are four groups and each participant is in only one group, the results can be analyzed with a one-way between-groups ANOVA.

Courtesy of Dr. David Tracer

This research design would be analyzed with a one-way between-groups ANOVA that uses the fairness measure, proportion of money given to the second player, as the dependent variable. There is one independent variable (type of society) and it has four levels (foraging, farming, natural resources, and industrial). It is a between-groups design because each player lived in one and only one of those societies. It is an ANOVA because it analyzes variance by estimating the variability among the different types of societies and dividing it by the variability within the types of societies. The fairness scores below are from 13 fictional people, but the groups have almost the same mean fairness scores that the researchers observed in their actual (much larger) data set.

Foraging: 28, 36, 38, 31
Farming: 32, 33, 40
Natural resources: 47, 43, 52
Industrial: 40, 47, 45

Let's begin by applying a familiar framework: the six steps of hypothesis testing. We will learn the calculations in the next section.

STEP 1: Identify the populations, distribution, and assumptions.

The first step of hypothesis testing is to identify the populations to be compared, the comparison distribution, the appropriate test, and the assumptions of the test. Let's summarize the fairness study with respect to this first step of hypothesis testing.

Summary: *The populations to be compared:* Population 1: All people living in foraging societies. Population 2: All people living in farming societies. Population 3: All people living in societies that extract natural resources. Population 4: All people living in industrial societies.

The comparison distribution and hypothesis test: The comparison distribution will be an F distribution. The hypothesis test will be a one-way between-groups ANOVA.

Assumptions: (1) The data are not selected randomly, so we must generalize only with caution. (2) We do not know if the underlying population distributions are normal, but the sample data do not indicate severe skew. (3) We will test homoscedasticity when we calculate the test statistics by checking whether the largest variance is not more than twice the smallest. (*Note:* Don't forget this step just because it comes later in the analysis.)

STEP 2: State the null and research hypotheses.

The second step is to state the null and research hypotheses. As usual, the null hypothesis posits no difference among the population means. The symbols are the same as before, but with more populations: $H_0: \mu_1 = \mu_2 = \mu_3 = \mu_4$. However, the research hypothesis is more complicated because we can reject the null hypothesis if only one group is different, on average, from the others. The research hypothesis that $\mu_1 \neq \mu_2 \neq \mu_3 \neq \mu_4$ does not include all possible outcomes, such as the hypothesis that groups 1 *and* 2 are greater than groups 3 and 4. The research hypothesis is that at least one population mean is different from at least one other population mean, so H_1 is that at least one μ is different from another μ.

Summary: Null hypothesis: People living in societies based on foraging, farming, the extraction of natural resources, and industry all exhibit, on average, the same fairness behaviors—$H_0: \mu_1 = \mu_2 = \mu_3 = \mu_4$. Research hypothesis: People living in societies based on foraging, farming, the extraction of natural resources, and industry do not all exhibit the same fairness behaviors, on average.

STEP 3: Determine the characteristics of the comparison distribution.

The third step is to explicitly state the relevant characteristics of the comparison distribution. This step is an easy one in ANOVA because most calculations are in step 5. Here we merely state that the comparison distribution is an F distribution and provide the appropriate degrees of freedom. As we discussed, the F statistic is a ratio of two independent estimates of the population variance, between-groups variance and within-groups variance (both of which we calculate in step 5). Each variance estimate has its own degrees of freedom. The sample between-groups variance estimates the population variance through the difference among the means of the samples, four in this case. The degrees of freedom for the between-groups variance estimate is the number of samples minus 1:

$$df_{between} = N_{groups} - 1 = 4 - 1 = 3$$

MASTERING THE FORMULA

11-1: The formula for the between-groups degrees of freedom is: $df_{between} = N_{groups} - 1$. We subtract 1 from the number of groups in the study.

Because there are four groups (foraging, farming, the extraction of natural resources, and industry), the between-groups degrees of freedom is 3.

The sample within-groups variance estimates the variance of the population by averaging the variances of the samples, without regard to differences among the sample means. We first must calculate the degrees of freedom for each sample. Because there are four participants in the first sample (farming), we would calculate:

$$df_1 = n_1 - 1 = 4 - 1 = 3$$

MASTERING THE FORMULA

11-2: The formula for the within-groups degrees of freedom for a one-way between-groups ANOVA conducted with four samples is: $df_{within} = df_1 + df_2 + df_3 + df_4$. We sum the degrees of freedom for each of the four groups. We calculate degrees of freedom for each group by subtracting 1 from the number of people in that sample. For example, for the first group, the formula is: $df_1 = n_1 - 1$.

n represents the number of participants in the particular sample. We would then do this for the remaining samples. For this example, there are four samples, so the formula would be:

$$df_{within} = df_1 + df_2 + df_3 + df_4$$

For this example, the calculations would be:

$$df_1 = 4 - 1 = 3$$
$$df_2 = 3 - 1 = 2$$
$$df_3 = 3 - 1 = 2$$
$$df_4 = 3 - 1 = 2$$
$$df_{within} = 3 + 2 + 2 + 2 = 9$$

Summary: We would use the F distribution with 3 and 9 degrees of freedom.

STEP 4: Determine the critical value, or cutoff.

The fourth step is to determine a critical value, or cutoff, indicating how extreme the data must be to reject the null hypothesis. For ANOVA, we use an F statistic, for which the critical value on an F distribution will always be positive (because the F is based on estimates of variance and variances are always positive). We determine the critical value by examining the F table in Appendix B (excerpted in Table 11-3). The between-groups degrees of freedom are found in a row across the top of the table. Notice that, in the full table, this row only goes up to 6, as it is rare to have more than seven conditions, or groups, in a study. The within-groups degrees of freedom are in a column along the left-hand side of the table. Because the number of participants in a study can range from a few to many, the column continues for several pages, with the same range of values of between-groups degrees of freedom on the top of each page.

TABLE 11-3. Excerpt from the *F* Table

We use the *F* table to determine a critical value for a given *p* level, based on the degrees of freedom in the numerator (between-groups degrees of freedom) and the degrees of freedom in the denominator (within-groups degrees of freedom). Note that critical values are in italics for 0.10, regular type for 0.05, and boldface for 0.01.

Within-Groups Degrees of Freedom: Denominator	*p* level	Between-Groups Degrees of Freedom: Numerator 1	2	3	4 . . .
. . .					
12	**0.01**	**9.33**	**6.93**	**5.95**	**5.41**
	0.05	4.75	3.88	3.49	3.26
	0.10	*3.18*	*2.81*	*2.61*	*2.48*
13	**0.01**	**9.07**	**6.70**	**5.74**	**5.20**
	0.05	4.67	3.80	3.41	3.18
	0.10	*3.14*	*2.76*	*2.56*	*2.43*
14	**0.01**	**8.86**	**6.51**	**5.56**	**5.03**
	0.05	4.60	3.74	3.34	3.11
	0.10	*3.10*	*2.73*	*2.52*	*2.39*
. . .					

Use the F table by first finding the appropriate within-groups degrees of freedom along the left-hand side of the page: 9. Then find the appropriate between-groups degrees of freedom along the top: 3. The place in the table where this row and this column intersect contains three numbers: p levels for 0.01, 0.05, and 0.10. Researchers usually use the middle one, 0.05, which for this study is 3.86 (Figure 11-2).

Summary: The cutoff, or critical value, for the F statistic for a p level of 0.05 is 3.86, as displayed in the curve in Figure 11-2.

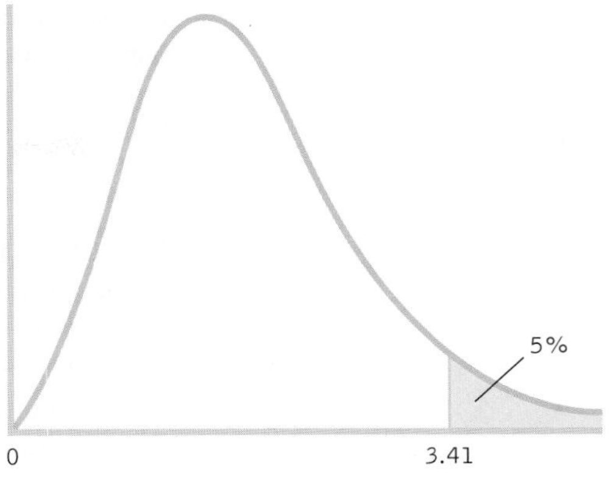

FIGURE 11-2
Determining Cutoffs for an *F* Distribution

We determine a single critical value on an F distribution. Because F is a squared version of a z or t in some circumstances, we have only one cutoff for a two-tailed test.

STEP 5: Calculate the test statistic. *In the fifth step, we calculate the test statistic.* We use the two estimates of the between-groups variance and the within-groups variance to calculate the F statistic. We compare the F statistic to the cutoff to determine whether to reject the null hypothesis. We will learn to do these calculations in the next section.

Summary: To be calculated in the next section.

STEP 6: Make a decision. *In the final step, we decide whether to reject or fail to reject the null hypothesis.* If the F statistic is beyond the critical value, then we know that it is in the most extreme 5% of possible test statistics *if* the null hypothesis is true. We can then reject the null hypothesis and conclude, "It seems that people exhibit different fairness behaviors, on average, depending on the type of society in which they live." ANOVA only tells us that at least one mean is significantly different from another; it does not tell us *which* societies are different.

If the test statistic is not beyond the critical value, then we must fail to reject the null hypothesis. The test statistic would not be very rare if the null hypothesis were true. In this circumstance, we report only that there is no evidence from the present study to support the research hypothesis.

Summary: We will be making an evidence-based decision, so we cannot make that decision until we complete step 5, in which we calculate the probabilities associated with that evidence. We will complete step 6 in the Making a Decision section. ■

◀ MASTERING THE CONCEPT

11.2: When conducting an ANOVA, we use the same six steps of hypothesis testing that we've already learned. One of the differences from what we've learned is that we calculate an F statistic, the ratio of between-groups variance to within-groups variance.

The Logic and Calculations of the *F* Statistic

In this section, we first review the logic behind ANOVA's use of between-groups variance and within-groups variance. Then we apply the same six steps of hypothesis testing we have used in previous statistical tests to make a data-driven decision about what story the data are trying to tell us. Our goal in performing the calculations of ANOVA is to understand the *sources* of all the variability in a study.

As we noted before, grown men, on average, are slightly taller than grown women, on average. We call that "between-groups variability." We also noted that not all women are the same height and not all men are the same height. We call that "within-groups variability." The F statistic is simply an

Gender Differences in Height Men, on average, are slightly taller than women (between-groups variance). However, neither men nor women are all the same height within their groups (within-groups variance). F is between-groups variability divided by within-groups variability.

estimate of between-groups variability (in the numerator) divided by an estimate of within-groups variability (in the denominator).

$$F = \frac{\text{between-groups variability}}{\text{within-groups variability}}$$

Quantifying Overlap with ANOVA Many women are taller than many men, so their distributions overlap. The amount of overlap is influenced by the distance between the means (between-groups variability) and the amount of spread (within-groups variability). In Figure 11-3a, there is a great deal of overlap; the means are close together and the distributions are spread out. Distributions with lots of overlap suggest that any differences among them are probably due to chance.

There is less overlap in the second set of distributions (b), but only because the means are farther apart; the within-groups variability remains the same. The *F* statistic

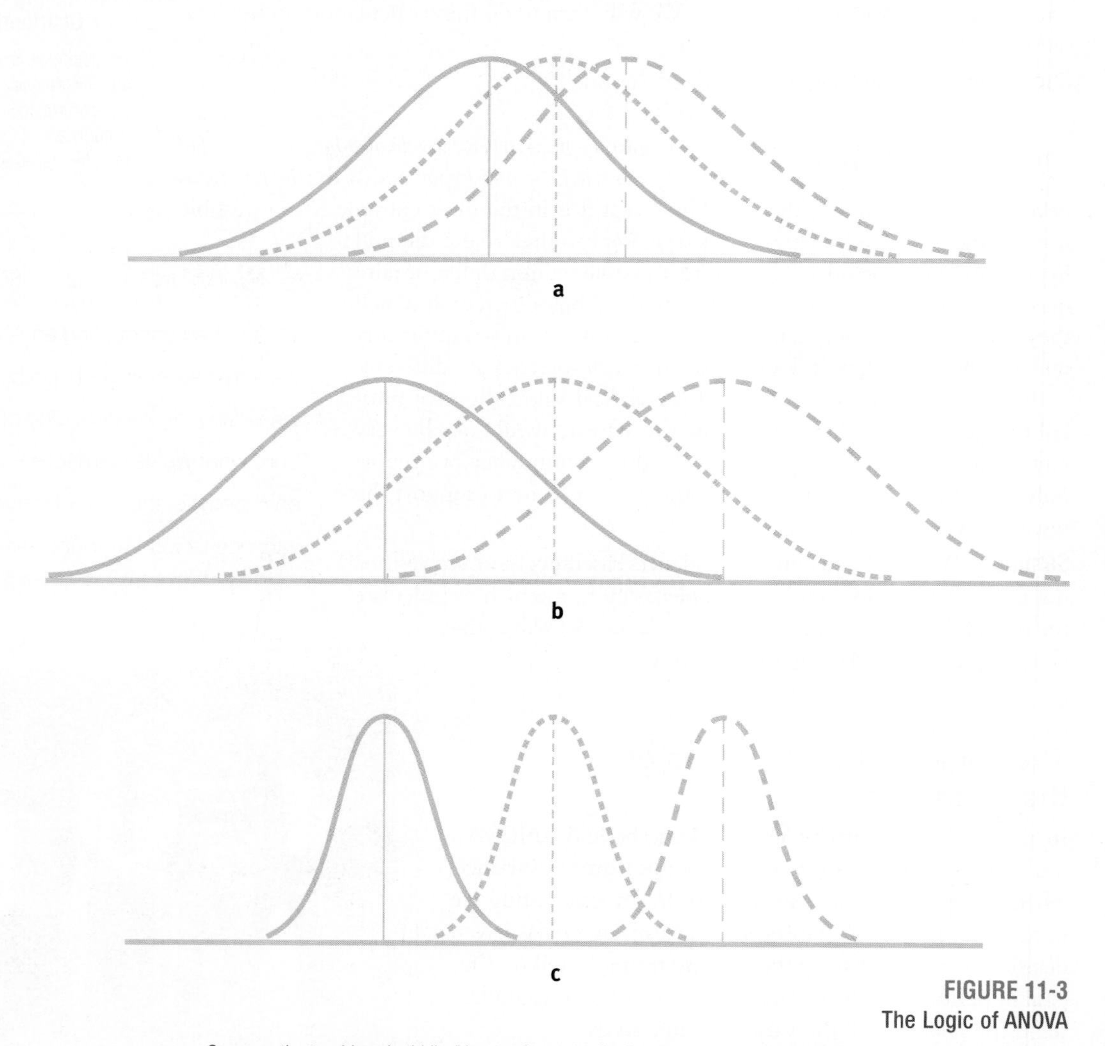

FIGURE 11-3
The Logic of ANOVA

Compare the top (a) and middle (b) sets of sample distributions. As the variability between means increases, the *F* statistic becomes larger. Compare the middle (b) and bottom (c) sets of sample distributions. As the variability within the samples themselves decreases, the *F* statistic becomes larger. The *F* statistic becomes larger as the curves overlap less. Both the increased spread among the sample means and the decreased spread within each sample contribute to this increase in the *F* statistic.

is larger because the numerator is larger; the denominator has not changed. Distributions with little overlap are less likely to be drawn from the same population. It is less likely that any differences among them are due to chance.

There is even less overlap in the third set of distributions (c) because the numerator representing the between-groups variance is still large *and* the denominator representing the within-groups variance has gotten smaller. Both changes contributed to a larger F statistic. Distributions with very little overlap suggest that any differences are probably not due to chance. It would be difficult to convince someone that these three samples were drawn by chance from the very same population.

Two Ways to Estimate Population Variance Between-groups variability and within-groups variability estimate two different kinds of variance in the population. If those two estimates are the same, then the F statistic will be 1.0. For example, if the estimate of the between-groups variance is 32 and the estimate of the within-groups variance is also 32, then the F statistic is $32/32 = 1.0$. This is a bit different from the z and t tests in which a z or t of 0 would mean no difference at all. Here, an F of 1 means no difference at all. As the sample means get farther apart, the between-groups variance (the numerator) increases, which means that the F statistic also increases.

Calculating the *F* Statistic with the Source Table The goal of any statistical analysis is to understand the *sources* of variability in a study. We achieve that in ANOVA by calculating a form of variance, the squared deviations from the mean and three sums of squares. We organize the results into *a* ***source table*** that *presents the important calculations and final results of an ANOVA in a consistent and easy-to-read format.* A source table is shown in Table 11-4; the symbols in this table would be replaced by numbers in an actual source table. We're going to explain the source table displayed in Table 11-4 by explaining column 1 first; we will then work backward from column 5 to column 4 to column 3 and finally to column 2.

▪ A **source table** presents the important calculations and final results of an ANOVA in a consistent and easy-to-read format.

Column 1: "*Source.*" One possible source of population variance comes from the spread *between* means; a second source comes from the spread *within* each sample. In this chapter, the row labeled "Total" allows us to check our calculations of the sum of squares (SS) and degrees of freedom (df). Now let's work backward through the source table to learn how it describes these two familiar sources of variability.

Column 5: "*F.*" We calculate F using simple division: between-groups variance divided by within-groups variance.

Column 4: "*MS.*" MS is the conventional symbol for variance in ANOVA. It stands for "mean square" because variance is the arithmetic mean of the squared deviations for between-groups variance ($MS_{between}$) and within-groups variance (MS_{within}). We divide $MS_{between}$ by MS_{within} to calculate F.

TABLE 11-4. The Source Table Organizes the ANOVA Calculations

A source table helps researchers organize the most important calculations necessary to conduct an ANOVA, as well as the final results. The numbers 1–5 in the first row are used in this particular table only to help you understand the format of source tables; they would not be included in an actual source table.

1 Source	2 *SS*	3 *df*	4 *MS*	5 *F*
Between	$SS_{between}$	$df_{between}$	$MS_{between}$	F
Within	SS_{within}	df_{within}	MS_{within}	
Total	SS_{total}	df_{total}		

MASTERING THE FORMULA

11-3: One formula for the total degrees of freedom for a one-way between-groups ANOVA is: $df_{total} = df_{between} + df_{within}$. We sum the between-groups degrees of freedom and the within-groups degrees of freedom. An alternate formula is: $df_{total} = N_{total} - 1$. We subtract 1 from the total number of people in the study—that is, from the number of people in all groups.

Column 3: "df." We calculate the between-groups degrees of freedom ($df_{between}$) and the within-groups degrees of freedom (df_{within}), and then add the two together to calculate the total degrees of freedom:

$$df_{total} = df_{between} + df_{within}$$

In our version of the fairness study, $df_{total} = 3 + 9 = 12$. A second way to calculate df_{total} is:

$$df_{total} = N_{total} - 1$$

N_{total} refers to the total number of people in the entire study. In our abbreviated version of the fairness study, there were four groups, with 4, 3, 3, and 3 participants in the groups, and $4 + 3 + 3 + 3 = 13$. We calculate total degrees of freedom for this study as $df_{total} = 13 - 1 = 12$. If we calculate degrees of freedom both ways and the answers don't match up, then we know we have to go back and check the calculations.

Column 2: "SS." We calculate three "sums of squares." One *SS* represents between-groups variability ($SS_{between}$), a second represents within-groups variability (SS_{within}), and a third represents total variability (SS_{total}). The first two sums of squares add up to the third; calculate all three to be sure they match.

The source table is a convenient summary because it describes everything we have learned about the sources of numerical variability. Once we calculate the sums of squares for between-groups variance and within-groups variance, there are just two steps.

Step 1: Divide each sum of squares by the appropriate degrees of freedom—the appropriate version of $(N - 1)$. We divide the $SS_{between}$ by the $df_{between}$ and the SS_{within} by the df_{within}. We then have the two variance estimates ($MS_{between}$ and MS_{within}).

Step 2: Calculate the ratio of $MS_{between}$ and MS_{within} to get the F statistic. Once we have the sums of squared deviations, the rest of the calculation is simple division.

Sums of Squared Deviations Language Alert! The term "deviations" is another word used to describe variability. ANOVA analyzes three different types of statistical deviations: (1) deviations between groups, (2) deviations within groups, and (3) total deviations. We begin by calculating the sum of squares for each type of deviation, or source of variability: between, within, and total.

It is easiest to start with the total sum of squares, SS_{total}. Organize all the scores and place them in a single column with a horizontal line dividing each sample from the next. Use the data (from our version of the fairness study) in the column labeled "X" of Table 11-5 as your model; X stands for each of the 13 individual scores listed below. Each set of scores is next to its sample; the means are underneath the names of each respective sample. (We have included subscripts on each mean in the first column—e.g., *for* for foraging, *nr* for natural resources—to indicate its sample.)

To calculate the total sum of squares, subtract the overall mean from each score, including everyone in the study, regardless of sample. The mean of all the scores is called the *grand mean,* and its symbol is *GM*. *The* ***grand mean*** *is the mean of every score in a study, regardless of which sample the score came from:*

MASTERING THE FORMULA

11-4: The grand mean is the mean score of all people in a study, regardless of which group they're in. The formula is: $GM = \frac{\Sigma(X)}{N_{total}}$. We add up everyone's score, then divide by the total number of people in the study.

$$GM = \frac{\Sigma(X)}{N_{total}}$$

The grand mean of these scores is 39.385. (As usual, we write each number to three decimal places until we get to the final answer, *F*. We report the final answer to two decimal places.)

TABLE 11-5. Calculating the Total Sum of Squares

The total sum of squares is calculated by subtracting the overall mean, called the *grand mean,* from every score to create deviations, then squaring the deviations and summing the squared deviations.

Sample	X	$(X - GM)$	$(X - GM)^2$
Foraging	28	−11.385	129.618
	36	−3.385	11.458
$M_{for} = 33.25$	38	−1.385	1.918
	31	−8.385	70.308
Farming	32	−7.385	54.538
	33	−6.385	40.768
$M_{farm} = 35.0$	40	0.615	0.378
Natural resources	47	7.615	57.988
	43	3.615	13.068
$M_{nr} = 47.333$	52	12.615	159.138
Industrial	40	0.615	0.378
	47	7.615	57.988
$M_{ind} = 44.0$	45	5.615	31.528
	$GM = 39.385$		$SS_{total} =$ **629.074**

The third column in Table 11-5 shows the deviation of each score from the grand mean. The fourth column shows the squares of these deviations. For example, for the first score, 28, we subtract the grand mean:

$$28 - 39.385 = -11.385$$

Then we square the deviation:

$$(-11.385)^2 = 129.618$$

Below the fourth column, we have summed the squared deviations: 629.074. This is the total sum of squares, SS_{total}. The formula for total sum of squares is:

$$SS_{total} = \Sigma(X - GM)^2$$

The model for calculating the within-groups sum of squares is shown in Table 11-6. This time the deviations are around the mean of each particular group (separated by horizontal lines) instead of around the grand mean. For the four scores in the first sample, we subtract their sample mean, 33.25. For example, the calculation for the first score is:

$$(28 - 33.25)^2 = 27.563$$

For the three scores in the second sample, we subtract their sample mean, 35.0. And so on for all four samples. (*Note:* Don't forget to switch means when you get to each new sample!)

Once we have all the deviations, we square them and sum them to calculate the within-groups sum of squares, 167.419, the number below the fourth column. Because we subtract the sample mean, rather than the grand mean, from each score, the formula is:

$$SS_{within} = \Sigma(X - M)^2$$

The **grand mean** is the mean of every score in a study, regardless of which sample the score came from.

MASTERING THE FORMULA

11-5: The total sum of squares in an ANOVA is calculated using the following formula: $SS_{total} = \Sigma(X - GM)^2$. We subtract the grand mean from every score, then square these deviations. We then sum all the squared deviations.

MASTERING THE FORMULA

11-6: The within-groups sum of squares in a one-way between-groups ANOVA is calculated using the following formula: $SS_{within} = \Sigma(X - M)^2$. From each score, we subtract its group mean. We then square these deviations. We sum all the squared deviations for everyone in all groups.

TABLE 11-6. Calculating the Within-Groups Sum of Squares

The within-groups sum of squares is calculated by taking each score and subtracting the mean of the sample from which it comes—not the grand mean—to create deviations, then squaring the deviations and summing the squared deviations.

Sample	X	$(X - M)$	$(X - M)^2$
Foraging	28	−5.25	27.563
	36	2.75	7.563
$M_{for} = 33.25$	38	4.75	22.563
	31	−2.25	5.063
Farming	32	−3.0	9.0
	33	−2.0	4.0
$M_{farm} = 35.0$	40	5.0	25.0
Natural resources	47	−0.333	0.111
	43	−4.333	18.775
$M_{nr} = 47.333$	52	4.667	21.781
Industrial	40	−4.0	16.0
	47	3.0	9.0
$M_{ind} = 44.0$	45	1.0	1.0
	$GM = 39.385$		$SS_{within} = \mathbf{167.419}$

Notice how the weighting for sample size is built into the calculation: The first sample has four scores and contributes four squared deviations to the total. The other samples have only three scores, so they only contribute three squared deviations.

Finally, we calculate the between-groups sum of squares. Remember, the goal for this step is to estimate how much each *group*—not each *individual participant*—deviates from the overall grand mean, so we use means rather than individual scores in the calculations. For each of the 13 people in this study, we subtract the grand mean from the mean of the group to which that individual belongs.

For example, the first person has a score of 28 and belongs to the group labeled "foraging," which has a mean score of 33.25. The grand mean is 39.385. We ignore this person's individual score and subtract 39.385 (the grand mean) from 33.25 (the group mean) to get the deviation score, −6.135. The next person, also in the group labeled "foraging," has a score of 36. The group mean of that sample is 33.25. Once again, we ignore that person's individual score and subtract 39.385 (the grand mean) from 33.25 (the group mean) to get the deviation score, also −6.135.

In fact, we subtract 39.385 from 33.25 for all four scores, as you can see in Table 11-7. When we get to the horizontal line between samples, we look for the next sample mean. For all three scores in the next sample, we subtract the grand mean, 39.385, from the sample mean, 35.0, and so on.

Notice that individual scores are *never* involved in the calculations, just sample means and the grand mean. Also notice that the first group (foraging) with four participants has more weight in the calculation than the other three groups that have only three participants. The third column of Table 11-7 includes the deviations and the fourth includes the squared deviations. The between-groups sum of squares, in bold under the fourth column, is 461.643. The formula for the between-groups sum of squares is:

MASTERING THE FORMULA

11-7: The between-groups sum of squares in an ANOVA is calculated using the following formula: $SS_{between} = \Sigma(M - GM)^2$. For each score, we subtract the grand mean from that score's group mean, and square this deviation. Note that we do not use the scores in any of these calculations. We sum all the squared deviations.

$$SS_{between} = \Sigma(M - GM)^2$$

TABLE 11-7. Calculating the Between-Groups Sum of Squares

The between-groups sum of squares is calculated by subtracting the grand mean from the sample mean for every score to create deviations, then squaring the deviations and summing the squared deviations. The individual scores themselves are not involved in any calculations.

Sample	X	$(M - GM)$	$(M - GM)^2$
Foraging	28	−6.135	37.638
	36	−6.135	37.638
$M_{for} = 33.25$	38	−6.135	37.638
	31	−6.135	37.638
Farming	32	−4.385	19.228
	33	−4.385	19.228
$M_{farm} = 35.0$	40	−4.385	19.228
Natural resources	47	7.948	63.171
	43	7.948	63.171
$M_{nr} = 47.333$	52	7.948	63.171
Industrial	40	4.615	21.298
	47	4.615	21.298
$M_{ind} = 44.0$	45	4.615	21.298
	$GM = 39.385$		$SS_{between} =$ **461.643**

Now is the moment of arithmetic truth. Were the calculations correct? To find out, we add the within-groups sum of squares (167.419) to the between-groups sum of squares (461.643) to see if they equal the total sum of squares (629.074). Here's the formula:

$$SS_{total} = SS_{within} + SS_{between} = 629.062 = 167.419 + 461.643$$

MASTERING THE FORMULA

11-8: We can also calculate the total sum of squares for a one-way between-groups ANOVA by adding the within-groups sum of squares and the between-groups sum of squares: $SS_{total} = SS_{within} + SS_{between}$. This is a useful check on the calculations.

Indeed, the total sum of squares, 629.074, is almost equal to the sum of the other two sums of squares, 167.419 and 461.643, which is 629.062. The slight difference is due to rounding decisions. So the calculations were correct.

To recap (Table 11-8), for the total sum of squares, we subtract the *grand mean* from each individual *score* to get the deviations. For the within-groups sum of squares, we subtract the appropriate *sample mean* from every *score* to get the deviations. And for the

TABLE 11-8. The Three Sums of Squares of ANOVA

The calculations in ANOVA are built on the foundation we learned in Chapter 4, sums of squared deviations. We calculate three types of sums of squares, one for between-groups variance, one for within-groups variance, and one for total variance. Once we have the three sums of squares, most of the remaining calculations involve simple division.

Sum of Squares	To calculate the deviations, subtract the . . .	Formula
Between-groups	Grand mean from the sample mean (for each score)	$SS_{between} = \Sigma(M - GM)^2$
Within-groups	Sample mean from each score	$SS_{within} = \Sigma(X - M)^2$
Total	Grand mean from each score	$SS_{total} = \Sigma(X - GM)^2$

MASTERING THE FORMULA

11-9: We calculate the mean squares from their associated sums of squares and degrees of freedom. For the between-groups mean square, we divide the between-groups sum of squares by the between-groups degrees of freedom: $MS_{between} = \frac{SS_{between}}{df_{between}}$. For the within-groups mean square, we divide the within-groups sum of squares by the within-groups degrees of freedom: $MS_{within} = \frac{SS_{within}}{df_{within}}$.

MASTERING THE FORMULA

11-10: The formula for the F statistic is: $F = \frac{MS_{between}}{MS_{within}}$. We divide the between-groups mean square by the within-groups mean square.

between-groups sum of squares, we subtract the *grand mean* from the appropriate *sample mean,* once for each score, to get the deviations; for the between-groups sum of squares, the actual scores are never involved in any calculations.

Now we insert these numbers into the source table to calculate the F statistic. See Table 11-9 for the source table that lists all the formulas and Table 11-10 for the completed source table. We divide the between-groups sum of squares and the within-groups sum of squares by their associated degrees of freedom to get the between-groups variance and the within-groups variance. The formulas are:

$$MS_{between} = \frac{SS_{between}}{df_{between}} = \frac{461.643}{3} = 153.881$$

$$MS_{within} = \frac{SS_{within}}{df_{within}} = \frac{167.419}{9} = 18.602$$

We then divide the between-groups variance by the within-groups variance to calculate the F statistic. The formula, in bold in Table 11-9, is:

$$F = \frac{MS_{between}}{MS_{within}} = \frac{153.881}{18.602} = 8.27$$

TABLE 11-9. A Source Table with Formulas

This table summarizes the formulas for calculating an F statistic.

Source	*SS*	*df*	*MS*	*F*
Between	$\Sigma(M - GM)^2$	$N_{groups} - 1$	$\frac{SS_{between}}{df_{between}}$	$\mathbf{\frac{MS_{between}}{MS_{within}}}$
Within	$\Sigma(X - M)^2$	$df_1 + df_2 + \ldots + df_{last}$	$\frac{SS_{within}}{df_{within}}$	
Total	$\Sigma(X - GM)^2$	$N_{total} - 1$		

[Expanded formula: $df_{within} = (N_1 - 1) + (N_2 - 1) + \ldots + (N_{last} - 1)$]

TABLE 11-10. A Completed Source Table

Once we calculate the sums of squares and the degrees of freedom, the rest is just simple division. We use the first two columns of numbers to calculate the variances and the F statistic. We divide the between-groups sum of squares and within-groups sum of squares by their associated degrees of freedom to get the between-groups variance and within-groups variance. Then we divide between-groups variance by within-groups variance to get the F statistic, 8.27.

Source	*SS*	*df*	*MS*	*F*
Between-groups	461.643	3	153.881	**8.27**
Within-groups	167.419	9	18.602	
Total	629.074	12		

TABLE 11-11. Calculating Sample Variances

We calculate the variances of the samples by dividing each sum of squares by the sample size minus 1 to check one of the assumptions of ANOVA. For unequal sample sizes, as we have here, we want the largest variance (20.917 in this case) to be no more than twice the smallest (13.0 in this case). Two times 13.0 is 26.0, so we meet this assumption.

Sample	Foraging	Farming	Natural Resources	Industrial
Squared deviations:	27.563	9.0	0.111	16.0
	7.563	4.0	18.775	9.0
	22.563	25.0	21.781	1.0
	5.063			
Sum of Squares:	62.752	38.0	40.667	26.0
***N* − 1:**	3	2	2	2
Variance:	**20.917**	**19.0**	**20.334**	**13.0**

Making a Decision

Now we have to come back to the six steps of hypothesis testing for ANOVA to fill in the gaps in steps 1 and 6. We finished steps 2 through 5 in the previous section.

Step 1: ANOVA assumes that participants were selected from populations with equal variances. Statistical software, such as SPSS, tests this assumption while analyzing the overall data. For now, we can use the within-groups variance column. Variance is computed by dividing the sum of squares by the sample size minus 1. We can add the squared deviations for each sample, then divide by the sample size minus 1. Table 11-11 shows the calculations for variance within each of the four samples. Because the largest variance, 20.917, is not more than twice the smallest variance, 13.0, we have met the assumption of equal variances.

Step 6: Now that we have the test statistic, we compare it with 3.86, the critical F value that we identified in step 4. The F statistic we calculated was 8.27, and Figure 11-4 demonstrates that the F statistic is beyond the critical value: We can reject the null hypothesis. It appears that people living in some types of societies are fairer, on average, than are people living in other types of societies. And congratulations on making your way through your first ANOVA! Statistical software will do all of these calculations for you, but understanding how the computer produced those numbers adds to your overall understanding.

The ANOVA, however, only allows us to conclude that at least one mean is different from at least one other mean. The next section describes how to determine which groups are different.

Summary: We reject the null hypothesis. It appears that mean fairness levels differ based on the type of society in which a person lives. In a scientific journal, these statistics are presented in a similar way to the z and t statistics but with separate degrees of freedom in parentheses for between-groups and within-groups: $F(3, 9) = 8.27$, $p < 0.05$. (*Note:* Use the actual p value when analyzing ANOVA with statistical software.)

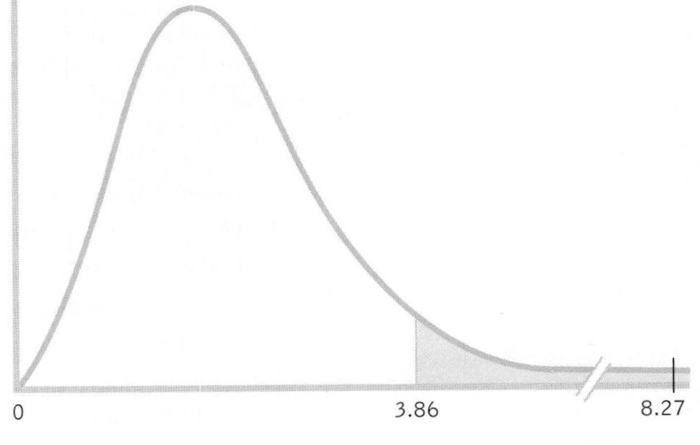

FIGURE 11-4
Making a Decision with an *F* Distribution

We compare the *F* statistic that we calculated for the samples to a single cutoff, or critical value, on the appropriate *F* distribution. We can reject the null hypothesis if the test statistic is beyond—more to the right than—the cutoff. Here, the *F* statistic of 8.27 is beyond the cutoff of 3.86, so we can reject the null hypothesis.

CHECK YOUR LEARNING

Reviewing the Concepts

> One-way between-groups ANOVA uses the same six steps of hypothesis testing that we learned in Chapter 7, but with a few minor changes in steps 3 and 5.

> In step 3, we merely state the comparison distribution and provide two different types of degrees of freedom, $df_{between}$ for the between-groups variance and df_{within} for the within-groups variance.

> In step 5, we complete the calculations, using a source table to organize the results. First, we estimate population variance by considering the differences among means (between-groups variance). Second, we estimate population variance by calculating a weighted average of the variances within each sample (within-groups variance).

> We divide between-groups variance by within-groups variance to calculate the F statistic.

> Before making a decision based on the F statistic, we check to see that the assumption of equal sample variances is met. This assumption is met when the largest sample variance is not more than twice the amount of the smallest variance.

Clarifying the Concepts

11-5 If the F statistic is beyond the cutoff, what does that tell us? What doesn't that tell us?

11-6 What is the primary subtraction that enters into the calculation of $SS_{between}$?

Calculating the Statistics

11-7 Calculate each type of degrees of freedom for the following data, assuming a between-groups design:

Group 1: 37, 30, 22, 29
Group 2: 49, 52, 41, 39
Group 3: 36, 49, 42

a. $df_{between} = N_{groups} - 1$

b. $df_{within} = df_1 + df_2 + \ldots + df_{last}$

c. $df_{total} = df_{between} + df_{within}$, or $df_{total} = N_{total} - 1$

11-8 Using the data in Check Your Learning 11-7, compute the grand mean.

11-9 Using the data in Check Your Learning 11-7, compute each type of sum of squares.

a. Total sum of squares

b. Within-groups sum of squares

c. Between-groups sum of squares

11-10 Using all of your calculations in Check Your Learning 11-7 to 11-9, perform the simple division to complete an entire between-groups ANOVA source table for these data.

Applying the Concepts

11-11 Let's create a context for the data provided above. Hollon, Thase, and Markowitz (2002) reviewed the efficacy of different treatments for depression, including medications, electroconvulsive therapy, psychotherapy, and placebo treatments. These data re-create some of the basic findings they present regarding psychotherapy. Each group is meant to represent people who received a different psychotherapy-based treatment, including psychodynamic therapy in group 1, interpersonal therapy in group 2, and cognitive-behavioral therapy in group 3. The scores presented here represent the extent to which someone responded to the treatment, with higher numbers indicating greater efficacy of treatment.

Group 1 (psychodynamic therapy): 37, 30, 22, 29
Group 2 (interpersonal therapy): 49, 52, 41, 39
Group 3 (cognitive-behavioral therapy): 36, 49, 42

a. Write hypotheses, in words, for this research.

b. Check the assumptions of ANOVA.

c. Determine the critical value for F. Using your calculations from Check Your Learning 11-10, make a decision about the null hypothesis for these treatment options.

Solutions to these Check Your Learning questions can be found in Appendix D.

Beyond Hypothesis Testing for the One-Way Between-Groups ANOVA

Do self-esteem techniques boost grades? Does the society in which one lives affect our sense of fairness? We need to move beyond hypothesis testing in two ways to fully answer such questions by (1) calculating effect size (as we did with *z* tests and *t* tests), and (2) conducting post hoc tests to determine exactly which groups are significantly different from each other.

▪ **R^2** is the proportion of variance in the dependent variable that is accounted for by the independent variable.

R^2, the Effect Size for ANOVA

In Chapter 8, we learned how to use Cohen's *d* to calculate effect size. However, Cohen's *d* only applies when subtracting one mean from another (as for a *z* test or a *t* test). With ANOVA, we calculate $\mathbf{R^2}$ (pronounced "r squared"), *the proportion of variance in the dependent variable that is accounted for by the independent variable.* We could also calculate a similar statistic called η^2 (pronounced "eta squared"). We can interpret η^2 exactly as we interpret R^2.

MASTERING THE CONCEPT

11.3: As with other hypothesis tests, it is recommended that we calculate an effect size in addition to conducting the hypothesis test. The most commonly reported effect size for ANOVA is R^2.

Like the *F* statistic, R^2 is a ratio. However, it calculates the proportion of variance accounted for by the independent variable out of all of the variance. Its numerator uses only the between-groups sum of squares, $SS_{between}$, to indicate variability among the means (ignoring the variability within each sample). The denominator uses total variability (both between-groups variance and within-groups variance) to calculate the total sums of squares: SS_{total}. The formula is:

$$R^2 = \frac{SS_{between}}{SS_{total}}$$

MASTERING THE FORMULA

11-11: The formula for the effect size we use with one-way between-groups ANOVA is $R^2 = \frac{SS_{between}}{SS_{total}}$. The calculation is a ratio, similar to the calculation for the *F* statistic. For R^2, we divide the between-groups sum of squares by the total sum of squares.

EXAMPLE 11.2

Let's apply this to the ANOVA we just conducted. We can use the statistics in the source table we created earlier to calculate R^2:

$$R^2 = \frac{SS_{between}}{SS_{total}} = \frac{461.643}{629.084} = 0.73$$

Table 11-12 displays the conventions for R^2 that, like Cohen's *d*, indicate whether the effect size is small, medium, or large. This R^2 of 0.73 is large. This is not surprising; if we can reject the null hypothesis when the sample size is small, then the effect size must be large.

We can also turn the proportion into the more familiar language of percentages by multiplying by 100. We can then say that a specific percentage of the variance in the dependent variable is accounted for by the independent variable. In this case, we could say that 73% of the variability in sharing is due to the type of society. ■

TABLE 11-12. Cohen's Conventions for Effect Sizes: R^2

The following guidelines, called *conventions* by statisticians, are meant to help researchers decide how important an effect is. These numbers are not cutoffs, merely rough guidelines to aid researchers in their interpretation of results.

Effect Size	Convention
Small	0.01
Medium	0.06
Large	0.14

Post Hoc Tests

The statistically significant *F* statistic means that some difference exists somewhere in the study. The R^2 tells us that the difference is large, but we still don't know which pairs of means are responsible for these effects. Here's an easy way to figure it out: Graph

> **MASTERING THE CONCEPT**
>
> **11.4:** ANOVA only tells us that there is a difference between at least two of the means in the study. We need a post hoc test to determine which pairs of means are statistically significantly different from each other.

the data. The picture will suggest which means are different, but those differences still need to be confirmed with a post hoc test. *A* ***post hoc test*** *is a statistical procedure frequently carried out after we reject the null hypothesis in an analysis of variance; it allows us to make multiple comparisons among several means.* The name of the test, post hoc, means "after this" in Latin; these tests are often referred to as *follow-up tests.* (Post hoc tests are not conducted if we fail to reject the null hypothesis, because we already know that there are no statistically significant differences among means.)

For example, the fairness study produced the following mean scores: foraging, 33.25; farming, 35.0; industrial, 44.0; and natural resources, 47.333. The ANOVA told us to reject the null hypothesis, so something is going on in this data set. The Pareto chart (organized by highest to lowest) and a post hoc test will tell us "where the action is" in this statistically significant ANOVA.

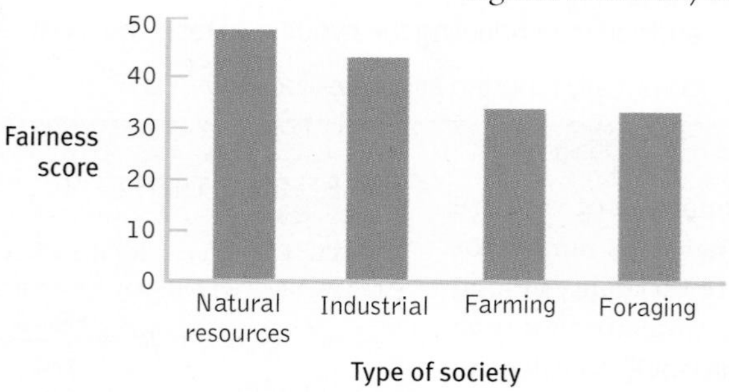

The graph in Figure 11-5 helps us think through the possibilities. For example, people in industrial societies and in societies that extract natural resources might exhibit higher levels of fairness, on average, than people in foraging or farming societies (groups 1 and 2 versus groups 3 and 4). Or people in societies that extract natural resources might be higher, on average, only compared with those in foraging societies (group 1 versus group 4). Maybe all four groups are different from one another, on average. There are so many possibilities that we need a post hoc test to reach a statistically valid conclusion. There are many post hoc tests and most are named for their founders, almost exclusively people with fabulous names—for example, Bonferroni, Scheffé (pronounced "sheff-ay"), and Tukey (pronounced "too-kee"). We will focus on the Tukey *HSD* test here.

Fairness score

50
40
30
20
10
0

Natural resources
Industrial
Farming
Foraging

Type of society

FIGURE 11-5
Which Types of Societies Are Different in Terms of Fairness?
This graph depicts the mean fairness scores of people living in each of four different types of societies. When we conduct an ANOVA and reject the null hypothesis, we only know that there is a difference somewhere; we do not know where the difference lies. We can see several possible combinations of differences by examining the means on this graph. A post hoc test will let us know which specific pairs of means are different from one another.

Tukey *HSD*

The ***Tukey HSD test*** *is a widely used post hoc test that determines the differences between means in terms of standard error; the HSD is compared to a critical value.* The Tukey *HSD* test (also called the *q test*) stands for "*h*onestly *s*ignificant *d*ifference" because it allows us to make multiple comparisons to identify differences that are "honestly" there.

The Tukey *HSD* test (1) calculates differences between each pair of means, (2) divides each difference by the standard error, and (3) compares the *HSD* for each pair of means to a critical value (a *q* value, found in Appendix B) to determine if the means are different enough to reject the null hypothesis. The formula for the Tukey *HSD* test is a variant of the *z* test and *t* tests for any two sample means:

$$HSD = \frac{(M_1 - M_2)}{s_M}$$

The formula for the standard error is:

$$s_M = \sqrt{\frac{MS_{within}}{N}}$$

N in this case is the sample size within each group, with the assumption that all samples have the same number of participants.

> **MASTERING THE FORMULA**
>
> **11-12:** To conduct a Tukey *HSD* test, we first calculate standard error: $s_M = \sqrt{\frac{MS_{within}}{N}}$. We divide the MS_{within} by the sample size and take the square root. We can then calculate the *HSD* for each pair of means: $HSD = \frac{(M_1 - M_2)}{s_M}$. For each pair of means, we subtract one from the other and divide by the standard error we calculated earlier.

When samples are different sizes, as in our example of societies, we have to calculate a weighted sample size, also known as a *harmonic mean*, N' (pronounced "N prime") before we can calculate standard error:

$$N' = \frac{N_{groups}}{\Sigma(1/N)}$$

MASTERING THE FORMULA

11-13: When we conduct an ANOVA with different-size samples, we have to calculate a harmonic mean, N': $N' = \frac{N_{groups}}{\Sigma(1/N)}$. To do that, we divide the number of groups in the study by the sum of 1 divided by the sample size for every group.

EXAMPLE 11.3

We calculate N' by dividing the number of groups (the numerator) by the sum of 1 divided by the sample size for every group (the denominator). For the example in which there were four participants in foraging societies and three in each of the other three types of societies, the formula is:

$$N' = \frac{4}{\left(\frac{1}{4} + \frac{1}{3} + \frac{1}{3} + \frac{1}{3}\right)} = \frac{4}{1.25} = 3.20$$

When sample sizes are not equal, we use a formula for s_M based on N' instead of N:

$$s_M = \sqrt{\frac{MS_{within}}{N'}} = \sqrt{\frac{18.602}{3.20}} = 2.411$$

MASTERING THE FORMULA

11-14: When we conduct an ANOVA with different-size samples, we have to calculate standard error using N': $s_M = \sqrt{\frac{MS_{within}}{N'}}$. To do that, we divide MS_{within} by N' and take the square root.

Now we use simple subtraction to calculate *HSD* for each pair of means. Which comes first doesn't matter; for example, we could subtract the mean for foraging societies from the mean for farming societies, or vice versa—subtract the mean for farming societies from the mean for foraging societies. We can ignore the sign of the answer because it is contingent on the arbitrary decision of which mean to subtract from the other.

Foraging (33.25) versus farming (35.0):

$$HSD = \frac{(33.25 - 35.0)}{2.411} = -0.73$$

Foraging (33.25) versus natural resources (47.333):

$$HSD = \frac{(33.25 - 47.333)}{2.411} = -5.84$$

Foraging (33.25) versus industrial (44.0):

$$HSD = \frac{(33.25 - 44.0)}{2.411} = -4.46$$

Farming (35.0) versus natural resources (47.333):

$$HSD = \frac{(35.0 - 47.333)}{2.411} = -5.12$$

- A **post hoc test** is a statistical procedure frequently carried out after we reject the null hypothesis in an analysis of variance; it allows us to make multiple comparisons among several means; often referred to as a *follow-up test.*
- The **Tukey *HSD* test** is a widely used post hoc test that determines the differences between means in terms of standard error; the *HSD* is compared to a critical value; sometimes called the *q test.*

Farming (35.0) versus industrial (44.0):

$$HSD = \frac{(35.0 - 44.0)}{2.411} = -3.73$$

Natural resources (47.333) versus industrial (44.0):

$$HSD = \frac{(47.333 - 44.0)}{2.411} = 1.38$$

Now all we need is a critical value from the q table in Appendix B (excerpted in Table 11-13) to which we can compare the *HSD*s. The numbers of means being compared (levels of the independent variable) are in a row along the top of the q table, and the within-groups degrees of freedom are in a column along the left-hand side. We first look up the within-groups degrees of freedom for the test, 9, along the left column. We then go across from 9 to the numbers below the number of means being compared, 4. For a p level of 0.05, the cutoff q is 4.41. Again, the sign of the *HSD* does not matter. This is a two-tailed test, and any *HSD* above 4.41 or below −4.41 would be considered statistically significant.

The q table indicates three statistically significant differences whose *HSD*s are beyond the critical value of −4.41: −5.84, −4.46, and −5.12. It appears that people in foraging societies are less fair, on average, than people in societies that depend on natural resources and people in industrial societies. In addition, people in farming societies are less fair, on average, than are people in societies that depend on natural resources. We

TABLE 11-13. Excerpt from the *q* Table

Like the *F* table, we use the *q* table to determine critical values for a given *p* level, based on the number of means being compared and the within-groups degrees of freedom. Note that critical values are in regular type for 0.05 and boldface for 0.01.

Within-Groups Degrees of Freedom	*p* level	*k* = Number of Treatments (levels) . . . 3	4	5 . . .
.				
.				
.				
8	.05	4.04	4.53	4.89
	.01	**5.64**	**6.20**	**6.62**
9	.05	3.95	4.41	4.76
	.01	**5.43**	**5.96**	**6.35**
10	.05	3.88	4.33	4.65
	.01	**5.27**	**5.77**	**6.14**
.				
.				
.				

have not rejected the null hypothesis for any other pairs, so we can only conclude that there is not enough evidence to determine whether their means are different.

What might explain these differences? The researchers observed that people who purchase food routinely interact with other people in an economic market. They concluded that higher levels of market integration are associated with higher levels of fairness (Henrich et al., 2010). Social norms of fairness may develop in market societies that require cooperative interactions between people who do not know each other.

How much faith can we have in these findings? Cautious confidence and replication are recommended; researchers could not randomly assign people to live in particular societies, so some third variable may explain the relation between market integration and fairness. ■

CHECK YOUR LEARNING

Reviewing the Concepts

> As with other hypothesis tests, it is recommended that we calculate a measure of effect size when we have conducted an ANOVA. The most commonly reported effect size for ANOVA is R^2.

> If we are able to reject the null hypothesis with ANOVA, we're not finished. We must conduct a post hoc test, such as a Tukey *HSD* test, to determine exactly which pairs of means are significantly different from one another.

Clarifying the Concepts

11-12 When do we conduct a posthoc test, such as a Tukey *HSD* test, and what does it tell us?

11-13 How is R^2 interpreted?

Calculating the Statistics

11-14 Assume that a researcher is interested in whether reaction time varies as a function of grade level. After measuring the reaction time of 10 children in fourth grade, 12 children in fifth grade, and 13 children in sixth grade, the researcher conducts an ANOVA and finds an $SS_{between}$ of 336.360 and an SS_{total} of 522.782.

a. Calculate R^2.

b. Write a sentence interpreting this R^2. Be sure to do so in terms of the independent and dependent variable described for this study.

11-15 If the researcher in Check Your Learning 11-14 rejected the null hypothesis after performing the ANOVA and intended to perform Tukey *HSD* post hoc comparisons, what would the critical value of the q statistic be for the comparisons?

Applying the Concepts

Solutions to these Check Your Learning questions can be found in Appendix D.

11-16 Perform Tukey *HSD* post hoc comparisons on the data you analyzed in Check Your Learning 11-10. For which comparisons do you reject the null hypothesis?

11-17 Calculate effect size for the data you analyzed in Check Your Learning 11-10 and interpret its meaning.

One-Way Within-Groups ANOVA

In the first part of this chapter, we learned how to analyze the multiple-group equivalent of an independent-samples t test—a one-way between-groups ANOVA—calculate its effect size, and conduct a post hoc test. Next, we learn how to analyze the multiple-group equivalent of a paired-samples t test, a one-way within-groups ANOVA (also called a *repeated-measures ANOVA*), calculate its effect size, and conduct a post hoc test.

EXAMPLE 11.4

Have you ever participated in a taste test? If you have, you were a participant in a within-groups experiment. About a decade ago, when pricier microbrew beers were becoming popular in North America, the journalist James Fallows found himself spending increasingly more on a bottle of beer and said to himself, "I love beer, but lately I've been wondering: Am I getting full value for my beer dollar?" He recruited 12 colleagues, all self-professed beer snobs, to participate in a taste test to see whether they really could tell whether a beer was expensive or cheap (Fallows, 1999).

Michael Newman/Photoedit

Taste Tests Are Within-Groups Experiments In a taste test, every person tries every flavor to determine a favorite. So it is an experiment in which every participant is in every group, or condition. If the order of the flavors is varied for each person, the researcher is using counterbalancing.

Fallows wanted to know whether his recruits could distinguish among widely available American beers that were categorized into three groups based on price—"high-end" beers like Sam Adams, "mid-range" beers like Budweiser, and "cheap" beers like Busch. All of these beers are lagers, a type of beer chosen because it can be found at every price point. Here are data—mean scores on a scale of 0–100 for each category of beer—for five of the participants. (*Note:* For teaching purposes, the means are slightly different and have been rounded to the nearest whole number; the take-home data story, however, remains the same.)

Participant	Cheap	Mid-Range	High-End
1	40	30	53
2	42	45	65
3	30	38	64
4	37	32	43
5	23	28	38

■

MASTERING THE CONCEPT

11.5: One-way within-groups ANOVA is used when we have one independent variable with at least three levels, a scale dependent variable, and participants who are in every group.

The Benefits of Within-Groups ANOVA

Fallows only reported his findings. If he had conducted hypothesis testing, then he would have used a one-way within-groups ANOVA, the appropriate statistic when you have one nominal or ordinal independent variable (type of beer), an independent variable with more than two levels (cheap, mid-range, and high-end), a scaled dependent variable (ratings of beers), and participants who experience every level of the independent variable (each participant tasted the beers in every category).

The beauty of the within-groups design is that it reduces errors due to differences between the groups because each group included exactly the same participants. The study could not be influenced by individual taste preferences, amount of alcohol typically consumed, tendency to be critical or lenient when rating, and so on. This enables us to reduce within-groups variability due to differences for the people in the study across groups. The lower within-groups variability means a smaller denominator for the F statistic and a larger F statistic that makes it is easier to reject the null hypothesis.

MASTERING THE CONCEPT

11.6: The calculations for a one-way within-groups ANOVA are similar to those for a one-way between-groups ANOVA, but we now calculate a subjects sum of squares in addition to the between-groups, within-groups, and total sum of squares. The subjects sum of squares reduces the within-groups sum of squares by removing variability associated with participants' differences across groups.

The Six Steps of Hypothesis Testing

EXAMPLE 11.5

We'll use the data from the beer taste test to walk through the same six steps of hypothesis testing that we have used for every other statistical test.

STEP 1: Identify the populations, distribution, and assumptions.

The one-way within-groups ANOVA requires an additional assumption compared to the one-way between-groups ANOVA: We must be careful to avoid order effects. In the beer study, order may have influenced participants' judgments because all participants tasted the beers in the same order: a mid-range beer, followed by a high-end beer, followed by a cheap beer, followed by another cheap beer, and so on. Perhaps the first sip of beer tastes the best, no matter what kind of beer is being tasted. Ideally, Fallows would have used counterbalancing, so that participants tasted the beers in different orders.

Summary: Population 1: People who drink cheap beer. Population 2: People who drink mid-range beer. Population 3: People who drink high-end beer.

The comparison distribution and hypothesis test: The comparison distribution is an F distribution. The hypothesis test is a one-way within-groups ANOVA.

Assumptions: (1) The participants were not selected randomly, so we must generalize with caution. (2) We do not know if the underlying population distributions are normal, but the sample data do not indicate severe skew. (3) We will test the homoscedasticity assumption by checking to see whether the largest variance is not more than twice the smallest, after we calculate the test statistic. (4) The experimenter did not counterbalance, so there may be order effects.

STEP 2: State the null and research hypotheses.

This step is identical to that for a one-way between-groups ANOVA.

Summary: Null hypothesis: People who drink cheap, mid-range, and high-end beer rate their beverages the same, on average—H_0: $\mu_1 = \mu_2 = \mu_3 = \mu_4$. Research hypothesis: People who drink cheap, mid-range, and high-end beer do not rate their beverages the same, on average—H_1 is that at least one μ is different from another μ.

STEP 3: Determine the characteristics of the comparison distribution.

We state that the comparison distribution is an F distribution and determine the degrees of freedom. Instead of three, we now calculate four kinds of degrees of freedom—between-groups, subjects, within-groups, and total. The subjects degrees of freedom corresponds to a sum of squares for differences

MASTERING THE FORMULA

11-15: The formula for the subjects degrees of freedom is: $df_{subjects} = n - 1$. We subtract one from the number of participants in the study. We use the lowercase n to indicate that this is the number of data points in a single sample, even though we know that every participant is in all three groups.

MASTERING THE FORMULA

11-16: The formula for the within-groups degrees of freedom for a one-way within-groups ANOVA is: $df_{within} = (df_{between})(df_{subjects})$. We multiply the between-groups degrees of freedom by the subjects degrees of freedom. This gives a lower number than we calculated for a one-way between-groups ANOVA because we want to exclude individual differences across groups from the within-groups sum of squares, and the degrees of freedom must reflect that.

MASTERING THE FORMULA

11-17: We can calculate the total degrees of freedom for a one-way within-groups ANOVA in two ways. We can sum all of the other degrees of freedom: $df_{total} = df_{between} + df_{subjects} + df_{within}$. Or we can subtract 1 from the total number of observations in the study: $df_{total} = N_{total} - 1$.

across participants: the *subjects sum of squares,* or $SS_{subjects}$. In a one-way within-groups ANOVA, we calculate between-groups degrees of freedom and subjects degrees of freedom first because we multiply these two together to calculate the within-groups degrees of freedom.

So we calculate the between-groups degrees of freedom exactly as before:

$$df_{between} = N_{groups} - 1 = 3 - 1 = 2$$

We next calculate the degrees of freedom that pairs with $SS_{subjects}$. Called $df_{subjects}$, it is calculated by subtracting 1 from the actual number of subjects, not data points. We use a lowercase n to indicate that this is the number of participants in a single sample (even though they're all in every sample). The formula is:

$$df_{subjects} = n - 1 = 5 - 1 = 4$$

Once we know the between-groups degrees of freedom and the subjects degrees of freedom, we calculate the within-groups degrees of freedom by multiplying the first two:

$$df_{within} = (df_{between})(df_{subjects}) = (2)(4) = 8$$

Note that the within-groups degrees of freedom is smaller than we would have calculated for a one-way between-groups ANOVA. For a one-way between-groups ANOVA, we would have subtracted 1 from each sample ($5 - 1 = 4$) and summed them to get 12. The within-groups degrees of freedom is smaller because we exclude variability related to differences among the participants from the within-groups sum of squares, and the degrees of freedom must reflect that.

Finally, we calculate total degrees of freedom using either method we learned earlier. We can sum the other degrees of freedom:

$$df_{total} = df_{between} + df_{subjects} + df_{within} = 2 + 4 + 8 = 14$$

Alternatively, we can use the second formula we learned before, treating the total number of participants as every data point, rather than every person. We know, of course, that there are just five participants and that they participate in all three levels of the independent variable, but for this step, we count the 15 total data points:

$$df_{total} = N_{total} - 1 = 15 - 1 = 14$$

We have calculated the four degrees of freedom that we will include in the source table. However, we only report the between-groups and within-groups degrees of freedom at this step.

Summary: We use the F distribution with 2 and 8 degrees of freedom.

STEP 4: Determine the critical values, or cutoffs.

The fourth step is identical to that for a one-way between-groups ANOVA. We use the between-groups degrees of freedom and within-groups degrees of freedom to look up a critical value on the F table in Appendix B.

Summary: The critical value for the F statistic for a p level of 0.05 and 2 and 8 degrees of freedom is 4.46.

STEP 5: Calculate the test statistic.

As before, we calculate the test statistic in the fifth step. To start, we calculate four sums of squares—one each for between-groups, subjects, within-groups, and total. For each sum of squares, we calculate deviations between two different types of means or scores, square the deviations, and then sum the squared differences. We calculate a squared deviation for *every* score; so for each sum of squares in this example, we sum 15 squared deviations.

As we did with the one-way between-groups ANOVA, let's start with the total sum of squares, SS_{total}. We calculate this exactly as we calculated it previously:

$$SS_{total} = \Sigma(X - GM)^2 = 2117.732$$

Type of Beer	Rating (X)	($X - GM$)	$(X - GM)^2$
Cheap	40	−0.533	0.284
Cheap	42	1.467	2.152
Cheap	30	−10.533	110.944
Cheap	37	−3.533	12.482
Cheap	23	−17.533	307.406
Mid-range	30	−10.533	110.944
Mid-range	45	4.467	19.954
Mid-range	38	−2.533	6.416
Mid-range	32	−8.533	72.812
Mid-range	28	−12.533	157.076
High-end	53	12.467	155.426
High-end	65	24.467	598.634
High-end	64	23.467	550.700
High-end	43	2.467	6.086
High-end	38	−2.533	6.416
	$GM = 40.533$		$\Sigma(X - GM)^2 = 2117.732$

Next, we calculate the between-groups sum of squares. It, too, is the same as for a one-way between-groups ANOVA:

$$SS_{between} = \Sigma(M - GM)^2 = 1092.130$$

Type of Beer	Rating (X)	Group Mean	($M - GM$)	$(M - GM)^2$
Cheap	40	34.4	−6.133	37.614
Cheap	42	34.4	−6.133	37.614
Cheap	30	34.4	−6.133	37.614
Cheap	37	34.4	−6.133	37.614
Cheap	23	34.4	−6.133	37.614
Mid-range	30	34.6	−5.933	35.200
Mid-range	45	34.6	−5.933	35.200
Mid-range	38	34.6	−5.933	35.200
Mid-range	32	34.6	−5.933	35.200
Mid-range	28	34.6	−5.933	35.200
High-end	53	52.6	12.067	145.612
High-end	65	52.6	12.067	145.612
High-end	64	52.6	12.067	145.612
High-end	43	52.6	12.067	145.612
High-end	38	52.6	12.067	145.612
	$GM = 40.533$			$\Sigma(M - GM)^2 = 1092.130$

So far, the calculations of the sums of squares for a one-way within-groups ANOVA have been the same as they were for a one-way between-groups ANOVA. We left the subjects sum of squares and within-groups sum of squares for last. Here is where we see some changes. We want to remove the variability due to participant differences from the estimate of variability across conditions. So we calculate the subjects sum of squares separately from the within-groups sum of squares. To do that, we subtract the grand mean from each participant's mean *for all of his scores.* We first have to calculate a mean for each participant across the three conditions. For example, the first participant had ratings of 40 for cheap beers, 30 for mid-range beers, and 53 for high-end beers. This participant's mean is 41.

So the formula for the subjects sum of squares is:

MASTERING THE FORMULA

11-18: The subjects sum of square in a one-way within-groups ANOVA is calculated using the following formula: $SS_{subjects} = \Sigma(M_{participant} - GM)^2$. For each score, we subtract the grand mean from that participant's mean for all of his or her scores and square this deviation. Note that we do not use the scores in any of these calculations. We sum all the squared deviations.

$$SS_{subjects} = \Sigma(M_{participant} - GM)^2 = 729.738$$

Participant	Type of Beer	Rating (X)	Participant Mean	$(M_{participant} - GM)$	$(M_{participant} - GM)^2$
1	Cheap	40	41	0.467	0.218
2	Cheap	42	50.667	10.134	102.698
3	Cheap	30	44	3.467	12.02
4	Cheap	37	37.333	−3.2	10.24
5	Cheap	23	29.667	−10.866	118.07
1	Mid-range	30	41	0.467	0.218
2	Mid-range	45	50.667	10.134	102.698
3	Mid-range	38	44	3.467	12.02
4	Mid-range	32	37.333	−3.2	10.24
5	Mid-range	28	29.667	−10.866	118.07
1	High-end	53	41	0.467	0.218
2	High-end	65	50.667	10.134	102.698
3	High-end	64	44	3.467	12.02
4	High-end	43	37.333	−3.2	10.24
5	High-end	38	29.667	−10.866	118.07
		$GM = 40.533$			$\Sigma(M_{participant} - GM)^2 = 729.738$

MASTERING THE FORMULA

11-19: The within-groups sum of squares for a one-way within-groups ANOVA is calculated using the following formula: $SS_{within} = SS_{total} - SS_{between} - SS_{subjects}$. We subtract the between-groups sum of squares and subjects sum of squares from the total sum of squares.

We only have one sum of squares left to go. To calculate the within-groups sum of squares from which we've removed the subjects sum of squares, we take the total sum of squares and subtract the two others that we've calculated so far—the between-groups sum of squares and the subjects sum of squares. The formula is:

$$SS_{within} = SS_{total} - SS_{between} - SS_{subjects}$$
$$= 2117.732 - 1092.130 - 729.738 = 295.864$$

MASTERING THE FORMULA

11-20: We calculate the subjects mean square by dividing its associated sum of squares by its associated degrees of freedom: $MS_{subjects} = \frac{SS_{subjects}}{df_{subjects}}$.

We now have enough information to fill in the first three columns of the source table—the source, *SS,* and *df* columns. We calculate the rest of the source table as we did for a one-way between-groups ANOVA. For each of the three sources—between-groups, subjects, and within-groups—we divide the sum of squares by the degrees of freedom to get its variance, *MS.*

$$MS_{subjects} = \frac{SS_{subjects}}{df_{subjects}} = \frac{729.738}{4} = 182.435$$

$$MS_{within} = \frac{SS_{within}}{df_{within}} = \frac{295.859}{8} = 36.982$$

We then calculate two F statistics—one for between-groups and one for subjects. For the between-groups F statistic, we divide its MS by the within-groups MS. For the subjects F statistic, we divide its MS by the within-groups MS.

$$F_{between} = \frac{MS_{between}}{MS_{within}} = \frac{546.068}{36.982} = 14.766$$

$$F_{subjects} = \frac{MS_{subjects}}{MS_{within}} = \frac{182.435}{36.982} = 4.933$$

MASTERING THE FORMULA

11-21: The formula for the subjects F statistic is: $F_{subjects} = \frac{MS_{subjects}}{MS_{within}}$. We divide the subjects mean square by the within-groups mean square.

The completed source table is shown here:

Source	*SS*	*df*	*MS*	*F*
Between-groups	1092.130	2	546.065	14.766
Subjects	729.738	4	182.435	4.933
Within-groups	295.859	8	36.982	
Total	2117.732	14		

Here is a recap of the formulas used to calculate a one-way within-groups ANOVA:

Source	*SS*	*df*	*MS*	*F*
Between-groups	$\Sigma(M - GM)^2$	$N_{groups} - 1$	$\frac{SS_{between}}{df_{between}}$	$\frac{MS_{between}}{MS_{within}}$
Subjects	$\Sigma(M_{participant} - GM)^2$	$df_{subjects} = n - 1$	$\frac{SS_{subjects}}{df_{subjects}}$	$\frac{MS_{subjects}}{MS_{within}}$
Within-groups	$SS_{total} - SS_{between} - SS_{subjects}$	$(df_{between})(df_{subjects})$	$\frac{SS_{within}}{df_{within}}$	
Total	$\Sigma(X - GM)^2$	$N_{total} - 1$		

We calculated two F statistics, but we're really only interested in the between-groups F statistic, 14.766, that tells us whether there is a statistically significant difference between groups.

Summary: The F statistic associated with the between-groups difference is 14.77.

STEP 6: Make a decision. This step is identical to that for the one-way between-groups ANOVA.

Summary: The F statistic, 14.77, is beyond the critical value, 4.46. We reject the null hypothesis. It appears that mean ratings of beers differ based on the type of beer in terms of price category, although we cannot yet know exactly which means differ. We report the statistics in a journal article as $F(2,8) = 14.77, p < 0.05$. (*Note:* If we used software, we would report the exact p value.) ■

CHECK YOUR LEARNING

Reviewing the Concepts

> We use one-way within-groups ANOVA when we have a nominal or ordinal independent variable with at least three levels, a scale dependent variable, and participants who experience all levels of the independent variable.

> Because all participants experience all levels of the independent variable, we reduce the within-groups variability by reducing individual differences; each person serves as a control for him or herself. A possible concern with this design is order effects.

> One-way within-groups ANOVA uses the same six steps of hypothesis testing that we used for one-way between-groups ANOVA—with one major exception. We calculate statistics for four sources, rather than three. In addition to between-groups, within-groups, and total, the fourth source is typically called "subjects."

Clarifying the Concepts

11-18 Why is the within-groups variability, or sum of squares, smaller for the within-groups ANOVA compared to the between-groups ANOVA?

Calculating the Statistics

11-19 Calculate the four degrees of freedom for the following groups, assuming a within-groups design:

	Participant 1	Participant 2	Participant 3
Group 1	7	9	8
Group 2	5	8	9
Group 3	6	4	6

a. $df_{between} = N_{groups} - 1$

b. $df_{subjects} = n - 1$

c. $df_{within} = (df_{between})(df_{subjects})$

d. $df_{total} = df_{between} + df_{subjects} + df_{within}$; or $df_{total} = N_{total} - 1$

11-20 Calculate the four sums of squares for the data listed in Check Your Learning 11-19:

a. $SS_{total} = \Sigma(X - GM)^2$

b. $SS_{between} = \Sigma(M - GM)^2$

c. $SS_{subjects} = \Sigma(M_{participant} - GM)^2$

d. $SS_{within} = SS_{total} - SS_{between} - SS_{subjects}$

11-21 Using all of your calculations in Check Your Learning 11-19 and 11-20, perform the simple division to complete an ANOVA source table for these data.

Applying the Concepts

11-22 Let's create a context for the data presented in Check Your Learning 11-19. Suppose a car dealer wants to sell a car by having people test drive it and two other cars in the same class (e.g., midsize sedans). The data from these three groups might represent driving-experience ratings (from 1, low quality, to 10, high quality) given by drivers after the test drives. Using the F values you calculated above, complete the following:

a. Write hypotheses, in words, for this study.

b. How might you conduct this research such that you would satisfy the fourth assumption of the within-groups ANOVA?

c. Determine the critical value for F and make a decision about the outcome of this research.

Solutions to these Check Your Learning questions can be found in Appendix D.

Beyond Hypothesis Testing for the One-Way Within-Groups ANOVA

Hypothesis testing with the one-way within-groups ANOVA can tell us the probability that people can distinguish between types of beer based on price category. Effect sizes help us figure out whether these differences are large enough to matter. The Tukey *HSD* test can tell us exactly which means are different from each other.

R^2, the Effect Size for ANOVA

The calculations for R^2 for a one-way within-groups ANOVA and a one-way between-groups ANOVA are similar. As before, the numerator is a measure of the variability that takes into account just the differences among means, $SS_{between}$. The denominator, however, takes into account the total variability, SS_{total}, but removes the variability due to differences among participants, $SS_{subjects}$. This enables us to determine the variability explained only by between-groups differences. The formula is:

$$R^2 = \frac{SS_{between}}{(SS_{total} - SS_{subjects})}$$

MASTERING THE FORMULA

11-22: The formula for effect size for a one-way within-groups ANOVA is: $R^2 = \frac{SS_{between}}{(SS_{total} - SS_{subjects})}$. We divide the between-groups sum of squares by the difference between the total sum of squares and the subjects sum of squares. We remove the subjects sum of squares so we can determine the variability explained only by between-groups differences.

EXAMPLE 11.6

Let's apply this to the ANOVA we just conducted. We can use the statistics in the source table shown on page 283 to calculate R^2:

$$R^2 = \frac{SS_{between}}{(SS_{total} - SS_{subject})} = \frac{1092.135}{(2117.732 - 729.738)} = 0.787$$

The conventions for R^2 are the same as those shown in Table 11-12 (see page 273). This effect size of 0.79 is a very large effect: 79% of the variability in ratings of beer is explained by price. ■

Tukey *HSD*

EXAMPLE 11.7

We use the same procedure that we used for a one-way between-groups ANOVA, the Tukey *HSD* test: We calculate an *HSD* for each pair of means by first calculating the standard error:

$$s_M = \sqrt{\frac{MS_{within}}{N}} = \sqrt{\frac{36.982}{5}} = 2.720$$

The standard error allows us to calculate *HSD* for each pair of means.

Cheap beer (34.4) versus mid-range beer (34.6):

$$HSD = \frac{(34.4 - 34.6)}{2.720} = -0.074$$

Cheap beer (34.4) versus high-end beer (52.6):

$$HSD = \frac{(34.4 - 52.6)}{2.720} = -6.691$$

Mid-range beer (34.6) versus high-end beer (52.6):

$$HSD = \frac{(34.6 - 52.6)}{2.720} = -6.618$$

Megan Maloy/Getty Images

Within-Groups Designs in Everyday Life We often use a within-groups design without even knowing it. A bride might use a within-groups design when she has all of her bridesmaids (the participants) try on several different possible dresses (the levels of the study). They would then choose the dress that is most flattering, on average, on the bridesmaids. We even have an innate understanding of order effects. A bride, for example, might ask her bridesmaids to try on the dress that she prefers either first or last (but not in the middle) so they'll remember it better and be more likely to prefer it!

Now we look up the critical value in the q table in Appendix B. For a comparison of three means with within-groups degrees of freedom of 8 and a p level of 0.05, the cutoff q is 4.04. As before, the sign of each *HSD* does not matter.

The q table indicates two statistically significant differences whose *HSD*s are beyond the critical values: -6.691 and -6.618. It appears that high-end beers elicit higher average ratings than cheap beers; high-end beers also elicit higher average ratings than mid-range beers. No statistically significant difference is found between cheap beers and mid-range beers.

What might explain these differences? It's not surprising that expensive beers came out ahead of cheap and mid-range beers, but Fallows was surprised that no observable average difference was found between cheap and mid-range beers, which led to this advice that he gave to his beer-drinking colleagues: Buy high-end beer "when [you] want an individual glass of lager to be as good as it can be," but buy cheap beer "at all other times, since it gives the maximum taste and social influence per dollar invested." The mid-range beers? Not worth the money.

How much faith can we have in these findings? As behavioral scientists, we critically examine the design and procedures. Did the darker color of Sam Adams (the beer that received the highest average ratings) give it away as a high-end beer? The beers were labeled with letters (Budweiser was labeled with F). Yet, in line with many academic grading systems, the letter A has a positive connotation and F has a negative one. Were there order effects? Did the testers get more lenient (or critical) with every swallow? The panel of tasters was mostly Microsoft employees and was all men. Would we get different results for non-tech employees or with female participants? Science is a slow but sure way of knowing that depends on replication of experiments. ■

CHECK YOUR LEARNING

Reviewing the Concepts

> It is recommended, as with other hypothesis tests, that we calculate a measure of effect size, R^2, for a one-way within-groups ANOVA.

> As with one-way between-groups ANOVA, if we are able to reject the null hypothesis with a one-way within-groups ANOVA, we're not finished. We must conduct a post hoc test, such as a Tukey *HSD* test, to determine exactly which pairs of means are significantly different from one another.

Clarifying the Concepts

11-23 How does the calculation of the effect size R^2 differ for the one-way within-groups ANOVA and the one-way between-groups ANOVA?

11-24 How does the calculation of the Tukey *HSD* differ for the one-way within-groups ANOVA and the one-way between-groups ANOVA?

Calculating the Statistics

11-25 A researcher measured the reaction time of six participants at three different times and found the mean reaction time at time 1 (M_1 = 155.833), time 2 (M_2 = 206.833), and time 3 (M_3 = 251.667). The researcher rejected the null hypothesis after performing a one-way within-groups ANOVA. For the ANOVA, $df_{between}$ = 2, df_{within} = 10, and MS_{within} = =771.256.

a. Calculate the *HSD* for each of the three mean comparisons.

b. What is the critical value of *q* for this Tukey *HSD* test?

c. For which comparisons do we reject the null hypothesis?

11-26 Use the following source table to calculate the effect size R^2 for the one-way within-groups ANOVA.

Source	*SS*	*df*	*MS*	*F*
Between	27,590.486	2	13,795.243	17.887
Subjects	16,812.189	5	3,362.438	4.360
Within	7,712.436	10	771.244	
Total	52,115.111	17		

Applying the Concepts

Solutions to these Check Your Learning questions can be found in Appendix D.

11-27 In Check Your Learning 11-21 and 11-22, we conducted an analysis of driver-experience ratings following test drives.

a. Calculate R^2 for this ANOVA.

b. What follow-up tests are needed for this ANOVA, if any?

REVIEW OF CONCEPTS

Using the *F* Distribution with Three or More Samples

The *F statistic* is used when we want to compare more than two means. As with the *z* and *t* statistics, the *F* statistic is calculated by dividing a measure of the differences among sample means (*between-groups variance*) by a measure of variability within the samples (*within-groups variance*). The hypothesis test based on the *F* statistic is called *analysis of variance (ANOVA)*.

ANOVA offers a solution to the problem of having to run multiple *t* tests because it allows for multiple comparisons in just one statistical analysis. There are several different types of ANOVA, and each has two descriptors. One indicates the number of independent variables, such as *one-way ANOVA* for one independent variable. The other indicates whether participants are in only one condition (*between-groups ANOVA*) or in every condition (*within-groups ANOVA*). The major assumptions for ANOVA are random selection of participants, normally distributed underlying populations, and *homoscedasticity,* which means that all populations have the same variance (versus *heteroscedasticity,* which means that the populations do not all have the same variance). As with previous statistical tests, most real-life analyses do not meet all of these assumptions.

One-Way Between-Groups ANOVA

The one-way between-groups ANOVA uses the six steps of hypothesis testing that we have already learned, but with some modifications, particularly to steps 3 and 5. Step 3 is simpler than with t tests; we only have to state that the comparison distribution is an F distribution and provide the degrees of freedom. In step 5, we calculate the F statistic; a *source table* helps us to keep track of the calculations. The F statistic is a ratio of two different estimates of population variance, both of distributions of scores rather than distributions of means. The denominator, within-groups variance, is similar to the pooled variance of the independent-samples t test; it's basically a weighted average of the variance within each sample. The numerator, between-groups variance, is an estimate based on the difference between the sample means, but it is then inflated to represent a distribution of scores rather than a distribution of means. As part of the calculations of between-groups variance and within-groups variance, we need to calculate a *grand mean,* the mean score of every participant in the study.

A large between-groups variance and a small within-groups variance indicate a small degree of overlap among samples and likely a small degree of overlap among populations. A large between-groups variance divided by a small within-groups variance produces a large F statistic. If the F statistic is beyond a prescribed cutoff, or critical value, then we can reject the null hypothesis.

Beyond Hypothesis Testing for the One-Way Between-Groups ANOVA

It is also recommended, as with other hypothesis tests, that we calculate an effect size—usually R^2—when we conduct an ANOVA. In addition, when we reject the null hypothesis in an ANOVA, we only know that at least one of the means is different from at least one other mean. But we do not know exactly where the differences lie until we conduct a *post hoc test* such as the *Tukey HSD test.*

One-Way Within-Groups ANOVA

We use a one-way within-groups ANOVA (also called a *repeated-measures ANOVA*) when we have one nominal or ordinal variable with at least three levels and a scale dependent variable, and every participant experiences every level of the independent variable. One-way within-groups ANOVA uses the same six steps of hypothesis testing that we used for one-way between-groups ANOVA, except that we calculate statistics for four sources instead of three. We still calculate statistics for the between-groups, within-groups, and total sources, but we also calculate statistics for a fourth source, "subjects."

Beyond Hypothesis Testing for the One-Way Within-Groups ANOVA

As we do with the one-way between-groups ANOVA, we calculate a measure of effect size, usually R^2, and we conduct a post hoc test, such as the Tukey *HSD* test, if we reject the null hypothesis.

SPSS®

In this chapter we conducted a one-way between-groups ANOVA to compare people in four different types of societies in terms of how fairly they behaved in a game, as assessed by the proportion of money that they gave to a second player in that game. The type of society was a nominal independent variable, and the proportion of money that they gave to the second player was a scale dependent variable. To conduct a one-way between-groups ANOVA using SPSS, we enter the data so that each participant has one row with all of her or his data. For example, a person would have a score in the first column indicating the type of society, perhaps a 1 for foraging or a 3 for natural resources, and a score in the second column indicating fairness level, the proportion of money he or she gave to a second player. The data as they should be entered are visible behind the output on the screenshot shown here.

Then we can instruct SPSS to conduct the ANOVA by selecting Analyze → Compare Means → One-Way ANOVA. Now select the variables. The independent variable, named "society" here, goes in the box marked "Factor," and the dependent variable, named "fairness" here, goes in the box labeled "Dependent List." To request a post hoc test to compare the means of the four groups, select "Post Hoc," then "Tukey," and then click "Continue." Click "OK" to run the ANOVA.

On the screenshot, we can see the source table near the top. Notice that the sums of squares, degrees of freedom, mean squares, and F statistic match the ones we calculated earlier. Any slight differences are due to differences in rounding decisions. The last column, titled "Sig.," says .006. This number indicates that the actual p value of this test statistic is just 0.006, which is less than the 0.05 p level typically used in hypothesis testing and an indication that we can reject the null hypothesis. Below the source table, we can see the output for the post hoc test. Mean differences with an asterisk are statistically significant at a p level of 0.05. The output here matches the post hoc test we conducted earlier.

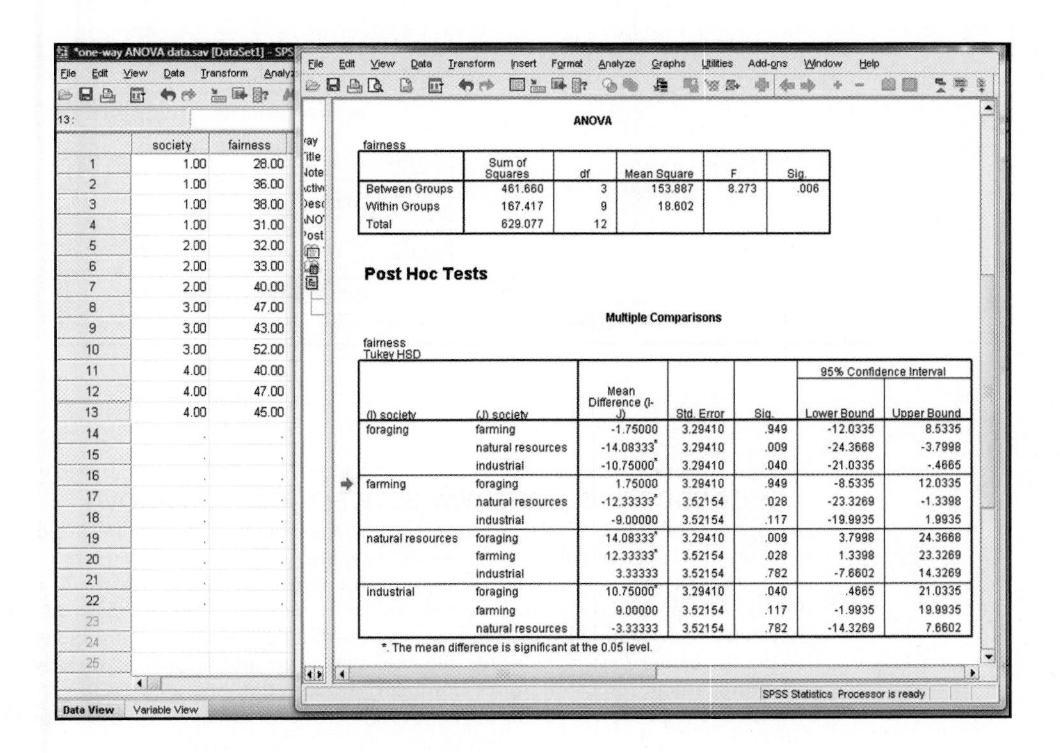

	society	fairness
1	1.00	28.00
2	1.00	36.00
3	1.00	38.00
4	1.00	31.00
5	2.00	32.00
6	2.00	33.00
7	2.00	40.00
8	3.00	47.00
9	3.00	43.00
10	3.00	52.00
11	4.00	40.00
12	4.00	47.00
13	4.00	45.00

ANOVA

fairness

	Sum of Squares	df	Mean Square	F	Sig.
Between Groups	461.660	3	153.887	8.273	.006
Within Groups	167.417	9	18.602		
Total	629.077	12			

Post Hoc Tests

Multiple Comparisons

fairness
Tukey HSD

(I) society	(J) society	Mean Difference (I-J)	Std. Error	Sig.	95% Confidence Interval Lower Bound	Upper Bound
foraging	farming	-1.75000	3.29410	.949	-12.0335	8.5335
	natural resources	-14.08333*	3.29410	.009	-24.3668	-3.7998
	industrial	-10.75000*	3.29410	.040	-21.0335	-.4665
farming	foraging	1.75000	3.29410	.949	-8.5335	12.0335
	natural resources	-12.33333*	3.52154	.028	-23.3269	-1.3398
	industrial	-9.00000	3.52154	.117	-19.9935	1.9935
natural resources	foraging	14.08333*	3.29410	.009	3.7998	24.3668
	farming	12.33333*	3.52154	.028	1.3398	23.3269
	industrial	3.33333	3.52154	.782	-7.6602	14.3269
industrial	foraging	10.75000*	3.29410	.040	.4665	21.0335
	farming	9.00000	3.52154	.117	-1.9935	19.9935
	natural resources	-3.33333	3.52154	.782	-14.3269	7.6602

*. The mean difference is significant at the 0.05 level.

To conduct a one-way within-groups ANOVA on SPSS using the beer-rating data from the chapter, we enter the data so that each participant has one row with all of his data. This results in a different format from that of the entered data for a one-way between-groups ANOVA. In that case, we had a score for each participant's level of the independent variable and a score for the dependent variable. For a within-groups ANOVA, each participant has multiple scores on the dependent variable. The levels of the independent variable are indicated in the titles for each of the three columns in SPSS. For example, as seen to the left of the SPSS screenshot on the next page, the first participant has scores of 40 for the cheap beer, 30 for the mid-range beer, and 53 for the high-end beer. We instruct SPSS to conduct the ANOVA by selecting Analyze → General Linear Model → Repeated Measures. Remember that *repeated measures* is another way to say "within groups" when describing ANOVA. Next, under "Within-Subject Factor Name" change the generic "factor1" to the actual name of the independent variable, such as "type_of_beer" (using underscores between words because SPSS doesn't recognize spaces in a variable name). Next to "Number of Levels," type "3" to represent the number of levels of the independent variable in this study. Now click "Add" followed by "Define." We define the levels by clicking each of the three levels followed by the arrow button, in turn. You can see the entered data, and the box in which we define the levels, in the second screenshot.

To see the results of the ANOVA, click "OK."

The output includes more tables than in the between-groups ANOVA SPSS output. The one we want to pay attention to is titled "Tests of Within-Subjects Effects." This table provides four F values and four "Sig." values (the actual p values). There are several more advanced considerations that play into deciding which one to use; for the purposes of this introduction to SPSS, we simply note that the F values are all the same, and all match the F of 14.77 that we calculated previously. Moreover, all of the p values are less than the cutoff of 0.05. As we could when we conducted this one-way within-groups ANOVA by hand, we can reject the null hypothesis.

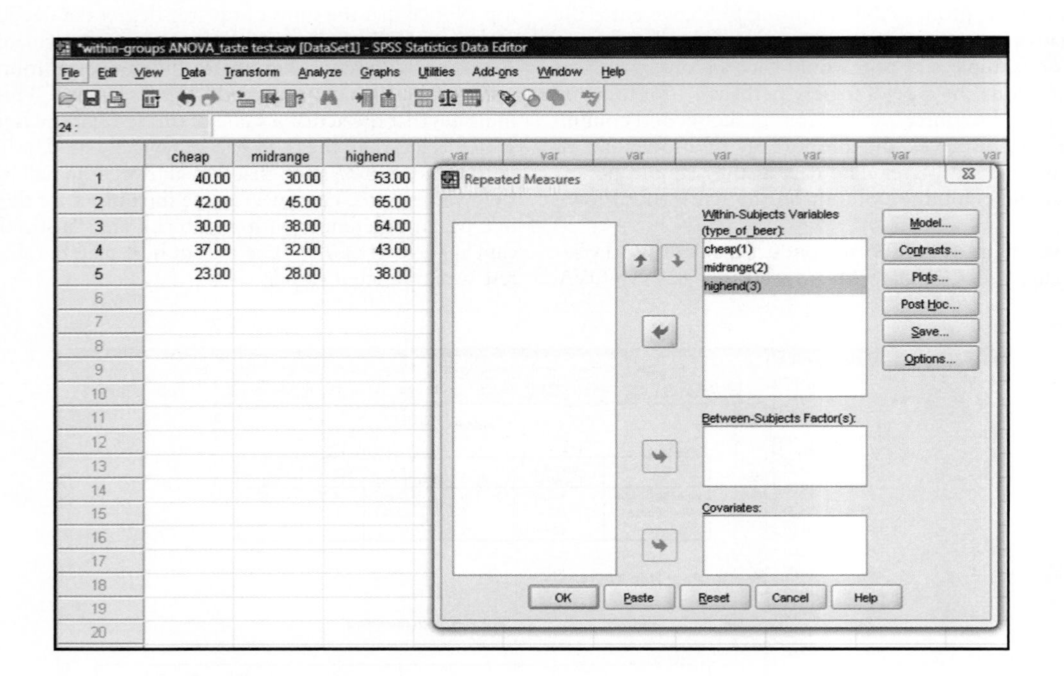

How It Works

11.1 CONDUCTING A ONE-WAY BETWEEN-GROUPS ANOVA

Irwin and colleagues (2004) are among a growing body of behavioral health researchers who are interested in adherence to medical regimens. These researchers studied adherence to an exercise regimen over one year in postmenopausal women, who are at increased risk for medical problems that may be reduced by exercise. Among the many factors that the research team examined was attendance at a monthly group education program that taught tactics to change exercise behavior; the researchers kept attendance and divided participants into three categories based on the number of sessions they attended. (*Note:* The researchers could have kept the data as numbers of sessions, a scale variable, rather than dividing them into categories based on numbers of sessions, an ordinal variable.)

Here is an abbreviated version of this study with fictional data points; the means of these data points, however, are the actual means of the study.

< 5 sessions: 155, 120, 130

5–8 sessions: 199, 160, 184

9–12 sessions: 230, 214, 195, 209

In this study, the independent variable was attendance, with three levels: <5 sessions, 5–8 sessions, and 9–12 sessions. The dependent variable was number of minutes of exercise per week. So we have one ordinal independent variable with three between-groups levels and one scale dependent variable. How can we conduct a one-way between-groups ANOVA?

Summary of Step 1

Population 1: Postmenopausal women who attended fewer than 5 sessions of a group exercise-education program. Population 2: Postmenopausal women who attended 5–8 sessions of a group exercise-education program. Population 3: Postmenopausal women who attended 9–12 sessions of a group exercise-education program.

The comparison distribution will be an F distribution. The hypothesis test will be a one-way between-groups ANOVA. The data were not selected randomly, so we must generalize only with caution. We do not know if the underlying population distributions are normal, but the sample data do not indicate severe skew. To see if we meet the homoscedasticity assumption, we will check to see if the largest variance is no greater than twice the smallest variance. From the calculations below, we see that the largest variance, 387, is not more than twice the smallest, 208.67, so we have met the homoscedasticity assumption. (The following information is taken from the calculation of SS_{within}.)

Sample	< 5	5–8	9–12
Squared deviations	400	324	324
	225	441	4
	25	9	289
	9		
Sum of squares	650	774	626
$N - 1$	2	2	3
Variance	**325**	**387**	**208.67**

Summary of Step 2

Null hypothesis: Postmenopausal women in different categories of attendance at a group exercise-education program exercise the same average number of minutes per week—H_0: $\mu_1 = \mu_2 = \mu_3$. Research hypothesis: Postmenopausal women in different categories of attendance at a group exercise-education program do not exercise the same average number of minutes per week—H_1 is that at least one μ is different from another μ.

Summary of Step 3

$$df_{between} = N_{groups} - 1 = 3 - 1 = 2$$

$$df_1 = 3 - 1 = 2; df_2 = 3 - 1 = 2; df_3 = 4 - 1 = 3$$

$$df_{within} = 2 + 2 + 3 = 7$$

The comparison distribution will be the F distribution with 2 and 7 degrees of freedom.

Summary of Step 4

The critical F statistic based on a p level of 0.05 is 4.74.

Summary of Step 5

$$df_{total} = 2 + 7 = 9 \text{ or } df_{total} = 10 - 1 = 9$$

$$SS_{total} = \Sigma(X - GM)^2 = 12{,}222.40$$

Sample	X	$(X - GM)$	$(X - GM)^2$
< 5	155	−24.6	605.16
$M_{<5} = 135$	120	−59.6	3552.16
	130	−49.6	2460.16
5–8	199	19.4	376.36
$M_{5-8} = 181$	160	−19.6	384.16
	184	4.4	19.36
9–12	230	50.4	2540.16
$M_{9-12} = 212$	214	34.4	1183.36
	195	15.4	237.16
	209	29.4	864.36
	$GM = 179.60$		$SS_{total} =$ **12,222.40**

$$SS_{within} = \Sigma(X - M)^2 = 2050.00$$

Sample	X	$(X - M)$	$(X - M)^2$
< 5	155	20	400
$M_{<5} = 135$	120	−15	225
	130	−5	25
5–8	199	18	324
$M_{5–8} = 181$	160	−21	441
	184	3	9
9–12	230	18	324
$M_{9–12} = 212$	214	2	4
	195	−17	289
	209	−3	9
	$GM = 179.60$		SS_{within} = **2050.00**

$$SS_{between} = \Sigma(M - GM)^2 = 10{,}172.40$$

Sample	X	$(M - GM)$	$(M - GM)^2$
< 5	155	−44.6	1989.16
$M_{<5} = 135$	120	−44.6	1989.16
	130	−44.6	1989.16
5–8	199	1.4	1.96
$M_{5–8} = 181$	160	1.4	1.96
	184	1.4	1.96
9–12	230	32.4	1049.76
$M_{9–12} = 212$	214	32.4	1049.76
	195	32.4	1049.76
	209	32.4	1049.76
	$GM = 179.60$		$SS_{between}$ = **10,172.40**

$$SS_{total} = SS_{within} + SS_{between} = 12{,}222.40 = 2050.00 + 10{,}172.40$$

$$MS_{between} = \frac{SS_{between}}{df_{between}} = \frac{10{,}172.40}{2} = 4086.20$$

$$MS_{within} = \frac{SS_{within}}{df_{within}} = \frac{2050.00}{7} = 292.857$$

$$F = \frac{MS_{between}}{MS_{within}} = \frac{5086.20}{292.857} = 17.37$$

Source	SS	df	MS	F
Between	10,172.40	2	5086.200	17.37
Within	2050.00	7	292.857	
Total	12,222.40	9		

Summary of Step 6

The F statistic, 17.37, is beyond the cutoff of 4.74. We can reject the null hypothesis. It appears that postmenopausal women in different categories of attendance at a group exercise-education program do exercise a different average number of minutes per week. However, the results from this ANOVA do not tell us where specific differences lie. The ANOVA tells us only that there is at least one difference between means. We must calculate a post hoc test to determine exactly which pairs of means are different.

11.2 CONDUCTING A ONE-WAY WITHIN-GROUPS ANOVA

Researchers followed the progress of 42 people undergoing inpatient rehabilitation following a spinal cord injury (White, Driver, & Warren, 2010). They assessed the patients on a variety of measures on three separate occasions—when they were admitted to the rehabilitation facility, 3 weeks later, and at discharge. Below are data that reflect patients' symptoms of depression on the Patient Health Questionnaire-9 (PHQ-9). (The data for these three fictional patients have the same means as the actual larger data set, as well as the same outcome in terms of the decision in step 6 of the ANOVA below.)

	Admission	Three Weeks	Discharge
Patient 1	6.1	5.5	5.3
Patient 2	6.9	5.7	4.2
Patient 3	7.4	6.5	4.9

How can we use one-way within-groups ANOVA to determine if depression levels changed as patients went through rehabilitation for spinal cord injury? We'll walk through all six steps of hypothesis testing for a one-way within-groups ANOVA.

Step 1: Population 1: People just admitted to an inpatient rehabilitation facility following a spinal cord injury. Population 2: People three weeks after they were admitted to an inpatient rehabilitation facility following a spinal cord injury. Population 3: People being discharged from an inpatient rehabilitation facility following spinal cord injury.

The comparison distribution will be an F distribution. The hypothesis test will be a one-way within-groups ANOVA. Regarding the assumptions: (1) The patients were not selected randomly (all were from the same hospital), so we must generalize with caution. (2) We do not know if the underlying population distributions are normal, but the sample data do not indicate severe skew. (3) To see if we meet the homoscedasticity assumption, we will check to see if the variances are similar (typically, when the largest variance is not more than twice the smallest) when we calculate the test statistic. (4) The experimenter could not counterbalance, so order effects might be present. With different levels of a time-related variable, it is not possible to assign someone to be measured at, for example, the final time point before the first time point.

Step 2: Null hypothesis: People in an inpatient rehabilitation hospital for a spinal cord injury have the same levels of depression, on average, at admission, 3 weeks later, and at discharge—H_0: $\mu_1 = \mu_2 = \mu_3$. Research hypothesis: People in an inpatient rehabilitation hospital for a spinal cord injury do not have the same levels of depression, on average, at admission, 3 weeks later, and at discharge—H_1 is that at least one μ is different from another μ.

Step 3: We use an F distribution with 2 and 4 degrees of freedom.

$$df_{between} = N_{groups} - 1 = 3 - 1 = 2$$

$$df_{subjects} = n - 1 = 3 - 1 = 2$$

$$df_{within} = (df_{between})(df_{subjects}) = (2)(2) = 4$$

$$df_{total} = df_{between} + df_{subjects} + df_{within} = 2 + 2 + 4 = 8 \text{ (or } df_{total} = N_{total} - 1 = 9 - 1 = 8)$$

Step 4: The critical value for the F statistic for a p level of 0.05 and 2 and 4 degrees of freedom is 6.95.

Step 5: $SS_{total} = \Sigma(X - GM)^2 = 8.059$.

Time	X	$X - GM$	$(X - GM)^2$
Admission	6.1	0.267	0.071
Admission	6.9	1.067	1.138
Admission	7.4	1.567	2.455
Three weeks	5.5	−0.333	0.111
Three weeks	5.7	−0.133	0.018
Three weeks	6.5	0.667	0.445
Discharge	5.3	−0.533	0.284
Discharge	4.2	−1.633	2.667
Discharge	4.9	−0.933	0.87
	$GM = 5.833$		$\Sigma(X - GM)^2 = 8.059$

$SS_{between} = \Sigma(M - GM)^2 = 6.018$

Time	X	Group Mean	$M - GM$	$(M - GM)^2$
Admission	6.1	6.8	0.967	0.935
Admission	6.9	6.8	0.967	0.935
Admission	7.4	6.8	0.967	0.935
Three weeks	5.5	5.9	0.067	0.004
Three weeks	5.7	5.9	0.067	0.004
Three weeks	6.5	5.9	0.067	0.004
Discharge	5.3	4.8	−1.033	1.067
Discharge	4.2	4.8	−1.033	1.067
Discharge	4.9	4.8	−1.033	1.067
	$GM = 5.833$		$\Sigma(M - GM)^2 = 6.018$	

$SS_{subjects} = \Sigma(M_{participant} - GM)^2 = 0.846$

Participant	Time	X	Participant Mean	$M_{participant} - GM$	$(M_{participant} - GM)^2$
1	Admission	6.1	5.633	−0.2	0.040
2	Admission	6.9	5.6	−0.233	0.054
3	Admission	7.4	6.267	0.434	0.188
1	Three weeks	5.5	5.633	−0.2	0.040
2	Three weeks	5.7	5.6	−0.233	0.054
3	Three weeks	6.5	6.267	0.434	0.188
1	Discharge	5.3	5.633	−0.2	0.040
2	Discharge	4.2	5.6	−0.233	0.054
3	Discharge	4.9	6.267	0.434	0.188
	$GM = 5.833$			$\Sigma(M_{participant} - GM)^2 = 0.846$	

$SS_{within} = SS_{total} - SS_{between} - SS_{subjects} = 8.059 - 6.018 - 0.846 = 1.195$

We now have enough information to fill in the first three columns of the source table—the source, *SS,* and *df* columns—and to divide each sum of squares by the degrees of freedom to get variance, *MS.*

$$MS_{between} = \frac{SS_{between}}{df_{between}} = \frac{6.018}{2} = 3.009$$

$$MS_{subjects} = \frac{SS_{subjects}}{df_{subjects}} = \frac{0.846}{2} = 0.423$$

$$MS_{within} = \frac{SS_{within}}{df_{within}} = \frac{1.195}{4} = 0.299$$

We then calculate two *F* statistics—one for between-groups and one for subjects—by dividing each *MS* by the within-groups *MS.*

$$F_{between} = \frac{MS_{between}}{MS_{within}} = \frac{3.009}{0.299} = 10.06$$

$$F_{subjects} = \frac{MS_{subjects}}{MS_{within}} = \frac{0.423}{0.299} = 1.41$$

The completed source table is:

Source	*SS*	*df*	*MS*	*F*
Between	6.018	2	3.009	10.06
Subjects	0.846	2	0.423	1.41
Within	1.195	4	0.299	
Total	8.059	8		

We want to know if there's a statistically significant difference between groups, so we'll look at the between-groups F statistic, 10.06.

Step 6: The F statistic, 10.06, is beyond the critical value, 6.95. We can reject the null hypothesis. It appears that depression scores differ based on the time point during rehabilitation. A post hoc test is necessary to know exactly which pairs of means are significantly different.

Exercises

Clarifying the Concepts

11.1 What is an ANOVA?

11.2 What do the F distributions allow us to do that the t distributions do not?

11.3 The F statistic is a ratio of between-groups variance and within-groups variance. What are these two types of variance?

11.4 What is the difference between a within-groups (repeated-measures) ANOVA and a between-groups ANOVA?

11.5 What are the three assumptions for a between-groups ANOVA?

11.6 The null hypothesis for ANOVA posits no difference among population means, as in other hypothesis tests, but the research hypothesis in this case is a bit different. Why?

11.7 Why is the F statistic always positive?

11.8 In your own words, define the word *source* as you would use it in everyday conversation. Provide at least two different meanings that might be used. Then define the word as a statistician would use it.

11.9 Explain the concept of *sum of squares.*

11.10 The total sum of squares for a between-groups ANOVA is found by adding which two statistics together?

11.11 What is the grand mean?

11.12 How do we calculate the between-groups sum of squares?

11.13 We typically measure effect size with ________ for a z test or a t test and with ________ for an ANOVA.

11.14 What are Cohen's conventions for interpreting effect size using R^2?

11.15 What does *post hoc* mean, and when are these tests needed with ANOVA?

11.16 Define the symbols in the following formula: $N' = \frac{N_{groups}}{\Sigma(1/N)}$.

11.17 Find the error in the statistics language in each of the following statements about z, t, or F distributions or their related tests. Explain why it is incorrect and provide the correct word.

a. The professor reported the mean and standard error for the final exam in the statistics class.

b. Before we can calculate a t statistic, we must know the population mean and the population standard deviation.

c. The researcher calculated the parameters for her three samples so that she could calculate an F statistic and conduct an ANOVA.

d. For her honors project, Evelyn calculated a z statistic so that she could compare a sample of students who had ingested caffeine and a sample of students who had not ingested caffeine on their video game performance mean scores.

11.18 Find the incorrectly used symbol or symbols in each of the following statements or formulas. For each statement or formula, (i) state which symbol(s) is/are used incorrectly, (ii) explain why the symbol(s) in the original statement is/are incorrect, and (iii) state what symbol(s) *should* be used.

a. When calculating an F statistic, the numerator includes the estimate for the between-groups variance, s.

b. $SS_{between} = (X - GM)^2$

c. $SS_{within} = (X - M)$

d. $F = \sqrt{t}$

11.19 What are the four assumptions for a within-groups ANOVA?

11.20 What are order effects?

11.21 Explain the source of variability called "subjects."

11.22 What is the advantage of the design of the within-groups ANOVA over that of the between-groups ANOVA?

11.23 What is counterbalancing?

11.24 Why is it appropriate to counterbalance when using a within-groups design?

11.25 How do we calculate the sum of squares for subjects?

11.26 How is the calculation of df_{within} different in a between-groups ANOVA from the calculation in a within-groups ANOVA?

11.27 How could we turn a between-groups study into a within-groups study?

11.28 What are some situations in which it might be impossible—or not make sense—to turn a between-groups study into a within-groups study?

11.29 How is the calculation of effect size different for a one-way between-groups ANOVA versus a one-way within-groups ANOVA?

Calculating the Statistics

11.30 For the following data, assuming a between-groups design, determine:

Group 1: 11, 17, 22, 15
Group 2: 21, 15, 16
Group 3: 7, 8, 3, 10, 6, 4
Group 4: 13, 6, 17, 27, 20

a. $df_{between}$
b. df_{within}
c. df_{total}
d. The critical value, assuming a p value of 0.05
e. The mean for each group and the grand mean
f. The total sum of squares
g. The within-groups sum of squares
h. The between-groups sum of squares
i. The rest of the ANOVA source table for these data
j. Tukey *HSD* values

11.31 For the following data, assuming a between-groups design, determine:

1970: 45, 211, 158, 74
1980: 92, 128, 382
1990: 273, 396, 178, 248, 374

a. $df_{between}$
b. df_{within}
c. df_{total}
d. The critical value, assuming a p value of 0.05
e. The mean for each group and the grand mean
f. The total sum of squares
g. The within-groups sum of squares
h. The between-groups sum of squares
i. The rest of the ANOVA source table for these data
j. The effect size

11.32 Calculate the F statistic, writing the ratio accurately, for each of the following cases:

a. Between-groups variance is 29.4 and within-groups variance is 19.1
b. Within-groups variance is 0.27 and between-groups variance is 1.56
c. Between-groups variance is 4595 and within-groups variance is 3972

11.33 Calculate the F statistic, writing the ratio accurately, for each of the following cases:

a. Between-groups variance is 321.83 and within-groups variance is 177.24
b. Between-groups variance is 2.79 and within-groups variance is 2.20
c. Within-groups variance is 41.60 and between-groups variance is 34.45

11.34 An incomplete one-way between-groups ANOVA source table is shown below. Compute the missing values.

Source	*SS*	*df*	*MS*	*F*
Between	191.450	—	47.863	—
Within	104.720	32	—	
Total	—	36		

11.35 An incomplete one-way between-groups ANOVA source table is shown below. Compute the missing values.

Source	*SS*	*df*	*MS*	*F*
Between	—	2	—	—
Within	89	11	—	
Total	132	—		

11.36 Each of the following is a calculated F statistic with its degrees of freedom. Using the F table, estimate the level of significance for each. You can do this by indicating whether its likelihood of occurring is greater than or less than a p level shown on the table.

a. $F = 4.11$, with 3 $df_{between}$ and 30 df_{within}
b. $F = 1.12$, with 5 $df_{between}$ and 83 df_{within}
c. $F = 2.28$, with 4 $df_{between}$ and 42 df_{within}

11.37 A researcher designs an experiment in which the single independent variable has four levels. If the researcher performed an ANOVA and rejected the null hypothesis, how many post hoc comparisons would the researcher make (assuming she was making all possible comparisons)?

11.38 A researcher designs an experiment in which the single independent variable has five levels. If the researcher performed an ANOVA and rejected the null hypothesis, how many post hoc comparisons would the researcher make (assuming he was making all possible comparisons)?

11.39 For the following data, assuming a within-groups design, determine:

	Person			
	1	2	3	4
Level 1 of the IV	7	16	3	9
Level 2 of the IV	15	18	18	13
Level 3 of the IV	22	28	26	29

a. $df_{between} = N_{groups} - 1$
b. $df_{subjects} = n - 1$
c. $df_{within} = (df_{between})(df_{subjects})$
d. $df_{total} = df_{between} + df_{subjects} + df_{within}$, or $df_{total} = N_{total} - 1$
e. $SS_{total} = \Sigma(X - GM)^2$
f. $SS_{between} = \Sigma(M - GM)^2$
g. $SS_{subjects} = \Sigma(M_{participant} - GM)^2$
h. $SS_{within} = SS_{total} - SS_{between} - SS_{subjects}$
i. The rest of the ANOVA source table for these data
j. The effect size
k. The Tukey *HSD* statistic for the comparisons between level 1 and level 3
l. The critical *q* value; then, make a decision for each comparison in part (k)
m. The effect size

11.40 For the following data, assuming a within-groups design, determine:

	Person					
	1	2	3	4	5	6
Level 1	5	6	3	4	2	5
Level 2	6	8	4	7	3	7
Level 3	4	5	2	4	0	4

a. $df_{between} = N_{groups} - 1$
b. $df_{subjects} = n - 1$
c. $df_{within} = (df_{between})(df_{subjects})$
d. $df_{total} = df_{between} + df_{subjects} + df_{within}$, or $df_{total} = N_{total} - 1$
e. $SS_{total} = \Sigma(X - GM)^2$
f. $SS_{between} = \Sigma(M - GM)^2$
g. $SS_{subjects} = \Sigma(M_{participant} - GM)^2$
h. $SS_{within} = SS_{total} - SS_{between} - SS_{subjects}$
i. The rest of the ANOVA source table for these data
j. The critical *F* value
k. If appropriate, the Tukey *HSD* statistic for all possible mean comparisons

11.41 For the following incomplete source table below for a one-way within-groups ANOVA:

Source	*SS*	*df*	*MS*	*F*
Between	941.102	2	—	—
Subjects	3807.322	—	—	—
Within	—	20	—	
Total	5674.502	—		

a. Complete the missing information.
b. Calculate R^2.

11.42 Assume that a researcher had 14 individuals participate in all three conditions of her experiment. Use this information to complete the source table below.

Source	*SS*	*df*	*MS*	*F*
Between	60	—	—	—
Subjects	—	—	—	—
Within	50	—	—	
Total	136	—		

Applying the Concepts

11.43 Comedy versus news and hypothesis testing: Focusing on coverage of the 2004 U.S. presidential election, Julia R. Fox, a telecommunications professor at Indiana University, wondered whether *The Daily Show,* despite its comedy format, was a valid source of news. She coded a number of half-hour episodes of *The Daily Show* as well as a number of half-hour episodes of the network news (Indiana University Media Relations, 2006). Fox reported that the average amounts of "video and audio substance" were not statistically significantly different between the two types of shows. Her analyses are described as "second-by-second," so, for this exercise, assume that all outcome variables are measures of time.

a. As the study is described, what are the independent and dependent variables? For nominal variables, state the levels.
b. As the study is described, what type of hypothesis test would Fox use?
c. Now imagine that Fox added a third category, a cable news channel such as CNN. Based on this new information, state the independent variable or variables and the levels of any nominal independent variables. What hypothesis test would she use?

11.44 The comparison distribution: For each of the following situations, state whether the distribution of interest is a z distribution, a t distribution, or an F distribution. Explain your answer.

a. A city employee locates a U.S. Census report that includes the mean and standard deviation for income in the state of Wyoming and then takes a random sample of 100 residents of the city of Cheyenne. He wonders whether residents of Cheyenne earn more, on average, than Wyoming residents as a whole.

b. A researcher studies the effect of different contexts on work interruptions. Using discreet video cameras, she observes employees working in enclosed offices in the workplace, in open cubicles in the workplace, and in home offices.

c. An honors student wondered whether an education in statistics reduced the tendency to believe advertising that cited data. He compared social science majors who had taken statistics and social science majors who had not taken statistics with respect to their responses to an interactive advertising assessment.

11.45 The comparison distribution: For each of the following situations, state whether the distribution of interest is a z distribution, a t distribution, or an F distribution. Explain your answer.

a. A student reads in her *Introduction to Psychology* textbook that the mean IQ is 100. She asks 10 friends what their IQ scores are (they attend a university that assesses everyone's IQ score) to determine whether her friends are smarter than average.

b. Is the presence of books in the home a marker of a stable family? A social worker counted the number of books on view in the living rooms of all the families he visited over the course of one year. He categorized families into four groups: no books visible, only children's books visible, only adult books visible, and both children's and adult books visible. The department for which he worked had stability ratings for each family based on a number of measures.

c. Which television show leads to more learning? A researcher assessed the vocabularies of a sample of children randomly assigned to watch *Sesame Street* as much as they wanted for a year but to not watch *The Wiggles.* She also assessed the vocabularies of a sample of children randomly assigned to watch *The Wiggles* as much as they wanted for a year but not to watch *Sesame Street*. She compared the average vocabulary scores for the two groups.

11.46 Links among distributions: The z, t, and F distributions are closely linked. In fact, it is possible to use an F distribution in all cases in which a t or a z could be used.

a. If you calculated an F statistic of 4.22 but you could have used a t statistic (i.e., the situation met all criteria for using a t statistic), what would the t statistic have been? Explain your answer.

b. If you calculated an F statistic of 4.22 but you could have used a z statistic, what would the z statistic have been? Explain your answer.

c. If you calculated a t statistic of 0.67 but you could have used a z statistic, what would the z statistic have been? Explain your answer.

d. Cite two reasons that all three types of distributions (i.e., z, t, and F) are still in use when we really only need an F distribution.

11.47 International students and type of ANOVA: Catherine Ruby (2006), a doctoral student at New York University, conducted an online survey to ascertain the reasons that international students chose to attend graduate school in the United States. One of several dependent variables that she considered was reputation; students were asked to rate the importance in their decision of factors such as the reputation of the institution, the institution and program's academic accreditations, and the reputation of the faculty. Students rated factors on a 1–5 scale, and then all reputation ratings were averaged to form a summary score for each respondent. For each of the following scenarios, state the independent variable with its levels (the dependent variable is reputation in all cases). Then state what kind of an ANOVA she would use.

a. Ruby compared the importance of reputation among graduate students in different types of programs: arts and sciences, education, law, and business.

b. Imagine that Ruby followed these graduate students for 3 years and assessed their rating of reputation once a year.

c. Ruby compared international students working toward a master's, a doctorate, or a professional degree (e.g., MBA) on reputation.

d. Imagine that Ruby followed international students from their master's program to their doctoral program to their postdoctoral fellowship, assessing their ratings of reputation once at each level of their training.

11.48 Type of ANOVA in study of remembering names: Do people remember names better under different circumstances? In a fictional study, a cognitive psychologist studied memory for names after a group activity that lasted 20 minutes. Participants were not told that this was a study of memory. After the group activity, participants were asked to name the other group members. The researcher randomly assigned 120 participants to one of three conditions: (1) group members introduced themselves once (one introduction only), (2) group members were introduced by the experimenter and by themselves (two introductions), and (3) group members were introduced by the experimenter and themselves and also wore nametags throughout the group activity (two introductions and nametags).

a. Identify the type of ANOVA that should be used to analyze the data from this study.

b. State what the researcher could do to redesign this study so it would be analyzed with a one-way within-groups ANOVA. Be specific.

11.49 Political party and ANOVA: Researchers asked 180 U.S. students to identify their political viewpoint as most similar to that of the Republicans, most similar to that of the Democrats, or neither. All three groups then completed a religiosity scale. The researchers wondered whether political orientation affected levels of religiosity, a measure that assesses how religious one is, regardless of the specific religion with which a person identifies.

a. What is the independent variable, and what are its levels?

b. What is the dependent variable?

c. What are the populations and what are the samples?

d. Would you use a between-groups or within-groups ANOVA? Explain.

e. Using this example, explain how you would calculate the F statistic.

11.50 Exercise and the Tukey *HSD* test: In How It Works 11.1, we conducted a one-way between-groups ANOVA on an abbreviated data set from research by Irwin and colleagues (2004) on adherence to an exercise regimen. Participants were asked to attend a monthly group education program to help them change their exercise behavior. Attendance was taken and participants were divided into three categories: those who attended fewer than 5 sessions, those who attended between 5 and 8 sessions, and those who attended between 9 and 12 sessions. The dependent variable was number of minutes of exercise per week. Here are the data once again:

< 5 sessions: 155, 120, 130

5–8 sessions: 199, 160, 184

9–12 sessions: 230, 214, 195, 209

a. What conclusion did we draw in step 6 of the ANOVA? Why could you not be more specific in your conclusion? That is, why is an additional test necessary when the ANOVA is statistically significant?

b. Conduct a Tukey *HSD* test for this example. State your conclusions based on this test. Show all calculations.

c. If we did not reject the null hypothesis for a particular pair of means, then why can't we conclude that the two means are the same?

11.51 Grade point average and comparing the t and F distributions: Based on your knowledge of the relation of the t and F distributions, complete the accompanying software output tables. The table for the independent-samples t test and the table for the one-way between-groups ANOVA were calculated using the identical fictional data comparing grade point averages (GPAs).

a. What is the F statistic? Show your calculations. (*Hint:* The "Mean Square" column includes the two estimates of variance used to calculate the F statistic.)

b. What is the t statistic? Show your calculations. (*Hint:* Use the F statistic that you calculated in part (a).)

c. In statistical software output, "Sig." refers to the actual p level of the statistic. We can compare the actual p level to a cutoff p level such as 0.05 to decide whether to reject the null hypothesis. For the t test, what is the "Sig."? Explain how you determined this. (*Hint:* Would we expect the "Sig." for the independent-samples t test to be the same as or different from that for the one-way between-groups ANOVA?)

Independent Samples Test

	t-test for Equality of Means						
	t	*df*	Sig. (2-tailed)	Mean Difference	Standard Error Difference	95% Confidence Interval of the Difference	
						Lower	Upper
GPA		82		−.28251	.12194	−.52508	−.03993

ANOVA

GPA	Sum of Squares	*df*	Mean Square	*F*	Sig.
Between groups	4.623	1	4.623		.005
Within groups	42.804	82	.522		
Total	47.427	83			

11.52 Consideration of Future Consequences and two kinds of hypothesis testing: Two samples of students, one comprised of social science majors and one comprised of students with other majors, completed the Consideration of Future Consequences scale (CFC). The accompanying tables include the output from software for an independent-samples *t* test and a one-way between-groups ANOVA on these data.

Independent Samples Test

	t	*df*	Sig. (2-tailed)	Mean Difference	Standard Error Difference	95% Confidence Interval of the Difference	
						Lower	Upper
CFC scores	−.650	28	.521	−.17500	.26930	−.72664	.37664

ANOVA

CFC Scores

	Sum of Squares	*df*	Mean Square	*F*	Sig.
Between Groups	.204	1	.204	.422	.521
Within Groups	13.538	28	.483		
Total	13.742	29			

Group Statistics

Major		*N*	Mean	Standard Deviation	Standard Error Mean
CFC Scores	Other	10	3.2000	.88819	.28087
	Social Science	20	3.3750	.58208	.13016

a. Demonstrate that the results of the independent-samples *t* test and the one-way between-groups ANOVA are the same. (*Hint:* Find the *t* statistic for the *t* test and the *F* statistic for the ANOVA.)

b. In statistical software output, "Sig." refers to the actual *p* level of the statistic. We can compare the actual *p* level to a cutoff *p* level such as 0.05 to decide whether to reject the null hypothesis. What are the "Sig." levels for the two tests here—the independent-samples *t* test and the one-way between-groups ANOVA? Are they the same or different? Explain why this is the case.

c. In the CFC ANOVA, the column titled "Mean Square" includes the estimates of variance. Show how the *F* statistic was calculated from two types of variance. (*Hint:* Look at the far left column to determine which estimate of variance is which.)

d. Looking at the table titled "Group Statistics," how many participants were in each sample?

e. Looking at the table titled "Group Statistics," what is the mean CFC score for the social science majors?

11.53 Imagine a researcher wanted to assess people's fear of dogs as a function of the size of the dog. He assessed fear among people who indicated they were afraid of dogs, using a 30-point scale from 0 (no fear) to 30 (extreme fear). The researcher exposed each participant to three different dogs, a small dog weighing 20 pounds, a medium-sized dog weighing 55 pounds, and a large dog weighing 110 pounds, assessing the fear level after each exposure. Here are some hypothetical data; note that these are the data from Exercise 11.39, on which you have already calculated numerous statistics:

	Person			
	1	2	3	4
Small dog	7	16	3	9
Medium dog	15	18	18	13
Large dog	22	28	26	29

a. State the null and research hypotheses.

b. Consider whether the assumptions of random selection and order effects were met.

c. In Exercise 11.39 you calculated the effect size for these data. What does this statistic tell us about the effect of size of dog on fear levels?

d. In Exercise 11.39, you calculated a Tukey *HSD* test for these data. What can you conclude about the effect of size of dog on fear levels based on this statistic?

11.54 Chewing-gum commercials and one-way within-groups ANOVA: Commercials for chewing gum make claims about how long the flavor will last. In fact, some commercials claim that the flavor lasts too long, affecting sales and profit. Let's put these claims to a test. Imagine a student decides to compare four different gums using five participants. Each randomly selected participant was asked to chew a different piece of gum each day for 4 days, such that at the end of the 4 days, each participant had chewed all four types of gum. The order of the gums was randomly determined for each participant. After 2 hours of chewing, participants recorded the intensity of flavor from 1 (not intense) to 9 (very intense). Here are some hypothetical data:

	Person				
	1	2	3	4	5
Gum 1	4	6	3	4	4
Gum 2	8	6	9	9	8
Gum 3	5	6	7	4	5
Gum 4	2	2	3	2	1

a. Conduct all six steps of the hypothesis test.

b. Are any additional tests warranted? Explain your answer.

11.55 Pessimism and one-way within-groups ANOVA: Researchers Busseri, Choma, and Sadava (2009) asked a sample of individuals who scored as pessimists on a measure of life orientation about past, present, and projected future satisfaction with their lives. Higher scores on the life satisfaction measure indicate higher satisfaction. The data below reproduce the pattern of means that the researchers observed in self-reported life satisfaction of the sample of pessimists for the three time points. Do pessimists predict a gloomy future for themselves?

	Person				
	1	2	3	4	5
Past	18	17.5	19	16	20
Present	18.5	19.5	20	17	18
Future	22	24	20	23.5	21

a. Perform steps 5 and 6 of hypothesis testing. Be sure to complete the source table when calculating the *F* ratio for step 5.

b. If appropriate, calculate the Tukey *HSD* for all possible mean comparisons. Find the critical value of *q* and make a decision regarding the null hypothesis for each of the mean comparisons.

c. Calculate the R^2 measure of effect size for this ANOVA.

11.56 Pessimism and one-way within-groups ANOVA: Exercise 11.55 describes a study conducted by Busseri and colleagues (2009) using a group of pessimists. These researchers asked the same question of a group of optimists: optimists rated their past, present, and projected future satisfaction with their lives. Higher scores on the life satisfaction measure indicate higher satisfaction. The data below reproduce the pattern of means that the researchers observed in self-reported life satisfaction of the sample of optimists for the three time points. Do optimists see a rosy future ahead?

	Person				
	1	2	3	4	5
Past	22	23	25	24	26
Present	25	26	27	28	29
Future	24	27	26	28	29

a. Perform steps 5 and 6 of hypothesis testing. Be sure to complete the source table when calculating the *F* ratio for step 5.

b. If appropriate, calculate the Tukey *HSD* for all possible mean comparisons. Find the critical value of *q* and make a decision regarding the null hypothesis for each of the mean comparisons.

c. Calculate the R^2 measure of effect size for this ANOVA.

11.57 Wagging tails and one-way within-groups ANOVA: How does a dog's tail wag in response to seeing different people and other pets? Quaranta, Siniscalchi, and Vallortigara (2007) investigated the amplitude and direction of a dog's tail wagging in response to seeing its owner, an unfamiliar cat, and an unfamiliar dog. The fictional data below are measures of amplitude. These data reproduce the pattern of results in the study, averaging left tail wags and right tail wags. Use these data to construct the source table for a one-way within-groups ANOVA.

Dog Participant	Owner	Cat	Other Dog
1	69	28	45
2	72	32	43
3	65	30	47
4	75	29	45
5	70	31	44

11.58 Memory, post hoc tests, and effect size: Luo, Hendriks, and Craik (2007) were interested in whether lists of words might be better remembered if they were paired with either pictures or sound effects. They asked participants to memorize lists of words under three different learning conditions. In the first condition, participants just saw a list of nouns that they were to remember (word-alone condition). In the second condition, the words were also accompanied by a picture of the object (picture condition). In the third condition, the words were accompanied by a sound effect matching the object (sound effect condition). The researchers measured the proportion of words participants got correct in a later recognition test. Fictional data from four participants produce results similar to those of the original study. The average proportion of words recognized was $M = 0.54$ in the word-alone condition, $M = 0.69$ in the picture condition, and $M = 0.838$ in the sound effect condition. The source table below depicts the results of the ANOVA on the data from the four fictional participants.

Source	*SS*	*df*	*MS*	*F*
Between	0.177	2	0.089	8.900
Subjects	0.002	3	0.001	0.100
Within	0.059	6	0.010	
Total	0.238	11		

a. Is it appropriate to perform post hoc comparisons on the data? Why or why not?

b. Use the information provided in the ANOVA table to calculate R^2. Interpret the effect size using Cohen's conventions. State what this R^2 means in terms of the independent and dependent variables used in this study.

11.59 Wagging tails, hypothesis test decision-making, and post hoc tests: Assume that we recruited a different sample of five dogs and attempted to replicate the Quaranta and colleagues (2007) study described in Exercise 11.57. The source table for our fictional replication appears below. Find the critical F value and make a decision regarding the null hypothesis. Based on this decision, is it appropriate to conduct post hoc comparisons? Why or why not?

Source	*SS*	*df*	*MS*	*F*
Between	58.133	2	29.067	0.066
Subjects	642.267	4	160.567	0.364
Within	3532.533	8	441.567	
Total	4232.933	14		

11.60 Post hoc tests and p values: The most recent version of the *Publication Manual of the American Psychological Association* recommends reporting the exact p values for all statistical tests to three decimal places (previously, it recommended reporting $p < 0.05$ or $p > 0.05$). Explain how the new reporting format allows a reader to more critically interpret the results of post hoc comparisons reported by an author.

Putting It All Together

11.61 Trust in leadership and one-way between-groups ANOVA: In Chapter 10, we introduced a study by Steele and Pinto (2006) that examined whether people's level of trust in their direct supervisor was related to their level of agreement with a policy supported by that leader. Steele and Pinto found that the extent to which subordinates agreed with their supervisor was related to trust and showed no relation to gender, age, time on the job, or length of time working with the supervisor. Let's assume we used a scale that sorted employees into three groups: low trust, moderate trust, and high trust in supervisors. We have presented fictional data regarding level of agreement with a leader's decision for these three groups. The scores presented are the level of agreement with a decision made by a leader, from 1, the least agreement, to 40, the highest level of agreement. *Note:* These fictional data are different from those presented in Chapter 11.

Employees with low trust in their leader: 9, 14, 11, 18

Employees with moderate trust in their leader: 14, 35, 23

Employees with high trust in their leader: 27, 33, 21, 34

a. What is the independent variable? What are its levels?

b. What is the dependent variable?

c. Conduct all six steps of hypothesis testing for a one-way between-groups ANOVA.

d. How would you report the statistics in a journal article?

e. Conduct a Tukey *HSD* test. What did you learn?

f. Why is it not possible to conduct a t test in this situation?

g. Why is it not possible to use a within-groups design for this study?

11.62 Orthodontics and one-way between-groups ANOVA: Iranian researchers studied factors affecting patients' likelihood of wearing orthodontic appliances, noting that orthodontics is perhaps the area of health care with the highest need for patient cooperation (Behenam & Pooya, 2007). Among their analyses, they compared students in primary school, junior high school, and high school. The data that follow have almost exactly the same means as they found in their study, but with far smaller samples. The score for each student is his or her daily hours of wearing the orthodontic appliance.

Primary school: 16, 13, 18

Junior high school: 8, 13, 14, 12

High school: 20, 15, 16, 18

a. What is the independent variable? What are its levels?

b. What is the dependent variable?

c. Conduct all six steps of hypothesis testing for a one-way between-groups ANOVA.

d. How would you report the statistics in a journal article?

e. Conduct a Tukey *HSD* test. What did you learn?

f. Calculate the appropriate measure of effect size for this sample.

g. Based on Cohen's conventions, is this a small, medium, or large effect size?

h. Why is it useful to know the effect size in addition to the results of a hypothesis test?

i. How could this study be conducted using a within-groups design?

11.63 Eye glare, football, and one-way within-groups ANOVA: Does the black grease beneath football players' eyes really reduce glare or does it just make them look intimidating? In a variation of a study actually conducted at Yale University, 46 participants placed one of three substances below their eyes: black grease, black antiglare stickers, or petroleum jelly. The researchers assessed eye glare using a contrast chart that gives a value for each participant on a scale measure. Every participant was assessed with each of the three substances, one at a time. Black grease led to a reduction in glare compared with the two other conditions, antiglare stickers or petroleum jelly (DeBroff & Pahk, 2003).

Person	Black Grease	Antiglare Stickers	Petroleum Jelly
1	19.8	17.1	15.9
2	18.2	17.2	16.3
3	19.2	18.0	16.2
4	18.7	17.9	17.0

a. What is the independent variable? What are its levels?

b. What is the dependent variable?

c. What kind of ANOVA is this?

d. What is the first assumption for ANOVA? Is it likely that the researchers met this assumption? Explain your answer.

e. What is the second assumption for ANOVA? How could the researchers check to see if they had met this assumption? Be specific.

f. What is the third assumption for ANOVA? How could the researchers check to see if they had met this assumption? Be specific.

g. What is the fourth assumption specific to the within-groups ANOVA? What would the researchers need to do to ensure that they meet this assumption?

h. Perform steps 5 and 6 of hypothesis testing. Be sure to complete the source table when calculating the *F* ratio for step 5.

i. If appropriate, calculate the Tukey *HSD* for all possible mean comparisons. Find the critical value of *q* and make a decision regarding the null hypothesis for each of the mean comparisons.

j. Calculate the R^2 measure of effect size for this ANOVA.

k. How could this study be conducted using a between-groups design?

Terms

analysis of variance (ANOVA) (p. 256)
F statistic (p. 256)
between-groups variance (p. 256)
within-groups variance (p. 256)
one-way ANOVA (p. 257)
between-groups ANOVA (p. 257)
within-groups ANOVA (p. 258)
homoscedastic (p. 258)
heteroscedastic (p. 258)
source table (p. 265)
grand mean (p. 266)
R^2 (p. 273)
post hoc test (p. 274)
Tukey *HSD* test (p. 274)

Formulas

$df_{between} = N_{groups} - 1$ (p. 261)

$df_{within} = df_1 + df_2 + \ldots + df_{last}$
(in which df_1, df_2, etc., are the degrees of freedom, $N - 1$, for each sample) [formula for a one-way between-groups ANOVA] (p. 261)

$df_{total} = df_{between} + df_{within}$ [formula for a one-way between-groups ANOVA] (p. 266)

$df_{total} = N_{total} - 1$ (p. 266)

$GM = \frac{\Sigma(X)}{N_{total}}$ (p. 266)

$SS_{total} = \Sigma(X - GM)^2$ (p. 267)

$SS_{within} = \Sigma(X - M)^2$ [formula for a one-way between-groups ANOVA] (p. 267)

$SS_{between} = \Sigma(M - GM)^2$ (p. 268)

$SS_{total} = SS_{within} + SS_{between}$ [alternate formula for a one-way between-groups ANOVA] (p. 269)

$MS_{between} = \frac{SS_{between}}{df_{between}}$ (p. 270)

$MS_{within} = \frac{SS_{within}}{df_{within}}$ (p. 270)

$F = \frac{MS_{between}}{MS_{within}}$ (p. 270)

$R^2 = \frac{SS_{between}}{SS_{total}}$ [formula for a one-way between-groups ANOVA] (p. 273)

$HSD = \frac{(M_1 - M_2)}{s_M}$, for any two sample means (p. 274)

$s_M = \sqrt{\frac{MS_{within}}{N}}$, if equal sample sizes (p. 274)

$N' = \frac{N_{groups}}{\Sigma(1/N)}$ (p. 275)

$s_M = \sqrt{\frac{MS_{within}}{N'}}$, if unequal sample sizes (p. 275)

$df_{subjects} = n - 1$ (p. 280)

$df_{within} = (df_{between})(df_{subjects})$ [formula for a one-way within-groups ANOVA] (p. 280)

$df_{total} = df_{between} + df_{subjects} + df_{within}$ [formula for a one-way within-groups ANOVA] (p. 280)

$SS_{subjects} = \Sigma(M_{participant} - GM)^2$ (p. 282)

$SS_{within} = SS_{total} - SS_{between} - SS_{subjects}$ [formula for a one-way within-groups ANOVA] (p. 282)

$MS_{subjects} = \frac{SS_{subjects}}{df_{subjects}}$ (p. 282)

$F_{subjects} = \frac{MS_{subjects}}{MS_{within}}$ (p. 283)

$R^2 = \frac{SS_{between}}{(SS_{total} - SS_{subjects})}$ [formula for a one-way within-groups ANOVA] (p. 285)

Symbols

F (p. 254)
$df_{between}$ (p. 261)
df_{within} (p. 261)
$MS_{between}$ (p. 265)
MS_{within} (p. 265)
df_{total} (p. 266)
$SS_{between}$ (p. 266)
SS_{within} (p. 266)
SS_{total} (p. 266)
GM (p. 266)
R^2 (p. 273)
HSD (p. 274)
N' (p. 275)
$df_{subjects}$ (p. 280)
$SS_{subjects}$ (p. 280)
$MS_{subjects}$ (p. 280)
$F_{subjects}$ (p. 283)

CHAPTER 12

Two-Way Between-Groups ANOVA

BEFORE YOU GO ON

- You should understand the six steps of hypothesis testing (Chapter 7).
- You should be able to conduct and interpret a one-way between-groups ANOVA (Chapter 11).
- You should understand the concept of effect size (Chapter 8) and the measure of effect size for ANOVA, R^2 (Chapter 11).

FIGURE 12-1
Ski Bums Beware!
This bar graph tells a useful story for consumers about a statistically significant interaction. Ski resorts tend to exaggerate about the amount of snowfall, especially snowfall on weekends.

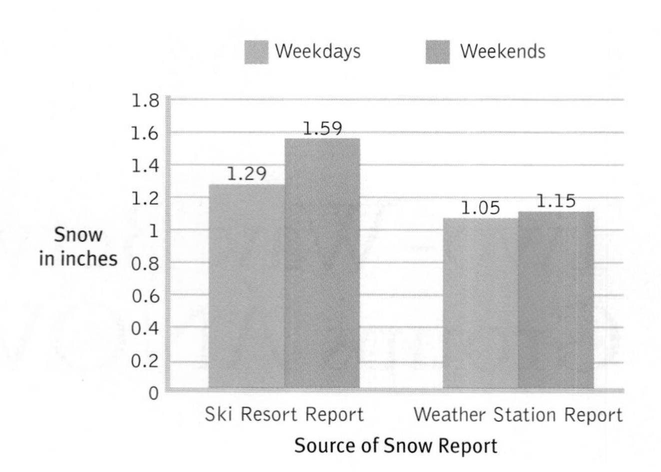

Jonathan Zinman and Eric Zitzewitz are social science researchers and ski bums who live in New Hampshire. They had become annoyed when they hit the slopes only to find that snowfall reports had clearly been exaggerated. So they collected snowfall data both from ski resorts and the weather station for weekdays (when fewer people go skiing or snowboarding) and for the more profitable weekends. Figure 12-1 summarizes what they found: The ski resorts reported greater snowfall than weather stations did, an effect that was even more pronounced on weekends, when such an exaggeration was likely to draw more paying customers.

In the study by Zinman and Zitzewitz (2009), the dependent variable is the amount of snow reported, in inches. But there are two independent variables (source of the snow report and time of the week), and each one seems to have its own effect on the amount of snow reported. When we ignore whether it was a weekday or a weekend, ski resorts reported greater snowfall, on average, than weather stations did. This is called a *main effect* of the source of the snow report. Figure 12-1 also shows that snow reports are higher on weekends than on weekdays, when we ignore the source of the snow report. This is called a *main effect* of time of the week.

But there is a third finding, and it's the most interesting one: Ski resorts bumped up their estimates even more on the weekend than they did during the week! So the source of the report variable interacted with the time of week variable to have its own unique impact on the reported amount of snowfall. This is called an *interaction* between the source of the snow report and the time of week.

Armed with empirical evidence of biased reporting, these researchers concluded that skiers and snowboarders could fight back. Zinman and Zitzewitz (2009) wrote: "Near the end of our sample period, a new iPhone application feature makes it easier for skiers to share information on ski conditions in real time. Exaggeration falls sharply, especially at resorts with better iPhone reception." It's a great reminder that we can question reported data, create new data, and take advantage of social media to share information in a meaningful way.

With two independent variables and one dependent variable, the experiment contains three comparisons that might influence a consumer's product evaluation: (1) the source of the snowfall report (ski resort versus weather station); (2) the time of week (weekday versus weekend); and (3) the combined effects of the source of snowfall report *and* the time of week. In this study, researchers discovered that the combined effects were the most important of the influences on the amount of snowfall reported. The effect of the source of the snowfall report on the amount of snow reported depended on whether it was a weekday or a weekend.

In this chapter, we examine a hypothesis test that checks for the presence of combined effects, also called *interactions. A statistical* ***interaction*** *occurs in a factorial design when two or more independent variables have an effect in combination that we do not see when we examine each independent variable on its own.* We learn about interactions in relation to a research design that has *two* nominal (or sometimes ordinal) independent variables, one scale dependent variable, and a between-groups design. We learn how the six steps of hypothesis testing apply to this statistical test, and how to calculate effect sizes for this test.

- A statistical **interaction** occurs in a factorial design when two or more independent variables have an effect in combination that we do not see when we examine each independent variable on its own.
- A **two-way ANOVA** is a hypothesis test that includes two nominal independent variables, regardless of their numbers of levels, and a scale dependent variable.
- A **factorial ANOVA** is a statistical analysis used with one scale dependent variable and at least two nominal independent variables (also called *factors*); also called a *multifactorial ANOVA.*
- **Factor** is a term used to describe an independent variable in a study with more than one independent variable.

Two-Way ANOVA

Our decisions about skiing and many other behaviors are routinely influenced by multiple variables, so we need a way to measure the interactive effects of multiple variables.

A two-way ANOVA (analysis of variance) allows us to compare levels from two independent variables plus the joint effects of those two variables. *A* ***two-way ANOVA*** *is a hypothesis test that includes two nominal independent variables, regardless of their numbers of levels, and a scale dependent variable.* We can also have ANOVAs with more than two independent variables. As the number of independent variables increases, the number increases in the name of the ANOVA—three-way, four-way, five-way, and so on. Table 12-1 shows a range of possibilities for naming ANOVAs.

Regardless of the number of independent variables, we can have different research designs. As with other hypothesis tests, a between-groups design is one in which every participant is in only one condition, and a within-groups design is one in which every participant is in all conditions. A mixed design is one in which one of the independent variables is between-groups and one is within-groups. In this chapter, we focus on the ANOVA that uses the second adjective from column 1 and the first adjective from column 2: the two-way between-groups ANOVA.

Language Alert! There is a catchall phrase for two-way, three-way, and higher-order ANOVAs; any ANOVA with at least two independent variables can be called a ***factorial ANOVA****, a statistical analysis used with one scale dependent variable and at least two nominal independent variables* (also called *factors*). This is also called a *multifactorial ANOVA.* ***Factor*** *is another word used to describe an independent variable in a study with more than one independent variable.*

In this section, we learn more about the situations in which we use a two-way ANOVA, as well as the language that is used in reference to this type of hypothesis test. Then we talk about the three outcomes we can examine with a two-way ANOVA.

TABLE 12-1. How to Name an ANOVA

ANOVAs are typically described by two adjectives, one from the first column and one from the second. We always have one descriptor from each column. So, we could have a one-way between-groups ANOVA or a one-way within-groups ANOVA, a two-way between-groups ANOVA or a two-way within-groups ANOVA, and so on. If at least one independent variable is between-groups and at least one is within-groups, it is a mixed-design ANOVA.

Number of Independent Variables: Pick One	Participants in One or All Samples: Pick One	Always Follows Descriptors
One-way	Between-groups	ANOVA
Two-way	Within-groups	
Three-way	Mixed-design	

Why We Use Two-Way ANOVA

To understand the benefits of a two-way ANOVA, let's consider a specific example. Since the mid-1990s, numerous studies (e.g., Bailey & Dresser, 2004; Mitchell, 1999) have documented the potential for grapefruit juice to increase the blood levels of certain medications, sometimes to toxic levels, by boosting the absorption of one or more of the active ingredients. Even scarier, this potentially life-threatening increase cannot be predicted for a given individual. For that reason, many physicians suggest that patients who take a wide range of medications (from some blood pressure drugs to many antidepressants) avoid grapefruit juice entirely. One commonly used anticholesterol drug whose effect is moderately boosted, sometimes dangerously, by the consumption of grapefruit juice is Lipitor (e.g., Bellosta, Paoletti, & Corsini, 2004). Let's use this particular interaction to understand how a two-way ANOVA gives us much more information with far less effort and expense.

EXAMPLE 12.1

Let's say that an investigator, Dr. Goldstein, wanted to know how to treat cholesterol but only knew how to analyze hypothesis tests that used one independent variable, the one-way between-groups ANOVA. She would conduct one study to compare the effect of Lipitor on cholesterol levels with the effect of another drug or a placebo. Then she would conduct a second study to compare the effect of grapefruit juice on cholesterol levels with that of another beverage or with no beverage, a study that might not even make much sense; after all, no one is predicting grapefruit juice on its own to be a treatment for high cholesterol. So how could she discover whether Lipitor works differently when combined with grapefruit juice?

Lew Robertson/Corbis

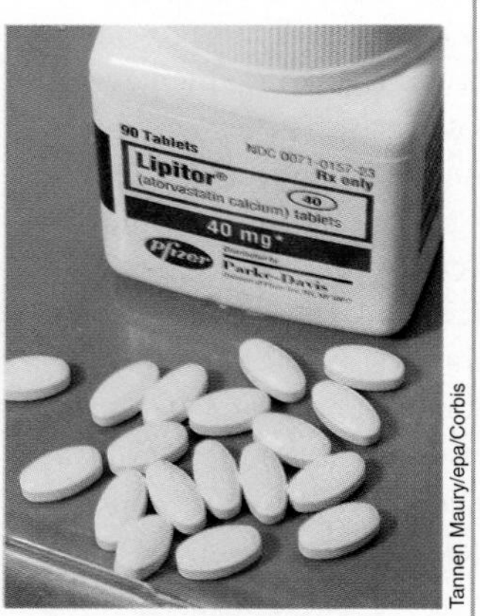

Tannen Maury/epa/Corbis

The Perils of Grapefruit Juice Studies have demonstrated that grapefruit juice (a level of one independent variable) can interact with many common medications (levels of a second independent variable) to cause higher levels of active ingredients (the dependent variable) to be absorbed into the bloodstream. The medical journals that physicians read report the results of such two-way ANOVAs because interactions have the potential to be toxic. This is why some physicians recommend that patients who take certain medications avoid grapefruit juice entirely.

A single study simultaneously examining medications like Lipitor *and* beverages like grapefruit juice is more efficient than two studies examining each independent variable separately. Two-way ANOVAs allow researchers to examine both hypotheses with the resources, time, and energy of a single study. But a two-way ANOVA yields even more information than two separate experiments.

Specifically, a two-way ANOVA allows researchers to explore exactly what Dr. Goldstein wanted to explore: interactions. Does the effect of some medications, but not others, depend on the particular levels of another independent variable, the beverages that accompany them? A two-way ANOVA can examine (1) the effect of Lipitor versus other medications, (2) the effect of grapefruit juice versus other beverages, *and* (3) the ways in which a drug and a juice might combine to create some entirely new, and often unexpected, effect. ■

The More Specific Vocabulary of Two-Way ANOVA

Every ANOVA, we learned, has two descriptors, one indicating the number of independent variables and one indicating the research design. Many researchers expand the first descriptor to provide even more information about the independent variables. Let's consider these expanded descriptors in the context of Dr. Goldstein's research.

TABLE 12-2. Interactions with Grapefruit Juice

A two-way ANOVA allows researchers to examine two independent variables, as well as the ways in which they might interact, simultaneously.

	Lipitor (L)	Zocor (Z)	Placebo (P)
Grapefruit Juice (G)	L & G	Z & G	P & G
Water (W)	L & W	Z & W	P & W

Were she to conduct just one study that examined both medication and beverage, she'd assign each participant to one level of medication (perhaps Lipitor, another cholesterol medication such as Zocor, or a placebo) *and* to one level of beverage (perhaps grapefruit juice or water). This research design is shown in Table 12-2.

When we draw the design of a study, such as in Table 12-2, we call each box of the design *a* ***cell****, a box that depicts one unique combination of levels of the independent variables in a factorial design*. When cells contain numbers, they are usually means of the scores of all participants who were assigned to that combination of levels. In Dr. Goldstein's study, participants are assigned to one of the six cells. Each participant is randomly assigned to one of the three levels of the variable medication listed in the columns of the table of cells.

Each participant is *also* assigned to one of the two levels of the variable beverage listed in the rows of the table of cells. A participant might be assigned to Lipitor and grapefruit juice (upper-left cell), placebo and water (lower-right cell), or any of the other four combinations.

Language Alert! This leads us to the new ANOVA vocabulary. Instead of the descriptor *two-way,* many researchers refer to an ANOVA with this arrangement of cells as a *3 × 2 ANOVA* (pronounced "three by two," not "three times two"). As with the *two-way* descriptor, the ANOVA is described with a second modifier—usually *between-groups* or *within-groups.* Because participants would receive only one medication and only one beverage, the hypothesis test could be called either a *two-way between-groups ANOVA* or a *3 × 2 between-groups ANOVA.* (An added benefit to the method of naming ANOVAs by the numbers of levels in each independent variable is the ease of calculating the total number of cells. Simply multiply the levels of the independent variables. In this case, the 3 × 2 ANOVA would have (3 × 2) = 6 cells.)

Two Main Effects and an Interaction

Two-way ANOVAs produce *three* F statistics: one for the first independent variable, one for the second independent variable, and one for the interaction between the two independent variables. The F statistics for each of the two independent variables describe *main effects. A* ***main effect*** *occurs in a factorial design when one of the independent variables has an influence on the dependent variable.* We evaluate whether there is a main effect by disregarding the influence of any other independent variables in the study—we temporarily pretend that the other variable doesn't exist.

So, with two independent variables, Dr. Goldstein would have two possibilities for a main effect. For example, after testing her participants in a two-way ANOVA, she might find a main effect of "type of medication," temporarily pretending that the variable beverage hasn't even been included in the study. For example, Lipitor and Zocor might both work better than the placebo at lowering cholesterol. That's the first F statistic. She also might find a main effect of "beverage," temporarily pretending that the variable type of medication hasn't even been included in the study. For example, drinking

- A **cell** is a box that depicts one unique combination of levels of the independent variables in a factorial design.
- A **main effect** occurs in a factorial design when one of the independent variables has an influence on the dependent variable.

grapefruit juice may reduce cholesterol, at least as compared to water. That's the second *F* statistic.

The third *F* statistic in a two-way ANOVA has the potential to be the most interesting because it is complicated by multiple, interacting variables. For example, Dr. Goldstein might find that both Lipitor and Zocor (but not the placebo) have more extreme effects on cholesterol when taken in combination with grapefruit juice versus water. In other words, the presence of the grapefruit juice changes the effects of Lipitor and Zocor, but not that of the placebo.

▶ MASTERING THE CONCEPT

12.1: In a two-way ANOVA, we test three different effects—two main effects, one for each independent variable, and one interaction, the joint effect of the two independent variables.

Each of the three *F* statistics has its own between-groups sum of squares (*SS*), degrees of freedom (*df*), mean square (*MS*), and critical value, but they all share a within-groups mean square (MS_{within}). The source table is shown in Table 12-3. The symbols in the body of the table are replaced by the specific values of these statistics in an actual source table.

TABLE 12-3. An Expanded Source Table

This source table is the framework into which we place the calculations for the two-way between-groups ANOVA with independent variables of medication and beverage. It tells three stories: the two main effects are listed first, then the interaction.

Source	*SS*	*df*	*MS*	*F*
Medication	$SS_{medication}$	$df_{medication}$	$MS_{medication}$	$F_{medication}$
Beverage	$SS_{beverage}$	$df_{beverage}$	$MS_{beverage}$	$F_{beverage}$
Medication × beverage	$SS_{medication \times beverage}$	$df_{medication \times beverage}$	$MS_{medication \times beverage}$	$F_{medication \times beverage}$
Within	SS_{within}	df_{within}	MS_{within}	
Total	SS_{total}	df_{total}		

CHECK YOUR LEARNING

Reviewing the Concepts

> Factorial ANOVAs are used with multiple independent variables because they allow us to examine several hypotheses in a single study and explore interactions.

> Factorial ANOVAs are often referred to by the levels of their independent variables (e.g., 2 × 2) rather than the number of independent variables (e.g., two-way). Sometimes the independent variables are called *factors.*

> A two-way ANOVA can have two main effects (one for each independent variable) and one interaction (the combined influence of both variables). Each effect and interaction has its own set of statistics, including its own *F* statistic, displayed in an expanded source table.

Clarifying the Concepts

12-1 What is a factorial ANOVA?

12-2 What is an interaction?

Calculating the Statistics

12-3 Determine how many factors are in each of the following designs:

a. The effect of three diet programs and two exercise programs on weight loss.

b. The effect of three diet programs, two exercise programs, and three different personal metabolism types on weight loss.

c. The effect of gift certificate value ($15, $25, $50, and $100) on the amount people spend over that value.

d. The effect of gift certificate value ($15, $25, $50, and $100) and store quality (low end versus high end) on consumer overspending.

Applying the Concepts

12-4 Adam Alter, a graduate student at Princeton University, and his advisor, Daniel Oppenheimer, studied whether names of stocks affected selling prices (Alter & Oppenheimer, 2006). They found that stocks with pronounceable ticker-code names, like "BAL," tended to sell at higher prices than did stocks with unpronounceable names, like "BDL." They examined this effect 1 day, 1 week, 6 months, and 1 year after the stock was offered for sale. The effect was strongest 1 day after the stocks were initially offered.

a. What are the "participants" in this study?

b. What are the independent variables and what are their levels?

c. What is the dependent variable?

d. Using the descriptors from Chapter 11, what would you call the hypothesis test that would be used?

e. Using the new descriptors from *this* chapter, what would you call the hypothesis test that would be used?

f. How many cells are there? Explain how you calculated this answer.

Solutions to these Check Your Learning questions can be found in Appendix D.

Understanding Interactions in ANOVA

Back in the 1970s, media attention about NASA's mission to Mars helped sell more Mars candy bars. Media attention primed the word *Mars,* which made Mars candy bars come to mind more easily. Priming, of course, would not increase sales for everybody—there would be exceptions to the rule. For example, diabetics who never eat chocolate and people who hate Mars bars would represent interacting exceptions to the rule that priming increases sales of Mars bars. In other words, any subgroup that represents a significant exception in any direction to the general trend in the data might indicate a statistical interaction. An interaction occurs when the effect of one independent variable depends on the specific level of another independent variable.

In this section, we explore the concept of an interaction in a two-way ANOVA in more depth. We look at a real-life example of an interaction, then introduce two different types of interactions—quantitative and qualitative.

Interactions and Public Policy

Hurricane Katrina demonstrates the importance of understanding interactions. First, the 2005 hurricane itself was an interaction among several weather variables. The devastating effects of the hurricane depended on particular levels of other variables, such as where it made landfall and the speed of its movement across the Gulf of Mexico.

Interactions were relevant for the people affected by Hurricane Katrina as well. For example, one would think that Hurricane Katrina would have been universally bad

AP Photo/Haraz N. Ghanbari

Disaster Relief and Pregnant Women This refugee from Hurricane Katrina and her newborn baby received care in a Baton Rouge, Louisiana, shelter that focused on pregnant women and newborn infants and their parents. Massive relief efforts sometimes mean that access to care for pregnant women is actually improved in the aftermath of a disaster. Of course, the quality of health care certainly doesn't improve for everybody, which means that an interaction is involved.

- A **quantitative interaction** is an interaction in which the effect of one independent variable is strengthened or weakened at one or more levels of the other independent variable, but the direction of the initial effect does not change.
- A **qualitative interaction** is a particular type of quantitative interaction of two (or more) independent variables in which one independent variable reverses its effect depending on the level of the other independent variable.
- A **marginal mean** is the mean of a row or a column in a table that shows the cells of a study with a two-way ANOVA design.

for the health of all those displaced people—a main effect of a hurricane on health care. However, there were exceptions to that rule, and three researchers from the Tulane University School of Public Health and Tropical Diseases in New Orleans proposed a startling interaction regarding the effects of the hurricane on health care for pregnant women (Buekens, Xiong, & Harville, 2006).

Some women gave birth in the squalor of the public shelter in New Orleans' Superdome or in alleys while waiting for rescuers. When it comes to pregnant women, the first priority of disaster relief agencies is to provide obstetric and neonatal care. Massive relief efforts sometimes mean that access to care for pregnant women is actually improved in the aftermath of a disaster. Of course, the quality of health care certainly doesn't improve for everybody, which means that an interaction is involved. So the quality of health care improved for pregnant women in the aftermath of the disaster, whereas it became worse for almost everyone else. In the language of two-way ANOVA, the effect of a disaster (one independent variable with two levels: disaster versus no disaster) on the quality of health care (the dependent variable) depends on the type of health care needed (the second independent variable, also with two levels: obstetric/neonatal versus all other types of health care).

Interpreting Interactions

The two-way between-groups ANOVA allows us to separate between-groups variance into three finer categories: the two main effects and an interaction effect. The interaction effect is a blended effect resulting from the interaction between the two independent variables; it is not a separate individual variable. The interaction effect is like mixing chocolate syrup into a glass of milk; the two foods blend into something familiar yet new.

Quantitative Interactions Two terms often used to describe interactions are *quantitative* and *qualitative* (e.g., Newton & Rudestam, 1999). *A **quantitative interaction** is an interaction in which the effect of one independent variable is strengthened or weakened at one or more levels of the other independent variable, but the direction of the initial effect does not change.* The researcher ski bums, Jonathan Zinman and Eric Zitzewitz, found a quantitative interaction when they discovered that resort snowfall reports were always exaggerated compared to weather reports, and especially on the weekends. More specifically, the effect of one independent variable is modified in the presence of another independent variable.

*A **qualitative interaction** is a particular type of quantitative interaction of two (or more) independent variables in which one independent variable reverses its effect depending on the level of the other independent variable.* In a qualitative interaction, the effect of one variable doesn't just become stronger or weaker; it actually reverses direction in the presence of another variable. Let's first examine the quantitative interaction.

> **MASTERING THE CONCEPT**
>
> **12.2:** Researchers often describe interactions with one of two terms—*quantitative* or *qualitative.* In a quantitative interaction, the effect of one independent variable is strengthened or weakened at one or more levels of the other independent variable, but the direction of the initial effect does not change. In a qualitative interaction, one independent variable actually reverses its effect depending on the level of the other independent variable.

EXAMPLE 12.2

The grapefruit juice example is a helpful illustration of a quantitative interaction. Lipitor and Zocor lead to elevations of some liver enzymes in combination with water, but the absorption levels are even higher with grapefruit juice. This effect is not seen with a placebo, which has an equal effect regardless of beverage. Let's invent some numbers to demonstrate this. The numbers in the cells in Table 12-4 don't represent actual absorption levels; rather, they are numbers that are easy for us to work with in our

TABLE 12-4. A Table of Means

We use a table to display the cell and marginal means so that we can interpret any main effects.

	Lipitor	Zocor	Placebo	
Grapefruit Juice	60	60	3	41
Water	30	30	3	21
	45	45	3	

understanding of interactions. For this exercise, we will consider every difference between numbers to be statistically significant. (Of course, if we really conducted this study, we would conduct the two-way ANOVA to determine exactly which effects were statistically significant.)

First, we consider main effects; then we consider the overall pattern that constitutes the interaction. If there is a significant interaction, we ignore any significant main effects. The significant interaction supersedes any significant main effects.

Table 12-4 includes mean absorption levels for the six cells of the study. It also includes numbers in the margins of the table, to the right of and below the cells; these numbers are also means, but for every participant in a given row or in a given column. Each of these is called a ***marginal mean***, *the mean of a row or a column in a table that shows the cells of a study with a two-way ANOVA design.* In Table 12-4, for example, the mean across from the row for grapefruit juice, 41, is the mean absorption level of every participant who was assigned to drink grapefruit juice, regardless of the assigned medication. The mean below the column for placebo, 3, is the mean absorption level of every participant who took the placebo, regardless of the assigned beverage. (Although we wouldn't expect any absorption with the placebo, we gave it a small value, 3, to facilitate the explanation of interactions.)

The easiest way to understand the main effects is to make a smaller table for each, with only the appropriate marginal means. Separate tables let us focus on one main effect at a time without being distracted by the means in the cells. For the main effect of beverage, we construct a table with two cells, as shown in Table 12-5. The table makes it easy to see that the absorption level was higher, on average, for grapefruit juice than for water.

Let's now consider the second main effect, that for medication. As before, we construct a table (such as Table 12-6) that shows only the means for medication, as if beverage were not included in the study. We kept the means for beverage in rows and for medication in columns, just as they were in the original table. You may, however, arrange them either way, using whichever makes sense to you. Table 12-6 shows that the absorption levels for Lipitor and Zocor were higher, on average, than for placebo, which led to almost no absorption. Both results would need to be verified with a hypothesis test, but we seem to have two main effects: (1) a main effect of beverage

TABLE 12-5. The Main Effect of Beverage

This table shows only the marginal means that demonstrate the main effect of beverage. Because we have isolated these marginal means, we cannot get distracted or confused by the other means in the table.

Grapefruit juice	41
Water	21

TABLE 12-6. The Main Effect of Medication

This table shows only the marginal means that demonstrate the main effect of medication. Because we have isolated these marginal means, we cannot get distracted or confused by the other means in the table.

Lipitor	Zocor	Placebo
45	45	3

TABLE 12-7. Examining the Overall Pattern of Means

A first step in understanding an interaction is examining the overall pattern of means in the cells.

	Lipitor	Zocor	Placebo
Grapefruit Juice	60	60	3
Water	30	30	3

(grapefruit juice leads to higher absorption, on average, than water does), and (2) a main effect of medication (Lipitor and Zocor lead to higher absorption, on average, than placebo does).

But that's not the whole story. Grapefruit juice, for example, does not lead to higher absorption, on average, among placebo users. Here's where the interaction comes in. Now we ignore the marginal means and get back to the means in the cells themselves, seen again in Table 12-7. Here we can see the overall pattern by framing it in two different ways. We can start by considering beverage. Does grapefruit juice boost mean absorption levels as compared to water? It depends. It depends on the level of the other independent variable, medication; specifically, it depends on whether the patient is taking one of the two medications or a placebo. We can also frame the question by starting with medication. Do Lipitor and Zocor boost mean absorption levels as compared to a placebo? It depends. They do anyway, even when just drinking water, but they do so to a far greater degree when drinking grapefruit juice. This is a quantitative interaction because the *strength* of the effect varies under certain conditions, but the *direction* does not.

People sometimes perceive an interaction where there is none. If Lipitor, Zocor, and a placebo all had higher mean absorption rates when drinking grapefruit juice (versus water), there would be no interaction. Lipitor and Zocor would *always* lead to a particular increase in average absorption levels versus a placebo—this would occur in the presence of any beverage. And grapefruit juice would *always* lead to a particular increase in average absorption levels versus water—this would occur in the presence of any medication. On the other hand, there is an interaction in the example we have been considering because grapefruit juice has a special effect with the two medications that it does not have with the placebo. The tendency to see an interaction when there is none can be diminished by constructing a bar graph, as in Figure 12-2.

The bar graph helps us to see the overall pattern, but one more step is necessary: to connect each set of bars with a line. We have two choices that match the two ways

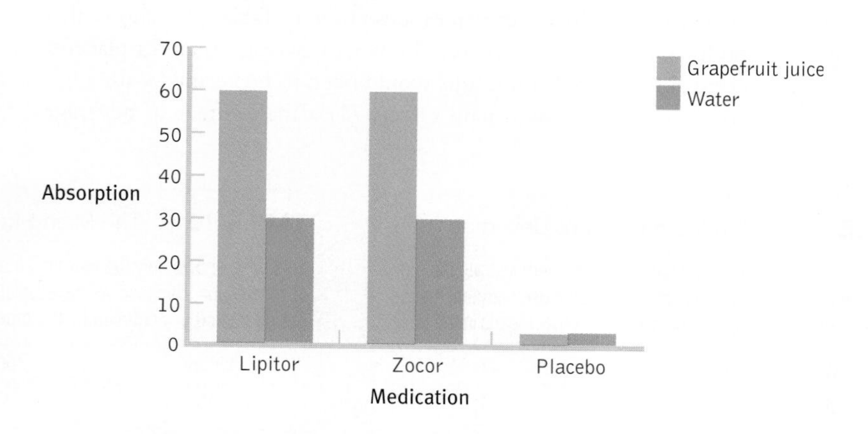

FIGURE 12-2
Bar Graphs and Interactions

Bar graphs help us determine if there really is an interaction. The bars in this graph help us to see that, among those taking the placebo, absorption is the same whether the placebo is accompanied by grapefruit juice or water, whereas among those taking Lipitor or Zocor, absorption is higher when accompanied by grapefruit juice than when accompanied by water.

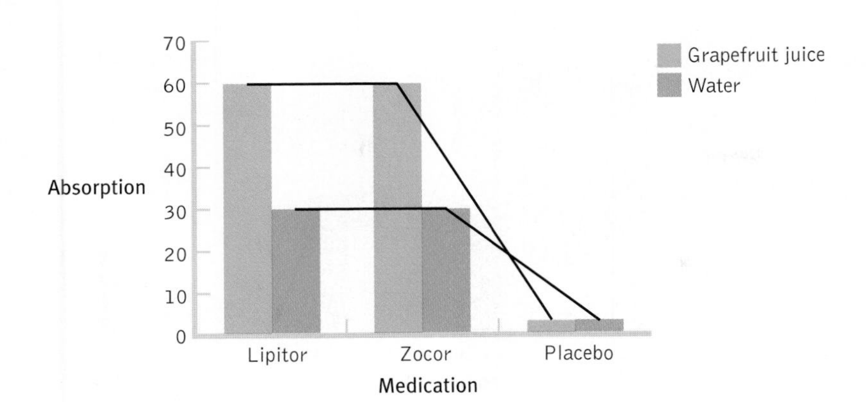

FIGURE 12-3
Are the Lines Parallel? Part I

We add lines to bar graphs to help us determine whether there really is an interaction. We draw a line connecting the three medications under the grapefruit juice condition. We then draw a line connecting the three medications under the water condition. These two lines intersect, an indication of an interaction that can be confirmed by conducting a hypothesis test.

we framed the interaction in words above: (1) As in Figure 12-3, we could connect the bars for the first independent variable, medication. We would connect the three medications for the grapefruit condition, and then we would connect the three medications for the water condition. (2) Alternatively, as in Figure 12-4, we could connect the bars for the two beverages. We would connect the two beverages for Lipitor, for Zocor, and for placebo.

In Figure 12-4, notice that the lines do not intersect, but they're not all parallel to one another either. If the lines were extended far enough, eventually the lines connecting the two bars for each medication would intersect the line connecting the bar for placebo. Perfectly parallel lines indicate the likely absence of an interaction, but we almost never see perfectly parallel lines emerging from real-life data sets; real-life data are usually messy. Nonparallel lines may indicate a statistically significant interaction, but we have to conduct an ANOVA to be sure. Only if the lines are significantly different from parallel can we reject the null hypothesis that there is no interaction; and we only want to interpret an interaction if we reject the null hypothesis.

Some social scientists refer to an interaction as a significant difference in differences. In the context of the grapefruit juice study, the mean difference between grapefruit juice and water is larger when participants are taking one of the medications than when they are taking the placebo. In fact, for those taking the placebo, there is no mean difference between grapefruit juice and water. This is an example of a significant difference between differences. This interaction is represented graphically whenever the lines connecting the bars are significantly different from parallel.

However, if grapefruit juice also led to an increase in mean absorption levels among those taking the placebo, the graph would look like the one in Figure 12-5. In this case,

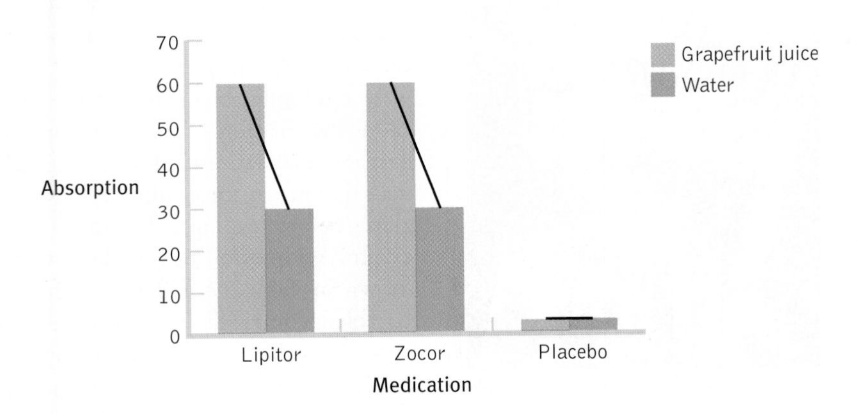

FIGURE 12-4
Are the Lines Parallel? Part II

There are always two ways to examine the pattern of the bar graphs. Here, we have drawn three lines, one connecting the two beverages under each of the three medication conditions. Were the three lines to continue, the two medication lines would eventually intersect with the horizontal placebo line, an indication of an interaction that can be confirmed by conducting a hypothesis test.

FIGURE 12-5
Parallel Lines
The three lines are exactly parallel. Were they to continue indefinitely, they would never intersect. Were this true among the population (not just this sample), there would be no interaction.

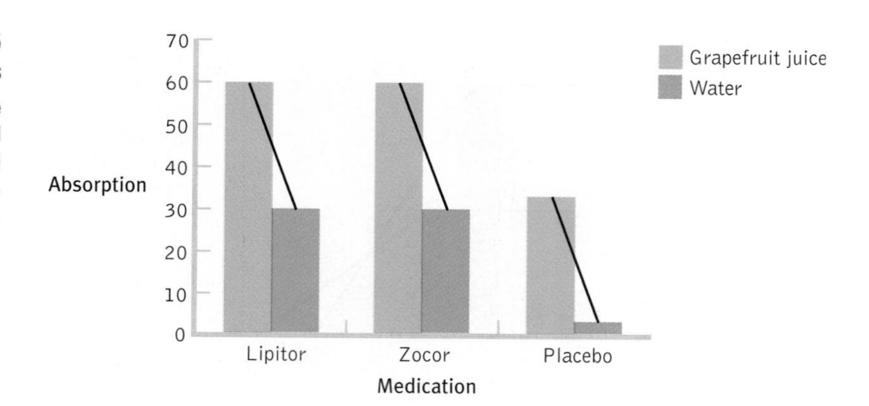

the mean absorption levels of Lipitor and Zocor do increase with grapefruit juice, but so does the mean absorption level of the placebo, and there is likely no interaction. Grapefruit juice has the same effect, regardless of the level of the other independent variable of medication. When in doubt about whether there is an interaction or just two main effects that add up to a greater effect, draw a graph and connect the bars with lines. ■

Qualitative Interactions Let's recall the definition of a qualitative interaction: a particular type of quantitative interaction in which the effect of one independent variable reverses its effect depending on the level of the other independent variable.

EXAMPLE 12.3

Do you think that, on average, people make better decisions when they consciously focus on the decision? Or do they make better decisions when the decision-making process is unconscious (i.e., making the decision after being distracted by other tasks)?

Researchers in the Netherlands conducted a series of studies in which participants were asked to decide between two options following either conscious or unconscious thinking about the choice. The studies were analyzed with two-way ANOVAs (Dijksterhuis, Bos, Nordgren, & van Baaren, 2006).

Choosing the Best Car When making decisions, such as which car to buy, do we make better choices after conscious or unconscious deliberation? Research by Dijksterhuis and colleagues (2006) suggests that less complex decisions are typically better after conscious deliberation, whereas more complex decisions are typically better after unconscious deliberation.

In one study, participants were asked to choose one of four cars. One car was objectively the best of the four, and one was objectively the worst. Some participants made a less complex decision; they were told 4 characteristics of each car. Some participants made a more complex decision; they were told 14 characteristics of each car. After learning about the cars, half the participants in each group were randomly assigned to think consciously about the cars for 4 minutes before making a decision. Half were randomly assigned to distract themselves for 4 minutes by solving anagrams before making a decision. The research design, with two independent variables, is shown in Table 12-8. The first independent variable is complexity, with two levels: less complex (4 attributes) and more complex (14 attributes). The second independent variable is

TABLE 12-8. A Two-Way Between-Groups ANOVA

Dutch researchers designed a study to examine what style of decision making led to the best choices in less complex and more complex situations. Would you predict an interaction? In other words, would the lines connecting bars on a graph be different from parallel? And if they are different from parallel, how are they different? If they are different just in strength, we are predicting a quantitative interaction. If the direction of effect actually reverses, we are predicting a qualitative interaction.

	Conscious Thought	Unconscious Thought (Distraction)
Less Complex (4 Attributes of Each Car)	less complex; conscious	less complex; unconscious
More Complex (12 Attributes of Each Car)	more complex; conscious	more complex; unconscious

type of decision making, with two levels: conscious thought and unconscious thought (distraction).

The researchers calculated a score for each participant that reflected his or her ability to differentiate between the objectively best and objectively worst cars in the group. This score represents the dependent variable, and higher numbers indicate a better ability to differentiate between the best and worst cars. Let's look at Table 12-9, which presents cell means and marginal means for this experiment. Note that the means are approximate and that the marginal means are created by assuming the same number of participants in each cell. As we consider these findings, we will assume that all differences are statistically significant. (In a real research situation, we would conduct an ANOVA to determine statistical significance.)

Because there was an overall pattern—an interaction—the researchers did not pay attention to the main effects in this study; an interaction trumps any main effects. However, let's examine the main effects to get some practice. We'll create tables for each of the two main effects so that we can examine them independently (Tables 12-10 and 12-11). The marginal means indicate that when type of decision making is ignored

TABLE 12-9. Decision-Making Tactics

To understand the main effects and overall pattern of a two-way ANOVA, we start by examining the cell means and marginal means.

	Conscious Thought	Unconscious Thought (Distraction)	
Less Complex	5.5	2.3	3.9
More Complex	0.6	5.0	2.8
	3.05	3.65	

TABLE 12-10. Main Effect of Complexity of Decision

These marginal means suggest that, overall, participants are better at making less complex decisions.

Less complex	3.9
More complex	2.8

TABLE 12-11. Main Effect of Type of Decision Making

These marginal means suggest that, overall, participants are better at making decisions when the decision making is unconscious—that is, when they are distracted.

Conscious	Unconscious
3.05	3.65

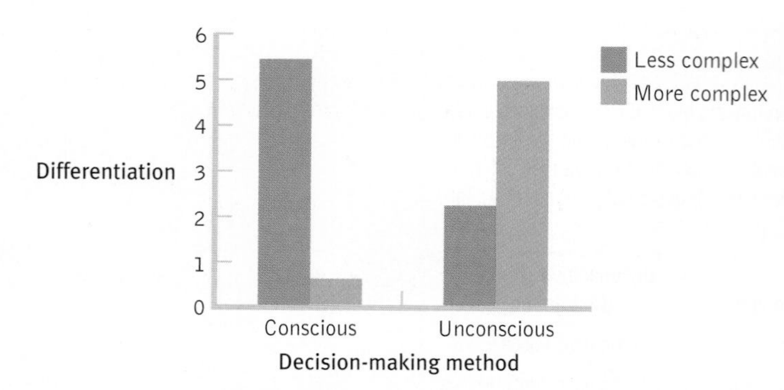

FIGURE 12-6
Graphing Decision-Making Methods

This bar graph displays the interaction far better than a table or words. We can see that it is a qualitative interaction; there is an actual reversal of direction of the effect of decision-making method in less complex versus more complex situations.

entirely, people make better decisions, on average, in less complex situations than in more complex situations. Further, the marginal means also suggest that, when complexity of decision is ignored, people make better decisions, on average, when the decision-making process is unconscious than when it is conscious.

However, if there is also a significant interaction, these main effects don't tell the whole story. The overall pattern of cell means renders this knowledge misleading, even inaccurate, under certain conditions. The interaction demonstrates that the effect of the decision-making method *depends* on the complexity of the decision. Conscious decision making tends to be better than unconscious decision making in less complex situations, but unconscious decision making tends to be better than conscious decision making in more complex situations. This reversal of direction is what makes this a qualitative interaction. It's not just the strength of the effect that changes, but the actual direction!

A bar graph, shown in Figure 12-6, makes the pattern of the data far clearer. We can actually see the qualitative interaction.

As with a quantitative interaction, we add lines to determine whether they are parallel (no matter how long the lines are) or intersect (or would do so if extended far enough), as in Figure 12-7. Here we see that the lines intersect without even having to extend them beyond the graph. This is likely an interaction. Type of decision making has an effect on differentiation between best and worst cars, but it depends on the complexity of the decision. Those making a less complex decision tend to make better choices if they use conscious thought. Those making a more complex decision tend to make better choices if they use unconscious thought. We would, as usual, verify this finding by conducting a hypothesis test before rejecting the null hypothesis that there is no interaction.

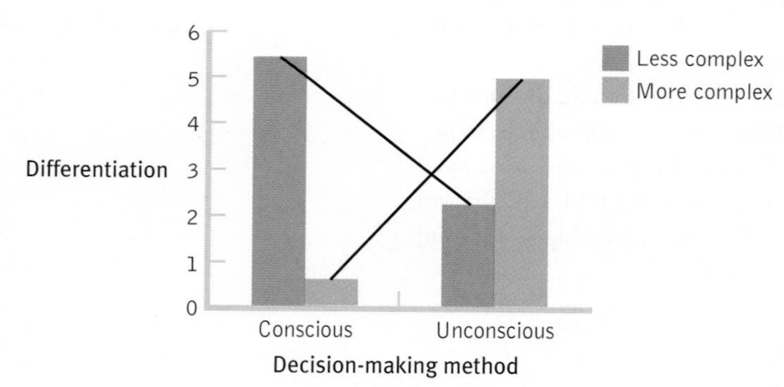

FIGURE 12-7
The Intersecting Lines of a Qualitative Interaction

When we draw two lines, one for the two bars that represent less complex situations and one for the two bars that represent more complex situations, we can easily see that they intersect. Lines that intersect, or would intersect if we extended them, indicate a possible interaction.

The qualitative interaction of decision-making method and complexity of situation was not likely to have been predicted by common sense. In such instances, we should be cautious before generalizing the findings. In this case, the research was carefully conducted and the researchers replicated their findings across several situations. For example, the researchers found similar effects in a real-life context when the less complex situation was shopping at a department store that sold clothing and kitchen products, and the more complex situation was purchasing furniture at IKEA. Such an intriguing finding would not have been possible without the inclusion of two independent variables in one study, which required that researchers use a two-way ANOVA capable of testing for an interaction.

So what do these findings mean for us as we approach the decisions we face every day? Which sunblock should we buy to best protect against UV rays? Should we go to graduate school or get a job following graduation? Should we consciously consider characteristics of sunblocks but "sleep on" graduate school–related factors? Research would suggest that the answer to the last question is yes (Dijksterhuis et al., 2006). Yet if the history of social science research is any indication, there are other factors not included in these studies that likely could affect the quality of our decisions. And so the research process continues. ■

CHECK YOUR LEARNING

Reviewing the Concepts

- A two-way ANOVA is represented by a grid in which cells represent each unique combination of independent variables. Means are calculated for cells, called *cell means.* Means are also computed for each level of an independent variable, by itself, regardless of the levels of the other independent variable. These means, found in the margins of the grid, are called *marginal means.*
- When there is a statistically significant interaction, the main effects are considered to be modified by an interaction. As a result, we focus only on the overall pattern of cell means that reveals the interaction.
- Two categories of interactions describe the overall pattern of cell means—*quantitative* and *qualitative* interactions.
- The most common interaction is a quantitative interaction in which the effect of the first independent variable depends on the levels of the second independent variable, but the differences at each level vary only in the strength of the effect.
- Qualitative interactions are those in which the effect of the first independent variable depends on the levels of the second independent variable, but the direction of the effect actually reverses.
- There are three ways to identify a statistically significant interaction: (1) visually, whenever the lines connecting the means of each group are significantly different from parallel; (2) conceptually, when you need to use the idea of "it depends" to tell the data's story; and (3) statistically, when the p value associated with an interaction in a source table is < 0.05, as with other hypothesis tests. This last option, the statistical analysis, is the only objective way to assess the interaction.

Clarifying the Concepts

12-5 What is the difference between a quantitative interaction and a qualitative interaction?

12-6 Why are main effects ignored when there is an interaction? (We often say they are *trumped* by an interaction.)

Calculating the Statistics

12-7 Data are presented here for two hypothetical independent variables (IVs) and their combinations.

IV 1, level A; IV 2, level A: 2, 1, 1, 3
IV 1, level B; IV 2, level A: 5, 4, 3, 4
IV 1, level A; IV 2, level B: 2, 3, 3, 3
IV 1, level B; IV 2, level B: 3, 2, 2, 3

a. Figure out how many cells are in this study's table, and draw a grid to represent them.

b. Calculate cell means and write them in the cells of the grid.

c. Calculate marginal means and write them in the margins of the grid.

d. Draw a bar graph of these data.

Applying the Concepts

12-8 For each of the following situations involving a real-life interaction, (i) state the independent variables, (ii) state the likely dependent variable, (iii) construct a table showing the cells, and (iv) explain whether the interaction is qualitative or quantitative.

a. Caroline and Mira are both really smart and do equally well in their psychology class, but something happens to Caroline when she goes to their philosophy class. She just can't keep up, whereas Mira does even better.

b. Our college baseball team has had a great few years. The team plays especially well at home versus away if playing teams in its own conference. However, it plays especially well at away games (versus home games) if playing teams from another conference.

c. Caffeinated drinks get me wired and make it somewhat difficult to sleep. So does working out in the evenings. When I do both, I'm so wired that I might as well stay up all night.

Solutions to these Check Your Learning questions can be found in Appendix D.

Conducting a Two-Way Between-Groups ANOVA

Advertising agencies understand that interactions can help them target their advertising campaigns. For example, researchers demonstrated that an increased exposure to *dogs* (linked in our memories to *cats* through familiar phrases such as "it's raining cats and dogs") positively influenced people's evaluations of Puma sneakers (a brand whose name refers to a cat), but only for people who recognized the Puma logo (Berger & Fitzsimons, 2008). The interaction between frequency of exposure to dogs (one independent variable) and whether or not someone could recognize the Puma logo (a second independent variable) *combined* to create a more positive evaluation of Puma sneakers (the dependent variable). Once again, both independent variables were needed to produce an interaction.

Behavioral scientists explore interactions by using two-way ANOVAs. Fortunately, hypothesis testing for a two-way between-groups ANOVA uses the same logic as for a one-way between-groups ANOVA. For example, the null hypothesis is exactly the same: There are no mean differences between groups. Type I and Type II errors still pose the same threats to decision making. We compare an F statistic to a critical F value to make a decision. The main way that a two-way ANOVA differs from a one-way ANOVA is that three ideas are being tested and each idea is a separate source of variability.

The three ideas being tested in a two-way between-groups ANOVA are the main effect of the first independent variable, the main effect of the second independent variable, and the interaction effect of the two independent variables. A fourth source of variability in a two-way ANOVA is within-groups variance. Let's learn how to separate and measure these four sources of variance by evaluating a commonly used educational method to improve public health: myth busting.

The Six Steps of Two-Way ANOVA

Two-way ANOVAs use the same six hypothesis-testing steps that you already know. The main difference is that you are essentially doing most of the steps three times—once for each main effect and once for the interaction. Let's look at an example.

EXAMPLE 12.4

Does myth busting really improve public health? Here are some myths and facts.

From the Web site of the Headquarters Counseling Center (2005) in Lawrence, Kansas:

> *Myth:* "Suicide happens without warning."
>
> *Fact:* "Most suicidal persons talk about and/or give behavioral clues about their suicidal feelings, thoughts, and intentions."

From the Web site for the World Health Organization (2007):

> *Myth:* "Disasters bring out the worst in human behavior."
>
> *Fact:* "Although isolated cases of antisocial behavior exist, the majority of people respond spontaneously and generously."

A group of Canadian researchers examined the effectiveness of myth busting (Skurnik, Yoon, Park, & Schwarz, 2005). They wondered whether the effectiveness of debunking false medical claims depends on the age of the person targeted by the message. In one study, they compared two groups of adults: younger adults, ages 18–

Don Smetzer/Getty Images

Do We Remember the Medical Myth or the Fact? Skurnik and colleagues (2005) studied the factors that influence the misremembering of false medical claims as facts. They asked: When a physician tells a patient a false claim, then debunks it with the facts, does the patient remember the false claim or the facts? A source table examines each factor in the study and tell us how much of the variability in the dependent variable is explained by that factor.

25, and older adults, ages 71–86. Participants were presented with a series of claims and were told that each claim was either true or false. (In reality, all claims were true, partly because researchers did not want to run the risk that participants would misremember false claims as being true.) In some cases, the claim was presented once, and in other cases, it was repeated three times. In either case, the accurate information was presented after each "false" statement. (Note that we have altered the study's design somewhat to make the study simpler for our purposes.)

The two independent variables in this study were age, with two levels (younger, older), and number of repetitions, with two levels (once, three times). The dependent variable, proportion of responses that were wrong after a 3-day delay, was calculated for each participant. This was a two-way between-groups ANOVA—more specifically, a 2 × 2 between-groups ANOVA. From this name, we know that the table has four cells: (2 × 2) = 4. There were 64 participants—16 in each cell. But here, we use an example with 12 participants—3 in each cell. Here are the data that we'll use; they have similar means to those in the actual study, and the F statistics are similar as well.

Experimental Conditions	Proportion of Responses That Were Wrong	Mean
Younger, one repetition	0.25, 0.21, 0.14	0.20
Younger, three repetitions	0.07, 0.13, 0.16	0.12
Older, one repetition	0.27, 0.22, 0.17	0.22
Older, three repetitions	0.33, 0.31, 0.26	0.30

Let's consider the steps of hypothesis testing for a two-way between-groups ANOVA in the context of this example.

STEP 1: Identify the populations, distribution, and assumptions.

The first step of hypothesis testing for a two-way between-groups ANOVA is very similar to that for a one-way between-groups ANOVA. First, we state the populations, but we specify that they are broken down into more than one category. In the current example, there are four populations, so there are four cells (as shown in Table 12-12). As we do the calculations, the first independent

TABLE 12-12. Studying the Memory of False Claims Using a Two-Way ANOVA

The study of memory for false claims has two independent variables: age (younger, older) and number of repetitions (one, three).

	One Repetition (1)	Three Repetitions (3)
Younger (Y)	Y; 1	Y; 3
Older (O)	O; 1	O; 3

variable, age, appears in the rows of the table, and the second independent variable, number of repetitions, appears in the columns of the tables.

There are four populations, each with labels representing the levels of the two independent variables to which they belong.

Population 1 (Y; 1): Younger adults who hear one repetition of a false claim.

Population 2 (Y; 3): Younger adults who hear three repetitions of a false claim.

Population 3 (O; 1): Older adults who hear one repetition of a false claim.

Population 4 (O; 3): Older adults who hear three repetitions of a false claim.

We next consider the characteristics of the data to determine the distributions to which we compare the sample. We have more than two groups, so we need to consider variances to analyze differences among means. Therefore, we use *F* distributions. Finally, we list the hypothesis test that we use for those distributions and check the assumptions for that test. For *F* distributions, we use ANOVA—in this case, two-way between-groups ANOVA.

The assumptions are the same for all types of ANOVA. The sample should be selected randomly; the populations should be distributed normally; and the population variances should be equal. Let's explore that a bit further.

(1) These data were not randomly selected. Younger adults were recruited from a university, and older adults were recruited from the local community. Because random sampling was not used, we must be cautious when generalizing from these samples. (2) The researchers did not report whether they investigated the shapes of the distributions of their samples to assess the shapes of the underlying populations. (3) The researchers did not provide standard deviations of the samples as an indication of whether the population spreads might be approximately equal, a condition known as *homoscedasticity*. We typically explore these assumptions using the sample data.

Summary: Population 1 (Y; 1): Younger adults who hear one repetition of a false claim. Population 2 (Y; 3): Younger adults who hear three repetitions of a false claim. Population 3 (O; 1): Older adults who hear one repetition of a false claim. Population 4 (O; 3): Older adults who hear three repetitions of a false claim.

The comparison distributions will be *F* distributions. The hypothesis test will be a two-way between-groups ANOVA. Assumptions: (1) The data are not from random samples, so we must generalize only with caution. (2) From the published research report, we do not know if the underlying population distributions are normal. (3) We do not know if the population variances are approximately equal (homoscedasticity).

STEP 2: State the null and research hypotheses.

The second step, to state the null and research hypotheses, is similar to that for a one-way between-groups ANOVA, except that we now have three sets of hypotheses, one for each main effect and one for the interaction. Those for the two main effects are the same as those for the one effect of a one-way between-groups ANOVA (see the summary below). If there are only two levels, then we can simply say that the two levels are not equal; if there are only two levels and there is a statistically significant difference, the difference must be between those two levels. Note that because there are two independent variables, we clarify which variable we are referring to by using initial letters or abbreviations for the levels of each (e.g., *Y* for younger and *O* for older). If an independent variable has more than two levels, the research hypothesis would be that any two levels of the independent variable are not equal.

The hypotheses for the interaction are typically stated in words but not in symbols. The null hypothesis is that the effect of one independent variable is not dependent on the levels of the other independent variable. The research hypothesis is that the effect of one independent variable depends on the levels of the other independent variable. It does not matter which independent variable we list first (e.g., "the effect of age is not dependent. . . ." or "the effect of number of repetitions is not dependent. . . ."). Write the hypotheses in the way that makes the most sense to you.

Summary: The hypotheses for the main effect of the first independent variable, age, are as follows. Null hypothesis: On average, younger adults have the same proportion of responses that are wrong when remembering which claims are myths compared with older adults—$H_0: \mu_Y = \mu_O$. Research hypothesis: On average, younger adults have a different proportion of responses that are wrong when remembering which claims are myths compared with older adults—$H_1: \mu_Y \neq \mu_O$.

The hypotheses for the main effect of the second independent variable, number of repetitions, are as follows. Null hypothesis: On average, those who hear one repetition have the same proportion of responses that are wrong when remembering which claims are myths compared with those who hear three repetitions—$H_0: \mu_1 = \mu_3$. Research hypothesis: On average, those who hear one repetition have a different proportion of responses that are wrong when remembering which claims are myths compared with those who hear three repetitions—$H_1: \mu_1 \neq \mu_3$.

The hypotheses for the interaction of age and number of repetitions are as follows. Null hypothesis: The effect of number of repetitions is not dependent on the levels of age. Research hypothesis: The effect of number of repetitions depends on the levels of age.

STEP 3: Determine the characteristics of the comparison distribution.

The third step is similar to that of a one-way between-groups ANOVA, except that there are three comparison distributions, all of them *F* distributions. We need to provide the appropriate degrees of freedom for each of these: two main effects and one interaction. As before, each *F* statistic is a ratio of between-groups variance and within-groups variance. Because there are three effects, there are three between-groups variance estimates, each with its own degrees of freedom. There is only one within-groups variance estimate, with its degrees of freedom for all three.

For each main effect, the between-groups degrees of freedom is calculated as for a one-way ANOVA: the number of groups minus 1. The first independent variable, age, is in the rows of the table of cells, so the between-groups degrees of freedom is:

$$df_{rows(age)} = N_{rows} - 1 = 2 - 1 = 1$$

MASTERING THE FORMULA

12-1: The formula for the between-groups degrees of freedom for the independent variable in the rows of the table of cells is: $df_{rows} = N_{rows} - 1$. We subtract 1 from the number of rows, representing levels, for that variable.

MASTERING THE FORMULA

12-2: The formula for the between-groups degrees of freedom for the independent variable in the columns of the table of cells is: $df_{columns} = N_{columns} - 1$. We subtract 1 from the number of columns, representing levels, for that variable.

MASTERING THE FORMULA

12-3: The formula for the between-groups degrees of freedom for the interaction is: $df_{interaction} = (df_{rows})(df_{columns})$. We multiply the degrees of freedom for each of the independent variables.

MASTERING THE FORMULA

12-4: To calculate the within-groups degrees of freedom, we first calculate degrees of freedom for each cell. That is, we subtract one from the number of participants in each cell. We then sum the degrees of freedom for each cell. For a study in which there are four cells, we'd use this formula: $df_{within} = df_{cell\ 1} + df_{cell\ 2} + df_{cell\ 3} + df_{cell\ 4}$.

MASTERING THE FORMULA

12-5: There are two ways to calculate the total degrees of freedom. We can subtract 1 from the total number of participants in the entire study: $df_{total} = N_{total} - 1$. We can also add the three between-groups degrees of freedom and the within-groups degrees of freedom. It's a good idea to calculate it both ways as a check on our work.

The second independent variable, number of repetitions, is in the columns of the table of cells, so the between-groups degrees of freedom is:

$$df_{columns(reps)} = N_{columns} - 1 = 2 - 1 = 1$$

We now need a between-groups degrees of freedom for the interaction, which is calculated by multiplying the degrees of freedom for the two main effects:

$$df_{interaction} = (df_{rows(age)})(df_{columns(reps)}) = (1)(1) = 1$$

The within-groups degrees of freedom is calculated like that for a one-way between-groups ANOVA, the sum of the degrees of freedom in each of the cells. In the current example, there are three participants in each cell, so the within-groups degrees of freedom is calculated as follows, with N representing the number in each cell:

$$df_{Y,1} = N - 1 = 3 - 1 = 2$$
$$df_{Y,3} = N - 1 = 3 - 1 = 2$$
$$df_{O,1} = N - 1 = 3 - 1 = 2$$
$$df_{O,3} = N - 1 = 3 - 1 = 2$$
$$df_{within} = df_{Y,1} + df_{Y,3} + df_{O,1} + df_{O,3} = 2 + 2 + 2 + 2 = 8$$

For a check on our work, we calculate the total degrees of freedom just as we did for the one-way between-groups ANOVA. We subtract 1 from the total number of participants:

$$df_{total} = N_{total} - 1 = 12 - 1 = 11$$

We now add up the three between-groups degrees of freedom and the within-groups degrees of freedom to see if they equal 11. In this case, they match:

$$11 = 1 + 1 + 1 + 8$$

Finally, for this step, we list the distributions with their degrees of freedom for the three effects. Note that, although the between-groups degrees of freedom for the three effects are the same in this case, they are often different. For example, if one independent variable had three levels and the other had four, the between-groups degrees of freedom for the main effects would be 2 and 3, respectively, and the between-groups degrees of freedom for the interaction would be 6.

Summary: Main effect of age: F distribution with 1 and 8 degrees of freedom. Main effect of number of repetitions: F distribution with 1 and 8 degrees of freedom. Interaction of age and number of repetitions: F distribution with 1 and 8 degrees of freedom. (*Note:* It is helpful to include all degrees of freedom calculations in this step.)

STEP 4: Determine the critical values, or cutoffs.

Again, this step for the two-way between-groups ANOVA is just an expansion of that for the one-way version. We now need three critical values but they're determined just as we determined them before. We use the F table in Appendix B.

For each main effect and for the interaction, we look up the within-groups degrees of freedom, which is always the same for each effect, along the left-hand side and the appropriate between-groups degrees of freedom across the top of the table. The place on the grid where this row and this column intersect contains three numbers. From top to bottom, the table provides cutoffs for p levels of 0.01, 0.05, and 0.10. As usual, we typically use 0.05. In this instance, it happens that the critical value is the same for all three effects because the between-groups degrees of freedom is the same for all three. But when the between-groups degrees of freedom are different, there are different critical values. Here, we look up the between-groups degrees of freedom of 1, within-groups degrees of freedom of 8, and p level of 0.05. The cutoff for all three is 5.32, as seen in Figure 12-8.

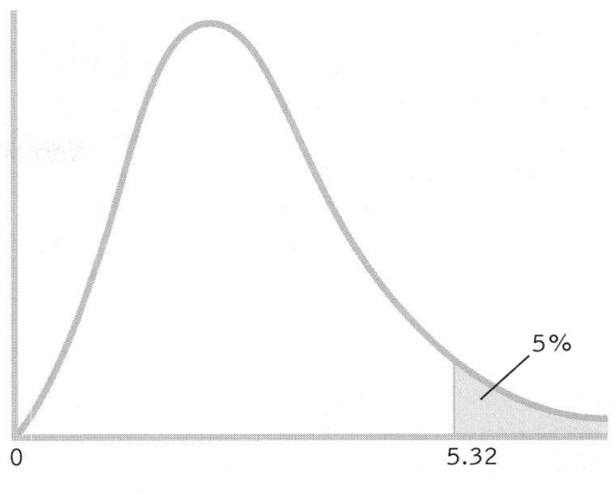

FIGURE 12-8
Determining Cutoffs for an *F* Distribution
We determine the critical values for an *F* distribution for a two-way between-groups ANOVA just as we did for a one-way between-groups ANOVA, except that we calculate three cutoffs, one for each main effect and one for the interaction. In this case, the between-groups degrees of freedom are the same for all three, so the cutoffs are the same.

Summary: There are three critical values (which in this case are all the same), as seen in the curve in Figure 12-8. The critical F value for the main effect of age is 5.32. The critical F value for the main effect of number of repetitions is 5.32. The critical F value for the interaction of age and number of repetitions is 5.32.

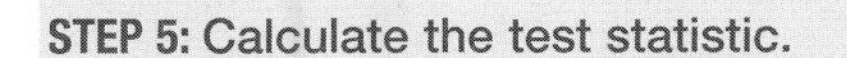
STEP 5: Calculate the test statistic. As with the one-way between-groups ANOVA, the fifth step for the two-way between-groups ANOVA is the most time consuming. As you might guess, it's similar to what we already learned, but we have to calculate three F statistics instead of one. We learn the logic and the specific calculations for this step in the next section.

STEP 6: Make a decision. This step is the same as for a one-way between-groups ANOVA, except that we compare each of the three F statistics to its appropriate cutoff F statistic. If the F statistic is beyond the critical value, then we know that it is in the most extreme 5% of possible test statistics *if* the null hypothesis is true. After making a decision for each F statistic, we present the results in one of three ways.

First, if we are able to reject the null hypothesis for the interaction, then we draw a specific conclusion with the help of a table and graph. Because we have more than two groups, we use a post hoc test, such as the one that we learned in Chapter 11 Because there are three effects, post hoc tests are typically implemented separately for each main effect and for the interaction (Hays, 1994). If the interaction is statistically significant, then it might not matter whether the main effects are also significant; if they are also significant, then those findings are usually qualified by the interaction, and are not described separately. The overall pattern of cell means tells the whole story.

Second, if we are not able to reject the null hypothesis for the interaction, then we focus on any significant main effects, drawing a specific directional conclusion for each. In this study, each independent variable has only two levels, so there is no need for a post hoc test. If there were three or more levels, however, then each significant main effect would require a post hoc test to determine exactly where the differences lie. Third, if we do not reject the null hypothesis for either main effect or the interaction, then we can only conclude that there is insufficient evidence from this study to support the research hypotheses. We will complete step 6 of hypothesis testing for this study in the next section, after we consider the calculations of the source table for a two-way between-groups ANOVA. ■

Identifying Four Sources of Variability in a Two-Way ANOVA

In this section, we complete step 5 for a two-way between-groups ANOVA. The calculations are similar to those for a one-way between-groups ANOVA, except that we calculate three F statistics. We use a source table with elements like those shown in Table 12-18, on page 329.

TABLE 12-13. Calculating the Total Sum of Squares

The total sum of squares is calculated by subtracting the overall mean, called the *grand mean,* from every score to create deviations, then squaring the deviations and summing them: $\Sigma(X - GM)^2 = 0.0672$.

	X	$(X - GM)$	$(X - GM)^2$
Y, 1	0.25	(0.25 − 0.21) = 0.04	0.0016
	0.21	(0.21 − 0.21) = 0.00	0.0000
	0.14	(0.14 − 0.21) = −0.07	0.0049
Y, 3	0.07	(0.07 − 0.21) = −0.14	0.0196
	0.13	(0.13 − 0.21) = −0.08	0.0064
	0.16	(0.16 − 0.21) = −0.05	0.0025
0, 1	0.27	(0.27 − 0.21) = 0.06	0.0036
	0.22	(0.22 − 0.21) = 0.01	0.0001
	0.17	(0.17 − 0.21) = −0.04	0.0016
0, 3	0.33	(0.33 − 0.21) = 0.12	0.0144
	0.31	(0.31 − 0.21) = 0.10	0.0100
	0.26	(0.26 − 0.21) = 0.05	0.0025

First, we calculate the total sum of squares (Table 12-13). We calculate this number in exactly the same way as for a one-way ANOVA. We subtract the grand mean, 0.21, from every score to create deviations, then square the deviations, and finally sum the squared deviations:

MASTERING THE FORMULA

12-6: We calculate the total sum of squares using the following formula: $SS_{total} = \Sigma(X - GM)^2$. We subtract the grand mean from every score to create deviations, then square the deviations, and finally sum the squared deviations.

$$SS_{total} = \Sigma(X - GM)^2 = 0.0672$$

We now calculate the between-groups sums of squares for the two main effects. Both are calculated similarly to the between-groups sum of squares for a one-way between-groups ANOVA. The table with the cell means, marginal means, and grand mean is shown in Table 12-14. The between-groups sum of squares for the main effect of the independent variable age would be the sum, for every score, of the marginal mean minus the grand mean, squared. We list all 12 scores in Table 12-15, marking the divisions among the cells. For each of the 6 younger participants, those in the top 6 rows of Table 12-15, we subtract the grand mean, 0.21, from the marginal mean, 0.16. For the 6 older participants, those in the bottom 6 rows, we subtract 0.21 from the marginal mean, 0.26. We square all of these deviations and

TABLE 12-14. Means for False Medical Claims Study

The study of the misremembering of false medical claims as true had two independent variables, age and number of repetitions. The cell means and marginal means for error rates are shown in the table. The grand mean is 0.21.

	One Repetition (1)	Three Repetitions (3)	
Younger (Y)	0.20	0.12	0.16
Older (O)	0.22	0.30	0.26
	0.21	0.21	0.21

TABLE 12-15. Calculating the Sum of Squares for the First Independent Variable

The sum of squares for the first independent variable is calculated by subtracting the overall mean (the grand mean) from the mean for each level of that variable—in this case, age—to create deviations, then squaring the deviations and summing them: $\Sigma(M_{row(age)} - GM)^2 = 0.03$.

	X	$(M_{row(age)} - GM)$	$(M_{row(age)} - GM)^2$
Y, 1	0.25	(0.16 − 0.21) = −0.05	0.0025
	0.21	(0.16 − 0.21) = −0.05	0.0025
	0.14	(0.16 − 0.21) = −0.05	0.0025
Y, 3	0.07	(0.16 − 0.21) = −0.05	0.0025
	0.13	(0.16 − 0.21) = −0.05	0.0025
	0.16	(0.16 − 0.21) = −0.05	0.0025
O, 1	0.27	(0.26 − 0.21) = 0.05	0.0025
	0.22	(0.26 − 0.21) = 0.05	0.0025
	0.17	(0.26 − 0.21) = 0.05	0.0025
O, 3	0.33	(0.26 − 0.21) = 0.05	0.0025
	0.31	(0.26 − 0.21) = 0.05	0.0025
	0.26	(0.26 − 0.21) = 0.05	0.0025

then add them to calculate the sum of squares for the rows, the independent variable of age:

$$SS_{between(rows)} = \Sigma(M_{row(age)} - GM)^2 = 0.03$$

We repeat this process for the second possible main effect, that of the independent variable in the columns (Table 12-16). The between-groups sum of squares for number

MASTERING THE FORMULA

12-7: We calculate the between-groups sum of squares for the first independent variable, that in the rows of the table of cells, using the following formula: $SS_{between(rows)} = \Sigma(M_{row} - GM)^2$. For every participant, we subtract the grand mean from the marginal mean for the appropriate row for that participant. We square these deviations, and sum the squared deviations.

TABLE 12-16. Calculating the Sum of Squares for the Second Independent Variable

The sum of squares for the second independent variable is calculated by subtracting the overall mean (the grand mean) from the mean for each level of that variable—in this case, number of repetitions—to create deviations, then squaring the deviations and summing them: $\Sigma(M_{column(reps)} - GM)^2 = 0$.

	X	$(M_{column(reps)} - GM)$	$(M_{column(reps)} - GM)^2$
Y, 1	0.25	(0.21 − 0.21) = 0	0
	0.21	(0.21 − 0.21) = 0	0
	0.14	(0.21 − 0.21) = 0	0
Y, 3	0.07	(0.21 − 0.21) = 0	0
	0.13	(0.21 − 0.21) = 0	0
	0.16	(0.21 − 0.21) = 0	0
O, 1	0.27	(0.21 − 0.21) = 0	0
	0.22	(0.21 − 0.21) = 0	0
	0.17	(0.21 − 0.21) = 0	0
O, 3	0.33	(0.21 − 0.21) = 0	0
	0.31	(0.21 − 0.21) = 0	0
	0.26	(0.21 − 0.21) = 0	0

of repetitions, then, would be the sum, for every score, of the marginal mean minus the grand mean, squared. We again list all 12 scores, marking the divisions among the cells. For each of the 6 participants who had one repetition, those in the left-hand column of Table 12-14 and in rows 1–3 and 7–9 of Table 12-16, we subtract the grand mean, 0.21, from the marginal mean, 0.21. For each of the 6 participants who had three repetitions, those in the right-hand column of Table 12-14 and in rows 4–6 and 10–12 of Table 12-16, we subtract 0.21 from the marginal mean, 0.21. (*Note:* It is a coincidence that in this case the marginal means are exactly the same.) We square all of these deviations and add them to calculate the between-groups sum of squares for the columns, the independent variable of number of repetitions. Again, the calculations for the between-groups sum of squares for each main effect are just like the calculations for a one-way between-groups ANOVA:

MASTERING THE FORMULA

12-8: We calculate the between-groups sum of squares for the second independent variable, that in the columns of the table of cells, using the following formula: $SS_{between(columns)} = \Sigma(M_{column} - GM)^2$. For every participant, we subtract the grand mean from the marginal mean for the appropriate column for that participant. We square these deviations, and then sum the squared deviations.

$$SS_{between(columns)} = \Sigma(M_{column(reps)} - GM)^2 = 0$$

The within-groups sum of squares is calculated in exactly the same way as for the one-way between-groups ANOVA (Table 12-17). The cell mean is subtracted from each of the 12 scores. The deviations are squared and summed:

MASTERING THE FORMULA

12-9: We calculate the within-groups sum of squares using the following formula: $SS_{within} = \Sigma(X - M_{cell})^2$. For every participant, we subtract the appropriate cell mean from that participant's score. We square these deviations, and sum the squared deviations.

$$SS_{within} = \Sigma(X - M_{cell})^2 = 0.018$$

All we need now is the between-groups sum of squares for the interaction. We calculate this by subtracting the other between-groups sums of squares (those for the two main effects) and the within-groups sum of squares from the total sum of squares. The between-groups sum of squares for the interaction is essentially what is left over when the main effects are accounted for. Mathematically, any variability that is predicted by

TABLE 12-17. Calculating the Within-Groups Sum of Squares

The within-groups sum of squares is calculated the same way for a two-way ANOVA as for a one-way ANOVA. We take each score and subtract the mean of the cell from which it comes—not the grand mean—to create deviations; then we square the deviations and sum them: $\Sigma(X - M_{cell})^2 = 0.018$.

	X	$\Sigma(X - M_{cell})$	$\Sigma(X - M_{cell})^2$
Y, 1	0.25	(0.25 − 0.20) = 0.05	0.0025
	0.21	(0.21 − 0.20) = 0.01	0.0001
	0.14	(0.14 − 0.20) = −0.06	0.0036
Y, 3	0.07	(0.07 − 0.12) = −0.05	0.0025
	0.13	(0.13 − 0.12) = 0.01	0.0001
	0.16	(0.16 − 0.12) = 0.04	0.0016
0, 1	0.27	(0.27 − 0.22) = 0.05	0.0025
	0.22	(0.22 − 0.22) = 0.00	0.0000
	0.17	(0.17 − 0.22) = −0.05	0.0025
0, 3	0.33	(0.33 − 0.30) = 0.03	0.0009
	0.31	(0.31 − 0.30) = 0.01	0.0001
	0.26	(0.26 − 0.30) = −0.04	0.0016

these variables, but is not directly predicted by either independent variable on its own, is attributed to the interaction. The formula is:

$$SS_{between(interaction)} = SS_{total} - (SS_{between(rows)} + SS_{between(columns)} + SS_{within})$$

The calculations are:

$$SS_{between(interaction)} = 0.0672 - (0.03 + 0 + 0.018) = 0.0192$$

Now we complete step 6 of hypothesis testing by calculating the F statistics using the formulas in Table 12-18. The results are in the source table (Table 12-19). The main effect of age is statistically significant because the F statistic, 13.04, is larger than the critical value of 5.32. The means tell us that older participants tend to make more mistakes, remembering more medical myths as true, than do younger participants. The main effect of number of repetitions is not statistically significant, however, because the F statistic of 0.00 is not larger than the cutoff of 5.32. It is unusual to have

MASTERING THE FORMULA

12-10: To calculate the between-groups sum of squares for the interaction, we subtract the two between-groups sums of squares for the independent variables and the within-groups sum of squares from the total sum of squares. The formula is: $SS_{between(interaction)} = SS_{total} - (SS_{between(rows)} + SS_{between(columns)} + SS_{within})$.

TABLE 12-18. The Expanded Source Table and the Formulas

This source table includes all of the formulas for the calculations necessary to conduct a two-way between-groups ANOVA.

Source	*SS*	*df*	*MS*	*F*
Age (between/rows)	$\Sigma(M_{between(rows)} - GM)^2$	$N_{rows} - 1$	$\frac{SS_{between(rows)}}{df_{between(rows)}}$	$\frac{MS_{between(rows)}}{MS_{within}}$
Repetitions (between/columns)	$\Sigma(M_{between(columns)} - GM)^2$	$N_{columns} - 1$	$\frac{SS_{between(columns)}}{df_{between(columns)}}$	$\frac{MS_{between(columns)}}{MS_{within}}$
Age × Repetitions (between/interaction)	$SS_{total} - (SS_{between(rows)} + SS_{between(columns)} + SS_{within})$	$(df_{rows})(df_{columns})$	$\frac{SS_{interaction}}{df_{interaction}}$	$\frac{MS_{interaction}}{MS_{within}}$
Within	$\Sigma(X - M_{cell})^2$	$df_{cell1} + df_{cell2} + df_{cell3} + df_{cell4}$ (and so on for any additonal cells)	$\frac{SS_{within}}{df_{within}}$	
Total	$\Sigma(X - GM)^2$	$N_{total} - 1$		

MASTERING THE FORMULA

12-11: The formulas to calculate the four mean squares are in Table 12-18. There are three between-groups mean squares—one for each main effect and one for the interaction—and one within-groups mean square. For each mean square, we divide the appropriate sum of squares by its related degrees of freedom. The formulas for the three F statistics, one for each main effect and one for the interaction, are also in Table 12-18. For each of the three effects, we divide the appropriate between-groups mean square by the within-group mean square. The denominator is the same in all three cases.

TABLE 12-19. The Expanded Source Table and False Medical Claims

This expanded source table shows the actual sums of squares, degrees of freedom, mean squares, and F statistics for the study on false medical claims.

Source	*SS*	*df*	*MS*	*F*
Age (A)	0.0300	1	0.0300	13.04
Repetitions (R)	0.0000	1	0.0000	0.00
A × R	0.0192	1	0.0190	8.26
Within	0.0180	8	0.0023	
Total	0.0672	11		

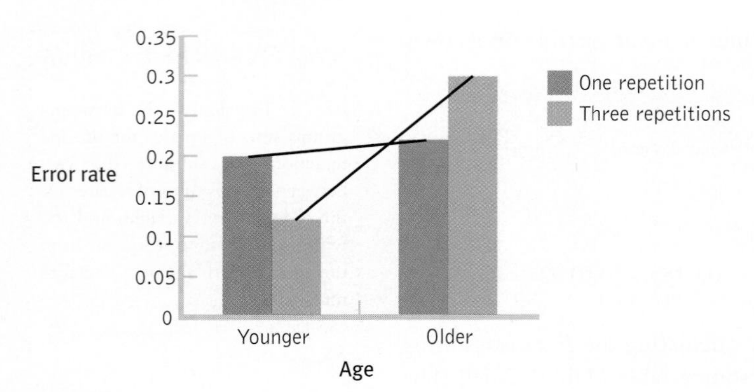

FIGURE 12-9
Interpreting the Interaction
The nonparallel lines demonstrate the interaction. The bars tell us that, on average, repetition decreases errors for younger people but increases them for older people. Because the direction reverses, this is a qualitative interaction.

an F statistic of 0.00. Even when there is no statistically significant effect, there is usually some difference among means due to random sampling. The interaction is also statistically significant because the F statistic of 8.35 is larger than the cutoff of 5.32. Therefore, we construct a bar graph of the cell means, as seen in Figure 12-9, to interpret the interaction.

In Figure 12-9, the lines are not parallel; in fact, they intersect without even having to extend them beyond the graph. We see that among younger participants, the proportion of responses that were incorrect was *lower,* on average, with three repetitions than with one repetition. Among older participants, the proportion of responses that were incorrect was *higher,* on average, with three repetitions than with one repetition. Does repetition help? It depends. It helps for younger people but is detrimental for older people. Specifically, repetition tends to help younger people distinguish between myth and fact. But the mere repetition of a medical myth tends to lead older people to be more likely to view it as fact. The researchers speculate that older people remember that they are familiar with a statement but forget the context in which they heard it. Because the direction of the effect of repetition reverses from one age group to another, this is a qualitative interaction.

Effect Size for Two-Way ANOVA

With a two-way ANOVA, as with a one-way ANOVA, we calculate R^2 as the measure of effect size. As before, we use sums of squares as indicators of variability. For each of the three effects—the two main effects and the interaction—we divide the appropriate between-groups sum of squares by the total sum of squares minus the sums of squares for both of the other effects. We subtract the sums of squares for the other two effects from the total so that we isolate the effect size for a single effect at a time. For example, if we want to determine effect size for the main effect in the rows, we divide the sum of squares for the rows by the total sum of squares minus the sum of squares for the column and the sum of squares for the interaction.

For the first main effect, the one in the rows of the table of cells, the formula is:

MASTERING THE FORMULA

12-12: To calculate effect size for two-way ANOVA, we make three R^2 calculations, one for each main effect and one for the interaction. In each case, we divide the appropriate between-groups sum of squares by the total sum of squares minus the sums of squares for the other two effects. For example, the effect size for the interaction is calculated using this formula: $R^2_{interaction} = \frac{SS_{interaction}}{(SS_{total} - SS_{rows} - SS_{columns})}$.

$$R^2_{rows} = \frac{SS_{rows}}{(SS_{total} - SS_{columns} - SS_{interaction})}$$

For the second main effect, the one in the columns of the table of cells, the formula is:

$$R^2_{columns} = \frac{SS_{columns}}{(SS_{total} - SS_{rows} - SS_{interaction})}$$

For the interaction, the formula is:

$$R^2_{interaction} = \frac{SS_{interaction}}{(SS_{total} - SS_{rows} - SS_{columns})}$$

EXAMPLE 12.5

Let's apply this to the ANOVA we just conducted. We use the statistics in the source table shown in Table 12-19 to calculate R^2 for each main effect and the interaction. Here are the calculations for the main effect for age:

$$R^2_{rows(age)} = \frac{SS_{rows(age)}}{(SS_{total} - SS_{columns(repetition)} - SS_{interaction})}$$

$$= \frac{0.0300}{(0.0672 - 0.000 - 0.0192)} = 0.625$$

Here are the calculations for the main effect for repetitions:

$$R^2_{columns(repetitions)} = \frac{SS_{columns(repetitions)}}{(SS_{total} - SS_{rows(age)} - SS_{interaction})}$$

$$= \frac{0.000}{(0.0672 - 0.0300 - 0.0192)} = 0.000$$

Here are the calculations for the interaction:

$$R^2_{interaction} = \frac{SS_{interaction}}{(SS_{total} - SS_{rows(age)} - SS_{columns(repetitions)})}$$

$$= \frac{0.0192}{(0.0672 - 0.0300 - 0.000)} = 0.516$$

The conventions are the same as those presented in Chapter 11, shown again here in Table 12-20. From this table, we can see that the R^2 of 0.63 for the main effect of age and 0.52 for the interaction are very large. The R^2 of 0.00 for the main effect of repetitions indicates that there is no observable effect in this study.

TABLE 12-20. Cohen's Conventions for Effect Sizes: R^2

The following guidelines, called *conventions* by statisticians, are meant to help researchers decide how important an effect is. These numbers are not cutoffs, merely rough guidelines to aid researchers in their interpretation of results.

Effect Size	Convention
Small	0.01
Medium	0.06
Large	0.14

CHECK YOUR LEARNING

Reviewing the Concepts

> The six steps of hypothesis testing for a two-way between-groups ANOVA are similar to those for a one-way between-groups ANOVA.

> Because we have the possibility of two main effects and an interaction, each step is broken down into three parts, with three sets of hypotheses, comparison distributions, critical F values, F statistics, and conclusions.

> An expanded source table helps us to keep track of the calculations.

> Significant F statistics require post hoc tests to determine where differences lie when there are more than two groups.

> We calculate a measure of effect size, R^2, for each main effect and for the interaction.

Clarifying the Concepts

12-9 What is the basic difference between the six steps of hypothesis testing for a two-way between-groups ANOVA and a one-way between-groups ANOVA?

12-10 What are the four sources of variability in a two-way ANOVA?

Calculating the Statistics

12-11 Compute the three between-groups degrees of freedom (both main effects and the interaction), the within-groups degrees of freedom, and the total degrees of freedom for the following data:

IV 1, level A; IV 2, level A: 2, 1, 1, 3
IV 1, level B; IV 2, level A: 5, 4, 3, 4
IV 1, level A; IV 2, level B: 2, 3, 3, 3
IV 1, level B; IV 2, level B: 3, 2, 2, 3

12-12 Using the degrees of freedom you calculated in Check Your Learning 12-11, determine critical values, or cutoffs, using a p level of 0.05, for the F statistics of the two main effects and the interaction.

Applying the Concepts

12-13 Researchers studied the effect of e-mail messages on students' final exam grades (Forsyth & Kerr, 1999; Forsyth et al., 2007). To test for possible interactions, participants included students whose first exam grade was either (1) a C, or (2) a D or an F. Participants were randomly assigned to receive several e-mails in one of three conditions: e-mails intended to bolster their self-esteem, e-mails intended to enhance their sense of control over their grades, and e-mails that just included review questions (control group). The accompanying table shows the cell means for the final exam grades (note that some of these are approximate, but all represent actual findings). For simplicity, assume there were 84 participants in the study evenly divided among cells.

	Self-Esteem (SE)	Take Responsibility (TR)	Control Group (CG)
C	67.31	69.83	71.12
D/F	47.83	60.98	62.13

a. From step 1 of hypothesis testing, list the populations for this study.
b. Conduct step 2 of hypothesis testing.
c. Conduct step 3 of hypothesis testing.
d. Conduct step 4 of hypothesis testing.
e. The F statistics are 20.84 for the main effect of the independent variable of initial grade, 1.69 for the main effect of the independent variable of type of e-mail, and 3.02 for the interaction. Conduct step 6 of hypothesis testing.

Solutions to these Check Your Learning questions can be found in Appendix D.

REVIEW OF CONCEPTS

Two-Way ANOVA

Factorial ANOVAs (also called *multifactorial ANOVAs*), those with more than one independent variable (or *factor*), permit us to test more than one hypothesis in a single study. They also allow us to examine *interactions* between independent variables. Factorial ANOVAs are often named by referring to the levels of their independent variables (e.g., 2 × 2) rather than the number of independent variables (e.g., two-way). With a *two-way ANOVA,* we can examine two *main effects,* one for each independent variable, and one interaction, the way in which the two variables might work together to influence the dependent variable. Because we are examining three hypotheses, we calculate three sets of statistics for a two-way ANOVA.

Understanding Interactions in ANOVA

Researchers typically interpret interactions by examining the overall pattern of cell means. A *cell* is one condition in a study. We typically write the mean of a group in its cell. We write the *marginal means* for each row to the right of the cells and the marginal means for each column below the cells. If the main effect of one independent variable is stronger under certain conditions of the second independent variable, there is a *quantitative interaction.* If the direction of the main effect actually reverses under certain conditions of the second independent variable, there is a *qualitative interaction.*

Conducting a Two-Way Between-Groups ANOVA

A two-way between-groups ANOVA uses the same six steps of hypothesis testing that we used previously, with minor changes. Because we test for two main effects and one interaction, each step is broken down into three parts. Specifically, we have three sets of hypotheses, three comparison distributions, three critical F values, three F statistics, and three conclusions. We use an expanded source table to aid in the calculations of the F statistics. We also calculate a measure of effect size, R^2, for each of the main effects and for the interaction.

SPSS®

Let's use SPSS to conduct a two-way ANOVA for the data on myth busting that we used in this chapter. We enter the data in three columns—one for each participant's scores on each independent variable (age and number of repetitions) and one for each participant's score on the dependent variable (false memory).

We instruct SPSS to conduct the ANOVA by selecting: Analyze → General Linear Model → Univariate and selecting the variables. We select the dependent variable, false memory, by highlighting it and clicking the arrow next to "Dependent Variable." We select the independent variables (called "fixed factors" here), age and repetitions, by clicking each of them, then clicking the arrow next to "Fixed Factor(s)." We can include specific descriptive statistics, as well as a measure of effect size, by selecting "Options," then selecting "Descriptive statistics" and "Estimates of effect size."

The screenshot shown here includes the same statistics that we calculated earlier. The small differences are due only to rounding decisions. For example, we see that the F statistic for the main effect of age is 13.333. Its p value is found in the column headed "Sig." and is .006. This is well below the typical p level of 0.05, which tells us that this is a statistically significant effect. The effect size is found in the final column, "Partial Eta Squared," which can be interpreted as we learned to interpret R^2. The effect size of .625 indicates that this is a very large effect.

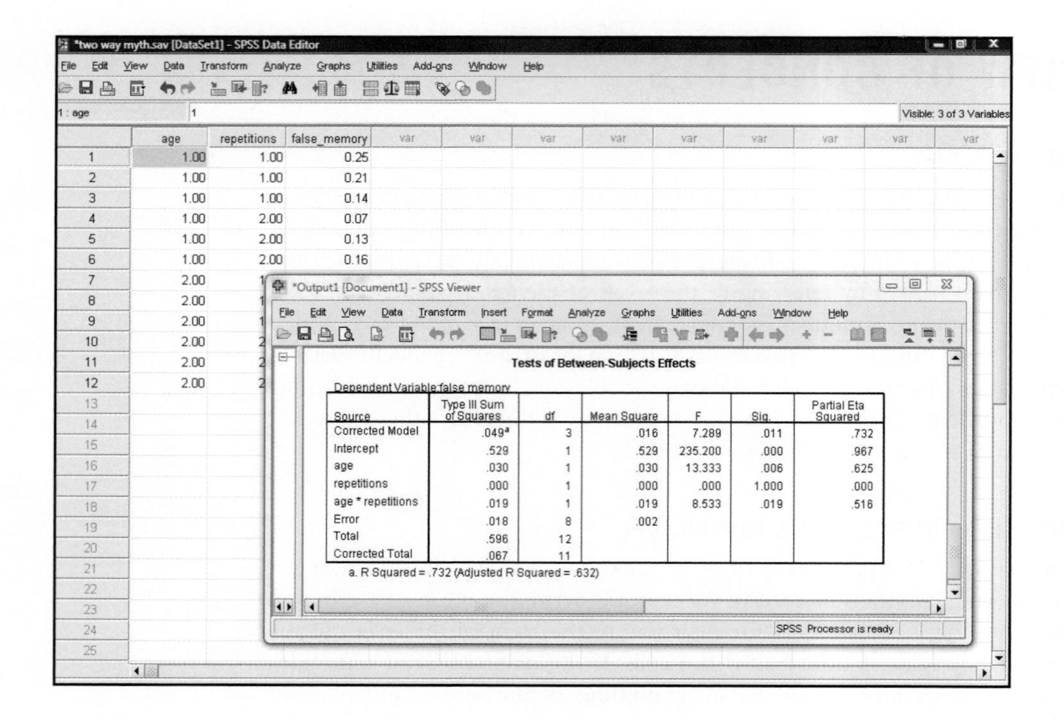

How It Works

12.1 CONDUCTING A TWO-WAY BETWEEN-GROUPS ANOVA

The online dating Web site Match.com allows users to post personal ads to meet others. Each person is asked to specify a range from the youngest age that would be acceptable in a dating partner to the oldest acceptable age. The following data were randomly selected from the ads of 25-year-old people living in the New York City area. The scores represent the youngest acceptable ages listed by those in the sample. So, in the first line, the first of the five 25-year-old women who are seeking men states that she will not date a man younger than 26 years old.

25-year-old women seeking men: 26, 24, 25, 24, 25

25-year-old men seeking women: 18, 21, 22, 22, 18

25-year-old women seeking women: 22, 25, 22, 25, 25

25-year-old men seeking men: 23, 25, 24, 22, 20

There are two independent variables and one dependent variable. The first independent variable is gender of the seeker, and its levels are male and female. The second independent variable is gender of the person being sought, and its levels are men and women. The dependent variable is the youngest acceptable age of the person being sought. Based on these variables, how can we conduct a two-way between-groups ANOVA on these data? The cell means are:

	Female seekers	Male seekers
Men Sought	24.8	22.8
Women Sought	23.8	20.2

Here are the six steps of hypothesis testing for this example.

Step 1: Population 1 (female, men): Women seeking men. Population 2 (male, women): Men seeking women. Population 3 (female, women): Women seeking women. Population 4 (male, men): Men seeking men.

The comparison distributions will be F distributions. The hypothesis test will be a two-way between-groups ANOVA. Assumptions: The data are not from random samples, so we must generalize with caution. The homogeneity of variance assumption is violated because the largest variance (3.70) is more than five times as large as the smallest variance (0.70). For the purposes of demonstration, we will proceed anyway.

Step 2: The hypotheses for the main effect of the first independent variable, gender of seeker, is as follows: Null hypothesis: On average, male and female seekers report the same youngest acceptable ages for their partners—H_0: $\mu_M = \mu_F$. Research hypothesis: On average, male and female seekers report different youngest acceptable ages for their partners—H_0: $\mu_M \neq \mu_F$.

The hypotheses for the main effect of the second independent variable, gender of person sought, is as follows: Null hypothesis: On average, those seeking men and those seeking women report the same youngest acceptable ages for their partners— H_0: $\mu_M = \mu_F$. Research hypothesis: On average, those seeking men and those seeking women report different youngest acceptable ages for their partners—H_0: $\mu_M \neq \mu_F$.

The hypotheses for the interaction of gender of seeker and gender of person sought are as follows: Null hypothesis: The effect of the gender of the seeker on youngest acceptable ages for their partners does not depend on the gender of the person sought. Research hypothesis: The effect of the gender of the seeker on youngest acceptable ages for their partners does depend on the gender of the person sought.

Step 3: $df_{columns(seeker)} = 2 - 1 = 1$

$df_{rows(sought)} = 2 - 1 = 1$

$df_{interaction} = (1)(1) = 1$

$df_{within} = df_{W,M} + df_{M,W} + df_{W,W} + df_{M,M} = 4 + 4 + 4 + 4 = 16$

Main effect of gender of seeker: F distribution with 1 and 16 degrees of freedom
Main effect of gender of sought: F distribution with 1 and 16 degrees of freedom
Interaction of seeker and sought: F distribution with 1 and 16 degrees of freedom

Step 4: Cutoff F for main effect of seeker: 4.49
Cutoff F for main effect of sought: 4.49
Cutoff F for interaction of seeker and sought: 4.49

Step 5: $SS_{total} = \Sigma(X - GM)^2 = 103.800$

$SS_{column(seeker)} = \Sigma(M_{column(seeker)} - GM)^2 = 39.200$

$SS_{row(sought)} = \Sigma(M_{row(sought)} - GM)^2 = 16.200$

$SS_{within} = \Sigma(X - M_{cell})^2 = 45.200$

$SS_{interaction} = SS_{total} - (SS_{row} + SS_{column} + SS_{within}) = 3.200$

Source	*SS*	*df*	*MS*	*F*
Seeker gender	39.200	1	39.200	13.876
Sought gender	16.200	1	16.200	5.736
Seeker × sought	3.200	1	3.200	1.133
Within	45.200	16	2.825	
Total	103.800	19		

Step 6: There is a significant main effect of gender of the seeker and a significant main effect of gender of the person being sought. We can reject the null hypotheses for both of these main effects. Male seekers are willing to accept younger partners, on average, than are female seekers. Those seeking women are willing to accept younger partners, on average, than are those seeking men. We cannot reject the null hypothesis for the interaction; we can only conclude that there is not sufficient evidence that the effect of the gender of the seeker on youngest acceptable age depends on the gender of the person sought.

12.2 CALCULATING EFFECT SIZE FOR A TWO-WAY BETWEEN-GROUPS ANOVA

How can we compute and interpret the effect sizes, R^2, for each main effect and the interaction for the ANOVA we conducted in How It Works 12.1? Here are the effect size calculations and interpretations, according to Cohen's conventions, for each of the three effects.

For the main effect of seeker gender:

$$R^2_{rows} = \frac{SS_{rows}}{(SS_{total} - SS_{columns} - SS_{interaction})} = \frac{39.2}{(103.8 - 16.2 - 3.2)} = 0.46$$

This is a large effect size.

For the main effect of sought gender:

$$R^2_{columns} = \frac{SS_{columns}}{(SS_{total} - SS_{rows} - SS_{interaction})} = \frac{16.2}{(103.8 - 39.2 - 3.2)} = 0.26$$

This is a large effect size.

For the interaction:

$$R^2_{interaction} = \frac{SS_{interaction}}{(SS_{total} - SS_{rows} - SS_{columns})} = \frac{3.2}{(103.8 - 39.2 - 16.2)} = 0.07$$

This is a medium effect size.

Exercises

Clarifying the Concepts

12.1 What is a two-way ANOVA?

12.2 What is a factor?

12.3 In your own words, define the word *cell,* first as you would use it in everyday conversation and then as a statistician would use it.

12.4 What is a four-way within-groups ANOVA?

12.5 What is the difference in information provided when we say *two-way ANOVA* versus *2 × 3 ANOVA*?

12.6 What are the three different F statistics in a two-way ANOVA?

12.7 What is a marginal mean?

12.8 What are the three ways to identify a statistically significant interaction?

12.9 How do bar graphs help us identify and interpret interactions? Explain how adding lines to the bar graph can help.

12.10 How do we calculate the between-groups degrees of freedom for an interaction?

12.11 In step 6 of hypothesis testing for a two-way between-groups ANOVA, we make a decision for each F statistic. What are the three possible outcomes with respect to the overall pattern of results?

12.12 When are post hoc tests needed for a two-way between-groups ANOVA?

12.13 Explain the following formula in your own words: $SS_{interaction} = SS_{total} - (SS_{rows} + SS_{columns} + SS_{within})$.

12.14 In your own words, define the word *interaction,* first as you would use it in everyday conversation and then as a statistician would use it.

12.15 What effect-size measure is used with two-way ANOVA?

Calculating the Statistics

12.16 For each of the following scenarios, what are two names for the ANOVA that would be conducted to analyze the data?

a. A researcher examined the effect of gender and pet ownership (no pets, one pet, more than one pet) on a measure of loneliness.

b. In a study on memory, participants completed a memory task once each week for 4 weeks—twice after sleeping 8 hours and twice after sleeping 4 hours. In each sleep condition, the participants completed the task after ingesting a caffeinated beverage and again, on another day, after ingesting a "placebo" beverage that they were told contained caffeine.

c. A study examined the impact of students' Facebook profiles on numbers of Facebook friends. The researchers were interested in the effect of the profile photo—either an identifiable photo of the student or a photo of someone or something else—and the effect of relationship status—whether it indicates the student is single or in a relationship.

12.17 Identify the factors and their levels in the following research designs.

a. Men and women's enjoyment of two different sporting events, Sport 1 and Sport 2, are compared using a 20-point enjoyment scale.

b. The amount of underage drinking, as documented in formal incident reports, is compared at "dry" college campuses (no alcohol at all) and "wet" campuses (those that enforce the legal age for possession of alcohol). Three different types of colleges are considered: state institutions, private schools, and schools with a religious affiliation.

c. The extent of contact with juvenile authorities is compared for youth across three age groups (12–13, 14–15, 16–17), considering both gender and family composition (two parents, single parent, or no identified authority figure).

12.18 Create an empty grid to represent the cells in each of these studies.

a. Men and women's enjoyment of two different sporting events, Sport 1 and Sport 2, are compared using a 20-point enjoyment scale.

b. The amount of underage drinking, as documented in formal incident reports, is compared at "dry" college campuses (no alcohol at all) and "wet" campuses (those that enforce the legal age for possession of alcohol). Three different types of colleges are considered: state institutions, private schools, and schools with a religious affiliation.

c. The extent of contact with juvenile authorities is compared for youth across three age groups (12–13, 14–15, 16–17), considering both gender and family composition (two parents, single parent, or no identified authority figure).

12.19 Use these "enjoyment" data to perform the following:

	Ice Hockey	Figure Skating
Men	19, 17, 18, 17	6, 4, 8, 3
Women	13, 14, 18, 8	11, 7, 4, 14

a. Calculate the cell and marginal means.

b. Draw a bar graph.

c. Calculate the five different degrees of freedom, and indicate the critical *F* value based on each set of degrees of freedom, assuming the *p* level is 0.01.

d. Calculate the total sum of squares.

e. Calculate the between-groups sum of squares for the independent variable gender.

f. Calculate the between-groups sum of squares for the independent variable sporting event.

g. Calculate the within-groups sum of squares.

h. Calculate the sum of squares for the interaction.

i. Create a source table.

12.20 Use these data—incidents of reports of underage drinking—to perform the following:

"Dry" campus, state school: 47, 52, 27, 50

"Dry" campus, private school: 25, 33, 31

"Wet" campus, state school: 77, 61, 55, 48

"Wet" campus, private school: 52, 68, 60

a. Calculate the cell and marginal means. Notice the unequal *N*s.

b. Draw a bar graph.

c. Calculate the five different degrees of freedom, and indicate the critical *F* value based on each set of degrees of freedom, assuming the *p* level is 0.05.

d. Calculate the total sum of squares.

e. Calculate the between-groups sum of squares for the independent variable campus.

f. Calculate the between-groups sum of squares for the independent variable school.

g. Calculate the within-groups sum of squares.

h. Calculate the sum of squares for the interaction.

i. Create a source table.

12.21 Using what you know about the expanded source table, fill in the missing values in the table shown here:

Source	*SS*	*df*	*MS*	*F*
Gender	248.25	1		
Parenting style	84.34	3		
Gender × style	33.60			
Within	1107.2	36		
Total				

12.22 Using the information in the source table provided on the next page, compute R^2 values for each effect. Using Cohen's conventions, explain what these values mean.

Source	*SS*	*df*	*MS*	*F*
A (rows)	0.267	1	0.267	0.004
B (columns)	3534.008	2	1767.004	24.432
A × B	5.371	2	2.686	0.037
Within	1157.167	16	72.323	
Total	4696.813	21		

12.23 Using the information in the source table provided here, compute R^2 values for each effect. Using Cohen's conventions, explain what these values mean.

Source	*SS*	*df*	*MS*	*F*
A (rows)	30.006	1	30.006	0.511
B (columns)	33.482	1	33.482	0.570
A × B	1.720	1	1.720	0.029
Within	587.083	10	58.708	
Total	652.291	13		

Applying the Concepts

12.24 Football, eye glare, and ANOVA: In Exercise 11.63 (page 303), we described a Yale University study in which researchers randomly assigned 46 participants to place one of three substances below their eyes: black grease, black antiglare stickers, or petroleum jelly. They assessed eye glare using a contrast chart that gives a value for each participant, a scale measure. Black grease led to a reduction in glare compared with the two other conditions, antiglare stickers or petroleum jelly (DeBroff & Pahk, 2003). Imagine that every participant was tested twice, once in broad daylight and again under the artificial lights used at night.

a. What are the independent variables and their levels?

b. What kind of ANOVA would we use?

12.25 Health-related myths and the type of ANOVA: Consider the study we used as an example for a two-way between-groups ANOVA. Older and younger people were randomly assigned to hear either one repetition or three repetitions of a health-related myth, accompanied by the accurate information that "busted" the myth.

a. Explain why this study would be analyzed with a between-groups ANOVA.

b. How could this study be redesigned to use a within-groups ANOVA? (*Hint:* Think long term.)

12.26 Memory for names and choosing the type of ANOVA: In a fictional study, a cognitive psychologist studied memory for names after a group activity. The researcher randomly assigned 120 participants to one of three conditions: (1) group members introduced themselves once, (2) group members were introduced by the experimenter and by themselves, and (3) group members were introduced by the experimenter and themselves, and they wore name tags throughout the group activity.

a. How could the researcher redesign this study so it would be analyzed with a two-way between-groups ANOVA? Be specific. (*Note:* There are several possible ways that the researcher could do this.)

b. How could the researcher redesign this study so it would be analyzed with a two-way mixed-design ANOVA? Be specific. (*Note:* There are several possible ways the researcher could do this.)

12.27 Age, online dating, and choosing the type of ANOVA: A researcher wondered about the degree to which age was a factor for those posting personal ads on Match.com. He randomly selected 200 ads and examined data about the posters (the people who posted the ads). Specifically, for each ad, he calculated the difference between the poster's age and the oldest acceptable age in a romantic prospect. So, if someone was 23 years old and would date someone as old as 30, his or her score would be 7; if someone was 25 and would date someone as old as 23, his or her score would be −2. The researcher then categorized the scores into male versus female and seeking a same-sex date versus seeking an opposite-sex date.

a. List any independent variables, along with the levels.

b. What is the dependent variable?

c. What kind of ANOVA would he use?

d. Now name the ANOVA using the more specific language that enumerates the numbers of levels.

e. Use your answer to part (d) to calculate the number of cells. Explain how you made this calculation.

f. Draw a table that depicts the cells of this ANOVA.

12.28 Racism, juries, and interactions: In a study of racism, Nail, Harton, and Decker (2003) had participants read a scenario in which a police officer assaulted a motorist. Half the participants read about an African American officer who assaulted a European American motorist, and half read about a European American officer who assaulted an African American motorist. Participants were categorized based on political orientation: liberal, moderate, or conservative. Participants were told that the officer was acquitted of assault charges in state court but was found guilty of violating the motorist's rights in federal court. Double jeopardy occurs when an individual is tried twice for the same crime. Participants were asked to rate, on a scale of 1–7, the degree to which the officer had been placed in double jeopardy by the second trial.

The researchers reported the interaction as $F(2, 58) = 10.93, p < 0.0001$. The means for the *liberal* participants were 3.18 for those who read about the African American officer and 1.91 for those who read about the European American officer. The means for the *moderate* participants were 3.50 for those who read about the African American officer and 3.33 for those who read about the European American officer. The means for the *conservative* participants were 1.25 for those who read about the African American officer and 4.62 for those who read about the European American officer.

a. Draw a table of cell means that includes the actual means for this study.
b. Do the reported statistics indicate that there is a significant interaction? If yes, describe the interaction in your own words.
c. Draw a bar graph that depicts the interaction. Include lines that connect the tops of the bars and show the pattern of the interaction.
d. Is this a quantitative or qualitative interaction? Explain.
e. Change the cell mean for the conservative participants who read about an African American officer so that this is now a quantitative interaction.
f. Draw a bar graph that depicts the pattern that includes the new cell means.
g. Change the cell means for the moderate and conservative participants who read about an African American officer so that there is now no interaction.
h. Draw a bar graph that depicts the pattern that includes the new cell means.

12.29 Self-interest, ANOVA, and interactions: Ratner and Miller (2001) wondered whether people are uncomfortable when they act in a way that's not obviously in their own self-interest. They randomly assigned 33 women and 32 men to read a fictional passage saying that federal funding would soon be cut for research into a gastrointestinal illness that mostly affected either (1) women or (2) men. They were then asked to rate, on a 1–7 scale, how comfortable they would be "attending a meeting of concerned citizens who share your position" on this cause (p. 11). A higher rating indicates a greater degree of comfort. The journal article reported the statistics for the interaction as $F(1, 58) = 9.83, p < 0.01$. Women who read about women had a mean of 4.88, whereas those who read about men had a mean of 3.56. Men who read about women had a mean of 3.29, whereas those who read about men had a mean of 4.67.

a. What are the independent variables and their levels? What is the dependent variable?
b. What kind of ANOVA did the researchers conduct?
c. Do the reported statistics indicate that there is a significant interaction? Explain your answer.
d. Draw a table that includes the cells of the study. Include the cell means.
e. Draw a bar graph that depicts these findings.
f. Describe the pattern of the interaction in words. Is this a qualitative or a quantitative interaction? Explain your answer.
g. Draw a new table of cells, but change the means for male participants reading about women so that there is now a quantitative, rather than a qualitative, interaction.
h. Draw a bar graph of the means in part (g).
i. Draw a new table of cells, but change the means for male participants reading about women so that there is no interaction.

12.30 Exercise, well-being, and type of ANOVA: Cox, Thomas, Hinton, and Donahue (2006) studied the effects of exercise on well-being. There were three independent variables: age (18–20 years old, 35–45 years old), intensity of exercise (low, moderate, high), and time point (15, 20, 25, and 30 minutes). The dependent variable was positive well-being. Every participant was assessed at all intensity levels and all time points. (Generally, moderate-intensity exercise and high-intensity exercise led to higher levels of positive well-being than did low-intensity exercise.) What type of ANOVA would the researchers conduct?

12.31 The cross-race effect, main effects, and interactions: Hugenberg, Miller, and Claypool (2007) conducted a study to better understand the cross-race effect, in which people have a difficult time recognizing members of different racial groups—colloquially known as the "they all look the same to me" effect. In a variation on this study, white participants viewed either 20 black faces or 20 white faces for 3 seconds each. Half the participants were told to pay particular attention to distinguishing features of the faces. Later, participants were shown 40 black faces or 40 white faces (the same race that they were shown in the prior stage of the experiment), 20 of which were new. Each participant received a score that measured his or her recognition accuracy.

The researchers reported two effects, one for the race of the people in the pictures, $F(1, 136) = 23.06, p < 0.001$, and one for the interaction of the race of the people in the pictures and the instructions, $F(1, 136) = 5.27, p < 0.05$. When given no instructions, the mean recognition scores were 1.46 for white faces and 1.04 for black faces. When given instructions to pay attention to distinguishing features, the mean recognition scores were 1.38 for white faces and 1.23 for black faces.

a. What are the independent variables and their levels? What is the dependent variable?

b. What kind of ANOVA did the researchers conduct?

c. Do the reported statistics indicate that there is a significant main effect? If yes, describe it.

d. Why is the main effect not sufficient in this situation to understand the findings? Be specific about why the main effect is misleading by itself.

e. Do the reported statistics indicate that there is a significant interaction? Explain your answer.

f. Draw a table that includes the cells of the study and the cell means.

g. Draw a bar graph that depicts these findings.

h. Describe the pattern of the interaction in words. Is this a qualitative or a quantitative interaction? Explain your answer.

12.32 Grade point average, fraternities, sororities, and two-way between-groups ANOVA: A sample of students from our statistics classes reported their GPAs, indicated their genders, and stated whether they were in the university's Greek system (i.e., in a fraternity or sorority). Following are the GPAs for the different groups of students:

Men in a fraternity: 2.6, 2.4, 2.9, 3.0

Men not in a fraternity: 3.0, 2.9, 3.4, 3.7, 3.0

Women in a sorority: 3.1, 3.0, 3.2, 2.9

Women not in a sorority: 3.4, 3.0, 3.1, 3.1

a. What are the independent variables and their levels? What is the dependent variable?

b. Draw a table that lists the cells of the study design. Include the cell means.

c. Conduct all six steps of hypothesis testing.

d. Draw a bar graph for all significant effects.

e. Is there a significant interaction? If yes, describe it in words and indicate whether it is a qualitative or a quantitative interaction. Explain.

f. Compute the effect sizes, R^2, for the main effects and interaction. Using Cohen's conventions, interpret the effect-size values.

12.33 Age, online dating, and two-way between-groups ANOVA: The data below were from the same 25-year-old participants described in How It Works 12.1, but now the scores represent the oldest age that would be acceptable in a dating partner.

25-year-old women seeking men: 40, 35, 29, 35, 35

25-year-old men seeking women: 26, 26, 28, 28, 28

25-year-old women seeking women: 35, 35, 30, 35, 45

25-year-old men seeking men: 33, 35, 35, 36, 38

a. What are the independent variables and their levels? What is the dependent variable?

b. Draw a table that lists the cells of the study design. Include the cell means.

c. Conduct all six steps of hypothesis testing.

d. Is there a significant interaction? If yes, describe it in words, indicate whether it is a quantitative or a qualitative interaction, and draw a bar graph.

e. Compute the effect sizes, R^2, for the main effects and interaction. Using Cohen's conventions, interpret the effect-size values.

12.34 Helping, payment, and two-way between-groups ANOVA: Heyman and Ariely (2004) were interested in whether effort and willingness to help were affected by the form and amount of payment offered in return for effort. They predicted that when money was used as payment, in what is called a *money market,* effort would increase as a function of payment level. On the other hand, if effort was performed out of altruism, in what is called a *social market,* the level of effort would be consistently high and unaffected by level of payment. In one of their studies, college students were asked to estimate another student's willingness to help load a sofa into a van in return for a cash payment or candy of equivalent value. Willingness to help was assessed using an 11-point scale ranging from "Not at all likely to help" to "Will help for sure." Data are presented here to re-create some of their findings.

Cash payment, low amount of $0.50: 4, 5, 6, 4

Cash payment, moderate amount of $5.00: 7, 8, 8, 7

Candy payment, low amount valued at $0.50: 6, 5, 7, 7

Candy payment, moderate amount valued at $5.00: 8, 6, 5, 5

a. What are the independent variables and their levels?

b. What is the dependent variable?

c. Draw a table that lists the cells of the study design. Include the cell and marginal means.

d. Create a bar graph.

e. Using this graph and the table of cell means, describe what effects you see in the pattern of the data.

f. Write the null and research hypotheses.

g. Complete all of the calculations, and construct a full source table for these data.

h. Determine the critical value for each effect at a p level of 0.05.

i. Make your conclusions. Is there a significant interaction? If yes, describe it in words and indicate whether it is a qualitative or a quantitative interaction. Explain.

j. Compute the effect sizes, R^2, for the main effects and interaction. Using Cohen's conventions, interpret the effect-size values.

12.35 Helping, payment, and interactions: Expanding on the work of Heyman and Ariely (2004) as described in Exercise 12.34, let's assume a higher level of payment was included and the following data were collected. (Notice that all data are the same as earlier, with the addition of new data under a high payment amount.)

Cash payment, low amount of \$0.50: 4, 5, 6, 4

Cash payment, moderate amount of \$5.00: 7, 8, 8, 7

Cash payment, high amount of \$50.00: 9, 8, 7, 8

Candy payment, low amount, valued at \$0.50: 6, 5, 7, 7

Candy payment, moderate amount, valued at \$5.00: 8, 6, 5, 5

Candy payment, high amount, valued at \$50.00: 6, 7, 7, 6

a. What are the independent variables and their levels? What is the dependent variable?

b. Draw a table that lists the cells of the study design. Include the cell and marginal means.

c. Create a new bar graph of these data.

d. Do you think there is a significant interaction? If yes, describe it in words.

e. Now that one independent variable has three levels, what additional analyses are needed? Explain what you would do and why. Based on the graph you created, where do you think there would be significant differences?

Putting It All Together

12.36 Skepticism, self-interest, and two-way ANOVA: A study on motivated skepticism examined whether participants were more likely to be skeptical when it served their self-interest (Ditto & Lopez, 1992). Ninety-three participants completed a fictitious medical test that told them they had high levels of a certain enzyme, TAA. Participants were randomly assigned to be told either that high levels of TAA had potentially unhealthy consequences or potentially healthy consequences. They were also randomly assigned to complete a dependent measure before or after the TAA test. The dependent measure assessed their perception of the accuracy of the TAA test on a scale of 1 (very inaccurate) to 9 (very accurate). Ditto and Lopez found the following means for those who completed the dependent measure before taking the TAA test: unhealthy result, 6.6; healthy result, 6.9. They found the following means for those who completed the dependent measure after taking the TAA test: unhealthy result, 5.6; healthy result, 7.3. From their ANOVA, they reported statistics for two findings. For the main effect of test outcome, they reported the following statistic: $F(1,73) = 7.74, p < 0.01$. For the interaction of test outcome and timing of the dependent measure, they reported the following statistic: $F(1, 73) = 4.01, p < 0.05$.

a. State the independent variables and their levels. State the dependent variable.

b. What kind of ANOVA would be used to analyze these data? State the name using the original language as well as the more specific language.

c. Use the more specific language of ANOVA to calculate the number of cells in this research design.

d. Draw a table of cell means, marginal means, and the grand mean. Assume that equal numbers of participants were assigned to each cell (even though this was not the case in the actual study).

e. Describe the significant main effect in your own words.

f. Draw a bar graph that depicts the main effect.

g. Why is the main effect misleading by itself?

h. Is the main effect qualified by a statistically significant interaction? Explain. Describe the interaction in your own words.

i. Draw a bar graph that depicts the interaction. Include lines that connect the tops of the bars and show the pattern of the interaction.

j. Is this a quantitative or qualitative interaction? Explain.

k. Change the cell mean for the participants who had a healthy test outcome and completed the dependent measure before the TAA test so that this is now a qualitative interaction.

l. Draw a bar graph depicting the pattern that includes the new cell mean.

m. Change the cell mean for the participants who had a healthy test outcome and completed the dependent measure before the TAA test so that there is now no interaction.

n. Draw a bar graph that depicts the pattern that includes the new cell mean.

Terms

interaction (p. 307)
two-way ANOVA (p. 307)
factorial ANOVA (p. 307)
factor (p. 307)
cell (p. 309)
main effect (p. 309)
quantitative interaction (p. 312)
qualitative interaction (p. 312)
marginal mean (p. 313)

Formulas

$df_{rows} = N_{rows} - 1$ (p. 323)

$df_{columns} = N_{columns} - 1$ (p. 324)

$df_{interaction} = (df_{rows})(df_{columns})$ (p. 324)

$df_{within} = df_{cell\ 1} + df_{cell\ 2} + df_{cell\ 3} + df_{cell\ 4}$ (p. 324)

$df_{total} = N_{total} - 1$ (p. 324)

$SS_{total} = \Sigma(X - GM)^2$ for each score (p. 326)

$SS_{between(rows)} = \Sigma(M_{row} - GM)^2$ for each score (p. 327)

$SS_{between(columns)} = \Sigma(M_{column} - GM)^2$ for each score (p. 328)

$SS_{within} = \Sigma(X - M_{cell})^2$ for each score (p. 328)

$SS_{between(interaction)} = SS_{total} - (SS_{between(rows)} + SS_{between(columns)} + SS_{within})$ (p. 329)

$$R^2_{rows} = \frac{SS_{rows}}{(SS_{total} - SS_{columns} - SS_{interaction})}$$ (p. 330)

$$R^2_{columns} = \frac{SS_{columns}}{(SS_{total} - SS_{rows} - SS_{interaction})}$$ (p. 330)

$$R^2_{interaction} = \frac{SS_{interaction}}{(SS_{total} - SS_{rows} - SS_{columns})}$$ (p. 330)

CHAPTER 13

Correlation

BEFORE YOU GO ON

- You should know the difference between correlational research and experimental research (Chapter 1).
- You should understand how to calculate the deviations of scores from a mean (Chapter 4).
- You should understand the concept of sum of squares (Chapter 4).
- You should understand the six steps of hypothesis testing (Chapter 7).
- You should understand the concept of effect size (Chapter 8).

A Correlation Between Cheating and Grades Does cheating lead to poorer grades? Researchers can't say whether cheating causes lower grades, but there is a negative correlation between cheating and final exam grade.

John Snow's map of the London cholera epidemic (Chapter 1) revealed a *correlation,* a systematic association (or relation) between two variables. Researchers from the Massachusetts Institute of Technology (MIT) also used a correlation to reveal the dangers of academic cheating (Palazzo, Lee, Warnakulasooriya, & Pritchard, 2010). They compared the final exam scores of 428 physics students with how frequently they had copied 3 days of computerized homework assignments from classmates. They assumed that the minority of students (< 10%) who finished their homework within a ridiculously short time had cheated. You might expect cheating to boost grades—after all, that is the point of cheating, isn't it? But you'd be wrong. The more a student cheated during the semester (variable 1), the lower his or her final exam grade (variable 2).

What might be going on here? Students' preexisting ability in math and physics was not correlated with cheating, so these were not weak students who were also incompetent at cheating. Their computerized homework revealed a correlation that suggested another possible answer: Cheaters procrastinated by starting their homework too late to complete it without cheating. Did they have outside jobs, suffer from anxiety, or have other commitments that cut into homework time? *Correlations can't tell us which explanation is right,* but they can force us to think about the possible explanations.

- A **correlation coefficient** is a statistic that quantifies a relation between two variables.
- A **positive correlation** is an association between two variables such that participants with high scores on one variable tend to have high scores on the other variable as well, and those with low scores on one variable tend to have low scores on the other variable.
- A **negative correlation** is an association between two variables in which participants with high scores on one variable tend to have low scores on the other variable.

This chapter demonstrates how to (a) assess the direction and size of a correlation, (b) identify limitations of correlation, and (c) calculate the most common form of correlation: *r,* the Pearson correlation coefficient. We then use the six steps of hypothesis testing to determine whether a correlation is statistically significant.

The Meaning of Correlation

A correlation is exactly what its name suggests: a co-relation between two variables. Lots of everyday observations are co-related: junk food eaten and body fat, miles driven and the wear on tires, air conditioner usage and your electric bill. If you can measure any two variables on a scale, you can calculate the degree to which they are related.

The Characteristics of Correlation

A ***correlation coefficient*** *is a statistic that quantifies a relation between two variables.* In this chapter, we learn how to quantify a relation—that is, we learn to calculate a correlation coefficient—when the data are linearly related. A linear relation means that the data form an overall pattern through which it would make sense to draw a straight line—that is, the dots on a scatterplot are roughly clustered around a line, rather than, say, a curve. You can actually see—and understand—the data story with just a glance. There are three main characteristics of the correlation coefficient.

> **MASTERING THE CONCEPT**
>
> **13.1:** A correlation coefficient always falls between −1.00 and 1.00. The size of the coefficient, not its sign, indicates how large it is.

1. The correlation coefficient can be either positive or negative.
2. The correlation coefficient always falls between −1.00 and 1.00.
3. It is the strength (also called the *magnitude*) of the coefficient, not its sign, that indicates how large it is.

The first important characteristic of the correlation coefficient is that it can be either positive or negative. A positive correlation has a positive sign (e.g., 0.32), and a negative correlation has a negative sign (e.g., −0.32). *A* ***positive correlation*** *is an association between two variables such that participants with high scores on one variable tend to have high scores on the other variable as well, and those with low scores on one variable tend to have low scores on the other variable.*

Contrary to what some people think, when participants with low scores on one variable tend to have low scores on the other, it is *not* a negative correlation. A positive correlation describes a situation in which participants tend to have *similar* scores, with respect to the mean and spread, on both variables—whether the scores are low, medium, or high. The line that summarizes a scatterplot with a positive correlation slopes upward and to the right.

EXAMPLE 13.1

The scatterplot in Figure 13-1 shows a positive correlation between Scholastic Aptitude Test (SAT) score and college grade point average (GPA). For example, the second dot from the left is for a person with a 980 on the SAT and a 2.2 GPA; this person is lower than average on both scores. The upper-right dot is for a person with a 1360 on the SAT and a 3.8 GPA; this person is higher than average on both scores. This makes sense, because we would expect people with higher SAT scores to get better grades, on average.

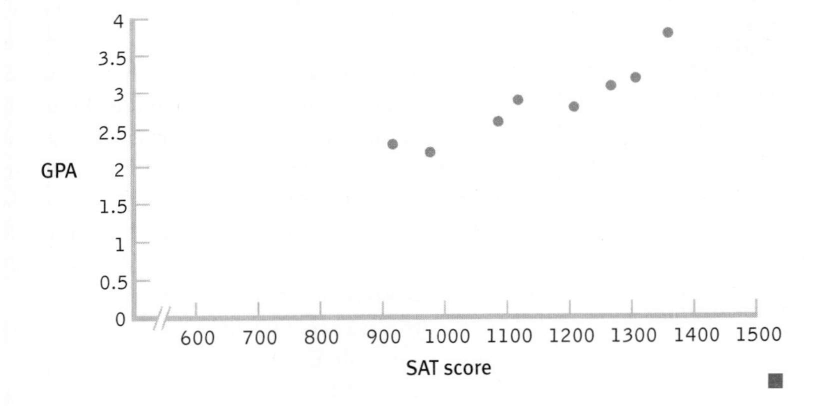

FIGURE 13-1
A Positive Correlation

These data points depict a positive correlation between SAT score and college GPA. Those with higher SAT scores tend to have higher GPAs, and those with lower SAT scores tend to have lower GPAs.

EXAMPLE 13.2

A ***negative correlation*** *is an association between two variables in which participants with high scores on one variable tend to have low scores on the other variable.* The line that summarizes a scatterplot with a negative correlation slopes downward and to the right.

The scatterplot in Figure 13-2 shows a negative correlation of −0.43 between cheating and final exam grade. Each dot represents one person's values on both variables. The proportion of homework copied during the semester is on the horizontal *x*-axis, and the final exam grade (converted to standardized *z* scores) is on the vertical *y*-axis. For example, the dot in the green diamond indicates a student who copied less than 0.2, or 20%, of the homework, and scored almost two standard deviations above the mean on the final exam. The dot in the red diamond indicates a student who copied almost 80% of the homework and scored more than three standard deviations below the mean on the final exam. Notice that most dots do not fit the pattern of the two students we just described. However, the overall trend is for students who copied more to perform more poorly on the final—a linear relation.

◀ MASTERING THE CONCEPT

13.2: The sign indicates the direction of the correlation, positive or negative. A positive correlation occurs when people who are high on one variable tend to be high on the other as well, and people who are low on one variable tend to be low on the other. A negative correlation occurs when people who are high on one variable tend to be low on the other.

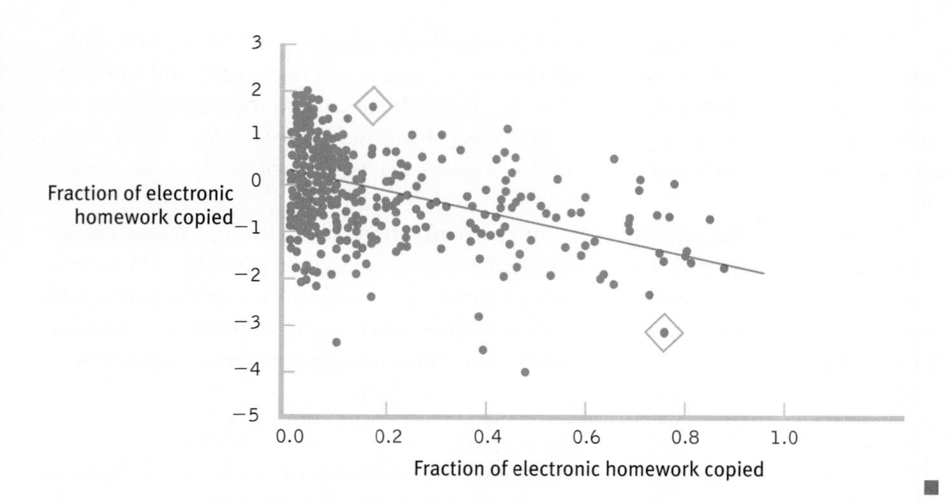

FIGURE 13-2
A Negative Correlation
These data points depict a negative correlation between cheating on homework and final exam grades. Those who cheat more tend to have a lower final exam grade, whereas those who cheat less tend to have a higher final exam grade.

A second important characteristic of the correlation coefficient is that it always falls between −1.00 and 1.00. Both −1.00 and 1.00 are perfect correlations. If we calculate a coefficient that is outside this range, we have made a mistake in the calculations. A correlation coefficient of 1.00 indicates a perfect positive correlation; every point on the scatterplot falls on one line, as seen in the imaginary relation between absences and exam grades depicted in Figure 13-3. Higher scores on one variable are associated with higher scores on the other, and lower scores on one variable are associated with lower scores on the other. When a correlation coefficient is either −1.00 or 1.00, knowing somebody's score on one variable tells you exactly what that person's score is on the other variable. They are perfectly related.

A correlation coefficient of −1.00 indicates a perfect negative correlation. Every point on the scatterplot falls on one line, as seen in the imaginary relation between absences and exam grades depicted in Figure 13-4, but now higher scores on one variable go with lower scores on the other variable. As with a perfect positive correlation, knowing somebody's score on one variable tells you that person's exact score on the other variable. A correlation of 0.00 falls right in the middle of the two extremes and indicates no correlation—no association between the two variables.

The third useful characteristic of the correlation coefficient is that its sign—positive or negative—indicates only the direction of the association, not the strength or size of the association. So a correlation coefficient of −0.35 is the same size as one of 0.35. A correlation coefficient of −0.67 is larger than one of 0.55. Don't be fooled by a negative sign; the sign indicates the direction of the relation, not the strength.

The strength of the correlation is determined by how close to "perfect" the data points are. The closer the data points are to the imaginary line that one could draw through them, the closer the correlation is to being perfect (either

FIGURE 13-3
A Perfect Positive Correlation
When every pair of scores falls on the same line on a scatterplot, with higher scores on one variable associated with higher scores on the other (and lower scores with lower scores), there is a perfect positive correlation of 1.00, a situation that almost never occurs in real life. Also, we would not predict that the number of absences would be positively correlated with exam grade!

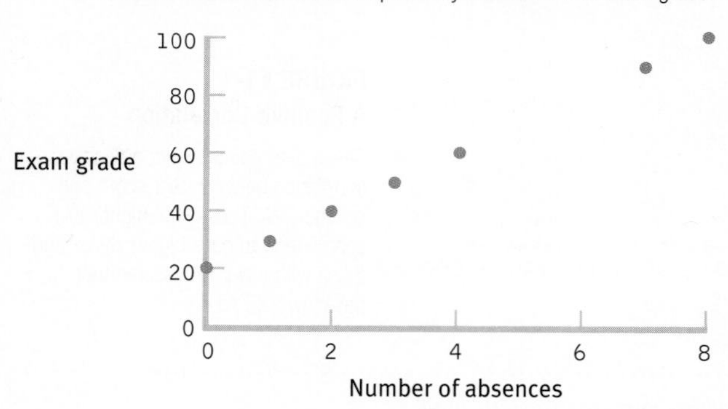

FIGURE 13-4
A Perfect Negative Correlation
When every pair of scores falls on the same line on a scatterplot and higher scores on one variable are associated with lower scores on the other variable, there is a perfect negative correlation of −1.00, a situation that almost never occurs in real life.

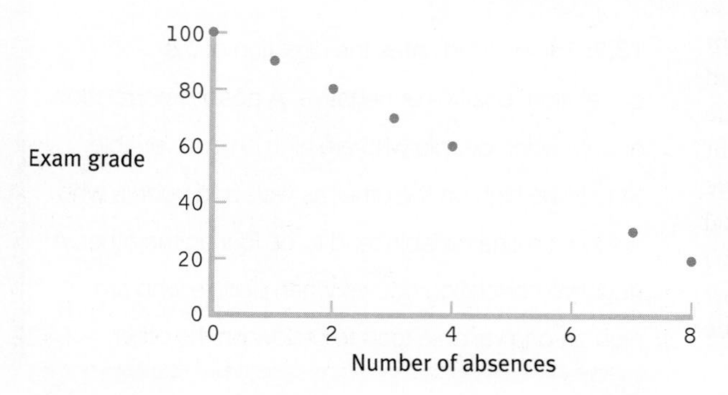

−1.00 or 1.00), and the stronger the relation between the two variables. The farther the points are from this imaginary line, the farther the correlation is from being perfect (so, closer to 0.00), and the weaker the relation between the two variables.

The scores in a positive correlation move up and down together, the same way the mercury rises or falls in a thermometer as the temperature goes up or down. The scores in a negative correlation move up and down in opposition to each other, like a teeter-totter. Knowing the direction of a correlation allows us to use a person's score on one variable to predict his or her score on another variable. Fortunately, the correlation statistic lets us identify both the direction and the strength of the relation between two variables.

Ole Graf/zefa/Corbis

The Teeter-Tottering Negative Correlation When two variables are negatively correlated, a high score on one variable indicates a likely low score on the other variable—just like children on a teeter-totter.

How big does a correlation coefficient have to be to be considered important? As he did for effect sizes, Cohen (1988) published standards, shown in Table 13-1, to help us interpret the correlation coefficient. Very few findings in the behavioral sciences have correlation coefficients of 0.50 or larger because a correlation is influenced by many variables. A student's exam grade, for example, is influenced by absences from class, attention level, hours of studying, interest in the subject matter, IQ, and many other variables. So, the correlation of −0.43 between cheating and exam grades found among MIT students is a large correlation for the behavioral sciences.

TABLE 13-1. How Strong Is an Association?

Cohen (1988) published guidelines to help researchers determine the strength of a correlation from the correlation coefficient. In behavioral science research, however, it is extremely unusual to have a correlation as high as 0.50, and some researchers have disputed the utility of Cohen's conventions for many behavioral science contexts.

Size of the Correlation	Correlation Coefficient
Small	0.10
Medium	0.30
Large	0.50

Correlation Is Not Causation

You need to understand what correlations do *not* reveal about the relation between variables. Correlations *only* provide clues to causality; they do *not* demonstrate or test for causality; they *only* quantify the strength and direction of the relation between variables. Your appreciation for what correlations do *not* reveal suggests that you are thinking scientifically. For example, we know that there was a strong negative correlation in the MIT study between cheating and final exam grade, and it is not unreasonable to think that cheating causes bad grades. However, there are three possible reasons for this observed correlation.

First, variable A (cheating) could cause variable B (poor grades). Second, variable B (poor grades) could cause variable A (cheating). Third, variable C (some other influence) could be causing the correlation between variable A (cheating) and variable B (poor grades). You can think of these three possibilities as the A-B-C model (Figure 13-5).

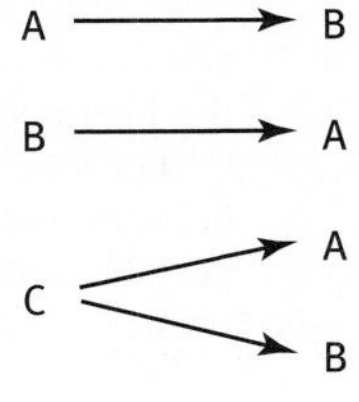

FIGURE 13-5
Three Possible Causal Explanations for a Correlation
Any correlation can be explained in one of several ways. The first variable (A) might cause the second variable (B). Or the reverse could be true—the second variable (B) could cause the first variable (A). Finally, a third variable, C, could cause both A and B. In fact, there could be many third variables.

> **MASTERING THE CONCEPT**
>
> **13.3:** Just because two variables are related doesn't mean one causes the other. It could be that one causes the second, the second causes the first, or a third variable causes both. Correlation does not indicate causation.

Knowing that correlation does *not* imply causation coaxes our brains into thinking of alternate explanations. The researchers found that physics and math ability did not correlate with cheating; so that's an unlikely answer. But we also mentioned working, anxiety, and other time commitments. You can probably think of even more possibilities. Never confuse correlation with causation.

CHECK YOUR LEARNING

Reviewing the Concepts

- A correlation coefficient is a statistic that quantifies a relation between two variables.
- The correlation coefficient always falls between −1.00 and 1.00.
- When two variables are related such that people with high scores on one tend to have high scores on the other and people with low scores on one tend to have low scores on the other, we describe the variables as positively correlated.
- When two variables are related such that people with high scores on one tend to have low scores on the other, we describe the variables as negatively correlated.
- When two variables are not related, there is no correlation and they have a correlation coefficient close to 0.
- The strength of the correlation, captured by the number value of the coefficient, is independent of its sign. Cohen established standards for evaluating the strength of association.
- Correlation is not equivalent to causation. In fact, a correlation does not help us decide the merits of different causal explanations.
- When two variables are correlated, this association might occur because the first variable (A) causes the second (B) or because the second variable (B) causes the first (A). In addition, a third variable (C) could cause both of the correlated variables (A and B).

Clarifying the Concepts

13-1 There are three main characteristics of the correlation coefficient. What are they?

13-2 Why doesn't correlation indicate causation?

Calculating the Statistics

13-3 Use Cohen's guidelines to describe the strength of the following coefficients:

a. −0.60

b. 0.35

c. 0.04

13-4 Draw a hypothetical scatterplot to depict the following correlation coefficients:

a. −0.60

b. 0.35

c. 0.04

Applying the Concepts

13-5 A writer for *Runner's World* magazine debated the merits of running while listening to music (Seymour, 2006). The writer, an avid iPod user, interviewed a clinical psychologist, whose response to the debate about whether to listen to music while running was: "I like to do what the great ones do and try to emulate that. What are the Kenyans doing?"

Let's say a researcher conducted a study in which he determined the correlation between the percentage of a country's marathon runners who train while using a portable music device and the average marathon finishing time for that country's runners. (Note that in this case the participants are countries, not people.) Let's say the researcher finds a strong positive correlation. That is, the more of a country's runners who train with music, the longer the average marathon finishing time. Remember, in a marathon, a longer time is bad. So this fictional finding is that training with music is associated with slower marathon finishing times; the United States, for example, would have a higher percentage of music use and higher (slower) finishing times than Kenya.

Using the A-B-C model, provide three possible explanations for this finding.

Solutions to these Check Your Learning questions can be found in Appendix D.

The Pearson Correlation Coefficient

The most widely used correlation coefficient is the ***Pearson correlation coefficient****, a statistic that quantifies a linear relation between two scale variables.* In other words, a single number is used to describe the direction and strength of the relation between two variables when their overall pattern indicates a straight-line relation. The Pearson correlation coefficient is symbolized by the italic letter r when it is a statistic based on sample data and by the Greek letter ρ (written as "rho" and pronounced "row," even though it looks a bit like the Latin letter p). We use ρ when we're referring to the population parameter for the correlation coefficient, such as when we're writing the hypotheses for significance testing.

Calculating the Pearson Correlation Coefficient

The correlation coefficient can be used as a descriptive statistic that describes the direction and strength of an association between two variables. However, it can also be used as an inferential statistic that relies on a hypothesis test to determine whether the correlation coefficient is significantly different from 0 (no correlation). In this section, we construct a scatterplot from the data and learn how to calculate the correlation coefficient. Then we walk through the steps of hypothesis testing.

EXAMPLE 13.3

Every couple of semesters, we have a student who avows that she does not have to attend statistics classes regularly to do well because she can learn it all from the book. What do you think? Table 13-2 displays the data for 10 students in one of our recent statistics classes. The second column shows the number of absences over the semester

TABLE 13-2. Is Skipping Class Related to Statistics Exam Grades?

Here are the scores for 10 students on two scale variables: number of absences from class in one semester and exam grade.

Student	Absences	Exam Grade
1	4	82
2	2	98
3	2	76
4	3	68
5	1	84
6	0	99
7	4	67
8	8	58
9	7	50
10	3	78

▪ The **Pearson correlation coefficient** is a statistic that quantifies a linear relation between two scale variables.

(out of 29 classes total) for each student, and the third column shows each student's final exam grade.

> **MASTERING THE CONCEPT**
>
> **13.4:** A scatterplot can indicate whether two variables are linearly related. It can also give us a sense of the direction and strength of the relation between the two variables.

Let's begin with a visual exploration of the scatterplot in Figure 13-6. The data, overall, have a pattern through which we could imagine drawing a straight line, so it makes sense to use the Pearson correlation coefficient. Look more closely at the scatterplot. Are the dots clustered closely around the imaginary line? If they are, then the correlation is probably close to 1.00 or −1.00; if they are not, then the correlation is probably closer to 0.00.

A positive correlation results when a high score (above the mean) on one variable tends to indicate a high score (also above the mean) on the other variable. A negative correlation results when a high score on one variable (above the mean) tends to indicate a low score (below the mean) on the other variable. We can determine whether an individual falls above or below the mean by calculating deviations from the mean for each score. If participants tend to have two positive deviations (both scores above the mean) or two negative deviations (both scores below the mean), then the two variables are likely to be positively correlated. If participants tend to have one positive deviation (above the mean) and one negative deviation (below the mean), then the two variables are likely to be negatively correlated. That's a big part of how the formula for the correlation coefficient does its work.

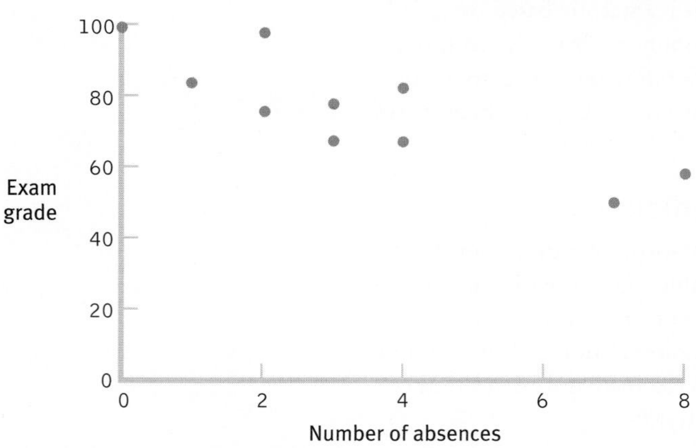

FIGURE 13-6
Always Start with a Scatterplot
Before calculating a correlation coefficient for the relation between number of absences from class and exam grade, we construct a scatterplot. If the relation between the variables appears to be roughly linear, we can calculate a Pearson correlation coefficient. We can also use the scatterplot to make a guess about what we expect from the correlation. Here, the correlation appears to be negative. The pattern of data goes down and to the right, and high scores on one variable are associated with low scores on the other. In addition, we would expect a somewhat large correlation—that is, fairly close to −1.00—because the data are fairly close to forming a straight line.

Think about why calculating deviations from the mean makes sense. With a positive correlation, high scores are above the mean and so would have positive deviations. The product of a pair of high scores would be positive. Low scores are below the mean and would have negative deviations. The product of a pair of low scores would also be positive. When we calculate a correlation coefficient, part of the process involves adding up the products of the deviations. If most of these are positive, we get a positive correlation coefficient.

Let's consider a negative correlation. High scores, which are above the mean, would have positive deviations. Low scores, which are below the mean, would have negative deviations. The product of one positive deviation and one negative deviation would be negative. If most of the products of the deviations are negative, we would get a negative correlation coefficient.

The process we just described is the calculation of the numerator of the correlation coefficient. Table 13-3 shows us the calculations. The first column has the number of absences for each student. The second column shows the deviations from the mean, 3.40. The third column has the exam grade for each student. The fourth column shows the deviations from the mean for that variable, 76.00. The fifth column shows the products of the deviations. Below the fifth column, we see the sum of the products of the deviations, −304.0.

As we see in Table 13-3, the pairs of scores tend to fall on either side of the mean—that is, for each student, a negative deviation on one score tends to indicate a positive deviation on the other score. For example, student 6 was never absent, so he has a score of 0, which is well below the mean, and he got a 99 on the exam, well above the mean. On the other hand, student 9 was absent 7 times, well above the mean, and she only got a 50 on the exam, well below the mean. So most of the products of the deviations are negative, and when we sum the products, we get a negative total. This indicates a negative correlation.

TABLE 13-3. Calculating the Numerator of the Correlation Coefficient

Absences (X)	$(X - M_X)$	Exam Grade (Y)	$(Y - M_Y)$	$(X - M_X)(Y - M_Y)$
4	0.6	82	6	3.6
2	−1.4	98	22	−30.8
2	−1.4	76	0	0.0
3	−0.4	68	−8	3.2
1	−2.4	84	8	−19.2
0	−3.4	99	23	−78.2
4	0.6	67	−9	−5.4
8	4.6	58	−18	−82.8
7	3.6	50	−26	−93.6
3	−0.4	78	2	−0.8
$M_X = 3.400$		$M_Y = 76.000$		$\Sigma[(X - M_X)(Y - M_Y)] = -304.0$

You might have noticed that this number, −304.0, is not between −1.00 and 1.00. The problem is that this number is influenced by two factors—sample size and variability. First, the more people in the sample, the more deviations there are to contribute to the sum. Second, if the scores in the study were more variable, the deviations would be larger and so would the sum of the products. So we have to correct for these two factors in the denominator.

It makes sense that we would have to correct for variability. In Chapter 6, we learned that z scores provide an important function in statistics by allowing us to standardize. You may remember that the formula for the z score that we first learned was $z = \frac{(X - M)}{SD}$. In the calculations in the numerator for correlation, we already subtracted the mean from the scores when we created deviations, but we didn't divide by the standard deviation. If we correct for variability in the denominator, that takes care of one of the two factors for which we have to correct.

But we also have to correct for sample size. You may remember that when we calculate standard deviation, the last two steps are (1) dividing the sum of squared deviations by the sample size, N, to remove the influence of the sample size and to calculate variance; and (2) taking the square root of the variance to get the standard deviation. So to factor in sample size along with standard deviation (which we just mentioned allows us to factor in variability), we can go backward in the calculations. If we multiply variance by sample size, we get the sum of squared deviations, or sum of squares. Because of this, the denominator of the correlation coefficient is based on the sums of squares for both variables. To make the denominator match the numerator, we multiply the two sums of squares together, and then we take their square root, as we would with standard deviation. Table 13-4 shows the calculations for the sum of squares for the two variables, absences and exam grades.

We now have all of the ingredients necessary to calculate the correlation coefficient. Here's the formula:

$$r = \frac{\Sigma[(X - M_X)(Y - M_Y)]}{\sqrt{(SS_X)(SS_Y)}}$$

The numerator is the sum of the products of the deviations for each variable (see Table 13-3).

MASTERING THE FORMULA

13-1: The formula for the correlation coefficient is: $r = \frac{\Sigma[(X - M_X)(Y - M_Y)]}{\sqrt{(SS_X)(SS_Y)}}$. We divide the sum of the products of the deviations for each variable by the square root of the products of the sums of squares for each variable. This calculation has a built-in standardization procedure: It subtracts a mean from each score and divides by some kind of variability. By using sums of squares in the denominator, it also takes sample size into account.

TABLE 13-4. Calculating the Denominator of the Correlation Coefficient

Absences (X)	$(X - M_X)$	$(X - M_X)^2$	Exam Grade (Y)	$(Y - M_Y)$	$(Y - M_Y)^2$
4	0.6	0.36	82	6	36
2	−1.4	1.96	98	22	484
2	−1.4	1.96	76	0	0
3	−0.4	0.16	68	−8	64
1	−2.4	5.76	84	8	64
0	−3.4	11.56	99	23	529
4	0.6	0.36	67	−9	81
8	4.6	21.16	58	−18	324
7	3.6	12.96	50	−26	676
3	−0.4	0.16	78	2	4
		$\Sigma(X - M_X)^2 = 56.4$			$\Sigma(Y - M_Y)^2 = 2262$

STEP 1: For each score, we calculate the deviation from its mean.

STEP 2: For each participant, we multiply the deviations for his or her two scores.

STEP 3: We sum the products of the deviations.

The denominator is the square root of the product of the two sums of squares. The sums of squares calculations are in Table 13-4.

STEP 1: We calculate a sum of squares for each variable.

STEP 2: We multiply the two sums of squares.

STEP 3: We take the square root of the product of the sums of squares.

Let's apply the formula for the correlation coefficient to the data:

$$r = \frac{\Sigma[(X - M_X)(Y - M_Y)]}{\sqrt{(SS_X)(SS_Y)}} = \frac{-304.0}{\sqrt{(56.4)(2262.0)}} = \frac{-304.0}{357.179} = -0.851$$

So the Pearson correlation coefficient, r, is -0.85. This is a very strong negative correlation. If we examine the scatterplot in Figure 13-6 carefully, we will notice that there aren't any glaring individual exceptions to this rule. The data tell a consistent story. So what should our students learn from this result? Go to class! ■

Hypothesis Testing with the Pearson Correlation Coefficient

We said earlier that correlation could be used as a descriptive statistic to simply describe a relation between two variables, and as an inferential statistic.

EXAMPLE 13.4

Here we outline the six steps for hypothesis testing with a correlation coefficient. Usually, when we conduct hypothesis testing with correlation, we want to test whether a correlation is statistically significantly different from no correlation—an *r* of 0.

MASTERING THE CONCEPT

13.5: As with other statistics, we can conduct hypothesis testing with the correlation coefficient. We compare the correlation coefficient to critical values on the *r* distribution.

STEP 1: Identify the populations, distribution, and assumptions. Population 1: Students like those whom we studied in Example 13.3. Population 2: Students for whom there is no correlation between number of absences and exam grade.

The comparison distribution is a distribution of correlations taken from the population, but with the characteristics of our study, such as a sample size of 10. In this case, it is a distribution of all possible correlations between the numbers of absences and exam grades when 10 students are considered.

The first two assumptions are like those for other parametric tests. (1) The data must be randomly selected, or external validity will be limited. In this case, we do not know how the data were selected, so we should generalize with caution. (2) The underlying population distributions for the two variables must be approximately normal. In our study, it's difficult to tell if the distribution is normal because we have so few data points.

The third assumption is specific to correlation: Each variable should vary equally, no matter the magnitude of the other variable. That is, number of absences should show the same amount of variability at each level of exam grade; conversely, exam grade should show the same amount of variability at each number of absences. You can get a sense of this by looking at the scatterplot in Figure 13-7. In our study, it's hard to determine whether the amount of variability is the same for each variable across all levels of the other variable because we have so few data points. But it seems as if there's variability of between 10 and 20 points on exam grade at each number of absences. The center of that variability decreases as we increase in number of absences, but the range stays roughly the same. It also seems that there's variability of between 2 and 3 absences at each exam grade. Again, the center of that variability decreases as exam grade increases, but the range stays roughly the same.

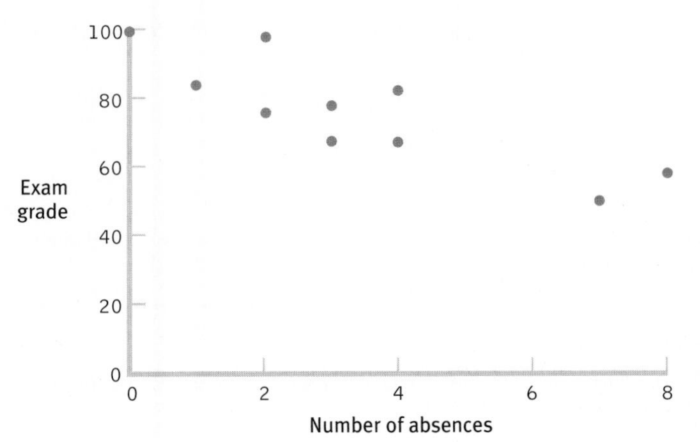

FIGURE 13-7
Using a Scatterplot to Examine the Assumptions
We can use a scatterplot to see whether one variable varies equally at each level of the other variable. With only 10 data points, we can't be certain. But this scatterplot suggests a variability of between 10 and 20 points on exam grade at each number of absences and a variability of between 2 and 3 absences at each exam grade.

STEP 2: State the null and research hypotheses. Null hypothesis: There is no correlation between number of absences and exam grade—H_0: $\rho = 0$. Research hypothesis: There is a correlation between number of absences and exam grade—H_1: $\rho \neq 0$. (*Note:* We use the Greek letter rho because hypotheses are about population parameters.)

STEP 3: Determine the characteristics of the comparison distribution. The comparison distribution is an *r* distribution with degrees of freedom calculated by subtracting 2 from the sample size, which in Pearson correlation is the number of participants rather than the number of scores:

MASTERING THE FORMULA

13-2: When conducting hypothesis testing for the Pearson correlation coefficient, *r*, we calculate degrees of freedom by subtracting 2 from the sample size. In Pearson correlation, the sample size is the number of participants, not the number of scores. The formula is: $df_r = N - 2$.

$$df_r = N - 2$$

In our study, degrees of freedom are calculated as follows:

$$df_r = N - 2 = 10 - 2 = 8$$

So the comparison distribution is an *r* distribution with 8 degrees of freedom.

STEP 4: Determine the critical values, or cutoffs. Now we can look up the critical values in the *r* table in Appendix B. Like the *z* table and the *t* table, the *r* table includes only positive values. For a two-tailed test, we take the negative and positive versions of the critical test statistic indicated in the table. So the critical values for an *r* distribution with 8 degrees of freedom for a two-tailed test with a *p* level of 0.05 are −0.632 and 0.632.

STEP 5: Calculate the test statistic. We already calculated the test statistic, *r*, in the preceding section. It is −0.85.

STEP 6: Make a decision. The test statistic, $r = -0.85$, is larger in magnitude than the critical value of −0.632. We can reject the null hypothesis and conclude that number of absences and exam grade seem to be negatively correlated. ■

CHECK YOUR LEARNING

Reviewing the Concepts

> The Pearson correlation coefficient allows us to quantify the relations that we observe.

> Before we calculate a Pearson correlation coefficient, we must always construct a scatterplot to be sure the two variables are linearly related.

> The Pearson correlation coefficient is calculated in three basic steps. (1) We calculate the deviation of each score from its mean, multiply the two deviations for each person, and sum the products of the deviations. (2) We calculate a sum of squares for each variable, multiply the sums of squares, and take the square root. (3) We divide the sum from step 1 by the square root in step 2.

> We use the six steps of hypothesis testing to determine whether the correlation coefficient is statistically significantly different from 0 on an *r* distribution.

Clarifying the Concepts

13-6 Define the Pearson correlation coefficient.

13-7 The denominator of the correlation equation corrects for which two issues present in the calculation of the numerator?

Calculating the Statistics

13-8 Create a scatterplot for the following data:

Variable A	Variable B
8.0	14.0
7.0	13.0
6.0	10.0
5.0	9.5
4.0	8.0
5.5	9.0
6.0	12.0
8.0	11.0

13-9 Calculate the correlation coefficient for the data provided in Check Your Learning 13-8.

Applying the Concepts

13-10 According to social learning theory, children exposed to aggressive behavior, including family violence, are more likely to engage in aggressive behavior than children who do not witness such violence. Let's assume the data you worked with in Check Your Learning 13-8 and 13-9 represent exposure to violence as the first variable (A) and aggressive behavior as the second variable (B). For both variables, higher values indicate higher levels, either of exposure to violence or of incidents of aggressive behavior. You computed the correlation coefficient, step 5 in hypothesis testing, in Check Your Learning 13-9. Now, complete steps 1, 2, 3, 4, and 6 of hypothesis testing.

Solutions to these Check Your Learning questions can be found in Appendix D.

Applying Correlation in Psychometrics

Here's an in-demand career available to students of the behavioral sciences: ***Psychometrics*** *is the branch of statistics used in the development of tests and measures.* Not surprisingly, *the statisticians and psychologists who develop tests and measures are called* ***psychometricians.*** Psychometricians use the statistical procedures referred to in this textbook, particularly those for which correlation forms the mathematical backbone. Psychometricians make sure that elections are fair, test for cultural biases in standardized tests, identify high-achieving employees, and make a wide range of social contributions—and we don't have nearly enough of them. The *New York Times* reported (Herszenhorn, 2006) a "critical shortage" of such experts and intense competition for the few who are available—offered U.S. salaries as high as $200,000 a year! Psychometricians use correlation to examine two important aspects of the development of measures—reliability and validity.

- **Psychometrics** is the branch of statistics used in the development of tests and measures.
- **Psychometricians** are the statisticians and psychologists who develop tests and measures.
- **Test–retest reliability** refers to whether the scale being used provides consistent information every time the test is taken.

Reliability

In Chapter 1, we defined a reliable measure as one that is consistent. For example, if we measure shyness, then a reliable measure leads to nearly the same score every time a person takes the shyness test. One particular type of reliability is test–retest reliability. ***Test–retest reliability*** *refers to whether the scale being used provides consistent information every time the test is taken.* To calculate a measure's test–retest reliability, the measure is given twice

AP Photo/Michael Manning

Correlation and Reliability Correlation is used by psychometricians to help professional sports teams assess the reliability of athletic performance, such as how fast a pitcher can throw a baseball.

to the *same sample,* typically with a delay between tests. The participants' scores for the first time they complete the measure are correlated with their scores for the second time they complete the measure. A large correlation indicates that the measure yields the same results consistently over time—that is, good test–retest reliability (Cortina, 1993).

Another way to measure the reliability of a test is by assessing its internal consistency in order to verify that all the items were measuring the same idea (DeVellis, 1991). Initially, researchers measured internal consistency via "split-half" reliability, correlating the odd-numbered items (1, 3, 5, etc.) with the even-numbered items (2, 4, 6, etc.). If this correlation coefficient is large, then the test has high internal consistency. The odd–even approach is easy to understand, but computers now allow researchers to take a more sophisticated approach. The computer can calculate the average of *every possible* split-half reliability.

Consider a 10-item measure. The computer can calculate correlations between the odd-numbered items and even-numbered items, between the first 5 items and the last 5 items, between items 1, 2, 4, 8, 10 and items 3, 5, 6, 7, 9, and so on for every combination of two groups of 5 items. The computer can then calculate what is essentially (although not always exactly) the average of all possible split-half correlations (Cortina, 1993). The average of these is called *coefficient alpha* (or *Cronbach's alpha* in honor of the statistician who developed it). ***Coefficient alpha*** *(symbolized as α) is a commonly used estimate of a test or measure's reliability and is calculated by taking the average of all possible split-half correlations.* Coefficient alpha is commonly used across a wide range of fields, including psychology, education, sociology, political science, medicine, economics, criminology, and anthropology (Cortina, 1993). (Note that this alpha is different from the *p* level.)

▶ MASTERING THE CONCEPT

13.6: Correlation is used to calculate reliability either through test–retest reliability or through a measure of internal consistency such as coefficient alpha.

When developing a new scale or measure, how high should its reliability be? It would not be worth using a scale in our research if its coefficient alpha is less than 0.80. However, if we are using a scale to make decisions about individuals—for example, if we are using the SAT or a diagnostic tool—we should aim for a coefficient alpha of 0.90 or even of 0.95 (Nunnally & Bernstein, 1994). We want high reliability when using a test that directly affects people's lives—but the test also needs to be valid.

Validity

In Chapter 1, we defined a valid measure as one that measures what it was designed or intended to measure. Many researchers consider validity to be the most important concept in the field of psychometrics (e.g., Nunnally & Bernstein, 1994). It can be a great deal more work to measure validity than reliability, however, so that work is not always done. In fact, it is quite possible to have a reliable test, one that measures a variable, such as shyness, consistently over time and is internally consistent but is still not valid. Just because the items on a test all measure the same thing doesn't mean that they're measuring what we want them to measure or what we think they are measuring.

▶ MASTERING THE CONCEPT

13.7: Correlation is used to calculate validity, often by correlating a new measure with existing measures known to assess the variable of interest.

For example, *Cosmopolitan* magazine often has quizzes that claim to assess readers' relationships with their boyfriends. If you've ever taken one of these quizzes, you might wonder whether some of the quiz items actually measure what the quiz suggests. One quiz, titled "Is He Devoted to You?," asks "Be honest: Do you ever worry that he might cheat on you?" Does this item assess a man's devotion or a woman's jealousy? Another item asks: "When you introduced him to your closest friends, he said:" and then offers three options—(1) "I've heard so much about all of you! So, how'd you become friends?" (2) "'Hi,' then silence—he looked a bit bored." (3) "'Nice to meet you' with a big smile." Does this measure his devotion or his social skills? Such a quiz might be reliable (you'd consistently get the same score), but it might not be a valid measure of a man's devotion to his girlfriend. Devotion, jealousy, and social skills are different concepts.

Spencer Grant/PhotoEdit

Validity and Personality Quizzes Correlation can also be used to establish the validity of a personality test. Establishing validity is usually much more difficult than establishing reliability. Moreover, most magazines and newspapers never examine the psychometric properties of the quizzes that they publish. Think of most of them as mere entertainment.

It takes a psychometrician who understands correlation to test the validity of such measures. Typically, a psychometrician finds other measures with which to correlate the new measure. So, a new scale to measure anxiety might be correlated with an existing measure known to be valid or with physiological measures of anxiety such as heart rate. If the new anxiety measure correlates with other measures, this is evidence of its validity.

Here's another example concerning validity. In a groundbreaking study on affirmative action in higher education, researchers studied the success of over 35,000 black and white students who attended 1 of 28 highly selective universities (Bowen & Bok, 2000). When determining validity, it is important that we consider how we will operationalize the variable of interest—here, success.

In this study, the researchers first considered the obvious criteria to operationalize success: these students' future graduate education and career achievement. Their findings debunked the myth that black graduates of such institutions did not achieve the successes of their white counterparts. The researchers then went a step further and assessed a success-related criterion very important to the social fabric of a society: graduates' levels of civic and community participation, including political involvement and community service. They found that significantly more black graduates than white graduates of these top institutions were actively involved in their communities. This research changed the nature of the debate on affirmative action through validity—by widening the pool of criteria by which we operationalize success.

- **Coefficient alpha**, symbolized as α, is a commonly used estimate of a test or measure's reliability and is calculated by taking the average of all possible split-half correlations; sometimes called *Cronbach's alpha.*

CHECK YOUR LEARNING

Reviewing the Concepts

> Correlation is a central part of psychometrics, the statistics of the construction of tests and measures.

> Psychometricians, the statisticians who practice psychometrics, use correlation to establish the reliability and the validity of a test.

> Test–retest reliability can be estimated by correlating the same participants' scores on the same test at two different time points.

> Coefficient alpha, now widely used to establish reliability, is essentially calculated by taking the average of all possible split-half correlations (i.e., not just the odds vs. the evens).

Clarifying the Concepts

13-11 How does the field of psychometrics make use of correlation?

13-12 What does coefficient alpha measure and how is it calculated?

continued on next page

Calculating the Statistics

13-13 A researcher is assessing a diagnostic tool for determining whether students should be placed in a remedial reading program. The researcher calculates coefficient alpha and finds that it is 0.85.

a. Does the test have sufficient reliability to be used as a diagnostic tool? Why or why not?

b. Does the test have sufficient validity to be used as a diagnostic tool? Why or why not?

c. What information would we need to appropriately assess the validity of the test?

Applying the Concepts

13-14 Remember the *Cosmopolitan* devotion quiz we referred to when discussing validity? Imagine that the magazine hired a psychometrician to assess the reliability and validity of its quizzes, and she administered this 10-item quiz to 100 female readers of that magazine who had boyfriends.

a. How could the psychometrician establish the reliability of the quiz? That is, which of the methods introduced above could she use in this case? Be specific, and cite at least two ways.

b. How could the psychometrician establish the validity of the quiz? Be specific, and cite at least two ways.

c. Choose one of your criteria from part (b) and explain why it might not actually measure the underlying variable of interest. That is, explain how your criterion itself might not be valid.

Solutions to these Check Your Learning Questions can be found in Appendix D.

REVIEW OF CONCEPTS

Correlation

Correlation is an association between two variables and is quantified by a *correlation coefficient*. A *positive correlation* indicates that a participant who has a high score on one variable is likely to have a high score on the other, and someone with a low score on one variable is likely to have a low score on the other. A *negative correlation* indicates that someone with a high score on one variable is likely to have a low score on the other. All correlation coefficients must fall between −1.00 and 1.00. The strength of the correlation is independent of its sign.

Correlation coefficients are useful, but they can be misleading. When interpreting a correlation coefficient, we must be certain not to confuse correlation with causation. We cannot know the causal direction in which two variables are related from a correlation coefficient, nor can we know if there is a hidden third variable that causes the apparent relation.

The Pearson Correlation Coefficient

The *Pearson correlation coefficient* is used when two scale variables are linearly related, as determined from a scatterplot. Calculating a correlation coefficient involves three steps. (1) We calculate the deviation of each score from its mean, multiply the deviations on each variable for each participant, and sum the products of the deviations. (2) We multiply the sums of squares for each variable, then take the square root of the product.

(3) We divide the sum of the products of the deviations (from step 1) by the square root of the product of the sums of squares (from step 2). We use the six steps of hypothesis testing to determine whether the correlation coefficient is statistically significantly different from 0 on the r distribution.

Applying Correlation in Psychometrics

Psychometrics is the statistics of the development of tests and measures. *Psychometricians* assess the reliability and validity of a test. Reliability is sometimes measured by *test–retest reliability,* whereby participants' scores on the same measure at two different points in time are correlated. With *coefficient alpha,* the computer essentially calculates the average of all possible split-half correlations (e.g., odd and even items, first and second halves of items). Validity is sometimes assessed by correlating a new measure with existing measures that have been shown to be valid.

SPSS®

Enter the data for the example used to calculate the correlation coefficient in this chapter: numbers of absences and exam grades. Be sure to put each student's two scores on the same row.

To view a scatterplot, select: Graphs → Chart Builder → Gallery → Scatter/Dot. Drag the upper-left sample scatterplot to the large box on top. Then select the variables to be included in the scatterplot by dragging the independent variable, absences, to the x-axis and the dependent variable, grade, to the y-axis. Click "OK."

If the scatterplot indicates that we meet the assumptions for a Pearson correlation coefficient, we can analyze the data. Select: Analyze → Correlate → Bivariate. Then select the two variables to be analyzed, absences and grade. "Pearson" will already be checked as the type of correlation coefficient to be calculated. (*Note:* If more than two variables are selected, SPSS will build a correlation matrix of all possible pairs of variables.) Click "OK" to see the Output screen. The screenshot here shows the output for the Pearson correlation coefficient. Notice that the correlation coefficient is −0.851, the same as the coefficient that we calculated by hand earlier. The two asterisks indicate that it is statistically significant at a p level of 0.01.

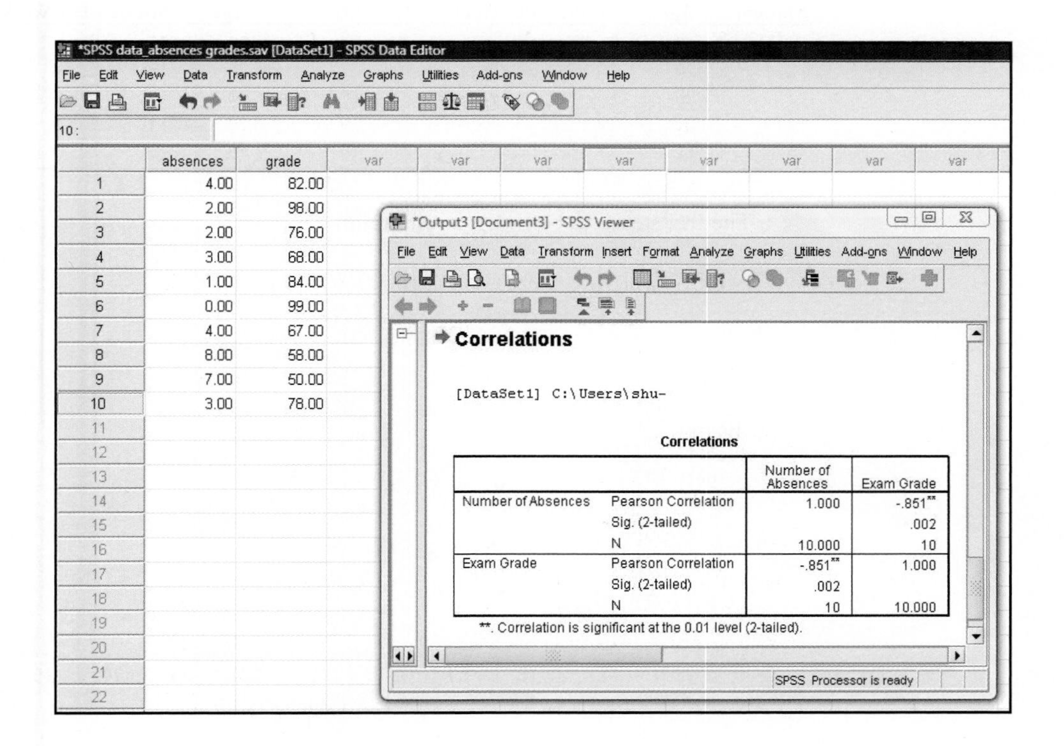

How It Works

13.1 UNDERSTANDING CORRELATION COEFFICIENTS

A researcher gathered data on psychology students' ratings of their likelihood of attending graduate school and the numbers of credits they had completed in their psychology major (Rajecki, Lauer, & Metzner, 1998). Imagine that each of the following numbers represents the Pearson correlation coefficient that quantifies the relation between these two variables. From each coefficient, what do we know about the relation between the two variables?

1. 1.00: This correlation coefficient reflects a perfect positive relation between students' ratings of the likelihood of attending graduate school and the number of psychology credits they completed. This correlation is the strongest correlation of the six options.
2. −0.001: This correlation coefficient reflects a lack of relation between students' ratings and the number of psychology credits they completed. This is the weakest correlation of the six options.
3. 0.56: This correlation coefficient reflects a large positive relation between students' ratings and the number of completed psychology credits.
4. −0.27: This coefficient reflects a medium negative relation between students' ratings and the number of completed psychology credits. (*Note:* This is the actual correlation between these variables found in the study.)
5. −0.98: This coefficient reflects a large (close to perfect) negative relation between students' ratings and the number of psychology credits they have completed.
6. 0.09: This coefficient reflects a small positive relation between students' ratings and the number of completed psychology credits.

13.2 CALCULATING THE PEARSON CORRELATION COEFFICIENT

Is age associated with how much people study? How can we calculate the Pearson correlation coefficient for the accompanying data (taken from students in some of our statistics classes)?

Student	Age	Number of Hours Studied Per Week	Student	Age	Number of Hours Studied Per Week
1	19	5	6	23	25
2	20	20	7	22	15
3	20	8	8	20	10
4	21	12	9	19	14
5	21	18	10	25	15

1. The first step is to construct a scatterplot:

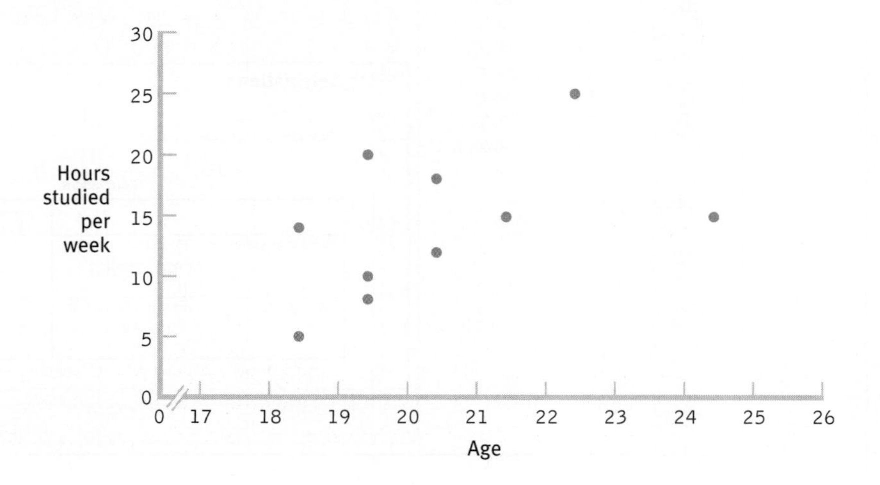

We see from the scatterplot that the data, overall, have a pattern through which we could imagine drawing a straight line. So, it is safe to calculate the Pearson correlation coefficient.

2. The next step is to calculate the numerator of the Pearson correlation coefficient. The numerator is the sum of the product of the deviations for each variable. The mean for age is 21, and the mean for hours studied is 14.2. We use these means to calculate each score's deviation from its mean. We then multiply the deviations for each student's two scores and sum the products of the deviations. Here are the calculations:

Age (X)	$(X - M_X)$	Hours Studied (Y)	$(Y - M_Y)$	$(X - M_X)(Y - M_Y)$
19	−2	5	−9.2	18.4
20	−1	20	5.8	−5.8
20	−1	8	−6.2	6.2
21	0	12	−2.2	0
21	0	18	3.8	0
23	2	25	10.8	21.6
22	1	15	0.8	0.8
20	−1	10	−4.2	4.2
19	−2	14	−0.2	0.4
25	4	15	0.8	3.2
$M_X = 21$		$M_Y = 14.2$		$\Sigma[(X - M_X)(Y - M_Y)] = 49$

The numerator is 49.

3. The next step is to calculate the denominator of the Pearson correlation coefficient. The denominator is the square root of the product of the two sums of squares. We first calculate a sum of squares for each variable. The calculations are here:

Age (X)	$(X - M_X)$	$(X - M_X)^2$	Hours Studied (Y)	$(Y - M_Y)$	$(Y - M_Y)^2$
19	−2	4	5	−9.2	84.64
20	−1	1	20	5.8	33.64
20	−1	1	8	−6.2	38.44
21	0	0	12	−2.2	4.84
21	0	0	18	3.8	14.44
23	2	4	25	10.8	116.64
22	1	1	15	0.8	0.64
20	−1	1	10	−4.2	17.64
19	−2	4	14	−0.2	0.04
25	4	16	15	0.8	0.64
$M_X = 21$	$\Sigma(X - M_X)^2 = 32$		$M_Y = 14.2$		$\Sigma(Y - M_Y)^2 = 311.6$

We now multiply the two sums of squares, then take the square root of the product of the sums of squares.

$$\sqrt{(SS_X)(SS_Y)} = \sqrt{(32)(311.6)} = 99.856$$

4. Finally, we can put the numerator and denominator together to calculate the Pearson correlation coefficient:

$$r = \frac{\Sigma[(X - M_X)(Y - M_Y)]}{\sqrt{(SS_X)(SS_Y)}} = \frac{49}{99.856} = 0.49$$

5. Now that we have calculated the Pearson correlation coefficient (0.49), we determine what the statistic tells us about the direction and the strength of the association between the two variables (age and number of hours studied). This is a positive correlation. Higher ages tend to be associated with longer hours spent studying, and lower ages tend to be associated with fewer hours spent studying.

Exercises

Clarifying the Concepts

13.1 What is a correlation coefficient?

13.2 What is a linear relation?

13.3 Describe a perfect correlation, including its possible coefficients.

13.4 What is the difference between a *positive correlation* and a *negative correlation*?

13.5 What *magnitude* of a correlation coefficient is large enough to be considered important, or worth talking about?

13.6 When we have a straight-line relation between two variables, we use a Pearson correlation coefficient. What does this coefficient describe?

13.7 Explain how the correlation coefficient can be used as a descriptive or an inferential statistic.

13.8 How are deviation scores used in assessing the relation between variables?

13.9 Explain how the sum of the product of deviations determines the sign of the correlation.

13.10 What are the null and research hypotheses for correlations?

13.11 What are the three basic steps to calculate the Pearson correlation coefficient?

13.12 Describe the third assumption of hypothesis testing with correlation.

13.13 What is the difference between test–retest reliability and coefficient alpha?

13.14 Why is a correlation coefficient never greater than 1 (or less than −1)?

Calculating the Statistics

13.15 Determine whether the data in each of the graphs provided would result in a negative or positive correlation coefficient.

a.

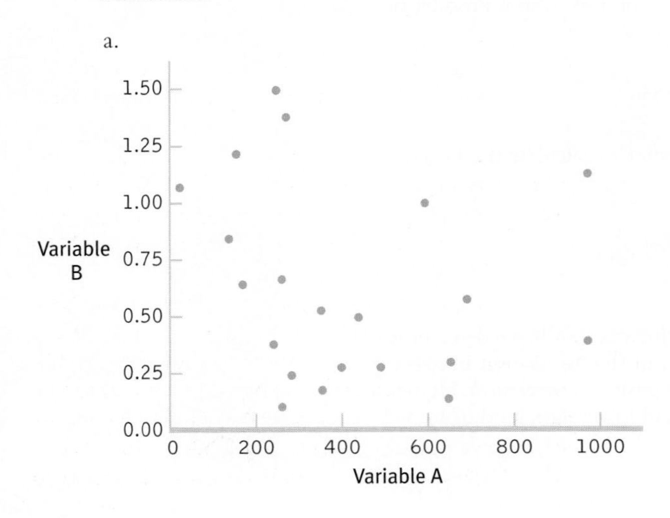

b.

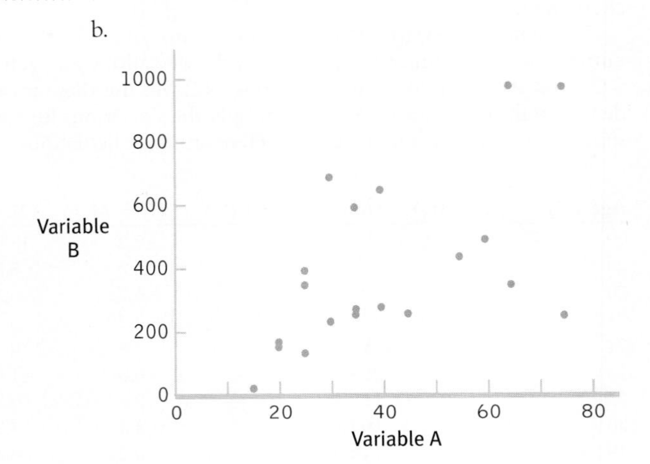

c.

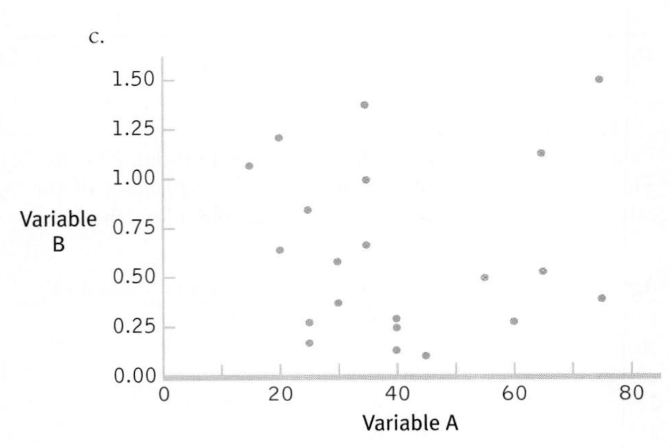

13.16 Decide which of the three correlation coefficient values below goes with each of the scatterplots presented in Exercise 13.15 above.

a. 0.545

b. 0.018

c. −0.20

13.17 Use Cohen's guidelines to describe the strength of the following correlation coefficients:

a. −0.28

b. 0.79

c. 1.0

d. −0.015

13.18 For each of the pairs of correlation coefficients provided, determine which one indicates a stronger relation between variables:

a. −0.28 and −0.31

b. 0.79 and 0.61

c. 1.0 and −1.0

d. −0.15 and 0.13

13.19 Using the following data:

X	Y	X	Y
0.13	645	0.57	689
0.27	486	0.84	137
0.49	435	0.64	167

a. Create a scatterplot.

b. Calculate deviation scores and products of the deviations for each individual, and then sum all products. This is the numerator of the correlation coefficient equation.

c. Calculate the sum of squares for each variable. Then compute the square root of the product of the sums of squares. This is the denominator of the correlation coefficient equation.

d. Divide the numerator by the denominator to compute the coefficient, r.

e. Calculate degrees of freedom.

f. Determine the critical values, or cutoffs, assuming a two-tailed test with a p level of 0.05.

13.20 Using the following data:

X	Y	X	Y
394	25	276	40
972	75	254	45
349	25	156	20
349	65	248	75
593	35		

a. Create a scatterplot.

b. Calculate deviation scores and products of the deviations for each individual, and then sum all products. This is the numerator of the correlation coefficient equation.

c. Calculate the sum of squares for each variable. Then compute the square root of the product of the sums of squares. This is the denominator of the correlation coefficient equation.

d. Divide the numerator by the denominator to compute the coefficient, r.

e. Calculate degrees of freedom.

f. Determine the critical values, or cutoffs, assuming a two-tailed test with a p level of 0.05.

13.21 Using the following data:

X	Y	X	Y
40	60	15	20
45	55	35	40
20	30	65	30
75	25		

a. Create a scatterplot.

b. Calculate deviation scores and products of the deviations for each individual, and then sum all products. This is the numerator of the correlation coefficient equation.

c. Calculate the sum of squares for each variable. Then compute the square root of the product of the sums of squares. This is the denominator of the correlation coefficient equation.

d. Divide the numerator by the denominator to compute the coefficient, r.

e. Calculate degrees of freedom.

f. Determine the critical values, or cutoffs, assuming a two-tailed test with a p level of 0.05.

13.22 Calculate degrees of freedom and the critical values, or cutoffs, assuming a two-tailed test with a p level of 0.05, for each of the following designs:

a. Forty students were recruited for a study about the relation between knowledge regarding academic integrity and values held by students, with the idea that students with less knowledge would care less about the issue than students with greater amounts of knowledge.

b. Twenty-seven couples are surveyed regarding their years together and their relationship satisfaction.

c. Data are collected to examine the relation between size of dog and rate of bone and joint health issues. Veterinarians from around the country contributed data on 3113 dogs.

d. Hours spent studying per week was correlated with credit-hour load for 72 students.

13.23 A researcher is deciding among three diagnostic tools. The first has a coefficient alpha of 0.82, the second has one of 0.95, and the third has one of 0.91. Based on this information, which tool would you suggest she use and why?

13.24 Which of the following is not a possible coefficient alpha: 1.67, 0.12, −0.88? Explain your answer.

Applying the Concepts

13.25 **Debunking astrology with correlation:** The *New York Times* reported that an officer of the International Society for Astrological Research, Anne Massey, stated that a certain phase of the planet Mercury, the retrograde phase, leads to breakdowns in areas as wide-ranging as communication and travel (Newman, 2006). The *Times* reporter, Andy Newman, documented the likelihood of breakdown on a number of variables, and discovered that, contrary to Massey's hypothesis, New Jersey Transit commuter trains were less likely to be late, although by just 0.4%, during the retrograde phase. On the other hand, consistent with Massey's hypothesis, the rate of baggage complaints at LaGuardia airport increased a tiny amount during retrograde periods. Newman's findings were contradictory across all examined variables—rates of theft, computer crashes, traffic disruptions, delayed plane arrivals—with some variables backing Massey and others not. Transportation statistics expert Bruce Schaller said, "If all of this is due to randomness, that's the result you'd expect." Astrologer Massey counters that the pattern she predicts would only emerge across thousands of years of data.

a. Do reporter Newman's data suggest a correlation between Mercury's phase and breakdowns?
b. Why might astrologer Massey believe there is a correlation? Discuss the confirmation bias and illusory correlations (Chapter 5) in your answer.
c. How do transportation expert Schaller's statement and Newman's contradictory results relate to what you learned about probability in Chapter 5? Discuss expected relative-frequency probability in your answer.
d. If there were indeed a small correlation that one could observe only across thousands of years of data, how useful would that knowledge be in terms of predicting events in your own life?
e. Write a brief response to Massey's contention of a correlation between Mercury's phases and breakdowns in aspects of day-to-day living.

13.26 Obesity, age at death, and correlation: In a newspaper column, Paul Krugman (2006) mentioned obesity (as measured by body mass index) as a possible correlate of age at death.

a. Describe the likely correlation between these variables. Is it likely to be positive or negative? Explain.
b. Draw a scatterplot that depicts the correlation you described in part (a).

13.27 Exercise, number of friends, and correlation: Does the amount that people exercise correlate with the number of friends they have? The accompanying table contains data collected in some of our statistics classes. The first and third columns show hours exercised per week and the second and fourth columns show the number of close friends reported by each participant.

Exercise	Friends	Exercise	Friends
1	4	8	4
0	3	2	4
1	2	10	4
6	6	5	7
1	3	4	5
6	5	2	6
2	4	7	5
3	5	1	5
5	6		

a. Create a scatterplot of these data. Be sure to label both axes.
b. What does the scatterplot suggest about the relation between these two variables?
c. Would it be appropriate to calculate a Pearson correlation coefficient? Explain your answer.

13.28 Externalizing behavior, anxiety, and correlation: A study on the relation between rejection and depression in adolescents (Nolan, Flynn, & Garber, 2003) also collected data on externalizing behaviors (e.g., acting out in negative ways, such as causing fights) and anxiety. They wondered whether externalizing behaviors were related to feelings of anxiety. Some of the data are presented in the accompanying table.

Externalizing	Anxiety	Externalizing	Anxiety
9	37	6	33
7	23	2	26
7	26	6	35
3	21	6	23
11	42	9	28

a. Create a scatterplot of these data. Be sure to label both axes.
b. What does the scatterplot suggest about the relation between these two variables?
c. Would it be appropriate to calculate a Pearson correlation coefficient? Explain your answer.
d. Construct a second scatterplot, but this time add a participant who scored 1 on externalizing and 45 on anxiety. Would you expect the correlation coefficient to be positive or negative now? Small in magnitude or large in magnitude?
e. The Pearson correlation coefficient for the first set of data is 0.65; for the second set of data it is 0.12. Explain why the correlation changed so much with the addition of just one participant.

13.29 Externalizing behavior, anxiety, and hypothesis testing for correlation: Using the data in Exercise 13.28, perform all six steps of hypothesis testing to explore the relation between externalizing and anxiety.

13.30 Direction of a correlation: For each of the following pairs of variables, would you expect a positive correlation or a negative correlation between the two variables? Explain your answer.

a. How hard the rain is falling and your commuting time
b. How often you say no to dessert and your body fat
c. The amount of wine you consume with dinner and your alertness after dinner

13.31 Cats, mental health problems, and the direction of a correlation: You may be aware of the stereotype about the "crazy" person who owns a lot of cats. Have you wondered whether the stereotype is true? As a researcher, you decide to assess 100 people on two variables: (1) the number of cats they own, and (2) their level of mental health problems (a higher score indicates more problems).

a. Imagine that you found a positive relation between these two variables. What might you expect for someone who owns a lot of cats? Explain.
b. Imagine that you found a positive relation between these two variables. What might you expect for someone who owns no cats or just one cat? Explain.
c. Imagine that you found a negative relation between these two variables. What might you expect for someone who owns a lot of cats? Explain.
d. Imagine that you found a negative relation between these two variables. What might you expect for someone who owns no cats or just one cat? Explain.

13.32 Cats, mental health problems, and scatterplots: Consider the scenario in Exercise 13.31 again. The two variables under consideration were (1) number of cats owned, and (2) level of mental health problems (with a higher score indicating more problems). Each possible relation between these variables would be represented by a different scatterplot. Using data for approximately 10 participants, draw a scatterplot that depicts a correlation between these variables for each of the following:

a. A weak positive correlation
b. A strong positive correlation
c. A perfect positive correlation
d. A weak negative correlation
e. A strong negative correlation
f. A perfect negative correlation
g. No (or almost no) correlation

13.33 Trauma, femininity, and correlation: Graduate student Angela Holiday (2007) conducted a study examining perceptions of combat veterans suffering from mental illness. Participants read a description of either a male or female soldier who had recently returned from combat in Iraq and who was suffering from depression. Participants rated the situation (combat in Iraq) with respect to how traumatic they believed it was; they also rated the combat veterans on a range of variables, including scales that assessed how masculine and how feminine they perceived the person to be. Among other analyses, Holiday examined the relation between the perception of the situation as traumatic and the perception of the veteran as being masculine or feminine. When the person was male, the perception of the situation as traumatic was strongly positively correlated with the perception of the man as feminine but was only weakly positively correlated with the perception of the man as masculine. What would you expect when the person was female? The accompanying table presents some of the data for the perception of the situation as traumatic (on a scale of 1–10, with 10 being the most traumatic) and the perception of the woman as feminine (on a scale of 1–10, with 10 being the most feminine).

Traumatic	Feminine	Traumatic	Feminine
5	6	5	6
6	5	7	4
4	6	8	5

a. Draw a scatterplot for these data. Does the scatterplot suggest that it is appropriate to calculate a Pearson correlation coefficient? Explain.
b. Calculate the Pearson correlation coefficient.
c. State what the Pearson correlation coefficient tells us about the relation between these two variables.
d. Explain why the pattern of pairs of deviation scores enables us to understand the relation between the two variables. (That is, consider whether pairs of deviations tend to have the same sign or opposite signs.)

13.34 Trauma, femininity, and hypothesis testing for correlation: Using the data and your work in Exercise 13.33, perform the remaining five steps of hypothesis testing to explore the relation between trauma and femininity. In step 6, be sure to evaluate the size of the correlation using Cohen's guidelines. [You completed step 5, the calculation of the correlation coefficient, in 13.33(b).]

13.35 Trauma, masculinity, and correlation: See the description of Holiday's experiment in Exercise 13.33. We calculated the correlation coefficient for the relation between the perception of a situation as traumatic and the perception of a woman's femininity. Now let's look at data to examine the relation between the perception of a situation as traumatic and the perception of a woman's masculinity.

Traumatic	Feminine	Traumatic	Feminine
5	3	5	2
6	3	7	4
4	2	8	3

a. Draw a scatterplot for these data. Does the scatterplot suggest that it is appropriate to calculate a Pearson correlation coefficient? Explain.
b. Calculate the Pearson correlation coefficient.
c. State what the Pearson correlation coefficient tells us about the relation between these two variables.

d. Explain why the pattern of pairs of deviation scores enables us to understand the relation between the two variables. (That is, consider whether pairs of deviation scores tend to share the same sign or to have opposite signs.)

e. Explain how the relations between the perception of a situation as traumatic and the perception of a woman as either masculine or feminine differ from those same relations with respect to men.

13.36 Trauma, masculinity, and hypothesis testing for correlation: Using the data and your work in Exercise 13.35, perform the remaining five steps of hypothesis testing to explore the relation between trauma and masculinity. In step 6, be sure to evaluate the size of the correlation using Cohen's guidelines. [You completed step 5, the calculation of the correlation coefficient, in 13.35 (b).]

13.37 Traffic, running late, and bias: A friend tells you that there is a correlation between how late she's running and the amount of traffic. Whenever she's going somewhere and she's behind schedule, there's a lot of traffic. And when she has plenty of time, the traffic is sparser. She tells you that this happens no matter what time of day she's traveling or where she's going. She concludes that she's cursed with respect to traffic.

a. Explain to your friend how other phenomena, such as coincidence, superstition, and the confirmation bias (Chapter 5), might explain her conclusion.

b. How could she quantify the relation between these two variables: the degree to which she is late and the amount of traffic? In your answer, be sure to explain how you might operationalize these variables. Of course, these could be operationalized in many different ways.

13.38 IQ-boosting water and illusory correlation: The trashy tabloid *Weekly World News* published an article—"Water from Mountain Falls Can Make You a Genius"—stating that drinking water from a special waterfall in a secret location in Switzerland "boosts IQ by 14 points—in the blink of an eye!" (exclamation point in the original). Hans and Inger Thurlemann, two hikers lost in the woods, drank some of the water, noticed an improvement in their thinking, and instantly found their way out of the woods. The more water they drank, the smarter they seemed to get. They credited the "miracle water" with enhancing their IQs. They brought some of the water home to their friends, who also claimed to notice an improvement in their thinking. Explain how a reliance on anecdotes led the Thurlemanns to perceive an illusory correlation (Chapter 5).

13.39 Driving a convertible, correlation, and causality: How safe are convertibles? *USA Today* (Healey, 2006) examined the pros and cons of convertible automobiles. The Insurance Institute for Highway Safety, the newspaper reported, determined that, depending on the model, 52 to 99 drivers of 1 million registered convertibles died in a car crash. The average rate of deaths for all passenger cars was 87. "Counter to conventional wisdom," the reporter wrote, "convertibles generally aren't unsafe."

a. What does the reporter suggest about the safety of convertibles?

b. Can you think of another explanation for the fairly low fatality rates? (*Hint:* The same article reported that convertibles "are often second or third cars.")

c. Given your explanation in part (b), suggest data that might make for a more appropriate comparison.

13.40 Standardized tests, correlation, and causality: A *New York Times* editorial ("Public vs. Private Schools," 2006) cited a finding by the U.S. Department of Education that standardized test scores were significantly higher among students in private schools than among students in public schools.

a. What are the researchers suggesting with respect to causality?

b. How could this correlation be explained by reversing the direction of hypothesized causality? Be specific.

c. How might a third variable account for this correlation? Be specific. Note that there are many possible third variables.
(*Note:* In the actual study, the difference between types of school disappeared when the researchers statistically controlled for related third variables including race, gender, parents' education, and family income.)

13.41 Athletes' grades, scatterplots, and correlation: At the university level, the stereotype of the "dumb jock" might be strong and ever present; however, a fair amount of research shows that athletes maintain decent grades and competitive graduation rates when compared to nonathletes. Let's play with some data to explore the relation between grade point average (GPA) and participation in athletics. Data are presented here for a hypothetical basketball team, including the GPA on a scale of 0.00 to 4.00 for each athlete and the average number of minutes played per game.

Minutes	GPA	Minutes	GPA
29.70	3.20	6.88	2.89
32.14	2.88	6.38	2.24
32.72	2.78	15.83	3.35
21.76	3.18	2.50	3.00
18.56	3.46	4.17	2.18
16.23	2.12	16.36	3.50
11.80	2.36		

a. Create a scatterplot of these data and describe your impression of the relation between these variables based on the scatterplot.

b. Compute the Pearson correlation coefficient for these data.

c. Explain why the correlation coefficient you just computed is a descriptive statistic, not an inferential statistic. What would you need to do to make this an inferential statistic?

d. Perform the six steps of hypothesis testing.

e. What limitations are there to the conclusions you can draw based on this correlation?

f. How else could you have studied this phenomenon such that you might have been able to draw a more sound, causal conclusion?

13.42 Romantic love, brain activation, and reliability: Aron and colleagues (2005) found a correlation between intense romantic love [as assessed by the Passionate Love Scale (PLS)] and activation in a specific region of the brain [as assessed by functional magnetic resonance imaging (fMRI)]. The PLS (Hatfield & Sprecher, 1986) assessed the intensity of romantic love by asking people in romantic relationships to respond to a series of questions, such as "I want ______ physically, emotionally, and mentally" and "Sometimes I can't control my thoughts; they are obsessively on ______," replacing the blanks with the name of their partner.

a. How might we examine the reliability of this measure using test–retest reliability techniques? Be specific and explain the role of correlation.

b. Would test–retest reliability be appropriate for this measure? That is, is there likely to be a practice effect? Explain.

c. How could we examine the reliability of this measure using coefficient alpha? Be specific and explain the role of correlation.

d. Coefficient alpha in this study was 0.81. Based on coefficient alpha, was the use of this scale in this study warranted? Explain.

e. What is the idea that this measure is trying to assess?

f. What would it mean for this measure to be valid? Be specific.

13.43 A biased exam question, validity, and correlation: New York State's fourth-grade English exam led to an outcry from parents because of a question that was perceived to be an unfair measure of fourth graders' performance. Students read a story, "Why the Rooster Crows at Dawn," that described an arrogant rooster who claims to be king, and Brownie, "the kindest of all the cows," who eventually acts in a mean way toward the rooster. In the beginning the rooster does whatever he wants, but by the end, the cows, led by Brownie, have convinced him that as self-proclaimed king, he must be the first to wake up in the morning and the last to go to sleep. To the cows' delight, the arrogant rooster complies. Students were then asked to respond to several questions about the story, including one that asked: "What causes Brownie's behavior to change?" Several parents started a Web site, http://browniethecow.org, to point out problems with the test, particularly with this question. Students, they argued, were confused because it seemed that it was the rooster's behavior, not the cow's behavior, that changed. The correct answer, according to a quote on the Web site from an unnamed state official, was that the cow started out kind and ended up mean.

a. This test item was supposed to evaluate writing skill. According to the Web site, test items should lead to good student writing; be unambiguous; test for writing, not another skill; and allow for objective, reliable scoring. If students were marked down for talking about the rooster rather than the cow, as alleged by the Web site, would it meet these criteria? Explain. Does this seem to be a valid question? Explain.

b. The Web site states that New York City schools use the tests to, among other things, evaluate teachers and principals. The logic behind this, ostensibly, is that good teachers and administrators cause higher test performance. List at least two possible third variables that might lead to better performance in some schools than in other schools, other than the presence of good teachers and administrators.

13.44 Holiday weight gain, reliability, and validity: The *Wall Street Journal* reported on a study of holiday weight gain. Researchers assessed weight gain by asking people how much weight they typically gain in the fall and winter (Parker-Pope, 2005). The average answer was 2.3 kilograms. But a study of actual weight gain over this period found that people gained, on average, 0.48 kilogram.

a. Is the method of asking people about their weight gain likely to be reliable? Explain.

b. Is this method of asking people about their weight gain likely to be valid? Explain.

Putting It All Together

13.45 Health care spending, longevity, and correlation: *New York Times* columnist Paul Krugman (2006) used the idea of correlation in a newspaper column when he asked, "Is being an American bad for your health?" Krugman explained that the United States has higher per capita spending on health care than any country in the world and yet is surpassed by many countries in life expectancy (Krugman cited a study published in the *Journal of the American Medical Association:* Banks, Marmot, Oldfield, & Smith, 2006).

a. Name the "participants" in this study.

b. What are the two scale variables being studied, and how was each of them operationalized? Suggest at least one alternate way, other than life expectancy, to operationalize health.

c. What was the study finding, and why might this finding be surprising? If the finding described above holds true across countries, would this be a negative correlation or a positive correlation? Explain.

d. Some people thought race or income might be a third variable related to higher spending and lower life expectancy. But Krugman further reported that a comparison of non-Hispanic white people from America and from England (thus taking race out of the equation) yielded a surprising finding: The wealthiest third of Americans have poorer health than do even the *least* wealthy third of the English. What are some other possible third variables that might affect both of the variables in this study?

e. Why is this research considered a correlational study rather than a true experiment?

f. Why would it not be possible to conduct a true experiment to determine whether the amount of health care spending causes changes in health?

13.46 Availability of food, amount eaten, and correlation: Did you know that sometimes you eat more just because the food is in front of you? Geier, Rozin, and Doros (2006) studied how portion size affected the amount people consumed. They discovered interesting things such as that people eat more M&M's when the candies are dispensed using a big spoon as compared with when a small spoon is used. They investigated whether people eat more when more food is available. Hypothetical data are presented below for the amount of candy presented in a bowl for customers to take and the amount of candy taken by the end of each day of the study:

Number of Pieces Presented	Number of Pieces Taken
10	3
25	14
50	26
75	44
100	36
125	57
150	41

a. Create a scatterplot of these data.

b. Describe your impression of the relation between these variables based on the scatterplot.

c. Compute the Pearson correlation coefficient for these data.

d. Summarize your findings using Cohen's guidelines.

e. Perform the remainder of the six steps of hypothesis testing.

f. What limitations are there to the conclusions you can draw based on this correlation?

g. Use the A-B-C model to explain possible causes for the relation between these variables.

Terms

correlation coefficient (p. 344)
positive correlation (p. 345)
negative correlation (p. 345)
Pearson correlation coefficient (p. 349)
psychometrics (p. 355)
psychometricians (p. 355)
test–retest reliability (p. 355)
coefficient alpha (p. 356)

Formulas

$$r = \frac{\Sigma[(X - M_X)(Y - M_Y)]}{\sqrt{(SS_X)(SS_Y)}}$$ (p. 351)

$$df_r = N - 2$$ (p. 354)

Symbols

r (p. 344)
ρ (p. 349)
α (p. 356)

CHAPTER 14

Regression

BEFORE YOU GO ON

- You should understand the six steps of hypothesis testing (Chapter 7).
- You should understand the concept of effect size (Chapter 8).
- You should understand the concept of correlation (Chapter 13).
- You should be able to explain the limitations of correlation (Chapter 13).

Ian Dagnall/Alamy

Facebook and Social Capital The prediction tools introduced in this chapter helped researchers determine that increased use of Facebook predicts higher levels of social capital.

In 2004, college student Mark Zuckerberg created the social networking site Facebook, which soon exploded in popularity across college campuses. By 2012, Facebook.com reported having almost 1 billion users, and it raised $16 billion in a public offering of its stock. As Facebook use ballooned, researchers at Michigan State University (MSU) (Ellison, Steinfeld, & Lampe, 2007) wanted to understand what college students were getting out of their Facebook relationships—that is, what they were gaining in *social capital.*

To find out, Ellison and colleagues' study focused on the idea of "bridging" social capital, the loose social connections we think of as acquaintances rather than friends. The researchers' hypothesis was that greater use of Facebook would predict more of this type of social capital. Researchers measured this by asking students to rate several items, such as "I feel I am part of the MSU community" and "At MSU, I come into contact with new people all the time."

Obviously, many influences determined how much social capital students get from Facebook, including the amount of time they spend on the site. Moreover, answers to the research question were complicated by gender, ethnicity, location of residence, and many other factors. In other words, to find out what students were getting out of their Facebook relationships, researchers had to account for the influence of many variables. The Michigan State University researchers controlled for all of these variables in their study and found that the more students used Facebook, the higher they tended to score on a measure of social capital.

The analytical methods we learn in this chapter build on correlation to help us to create prediction tools. We learn how to use one scale variable to predict outcome on a second scale variable. Then we discuss the limitations of this method—limitations that are similar to those we encountered with correlation. Finally, we expand this analytical method to allow us to use multiple scale variables to predict outcome on another scale variable.

Simple Linear Regression

Correlation is a marvelous tool that allows us to know the direction and strength of a relation between two variables. We can also use a correlation coefficient to develop a prediction tool—an equation to predict a person's score on a scale dependent variable from his or her score on a scale independent variable. For instance, the research team at Michigan State could predict a high score on a measure of social capital for a student who spends a lot of time on Facebook.

▪ **Simple linear regression** is a statistical tool that lets us predict a person's score on the dependent variable from his or her score on one independent variable.

Statistical prediction is powerful stuff! Many universities use variables such as high school grade point average (GPA) and Scholastic Aptitude Test (SAT) score to predict the success of prospective students. Similarly, insurance companies input demographic data into an equation to predict the likelihood of a class of people (such as young male

Jeff Greenberg/Photo Edit

Prediction and Car Insurance
When you call an insurance company for a car insurance estimate, the salesperson asks a number of questions about you (e.g., age, gender, marital status) and about your car (e.g., make, model, year, color). These characteristics are input into a type of statistical equation; the output is your quote. A flashy, expensive car driven by a young, unmarried man leads to a higher quote than a basic sedan driven by a married 50-year-old woman.

drivers) to submit a claim. Mark Zuckerberg, the founder of Facebook, is even alleged to have used data from Facebook users to predict breakups of romantic relationships! He used independent variables, such as the amount of time looking at others' Facebook profiles, changes in postings to others' Facebook walls, and photo-tagging patterns, to predict the dependent variable of the end of a relationship as evidenced by the user's Facebook relationship status. He was right one-third of the time ("Can Facebook Predict Your Breakup?," 2010).

Prediction versus Relation

The name for the prediction tool that we've been discussing is *regression,* a statistical technique that can provide specific quantitative information that predicts relations between variables. More specifically, ***simple linear regression*** *is a statistical tool that lets us predict a person's score on a dependent variable from his or her score on one independent variable.*

> **MASTERING THE CONCEPT**
>
> **14.1:** Simple linear regression allows us to determine an equation for a straight line that predicts a person's score on a dependent variable from his or her score on the independent variable. We can only use it when the data are approximately linearly related.

Simple linear regression works by calculating the equation for a straight line. Once we have a line, we can look at any point on the x-axis and find its corresponding point on the y-axis. That corresponding point is what we predict for y. (*Note:* Like the Pearson correlation coefficient, simple linear regression would not be used if the data do not form the pattern of a straight line.) Let's consider an example of research that uses regression techniques, and then walk through the steps to develop a regression equation.

Christopher Ruhm, an economist, often uses regression in his research. In one study, he wanted to explore the reasons for his finding (Ruhm, 2000) that the death rate *decreases* when unemployment goes up—a surprising negative relation between the death rate and an economic indicator. He took this relation a step further, into the realm of prediction: He found that an increase of 1% in unemployment predicted a decrease in the death rate of 0.5%, on average. A *poorer* economy predicted *better* health!

To explore the reasons for this surprising finding, Ruhm (2006) conducted regression analyses for independent variables related to health (smoking, obesity, and physical activity) and dependent variables related to the economy (income, unemployment, and the length of the workweek). He analyzed data from a sample of nearly 1.5 million participants collected from telephone surveys between 1987 and 2000. Among other things, Ruhm found that a decrease in working hours predicted decreases in smoking, obesity, and physical inactivity.

Regression can take us a step beyond correlation. Regression can provide specific quantitative predictions that more precisely explain relations among variables. For example, Ruhm reported that a decrease in the workweek of just 1 hour predicted a 1% decrease in physical inactivity. Ruhm suggested that shorter working hours free up time for physical activity—something he might not have thought of without the more specific quantitative information provided by regression. Let's now conduct a simple linear regression analysis using information that we're already familiar with: z scores.

Regression with *z* Scores

In Chapter 13 we calculated a Pearson correlation coefficient to quantify the relation between students' numbers of absences from statistics class and their statistics final exam grades; the data for the 10 students in the sample are shown in Table 14-1. Remember, the mean number of absences was 3.400; the standard deviation for these data is 2.375. The mean exam grade was 76.000; the standard deviation for these data is 15.040. The Pearson correlation coefficient that we calculated in Chapter 13 was −0.85, but simple linear regression can take us a step further. We can develop an equation to predict students' final exam grades from their numbers of absences.

Let's say that a student (let's call him Skip) announces on the first day of class that he intends to skip five classes during the semester. We can refer to the size and direction of the correlation (−0.85) as a benchmark to predict his final exam grade. To predict his grade, we unite regression with a statistic we are more familiar with: z scores. If we know Skip's z score on one variable, we can multiply by the correlation coefficient to calculate his predicted z score on the second variable. Remember that z scores indicate how far a participant falls from the mean in terms of standard deviations. The formula, called the *standardized regression equation* because it uses z scores, is:

MASTERING THE FORMULA

14-1: The standardized regression equation predicts the z score of a dependent variable, Y, from the z score of an independent variable, X. We simply multiply the independent variable's z score by the Pearson correlation coefficient to get the predicted z score on the dependent variable: $z_{\hat{Y}} = (r_{XY})(z_X)$.

$$z_{\hat{Y}} = (r_{XY})(z_X)$$

The subscripts in the formula indicate that the first z score is for the dependent variable, Y, and that the second z score is for the independent variable, X. The ^ symbol over

TABLE 14-1. Is Skipping Class Related to Exam Grades?

Here are the scores for 10 students on two scale variables, number of absences from class in one semester and the final exam grade for that semester. The correlation between these variables is −0.85, but regression can take us a step further. We can develop a regression equation to assist with prediction.

Student	Absences	Exam Grade
1	4	82
2	2	98
3	2	76
4	3	68
5	1	84
6	0	99
7	4	67
8	8	58
9	7	50
10	3	78

the subscript Y, called a "hat" by statisticians, refers to the fact that this variable is predicted. This is the z score for "Y hat"—the z score for the *predicted* score on the dependent variable, not the actual score. We cannot, of course, predict the actual score, and the "hat" reminds us of this. When we refer to this score, we can either say "the predicted score for Y" (with no hat, because we have specified with words that it is predicted) or we can use the hat, $\hat{Y}$, to indicate that it is predicted. (We would not use both expressions because that would be redundant.) The subscripts X and Y for the Pearson correlation coefficient, r, indicate that this is the correlation between variables X and Y.

EXAMPLE 14.1

If Skip's projected number of absences was identical to the mean number of absences for the entire class, then he'd have a z score of 0. If we multiply that by the correlation coefficient, then he'd have a predicted z score of 0 for final exam grade:

$$z_{\hat{Y}} = (-0.85)(0) = 0$$

So if he's right at the mean on the independent variable, then we'd predict that he'd be right at the mean on the dependent variable.

If Skip missed more classes than average and had a z score of 1.0 on the independent variable (1 standard deviation above the mean), then his predicted score on the dependent variable would be -0.85 (that is, 0.85 standard deviation below the mean):

$$z_{\hat{Y}} = (-0.85)(1) = -0.85$$

If his z score was -2 (that is, if it was 2 standard deviations below the mean), his predicted z score on the dependent variable would be 1.7 (that is, 1.7 standard deviations above the mean):

$$z_{\hat{Y}} = (-0.85)(-2) = 1.7$$

Notice two things: First, because this is a negative correlation, a score above the mean on absences predicts a score below the mean on grade, and vice versa. Second, the predicted z score on the dependent variable is closer to its mean than is the z score on the independent variable. Table 14-2 illustrates this for several z scores.

TABLE 14-2. Regression to the Mean

One reason that regression equations are so named is because they predict a z score on the dependent variable that is closer to the mean than is the z score on the independent variable. This phenomenon is often called *regression to the mean.* The following predicted z scores for the dependent variable, Y, were calculated by multiplying the z score for the independent variable, X, by the Pearson correlation coefficient of -0.85.

z Score for the Independent Variable, X	Predicted z Score for the Dependent Variable, Y
−2.0	1.70
−1.0	0.85
0.0	0.00
1.0	−0.85
2.0	−1.70

This regressing of the dependent variable—the fact that it is closer to its mean—is called ***regression to the mean***, *the tendency of scores that are particularly high or low to drift toward the mean over time.*

In the social sciences, many phenomena demonstrate regression to the mean. For example, parents who are very tall tend to have children who are somewhat shorter than they are, although probably still above average. And parents who are very short tend to have children who are somewhat taller than they are, although probably still below average. We explore this concept in more detail later in this chapter.

When we don't have a person's z score on the independent variable, we have to perform the additional step of converting his or her raw score to a z score. In addition, when we calculate a predicted z score on the dependent variable, we can use the formula that determines a raw score from a z score. Let's try it with the skipping class and exam grade example using Skip as the subject. ■

EXAMPLE 14.2

We already know that Skip has announced his plans to skip five classes. What would we predict for his final exam grade?

STEP 1: Calculate the *z* score.

We first have to calculate Skip's z score on number of absences. Using the mean (3.400) and the standard deviation (2.375) that we calculated in Chapter 13, we calculate:

$$z_X = \frac{(X - M_X)}{SD_X} = \frac{(5 - 3.400)}{2.375} = 0.674$$

STEP 2: Multiply the *z* score by the correlation coefficient.

We multiply this z score by the correlation coefficient to get his predicted z score on the dependent variable, final exam grade:

$$z_{\hat{Y}} = (r_{XY})(z_X) = (-0.85)(0.674) = -0.573$$

STEP 3: Convert the *z* score to a raw score.

We convert from the predicted z score on Y, -0.573, to a predicted raw score for Y:

$$\hat{Y} = z_{\hat{Y}}(SD_Y) + M_Y = -0.573(15.040) + 76.000 = 67.38$$

If Skip skipped five classes, this number would reflect more classes than the typical student skipped, so we would expect him to earn a lower-than-average grade. And the formula makes this very prediction—that Skip's final exam grade would be 67.38, which is lower than the mean (76.00).

The admissions counselor, the insurance salesperson, and Mark Zuckerberg of Facebook, however, are unlikely to have the time or interest to do conversions from raw scores to z scores and back. So the z score regression equation is not useful in a practical sense for situations in which we must make ongoing predictions using the same variables. It is very useful, however, as a tool to help us develop a regression equation we can use with raw scores, a procedure we look at in the next section. ■

- **Regression to the mean** is the tendency of scores that are particularly high or low to drift toward the mean over time.
- The **intercept** is the predicted value for Y when X is equal to 0, which is the point at which the line crosses, or intercepts, the y-axis.
- The **slope** is the amount that Y is predicted to increase for an increase of 1 in X.

Determining the Regression Equation

You may remember the equation for a line that you learned in geometry class. The version you likely learned was: $y = m(x) + b$. (In this equation, b is the intercept and m is the slope.) In statistics, we use a slightly different version of this formula:

$$\hat{Y} = a + b(X)$$

MASTERING THE FORMULA

14-2: The simple linear regression equation uses the formula: $\hat{Y} = a + b(X)$. In this formula, X is the raw score on the independent variable and $\hat{Y}$ is the predicted raw score on the dependent variable. a is the intercept of the line, and b is its slope.

In the regression formula, *a is the* ***intercept****, the predicted value for Y when X is equal to 0, which is the point at which the line crosses, or intercepts, the y-axis.* In Figure 14-1, the intercept is 5. *b is the* ***slope****, the amount that Y is predicted to increase for an increase of 1 in X.* In Figure 14-1, the slope is 2. As X increases from 3 to 4, for example, we see an increase in what we predict for a Y of 2: from 11 to 13. The equation, therefore, is: $\hat{Y} = 5 + 2(X)$. If the score on X is 6, for example, the predicted score for Y is: $\hat{Y} = 5 + 2(6) = 5 + 12 = 17$. We can verify this on the line in Figure 14-1. Here, we were given the regression equation and regression line, but usually we have to determine these from the data. In this section, we learn the process of calculating a regression equation from data.

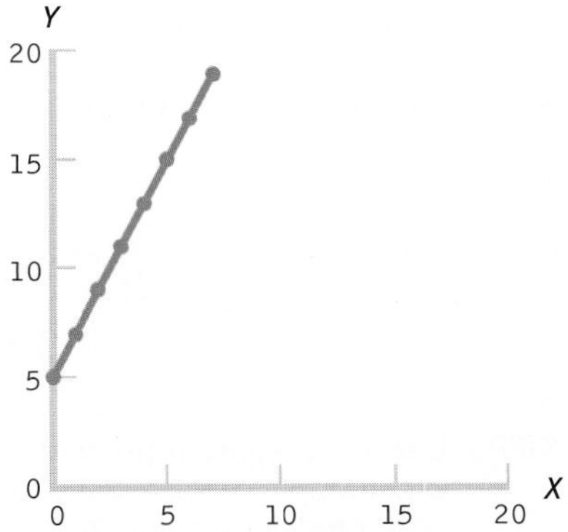

Once we have the equation for a line, it's easy to input any value for X to determine the predicted value for Y. Let's imagine that one of Skip's classmates, Allie, anticipates two absences this semester. If we had a regression equation, then we could input Allie's score of 2 on X and find her predicted score on Y. But first we have to develop the regression equation. Using the z score regression equation to find the intercept and slope enables us to "see" where these numbers come from in a way that makes sense (Aron & Aron, 2002). For this, we use the z score regression equation: $z_{\hat{Y}} = (r_{XY})(z_X)$.

FIGURE 14-1
The Equation for a Line

The equation for a line includes the intercept, the point at which the line crosses the *y*-axis; here the intercept is 5. It also includes the slope, the amount that $\hat{Y}$ increases for an increase of 1 in *X*. Here, the slope is 2. The equation, therefore, is: $\hat{Y} = 5 + 2(X)$.

EXAMPLE 14.3

We start by calculating *a*, the intercept, a process that takes three steps.

STEP 1: Find the *z* score for an *X* of 0.

We know that the intercept is the point at which the line crosses the *y*-axis when X is equal to 0. So we start by finding the z score for X using the formula: $z_X = \frac{(X - M_X)}{SD_X}$.

$$z_X = \frac{(X - M_X)}{SD_X} = \frac{(0 - 3.400)}{2.375} = -1.432$$

STEP 2: Use the *z* score regression equation to calculate the predicted *z* score on *Y*.

We use the z score regression equation, $z_{\hat{Y}} = (r_{XY})(z_X)$, to calculate the predicted raw score on Y.

$$z_{\hat{Y}} = (r_{XY})(z_X) = (-0.85)(-1.432) = 1.217$$

STEP 3: Convert the z score to its raw score. We convert the z score for $\hat{Y}$ to its raw score using the formula: $\hat{Y} = z_{\hat{Y}}(SD_Y) + M_Y$.

$$\hat{Y} = z_{\hat{Y}}(SD_Y) + M_Y = 1.217(15.040) + 76.000 = 94.30$$

We have the intercept! When X is 0, $\hat{Y}$ is 94.30. That is, we would predict that someone who never misses class would earn a final exam grade of 94.30. ■

EXAMPLE 14.4

Next, we calculate b, the slope, a process that is similar to the one for calculating the intercept, but calculating the slope takes four steps. We know that the slope is the amount that $\hat{Y}$ increases when X increases by 1. So all we need to do is calculate what we would predict for an X of 1. We can then compare the $\hat{Y}$ for an X of 0 to the $\hat{Y}$ for an X of 1. The difference between the two is the slope.

STEP 1: Find the z score for an X of 1. We find the z score for an X of 1, using the formula: $z_X = \frac{(X - M_X)}{SD_X}$.

$$z_X = \frac{(X - M_X)}{SD_X} = \frac{(1 - 3.400)}{2.375} = -1.011$$

STEP 2: Use the z score regression equation to calculate the predicted z score on Y. We use the z score regression equation, $z_{\hat{Y}} = (r_{XY})(z_X)$, to calculate the predicted score on Y.

$$z_{\hat{Y}} = (r_{XY})(z_X) = (-0.85)(-1.011) = 0.859$$

STEP 3: Convert the z score to its raw score. We convert the z score for $\hat{Y}$ to its raw score, using the formula: $\hat{Y} = z_{\hat{Y}}(SD_Y) + M_Y$.

$$\hat{Y} = z_{\hat{Y}}(SD_Y) + M_Y = 0.859(15.040) + 76.000 = 88.919$$

STEP 4: Determine the slope. The prediction is that a student who misses one class would achieve a final exam grade of 88.919. As X, number of absences, increased from 0 to 1, what happened to $\hat{Y}$? First, ask yourself if it increased or decreased. An increase would mean a positive slope, and a decrease would mean a negative slope. Here, we see a decrease in exam grade as the number of absences increased. Next, determine how much it increased or decreased. In this case, the decrease is 5.385 (calculated as 94.304 − 88.919 = 5.385). So the slope here is −5.39.

We now have the intercept and the slope and can put them into the equation: $\hat{Y} = a + b(X)$, which becomes $\hat{Y} = 94.30 - 5.39(X)$. We can use this equation to predict Allie's final exam grade based on her number of absences, two.

$$\hat{Y} = 94.30 - 5.39(X) = 94.30 - 5.39(2) = 83.52$$

Based on the data from our statistics classes, we predict that Allie would earn a final exam grade of 83.52 if she skips two classes. We could have predicted this same grade for Allie using the z score regression equation. The difference is that now we can input any score into the raw-score regression equation, and it does all the work of converting for us. The admissions counselor, insurance salesperson, or Facebook founder has an easy formula and doesn't have to know z scores.

We can also use the regression equation to draw the regression line and get a visual sense of what it looks like. We do this by calculating at least two points on the regression line, usually for one low score on X and one high score on X. We would always have $\hat{Y}$ for two scores, 0 and 1 (although in some cases these numbers won't make sense, such as for the variable of human body temperature; you'd never have a temperature that low!). Because these scores are low on the scale for number of absences, we would choose a high score as well; 8 is the highest score in the original data set, so we can use that:

$$\hat{Y} = 94.30 - 5.39(X) = \hat{Y} = 94.30 - 5.39(8) = 51.18$$

For someone who skipped eight classes, we predict a final exam grade of 51.18. We now have three points, as shown in Table 14-3. It's useful to have three points because the third point serves as a check on the other two. If the three points do not fall in a straight line, we have made an error.

We then draw a line through the dots, but it's not just any line. This line, which you can see in Figure 14-2, is the regression line, which has another name that is wonderfully intuitive: the line of best fit. If you have ever had some clothes tailored to fit your body, perhaps for a wedding or other special occasion, then you know that there really is such a thing as a "best fit."

TABLE 14-3. Drawing a Regression Line

We calculate at least two, and preferably three, pairs of scores for X and Y. Ideally, at least one is low on the scale for X and at least one is high.

X	$\hat{Y}$
0	94.30
1	88.92
8	51.18

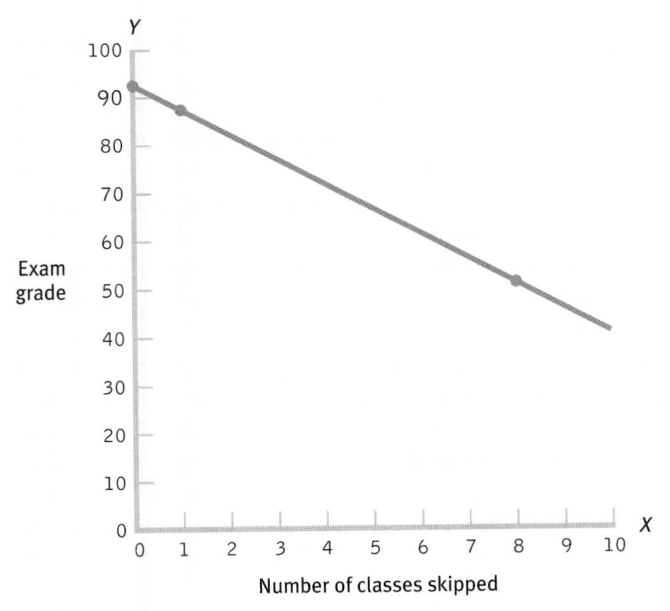

FIGURE 14-2
The Regression Line

To draw a regression line, we plot at least two, and preferably three, pairs of scores for X and $\hat{Y}$. We then draw a line through the dots.

The Line of Best Fit The line of best fit in regression has the same characteristics as tailored clothes; there is nothing we could do to that line that would make it fit the data any better.

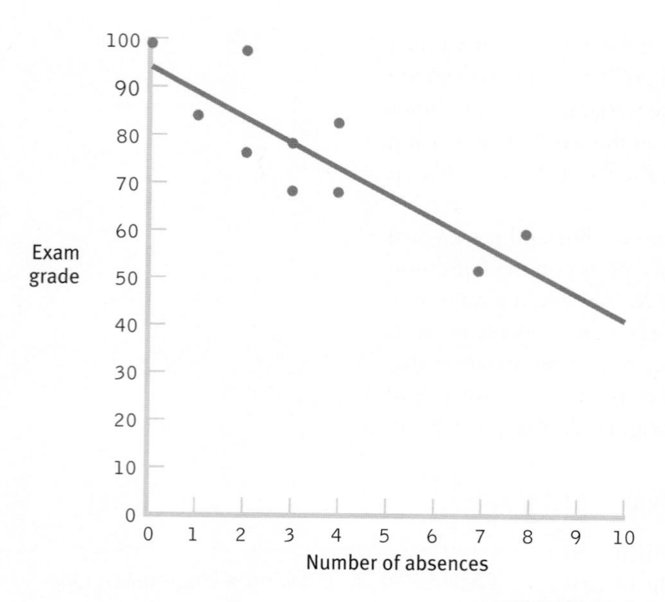

FIGURE 14-3
The Line of Best Fit
The regression line is the line that best fits the points on the scatterplot. Statistically, the regression line is the line that leads to the least amount of error in prediction.

In regression, the meaning of "the line of best fit" is the same as that characteristic in a tailored set of clothes. We couldn't make the line a little steeper, or raise or lower it, or manipulate it in any way that would make it represent those dots any better than it already does. When we look at the scatterplot around the line in Figure 14-3, we see that the line goes precisely through the middle of the data. Statistically, this is the line that leads to the least amount of error in prediction.

Language Alert! Notice that the line we just drew starts in the upper left of the graph and ends in the lower right, meaning that it has a negative slope. The word *slope* is often used when discussing, say, ski slopes. A negative slope means that the line looks like it's going downhill as we move from left to right. This makes sense because the calculations for the regression equation are based on the correlation coefficient, and the scatterplot associated with a negative correlation coefficient has dots that also go "downhill." If the slope was positive, the line would start in the lower left of the graph and end in the upper right. A positive slope means that the line looks like it's going uphill as we move from left to right. Again, this makes sense, because we base the calculations on a positive correlation coefficient, and the scatterplot associated with a positive correlation coefficient has dots that also go "uphill." ■

The Standardized Regression Coefficient and Hypothesis Testing with Regression

The steepness of the slope tells us the amount that the dependent variable changes as the independent variable increases by 1. So, for the skipping class and exam grades example, the slope of −5.39 tells us that for each additional class skipped, we can predict that the exam grade will be 5.39 points lower. Let's say that another professor uses skipped classes to predict the class grade on a GPA scale of 0–4. And let's say that we found a slope of −0.23 with these data. For each additional skipped class, we would predict that the grade, in terms of the 0–4 scale, would decrease by 0.23. The problem here is that we can't directly compare one professor's findings with another professor's findings. A decrease of 5.39 is larger than a decrease of 0.23, but they're not comparable because they're on different scales.

This problem might remind you of the problems we faced in comparing scores on different scales. To appropriately compare scores, we standardized them using the z statistic. We can standardize slopes in a similar way by calculating the standardized regression coefficient. *The* ***standardized regression coefficient****, a standardized version of the slope in a regression equation, is the predicted change in the dependent variable in terms of standard deviations for an increase of 1 standard deviation in the independent variable.* It is symbolized by β and is often called a *beta weight* because of its symbol (pronounced "beta"). It is calculated using the formula:

MASTERING THE FORMULA

14-3: The standardized regression coefficient, β, is calculated by multiplying the slope of the regression equation by the quotient of the square root of the sum of squares for the independent variable by the square root of the sum of squares for the dependent variable: $\beta = (b)\frac{\sqrt{SS_X}}{\sqrt{SS_Y}}$.

$$\beta = (b)\frac{\sqrt{SS_X}}{\sqrt{SS_Y}}$$

TABLE 14-4. The Denominator of the Correlation Coefficient: The Calculations for Sums of Squares

Absences (X)	$(X - M_X)$	$(X - M_X)^2$	Exam Grade (Y)	$(Y - M_Y)$	$(Y - M_Y)^2$
4	0.6	0.36	82	6	36
2	−1.4	1.96	98	22	484
2	−1.4	1.96	76	0	0
3	−0.4	0.16	68	−8	64
1	−2.4	5.76	84	8	64
0	−3.4	11.56	99	23	529
4	0.6	0.36	67	−9	81
8	4.6	21.16	58	−18	324
7	3.6	12.96	50	−26	676
3	−0.4	0.16	78	2	4
		$\Sigma(X - M_X)^2 = 56.4$			$\Sigma(Y - M_Y)^2 = 2262$

We calculated the slope, −5.39, earlier in this chapter. We calculated the sums of squares in Chapter 13. Table 14-4 repeats part of the calculations for the denominator of the correlation coefficient equation. At the bottom of the table, we can see that the sum of squares for the independent variable of classes skipped is 56.4 and the sum of squares for the dependent variable of exam grade is 2262. By inputting these numbers into the formula, we calculate:

▪ The **standardized regression coefficient**, a standardized version of the slope in a regression equation, is the predicted change in the dependent variable in terms of standard deviations for an increase of 1 standard deviation in the independent variable; symbolized by β and often called *beta weight.*

$$\beta = (b)\frac{\sqrt{SS_X}}{\sqrt{SS_Y}} = -5.39\frac{\sqrt{56.4}}{\sqrt{2262}} = -5.39\frac{7.510}{47.560} = -5.39(0.158) = -0.85$$

Notice that this result is the same as the Pearson correlation coefficient of −0.85. In fact, for simple linear regression, it is always exactly the same. Any difference would be due to rounding decisions for both calculations. Both the standardized regression coefficient and the correlation coefficient indicate the change in standard deviation that we expect when the independent variable increases by 1 standard deviation. Note that the correlation coefficient is *not* the same as the standardized regression coefficient when an equation includes more than one independent variable, a situation we'll encounter later in the section "Multiple Regression."

MASTERING THE CONCEPT

14.2: A standardized regression coefficient is the standardized version of a slope, much like a *z* statistic is a standardized version of a raw score. For simple linear regression, the standardized regression coefficient is identical to the correlation coefficient. This means that when we conduct hypothesis testing and conclude that a correlation coefficient is statistically significantly different from 0, we can draw the same conclusion about the standardized regression coefficient.

Because the standardized regression coefficient is the same as the correlation coefficient with simple linear regression, the outcome of hypothesis testing is also identical. The hypothesis-testing process that we used to test whether the correlation coefficient is statistically significantly different from 0 can also be used to test whether the standardized regression coefficient is statistically significantly different from 0. As you'll remember from Chapter 13, the Pearson correlation coefficient, $r = -0.85$, was larger in magnitude than the critical value of -0.632 (determined based on 8 degrees of freedom and a p level of 0.05). We rejected the null hypothesis and concluded that number of absences and exam grade seemed to be negatively correlated.

CHECK YOUR LEARNING

Reviewing the Concepts

> Regression builds on correlation, enabling us not only to quantify the relation between two variables but also to predict a score on a dependent variable from a score on an independent variable.

> With the standardized regression equation, we simply multiply a person's z score on an independent variable by the Pearson correlation coefficient to predict that person's z score on a dependent variable.

> The raw-score regression equation is easier to use in that the equation itself does the transformations from raw score to z score and back.

> We use the standardized regression equation to build the regression equation that can predict a raw score on a dependent variable from a raw score on an independent variable.

> We can graph the regression line, $\hat{Y} = a + b(X)$, based on values for the y intercept (a); the value on Y when X is zero; and the slope (b), which is the change in Y expected for a 1-unit increase in X.

> The slope, which captures the nature of the relation between the variables, can be standardized by calculating the standardized regression coefficient. The standardized regression coefficient tells us the predicted change in the dependent variable in terms of standard deviations for every increase of 1 standard deviation in the independent variable.

> With simple linear regression, the standardized regression coefficient is identical to the Pearson correlation coefficient. Because of this fact, hypothesis testing with simple linear regression gives us the same outcome as with correlation.

Clarifying the Concepts

14-1 What is simple linear regression?

14-2 What purpose does the regression line serve?

Calculating the Statistics

14-3 Let's assume we know that women's heights and weights are correlated and the Pearson coefficient is 0.28. Let's also assume that we know the following descriptive statistics: For women's height, the mean is 5 feet, 4 inches (64 inches), with a standard deviation of 2 inches; for women's weight, the mean is 155 pounds, with a standard deviation of 15 pounds. Sarah is 5 feet, 7 inches tall. How much would you predict she weighs? To answer this question, complete the following steps:

a. Transform the raw score for the independent variable into a z score.

b. Calculate the predicted z score for the dependent variable.

c. Transform the z score for the dependent variable back into a raw score.

14-4 Given the regression line $\hat{Y} = 12 + 0.67(X)$, make predictions for each of the following:

a. $X = 78$

b. $X = -14$

c. $X = 52$

Applying the Concepts

14-5 In Exercise 13.41, we explored the relation between athletic participation, measured by average minutes played by players on a basketball team, and academic achievement, as measured by GPA. We computed a correlation of 0.344 between these variables. The original, fictional data are presented below. The regression equation for these data is: $\hat{Y} = 2.586 + 0.016(X)$.

Minutes	GPA	Minutes	GPA
29.70	3.20	6.88	2.89
32.14	2.88	6.38	2.24
32.72	2.78	15.83	3.35
21.76	3.18	2.50	3.00
18.56	3.46	4.17	2.18
16.23	2.12	16.36	3.50
11.80	2.36		

a. Interpret both the y intercept and the slope in this regression equation.

b. Compute the standardized regression coefficient.

c. Explain how the strength of the correlation relates to the utility of the regression line.

d. What conclusion would you make if you performed a hypothesis test for this regression?

Solutions to these Check Your Learning questions can be found in Appendix D.

Interpretation and Prediction

In this section, we explore how the logic of regression is already a part of our everyday reasoning. Then we discuss why regression doesn't allow us to designate causation as we interpret data; for instance, MSU researchers could not say that spending more time on Facebook *caused* students to bridge more social capital with more online connections. This discussion of causation then leads us to a familiar caution about interpreting the meaning of regression, this time due to the process called *regression to the mean*. Finally, we learn how to calculate effect sizes so we can make interpretations about how well a regression equation predicts behavior.

Regression and Error

For many different reasons, predictions are full of errors, and that, too, is factored into the regression analysis. For example, we might predict that a student would get a certain grade based on how many classes she skipped, but we could be wrong in our prediction. Other factors, such as her intelligence, the amount of sleep she got the night before, and the number of related classes she's taken all are likely to affect her grade as well. The number of skipped classes is highly unlikely to be a perfect predictor.

Statistically speaking, errors in prediction lead directly back to variability, which is often assessed by standard deviation and standard error. This time, however, we are concerned with variability around the line of best fit rather than variability around the mean. A graph that includes a line of best fit can give us a sense of how much error there is in a regression equation. That is, as we can see in Figure 14-4, the arrangement of the dots around the line of best fit in a graph tells us something about the error that's likely to occur when we use the regression equation.

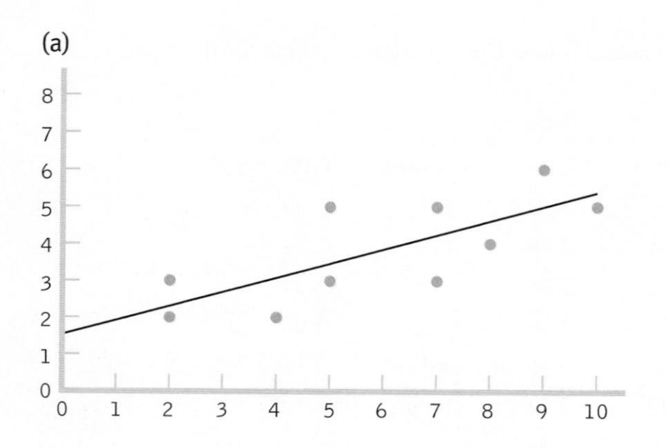

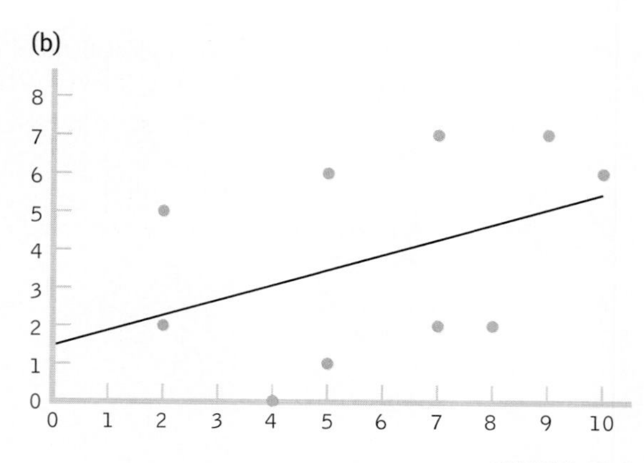

FIGURE 14-4
The Standard Error of the Estimate

Data points clustered closely around the line of best fit, as in graph (a), are described by a small standard error of the estimate. Data points clustered far away from the line of best fit, as in graph (b), are described by a large standard error of the estimate. We enjoy a high level of confidence in the predictive ability of the independent variable when the data points are tightly clustered around the line of best fit, as in (a). That is, there is much less error. And we have a low level of confidence in the predictive ability of the independent variable when the data points vary widely around the line of best fit, as in (b). That is, there is much more error.

However, we can go a step further by quantifying the amount of error. The number that describes how far away, on average, the data points are from the line of best fit is called *the* ***standard error of the estimate****, a statistic indicating the typical distance between a regression line and the actual data points.* The standard error of the estimate is essentially the standard deviation of the actual data points around the regression line. We usually get the standard error of the estimate using software, so its calculation is not covered here.

■ The **standard error of the estimate** is a statistic indicating the typical distance between a regression line and the actual data points.

Applying the Lessons of Correlation to Regression

In addition to understanding the ways in which regression can help us, it is important to understand the limitations associated with using regression. It is extremely rare that the data analyzed in a regression equation are from a true experiment (one that used randomization to assign participants to conditions). Typically, we cannot randomly assign participants to conditions when the independent variable is a scale variable (rather than a nominal variable), as is usually the case with regression. So, the results are subject to the same limitations in interpretation that we discussed with respect to correlation.

In Chapter 13, we introduced the A-B-C model of understanding correlation. We noted that the correlation between number of absences and exam grade could be explained if skipping class (A) harmed one's grade (B); if a bad grade (B) led one to skip class more often (A) because of frustration; or if a third variable (C)—such as intelligence—might lead both to the awareness that going to class is a good thing (A) and to good grades (B). When drawing conclusions from regression, we must consider the same set of possible confounding variables that limited our confidence in our findings following a correlation.

In fact, regression, like correlation, can be wildly inaccurate in its predictions. As with the Pearson correlation coefficient, a good statistician questions causality *after* the statistical analysis (to identify potential confounding variables). But one more source of error can affect fair-minded interpretations of regression analyses: regression to the mean.

Regression to the Mean

In the study that we considered earlier in this chapter (Ruhm, 2006), economic factors predicted several indicators of health. The study also reported that "the drop in tobacco use disproportionately occurs among heavy smokers, the fall in body weight among the severely obese, and the increase in exercise among those who were completely inactive" (p. 2). What Ruhm describes captures the meaning of the word *regression,* as defined by its early proponents. Those who were most extreme on a given variable regressed (toward the mean). In other words, they became somewhat less extreme on that variable.

MASTERING THE CONCEPT

14.3: Regression to the mean occurs because extreme scores tend to become less extreme—that is, they tend to regress toward the mean. Very tall parents do tend to have tall children, but usually not as tall as they are, whereas very short parents do tend to have short children, but usually not as short as they are.

Francis Galton (Darwin's cousin) was the first to describe the phenomenon of regression to the mean, and he did so in a number of contexts (Bernstein, 1996). For example, Galton asked nine people—including Darwin—to plant sweet pea seeds in the widely scattered locations in Britain where these people lived. Galton found that the variability among the seeds he sent out to be planted was larger than among the seeds that were produced by these plants. The largest seeds produced seeds smaller than they were. The smallest seeds produced seeds larger than they were.

Similarly, among people, Galton documented that, although tall parents tend to have taller-than-average children, their children tend to be a little shorter than they are. And although short parents tend to have shorter-than-average children, their children tend to be a little taller than they are. Galton noted that if regression to the mean did *not* occur, with tall people and large sweet peas producing offspring even taller or larger, and short people and small sweet peas producing offspring even shorter or smaller, "the world would consist of nothing but midgets and giants" (quoted in Bernstein, 1996, p. 167).

An understanding of regression to the mean can help us make better choices in our daily lives. For example, regression to the mean is a particularly important concept to remember when we begin to save for retirement and have to choose the specific allocations of our savings. Table 14-5 shows data from *Morningstar,* an investment publication. The percentages represent the increase in that investment vehicle over two 5-year periods: 1984–1989 and 1989–1994 (Bernstein, 1996). As most descriptions of mutual funds remind potential investors, previous performance is not necessarily

Jack Hollingsworth/Getty Images

Regression to the Mean Tall parents tend to have children who are taller than average but not as tall as they are. Similarly, short parents (like the older parents in this photograph) tend to have children who are shorter than average but not as short as they are. Francis Galton was the first to observe this phenomenon, which came to be called regression to the mean.

TABLE 14-5. Regression to the Mean: Investing

Bernstein (1996) presented these data from *Morningstar,* an investment publication, demonstrating regression to the mean in action. Notice that the category that showed the highest performances during the first time period (international stocks) had declined by the second time period, whereas the category with the poorest performances in the first time period (aggressive growth) had improved by the second time period.

5 Years to Objective	5 Years to March 1989	March 1994
International stocks	20.6%	9.4%
Income	14.3%	11.2%
Growth and income	14.2%	11.9%
Growth	13.3%	13.9%
Small company	10.3%	15.9%
Aggressive growth	8.9%	16.1%
Average	13.6%	13.1%

▪ The **proportionate reduction in error** is a statistic that quantifies how much more accurate predictions are when we use the regression line instead of the mean as a prediction tool; also called the *coefficient of determination.*

indicative of future performance. Consider regression to the mean in your own investment decisions. It might help you ride out a decrease in a mutual fund rather than panic and sell before the likely drift back toward the mean. And it might help you avoid buying into the fund that's been on top for several years, knowing that it stands a chance of sliding back toward the mean.

Proportionate Reduction in Error

In the previous section, we developed a regression equation to predict a final exam score from number of absences. Now we want to know: How good is this regression equation? Is it worth having students use this equation to predict their own final exam grades from the numbers of classes they plan to skip? To answer this question, we calculate a form of effect size, *the* ***proportionate reduction in error****—a statistic that quantifies how much more accurate predictions are when we use the regression line instead of the mean as a prediction tool.* (Note that the proportionate reduction in error is sometimes called the *coefficient of determination.*) More specifically, the proportionate reduction in error is a statistic that quantifies how much more accurate predictions are when we predict scores using a specific regression equation rather than just predicting the mean for everyone.

Earlier in this chapter, we noted that if we did not have a regression equation, the best we could do is predict the mean for everyone, regardless of number of absences. The average final exam grade for students in this sample is 76. With no further information, we could only tell our students that our best guess for their statistics grade is a 76. There would obviously be a great deal of error if we predicted the mean for everyone. Using the mean to estimate scores is a reasonable way to proceed if that's all the information we have. But the regression line provides a more precise picture of the relation between variables, so using a regression equation reduces error.

Less error is the same thing as having a smaller standard error of the estimate. And a smaller standard error of the estimate means that we'd be doing much better in our predictions than if we had a larger one; visually, this means that the actual scores are closer to the regression line. And with a larger standard error of the estimate, we'd be doing much worse in our predictions than if we had a smaller one; visually, the actual scores are farther away from the regression line.

But we can do more than just quantify the standard deviation around the regression line. We can determine how much better the regression equation is compared to the mean: We calculate the proportion of error that we can eliminate by using the regression equation, rather than the mean, to make a prediction. (In this next section, we learn the long way to calculate this proportion in order to understand exactly what the proportion represents. Then we learn a shortcut.)

EXAMPLE 14.5

Using a sample, we can calculate the amount of error from using the mean as a predictive tool. We quantify that error by determining how far off a person's score on the dependent variable (final exam grade) is from the mean, as seen in the column labeled "Error" in Table 14-6.

For example, for student 1, the error is $82 - 76 = 6$. We then square these errors for all 10 students and sum them. This is another type of sum of squares: the sum of squared errors. Here, the sum of squared errors is 2262 (the sum of the values in column 5). This is a measure of the error that would result if we predicted the mean for every person in the sample. We'll call this particular type of sum of squared errors the *sum of squares total,* SS_{total}, because it represents the worst-case scenario, the total error we would have if there were no regression equation. We can visualize this error on a graph that depicts a horizontal line for the mean, as seen in Figure 14-5. We can add the actual

TABLE 14-6. Calculating Error When We Predict the Mean for Everyone

If we do not have a regression equation, the best we can do is predict the mean for Y for every participant. When we do that, we will, of course, have some error, because not everyone will have exactly the mean value on Y. This table presents the squared errors for each participant when we predict the mean for each of them.

Student	Grade (Y)	Mean For Y	$(Y - M_Y)$ Error	Squared Error
1	82	76	6	36
2	98	76	22	484
3	76	76	0	0
4	68	76	−8	64
5	84	76	8	64
6	99	76	23	529
7	67	76	−9	81
8	58	76	−18	324
9	50	76	−26	676
10	78	76	2	4

points, as we would in a scatterplot, and draw vertical lines from each point to the mean. These vertical lines give us a visual sense of the error that results from predicting the mean for everyone.

The regression equation can't make the predictions any worse than they would be if we just predicted the mean for everyone. But it's not worth the time and effort to use a regression equation if it doesn't lead to a substantial improvement over just predicting the mean. As with the mean, we can calculate the amount of error from using the regression equation with the sample. We can then see how much better we do with the regression equation than with the mean.

First, we calculate what we would predict for each student if we used the regression equation. We do this by plugging each X into the regression equation. Here are the calculations using the equation $\hat{Y} = 94.30 - 5.39(X)$:

$$\hat{Y} = 94.30 - 5.39(4); \hat{Y} = 72.74$$

$$\hat{Y} = 94.30 - 5.39(2); \hat{Y} = 83.52$$

$$\hat{Y} = 94.30 - 5.39(2); \hat{Y} = 83.52$$

$$\hat{Y} = 94.30 - 5.39(3); \hat{Y} = 78.13$$

$$\hat{Y} = 94.30 - 5.39(1); \hat{Y} = 88.91$$

$$\hat{Y} = 94.30 - 5.39(0); \hat{Y} = 94.30$$

$$\hat{Y} = 94.30 - 5.39(4); \hat{Y} = 72.74$$

$$\hat{Y} = 94.30 - 5.39(8); \hat{Y} = 51.18$$

$$\hat{Y} = 94.30 - 5.39(7); \hat{Y} = 56.57$$

$$\hat{Y} = 94.30 - 5.39(3); \hat{Y} = 78.13$$

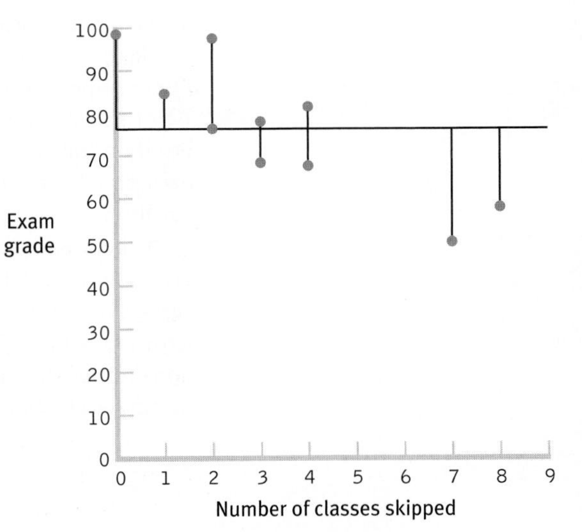

FIGURE 14-5
Visualizing Error

A graph with a horizontal line for the mean, 76, allows us to visualize the error that would result if we predicted the mean for everyone. We draw lines for each person's point on a scatterplot to the mean. Those lines are a visual representation of error.

TABLE 14-7. Calculating Error When We Use the Regression Equation to Predict

When we use a regression equation for prediction, as opposed to using the mean, we have less error. However, we still have some error because not every participant falls exactly on the regression line. This table presents the squared errors for each participant when we predict each one's score on Y using the regression equation.

Student	Absences (X)	Grade (Y)	Predicted ($\hat{Y}$)	$(Y - \hat{Y})$ Error	Squared Error
1	4	82	72.74	9.26	85.748
2	2	98	83.52	14.48	209.670
3	2	76	83.52	−7.52	56.550
4	3	68	78.13	−10.13	102.617
5	1	84	88.91	−4.91	24.108
6	0	99	94.30	4.70	22.090
7	4	67	72.74	−5.74	32.948
8	8	58	51.18	6.82	46.512
9	7	50	56.57	−6.57	43.165
10	3	78	78.13	−0.13	0.017

The $\hat{Y}$'s, or predicted scores for Y, that we just calculated are presented in Table 14-7, where the errors are calculated based on the predicted scores rather than the mean. For example, for student 1, the error is the actual score minus the predicted score: $82 - 72.74 = 9.26$. As before, we square the errors and sum them. The sum of squared errors based on the regression equation is 623.425. We call this the *sum of squared error*, SS_{error}, because it represents the error that we'd have if we predicted Y using the regression equation.

As before, we can visualize this error on a graph that includes the regression line, as seen in Figure 14-6. We again add the actual points, as in a scatterplot, and we draw vertical lines from each point to the regression line. These vertical lines give us a visual sense of the error that results from predicting Y for everyone using the regression equation. Notice that these vertical lines in Figure 14-6 tend to be shorter than those connecting each person's point with the mean in Figure 14-5.

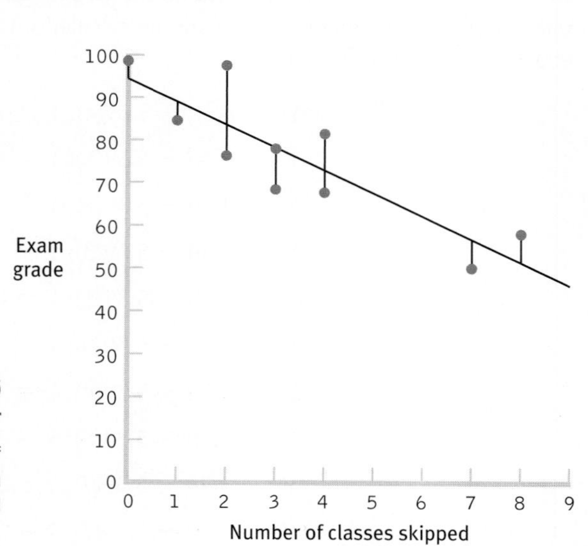

FIGURE 14-6
Visualizing Error

A graph that depicts the regression line allows us to visualize the error that would result if we predicted Y for everyone using the regression equation. We draw lines for each person's point on a scatterplot to the regression line. Those lines are a visual representation of error.

So how much better did we do? The error we predict by using the mean for everyone in this sample is 2262. The error we predict by using the regression equation for everyone in this sample is 623.425. Remember that the measure of how well the regression equation predicts is called the proportionate *reduction* in error. What we want to know is how much error we have gotten rid of—reduced—by using the regression equation instead of the mean. The amount of error we've reduced is 2262 − 623.425 = 1638.575. But the word *proportionate* indicates that we want a proportion of the total error that we have reduced, so we set up a ratio to determine this. We have reduced 1638.575 of the original 2262, or $\frac{1638.575}{2262} = 0.724$.

We have reduced 0.724, or 72.4%, of the original error by using the regression equation versus using the mean to predict *Y*. This ratio can be calculated using an equation that represents what we just calculated: the proportionate reduction in error, symbolized as:

$$r^2 = \frac{(SS_{total} - SS_{error})}{SS_{total}} = \frac{(2262 - 623.425)}{2262} = 0.724$$

MASTERING THE FORMULA

14-4: The proportionate reduction in error is calculated by subtracting the error generated using the regression equation as a prediction tool from the total error that would occur if we used the mean as everyone's predicted score. We then divide this difference by the total error: $r^2 = \frac{(SS_{total} - SS_{error})}{SS_{total}}$. We can interpret the proportionate reduction in error as we did the effect-size estimate for ANOVA. It represents the same statistic.

To recap, we simply have to do the following:

1. Determine the error associated with using the mean as the predictor.
2. Determine the error associated with using the regression equation as the predictor.
3. Subtract the error associated with the regression equation from the error associated with the mean.
4. Divide the difference (calculated in step 3) by the error associated with using the mean.

The proportionate reduction in error tells us how good the regression equation is. Here is another way to state it: The proportionate reduction in error is a measure of the amount of variance in the dependent variable that is explained by the independent variable. Did you notice the symbol for the proportionate reduction in error? The symbol is r^2. Perhaps you see the connection with another number we have calculated. Yes, we could simply square the correlation coefficient!

The longer calculations are necessary, however, to see the difference between the error in prediction from using the regression equation and the error in prediction from simply predicting the mean for everyone. Once you have calculated the proportionate reduction in error the long way a few times, you'll have a good sense of exactly what you're calculating. In addition to the relation of the proportionate reduction in error to the correlation coefficient, it also is the same as another number we've calculated—the effect size for ANOVA, R^2. In both cases, this number represents the proportion of variance in the dependent variable that is explained by the independent variable.

MASTERING THE CONCEPT

14.4: Proportionate reduction in error is the effect size used with regression. It is the same number we calculated as the effect size estimate for ANOVA. It tells us the proportion of error that is eliminated when we predict scores on the dependent variable using the regression equation versus simply predicting that everyone is at the mean on the dependent variable.

Because the proportionate reduction in error can be calculated by squaring the correlation coefficient, we can have a sense of the amount of error that would be reduced simply by looking at the correlation coefficient. A correlation coefficient that is high in magnitude, whether negative or positive, indicates a strong relation between two variables. If two variables are highly related, it makes sense that one of them is going to be a good predictor of the other. And it makes sense that when we use one variable to predict the other, we're going to reduce error. ■

CHECK YOUR LEARNING

Reviewing the Concepts

> Findings from regression analyses are subject to the same types of limitations as correlation. Regression, like correlation, does not tell us about causation.

> People with extreme scores at one point in time tend to have less extreme scores (scores closer to the mean) at a later point in time, a phenomenon called regression to the mean.

> Error based on the mean is referred to as the sum of squares total (SS_{total}), whereas error based on the regression equation is referred to as the sum of squared error (SS_{error}).

> Proportionate reduction in error, r^2, determines the amount of error we have eliminated by using a particular regression equation to predict a person's score on the dependent variable versus simply predicting the mean on the dependent variable for that person.

Clarifying the Concepts

14-6 Distinguish error of prediction when the mean is used from the standard error of the estimate.

14-7 Explain how the strength of the correlation is related to the proportionate reduction in error for regression.

Calculating the Statistics

14-8 Data are provided here with means, standard deviations, a correlation coefficient, and a regression equation: $r = -0.77$, $\hat{Y} = 7.846 - 0.431(X)$.

X	Y
5	6
6	5
4	6
5	6
7	4
8	5
$M_X = 5.833$	$M_Y = 5.333$
$SD_X = 1.344$	$SD_Y = 0.745$

a. Using this information, calculate the sum of squared error for the mean, SS_{total}.

b. Now, using the regression equation provided, calculate the sum of squared error for the regression equation, SS_{error}.

c. Using your work from parts (a) and (b), calculate the proportionate reduction in error for these data.

d. Check that this calculation of r^2 equals the square of the correlation coefficient.

Applying the Concepts

14-9 Many athletes and sports fans believe that an appearance on the cover of *Sports Illustrated* (*SI*) is a curse. The tendency for *SI* cover subjects to face imminent bad sporting luck is documented in the pages of (what else?) *Sports Illustrated* and even has a name, the "*SI* jinx" (Wolff, 2002). Players or teams, shortly after appearing on the cover, often have a particularly poor performance. In fact, of 2456 covers, *SI* counted 913 "victims." And their potential victims have noticed: After the New England Patriots football team won their league championship in 1996, their coach at the time, Bill Parcells, called his daughter, an *SI* staffer, and ordered: "No cover." Using your knowledge about the limitations of regression, what would you tell Coach Parcells?

Solutions to these Check Your Learning questions can be found in Appendix D.

Multiple Regression

In regression analysis, we explain more of the variability in the dependent variable if we can discover genuine predictors that are separate and distinct. This involves ***orthogonal variables****, independent variables that make separate and distinct contributions in the prediction of a dependent variable, as compared with the contributions of other variables.* Orthogonal variables do not overlap each other. For example, the study we discussed earlier explored whether the amount of Facebook use predicted social capital. It is likely that a person's personality also predicts social capital; for example, we would expect extroverted, or outgoing, people to be more likely to have this kind of social capital than introverted, or shy, people. It would be useful to separate the effects of the amount of Facebook use and extroversion on social capital.

The statistical technique we consider next is a way of quantifying (1) whether multiple pieces of evidence really are better than one, and (2) precisely how much better each additional piece of evidence actually is.

- An **orthogonal variable** is an independent variable that makes a separate and distinct contribution in the prediction of a dependent variable, as compared with another variable.
- **Multiple regression** is a statistical technique that includes two or more predictor variables in a prediction equation.

Understanding the Equation

Just as a regression equation using one independent variable is a better predictor than the mean, a regression equation using more than one independent variable is likely to be an even better predictor. This makes sense in the same way that knowing a baseball player's historical batting average *plus* knowing that the player continues to suffer from a serious injury is likely to change the prediction yet again. So it is not surprising that multiple regression is far more common than simple linear regression. ***Multiple regression*** *is a statistical technique that includes two or more predictor variables in a prediction equation.*

Let's examine an equation that might be used to predict a final exam grade from two variables, number of absences *and* score on the mathematics portion of the SAT. Table 14-8 repeats the data from Table 14-1, with the added variable of SAT score. (Note that although

MASTERING THE CONCEPT

14.5: Multiple regression predicts scores on a single dependent variable from scores on more than one independent variable. Because behavior tends to be influenced by many factors, multiple regression allows us to better predict a given outcome.

TABLE 14-8. Predicting Exam Grade from Two Variables

Multiple regression allows us to develop a regression equation that predicts a dependent variable from two or more independent variables. Here, we will use these data to develop a regression equation that predicts exam grade from number of absences *and* SAT score.

Student	Absences	SAT	Exam Grade
1	4	620	82
2	2	750	98
3	2	500	76
4	3	520	68
5	1	540	84
6	0	690	99
7	4	590	67
8	8	490	58
9	7	450	50
10	3	560	78

Coefficients(a)

Model		Unstandardized Coefficients		Standardized Coefficients	t	Sig.
		B	Std. Error	Beta		
1	(Constant)	33.422	13.584		2.460	.043
	number of absences	-3.340	.773	-.527	-4.320	.003
	SAT	.094	.021	.558	4.569	.003

a. Dependent Variable: mean exam grade

FIGURE 14-7
Software Output for Regression

Computer software provides the information necessary for the multiple regression equation. All necessary coefficients are in column B under "Unstandardized coefficients." The constant, 33.422, is the intercept; the number next to "Number of absences," −3.340, is the slope for that independent variable; and the number next to "SAT," .094, is the slope for that independent variable.

the scores on number of absences and final exam grade are real-life data from our statistics classes, the SAT scores are fictional.)

The computer gives us the printout seen in Figure 14-7. The column in which we're interested is the one labeled "B" under "Unstandardized coefficients." The first number, across from "(Constant)," is the intercept. The intercept is called "constant" because it does not change; it is not multiplied by any value of an independent variable. The intercept here is 33.422. The second number is the slope for the independent variable, number of absences. Number of absences is negatively correlated with final exam grade, so the slope, −3.340, is negative. The third number in this column is the slope for the independent variable of SAT score. As we might guess, SAT score and final exam grade are positively correlated; a student with a high SAT score tends to have a higher final exam grade. So the slope, 0.094, is positive. We can put these numbers into a regression equation:

$$\hat{Y} = 33.422 - 3.34(X_1) + 0.094(X_2)$$

Once we develop the multiple regression equation, we can input raw scores on number of absences and mathematics SAT score to determine a student's predicted score on Y. Imagine that our student, Allie, scored 600 on the mathematics portion of the SAT. We already know she planned to miss two classes this semester. What would we predict her final exam grade to be?

$$\begin{aligned}\hat{Y} &= 33.422 - 3.34(X_1) + 0.094(X_2)\\ &= 33.422 - 3.34(2) + 0.094(600)\\ &= 33.422 - 6.68 + 56.4 = 83.142\end{aligned}$$

Based on these two variables, we predict a final exam grade of 83.142 for Allie. How good is this multiple regression equation? From software, we calculated that the proportionate reduction in error for this equation is a whopping 0.93. By using a multiple regression equation with the independent variables of number of absences and SAT score, we have reduced 93% of the error that would result from predicting the mean of 76 for everyone.

When we calculate proportionate reduction in error for a multiple regression, the symbol changes slightly. The symbol is now R^2 instead of r^2. The capitalization of this statistic is an indication that the proportionate reduction in error is based on more than one independent variable.

Multiple Regression in Everyday Life

With the development of increasingly more powerful computers and the availability of ever-larger amounts of computerized data, tools based on multiple regression have proliferated. Now the general public can access many of them online (Darlin, 2006). Bing Travel (formerly Farecast.com) predicts the price of an airline ticket for specific routes, travel dates, and, most important, purchase dates. Using the same data available to travel agents, along with additional independent variables such as the weather and even which sports teams' fans might be traveling to a championship game, Bing Travel mimics the regression equations used by the airlines. Airlines predict how much money potential travelers are willing to pay on a given date for a given flight and use these predictions to adjust their fares so they can earn the most money.

Bing Travel is an attempt at an end run, using mathematical prediction tools, to help savvy airline consumers either beat or wait out the airlines' price hikes. In 2007, Bing Travel's precursor, Farecast.com, claimed a 74.5% accuracy rate for its predictions. Zillow.com does for real estate what Bing Travel does for airline tickets. Using archival land records, Zillow.com predicts U.S. housing prices and claims to be accurate to within 10% of the actual selling price of a given home.

Another company, Inrix, predicts the dependent variable, traffic, using the independent variables of the weather, traveling speeds of vehicles that have been outfitted with Global Positioning Systems (GPS), and information about events such as rock concerts. It even suggests, via cell phone or in-car navigation systems, alternative routes for gridlocked drivers. Like the future of visual displays of data, the future of the regression equation is limited only by the creativity of the rising generation of behavioral scientists and statisticians.

CHECK YOUR LEARNING

Reviewing the Concepts

> Multiple regression is used when we want to predict a dependent variable from more than one independent variable. Ideally, these variables are distinct from one another in such a way that they contribute uniquely to the predictions.

> We can develop a multiple regression equation and input specific scores for each independent variable to determine the predicted score on the dependent variable.

> Multiple regression is the backbone of many online tools that we can use for predicting everyday variables such as traffic or home prices.

Clarifying the Concepts

14-10 What is multiple regression, and what are its benefits over simple linear regression?

Calculating the Statistics

14-11 Write the equation for the line of prediction using the following output from a multiple regression analysis:

Coefficients[a]

Model		Unstandardized Coefficients		Standardized Coefficients	t	Sig.
		B	Std. Error	Beta		
1	(Constant)	5.251	4.084		1.286	.225
	Variable A	.060	.107	.168	.562	.585
	Variable B	1.105	.437	.758	2.531	.028

a. Dependent Variable: Outcome variable

continued on next page

14-12 Use the equation for the line you created in Check Your Learning 14-11 to make predictions for each of the following.

a. $X_1 = 40, X_2 = 14$

b. $X_1 = 101, X_2 = 39$

c. $X_1 = 76, X_2 = 20$

Applying the Concepts

14-13 The accompanying computer printout shows a regression equation that predicts GPA from three independent variables: hours slept per night, hours studied per week, and admiration for Pamela Anderson, the B-level actress whom many view as tacky. The data are from some of our statistics classes. (*Note:* Hypothesis testing shows that all three independent variables are statistically significant predictors of GPA!)

Coefficients(a)

Model		Unstandardized Coefficients		Standardized Coefficients		
		B	Std. Error	Beta	t	Sig.
1	(Constant)	2.695	.228		11.829	.000
	Hours Slept Per Night	.069	.032	.173	2.186	.030
	Hours Studied Per Week	.015	.006	.209	2.637	.009
	Level of Admiration for Pamela Anderson	-.072	.025	-.229	-2.882	.005

a. Dependent Variable: GPA

a. What is the regression equation based on these data?

b. If someone reports that he typically sleeps 6 hours a night, studies 20 hours per week, and has a Pamela Anderson admiration level of 4 (on a scale of 1–7, with 7 indicating the highest level of admiration), what would you predict for his GPA?

c. What does the negative sign in the slope for the independent variable, level of admiration for Pamela Anderson, tell you about this variable's predictive association with grade point average?

Solutions to these Check Your Learning questions can be found in Appendix D.

REVIEW OF CONCEPTS

Simple Linear Regression

Regression is an expansion of correlation in that it allows us not only to quantify a relation between two variables but also to quantify one variable's ability to predict another variable. We can predict a dependent variable's *z* score from an independent variable's *z* score, or we can do a bit more initial work and predict a dependent variable's raw score from an independent variable's raw score. The latter method uses the equation for a line with an *intercept* and a *slope.*

We use *simple linear regression* when we predict one dependent variable from one independent variable when the two variables are linearly related. We can graph this line using the regression equation, plugging in low and high values of *X* and plotting those values with their associated predicted values on Y, then connecting the dots to form the regression line.

Just as we can standardize a raw score by converting it to a z score, we can standardize a slope by converting it to a *standardized regression coefficient*. This number indicates the predicted change on the dependent variable in terms of standard deviation for every increase of 1 standard deviation in the independent variable. For simple linear regression, the standardized regression coefficient is the same as the Pearson correlation coefficient. Hypothesis testing that determines whether the correlation coefficient is statistically significantly different from 0 also indicates whether the standardized regression coefficient is statistically significantly different from 0.

When we use regression, we must also be aware of the phenomenon called *regression to the mean,* in which extreme values tend to become less extreme over time.

Interpretation and Prediction

A regression equation is rarely a perfect predictor of scores on the dependent variable. There is always some prediction error, which can be quantified by the *standard error of estimate,* the number that describes the typical amount that an observation falls from the regression line. In addition, regression suffers from the same drawbacks as correlation. For example, we cannot know if the predictive relation is causal; the posited direction could be the reverse (with Y causally predicting X), or there could be a third variable at work.

When we use regression, we must consider the degree to which an independent variable predicts a dependent variable. To do this, we can calculate the *proportionate reduction in error,* symbolized as r^2. The proportionate reduction in error tells us how much better our prediction is with the regression equation than with the mean as the only predictive tool.

Multiple Regression

We use *multiple regression* when we have more than one independent variable, as is usual in most research in the behavioral sciences. Multiple regression is particularly useful when we have *orthogonal variables,* independent variables that make separate contributions to the prediction of a dependent variable. Multiple regression has led to the development of many Web-based prediction tools that allow us to make educated guesses about such outcomes as airplane ticket prices.

SPSS®

The most common form of regression analysis in SPSS uses at least two scale variables: an independent variable (predictor) and a dependent variable (the variable being predicted). Let's use the number of absences and exam grade data as an example. Once again, begin by visualizing the data.

Request the scatterplot of the data by selecting: Graphs → Chart Builder → Gallery → Scatter/Dot. Drag the upper-left sample graph to the large box on top. Then select the variables to be included in the scatterplot by dragging the independent variable, number, to the x-axis, and the dependent variable, grade, to the y-axis. Click "OK." Then click on the graph to make changes. To add the regression line, click "Elements," then "Fit Line at Total." Choose "Linear," then click "Apply."

To analyze the linear regression, select: Analyze → Regression → Linear. Select "number" as the independent (predictor) variable and "grade" as the dependent variable being predicted.

As usual, click on "OK" to see the Output screen. Part of the output is shown in the screenshot here. In the box titled "Model Summary," we can see the correlation coefficient of .851 under "R" and the proportionate reduction of error, .724, under "R Square." In the box titled "Coefficients," we can look in the first column under "B" to determine the regression equation. The intercept, 94.326, is across from "(Constant)," and the slope, −5.390, is across from "Number of Absences." (Any slight differences from the numbers we calculated earlier are due to rounding decisions.)

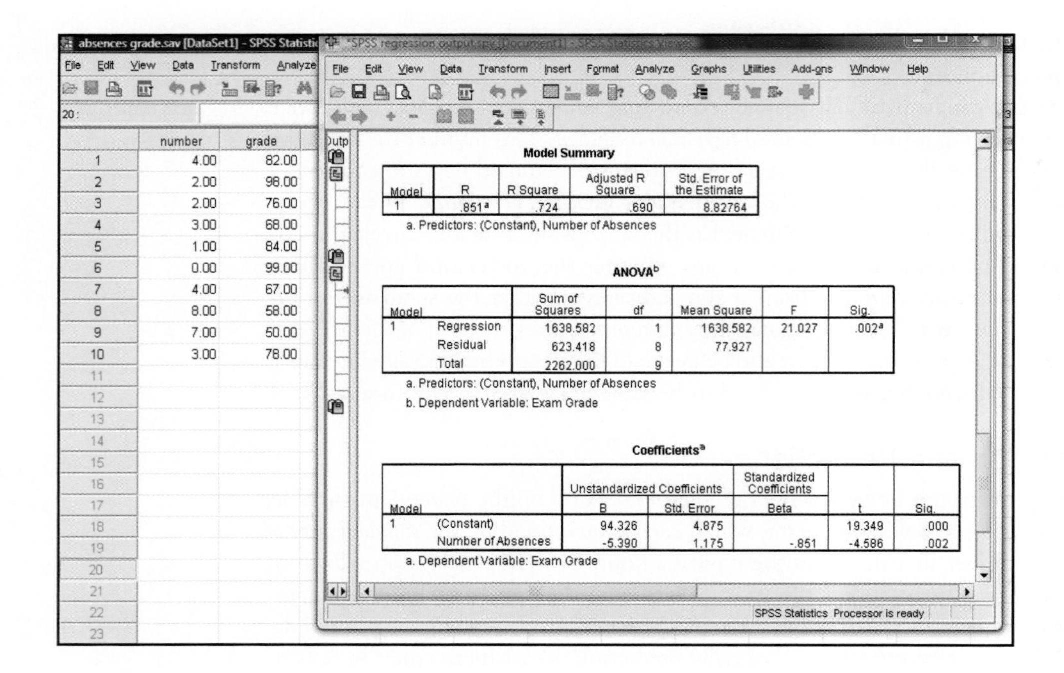

	number	grade
1	4.00	82.00
2	2.00	98.00
3	2.00	76.00
4	3.00	68.00
5	1.00	84.00
6	0.00	99.00
7	4.00	67.00
8	8.00	58.00
9	7.00	50.00
10	3.00	78.00

Model Summary

Model	R	R Square	Adjusted R Square	Std. Error of the Estimate
1	.851[a]	.724	.690	8.82764

a. Predictors: (Constant), Number of Absences

ANOVA[b]

Model		Sum of Squares	df	Mean Square	F	Sig.
1	Regression	1638.582	1	1638.582	21.027	.002[a]
	Residual	623.418	8	77.927		
	Total	2262.000	9			

a. Predictors: (Constant), Number of Absences

b. Dependent Variable: Exam Grade

Coefficients[a]

Model		Unstandardized Coefficients		Standardized Coefficients	t	Sig.
		B	Std. Error	Beta		
1	(Constant)	94.326	4.875		19.349	.000
	Number of Absences	-5.390	1.175	-.851	-4.586	.002

a. Dependent Variable: Exam Grade

How It Works

14.1 Regression with z Scores

Shannon Callahan, a former student in the experimental psychology master's program at Seton Hall University, conducted a study that examined evaluations of faculty members on Ratemyprofessor.com. She wondered if professors who were rated high on "clarity" were more likely to be viewed as "easy." Callahan found a significant correlation of 0.267 between the average easiness rating a professor garnered and the average rating he or she received with respect to clarity of teaching.

If we know that a professor's z score on clarity is 2.2 (an indication that she is very clear), how could we predict her z score on easiness?

$$z_{\hat{Y}} = (r_{XY})(z_X) = (0.267)(2.2) = 0.59$$

When there's a positive correlation, we predict a z score above the mean when her original z score is above the mean.

And if a professor's z score on clarity is -1.8 (an indication that she's not very clear), how could we predict her z score on easiness?

$$z_{\hat{Y}} = (r_{XY})(z_X) = (0.267)(-1.8) = -0.48$$

When there's a positive correlation, we predict a z score below the mean when her original z score is below the mean.

14.2 Regression with Raw Scores

Using Shannon Callahan's data, how can we develop the regression equation so that we can work directly with raw scores? To do this, we need a little more information. For this data set, the mean clarity score is 3.673, with a standard deviation of 0.890; the mean easiness score is 2.843, with a standard deviation of 0.701. As noted before, the correlation between these variables is 0.267.

To calculate the regression equation, we need to find the intercept and the slope. We determine the *intercept* by calculating what we predict for Y (easiness) when X (clarity) equals 0. Given the means, the standard deviations, and the correlation calculated above, we first find z_X:

$$z_X = \frac{(X - M_X)}{SD_X} = \frac{(0 - 3.673)}{0.890} = -4.127$$

We then calculate the predicted z score for easiness:

$$z_{\hat{Y}} = (r_{XY})(z_X) = (0.267)(-4.127) = -1.102$$

Finally, we transform the predicted easiness z score into the predicted easiness raw score:

$$\hat{Y} = z_{\hat{Y}}(SD_Y) + M_Y = -1.102(0.701) + 2.843 = 2.070$$

The intercept, therefore, is 2.070.

To determine the *slope,* we calculate what we would predict for Y (easiness) when X (clarity) equals 1, and determine how much that differs from what we would predict when X equals 0. The z score for X corresponding to the raw score of 1 is:

$$z_X = \frac{(X - M_X)}{SD_X} = \frac{(1 - 3.673)}{0.890} = -3.003$$

We then calculate the predicted z score for easiness:

$$z_{\hat{Y}} = (r_{XY})(z_X) = (0.267)(-3.003) = -0.802$$

Finally, we transform the predicted easiness z score into the predicted easiness raw score:

$$\hat{Y} = z_{\hat{Y}}(SD_Y) + M_Y = -0.802(0.701) + 2.843 = 2.281$$

The difference between the predicted Y when X equals 1 (2.281) and that when X equals 0 (2.070) yields the slope, which is $2.281 - 2.070 = 0.211$. So the regression equation is:

$$\hat{Y} = 2.07 + 0.21(X)$$

We can then use this regression equation to calculate a professor's predicted easiness score from his or her clarity score. Let's say a professor has a clarity score of 3.2. We would use the regression equation to predict her easiness score as follows:

$$\hat{Y} = 2.07 + 0.21(X) = 2.07 + 0.21(3.2) = 2.742$$

This result makes sense because she is below the mean on clarity, so, given that there is a positive correlation, we predict her score to fall below the mean on easiness.

Exercises

Clarifying the Concepts

14.1 What does regression add above and beyond what we learn from correlation?

14.2 How does the regression line relate to the correlation of the two variables?

14.3 Is there any difference between $\hat{Y}$ and a predicted score for Y?

14.4 What do each of the symbols stand for in the formula for the regression equation: $z_{\hat{Y}} = (r_{XY})(z_X)$?

14.5 The equation for a line is $\hat{Y} = a + b(X)$. Define the symbols a and b.

14.6 What are the three steps to calculate the intercept?

14.7 When is the intercept not meaningful or useful?

14.8 What does the slope tell us?

14.9 Why do we also call the regression line the line of best fit?

14.10 How are the sign of the correlation coefficient and the sign of the slope related?

14.11 What is the difference between a small standard error of the estimate and a large one?

14.12 Why are explanations of the causes behind relations explored with regression limited in the same way they are with correlation?

14.13 What is the connection between regression to the mean and the bell-shaped normal curve?

14.14 Explain why the regression equation is a better source of predictions than the mean.

14.15 What is the SS_{total}?

14.16 When drawing error lines between data points and the regression line, why is it important that these lines be perfectly vertical?

14.17 What are the basic steps to calculate the proportionate reduction in error?

14.18 What information does the proportionate reduction in error give us?

14.19 What is an orthogonal variable?

14.20 If you know the correlation coefficient, how can you determine the proportionate reduction in error?

Calculating the Statistics

14.21 Using the following information, make a prediction for *Y*, given an *X* score of 2.9:

Variable *X*: $M = 1.9$, $SD = 0.6$

Variable *Y*: $M = 10$, $SD = 3.2$

Pearson correlation of variables *X* and *Y* = 0.31

a. Transform the raw score for the independent variable to a *z* score.

b. Calculate the predicted *z* score for the dependent variable.

c. Transform the *z* score for the dependent variable back into a raw score.

14.22 Using the following information, make a prediction for *Y*, given an *X* score of 8:

Variable *X*: $M = 12$, $SD = 3$

Variable *Y*: $M = 74$, $SD = 18$

Pearson correlation of variables *X* and *Y* = 0.46

a. Transform the raw score for the independent variable to a *z* score.

b. Calculate the predicted *z* score for the dependent variable.

c. Transform the *z* score for the dependent variable back into a raw score.

d. Calculate the *y* intercept, *a*.

e. Calculate the slope, *b*.

f. Write the equation for the line.

g. Draw the line on an empty scatterplot, basing the line on predicted *Y* values for *X* values of 0, 1, and 18.

14.23 Let's assume we know that age is related to bone density, with a Pearson correlation coefficient of −0.19. (Notice that the correlation is negative, indicating that bone density tends to be lower at older ages than at younger ages.) Assume we also know the following descriptive statistics:

Age of people studied: 55 years on average, with a standard deviation of 12 years

Bone density of people studied: 1000 mg/cm^2 on average, with a standard deviation of 95 mg/cm^2

Virginia is 76 years old. What would you predict her bone density to be? To answer this question, complete the following steps:

a. Transform the raw score for the independent variable to a *z* score.

b. Calculate the predicted *z* score for the dependent variable.

c. Transform the *z* score for the dependent variable back into a raw score.

d. Calculate the *y* intercept, *a*.

e. Calculate the slope, *b*.

f. Write the equation for the line.

g. Draw the line on an empty scatterplot, basing the line on predicted *Y* values for *X* values of 0, 1, and 18.

14.24 Given the regression line $\hat{Y} = -6 + 0.41(X)$, make predictions for each of the following:

a. $X = 25$

b. $X = 50$

c. $X = 75$

14.25 Given the regression line $\hat{Y} = 49 - 0.18(X)$, make predictions for each of the following:

a. $X = -31$

b. $X = 65$

c. $X = 14$

14.26 Data are provided here with descriptive statistics, a correlation coefficient, and a regression equation: $r = 0.426$, $\hat{Y} = 219.974 + 186.595(X)$.

X	*Y*
0.13	200.00
0.27	98.00
0.49	543.00
0.57	385.00
0.84	420.00
1.12	312.00
$M_X = 0.57$	$M_Y = 326.333$
$SD_X = 0.333$	$SD_Y = 145.752$

Using this information, compute the following estimates of prediction error:

a. Calculate the sum of squared error for the mean, SS_{total}.

b. Now, using the regression equation provided, calculate the sum of squared error for the regression equation, SS_{error}.

c. Using your work, calculate the proportionate reduction in error for these data.

d. Check that this calculation of r^2 equals the square of the correlation coefficient.

e. Compute the standardized regression coefficient.

14.27 Data are provided here with descriptive statistics, a correlation coefficient, and a regression equation: $r = 0.52$, $\hat{Y} = 2.643 + 0.469(X)$.

X	Y
4.00	6.00
6.00	3.00
7.00	7.00
8.00	5.00
9.00	4.00
10.00	12.00
12.00	9.00
14.00	8.00
$M_X = 8.75$	$M_Y = 6.75$
$SD_X = 3.031$	$SD_Y = 2.727$

Using this information, compute the following estimates of prediction error:

a. Calculate the sum of squared error for the mean, SS_{total}.

b. Now, using the regression equation provided, calculate the sum of squared error for the regression equation, SS_{error}.

c. Using your work, calculate the proportionate reduction in error for these data.

d. Check that this calculation of r^2 equals the square of the correlation coefficient.

e. Compute the standardized regression coefficient.

14.28 Use this output from a multiple regression analysis to answer the following questions:

Coefficients[a]

Model		Unstandardized Coefficients		Standardized Coefficients		
		B	Std. Error	Beta	t	Sig.
1	(Constant)	3.977	1.193		3.333	.001
	Variable 1	.414	.096	.458	4.313	.000
	Variable 2	-.019	.011	-.181	-1.704	.093

a. Dependent Variable: Outcome (Y)

a. Write the equation for the line of prediction.

b. Use the equation for part (a) to make predictions for: Variable 1 = 6, variable 2 = 60

c. Use the equation for part (a) to make predictions for: Variable 1 = 9, variable 2 = 54.3

d. Use the equation for part (a) to make predictions for: Variable 1 = 13, variable 2 = 44.8

14.29 Use this output from a multiple regression analysis to answer the following questions:

Coefficients[a]

Model		Unstandardized Coefficients		Standardized Coefficients		
		B	Std. Error	Beta	t	Sig.
1	(Constant)	1.675	.563		2.972	.004
	SAT	.001	.000	.321	2.953	.004
	Rank	-.008	.003	-.279	-2.566	.012

a. Dependent Variable: GPA

a. Write the equation for the line of prediction.

b. Use the equation for part (a) to make predictions for: SAT = 1030, rank = 41

c. Use the equation for part (a) to make predictions for: SAT = 860, rank = 22

d. Use the equation for part (a) to make predictions for: SAT = 1060, rank = 8

Applying the Concepts

14.30 Weight, blood pressure, and regression: Several studies have found a correlation between weight and blood pressure.

a. Explain what is meant by a correlation between these two variables.

b. If you were to examine these two variables with simple linear regression instead of correlation, how would you frame the question? (*Hint:* The research question for correlation would be: Is weight related to blood pressure?)

c. What is the difference between simple linear regression and multiple regression?

d. If you were to conduct a multiple regression instead of a simple linear regression, what other independent variables might you include?

14.31 Temperature, hot chocolate sales, and prediction: Running a football stadium involves innumerable predictions. For example, when stocking up on food and beverages for sale at the game, it helps to have an idea of how much will be sold. In the football stadiums in colder climates, stadium managers use expected outdoor temperature to predict sales of hot chocolate.

a. What is the independent variable in this example?

b. What is the dependent variable?

c. As the value of the independent variable increases, what can we predict would happen to the value of the dependent variable?

d. What other variables might predict this dependent variable? Name at least three.

14.32 Age, hours studied, and prediction: In How It Works 13.2, we calculated the correlation coefficient between students' age and number of hours they study per week. The correlation between these two variables is 0.49.

a. Elif's z score for age is -0.82. What would we predict for the z score for the number of hours she studies per week?

b. John's z score for age is 1.2. What would we predict for the z score for the number of hours he studies per week?

c. Eugene's z score for age is 0. What would we predict for the z score for the number of hours he studies per week?

d. For part (c) explain why the concept of *regression* to the mean is not relevant (and why you didn't really need the formula).

14.33 Consideration of Future Consequences scale, z scores, and raw scores: A study of Consideration of Future Consequences (CFC) found a mean score of 3.51, with a standard deviation of 0.61, for the 664 students in the sample (Petrocelli, 2003).

a. Imagine that your z score on the CFC score was -1.2. What would your raw score be? Use symbolic notation and the formula. Explain why this answer makes sense.

b. Imagine that your z score on the CFC score was 0.66. What would your raw score be? Use symbolic notation and the formula. Explain why this answer makes sense.

14.34 The GRE, z scores, and raw scores: The verbal subtest of the Graduate Record Examination (GRE) has a population mean of 500 and a population standard deviation of 100 by design (the quantitative subtest has the same mean and standard deviation).

a. Convert the following z scores to raw scores *without* using a formula: (i) 1.5, (ii) -0.5, (iii) -2.0

b. Now convert the same z scores to raw scores using symbolic notation and the formula: (i) 1.5, (ii) -0.5, (iii) -2.0

14.35 Hours studied, grade, and regression: A regression analysis of data from some of our statistics classes yielded the following regression equation for the independent variable (hours studied) and the dependent variable (grade point average [GPA]): $\hat{Y} = 2.96 + 0.02(X)$.

a. If you plan to study 8 hours per week, what would you predict your GPA will be?

b. If you plan to study 10 hours per week, what would you predict your GPA will be?

c. If you plan to study 11 hours per week, what would you predict your GPA will be?

d. Create a graph and draw the regression line based on these three pairs of scores.

e. Do some algebra, and determine the number of hours you'd have to study to have a predicted GPA of the maximum possible, 4.0. Why is it misleading to make predictions for anyone who plans to study this many hours (or more)?

14.36 Precipitation, violence, and limitations of regression: Does the level of precipitation predict violence? Dubner and Levitt (2006b) reported on various studies that found links between rain and violence. They mentioned one study by Miguel, Satyanath, and Sergenti that found that decreased rain was linked with an increased likelihood of civil war across a number of African countries they examined. Referring to the study's authors, Dubner and Levitt state, "The causal effect of a drought, they argue, was frighteningly strong."

a. What is the independent variable in this study?

b. What is the dependent variable?

c. What possible third variables might play a role in this connection? That is, is it just the lack of rain that's causing violence, or is it something else? (*Hint:* Consider the likely economic base of many African countries.)

14.37 Cola consumption, bone mineral density, and limitations of regression: Does one's cola consumption predict one's bone mineral density? Using regression analyses, nutrition researchers found that older women who drank more cola (but not more of other carbonated drinks) tended to have lower bone mineral density, a risk factor for osteoporosis (Tucker, Morita, Qiao, Hannan, Cupples, & Kiel, 2006). Cola intake, therefore, does seem to predict bone mineral density.

a. Explain why we cannot conclude that cola intake causes a decrease in bone mineral density.

b. The researchers included a number of possible third variables in their regression analyses. Among the included variables were physical activity score, smoking, alcohol use, and calcium intake. They included the possible third variables first, and then added the bone density measure. Why would they have used multiple regression in this case? Explain.

c. How might physical activity play a role as a third variable? Discuss its possible relation to both bone density and cola consumption.

d. How might calcium intake play a role as a third variable? Discuss its possible relation to both bone density and cola consumption.

14.38 Tutoring, mathematics performance, and problems with regression: A researcher conducted a study in which children with problems learning mathematics were offered the opportunity to purchase time with special tutors. The number of weeks that children met with their tutors varied from 1 to 20. He found that the number of weeks of tutoring predicted these children's mathematics performance and recommended that parents of such children send them for tutoring.

a. List one problem with that interpretation. Explain your answer.

b. If you were to develop a study that uses a multiple regression equation instead of a simple linear regression equation, what additional variables might be good independent variables? List at least one variable that can be manipulated (e.g., weeks of tutoring) and at least one variable that cannot be manipulated (e.g., parents' years of education).

14.39 Anxiety, depression, and simple linear regression: We analyzed data from a larger data set that one of the authors used for previous research (Nolan, Flynn, & Garber, 2003). In the current analyses, we used regression to look at factors that predict anxiety over a 3-year period. Shown below is the output for the regression analysis examining whether depression at year 1 predicted anxiety at year 3.

Coefficients(a)

	Unstandardized Coefficients		Standardized Coefficients	t	Sig.
	B	Std. Error	Beta		
(Constant)	24.698	.566		43.665	.000
Depression Year 1	.161	.048	.235	3.333	.001

a. Dependent Variable: Anxiety Year 3

a. From this software output, write the regression equation.

b. As depression at year 1 increases by 1 point, what happens to the predicted anxiety level for year 3? Be specific.

c. If someone has a depression score of 10 at year 1, what would we predict for her anxiety score at year 3?

d. If someone has a depression score of 2 at year 1, what would we predict for his anxiety score at year 3?

14.40 Anxiety, depression, and multiple regression: We conducted a second regression analysis on the data from Exercise 14.39. In addition to depression at year 1, we included a second independent variable to predict anxiety at year 3. We also included anxiety at year 1. (We might expect that the best predictor of anxiety at a later point in time is one's anxiety at an earlier point in time.) Here is the output for that analysis.

Coefficients(a)

	Unstandardized Coefficients		Standardized Coefficients	t	Sig.
	B	Std. Error	Beta		
(Constant)	17.038	1.484		11.482	.000
Depression Year 1	-.013	.055	-.019	-.237	.813
Anxiety Year 1	.307	.056	.442	5.521	.000

a. Dependent Variable: Anxiety Year 3

a. From this software output, write the regression equation.

b. As the first independent variable, depression at year 1, increases by 1 point, what happens to the predicted score on anxiety at year 3?

c. As the second independent variable, anxiety at year 1, increases by 1 point, what happens to the predicted score on anxiety at year 3?

d. Compare the predictive utility of depression at year 1 using the regression equation in Exercise 14.39 and using the regression equation you just wrote in 14.40(a). In which regression equation is depression at year 1 a better predictor? Given that we're using the same sample, is depression at year 1 actually better at predicting anxiety at year 3 in one regression equation versus the other? Why do you think there's a difference?

e. The table on the next page is the correlation matrix for the three variables. As you can see, all three are highly correlated with one another. If we look at the intersection of each pair of variables, the number next to "Pearson correlation" is the correlation coefficient. For example, the correlation between "Anxiety year 1" and "Depression year 1" is .549. Which two variables show the strongest correlation? How might this explain the fact that depression at year 1 seems to be a better predictor when it's the only independent variable than when anxiety at year 1 also is included? What does this tell us about the importance of including third variables in the regression analyses when possible?

f. Let's say you want to add a fourth independent variable. You have to choose among three possible independent variables: (1) a variable highly correlated with both independent variables and the

Correlations

		Depression Year 1	Anxiety Year 1	Anxiety Year 3
Depression Year 1	Pearson Correlation	1	.549(**)	.235(**)
	Sig. (2-tailed)		.000	.001
	N	240	240	192
Anxiety Year 1	Pearson Correlation	.549(**)	1	.432(**)
	Sig. (2-tailed)	.000		.000
	N	240	240	192
Anxiety Year 3	Pearson Correlation	.235(**)	.432(**)	1
	Sig. (2-tailed)	.001	.000	
	N	192	192	192

** Correlation is significant at the 0.01 level (2-tailed).

dependent variable, (2) a variable highly correlated with the dependent variable but not correlated with either independent variable, and (3) a variable not correlated with either of the independent variables or with the dependent variable. Which of the three variables is likely to make the multiple regression equation better? That is, which is likely to increase the proportionate reduction in error? Explain.

14.41 Cohabitation, divorce, and prediction: A study by the Institute for Fiscal Studies (Goodman & Greaves, 2010) found that parents' marital status when a child was born predicted the likelihood of the relationship's demise. Parents who were cohabitating when their child was born had a 27% chance of breaking up by the time the child was 5, whereas those who were married when their child was born had a 9% chance of breaking up by the time the child was 5—a difference of 18%. The researchers, however, reported that cohabiting parents tended to be younger, less affluent, less likely to own a home, less educated, and more likely to have an unplanned pregnancy. When the researchers statistically controlled for these variables, they found that there was just a 2% difference between cohabitating and married parents.

a. What are the independent and dependent variables used in this study?

b. Were the researchers likely to have used simple linear regression or multiple regression for their analyses? Explain your answer.

c. In your own words, explain why the ability of marital status at the time of a child's birth to predict divorce within 5 years almost disappeared when other variables were considered.

d. Name at least one additional "third variable" that might have been at play in this situation.

Putting It All Together

14.42 Corporate political contributions, profits, and regression: Researchers studied whether corporate political contributions predicted profits (Cooper, Gulen, & Ovtchinnikov, 2007). From archival data, they determined how many political candidates each company supported with financial contributions, as well as each company's profit in terms of a percentage. The accompanying table shows data for five companies. (*Note:* The data points are hypothetical but are based on averages for companies falling in the 2nd, 4th, 6th, and 8th deciles in terms of candidates supported. A decile is a range of 10%, so the 2nd decile includes those with percentiles between 10 and 19.9.)

Number of Candidates Supported	Profit (%)
6	12.37
17	12.91
39	12.59
62	13.43
98	13.42

a. Create the scatterplot for these scores.

b. Calculate the mean and standard deviation for the variable "number of candidates supported."

c. Calculate the mean and standard deviation for the variable "profit."

d. Calculate the correlation between number of candidates supported and profit.

e. Calculate the regression equation for the prediction of profit from number of candidates supported.

f. Create a graph and draw the regression line.

g. What do these data suggest about the political process?

h. What third variables might be at play here?

i. Compute the standardized regression coefficient.

j. How does this coefficient relate to other information you know?

k. Draw a conclusion about your analysis based on what you know about hypothesis testing with simple linear regression.

14.43 Age, hours studied, and regression: In How It Works 13.2, we calculated the correlation coefficient between students' age and number of hours they study per week. The mean for age is 21, and the standard deviation is 1.789. The mean for hours studied is 14.2, and the standard deviation is 5.582. The correlation between these two variables is 0.49. Use the z score formula.

a. João is 24 years old. How many hours would we predict he studies per week?

b. Kimberly is 19 years old. How many hours would we predict she studies per week?

c. Seung is 45 years old. Why might it not be a good idea to predict how many hours per week he studies?

d. From a mathematical perspective, why is the word *regression* used? (*Hint:* Look at parts (a) and (b), and discuss the scores on the first variable with respect to their mean versus the predicted scores on the second variable with respect to their mean.)

e. Calculate the regression equation.

f. Use the regression equation to predict the number of hours studied for a 17-year-old student and for a 22-year-old student.

g. Using the four pairs of scores that you have (age and predicted hours studied from part (b), and the predicted scores for a score of 0 and 1 from calculating the regression equation), create a graph that includes the regression line.

h. Why is it misleading to include young ages such as 0 and 5 on the graph?

i. Construct a graph that includes both the scatterplot for these data and the regression line. Draw vertical lines to connect each dot on the scatterplot with the regression line.

j. Construct a second graph that includes both the scatterplot and a line for the mean for hours studied, 16.2. The line will be horizontal and will begin at 16.2 on the y-axis. Draw vertical lines to connect each dot on the scatterplot with the regression line.

k. Part (i) is a depiction of the error we make if we use the regression equation to predict hours studied. Part (j) is a depiction of the error we make if we use the mean to predict hours studied (i.e., if we predict that everyone has the mean of 16.2 on hours studied per week). Which one appears to have less error? Briefly explain why the error is less in one situation.

l. Calculate the proportionate reduction in error the long way.

m. Explain what the proportionate reduction in error that you calculated in part (a) tells us. Be specific about what it tells us about predicting using the regression equation versus predicting using the mean.

n. Demonstrate how the proportionate reduction in error could be calculated using the shortcut. Why does this make sense? That is, why does the correlation coefficient give us a sense of how useful the regression equation will be?

o. Compute the standardized regression coefficient.

p. How does this coefficient relate to other information you know?

q. Draw a conclusion about your analysis based on what you know about hypothesis testing with regression.

Terms

simple linear regression (p. 371)
regression to the mean (p. 374)
intercept (p. 375)
slope (p. 375)
standardized regression coefficient (p. 378)
standard error of the estimate (p. 382)
proportionate reduction in error (p. 384)
orthogonal variable (p. 389)
multiple regression (p. 389)

Formulas

$z_{\hat{Y}} = (r_{XY})(z_X)$ (p. 372)

$\hat{Y} = a + b(X)$ (p. 375)

$\beta = (b)\frac{\sqrt{SS_X}}{\sqrt{SS_Y}}$ (p. 378)

$r^2 = \frac{(SS_{total} - SS_{error})}{SS_{total}}$ (p. 387)

Symbols

$\hat{Y}$ (p. 373)
a (p. 375)
b (p. 375)
β (p. 378)
SS_{total} (p. 384)
SS_{error} (p. 386)
r^2 (p. 387)
R^2 (p. 387)

CHAPTER 15

Nonparametric Tests

BEFORE YOU GO ON

- You should be able to differentiate between a parametric and a nonparametric hypothesis test (Chapter 7).
- You should know the six steps of hypothesis testing (Chapter 7).
- You should understand the concept of effect size (Chapter 8).
- You should understand the concept of correlation (Chapters 1 and 13).
- You should know when to use an independent-samples *t* test (Chapter 10).

David Santiago/*El Nuevo Herald*/MCT via Getty Images

Does LeBron James have a "hot hand"? Studies of baseball, basketball, and many other sports tell the same story: You may feel as if you have a "hot hand," but the pattern is only a strongly felt illusion. In basketball, the success or failure of a previous shot does not influence the outcome of the next shot.

You can't believe everything you think—or feel. Statistically, the "hot hand" in basketball, the "hot seat" at the poker table, and the "Big Mo" (momentum) in football are not real; however, they feel real because of the hindsight bias and the confirmation bias (introduced in Chapter 5). With the possible exception of bowling, "feeling it" does not statistically predict that a "hot streak" will continue (Alter & Oppenheimer, 2006; Gilovich, 1991; Kida, 2006). Nevertheless, basketball players and fans have believed that a player's chance of hitting a shot is greater after a make than a miss (Gilovich, Vallone, & Tversky, 1985). However, shooting records of the Philadelphia 76ers, free-throw records from the Boston Celtics, and a controlled study of 14 men and 12 women from Cornell University's basketball teams all told the same data story. The Cornell athletes, for example, believed that the outcome of the previous shot influenced the next shot even though they did *not* perform that way (Gilovich et al., 1985).

Statistical inference helps us overcome confirmation bias by separating real patterns from chance events, even when the data do not have a scale dependent variable. Table 15-1 summarizes the parametric tests we've learned. This chapter helps us explore hypotheses about nonparametric data by teaching (a) when to use a nonparametric test, (b) two types of nonparametric tests used with nominal data, and (c) two types of nonparametric tests used with ordinal data.

Nonparametric Statistics

Listen carefully to most sports commentators. They will identify a player as "on fire" only *after* the player succeeds; they are not good at predicting success (Koehler & Conley, 2003). However, many field studies violate parametric assumptions for hypothesis testing (especially that the data be drawn from a normally distributed population). The solution? Use more conservative nonparametric statistics that are *not* based on critical assumptions about the population.

An Example of a Nonparametric Test

Nonparametric statistics enlarge the universe of things we can study. For example, a team of Israeli physician-researchers, led by Dr. Shevach Friedler, a trained mime as well as a physician (Ryan, 2006), found that live entertainment by clowns—yes, clowns!—was associated with higher rates of conception during in vitro fertilization (IVF) (Rockwell, 2006). Friedler had a professional clown entertain 93 women during the 15 minutes after embryo transfer (Brinn, 2006); a comparison group of 93 women did not receive entertainment by a clown during IVF. Thirty-three who were entertained by a clown (35%) conceived, compared with 19% in the comparison group. Is this real or just chance?

TABLE 15-1. A Summary of Research Designs

We have encountered several research designs so far, most of which fall in one of two categories. Some designs—those listed in category I—include at least one scale independent variable and a scale dependent variable. Other designs—those listed in category II—include a nominal (or sometimes ordinal) independent variable and a scale dependent variable. Until now, we have not encountered a research design with a nominal independent variable and a nominal dependent variable, or a research design with an ordinal dependent variable.

I. Scale Independent Variable and Scale Dependent Variable	II. Nominal Independent Variable and Scale Dependent Variable
Correlation	*z* test
Regression	All kinds of *t* tests
	All kinds of ANOVAs

The hypothesis is that whether a woman becomes pregnant depends on whether she receives clown therapy. The independent variable is type of post-IVF treatment, with two levels (clown therapy versus no clown therapy). The dependent variable is outcome, with two levels (becomes pregnant versus does not become pregnant). Pregnancy, of course, is not a scale variable. You can't be "just a little pregnant," so this is a new statistical situation: Both the independent variable and the dependent variable are nominal.

This new situation (two nominal variables) calls for a new statistic and a new hypothesis test: The chi-square statistic is symbolized as χ^2 (pronounced "kai square"—rhymes with *sky*) and relies on the chi-square distribution.

When to Use Nonparametric Tests

We use a nonparametric test when (1) the dependent variable is nominal, (2) the dependent variable is ordinal, or (3) the sample size is small and the population of interest may be skewed.

Situation 1 (a nominal dependent variable) occurs whenever we categorize the observations: pregnant or not pregnant, male or female, driver's license or no driver's license. It's not always perfect, but we often think of our world in terms of categories.

Situation 2 (an ordinal dependent variable) describes rank, such as in athletic competitions, class position, and preferred flavor of ice cream. Top 10 lists and favorite nephews are also ordinal observations.

Situation 3 (a small sample size, usually less than 30, and a potentially skewed population) occurs less frequently. It would be difficult to recruit enough participants to study the brain patterns among people who have won the Nobel Prize in literature, no matter how hard we tried or how much we paid people to participate.

> **MASTERING THE CONCEPT**
>
> **15.1:** We use nonparametric tests in three different statistical situations: when (1) the dependent variable is nominal, (2) the dependent variable is ordinal, or (3) the sample size is small and we suspect that the underlying population distribution is not normal.

Although nonparametric tests expand the range of variables available for research, they have two big problems: (1) confidence intervals and effect-size measures are not typically available for nominal or ordinal data; (2) nonparametric tests tend to have less statistical power than parametric tests. This increases the risk of a Type II error: We are less likely to reject the null hypothesis when we should reject it—that is, when there is a real difference between groups. Nonparametric tests are usually the backup plan, not the go-to statistical tests.

CHECK YOUR LEARNING

Reviewing the Concepts

> We use a nonparametric test when we cannot meet the assumptions of a parametric test, primarily the assumptions of having a scale dependent variable and a normally distributed population.

> The most common situations in which we use a nonparametric test are when we have a nominal or ordinal dependent variable or a small sample in which the data for the dependent variable suggest that the underlying population distribution might be skewed.

Clarifying the Concepts

15-1 Distinguish parametric tests from nonparametric tests.

15-2 When do we use nonparametric tests?

Calculating the Statistics

15-3 For each of the following situations, identify the independent and dependent variables and how they are measured (nominal, ordinal, or scale).

a. Bernstein (1996) reported that Francis Galton created a "beauty map" by recording the numbers of women he encountered in different cities in England who were either pretty or not so pretty. London women, he found, were the most likely to be pretty and Aberdeen women the least likely.

b. Imagine that Galton instead gave every woman a beauty score on a scale of 1–10 and then compared means for the women in each of five cities.

c. Galton was famous for discounting the intelligence of most women (Bernstein, 1996). Imagine that he assessed the intelligence of 50 women and then applied the

continued on next page

beauty scale mentioned in part (b). Let's say he found that women with higher intelligence were more likely to be pretty, whereas women with lower intelligence were less likely to be pretty.

d. Imagine that Galton now ranked 50 women on their beauty and on their intelligence.

Applying the Concepts

Solutions to these Check Your Learning questions can be found in Appendix D.

15-4 For each of the situations listed in Check Your Learning 15-3, state the category (I or II) from Table 15-1 from which you would choose the appropriate hypothesis test. If you would not choose a test from either category I or II, simply list category III—other. Explain why you chose I, II, or III.

Chi-Square Tests

Hot-hand research has moved from individual performance to team performance. For example, in 2005, Super Bowl coach Mike Holmgren was uncertain about whether to rest key players prior to the playoffs because he didn't "want to lose momentum." The reality, however, is that the data are not a good fit with Holmgren's belief in Big Mo (Vergin, 2000). A "good fit" describes what is being tested in two common kinds of chi-square statistical tests: (1) the ***chi-square test for goodness of fit,*** *a nonparametric hypothesis test that is used when there is one nominal variable;* (2) the ***chi-square test for independence***, *a nonparametric hypothesis test that is used when there are two nominal variables.* Both chi-square tests involve the by-now familiar six steps of hypothesis testing.

Both chi-square tests use the chi-square statistic: χ^2. The chi-square statistic is based on the chi-square distribution. As with t and F distributions, there are also several chi-square distributions, depending on the degrees of freedom. After we introduce chi-square tests, we'll introduce several ways of determining the size of a finding—by calculating an effect size, graphing the finding, or determining relative risk.

MASTERING THE CONCEPT

15.2: When we only have nominal variables, we use the chi-square statistic. Specifically, we use a chi-square test for goodness of fit when there is one nominal variable, and we use a chi-square test for independence when there are two nominal variables.

Chi-Square Test for Goodness of Fit

The chi-square test for goodness of fit calculates a statistic based on just one variable. There is no independent variable or dependent variable, just one categorical variable with two or more categories into which participants are placed. In fact, the chi-square test for goodness of fit received its name because it measures how good the fit is between the observed data in the various categories of a single nominal variable and the data we would expect according to the null hypothesis. If there's a really good fit with the null hypothesis, then we cannot reject the null hypothesis. If we hope to receive empirical support for the research hypothesis, then we're actually hoping for a *bad fit* between the observed data and what we expect according to the null hypothesis.

EXAMPLE 15.1

For example, researchers reported that the best youth soccer players in the world were more likely to have been born early in the year than later (Dubner & Levitt, 2006a). As one example, they reported that 52 elite youth players in Germany were born in January, February, or March, whereas only 4 players were born in October, November, or December. (Those born in other months were not included in this study.)

The null hypothesis predicts that when a person was born will not make any difference; the research hypothesis predicts that the month a person was born will matter when it comes to being an elite soccer player. Assuming that births in the general population are evenly distributed across months of the year, the null hypothesis posits equal numbers of elite soccer players born in the first 3 months and the last 3 months of the year. With 56 participants in the study (52 born in the first 3 months and 4 in the last 3 months), equal frequencies lead us to expect 28 players born in the first 3 months and 28 in the last 3 months just by chance. The birth months don't appear to be evenly distributed, but is this a real pattern, or just chance?

Like previous hypothesis tests, the chi-square goodness of fit test uses the six steps of hypothesis testing.

Are Elite Soccer Players Born in the Early Months of the Year? Researchers reported that elite soccer players are far more likely to be born in the first 3 months of the year than in the last 3 months (Dubner & Levitt, 2006a). Based on data for elite German youth soccer players, a chi-square test for goodness of fit showed a significant effect: Players were significantly more likely to be born in the first 3 months than in the last 3 months of the year.

STEP 1: Identify the populations, distribution, and assumptions.

In this case, there is a population of elite German youth soccer players with birth dates like those we observed and a population of elite German youth soccer players with birth dates like those in the general population. The comparison distribution is a chi-square distribution. There's just one nominal variable, birth months, so we'll conduct a chi-square test for goodness of fit.

The first assumption is that the variable (birth month) is nominal. The second assumption is that each observation is independent; no single participant can be in more than one category. The third assumption is that participants were randomly selected. If not, it may be unwise to confidently generalize beyond the sample. A fourth assumption is that there is a minimum number of expected participants in every category (also called a *cell*)—at least 5 and preferably more. An alternative guideline (Delucchi, 1983) is for at least five times as many participants as cells. In any case, the chi-square tests seem robust to violations of this last assumption.

Summary: Population 1: Elite German youth soccer players with birth dates like those we observed. Population 2: Elite German youth soccer players with birth dates like those in the general population.

The comparison distribution is a chi-square distribution. The hypothesis test will be a chi-square test for goodness of fit because we have one nominal variable only, birth months. This study meets three of the four assumptions: (1) The one variable is nominal. (2) Every participant is in only one cell (you can't be born in January and November). (3) This is not a randomly selected sample of all elite soccer players. The sample includes only German youth soccer players in the elite leagues. We must be cautious in generalizing beyond young German elite players. (4) There are more than five times as many participants as cells (the table has two cells, and $2 \times 5 = 10$). We have 56 participants, far more than the 10 necessary to meet this guideline.

- The **chi-square test for goodness of fit** is a nonparametric hypothesis test used with one nominal variable.
- The **chi-square test for independence** is a nonparametric hypothesis test used with two nominal variables.

STEP 2: State the null and research hypotheses.

For chi-square tests, it's easiest to state the hypotheses in words only, rather than in both words and symbols.

Summary: Null hypothesis: Elite German youth soccer players have the same pattern of birth months as those in the general population. Research hypothesis: Elite German youth soccer players have a different pattern of birth months than those in the general population.

STEP 3: Determine the characteristics of the comparison distribution. Our only task at this step is to determine the degrees of freedom. In most previous hypothesis tests, the degrees of freedom have been based on sample size. For the chi-square hypothesis tests, however, the degrees of freedom are based on the numbers of categories, or cells, in which participants can be counted. The degrees of freedom for a chi-square test for goodness of fit is the number of categories minus 1:

MASTERING THE FORMULA

15-1: We calculate the degrees of freedom for the chi-square test for goodness of fit by subtracting 1 from the number of categories, represented in the formula by k. The formula is: $df_{\chi^2} = k - 1$.

$$df_{\chi^2} = k - 1$$

Here, k is the symbol for the number of categories. The current example has only two categories: Each soccer player in this study was born in either the first 3 months of the year or the last 3 months of the year:

$$df_{\chi^2} = 2 - 1 = 1$$

Summary: The comparison distribution is a chi-square distribution, which has 1 degree of freedom: $df_{\chi^2} = 2 - 1 = 1$.

STEP 4: Determine the critical value, or cutoff. To determine the cutoff, or critical value, for the chi-square statistic, we use the chi-square table in Appendix B. χ^2 is based on squares and can never be negative, so there is just one critical value. An excerpt from Appendix B that applies to the soccer study is given in Table 15-2. We look under the p level, usually 0.05, and across from the appropriate degrees of freedom, in this case, 1. For this situation, the critical chi-square statistic is 3.841.

Summary: The critical χ^2, based on a p level of 0.05 and 1 degree of freedom, is 3.841, as seen in the curve in Figure 15-1.

STEP 5: Calculate the test statistic. To calculate a chi-square statistic, we determine the observed frequencies and the expected frequencies, as seen in the second and third columns of Table 15-3. The expected

TABLE 15-2. Excerpt from the χ^2 Table

We use the χ^2 table to determine critical values for a given p level, based on the degrees of freedom.

	Proportion in Critical Region		
df	0.10	0.05	0.01
1	2.706	3.841	6.635
2	4.605	5.992	9.211
3	6.252	7.815	11.345
.			
.			
.			

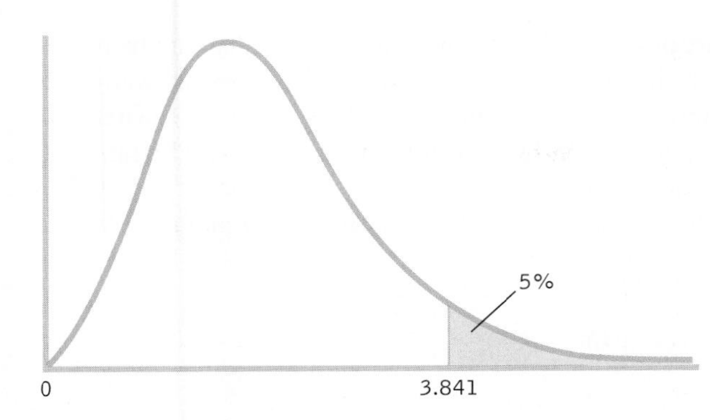

FIGURE 15-1
Determining the Cutoff for a Chi-Square Statistic

We look up the critical value for a chi-square statistic, based on a certain p level and degrees of freedom, in the chi-square table. Because the chi-square statistic is squared, it is never negative, so there is only one critical value.

frequencies are determined from the information we have about the general population. In this case, we estimate that, in the general population, about half of all births (only, of course, among those born in the first or last 3 months of the year) occur in the first 3 months of the year, a proportion of 0.50.

$$(0.50)(56) = 28$$

Of the 56 elite German youth soccer players in the study, we would expect to find that 28 were born in the first 3 months of the year (versus the last 3 months of the year) if these youth soccer players are no different from the general population with respect to birth date. Similarly, we would expect a proportion of $1 - 0.50 = 0.50$ of these soccer players to be born in the last 3 months of the year:

$$(0.50)(56) = 28$$

These numbers are identical only because the proportions are 0.50 and 0.50. If the proportion expected for the first 3 months of the year, based on the general population, was 0.60, then we would expect a proportion of $1 - 0.60 = 0.40$ for the last 3 months of the year.

The next step in calculating the chi-square statistic is to calculate a sort of sum of squared differences. We start by determining the difference between each observed frequency and its matching expected frequency. This is usually done in columns, so we use this format even though we have only two categories. The first three columns of Table 15-3 show us the categories, observed frequencies, and expected frequencies, respectively. The fourth column, using O for observed and E for expected, displays the differences. As in the past, if we sum the differences, we get 0; they cancel out because

TABLE 15-3. The Chi-Square Calculations

As with many other statistics, we calculate the chi-square statistic using columns to keep track of our work. We list the observed frequencies, then calculate the expected frequencies, the difference between the observed frequencies and the expected frequencies, square the differences, then divide each square by its appropriate expected frequency. Finally, we add up the numbers in the sixth column to find the chi-square statistic.

Column 1	2	3	4	5	6
Category	Observed (O)	Expected (E)	$O - E$	$(O - E)^2$	$\frac{(O-E)^2}{E}$
First 3 months	52	28	24	576	20.571
Last 3 months	4	28	−24	576	20.571

some are positive and some are negative. We solve this problem as we have in the past—by squaring the differences, as shown in the fifth column. Next, however, we have a step that we haven't seen before with squared differences. We divide each squared difference by the expected value for its cell, as seen in the sixth column. The numbers in the sixth column are the ones we sum.

As an example, here are the calculations for the category "first 3 months":

$$O - E = (52 - 28) = 24$$

$$(O - E)^2 = (24)^2 = 576$$

$$\frac{(O - E)^2}{E} = \frac{576}{28} = 20.571$$

Once we complete the table, the last step is easy. We just add up the numbers in the sixth column. In this case, the chi-square statistic is 20.571 + 20.571 = 41.14. We can finish the formula by adding a summation sign to the formula in the sixth column. Note that we don't have to divide this sum by anything, as we've done with other statistics. We already did the dividing before we summed. This sum is the chi-square statistic. Here is the formula:

MASTERING THE FORMULA

15-2: The formula for the chi-square statistic is: $\chi^2 = \Sigma\left[\frac{(O-E)^2}{E}\right]$. For each cell, we subtract the expected count, E, from the observed count, O. Then we square each difference and divide the square by the expected count. Finally, we sum the calculations for each of the cells.

$$\chi^2 = \Sigma\left[\frac{(O-E)^2}{E}\right]$$

Summary: $\chi^2 = \Sigma\left[\frac{(O-E)^2}{E}\right] = (20.571 + 20.571) = 41.14$

STEP 6: Make a decision. This last step is identical to that of previous hypothesis tests. We reject the null hypothesis if the test statistic is beyond the critical value, and we fail to reject the null hypothesis if the test statistic is not beyond the critical value. In this case, the test statistic, 41.14, is far beyond the cutoff, 3.841, as seen in Figure 15-2. We reject the null hypothesis. Because there are only two categories, it's clear where the difference lies. It appears that elite German youth soccer players are more likely to have been born in the first 3 months of the year, and less likely to have been born in the last 3 months of the year, than members of the general population. (If we had failed to reject the null hypothesis, we could only have concluded that these data did not provide sufficient evidence to show that elite German youth soccer players have a different likelihood of being born in the first, versus last, 3 months of the year than those in the general population.)

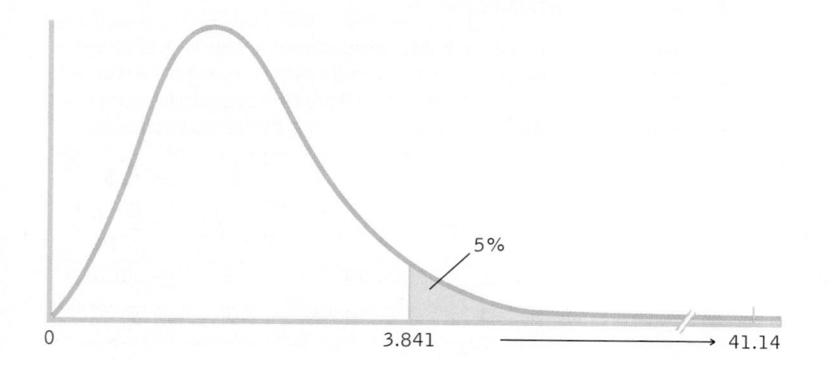

FIGURE 15-2
Making a Decision

As with other hypothesis tests, we make a decision with a chi-square test by comparing the test statistic to the cutoff, or critical value. We see here that 41.14 would be *far* to the right of 3.841.

Summary: Reject the null hypothesis; it appears that elite German youth soccer players are more likely to have been born in the first 3 months of the year, and less likely to have been born in the last 3 months of the year, than people in the general population.

We report these statistics in a journal article in almost the same format that we've seen previously. We report the degrees of freedom, the value of the test statistic, and whether the p value associated with the test statistic is less than or greater than the cutoff based on the p level of 0.05. (As usual, we would report the actual p level if we conducted this hypothesis test using software.) In addition, we report the sample size in parentheses with the degrees of freedom. In the current example, the statistics read:

$$\chi^2(1, N = 56) = 41.14, p < 0.05$$

The researchers who conducted this study imagined four possible explanations: "a) certain astrological signs confer superior soccer skills; b) winterborn babies tend to have higher oxygen capacity, which increases soccer stamina; c) soccer-mad parents are more likely to conceive children in springtime, at the annual peak of soccer mania; d) none of the above" (Dubner & Levitt, 2006a). What's your guess?

The researchers picked (d) and suggested another alternative (Dubner & Levitt, 2006a). Participation in youth soccer leagues has a strict cutoff date: December 31. Compared to those born in December, children born the previous January are likely to be more mature, perceived as more talented, chosen for the best leagues, and given better coaching—a self-fulfilling prophecy. All this from a simple chi-square test for goodness of fit! ■

Chi-Square Test for Independence

The chi-square test for goodness of fit analyzes just one nominal variable. The chi-square test for independence analyzes *two* nominal variables.

Like the correlation coefficient, the chi-square test for independence does not require that we identify independent and dependent variables. However, specifying an independent variable and a dependent variable can help us articulate hypotheses. The chi-square test for independence is so named because we are trying to determine whether the two variables—no matter which one is considered to be the independent variable—are independent of each other. Let's take a closer look at whether pregnancy rates are independent of (that is, depend on) whether one is entertained by a clown after IVF treatment.

Clown Therapy Israeli researchers tested whether entertainment by a clown led to higher pregnancy rates after in vitro fertilization treatment. Their study had two nominal variables—entertainment (clown, no clown) and pregnancy (pregnant, not pregnant)—and could have been analyzed with a chi-square test for independence.

EXAMPLE 15.2

In the clown study, as reported in the mass media (Ryan, 2006), 186 women were randomly assigned to receive IVF treatment only or to receive IVF treatment followed by 15 minutes of clown entertainment. Eighteen of the 93 who received only the IVF treatment became pregnant, whereas 33 of the 93 who received both IVF treatment and clown entertainment became pregnant. The cells for these observed frequencies can be seen in Table 15-4. The table of cells for a chi-square test for independence is called a *contingency table* because we are trying to see if the outcome of one variable (e.g., becoming pregnant versus not becoming pregnant) is contingent on the other variable (clown versus no clown). Let's implement the six steps of hypothesis testing for a chi-square test for independence.

STEP 1: Identify the populations, distribution, and assumptions. Population 1: Women receiving IVF treatment like the women we observed. Population 2: Women receiving IVF treatment for whom the presence of a clown is not associated with eventual pregnancy.

The comparison distribution is a chi-square distribution. The hypothesis test will be a chi-square test for independence because we have two nominal variables. This study meets three of the four assumptions: (1) The two variables are nominal. (2) Every participant is in only one cell. (3) The participants were not, however, randomly selected from the population of all women undergoing IVF treatment. We must be cautious in generalizing beyond the sample of Israeli women at this particular hospital. (4) There are more than five times as many participants as cells (186 participants and 4 cells—$4 \times 5 = 20$). We have far more participants, 186, than the 20 necessary to meet this guideline.

STEP 2: State the null and research hypotheses. Null hypothesis: Pregnancy rates are independent of whether one is entertained by a clown after IVF treatment. Research hypothesis: Pregnancy rates depend on whether one is entertained by a clown after IVF treatment.

STEP 3: Determine the characteristics of the comparison distribution. For a chi-square test for independence, we calculate degrees of freedom for each variable and then multiply the two to get the overall degrees of freedom. The degrees of freedom for the variable in the rows of the contingency table is:

$$df_{row} = k_{row} - 1$$

The degrees of freedom for the variable in the columns of the contingency table is:

$$df_{column} = k_{column} - 1$$

The overall degrees of freedom is:

$$df_{\chi^2} = (df_{row})(df_{column})$$

To expand this last formula, we write:

$$df_{\chi^2} = (k_{row} - 1)(k_{column} - 1)$$

The comparison distribution is a chi-square distribution, which has 1 degree of freedom:

$$df_{\chi^2} = (k_{row} - 1)(k_{column} - 1) = (2 - 1)(2 - 1) = 1$$

MASTERING THE FORMULA

15-3: To calculate the degrees of freedom for the chi-square test for independence, we first have to calculate the degrees of freedom for each variable. For the variable in the rows, we subtract 1 from the number of categories in the rows: $df_{row} = k_{row} - 1$. For the variable in the columns, we subtract 1 from the number of categories in the columns: $df_{column} = k_{column} - 1$. We multiply these two numbers to get the overall degrees of freedom: $df_{\chi^2} = (df_{row})(df_{column})$. To combine all the calculations, we can use the following formula instead: $df_{\chi^2} = (k_{row} - 1)(k_{column} - 1)$.

STEP 4: Determine the critical values, or cutoffs. The critical value, or cutoff, for the chi-square statistic based on a p level of 0.05 and 1 degree of freedom is 3.841 (Figure 15-3).

STEP 5: Calculate the test statistic. The next step, the determination of the appropriate expected frequencies, is the most important in the calculation of the chi-square test for independence. Errors are often made in this step, and if the wrong expected frequencies are used, the chi-square statistic derived from them will also be wrong. Many students want to divide the total number of participants (here, 186) by the number of cells (here, 4) and place equivalent frequencies in all cells for the expected data. Here, that would mean that the expected frequencies would be 46.5.

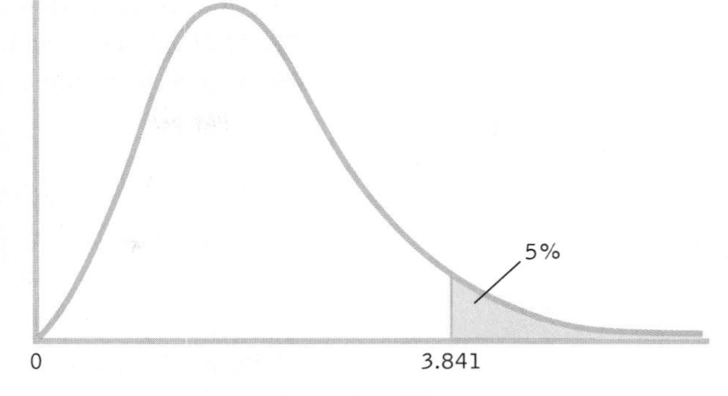

FIGURE 15-3
The Cutoff for a Chi-Square Test for Independence

The shaded region is beyond the critical value for a chi-square test for independence with a *p* level of 0.05 and 1 degree of freedom. If the test statistic falls within this shaded area, we will reject the null hypothesis.

But this would not make sense. Of the 186 women, only 51 became pregnant; 51/186 = 0.274, or 27.4%, of these women became pregnant. If pregnancy rates do not depend on clown entertainment, then we would expect the same percentage of successful pregnancies, 27.4%, regardless of exposure to clowns. If we have expected frequencies of 46.5 in all four cells, then we have a 50%, not a 27.4%, pregnancy rate. We must always consider the specifics of the situation.

In the current study, we already calculated that 27.4% of all women in the study became pregnant. If pregnancy rates are independent of whether a woman is entertained by a clown, then we would expect 27.4% of the women who were entertained by a clown to become pregnant and 27.4% of women who were not entertained by a clown to become pregnant. Based on this percentage, 100 − 27.4 = 72.6% of women in the study did not become pregnant. We would therefore expect 72.6% of women who were entertained by a clown to fail to become pregnant and 72.6% of women who were not entertained by a clown to fail to become pregnant. Again, we expect the same pregnancy and nonpregnancy rates in both groups—those who were and were not entertained by clowns.

Table 15-4 shows the observed data, but it also shows totals for each row, each column, and the whole table.

From Table 15-4, we see that 93 women were entertained by a clown after IVF treatment. As we calculated above, we would expect 27.4% of them to become pregnant:

$$(0.274)(93) = 25.482$$

Of the 93 women who were not entertained by a clown, we would expect 27.4% of them to become pregnant if clown entertainment is independent of pregnancy rates:

$$(0.274)(93) = 25.482$$

TABLE 15-4. Observed Frequencies with Totals

This table depicts the cells and their frequencies for the study on whether entertainment by a clown is associated with pregnancy rates among women undergoing in vitro fertilization. It also includes row totals (93, 93), column totals (51, 135), and the grand total for the whole table (186).

	Observed		
	Pregnant	Not Pregnant	
Clown	33	60	93
No clown	18	75	93
	51	135	186

We now repeat the same procedure for not becoming pregnant. We would expect 72.6% of women in both groups to fail to become pregnant. For the women who were entertained by a clown, we would expect 72.6% of them to fail to become pregnant:

$$(0.726)(93) = 67.518$$

For the women who were not entertained by a clown, we would expect 72.6% of them to fail to become pregnant:

$$(0.726)(93) = 67.518$$

(Note that the two expected frequencies for the first row are the same as the two expected frequencies for the second row, but only because the same number of people were in each clown condition, 93. If these two numbers were different, we would not see the same expected frequencies in the two rows.)

The method of calculating the expected frequencies that we described above is ideal because it is directly based on our own thinking about the frequencies in the rows and in the columns. Sometimes, however, our thinking can get muddled, particularly when the two (or more) row totals do not match and the two (or more) column totals do not match. For these situations, a simple set of rules leads to accurate expected frequencies. For each cell, we divide its column total ($Total_{column}$) by the grand total (N) and multiply that by the row total ($Total_{row}$):

MASTERING THE FORMULA

15-4: When conducting a chi-square test for independence, we can calculate the expected frequencies in each cell by taking the total for the column that the cell is in, dividing it by the total in the study, and then multiplying by the total for the row that the cell is in: $\frac{Total_{column}}{N}(Total_{row})$.

$$\frac{Total_{column}}{N}(Total_{row})$$

As an example, the observed frequency of those who became pregnant and were entertained by a clown is 33. The row total for this cell is 93. The column total is 51. The grand total, N, is 186. The expected frequency, therefore, is:

$$\frac{Total_{column}}{N}(Total_{row}) = \frac{51}{186}(93) = (0.274)(93) = 25.482$$

Notice that this result is identical to what we calculated without a formula. The middle step above shows that, even with the formula, we actually did calculate the pregnancy rate overall, by dividing the column total (51) by the grand total (186). We then calculated how many in that row of 93 participants we would expect to get pregnant using this overall rate:

$$(0.274)(93) = 25.482$$

The formula follows our logic, but it also keeps us on track when there are multiple calculations.

As a final check on the calculations, shown in Table 15-5, we can add up the frequencies to be sure that they still match the row, column, and grand totals. For example, if we add the two numbers in the first column, 25.482 and 25.482, we get 50.964 (different from 51 only because of rounding decisions). If we had made the mistake of dividing the 186 participants into cells by dividing by 4, we would have had 46.5 in each cell; then the total for the first column would have been 46.5 + 46.5 = 93, not a match with 51. This final check ensures that we have the appropriate expected frequencies in the cells.

The remainder of the fifth step is identical to that for a chi-square test for goodness of fit, as seen in Table 15-6. As before, we calculate the difference between each ob-

TABLE 15-5. Expected Frequencies with Totals

This table includes the expected frequencies for each of the four cells. The expected frequencies should still add up to the row totals (93, 93), column totals (51, 135), and the grand total for the whole table (186).

	Expected		
	Pregnant	Not Pregnant	
Clown	25.482	67.518	93
No clown	25.482	67.518	93
	51	135	186

TABLE 15-6. The Chi-Square Calculations

For the calculations for the chi-square test for independence, we use the same format as we did for the chi-square test for goodness of fit. We calculate the difference between each observed frequency and expected frequency, square the difference, then divide each square by its appropriate expected frequency. Finally, we add up the numbers in the last column, and that's the chi-square statistic.

Category	Observed (O)	Expected (E)	$O - E$	$(O - E)^2$	$\frac{(O-E)^2}{E}$
Clown; pregnant	33	25.482	7.518	56.520	2.218
Clown; not pregnant	60	67.518	−7.518	56.520	0.837
No clown; pregnant	18	25.482	−7.482	55.980	2.197
No clown; not pregnant	75	67.518	7.482	55.980	0.829

served frequency and its matching expected frequency, square these differences, and divide each squared difference by the appropriate expected frequency. We add up the numbers in the final column of the table to calculate the chi-square statistic:

$$\chi^2 = \Sigma\left[\frac{(O-E)^2}{E}\right] = (2.218 + 0.837 + 2.197 + 0.829) = 6.081$$

STEP 6: Make a decision. Reject the null hypothesis; it appears that pregnancy rates depend on whether a woman receives clown entertainment following IVF treatment (Figure 15-4).

The statistics, as reported in a journal article, would follow the format we learned for a chi-square test for goodness of fit as well as for other hypothesis tests in earlier

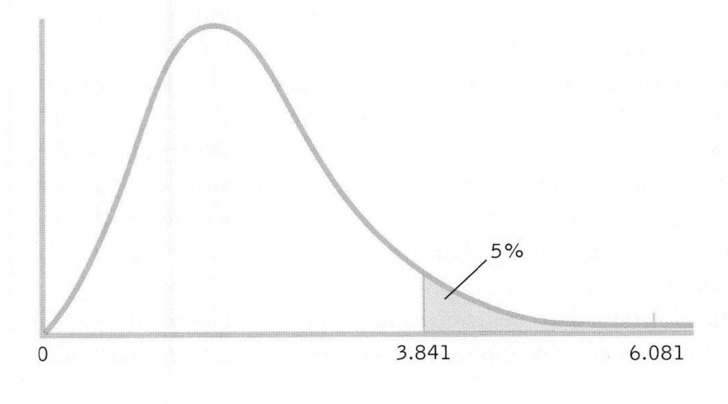

FIGURE 15-4
The Decision

Because the chi-square statistic, 6.081, is beyond the critical value, 3.841, we can reject the null hypothesis. It is unlikely that the pregnancy rates for those who received clown therapy versus those who did not were this different from each other just by chance.

Cramer's *V* is the standard effect size used with the chi-square test for independence; also called *Cramer's phi,* symbolized as φ.

chapters. We report the degrees of freedom and sample size, the value of the test statistic, and whether the *p* value associated with the test statistic is less than or greater than the critical value based on the *p* level of 0.05. (We would report the actual *p* level if we conducted this hypothesis test using software.) In the current example, the statistics would read:

$$\chi^2(1, N = 186) = 6.08, p < 0.05$$ ■

Cramer's *V*, the Effect Size for Chi Square

A hypothesis test tells us only that there is a likely effect—that it would be unlikely that the observed effect would have occurred merely by chance if the null hypothesis was true. But we have to calculate an additional statistic, an effect size, before we can make claims about the importance of a study's finding.

***Cramer's V** is the standard effect size used with the chi-square test for independence.* It is also called *Cramer's phi* (pronounced "fie"—rhymes with *fly*) and symbolized by φ. Once we have calculated the test statistic, it is easy to calculate Cramer's *V* by hand. The formula is:

MASTERING THE FORMULA

15-5: The formula for Cramer's *V*, the effect size typically used with the chi-square statistic, is: Cramer's $V = \sqrt{\frac{\chi^2}{(N)(df_{row/column})}}$. The numerator is the chi-square statistic, χ^2. The denominator is the product of the sample size, *N*, and either the degrees of freedom for the rows or the degrees of freedom for the columns, whichever is smaller. We take the square root of this quotient to get Cramer's *V*.

$$\text{Cramer's } V = \sqrt{\frac{\chi^2}{(N)(df_{row/column})}}$$

χ^2 is the test statistic we just calculated, *N* is the total number of participants in the study (the lower-right number in the contingency table), and $df_{row/column}$ is the degrees of freedom for either the category in the rows or the category in the columns, whichever is smaller.

EXAMPLE 15.3

For the clown example, we calculated a chi-square statistic of 6.081, there were 186 participants, and the degrees of freedom for both categories were 1. When neither degrees of freedom is smaller than the other, of course, it doesn't matter which one we choose. The effect size for the clown study, therefore, is:

$$\text{Cramer's } V = \sqrt{\frac{x^2}{(N)(df_{row/column})}} = \sqrt{\frac{6.081}{(186)(1)}} = \sqrt{0.033} = 0.181$$

Now that we have the effect size, what does it mean? As with other effect sizes, Jacob Cohen (1992) has developed guidelines, shown in Table 15-7, for determining whether a particular effect is small, medium, or large. The guidelines vary based on the size of the contingency table. When the smaller of the two degrees of freedom for the row and column is 1, we use the guidelines in the second column. When the smaller of the two degrees of freedom is 2, we use the guidelines in the third column. And when it is 3, we use the guidelines in the fourth column. As with the other guidelines for judging effect sizes, such as those for Cohen's *d,* the guidelines are not cutoffs. Rather, they are rough indicators to help researchers gauge a finding's importance.

The effect size for the clowning and pregnancy study was 0.18. The smaller of the two degrees of freedom, that for the row and that for the column, was 1 (in fact, both

TABLE 15-7. Conventions for Determining Effect Size Based on Cramer's *V*

Jacob Cohen (1992) developed guidelines to determine whether particular effect sizes should be considered small, medium, or large. The effect-size guidelines vary depending on the size of the contingency table. There are different guidelines based on whether the smaller of the two degrees of freedom (row or column) is 1, 2, or 3.

Effect Size	When $df_{row/column} = 1$	When $df_{row/column} = 2$	When $df_{row/column} = 3$
Small	0.10	0.07	0.06
Medium	0.30	0.21	0.17
Large	0.50	0.35	0.29

were 1). So we use the second column in Table 15-7. This Cramer's *V* falls about halfway between the effect-size guidelines for a small effect (0.10) and a medium effect (0.30). We would call this a small-to-medium effect. We can build on the report of the statistics by adding the Cramer's *V* to the end:

$$\chi^2(1, N = 186) = 6.08, p < 0.05, \text{Cramer's } V = 0.18$$ ■

Graphing Chi-Square Percentages

In addition to calculating Cramer's *V*, we can graph the data. A visual depiction of the pattern of results is an effective way to understand the size of the relation between two variables assessed using the chi-square statistic. We don't graph the frequencies, however. We graph proportions or percentages.

EXAMPLE 15.4

For the women entertained by a clown, we calculate the proportion who became pregnant and the proportion who did not. For the women not entertained by a clown, we again calculate the proportion who became pregnant and the proportion who did not. The calculations for the proportions are below.

In each case, we're dividing the number of a given outcome by the total number of women in that group. The proportions are called conditional proportions because we're not calculating the proportions out of all women in the study; we're calculating proportions for women in a certain condition. We calculate the proportion of women who became pregnant, for example, conditional on their having been entertained by a clown.

Entertained by a clown

Became pregnant: $33/93 = 0.355$

Did not become pregnant: $60/93 = 0.645$

Not entertained by a clown

Became pregnant: $18/93 = 0.194$

Did not become pregnant: $75/93 = 0.806$

We can put those proportions into a table (such as Table 15-8). For each category of entertainment (clown, no clown), the proportions should add up to 1.00; or if we used percentages, they should add up to 100%.

We can now graph the conditional proportions, as in Figure 15-5. Alternately, we could have simply graphed the two rates at which women got pregnant—0.355 and 0.194—given that the rates at which they did not become pregnant are based on these

TABLE 15-8. Conditional Proportions

To construct a graph depicting the results of a chi-square test for independence, we first calculate conditional proportions. For example, we calculate the proportions of women who got pregnant, conditional on having been entertained by a clown post-IVF: 33/93 = 0.355.

	Conditional Proportions		
	Pregnant	Not Pregnant	
Clown	0.355	0.645	1.00
No clown	0.194	0.806	1.00

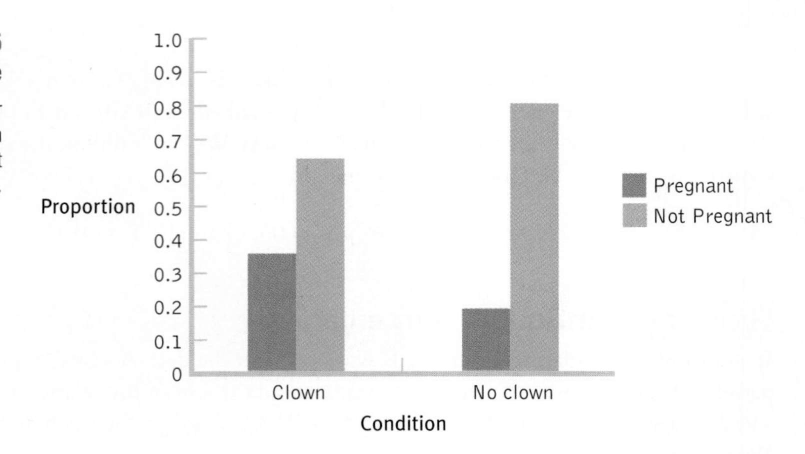

FIGURE 15-5
Graphing and Chi Square
When we graph the data for a chi-square test for independence, we graph conditional proportions rather than frequencies. The proportions allow us to compare the rates at which women became pregnant in the two conditions.

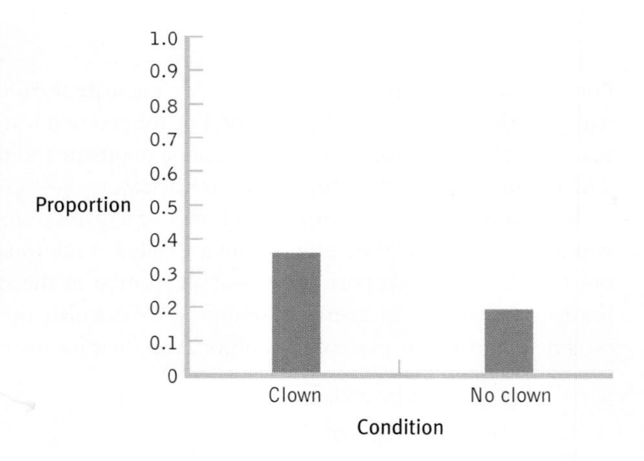

FIGURE 15-6
A Simpler Graph of Conditional Probabilities
Because the rates at which women did not become pregnant are based on the rates at which they did become pregnant, we can simply graph one set of rates. Here we see the rates at which women became pregnant in each of the two clown conditions.

rates. This graph is depicted in Figure 15-6. In both cases, we include the scale of proportions on the *y*-axis from 0 to 1.0 so that the graph will not mislead the viewer into thinking that rates are higher than they are. ■

Relative Risk

▪ **Relative risk** is a measure created by making a ratio of two conditional proportions; also called *relative likelihood* or *relative chance.*

Public health statisticians like John Snow (Chapter 1) are called epidemiologists. They often think about the size of an effect with chi square in terms of ***relative risk**, a measure created by making a ratio of two conditional proportions.* It is also called *relative likelihood* or *relative chance.*

EXAMPLE 15.5

As with Figure 15-5, we calculate the chance of getting pregnant with clown entertainment post-IVF by dividing the number of pregnancies in this group by the total number of women entertained by clowns:

$$33/93 = 0.355$$

We then calculate the chance of getting pregnant with no clown entertainment post-IVF by dividing the number of pregnancies in this group by the total number of women not entertained by clowns:

$$18/93 = 0.194$$

If we divide the chance of getting pregnant having been entertained by clowns by the chance of getting pregnant not having been entertained by clowns, then we get the relative likelihood:

$$0.355/0.194 = 1.830$$

Based on the relative risk calculation, the chance of getting pregnant when IVF is followed by clown entertainment is 1.83 times the chance of getting pregnant when IVF is not followed by clown entertainment. This matches the impression that we get from the graph.

Alternately, we can reverse the ratio, dividing the chance of becoming pregnant without clown entertainment, 0.194, by the chance of becoming pregnant following clown entertainment, 0.355. This is the relative likelihood for the reversed ratio:

$$0.194/0.355 = 0.546$$

This number gives us the same information in a different way. The chance of getting pregnant when IVF is followed by no entertainment is 0.55 (or about half) the chance of getting pregnant when IVF is followed by clown entertainment. Again, this matches the graph; one bar is about half that of the other.

(*Note:* When this calculation is made with respect to diseases, it's referred to as relative risk [rather than relative likelihood].) We should be careful when relative risks and relative likelihoods are reported, however. We must always be aware of base rates. If, for example, a certain disease occurs in just 0.01% of the population (that is, 1 in 10,000) and is twice as likely to occur among people who eat ice cream, then the rate is 0.02% (2 in 10,000) among those who eat ice cream. Relative risks and relative likelihoods can be used to scare the general public unnecessarily—which is one more reason why statistical reasoning is a healthy way to think. ■

MASTERING THE CONCEPT

15.3: We can quantify the size of an effect with chi square through relative risk, also called relative likelihood. By making a ratio of two conditional proportions, we can say, for example, that one group is twice as likely to show some outcome or, conversely, that the other group is one-half as likely to show that outcome.

CHECK YOUR LEARNING

Reviewing the Concepts

- The chi-square test for goodness of fit is used with one nominal variable.
- The chi-square test for independence is used with two nominal variables; usually one can be thought of as the independent variable and one as the dependent variable.
- Both chi-square hypothesis tests use the same six steps of hypothesis testing with which we are familiar.
- The appropriate effect-size measure for the chi-square test for independence is Cramer's *V.*

continued on next page

> We can depict the effect size visually by calculating and graphing conditional proportions.

> Another way to consider the size of an effect is through relative risk, a ratio of conditional proportions for each of two groups.

Clarifying the Concepts

15-5 When do we use chi-square tests?

15-6 What are observed frequencies and expected frequencies?

15-7 What is the effect-size measure for chi-square tests and how is it calculated?

Calculating the Statistics

15-8 Imagine a town that boasts clear blue skies 80% of the time. You get to work in that town one summer for 78 days and record the following data. (*Note:* For each day, you picked just one label.)

Clear blue skies: 59 days

Cloudy/hazy/gray skies: 19 days

a. Calculate degrees of freedom for this chi-square test for goodness of fit.

b. Determine the observed and expected frequencies.

c. Calculate the differences and squared differences between frequencies, and calculate the chi-square statistic. Use the six-column format provided here.

Category	Observed (O)	Expected (E)	$O - E$	$(O - E)^2$	$\frac{(O - E)^2}{E}$
Clear blue skies					
Unclear skies					

15-9 Assume you are interested in whether students with different majors tend to have different political affiliations. You ask U.S. psychology majors and business majors to indicate whether they are Democrats or Republicans. Of 67 psychology majors, 36 indicated that they were Republicans and 31 indicated that they were Democrats. Of 92 business majors, 54 indicated that they were Republicans and 38 indicated that they were Democrats. Calculate the relative likelihood of being a Republican given that a person is a business major as opposed to a psychology major.

Applying the Concepts

15-10 The Chicago Police Department conducted a study comparing two types of lineups for suspect identification: simultaneous lineups and sequential lineups (Mecklenburg, Malpass, & Ebbesen, 2006). In simultaneous lineups, witnesses see the suspects all at once. In sequential lineups, witnesses see the people in the lineup one at a time, saying yes or no to each. Over one year, three jurisdictions in Illinois compared the two types of lineups. Of 319 simultaneous lineups, 191 led to the identification of the suspect, 8 led to the identification of another person in the lineup, and 120 led to no identification. Of 229 sequential lineups, 102 led to the identification of the suspect, 20 led to the identification of another person in the lineup, and 107 led to no identification.

a. Who or what are the participants in this study? Identify the independent variable and its levels as well as the dependent variable and its levels.

b. Conduct all six steps of hypothesis testing.

c. Report the statistics as you would in a journal article.

d. Why is this study an example of the importance of using two-tailed rather than one-tailed hypothesis tests?

e. Calculate the appropriate measure of effect size for this study.

f. Create a graph of the conditional proportions for these data.

g. Calculate the relative likelihood of a suspect being accurately identified in the simultaneous versus the sequential lineups.

Solutions to these Check Your Learning questions can be found in Appendix D.

Ordinal Data and Correlation

The statistical tests we discuss in this section allow researchers to draw conclusions from data that do not meet the assumptions for a parametric test, such as when we have rank-ordered data. In this section, we learn how to convert scale data to ordinal data. Then we examine two tests that can be used with ordinal data.

When the Data Are Ordinal

A 2006 University of Chicago News Office press release proclaimed, "Americans and Venezuelans Lead the World in National Pride." Researchers from the University of Chicago's National Opinion Research Center (NORC) surveyed citizens of 33 countries (Smith & Kim, 2006) and developed two different kinds of national pride scores: pride in specific accomplishments of their nations, like science or sports (which they called domain-specific national pride), and a more general national pride in which citizens responded to items such as "People should support their country even if the country is in the wrong."

So, the researchers had two sets of national pride scores—accomplishment-related and general—for each country. They converted the scores to ranks, and when results on the two scales were merged, Venezuela and the United States were tied for first place. These findings suggest many hypotheses about what creates and inflates national pride. The authors noted that countries that were settled as colonies tend to rank higher than their "mother country," that ex-socialist countries tend to rank lower than other countries, and that countries in Asia tend to rank lower than those from other continents.

AP Photo/Leslie Mazoch

National Pride University of Chicago researchers ranked 33 countries in terms of national pride. Venezuela, along with the United States, came out on top. Ordinal data such as these are analyzed using nonparametric statistics.

We wondered about other possible precursors of national pride, such as competitiveness. Because the researchers provided ordinal data, the only way we can explore these interesting hypotheses is by using nonparametric statistics. The very nature of an ordinal variable means that it will not meet the assumptions of a scale dependent variable and a normally distributed population. As we can see in Figure 15-7, the shape of a distribution of ordinal variables is rectangular because every participant has a different rank.

Fortunately, the logic of many nonparametric statistics for ordinal data will be familiar to students. This is because many of the nonparametric statistical tests are specific alternatives to parametric statistical tests. The four most common of these tests (shown in Table 15-9) are: (1) a nonparametric equivalent for the Pearson correlation coefficient, the *Spearman rank-order correlation coefficient;* (2) a nonparametric equivalent for the paired-samples *t* test, the *Wilcoxon signed-rank test;* (3) a nonparametric equivalent

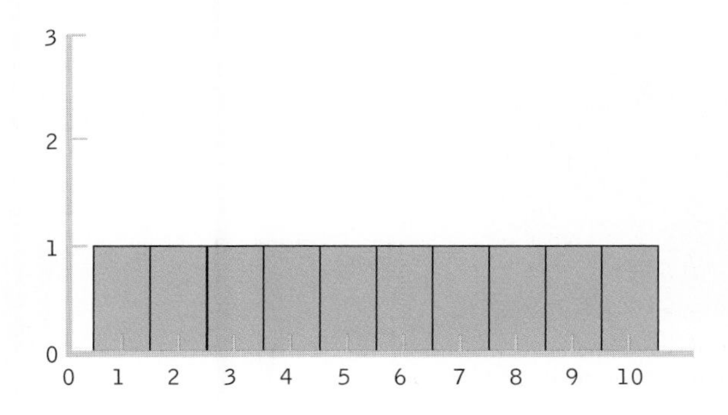

FIGURE 15-7
A Histogram of Ordinal Data

When ordinal data are graphed in a histogram, the resulting distribution is rectangular. These are data for ranks 1–10. For each rank, there is one individual. Ordinal data are never normally distributed.

TABLE 15-9. Parametric and Nonparametric Partners

Most parametric hypothesis tests have at least one equivalent nonparametric alternative. Here, all the parametric tests call for scale dependent variables, and their nonparametric counterparts all call for ordinal dependent variables.

Design	Parametric Test	Nonparametric Test
Association between two variables	Pearson correlation coefficient	Spearman rank-order correlation coefficient
Two groups; within-groups design	Paired-samples *t* test	Wilcoxon signed-rank test
Two groups; between-groups design	Independent-samples *t* test	Mann–Whitney *U* test
More than two groups; between-groups design	One-way between-groups ANOVA	Kruskal–Wallis *H* test

for the independent-samples *t* test, the *Mann–Whitney U test;* (4) a nonparametric equivalent for the one-way between-groups ANOVA, the *Kruskal–Wallis H test.* In this chapter, we'll consider the Spearman rank-order correlation coefficient and the Mann–Whitney *U* test.

EXAMPLE 15.6

Nonparametric tests for ordinal data are typically used in one of two situations. First and most obviously, we use nonparametric tests for ordinal data when the sample data are ordinal. Second, we use nonparametric tests when the dependent variable suggests that the underlying population distribution is greatly skewed, a situation that often develops when we have a small sample size. This second reason is likely why the national pride researchers converted their data to ranks (Smith & Kim, 2006). Figure 15-8 shows

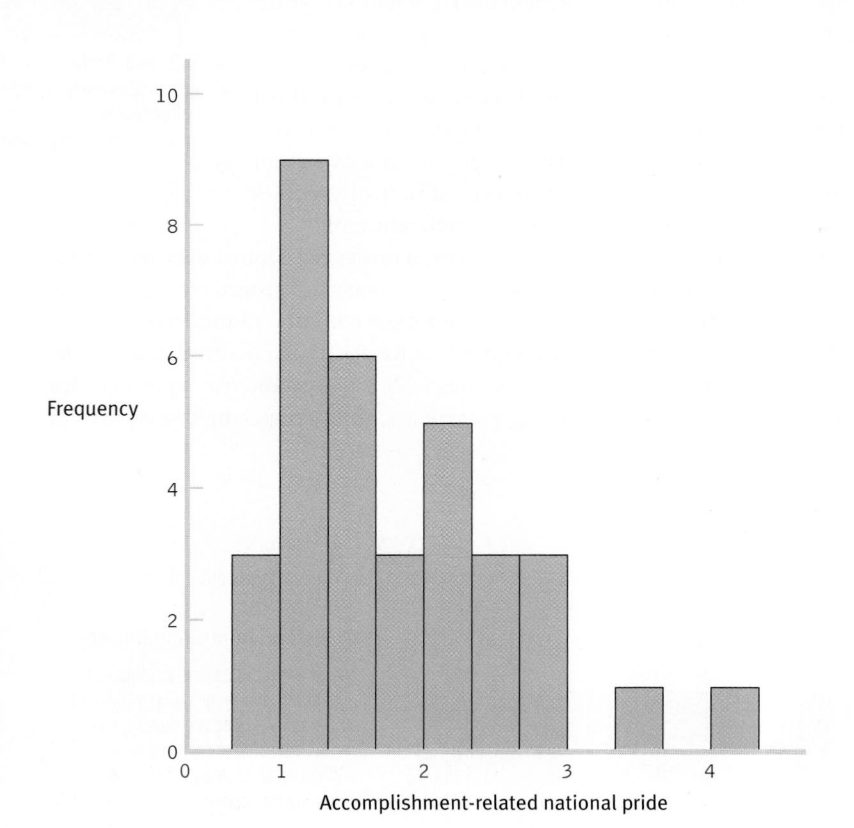

FIGURE 15-8
Skewed Data

The sample data for the variable, accomplishment-related national pride, are skewed. This indicates the possibility that the underlying population distribution is skewed. It is likely that the researchers chose to report their data as ranks for this reason (Smith & Kim, 2006).

a histogram of their full set of data for the variable accomplishment-related national pride—the variable that we will use for many examples in this chapter. The data appear to be positively skewed, most likely because two countries, Venezuela and the United States, appear to be outliers. Because of this, we have to transform the data from scale to ordinal.

Transforming scale data to ordinal data is not uncommon. This conversion can help in situations like that described above: when the data from a small sample are skewed. Look what happens to the following five data points for income when we change the data from scale to ordinal. In the first row, the one that includes the scale data, there is a severe outlier ($550,000) and the sample data suggest a skewed distribution. In the second row, the severe outlier merely becomes the last ranking. The ranked data do not have an outlier.

Scale: $24,000 $27,000 $35,000 $46,000 $550,000

Ordinal: 1 2 3 4 5

In the next section, we'll use this technique to convert scale data to ordinal data so that we can calculate the Spearman rank-order correlation coefficient. ■

Spearman Rank-Order Correlation Coefficient

Many daily observations represent rank-ordered data. For example, a person may prefer Chunky Monkey ice cream to Chubby Hubby ice cream but would not be able to specify that he liked it precisely twice as much. When we collect ranked data, we analyze it using nonparametric statistics. The ***Spearman rank-order correlation coefficient*** *is a nonparametric statistic that quantifies the association between two ordinal variables.*

MASTERING THE CONCEPT

15.4: We calculate a Spearman rank-order correlation coefficient to quantify the association between two ordinal variables. It is the nonparametric equivalent of the Pearson correlation coefficient.

EXAMPLE 15.7

To see how the Spearman rank-order correlation coefficient works, let's look at a study that uses two ordinal variables, one taken from the University of Chicago study on national pride (Smith & Kim, 2006). We wondered whether accomplishment-related national pride is related to the underlying trait of competitiveness. So we randomly selected 10 countries from this list and compiled their scores for accomplishment-related national pride. We also located rankings of competitiveness compiled by an international business school (IMD International, 2001).

A correlation between these variables, if found, would be evidence that countries' levels of accomplishment-related national pride are tied to levels of competitiveness. The competitiveness variable we borrowed from the business school rankings was already ordinal. However, the accomplishment-related national pride variable was initially a scale variable. When even one of the variables is ordinal, we use the Spearman rank-order correlation coefficient (often called just the Spearman correlation coefficient, or Spearman's rho). Its symbol is almost like the one for the Pearson correlation coefficient, but it has a subscript S to indicate that it is Spearman's correlation coefficient: r_S.

To convert scale data to ordinal data, we simply organize the data from highest to lowest (or lowest to highest if that makes more sense) and then rank them. Table 15-10 shows the conversion of accomplishment-related national pride from scale data to ordinal data. Sometimes, as seen for Austria and Canada, we have a tie. Both of these countries had an accomplishment-related national pride score of 2.40. When we rank

▪ The **Spearman rank-order correlation coefficient** is a nonparametric statistic that quantifies the association between two ordinal variables.

TABLE 15-10. Converting Pride Scores to Ranks

When we convert scale data to ordinal data, we simply arrange the data from highest to lowest (or lowest to highest if that makes more sense) and then rank them. These are the original data for accomplishment-related national pride. In cases of ties, we average the two ranks that these participants—countries, in this case—would hold.

Country	Pride Score	Pride Rank
United States	4.0	1
South Africa	2.7	2
Austria	2.4	3.5
Canada	2.4	3.5
Chile	2.3	5
Japan	1.8	6
Hungary	1.6	7
France	1.5	8
Norway	1.3	9
Slovenia	1.1	10

the data, these countries take the third and fourth positions, but they must have the same rank because their scores are the same. So we take the average of the two ranks they would hold if the scores were different: $(3 + 4)/2 = 3.5$. Both of these countries receive the rank of 3.5.

Now that we have the ranks, we can compute the Spearman correlation coefficient. We first need to include both sets of ranks in the same table, as in the second and third columns in Table 15-11. We then calculate the difference (D) between each pair of ranks, as in the fourth column. The differences always add up to 0, so we must square

TABLE 15-11. Calculating a Spearman Correlation Coefficient

The first step in calculating a Spearman correlation coefficient is creating a table that includes the ranks for all participants—countries, in this case—on both variables of interest (accomplishment-related national pride and competitiveness). We then calculate differences for each participant (country, here) and square each difference.

Country	Pride Rank	Competitiveness Rank	Difference (D)	Squared Difference (D^2)
United States	1	1	0	0
South Africa	2	10	−8	64
Austria	3.5	2	1.5	2.25
Canada	3.5	3	0.5	0.25
Chile	5	5	0	0
Japan	6	7	−1	1
Hungary	7	8	−1	1
France	8	6	2	4
Norway	9	4	5	25
Slovenia	10	9	1	1

the differences, as in the last column. As we have frequently done with squared differences in the past, we sum them—another variation on the concept of a sum of squares. The sum of these squared differences is:

$$\Sigma D^2 = (0 + 64 + 2.25 + 0.25 + 0 + 1 + 1 + 4 + 25 + 1) = 98.5$$

The formula for calculating the Spearman correlation coefficient includes the sum of the squared differences that we just calculated, 98.5. The formula is:

$$r_S = 1 - \frac{6(\Sigma D^2)}{N(N^2 - 1)}$$

MASTERING THE FORMULA

15-6: The formula for the Spearman correlation coefficient is: $r_S = 1 - \frac{6(\Sigma D^2)}{N(N^2 - 1)}$. The numerator includes a constant, 6, as well as the sum of the squared differences between ranks for each participant. The denominator is calculated by multiplying the sample size, *N*, by the square of the sample size minus 1.

In addition to the sum of squared differences, the only other information we need is the sample size, *N*, which is 10 in this example. (The number 6 is a constant; it is always included in the calculation of the Spearman correlation coefficient.) The Spearman correlation coefficient, therefore, is:

$$r_S = 1 - \frac{6(\Sigma D^2)}{N(N^2 - 1)} = 1 - \frac{6(98.5)}{10(10^2 - 1)} = 1 - \frac{591}{10(100 - 1)} = 1 - \frac{591}{10(99)} = 1 - \frac{591}{990}$$
$$= 1 - 0.597 = 0.403$$

The Spearman correlation coefficient is 0.40.

The interpretation of the Spearman correlation coefficient is identical to that for the Pearson correlation coefficient. The coefficient can range from −1, a perfect negative correlation, to 1, a perfect positive correlation. A correlation coefficient of 0 indicates no relation between the two variables. As with the Pearson correlation coefficient, it is not the sign of the Spearman correlation coefficient that indicates the strength of a relation. So, for example, a coefficient of −0.66 indicates a stronger association than does a coefficient of 0.23. Finally, as with the Pearson correlation coefficient, we can implement the six steps of hypothesis testing to determine whether the Spearman correlation coefficient is statistically significantly different from 0. If we do decide to conduct hypothesis testing, we can find the critical values for the Spearman correlation coefficient in Appendix B.7.

Like the Pearson correlation coefficient, the Spearman correlation coefficient does not tell us about causation. It is possible that there is a causal relation in one of two directions. The relation between competitiveness (variable A) and accomplishment-related national pride (variable B) is 0.40, a fairly strong positive correlation. It is possible that competitiveness (variable A) causes a country to feel prouder (variable B) of its accomplishments. On the other hand, it is also possible that accomplishment-related national pride (variable B) causes competitiveness (variable A). Finally, it is also possible that a third variable, C, causes both of the other two variables (A and B). For example, a high gross domestic product (variable C) might cause both a sense of competitiveness with other economic powerhouses (variable A) and a feeling of national pride at this economic accomplishment (variable B). A strong correlation indicates only a strong association; we can draw no conclusions about causation. ■

The Mann–Whitney *U* Test

As mentioned earlier, most parametric hypothesis tests have nonparametric equivalents. In this section, we learn how to conduct one of the most common of these tests—the Mann–Whitney *U* test, the nonparametric equivalent of the independent-samples *t* test.

▶ MASTERING THE CONCEPT

15.5: We conduct a Mann-Whitney *U* test to compare two independent groups with respect to an ordinal dependent variable. It is the nonparametric equivalent of the independent-samples *t* test.

The ***Mann–Whitney U test*** *is a nonparametric hypothesis test used when there are two groups, a between-groups design, and an ordinal dependent variable.* The test statistic is symbolized as *U.* Let's use this new statistic to test more hypotheses about the ranked data on national pride.

EXAMPLE 15.8

The researchers observed that countries with recent communist pasts tended to have lower ranks on national pride (Smith & Kim, 2006). Let's choose 10 European countries, 5 of which were communist during part of the twentieth century. The independent variable is type of country, with two levels: formerly communist and not formerly communist. The dependent variable is rank on accomplishment-related national pride. As in previous situations, we may start with ordinal data, or we may convert scale data to ordinal data because we were far from meeting the assumptions of a parametric test. Table 15-12 shows the scores for the 10 countries.

As noted earlier, nonparametric tests use the same six steps of hypothesis testing as parametric tests but are usually easier to calculate.

STEP 1: Identify the assumptions. There are three assumptions. (1) The data must be ordinal. (2) We should use random selection; otherwise, our ability to generalize will be limited. (3) Ideally, no ranks are tied. The Mann–Whitney *U* test is robust with respect to violations of the third assumption; if there are only a few ties, then it is usually safe to proceed.

TABLE 15-12. Comparing Two Groups

Here are the data for two samples: European countries that were recently communist and European countries that were not recently communist. The data in this table are scale; because we do not meet the assumptions for a parametric test, we have to convert the data from scale to ordinal as one step of the calculations.

Country	Pride Score
Noncommunist	
Ireland	2.9
Austria	2.4
Spain	1.6
Portugal	1.6
Sweden	1.2
Communist	
Hungary	1.6
Czech Republic	1.3
Slovenia	1.1
Slovakia	1.1
Poland	0.9

Summary: (1) We need to convert the data from scale to ordinal. (2) The researchers did not indicate whether they used random selection to choose the European countries in the sample, so we must be cautious when generalizing from these results. (3) There are some ties, but we will assume that there are not so many as to render the results of the test invalid.

▪ The **Mann–Whitney *U* test** is a nonparametric hypothesis test used when there are two groups, a between-groups design, and an ordinal dependent variable.

STEP 2: State the null and research hypotheses.

We state the null and research hypotheses only in words, not in symbols.

Summary: Null hypothesis: European countries with recent communist histories and those without recent communist histories do not tend to differ in accomplishment-related national pride. Research hypothesis: European countries with recent communist histories and those without recent communist histories tend to differ in accomplishment-related national pride.

STEP 3: Determine the characteristics of the comparison distribution.

The Mann–Whitney *U* test compares the two distributions—those represented by the two samples. There is no comparison distribution in the sense of a parametric test. To complete step 4 and find a cutoff, or critical value, we need two pieces of information: the sample size for the first group and the sample size for the second group.

Summary: There are five countries in the communist group and five countries in the noncommunist group.

STEP 4: Determine the critical values, or cutoffs.

There are two Mann–Whitney *U* tables. We use Table B.8A (for a one-tailed test) or Table B.8B (for a two-tailed test) from Appendix B to determine the cutoff, or critical value, for the Mann–Whitney *U* test. In the tables, we find the sample size for the first group across the top row and the sample size for the second group down the left-hand column. The table includes only critical values for a hypothesis test with a p level of 0.05. There are two differences between this critical value and those we considered with parametric tests. First, we calculate two test statistics, but we only compare the *smaller* one with the critical value. Second, we want the test statistic to be equal to or *smaller than* the critical value.

Summary: The cutoff, or critical value, for a Mann–Whitney *U* test with two groups of five participants (countries), a p level of 0.05, and a two-tailed test is 2. (*Note:* Remember that we want the *smaller* of the test statistics to be equal to or *smaller than* this critical value.)

STEP 5: Calculate the test statistic.

As noted above, we calculate two test statistics for a Mann–Whitney *U* test, one for each group. We start the calculations by organizing the data by raw score from highest to lowest in one single column and then by rank in the next column, as shown in Table 15-13. For the two sets of tied scores, we take the average of the ranks they would have held and apply that rank to the tied scores. For example, Spain, Portugal, and Hungary all received scores of 1.6; they would have been ranked 3, 4, and 5, but because they're tied, they all received the average of these three scores, 4. (We have chosen to give the highest score a rank of 1, as did the researchers.) We include the group membership of each participant (country) next to its score and rank: C indicates a formerly communist country and NC represents a noncommunist country. The final two columns separate the ranks by group; from these columns we can easily see that the noncommunist countries tend to hold the higher ranks and the communist countries, the lower ranks.

TABLE 15-13. Organizing Data for a Mann–Whitney *U* Test

To conduct a Mann–Whitney *U* test, we first organize the data. We organize the raw scores from highest to lowest in a single column, then rank them in an adjacent column. Notice that when scores are tied, we average the ranks of the two or three tied scores. The next column includes the group to which each country belongs—in this case, a recently communist country (C) or a noncommunist (NC) country. The last two columns separate the ranks by group.

Country	Pride Score	Pride Rank	Type of Country	NC Ranks	C Ranks
Ireland	2.9	1	NC	1	
Austria	2.4	2	NC	2	
Spain	1.6	4	NC	4	
Portugal	1.6	4	NC	4	
Hungary	1.6	4	C		4
Czech Republic	1.3	6	C		6
Sweden	1.2	7	NC	7	
Slovenia	1.1	8.5	C		8.5
Slovakia	1.1	8.5	C		8.5
Poland	0.9	10	C		10

Before we continue, we sum the ranks (R) for each group and add subscripts to indicate which group is which:

$$\Sigma R_{nc} = 1 + 2 + 4 + 4 + 7 = 18$$

$$\Sigma R_{c} = 4 + 6 + 8.5 + 8.5 + 10 = 37$$

The formula for the first group, with the n's referring to sample size in a particular group, is:

MASTERING THE FORMULA

15-7: The formula for the first group in a Mann-Whitney U test is: $U_1 = (n_1)(n_2) + \frac{n_1(n_1+1)}{2} - \Sigma R_1$. The formula for the second group is: $U_2 = (n_1)(n_2) + \frac{n_2(n_2+1)}{2} - \Sigma R_2$. The symbol n refers to the sample size for a particular group. In these formulas, the first group is labeled 1 and the second group is labeled 2. ΣR refers to the sum of the ranks for a particular group.

$$U_{nc} = (n_{nc})(n_c) + \frac{n_{nc}(n_{nc}+1)}{2} - \Sigma R_{nc} = (5)(5) + \frac{5(5+1)}{2} - 18 = 25 + 15 - 18 = 22$$

The formula for the second group is:

$$U_c = (n_{nc})(n_c) + \frac{n_c(n_c+1)}{2} - \Sigma R_c = (5)(5) + \frac{5(5+1)}{2} - 37 = 25 + 15 - 37 = 3$$

Summary: $U_{NC} = 22$; $U_C = 3$.

STEP 6: Make a decision. For a Mann–Whitney U test, we compare only the smaller test statistic, 3, with the critical value, 2. This test statistic is not smaller than the critical value, so we fail to reject the null hypothesis. We cannot conclude that the two groups are different with respect to accomplishment-related national pride rankings. The researchers concluded, however, that noncommunist countries tend to have more national pride, but remember, they used more countries in their analyses, so they had more statistical power (Smith & Kim, 2006). We selected just 10 of the European countries on their list. As with parametric tests, increased sample sizes lead to increased statistical power.

Summary: The test statistic, 3, is not smaller than the critical value, 2. We cannot reject the null hypothesis. We conclude only that insufficient evidence exists to show that the two groups are different with respect to accomplishment-related national pride.

After completing the hypothesis test, we want to present the primary statistical information in a report. In the write-up, we list the two groups and their sample sizes, but there are no degrees of freedom. In addition, we report the smaller test statistic; because this is the standard, we do not include a subscript. The statistics read:

$$U = 3, p > 0.05$$

(Note that if we conduct the Mann-Whitney U test using software, we report the actual p value associated with the test statistic.) ■

CHECK YOUR LEARNING

Reviewing the Concepts

> Nonparametric tests for ordinal data are used when the data are already ordinal or when it is clear that the assumptions are severely violated. In the latter case, the scale data must be converted to ordinal data.

> When we want to calculate a correlation between two ordinal variables, we calculate a Spearman rank-order correlation coefficient, which is interpreted in the same way as the Pearson correlation coefficient.

> As with the Pearson correlation coefficient, the Spearman correlation coefficient does not tell us about causation. It simply quantifies the magnitude and direction of association between two ordinal variables.

> There are nonparametric hypothesis tests that can be used to replace the various parametric hypothesis tests when it seems clear that there are severe violations of the assumptions.

> We use the Mann–Whitney U test in place of the independent-samples t test. Nonparametric hypothesis tests use the same six steps of hypothesis testing that are used for parametric tests, but the steps and the calculations in nonparametric hypothesis tests tend to be simpler.

Clarifying the Concepts

15-11 Describe a common situation in which we use nonparametric tests other than chi-square tests.

15-12 Why must scale data be transformed into ordinal data before we can perform any nonparametric tests on those data?

Calculating the Statistics

15-13 Convert the following scale data to ordinal or ranked data, starting with a rank of 1 for the smallest data point.

Observation	Variable 1	Variable 2
1	1.30	54.39
2	1.80	50.11
3	1.20	53.39
4	1.06	44.89
5	1.80	48.50

15-14 Compute the Spearman correlation coefficient for the data listed in Check Your Learning 15-13.

continued on next page

Applying the Concepts

15-15 Here are IQ scores for 10 people: 88, 90, 91, 99, 103, 103, 104, 112, 114, and 139.

a. Why might it be better to use a nonparametric test than a parametric test in this case?

b. Convert the scores for IQ (a scale variable) to ranks (an ordinal variable).

c. What happens to the outlier when the scores are converted from a scale measure to an ordinal measure?

15-16 Researchers provided accomplishment-related national pride scores for a number of countries (Smith & Kim, 2006). We selected seven countries for which English is the primary language and seven countries for which it is not. We wondered whether English-speaking countries would be different on the variable of accomplishment-related national pride from non-English-speaking countries. The data are in the accompanying table. Conduct a Mann–Whitney *U* test on these data. Remember to organize the data in one column before starting.

English-Speaking Countries	Pride Score	Non-English-Speaking Countries	Pride Score
United States	4.00	Chile	2.30
Australia	2.90	Japan	1.80
Ireland	2.90	France	1.50
South Africa	2.70	Czech Republic	1.30
New Zealand	2.60	Norway	1.30
Canada	2.40	Slovenia	1.10
Great Britain	2.20	South Korea	1.00

Solutions to these Check Your Learning questions can be found in Appendix D.

REVIEW OF CONCEPTS

Nonparametric Statistics

Nonparametric hypothesis tests are used when we do not meet the assumptions of a parametric test. This often occurs when we have a nominal or ordinal dependent variable, or a small sample in which the data suggest a skewed population distribution. Given the choice, we should use a parametric test because these tests tend to have more statistical power and because we can more frequently calculate confidence intervals and effect sizes for parametric hypothesis tests.

Chi-Square Tests

We use the *chi-square test for goodness of fit* when we have only one variable and it is nominal. We use the *chi-square test for independence* when we have two nominal variables; typically, for the purposes of articulating hypotheses, one variable is thought of as the independent variable and the other is thought of as the dependent variable. With both chi-square tests, we analyze whether the data that we observe match what we would expect according to the null hypothesis. Both tests use the same basic six steps of hypothesis testing that we learned previously. We usually calculate an effect size as well; the most commonly calculated effect size with chi square is *Cramer's V,* also called *Cramer's phi.* We can also create a graph that depicts the conditional proportions of an outcome for each group. Alternately, we can calculate *relative risk (relative likelihood/chance)* to more easily compare the rates of certain outcomes in each of two groups.

Ordinal Data and Correlation

The nonparametric parallel to the Pearson correlation coefficient is the *Spearman rank-order correlation coefficient,* a statistic that is interpreted just like its parametric cousin with respect to magnitude and direction.

The *Mann–Whitney U test* is the nonparametric parallel of the independent-samples t test. The same six steps of hypothesis testing are used for both parametric and nonparametric tests, but the steps and the calculations for the nonparametric tests tend to be simpler.

SPSS®

In SPSS, we conduct a chi-square test for independence by first entering the data. Each participant gets a score on each variable. For the pregnancy and clown entertainment data, there are two columns: one for a woman's status with respect to entertainment by a clown (yes or no) and another for her pregnancy status (yes or no). We can use the numbers 1 and 2 to represent the levels of these variables. Select: Analyze → Descriptive Statistics → Crosstabs (select a nominal variable for the row and a nominal variable for the column; we selected entertainment-by-clown status for the rows and pregnancy status for the columns, but it doesn't matter which we choose) → Statistics → Chi-Square and Phi & Cramer's V (for effect sizes) → Continue. Select "Cells" and then click "Row" under percentages to get us the percentage of women who got pregnant in each clown condition. Click "Continue," and then click "OK" to run the analysis.

Most of the output, along with a view of some of the data, can be seen in the accompanying screenshot. In the top box of the output, we can see the percentages of women who did or did not become pregnant in each condition. For example, 35.5% of women who were entertained by a clown became pregnant. We can also see that the chi-square statistic (in the box titled "Chi-Square Tests" in the row labeled "Pearson Chi-Square") is 6.078, the same as the one we calculated by hand earlier. In the box titled "Symmetric Measures," we can see the Cramer's V statistic of .181, also the same as we calculated earlier. (Any slight differences we see in this table versus what we calculated earlier are due to rounding decisions.)

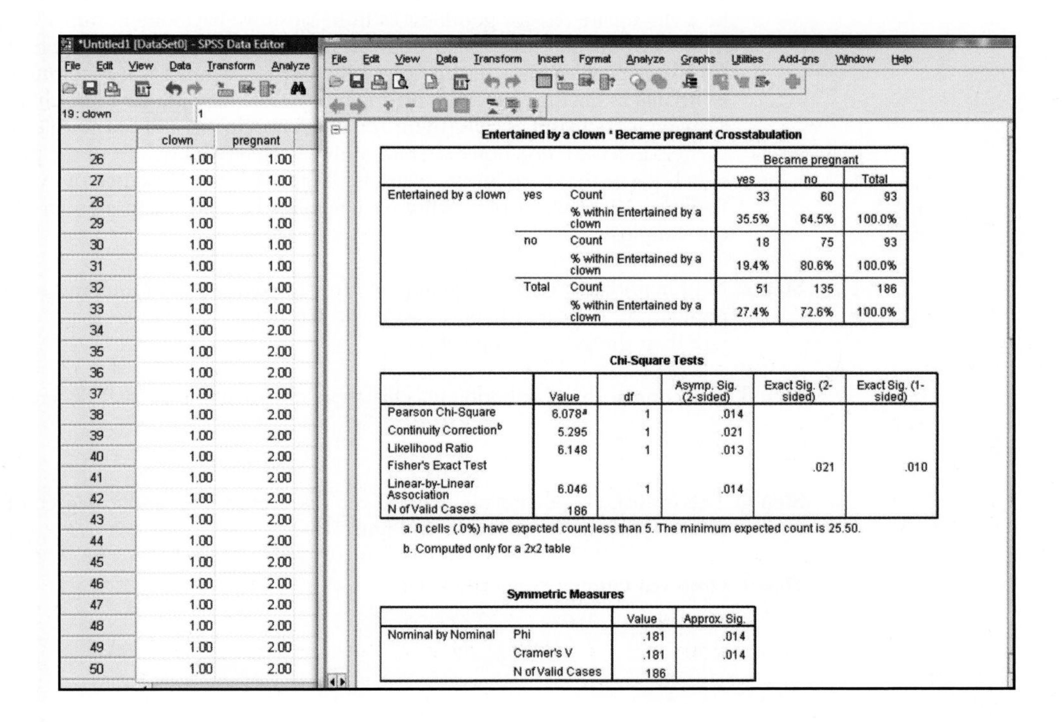

*Untitled1 [DataSet0] - SPSS Data Editor

19 : clown 1

	clown	pregnant
26	1.00	1.00
27	1.00	1.00
28	1.00	1.00
29	1.00	1.00
30	1.00	1.00
31	1.00	1.00
32	1.00	1.00
33	1.00	1.00
34	1.00	2.00
35	1.00	2.00
36	1.00	2.00
37	1.00	2.00
38	1.00	2.00
39	1.00	2.00
40	1.00	2.00
41	1.00	2.00
42	1.00	2.00
43	1.00	2.00
44	1.00	2.00
45	1.00	2.00
46	1.00	2.00
47	1.00	2.00
48	1.00	2.00
49	1.00	2.00
50	1.00	2.00

Entertained by a clown * Became pregnant Crosstabulation

			Became pregnant		
			yes	no	Total
Entertained by a clown	yes	Count	33	60	93
		% within Entertained by a clown	35.5%	64.5%	100.0%
	no	Count	18	75	93
		% within Entertained by a clown	19.4%	80.6%	100.0%
	Total	Count	51	135	186
		% within Entertained by a clown	27.4%	72.6%	100.0%

Chi-Square Tests

	Value	df	Asymp. Sig. (2-sided)	Exact Sig. (2-sided)	Exact Sig. (1-sided)
Pearson Chi-Square	6.078[a]	1	.014		
Continuity Correction[b]	5.295	1	.021		
Likelihood Ratio	6.148	1	.013		
Fisher's Exact Test				.021	.010
Linear-by-Linear Association	6.046	1	.014		
N of Valid Cases	186				

a. 0 cells (.0%) have expected count less than 5. The minimum expected count is 25.50.

b. Computed only for a 2x2 table

Symmetric Measures

		Value	Approx. Sig.
Nominal by Nominal	Phi	.181	.014
	Cramer's V	.181	.014
	N of Valid Cases	186	

SPSS®

Let's conduct a Mann–Whitney U test on the data we used when comparing communist and noncommunist countries on accomplishment-related pride. In SPSS, we conduct a Mann–Whitney U test by selecting: Analyze → Nonparametric Tests → 2 Independent-Samples Tests. Select "Mann–Whitney U" as the test type. Select "Descriptive" under "Options" if you want the descriptive data as well. The dependent variable, rankings on accomplishment-related national pride, goes under "Test Variable List," and the independent variable, political system (noncommunist versus communist), goes under "Grouping Variable." Be sure to define the groups by clicking "Define Groups" and telling SPSS what you have called each of the conditions (e.g., 1 and 2 for noncommunist and communist, respectively). The output will give us the same value for the Mann–Whitney U statistic, 3, that we calculated earlier. (*Note:* You may enter either scale or ordinal data as the dependent variable. SPSS automatically ranks the data. If the data are already ranked, no change is made to the values. If the data are scale, they are converted to ranks.)

How It Works

15.1 CONDUCTING A CHI-SQUARE TEST FOR GOODNESS OF FIT

Gary Steinman (2006), an obstetrician and gynecologist, studied whether a woman's diet could affect the likelihood that she would have twins. Insulin-like growth factor (IGF), often found in diets that include animal products like milk and beef, is hypothesized to lead to higher rates of twin births. Steinman wondered whether women who were vegans (those who eat neither meat nor dairy products) would have lower rates of twin births than would women who were vegetarians and consumed dairy products or women who ate meat. Steinman reported that, in the general population, 1.9% of births result in twins (without the aid of reproductive technologies). In Steinman's study of 1042 vegans who gave birth (without reproductive technologies), four sets of twins were born. How can we use Steinman's data to conduct the six steps of hypothesis testing for a chi-square test for goodness of fit?

Step 1: Population 1: Vegans who recently gave birth, like those whom we observed. Population 2: Vegans who recently gave birth who are like the general population of mostly nonvegans.

The comparison distribution is a chi-square distribution. The hypothesis test will be a chi-square test for goodness of fit because we have one nominal variable only. This study meets three of the four assumptions: (1) The one variable is nominal. (2) Every participant is in only one cell (a vegan woman is not counted as having twins *and* as having one child, or singleton). (3) There are far more than five times as many participants as cells (there are 1042 participants and only two cells). (4) The participants were not, however, randomly selected. We learn from the published research paper that participants were recruited with the assistance of "various vegan societies." This limits our ability to generalize beyond vegan women like those in the sample.

Step 2: Null hypothesis: Vegan women give birth to twins at the same rate as the general population. Research hypothesis: Vegan women give birth to twins at a different rate than the general population.

Step 3: The comparison distribution is a chi-square distribution that has 1 degree of freedom:

$$df_{\chi^2} = 2 - 1 = 1$$

Step 4: The critical chi-square value, based on a p level of 0.05 and 1 degree of freedom, is 3.841, as seen in the curve in Figure 15-3.

Step 5: **Observed (among vegan mothers)**

Singleton	Twins
1038	4

Expected (based on the 1.9% rate in the general population)

Singleton	Twins
1022.202	19.798

Category	Observed (O)	Expected (E)	$O - E$	$(O - E)^2$	$\frac{(O-E)^2}{E}$
Singleton	1038	1022.202	15.798	249.577	0.244
Twins	4	19.798	−15.798	249.577	12.606

$$\chi^2 = \Sigma\left(\frac{(O-E)^2}{E}\right) = 0.244 + 12.606 = 12.85$$

Step 6: Reject the null hypothesis; it appears that vegan mothers are less likely to have twins than are mothers in the general population.

The statistics, as reported in a journal article, would read:

$\chi^2(1, N = 1042) = 12.85, p < 0.05$

15.2 CONDUCTING A CHI-SQUARE TEST FOR INDEPENDENCE

Do people who move far from their hometown have a more exciting life? Since 1972, the General Social Survey (GSS) has asked approximately 40,000 adults in the United States numerous questions about their lives. During several years of the GSS, participants were asked, "In general, do you find life exciting, pretty routine, or dull?" (a variable called LIFE) and "When you were 16 years old, were you living in the same (city/town/country)?" (a variable called MOBILE16). How can we use these data to conduct the six steps of hypothesis testing for a chi-square test for independence?

In this case, there are two nominal variables. The independent variable is where a person lives relative to when he or she was 16 years old (same city, same state but different city, different state). The dependent variable is how the person finds life (exciting, routine, dull). Here are the data:

	Exciting	Routine	Dull
Same city	4890	6010	637
Same state/different city	3368	3488	337
Different state	4604	4139	434

Step 1: Population 1: People like those in this sample. Population 2: People from a population in which a person's characterization of life as exciting, routine, or dull does not depend on where that person is living relative to when he or she was 16 years old.

The comparison distribution is a chi-square distribution. The hypothesis test will be a chi-square test for independence because we have two nominal variables. This study meets all four assumptions: (1) The two variables are nominal. (2) Every participant is in only one cell. (3) There are more than five times as many participants as there are cells (there are 27,907 participants and 9 cells). (4) The GSS sample uses a form of random selection.

Step 2: Null hypothesis: The proportion of people who find life to be exciting, routine, or dull does not depend on where they live relative to where they lived when they were 16 years old. Research hypothesis: The proportion of people who find life exciting, routine, or dull differs depending on where they live relative to where they lived when they were 16 years old.

Step 3: The comparison distribution is a chi-square distribution with 4 degrees of freedom:

$$df_{\chi^2} = (k_{row} - 1)(k_{column} - 1) = (3 - 1)(3 - 1) = (2)(2) = 4$$

Step 4: The critical chi-square statistic, based on a p level of 0.05 and 4 degrees of freedom, is 9.488.

Step 5:

OBSERVED (EXPECTED IN PARENTHESES)

	Exciting	Routine	Dull	
Same city	4890 (5317.264)	6010 (5637.656)	637 (582.080)	11,537
Same state/different city	3368 (3315.167)	3488 (3514.923)	337 (362.911)	7193
Different state	4604 (4230.030)	4139 (4484.910)	434 (463.060)	9178
	12,862	13,637	1408	27,907

$$\frac{Total_{column}}{N}(Total_{row}) = \frac{12,862}{27,907}(11,537) = 5317.264$$

$$\frac{Total_{column}}{N}(Total_{row}) = \frac{12,862}{27,907}(7193) = 3315.167$$

$$\frac{Total_{column}}{N}(Total_{row}) = \frac{12,862}{27,907}(9178) = 4230.030$$

$$\frac{Total_{column}}{N}(Total_{row}) = \frac{13,637}{27,907}(11,537) = 5637.656$$

$$\frac{Total_{column}}{N}(Total_{row}) = \frac{13,637}{27,907}(7193) = 3514.923$$

$$\frac{Total_{column}}{N}(Total_{row}) = \frac{13,637}{27,907}(9178) = 4484.910$$

$$\frac{Total_{column}}{N}(Total_{row}) = \frac{1408}{27,907}(11,537) = 582.080$$

$$\frac{Total_{column}}{N}(Total_{row}) = \frac{1408}{27,907}(7193) = 362.911$$

$$\frac{Total_{column}}{N}(Total_{row}) = \frac{1408}{27,907}(9178) = 463.060$$

Category	$(O - E)^2$	$\frac{(O - E)^2}{E}$
Same city; exciting	182,554.530	34.332
Same city; routine	138,640.050	24.592
Same city; dull	3,016.206	5.182
Same state/different city; exciting	2,791.326	0.842
Same state/different city; routine	724.848	0.206
Same state/different city; dull	671.380	1.850
Different state; exciting	139,853.560	33.062
Different state; routine	119,653.730	26.679
Different state; dull	844.484	1.824

$$\chi^2 = \Sigma\left(\frac{(O - E)^2}{E}\right) = 128.569$$

Step 6: Reject the null hypothesis. The calculated chi-square statistic exceeds the critical value. How exciting a person finds life does appear to vary with where the person lives relative to where he or she lived when he or she was 16 years old.

We'd present these statistics in a journal article as: $\chi^2(4, N = 27,907) = 128.57$, $p < 0.05$.

15.3 CALCULATING CRAMER'S *V*

What is the effect size, Cramer's *V*, for the chi-square test for independence we conducted in How It Works 15.2?

$$V = \sqrt{\frac{\chi^2}{(N)(df_{row/column})}} = \sqrt{\frac{128.569}{(27,907)(2)}} = \sqrt{\frac{128.569}{55,814}} = 0.048$$

According to Cohen's conventions, this is a small effect size. With this piece of information, we'd present the statistics in a journal article as:

$\chi^2(4, N = 27,907) = 128.57, p < 0.05$, Cramer's $V = 0.05$

15.4 CALCULATING THE SPEARMAN RANK-ORDER CORRELATION COEFFICIENT

The accompanying table includes ranks for accomplishment-related national pride, along with numbers of medals won at the 2000 Sydney Olympics for 10 countries. (Of course, this might not be the best way to operationalize the variable of Olympic performance; perhaps we should be ranking Olympic medals per capita.) How can we calculate the Spearman correlation coefficient for these two variables—seen in the first two columns of the accompanying table—pride rank and Olympic medals?

First, we have to convert the numbers of Olympic medals to ranks. Then we can calculate the correlation coefficient.

Country	Pride Rank	Olympic Medals	Medals Rank	Difference (*D*)	Squared Difference (D^2)
United States	1	97	1	0	0
South Africa	2	5	7	−5	25
Austria	3	3	8	−5	25
Canada	4	14	5	−1	1
Chile	5	1	10	−5	25
Japan	6	18	3	3	9
Hungary	7	17	4	3	9
France	8	38	2	6	36
Norway	9	10	6	3	9
Slovenia	10	2	9	1	1

$$\Sigma D^2 = (0 + 25 + 25 + 1 + 25 + 9 + 9 + 36 + 9 + 1) = 140$$

$$r_S = 1 - \frac{6(\Sigma D^2)}{N(N^2 - 1)} = 1 - \frac{6(140)}{10(10^2 - 1)} = 1 - \frac{840}{10(100 - 1)} = 1 - \frac{840}{990} = 1 - 0.848 = 0.152$$

15.5 CONDUCTING THE MANN–WHITNEY *U* TEST

In which region do political science graduate programs tend to have the best rankings—on the East Coast (E) or in the Midwest (M)? Here are data from *U.S. News & World Report*'s 2005 online rankings of graduate schools. These are the top 13 doctoral programs in political science that are either on the East Coast or in the Midwest. Schools listed at the same rank are tied.

1	Harvard University (E)
2	University of Michigan, Ann Arbor (M)
3	Princeton University (E)
4	Yale University (E)
5.5	Duke University (E)
5.5	University of Chicago (M)
7.5	Columbia University (E)
7.5	Massachusetts Institute of Technology (E)
10	The Ohio State University (M)
10	University of North Carolina, Chapel Hill (E)
10	University of Rochester (E)
12.5	University of Wisconsin, Madison (M)
12.5	Washington University in St. Louis (M)

How can we conduct a Mann–Whitney *U* test for this example? The independent variable is region of the country, and its levels are East Coast and Midwest. The dependent variable is *U.S. News & World Report* ranking.

Step 1: This study meets the first and third of the three assumptions: (1) There is ordinal data after we convert the data from scale to ordinal. (2) The researchers did not use random selection, so our ability to generalize beyond this sample is limited. (3) There are some ties, but we will assume that there are not so many as to render the results of the test invalid.

Step 2: Null hypothesis: Political science programs on the East Coast and those in the Midwest do not differ in national ranking. Research hypothesis: Political science programs on the East Coast and those in the Midwest differ in national ranking.

Step 3: There are eight political science programs on the East Coast and five in the Midwest.

Step 4: The cutoff, or critical value, for a Mann–Whitney *U* test with one group of eight programs and one group of five programs, a p level of 0.05, and a two-tailed test is 6.

Step 5:

School	Rank	East Coast Rank	Midwest Rank
Harvard	1	1	
Michigan, Ann Arbor	2		2
Princeton	3	3	
Yale	4	4	
Duke	5.5		5.5
Chicago	5.5	5.5	
Columbia	7.5	7.5	
MIT	7.5	7.5	
Ohio State	10		10
North Carolina, Chapel Hill	10	10	
Rochester	10	10	
Wisconsin	12.5		12.5
Washington University in St. Louis	12.5		12.5

Before we continue, we sum the ranks for each group and add subscripts to indicate which group is which:

$$\Sigma R_E = 1 + 3 + 4 + 5.5 + 7.5 + 7.5 + 10 + 10 = 48.5$$

$$\Sigma R_M = 2 + 5.5 + 10 + 12.5 + 12.5 = 42.5$$

The formula for the first group is:

$$U_E = (n_E)(n_M) + \frac{n_E(n_E + 1)}{2} - \Sigma R_E = (8)(5) + \frac{8(8+1)}{2} - 48.5 = 40 + 36 - 48.5 = 27.5$$

The formula for the second group is:

$$U_M = (n_E)(n_M) + \frac{n_M(n_M + 1)}{2} - \Sigma R_M = (8)(5) + \frac{5(5+1)}{2} - 42.5 = 40 + 15 - 42.5 = 12.5$$

Step 6: For a Mann–Whitney *U* test, we compare only the smaller test statistic, 12.5, with the critical value, 6. This test statistic is not smaller than the critical value, so we fail to reject the null hypothesis. We cannot conclude that the two groups are different with respect to national rankings.

In a journal article, the statistics would read:

$U = 12.5, p > 0.05$

Exercises

Clarifying the Concepts

15.1 Distinguish nominal, ordinal, and scale data.

15.2 What are the three main situations in which we use a nonparametric test?

15.3 What is the difference between the chi-square test for goodness of fit and the chi-square test for independence?

15.4 What are the four assumptions for the chi-square tests?

15.5 List two ways in which statisticians use the word *independence* or *independent* with respect to concepts introduced earlier in this book. Then describe how *independence* is used by statisticians with respect to chi square.

15.6 What are the hypotheses when conducting the chi-square test for goodness of fit?

15.7 How are the degrees of freedom for the chi-square hypothesis tests different from those of most other hypothesis tests?

15.8 Why is there just one critical value for a chi-square test, even when the hypothesis is a two-tailed test?

15.9 What information is presented in a contingency table in the chi-square test for independence?

15.10 What measure of effect size is used with chi square?

15.11 Define the symbols in the following formula: $\chi^2 = \Sigma\left[\frac{(O-E)^2}{E}\right]$.

15.12 What is the formula $\frac{Total_{column}}{N}(Total_{row})$ used for?

15.13 What information does the measure of relative likelihood provide?

15.14 In order to calculate relative likelihood, what must be calculated first?

15.15 What is the difference between relative likelihood and relative risk?

15.16 Why might relative likelihood be easier to understand as an effect size than Cramer's V?

15.17 When do we convert scale data to ordinal data?

15.18 When the data on at least one variable are ordinal, the data on any scale variable must be converted from scale to ordinal. How do we convert a scale variable into an ordinal one?

15.19 How does the transformation of scale data to ordinal data solve the problem of outliers?

15.20 What does a histogram of rank-ordered data look like and why does it look that way?

15.21 Explain how the relation between ranks is the core of the Spearman rank-order correlation.

15.22 Define the symbols in the following term: $r_S = 1 - \frac{6(\Sigma D^2)}{N(N^2-1)}$.

15.23 What is the possible range of values for the Spearman rank-order correlation and how are these values interpreted?

15.24 If your data meet the assumptions of the parametric test, why is it preferable to use the parametric test rather than the nonparametric alternative?

15.25 When is it appropriate to use the Wilcoxon signed-rank test?

15.26 When do we use the Mann–Whitney U test?

15.27 What are the assumptions of the Mann–Whitney U test?

15.28 How is the critical value for the Mann–Whitney U test used differently than critical values for parametric tests?

15.29 When is it appropriate to use the Kruskal–Wallis H test?

Calculating the Statistics

15.30 For each of the following, (i) identify the incorrect symbol, (ii) state what the correct symbol should be, and (iii) explain why the initial symbol was incorrect.

a. For the chi-square test for goodness of fit: $df_{\chi^2} = N - 1$

b. For the chi-square test for independence: $df_{\chi^2} = (k_{row} - 1) + (k_{column} - 1)$

c. $\chi^2 = \Sigma\left[\frac{(M-E)^2}{E}\right]$

d. Cramer's $V = \sqrt{\frac{\chi^2}{(N)(k_{row/column})}}$

e. Expected frequency for each cell $= \frac{k_{column}}{N(k_{row})}$

15.31 For each of the following, identify the independent variable(s), the dependent variable(s), and the level of measurement (nominal, ordinal, scale).

a. The number of loads of laundry washed per month was tracked for women and men living in college dorms.

b. A researcher interested in people's need to maintain social image collected data on the number of miles on someone's car and his or her rank for "need for approval" out of the 183 people studied.

c. A professor of social science was interested in whether involvement in campus life is significantly impacted by whether a student lives on or off campus. Thirty-seven students living on campus and 37 students living off campus were asked whether they were an active member of a club.

15.32 Use this calculation table for the chi-square test for goodness of fit to complete this exercise.

Category	Observed (O)	Expected (E)	$O - E$	$(O - E)^2$	$\frac{(O-E)^2}{E}$
1	48	60			
2	46	30			
3	6	10			

a. Calculate degrees of freedom for this chi-square test for goodness of fit.
b. Perform all of the calculations to complete this table.
c. Compute the chi-square statistic.

15.33 Use this calculation table for the chi-square test for goodness of fit to complete this exercise.

a. Calculate degrees of freedom for this chi-square test for goodness of fit.
b. Perform all of the calculations to complete this table.
c. Compute the chi-square statistic.

Category	Observed (O)	Expected (E)	$O - E$	$(O - E)^2$	$\frac{(O-E)^2}{E}$
1	750	625			
2	650	625			
3	600	625			
4	500	625			

15.34 Below are some data to use in a chi-square test for independence.

Observed

	Accidents	No Accidents	
Rain	19	26	45
No Rain	20	71	91
	39	97	136

a. Calculate the degrees of freedom for this test.
b. Complete this table of expected frequencies.

Expected

	Accidents	No Accidents
Rain		
No Rain		

c. Calculate the test statistic.
d. Calculate the appropriate measure of effect size.
e. Calculate the relative likelihood of accidents, given that it is raining.

15.35 The data below are from a study of lung cancer patients in Turkey (Yilmaz et al., 2000). Use these data to calculate the relative likelihood of being a smoker, given that a person is female rather than male.

	Nonsmoker	Smoker
Female	186	13
Male	182	723

15.36 In order to compute statistics, we need to have working formulas. For the following, (i) identify the incorrect symbol, (ii) state what the correct symbol should be, and (iii) explain why the initial symbol was incorrect.

a. $U_1 = (n_1)(n_2) + \frac{n_1(n_1 + 1)}{2} - \Sigma R_1^2$

b. $r = 1 - \frac{6(\Sigma D^2)}{N(N^2 - 1)}$

15.37 Consider the following scale data.

Count	Variable X	Variable Y
1	134.5	64.00
2	186	60.00
3	157	61.50
4	129	66.25
5	147	65.50
6	133	62.00
7	141	62.50
8	147	62.00
9	136	63.00
10	147	65.50

a. Convert the data to ordinal or ranked data, starting with a rank of 1 for the smallest data point.
b. Compute the Spearman correlation coefficient.

15.38 Consider the following scale data.

Count	Variable X	Variable Y
1	$1250	25
2	$1400	21
3	$1100	32
4	$1450	54
5	$1600	38
6	$2100	62
7	$3750	43
8	$1300	32

a. Convert the data to ordinal or ranked data, starting with a rank of 1 for the smallest data point.

b. Compute the Spearman correlation coefficient.

15.39 The following fictional data represent the finishing place for runners of a 5-kilometer race and the number of hours they trained per week.

Race Rank	Hours Trained	Race Rank	Hours Trained
1	25	6	18
2	25	7	12
3	22	8	17
4	18	9	15
5	19	10	16

a. Calculate the Spearman correlation for this set of data.

b. Make a decision regarding the null hypothesis. Is there a significant correlation between a runner's finishing place and the amount the runner trained?

15.40 Assume a researcher compared the performance of two independent groups of participants on an ordinal variable using the Mann–Whitney *U* test. The first group had 8 participants and the second group had 11 participants.

a. Using a *p* level of 0.05 and a two-tailed test, determine the critical value.

b. Assume the researcher calculated $U_1 = 22$ and $U_2 = 17$. Make a decision regarding the null hypothesis and explain that decision.

c. Assume the researcher calculated $U_1 = 24$ and $U_2 = 30$. Make a decision regarding the null hypothesis and explain that decision.

d. Assume the researcher calculated $U_1 = 13$ and $U_2 = 9$. Make a decision regarding the null hypothesis and explain that decision.

15.41 Compute the Mann–Whitney *U* statistic for the following data. The numbers under the Group 1 and Group 2 columns are participant numbers.

Group 1	Ordinal Dependent Variable	Group 2	Ordinal Dependent Variable
1	1	1	11
2	2.5	2	9
3	8	3	2.5
4	4	4	5
5	6	5	7
6	10	6	12

15.42 Compute the Mann–Whitney *U* statistic for the following data. The numbers under the Group 1 and Group 2 columns are participant numbers.

Group 1	Scale Dependent Variable	Group 2	Scale Dependent Variable
1	8	9	3
2	5	10	4
3	5	11	2
4	7	12	1
5	10	13	1
6	14	14	5
7	9	15	6
8	11		

15.43 Are men or women more likely to be at the top of their class? The following table depicts fictional class standings for a group of men and women:

Student	Gender	Class Standing	Student	Gender	Class Standing
1	Male	98	7	Male	43
2	Female	72	8	Male	33
3	Male	15	9	Female	17
4	Female	3	10	Female	82
5	Female	102	11	Male	63
6	Female	8	12	Male	25

a. Compute the Mann–Whitney *U* test statistic.

b. Make a decision regarding the null hypothesis. Is there a significant difference in the class ranks of men and women?

Applying the Concepts

15.44 Gender, the Oscars, and nonparametric tests: In 2010, Sandra Bullock won an Academy Award for Best Actress. Shortly thereafter, she discovered her husband was cheating on her. Headlines erupted about a supposed Oscar curse that befalls women, and many in the media wondered whether ambitious women—whether actors or corporate leaders—are more likely than ambitious men to run the risk of ruining their family lives. Reporters breathlessly listed female actors who were divorced within a couple of years of winning an Oscar—Julia Roberts, Helen Hunt, Kate Winslet, Halle Berry, and Reese Witherspoon among them.

a. A good researcher always asks, "Compared to what?" In this case, what would be an appropriate comparison group to use to determine whether there really is a gender difference in likelihood of relationship breakups among Oscar winners? Explain your answer.

b. Kate Harding (2010) reported that many men—including Russell Crowe, William Hurt, Dustin Hoffman, Robert Duvall, and Clark Gable—experienced the same outcome. Indeed, Harding counted 15 Best Actor winners, compared with just 8 Best Actress winners, who divorced not long after winning an Oscar. If she wanted to conduct statistical analyses, what test would Harding use? Explain your answer.

c. Explain how an illusory correlation, bolstered by a confirmation bias, might have led to the headlines despite evidence to the contrary.

15.45 Parametric or nonparametric test? For each of the following research questions, state whether a parametric or nonparametric hypothesis test is more appropriate. Explain your answers.

a. Are women more or less likely than men to be economics majors?

b. At a small company with 15 staff and 1 top boss, do those with a college education tend to make a different amount of money than those without one?

c. At your high school, did athletes or nonathletes tend to have higher grade point averages?

d. At your high school, did athletes or nonathletes tend to have higher class ranks?

e. Compare car accidents in which the occupants were wearing seat belts with accidents in which the occupants were not wearing seat belts. Do seat belts seem to make a difference in the numbers of accidents that lead to no injuries, nonfatal injuries, and fatal injuries?

f. Compare car accidents in which the occupants were wearing seat belts with accidents in which the occupants were not wearing seat belts. Were those wearing seat belts driving at slower speeds, on average, than those not wearing seat belts?

15.46 Evaluating professors and types of variables: Weinberg, Fleisher, and Hashimoto (2007) studied almost 50,000 students' evaluations of their professors in almost 400 economics courses at The Ohio State University over a 10-year period. For each of their findings, outlined below, state (i) the independent variable or variables, and, where appropriate, their levels; (ii) the dependent variable(s); and (iii) which category of research design is being used:

I—Scale independent variable(s) and scale dependent variable

II—Nominal independent variable(s) and scale dependent variable

III—Only nominal variables

Explain your answer to part (iii).

a. The researchers found that students' ratings of their professors were predictive of grades in the class for which the professor was evaluated.

b. The researchers also found that students' ratings of their professors were not predictive of grades for other, related future classes. (The researchers stated that these first two findings suggest that student ratings of professors are tied to their current grades but not to learning—which would affect future grades.)

c. The researchers found that male professors received statistically significantly higher student ratings, on average, than did female professors.

d. The researchers reported, however, that average levels of students' learning (as assessed by grades in related future classes) were not statistically significantly different for those who had male and those who had female professors.

e. The researchers might have been interested in whether there were proportionally more female professors teaching upper-level than lower-level courses and proportionally more male professors teaching lower-level than upper-level courses (perhaps a reason for the lower average ratings of female professors).

f. The researchers found no statistically significant differences in average student evaluations among non-tenure-track lecturers, graduate student teaching associates, and tenure-track faculty members.

15.47 Grade inflation and types of variables: A *New York Times* article on grade inflation reported several findings related to a tendency for average grades to rise over the years and a tendency for the top-ranked institutions to give the highest average grades (Archibold, 1998). For each of the findings outlined below, state (i) the independent variable or variables, and, where appropriate, their levels; (ii) the dependent variable(s); and (iii) which category of research design is being used:

I—Scale independent variable(s) and scale dependent variable

II—Nominal independent variable(s) and scale dependent variable

III—Only nominal variables

Explain your answer to part (iii).

a. In 1969, 7% of all grades were A's; in 1994, 25% of all grades were A's.

b. The average GPA for the graduating students of elite schools is 3.2, the average GPA for graduating students at selective schools (the level below elite schools) is 3.04, and the average GPA for graduating students at state colleges is 2.95.

c. At Dartmouth College, an elite university, SAT scores of incoming students have increased along with their subsequent college GPAs (perhaps an explanation for grade inflation).

15.48 High school academic performance and types of variables: Here are three ways to assess one's performance in high school: (1) GPA at graduation, (2) whether one graduated with honors (as indicated by graduating with a GPA of at least 3.5), and (3) class rank at graduation. For example, Abdul had a 3.98 GPA, graduated with honors, and was ranked 10th in his class.

a. Which of these variables could be considered a nominal variable? Explain.

b. Which of these variables is most clearly an ordinal variable? Explain.

c. Which of these variables is a scale variable? Explain.

d. Which of these variables gives us the most information about Abdul's performance?

e. If we were to use one of these variables in an analysis, which variable (as the dependent variable) would lead to the lowest chance of a Type II error? Explain why.

15.49 Immigration, crime, and research design: "Do Immigrants Make Us Safer?" asked the title of a *New York Times Magazine* article (Press, 2006). The article reported findings from several U.S.-based studies, including several conducted by Harvard sociologist Robert Sampson in Chicago. For each of the following findings, draw the table of cells that would comprise the research design. Include the labels for each row and column.

a. Mexicans were more likely to be married (versus single) than either blacks or whites.

b. People living in immigrant neighborhoods were 15% less likely than were people living in nonimmigrant neighborhoods to commit crimes. This finding was true among both those living in households headed by a married couple and those living in households not headed by a married couple.

c. The crime rate was higher among second-generation than among first-generation immigrants; moreover, the crime rate was higher among third-generation than among second-generation immigrants.

15.50 Sex selection and hypothesis testing: Across all of India, there are only 933 girls for every 1000 boys (Lloyd, 2006), evidence of a bias that leads many parents to illegally select for boys or to kill their infant girls. (Note that this translates into a proportion of girls of 0.483.) In Punjab, a region of India in which residents tend to be more educated than in other regions, there are only 798 girls for every 1000 boys. Assume that you are a researcher interested in whether sex selection is more or less prevalent in educated regions of India and that 1798 children from Punjab constitute the entire sample. (*Hint:* You will use the proportions from the national database for comparison.)

a. How many variables are there in this study? What are the levels of any variable you identified?

b. Which hypothesis test would be used to analyze these data? Justify your answer.

c. Conduct the six steps of hypothesis testing for this example. (*Note:* Be sure to use the correct proportions for the expected values, not the actual numbers for the population.)

d. Report the statistics as you would in a journal article.

15.51 Gender, op-ed writers, and hypothesis testing: Richards (2006) reported data from a study by the *American Prospect* on the genders of op-ed writers who addressed the topic of abortion in the *New York Times.* Over a 2-year period, the *American Prospect* counted 124 articles that discussed abortion (from a wide range of political and ideological perspectives). Of these, just 21 were written by women.

a. How many variables are there in this study? What are the levels of any variable you identified?

b. Which hypothesis test would be used to analyze these data? Justify your answer.

c. Conduct the six steps of hypothesis testing for this example.

d. Report the statistics as you would in a journal article.

15.52 Romantic music, behavior, chi square, and effect size: Guéguen, Jacob, and Lamy (2010) investigated whether exposure to romantic music affects dating behavior. The participants, young, single French women, waited for the experiment to start in a room in which songs with either romantic lyrics or neutral lyrics were playing. After a few minutes, each woman who participated completed a marketing survey administered by a young male confederate. During a break, the confederate asked the participant for her phone number. Of the women who listened to romantic music, 52.2% (23 out of 44) gave him her phone number, whereas 27.9% (12 out of 43) of the women who listened to neutral music did so. The researchers conducted a chi-square test for independence, and found the following results, ($\chi^2(1, N = 83) = 5.37, p = .02$).

a. Calculate Cramer's *V.* What size effect is this?

b. Calculate the relative likelihood of providing her phone number for women listening to romantic music versus neutral music. Explain what we learn from this relative likelihood.

15.53 The General Social Survey, an exciting life, and relative risk: In How It Works 15.2, we walked through a chi-square test for independence using two items from the General Social Survey (GSS)—LIFE and MOBILE16. Use these data to answer the following questions.

a. Construct a table that shows only the appropriate conditional proportions for this example. For example, the percentage of people who find life exciting, given that they live in the same city, is 42.4. The proportion, therefore, is 0.424.

b. Construct a graph that displays these conditional proportions.

c. Calculate the relative risk (or relative likelihood) of finding life exciting if one lives in a different state compared to if one lives in the same city as one did at 16.

15.54 University students, cell phone bills, and ordinal data: Here are some monthly cell phone bills, in dollars, for university students:

100 60 35 50 50 50 60 65
0 75 100 55 50 40 80
200 30 50 108 500 100 45
40 45 50 40 40 100 80

a. Convert these data from scale to ordinal. (Don't forget to put them in order first.) What happens to an outlier when you convert these data to ordinal?

b. What approximate shape would the distribution of these data take? Would they likely be normally distributed? Explain why the distribution of ordinal data is never normal.

c. Why does it not matter if the ordinal variable is normally distributed? (*Hint:* Think about what kind of hypothesis test you would conduct.)

15.55 World cities, livability, and nonparametric hypothesis tests: CNN.com reported on a 2005 study that ranked the world's cities in terms of how livable they are (http://www.cnn.com/2005/WORLD/europe/10/04/eui.survey/), using a range of criteria related to stability, health care, culture and environment, education, and infrastructure. Vancouver came out on top. For each of the following research questions, state which nonparametric hypothesis test is appropriate: Spearman rank-order correlation coefficient, Wilcoxon signed-rank test, Mann–Whitney U test, or Kruskal–Wallis H test. Explain your answers.

a. Which cities tend to receive higher rankings—those north of the equator or those south of the equator?

b. Did the top 10 cities tend to change their rankings relative to their position in the previous study?

c. Are the livability rankings related to a city's economic status?

15.56 Fantasy baseball and the Spearman correlation coefficient: In fantasy baseball, groups of 12 league participants conduct a draft in which they can "buy" any baseball players from any teams across one of the leagues (i.e., the American League or the National League). These makeshift teams are compared on the basis of the combined statistics of the individual baseball players. Statistics such as home runs are awarded points, and each fantasy team receives a total score of all combined points for its baseball players, regardless of their real-life team. Many in the fantasy and real-life baseball worlds have wondered how success in fantasy leagues maps onto the real-life success of winning baseball games. Walker (2006) compared the fantasy league performances of the players for each American League team with their actual American League finishes for the 2004 season, the year the Boston Red Sox broke the legendary "curse" against them and won the World Series. The data, sorted from highest to lowest fantasy league score, are shown in the accompanying table.

Team	Fantasy League Points	Actual American League Finish
Boston	117.5	2
New York	109.5	1
Anaheim	108	3.5
Minnesota	97	3.5
Texas	85	6
Chicago	80	7
Cleveland	79	8
Oakland	77	5
Baltimore	74.5	9
Detroit	68.5	10
Seattle	51	13
Tampa Bay	47.5	11
Toronto	35.5	12
Kansas City	20	14

a. What are the two variables of interest? For each variable, state whether it's scale or ordinal.

b. Calculate the Spearman correlation coefficient for these two variables. Remember to convert any scale variables to ranks.

c. What does the coefficient tell us about the relation between these two variables?

d. Why couldn't we calculate a Pearson correlation coefficient for these data?

15.57 Test-taking speed, grade, and the Spearman correlation coefficient: Does speed in completing a test correlate with one's grade? Here are test scores for eight students in one of our statistics classes. They are arranged in order from the student who turned in the test first to the student who turned in the test last.

98 74 87 92 88 93 62 67

a. What are the two variables of interest? For each variable, state whether it's scale or ordinal.

b. Calculate the Spearman correlation coefficient for these two variables. Remember to convert any scale variables to ranks.

c. What does the coefficient tell us about the relation between these two variables?

d. Why couldn't we calculate a Pearson correlation coefficient for these data?

e. Does this Spearman correlation coefficient suggest that students should take their tests as quickly as possible? That is, does it indicate that taking the test quickly *causes* a good grade? Explain your answer.

f. What third variables might be responsible for this correlation? That is, what third variables might cause both speedy test-taking and a good test grade?

15.58 Test-taking speed, grade, and interpreting the Spearman correlation coefficient: Consider again the two variables described in Exercise 15.57, test grade and speed in taking the test. Imagine that each of the following numbers represents the Spearman correlation coefficient that quantifies the relation between test grade and speed in taking the test. Recall that test grade was converted to ranks such that the top grade of 98 is ranked 1, and for speed in taking the test, the fastest person was ranked 1. What does each coefficient suggest about the relation between the variables? Using the guidelines for the Pearson correlation coefficient, indicate whether each coefficient is roughly small (0.10), medium (0.30), or large (0.50). Specify which of these coefficients suggests the strongest relation between the two variables as well as which coefficient suggests the weakest relation between the two variables. [You calculated the actual correlation between these variables in Exercise 15.57(b).]

a. 1.00

b. −0.001

c. 0.52

d. −0.27

e. −0.98

f. 0.09

15.59 Choosing a graduate school and choosing a nonparametric test: You're applying to graduate school and have found a list of the top 50 PhD programs for your area of study. For each of the following scenarios, state which nonparametric hypothesis test is most appropriate: Spearman rank-order correlation coefficient, Wilcoxon signed-rank test, Mann–Whitney U test, or Kruskal–Wallis H test. Explain your answers.

a. You want to determine which institutions tend to be higher ranked: those that fund students primarily by offering fellowships, those that fund students primarily by offering teaching assistantships, or those that don't have full funding for most students.

b. You wonder whether rankings are related to the typical Graduate Record Examination (GRE) scores of incoming students.

c. You decide to compare the rankings of institutions within a 3-hour drive of your current home and those beyond a 3-hour drive.

15.60 Public versus private universities and the Mann–Whitney U test: Do public or private universities tend to have better sociology graduate programs? *U.S. News & World Report* publishes online rankings of graduate schools across a range of disciplines. Here is its 2005 list of the top 21 doctoral programs in sociology, along with an indication of whether the schools are public or private institutions. Schools listed at the same rank are tied.

1	University of Wisconsin, Madison (public)
2	University of California, Berkeley (public)
3	University of Michigan, Ann Arbor (public)
4.5	University of Chicago (private)
4.5	University of North Carolina (public)
6.5	Princeton University (private)
6.5	Stanford University (private)
8.5	Harvard University (private)
8.5	University of California, Los Angeles (public)
10	University of Pennsylvania (private)
12	Columbia University (private)
12	Indiana University, Bloomington (public)
12	Northwestern University (private)
15	Cornell University (private)
15	Duke University (private)
15	University of Texas, Austin (public)
18	Pennsylvania State University, University Park (public)
18	University of Arizona (public)
18	University of Washington (public)
20.5	The Ohio State University (public)
20.5	Yale University (private)

a. What is the independent variable, and what are its levels? What is the dependent variable?

b. Is this a between-groups or within-groups design? Explain.

c. Why do we have to use a nonparametric hypothesis test for these data?

d. Conduct all six steps of hypothesis testing for a Mann–Whitney *U* test.

e. How would you present these statistics in a journal article?

15.61 Political party, voter turnout, and the Mann–Whitney *U* test: Do red states (U.S. states whose residents tend to vote Republican) have different voter turnouts than blue states (U.S. states whose residents tend to vote Democratic)? The accompanying table shows voter turnouts (in percentages) for the 2004 presidential election for eight randomly selected red states and eight randomly selected blue states.

Red States	Voted in 2004 Election (%)	Blue States	Voted in 2004 Election (%)
Georgia	57.38	California	60.01
Idaho	64.89	Illinois	60.73
Indiana	55.69	Maine	73.40
Louisiana	60.78	New Jersey	64.54
Missouri	66.89	Oregon	70.50
Montana	64.36	Vermont	66.19
Texas	53.35	Washington	67.42
Virginia	61.50	Wisconsin	76.73

a. What is the independent variable, and what are its levels? What is the dependent variable?

b. Is this a between-groups or within-groups design? Explain.

c. Conduct all six steps of hypothesis testing for a Mann–Whitney *U* test.

d. How would you present these statistics in a journal article?

15.62 Gender, aggression, and the interpretation of a Mann–Whitney *U* test: Spanish researchers examining aggression in children's dreams reported the following: "Using the Mann–Whitney nonparametrical statistical test on the gender differences, we found a significant difference between boys and girls in Group 1 for overall [aggression] ($U = 44.00, p = 0.004$) and received aggression ($U = 48.00, p = 0.005$). So, in their dreams, younger boys not only had a higher level of general aggression but also received more *severe* aggressive acts than girls of the same age" (emphasis in original) (Oberst, Charles, & Chamarro, 2005, p. 175).

a. What is the independent variable, and what are its levels? What is the dependent variable?

b. Is this a between-groups or within-groups design?

c. Which hypothesis test did the researchers conduct? Why might they have chosen a nonparametric test? Why do you think they chose this particular nonparametric test?

d. Describe what they found in your own words.

e. Can we conclude that gender caused a difference in levels of aggression in dreams? Explain. Provide at least two reasons why gender might not cause certain levels of aggression in dreams even though these variables are associated.

15.63 Cell phone bill, hours studied, and the shapes of distributions: The following figures display data that depict the relation between students' monthly cell phone bills and the number of hours they report that they study per week.

a. What does the accompanying scatterplot suggest about the shape of the distribution for hours studied per week? What does it suggest about the shape of the distribution for monthly cell phone bill?

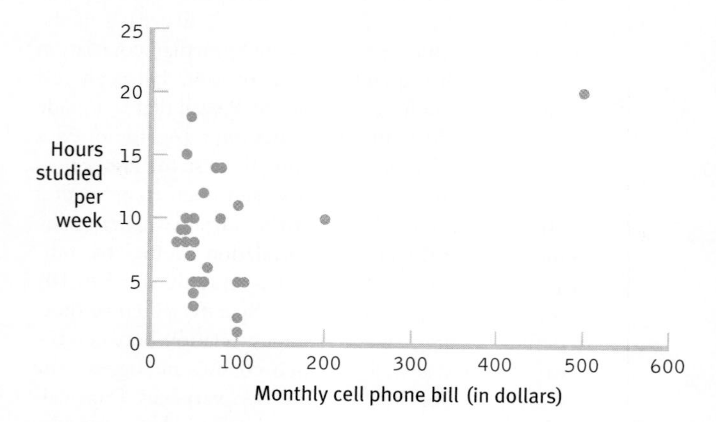

b. What does the accompanying grouped frequency histogram suggest about the shape of the distribution for monthly cell phone bill?

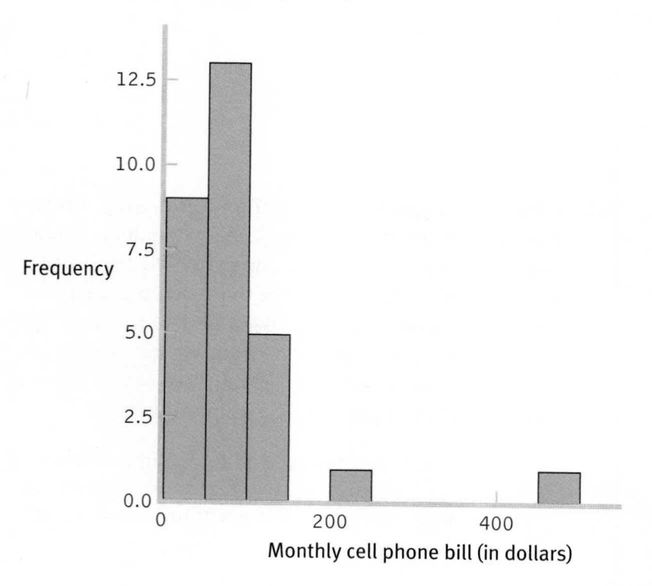

c. Is it a good idea to use a parametric hypothesis test for these data? Explain.

15.64 Stroke patients, treatment, and type of nonparametric test: A common situation faced by researchers working with special populations, such as neurologically impaired people or people with less common psychiatric conditions, is that the studies often have small sample sizes due to the relatively few numbers of patients. As a result, these researchers often turn to nonparametric statistical tests. For each of the following research descriptions, state which nonparametric hypothesis test is most appropriate: Spearman rank-order correlation coefficient, Wilcoxon signed-rank test, Mann–Whitney *U* test, or Kruskal–Wallis *H* test. Explain your answers.

a. People who have had a stroke often have whole or partial paralysis on the side of their body opposite the side of the brain damage. Leung, Ng, and Fong (2009) were interested in the effects of a treatment program for constrained movement on the recovery from paralysis. They compared the arm-movement ability of eight stroke patients before and after the treatment.

b. Leung and colleagues (2009) were also interested in whether the amount of improvement after the therapy was related to the number of months that had passed since the patient experienced the stroke.

c. Five of Leung and colleagues' (2009) patients were male and three were female. We could ask whether post-treatment movement performance was different between men and women.

Putting It All Together

15.65 Gender bias, poor growth, and hypothesis testing: Grimberg, Kutikov, and Cucchiara (2005) wondered whether gender biases were evident in referrals of children for poor growth. They believed that boys were more likely to be referred even when there was no problem—which is bad for boys because families of short boys might falsely view their height as a medical problem. They also believed that girls were less likely to be referred even when there was a problem—which is bad for girls because real problems might not be diagnosed and treated. They studied all new patients at The Children's Hospital of Philadelphia Diagnostic and Research Growth Center who were referred for potential problems related to short stature. Of the 182 boys who were referred, 27 had an underlying medical problem, 86 did not but were below norms for their age, and 69 were of normal height according to growth charts. Of the 96 girls who were referred, 39 had an underlying medical problem, 38 did not but were below norms for their age, and 19 were of normal height according to growth charts.

a. How many variables are there in this study? What are the levels of any variable you identified?

b. Which hypothesis test would be used to analyze these data? Justify your answer.

c. Conduct the six steps of hypothesis testing for this example.

d. Calculate the appropriate measure of effect size. According to Cohen's conventions, what size effect is this?

e. Report the statistics as you would in a journal article.

f. Draw a table that includes the conditional proportions for boys and for girls.

g. Create a graph with bars showing the proportions for all six conditions.

h. Among only children who are below height norms, calculate the relative risk of having an underlying medical condition if one is a boy as opposed to a girl. Show your calculations.

i. Explain what we learn from this relative risk.

j. Now calculate the relative risk of having an underlying medical condition if one is a girl. Show your calculations.

k. Explain what we learn from this relative risk.

l. Explain how the calculations in parts (h) and (j) provide us with the same information in two different ways.

15.66 The prisoner's dilemma, cross-cultural research, and hypothesis testing: In a classic prisoner's dilemma game with money for prizes, players who cooperate with each other both earn good prizes. If, however, your opposing player cooperates but you do not (the term used is *defect*), you receive an even bigger payout and your opponent receives nothing. If you cooperate but your opposing player defects, he or she receives that bigger payout and you receive nothing. If you both defect, you each get a small prize. Because of this, most players of such games choose to defect, knowing that if they cooperate but their partners don't, they won't win anything. The strategies of U.S. and Chinese students were compared. The researchers hypothesized that those from the market economy (United States) would cooperate less (i.e., would defect more often) than would those from the nonmarket economy (China).

	Defect	Cooperate
China	31	36
United States	41	14

a. How many variables are there in this study? What are the levels of any variables you identified?

b. Which hypothesis test would be used to analyze these data? Justify your answer.

c. Conduct the six steps of hypothesis testing for this example, using the above data.

d. Calculate the appropriate measure of effect size. According to Cohen's conventions, what size effect is this?

e. Report the statistics as you would in a journal article.

f. Draw a table that includes the conditional proportions for participants from China and from the United States.

g. Create a graph with bars showing the proportions for all four conditions.

h. Create a graph with two bars showing just the proportions for the defections for each country.

i. Calculate the relative risk (or relative likelihood) of defecting, given that one is from China versus the United States. Show your calculations.

j. Explain what we learn from this relative risk.

k. Now calculate the relative risk of defecting, given that one is from the United States versus China. Show your calculations.

l. Explain what we learn from this relative risk.

m. Explain how the calculations in parts (i) and (k) provide us with the same information in two different ways.

Terms

chi-square test for goodness of fit (p. 406)
chi-square test for independence (p. 406)
Cramer's *V* (p. 416)
relative risk (p. 418)
Spearman rank-order correlation coefficient (p. 423)
Mann–Whitney *U* test (p. 426)

Formulas

$df_{\chi^2} = k - 1$ (degrees of freedom for chi-square test for goodness of fit) (p. 408)

$$\chi^2 = \Sigma\left[\frac{(O - E)^2}{E}\right]$$ (p. 410)

$df_{\chi^2} = (k_{row} - 1)(k_{column} - 1)$ (degrees of freedom for chi-square test for independence) (p. 412)

Expected frequency for each cell $= \frac{Total_{column}}{N}(Total_{row})$, where we use the overall number of participants, N, along with the totals for the rows and columns for each particular cell (p. 414)

$$\text{Cramer's } V = \sqrt{\frac{\chi^2}{(N)(df_{row/column})}}$$ (p. 416)

$$r_S = 1 - \frac{6(\Sigma D^2)}{N(N^2 - 1)}$$ (p. 425)

$$U_1 = (n_1)(n_2) + \frac{n_1(n_1 + 1)}{2} - \Sigma R_1$$ (p. 428)

$$U_2 = (n_1)(n_2) + \frac{n_2(n_2 + 1)}{2} - \Sigma R_2$$ (p. 428)

Symbols

χ^2 (p. 405)
k (p. 408)
Cramer's V (p. 416)
Cramer's φ (p. 416)
r_S (p. 423)
U (p. 426)

Reference for Basic Mathematics

This appendix serves as a reference for the basic mathematical operations that are used in the book. We provide quick reference tables to help you with symbols and notation; instruction on the order of operations for equations with multiple operations; guidelines for converting fractions, decimals, and percentages; and examples of how to solve basic algebraic equations. Some of you will need a more extensive review than is presented in these pages. That review, which involves greater detail and instruction, can be found on the book's companion Web site. Most of you will be familiar with much of this material. However, the inclusion of this reference can help you to solve problems throughout this book, particularly when you come across material that appears unfamiliar.

We include a diagnostic quiz for you to assess your current comfort level with this material. Following the diagnostic test, we provide instruction and reference tables for each section so that you can review the concepts, apply the concepts through worked problems, and review your skills with a brief self-quiz.

A.1 Diagnostic Test: Skills Evaluation

This diagnostic test is divided into four parts that correspond to the sections of the basic mathematics review that follows. The purpose of the diagnostic test is to help understand which areas you need to review prior to completing work in this book. (Answers to each of the questions can be found at the end of the review on page A-6.)

SECTION 1 (Symbols and Notation: Arithmetic Operations)

1. $8 + 2 + 14 + 4 =$ ______________
2. $4 \times (-6) =$ ______________
3. $22 - (-4) + 3 =$ ______________
4. $8 \times 6 =$ ______________
5. $36 \div (-9) =$ ______________
6. $13 + (-2) + 8 =$ ______________
7. $44 \div 11 =$ ______________
8. $-6\ (-3) =$ ______________
9. $-6 - 8 =$ ______________
10. $-14\ /-2 =$ ______________

SECTION 2 (Order of Operations)

1. $3 \times (6 + 4) - 30 =$ ______________
2. $4 + 6(2 + 1) + 6 =$ ______________
3. $(3 - 6) \times 2 + 5 =$ ______________
4. $4 + 6 \times 2 =$ ______________
5. $16/2 + 6(3 - 1) =$ ______________
6. $2^2\ (12 - 8) =$ ______________
7. $5 - 3(4 - 1) =$ ______________
8. $7 \times 2 - (9 - 3) \times 2 =$ ______________
9. $15 \div 5 + (6 + 2)/2 =$ ______________
10. $15 - 3^2 + 5(2) =$ ______________

SECTION 3 (Proportions: Fractions, Decimals, and Percentages)

1. Convert 0.42 into a fraction ______________
2. Convert $^6/_{10}$ into a decimal ______________
3. Convert $^4/_5$ into a percentage ______________
4. $^6/_{13} + {}^4/_{13} =$ ______________
5. $0.8 \times 0.42 =$ ______________
6. 40% of 120 = ______________
7. $^2/_7 + {}^2/_5 =$ ______________
8. $^2/_5 \times 80 =$ ______________
9. $^1/_4 \div {}^1/_3 =$ ______________
10. $^4/_7 \times {}^5/_9 =$ ______________

SECTION 4 (Solving Equations with a Single Unknown Variable)

1. $5X - 13 = 7$ ______________
2. $3(X - 2) = 9$ ______________
3. $X/3 + 2 = 10$ ______________
4. $X(-3) + 2 = -16$ ______________
5. $X(6 - 4) + 3 = 15$ ______________
6. $X/4 + 3 = 6$ ______________
7. $3X + (-9)/(-3) = 24$ ______________
8. $9 + X/4 = 12$ ______________
9. $4X - 5 = 19$ ______________
10. $5 + (-2) + 3X = 9$ ______________

A.2 Symbols and Notation: Arithmetic Operations

SYMBOLS AND NOTATION

The basic mathematical symbols used throughout this book are located in Table A.1. These include the most common arithmetic operations, and most of you will find that you are familiar with them. However, it is worth your time to review the reference table and material that outline the operations using positive and negative numbers. For those of you who have spent little time solving math equations recently, familiarizing yourselves with this material can be quite helpful in avoiding common mistakes.

TABLE A.1 Symbols and Notations

Symbol	Operation	Example
$+$	Addition	$8 + 3 = 11$
$-$	Subtraction	$14 - 6 = 8$
$\times$, ()	Multiplication	$4 \times 3 = 12$, $4(3) = 12$
$\div$, /	Division	$12 \div 6 = 2$, $^{12}/_{6} = 2$
$>$	Greater than	$7 > 5$
$<$	Less than	$4 < 9$
$\geq$	Greater than or equal to	$7 \geq 5$, $4 \geq 4$
$\leq$	Less than or equal to	$5 \leq 9$, $6 \leq 6$
$\neq$	Not equal to	$5 \neq 3$

ARITHMETIC OPERATIONS: Worked Examples

Adding, Subtracting, Multiplying, and Dividing with Positive and Negative Numbers

1. Adding with positive numbers: Add the two (or series of) numbers to produce a sum.
   a. $4 + 7 = 11$
   b. $7 + 4 + 9 = 20$
   c. $4 + 6 + 7 + 2 = 19$

2. Adding with negative numbers: Sum the absolute values of each number and place a negative sign in front of the sum. (*Hint:* When a positive sign directly precedes a negative sign, change both signs to a single negative sign.)
   a. $-6 + (-4) = -10$
      $-6 - 4 = -10$
   b. $-3 + (-2) = -5$
      $-3 - 2 = -5$

3. Adding two numbers with opposite signs: Find the difference between the two numbers and assign the sign (positive or negative) of the larger number.
   a. $17 + (-9) = 8$
   b. $-16 + 10 = -6$

4. Subtracting one number from another number. (*Hint:* When subtracting a negative number from another number, two negative signs come in sequence, as in part (a). To solve the equations, change the two sequential negative signs into a single positive sign.)
   a. $5 - (-4) = 9$
      $5 + 4 = 9$
   b. $5 - 8 = -3$
   c. $-6 - 3 = -9$

5. Multiplying two positive numbers produces a positive result.
   a. $6 \times 9 = 54$
   b. $6(9) = 54$
   c. $4 \times 3 = 12$
   d. $4(3) = 12$

6. Multiplying two negative numbers produces a positive result.
   a. $-3 \times -9 = 27$
   b. $-3(-9) = 27$
   c. $-4 \times (-3) = 12$
   d. $-4(-3) = 12$

7. Multiplying one positive and one negative number produces a negative result.
   a. $-3 \times 9 = -27$
   b. $-3(9) = -27$
   c. $4 \times (-3) = -12$
   d. $4(-3) = -12$

8. Dividing two positive numbers produces a positive result.
   a. $12 \div 4 = 3$
   b. $12 / 4 = 3$
   c. $16 \div 8 = 2$
   d. $16 / 8 = 2$

9. Dividing two negative numbers produces a positive result.
   a. $-12 \div -4 = 3$
   b. $-12 / -4 = 3$
   c. $-16 \div (-8) = 2$
   d. $-16 / (-8) = 2$

10. Dividing a positive number by a negative number (or dividing a negative number by a positive number) produces a negative result.
    a. $-12 \div 4 = -3$
    b. $-12 / 4 = -3$
    c. $16 \div (-8) = -2$
    d. $16 / (-8) = -2$

SELF-QUIZ #1: Symbols and Notation: Arithmetic Operations

(Answers to this quiz can be found on page A-7.)

1. $4 \times 7 =$
2. $6 + 3 + 9 =$
3. $-6 - 3 =$
4. $-27 / 3 =$
5. $4(9) =$
6. $12 + (-5) =$
7. $16(-3) =$
8. $-24 / -3 =$
9. $75 \div 5 =$
10. $-7(-4) =$

A.3 Order of Operations

Equations and formulas often include a number of mathematical operations combining addition, subtraction, multiplication, and division. Some will also include exponents and square roots. In complex equations with more than one operation, it is important to perform the operations in a specific sequence. Deviating from this sequence can produce a wrong answer. Table A.2 lists the order of operations for quick reference.

TABLE A.2 Order of Operations

Rule of Operation	Example
1. Calculations within parentheses are completed first.	1a. $(6 + 2) - 4 \times 3 / 2^2 + 6 =$ 1b. $8 - 4 \times 3 / 2^2 + 6 =$
2. Squaring (or raising to another exponent) is completed second.	2a. $8 - 4 \times 3 / 2^2 + 6 =$ 2b. $8 - 4 \times 3 / 4 + 6 =$
3. From **left** to **right**, complete all multiplication and division operations. This may require multiple steps.	3a. $8 - 4 \times 3 / 4 + 6 =$ 3b. $8 - 12 / 4 + 6 =$ 3c. $8 - 12 / 4 + 6 =$ 3d. $8 - 3 + 6 =$
4. Last, complete all the addition and subtraction operations.	4a. $8 - 3 + 6 =$ 4b. $5 + 6 =$ 4c. $11 = 11$

ORDER OF OPERATIONS: Worked Examples

1.

$-3 + 6(4) - 7 =$	Multiplication
$-3 + 24 - 7 =$	Addition
$21 - 7 =$	Subtraction
$14 = 14$	Answer

2.

$2(8) + 6 / 3 \times 8 =$	Multiplication.
$16 + 6 / 3 \times 8 =$	Division
$16 + 2 \times 8 =$	Multiplication
$16 + 16 =$	Addition
$32 = 32$	Answer

3.

$3^2 + 6 / 3 - 12(2) =$	Square (raise exponent)
$9 + 6 / 3 - 12(2) =$	Division
$9 + 2 - 12(2) =$	Multiplication
$9 + 2 - 24 =$	Addition
$11 - 24 =$	Subtraction
$-13 = -13$	Answer

4.

$(10 + 6) - 6^2 / 4 + 3(10) =$	Within parentheses
$16 - 6^2 / 4 + 3(10) =$	Square (raise exponent)
$16 - 36 / 4 + 3(10) =$	Division
$16 - 9 + 3(10) =$	Multiplication
$16 - 9 + 30 =$	Subtraction
$7 + 30 =$	Addition
$37 = 37$	Answer

5.

$8 + (-4) + 3(12 - 8) =$	Within parentheses
$8 + (-4) + 3(4) =$	Multiplication
$8 + (-4) + 12 =$	Addition
$4 + 12 =$	Addition
$16 = 16$	Answer

SELF-QUIZ #2: Order of Operations

(Answers to this quiz can be found on page A-7.)

1. $3(7) - 12/3 + 2 =$
2. $4/2 + 6 - 2(3) =$
3. $-5(4) + 16 =$
4. $8 + (-16)/4 =$
5. $6 - 3 + 5 - 3(5) + 10 =$
6. $4^2/8 - 4(3) + (8 - 3) =$
7. $(14 - 6) + 72/9 + 4 =$
8. $(54 - 18)/4 + 7 \times 3 =$
9. $32 - 4(3 + 4) + 8 =$
10. $100 \times 3 - 87 =$

A.4 Proportions: Fractions, Decimals, and Percentages

A proportion is a part in relation to a whole. When we look at fractions, we understand the denominator (the bottom number) to be the number of equal parts that comprise the whole. The numerator represents the proportion of parts of that whole that are present. Fractions can be converted into decimals by dividing the numerator by the denominator. Decimals can then be converted into percentages by multiplying by 100 (Table A.3). It is important to use the percentage symbol (%) when differentiating decimals from percentages. Additionally, decimals are often rounded to the nearest hundredth before they are converted into a percentage.

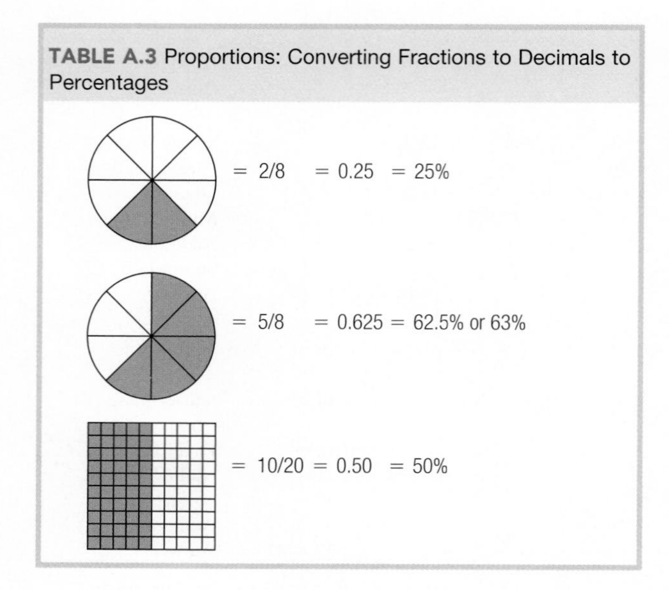

FRACTIONS

Equivalent Fractions

The same proportion can be expressed in a number of equivalent fractions. Equivalent fractions are found by multiplying both the numerator and the denominator by the same number.

$$^{1}/_{2} = {}^{2}/_{4} = {}^{6}/_{12} = {}^{30}/_{60}$$

In this case, we multiply each side of $^{1}/_{2}$ by 2 to reach the equivalent $^{2}/_{4}$, then by 3 to reach the equivalent $^{6}/_{12}$, then by 5 to reach the equivalent $^{30}/_{60}$. Or we could have multiplied the numerator and denominator of the original $^{1}/_{2}$ by 30 to reach our concluding $^{30}/_{60}$.

Fractions can also be reduced to a simpler form by dividing the numerator and denominator by the same number. Be sure to divide each by a number that will result in a whole number for both the numerator and the denominator.

$$^{25}/_{75} = {}^{5}/_{15} = {}^{1}/_{3}$$

By dividing each side by 5, the fraction was reduced from $^{25}/_{75}$ to $^{5}/_{15}$. By further dividing by 5, we reduce the fraction to its simplest form, $^{1}/_{3}$. Or we could have divided the numerator and denominator of the original $^{25}/_{75}$ by 25, resulting in the simplest expression of this fraction, $^{1}/_{3}$.

Adding and Subtracting Fractions (with the same denominator)

Finding equivalent fractions is essential to adding and subtracting two or more fractions. In order to add or subtract, each fraction must have the same denominator. If the two fractions already have the same denominator, add or subtract the numbers in the numerators only.

$$^{2}/_{7} + {}^{1}/_{7} = {}^{3}/_{7} \qquad {}^{4}/_{5} - {}^{3}/_{5} = {}^{1}/_{5}$$

In each of these instances, we are adding or subtracting from the same whole (or same pie, as in lines one and two of Table A.3). In the first equation, we are increasing our proportion of 2 by 1 to equal 3 pieces of the whole. In the second equation, we are reducing the number of proportions from 4 by 3 to equal just 1 piece of the whole.

Adding and Subtracting Fractions (with different denominators)

When adding or subtracting two proportions with different denominators, it is necessary to find a common denominator before performing the operation. It is often easiest to multiply each side (numerator and denominator) by the number equal to the denominator of the other fraction. This provides an easy route to finding a common denominator.

$$^{2}/_{5} + {}^{1}/_{6} =$$

Multiply the numerator and denominator of $^{2}/_{5}$ by 6, equaling $^{12}/_{30}$.

Multiply the numerator and denominator of $^{1}/_{6}$ by 5, equaling $^{5}/_{30}$.

$$^{12}/_{30} + {}^{5}/_{30} = {}^{17}/_{30}$$

Multiplying Fractions

When multiplying fractions, it is not necessary to find common denominators. Just multiply the two numerators in each fraction and the two denominators in each fraction.

$$^{4}/_{7} \times {}^{5}/_{8} = (4 \times 5)/(7 \times 8) = {}^{20}/_{56}$$

(*Note:* This fraction can be reduced to a simpler equivalent by dividing both the numerator and denominator by 4. The result is $^{5}/_{14}$.)

Dividing Fractions

When dividing a fraction by another fraction, invert the second fraction and multiply as above.

$$^{1}/_{3} \div {}^{2}/_{3} = {}^{1}/_{3} \times {}^{3}/_{2} = (1 \times 3)\ /\ (3 \times 2) = {}^{3}/_{6}$$

(*Note:* This can be reduced to a simpler equivalent by dividing both the numerator and denominator by 3. The result is one-half, $^{1}/_{2}$.)

SELF-QUIZ #3: Fractions

(Answers to this quiz can be found on page A-7.)

1. $^{2}/_{5} + {}^{1}/_{5} =$
2. $^{2}/_{7} \times {}^{4}/_{5} =$
3. $^{11}/_{15} - {}^{2}/_{5} =$
4. $^{3}/_{5} \div {}^{6}/_{8} =$
5. $^{3}/_{8} + {}^{1}/_{4} =$
6. $^{1}/_{8} \div {}^{4}/_{5} =$
7. $^{8}/_{9} - {}^{5}/_{9} + {}^{2}/_{9}$
8. $^{2}/_{7} + {}^{1}/_{3} =$
9. $^{4}/_{15} \times {}^{3}/_{5} =$
10. $^{6}/_{7} - {}^{3}/_{4} =$

DECIMALS

Converting Decimals to Fractions

Decimals represent proportions of a whole similar to fractions. Each decimal place represents a factor of 10. So the first decimal place represents a number over 10, the second decimal place represents a number over 100, the third decimal place represents a number over 1000, the fourth decimal place represents a number over 10,000, and so on.

To convert a decimal to a fraction, take the number as the numerator and place it over 10, 100, 1000, and so on based on how many numbers are to the right of the decimal point. For example:

$$0.6 = {}^{6}/_{10} \qquad 0.58 = {}^{58}/_{100}$$
$$0.926 = {}^{926}/_{1000} \qquad 0.7841 = {}^{7841}/_{10{,}000}$$

Adding and Subtracting Decimals

When adding or subtracting decimal points, it is necessary to keep the decimal points in a vertical line. Then add or subtract each vertical row as you normally would.

```
  3.83              4.4992
+1.358             −1.738
 5.188              2.7612
```

Multiplying Decimals

Multiplying decimals requires two basic steps. First, multiply the two decimals just as you would any numbers, paying no concern to where the decimal point is located. Once you have completed that operation, add the number of places to the right of the decimal in each number and count off that many decimal points in the solution line. That is your answer, which you may round up to three decimal places (two for the final answer).

```
    4.26 (two decimal places)         0.532 (three decimal places)
  ×0.398 (three decimal places)        ×0.8 (one decimal place)
    3408                              0.4256 (four decimal places)
   3834
  1278
 1.69548 (five decimal places)
```

Dividing Decimals

When dividing decimals, it is easiest to multiply each decimal by the factor of 10 associated with the number of places to the right of the decimal point. So, if one of the numbers has two numbers to the right of the decimal point and the other number has one, each number should be multiplied by 100. For example,

$$0.7 \div 1.32 = {}^{0.7}/_{1.32}$$

Then multiply each side by the factor of 10 associated with the most spaces to the right of the decimal point in either number. In this case, that is 2, so we multiply each side by 100.

$$0.7 \times 100 = 70$$
$$1.32 \times 100 = 132$$

The new fraction is ${}^{70}/_{132}$, which we can solve: $70 \div 132 = 0.530$

SELF-QUIZ #4: Decimals

(Answers to this quiz can be found on page A-7.)

1. $1.83 \times 0.68 =$
2. $2.637 + 4.2 =$
3. $1.894 - 0.62 =$
4. $0.35 \div 0.7 =$
5. $3.419 \times 0.12 =$
6. ${}^{0.82}/_{1.74} =$
7. $0.125 \div 0.625 =$
8. $0.44 \times 0.163 =$
9. $0.8 + 1.239 =$
10. $13.288 - 4.46 =$

PERCENTAGES

Converting Percentages to Fractions or Decimals

Convert a percentage into a fraction by removing the percentage symbol and placing the number over a denominator of 100.

$$82\% = {}^{82}/_{100} \text{ or } {}^{41}/_{50} \text{ or } 0.82$$
$$20\% = {}^{20}/_{100} \text{ or } {}^{1}/_{5} \text{ or } 0.2$$

Multiplying with Percentages

In statistics, it is often necessary to determine the percentage of a whole number when analyzing data. To multiply with a percentage, convert the percentage to a decimal (Table A.3) and solve the equation. To convert a percentage to a decimal, remove the percentage symbol and move the decimal point two places to the left.

$$80\% \text{ of } 45 = 80\% \times 45 = 0.80 \times 45 = 36$$
$$25\% \text{ of } 94 = 25\% \times 94 = 0.25 \times 94 = 23.5$$

SELF-QUIZ #5: Percentages

(Answers to this quiz can be found on page A-7.)

1. $45\% \times 100 =$
2. 22% of 80 =
3. 35% of 90 =
4. $80\% \times 23 =$
5. $58\% \times 60 =$
6. $32 \times 16\% =$
7. $125 \times 73\% =$
8. $24 \times 75\% =$
9. 69% of 224 =
10. $51\% \times 37 =$

A.5 Solving Equations with a Single Unknown Variable

When solving equations with an unknown variable, isolate the unknown variable on one side of the equation. By isolating the variable, you free up the other side of the equation so you can solve it to a single number, thus providing you with the value of the variable.

To isolate the variable, add, subtract, multiply, or divide each side of the equation to solve operations on the side of the equation that contains the variable (Table A.4).

TABLE A.4 Solving Equations with a Single Variable

Addition $X + 7 = 18$ $X + 7 - 7 = 18 - 7$ $X = 11$	Subtracting 7 from each side zeros the addition operation.
Subtraction $X - 13 = 27$ $X - 13 + 13 = 27 + 13$ $X = 40$	Adding 13 to each side zeros the subtraction operation.
Multiplication $X \times 5 = 20$ $X \times 5/5 = 20/5$ $X = 4$	Dividing each side by 5 zeros the multiplication operation.
Division $X/5 = 40$ $X/5 \times 5 = 40 \times 5$ $X = 200$	Multiplying each side by 5 zeros the division operation.
Multiple Operations $4X + 6 = 18$ $4X + 6 - 6 = 18 - 6$ $4X = 12$ $4X/4 = 12/4$ $X = 3$	When isolating a variable, work *backward* through the order of operations (see Table A.2). Isolate addition and subtraction operations first. Then isolate operations for multiplication and division.

SOLVING EQUATIONS WITH A SINGLE UNKNOWN VARIABLE: Worked Examples

1. $X + 12 = 42$
   $X + 12 - 12 = 42 - 12$
   $X = 30$

2. $X - 13 = -5$
   $X - 13 + 13 = -5 + 13$
   $X = 8$

3. $(X - 3)/6 = 2$
   $(X - 3)/6 \times 6 = 2 \times 6$
   $X - 3 = 12$
   $X - 3 + 3 = 12 + 3$
   $X = 15$

4. $(3X + 4)/2 = 8$
   $(3X + 4)/2 \times 2 = 8 \times 2$
   $3X + 4 = 16$
   $3X + 4 - 4 = 16 - 4$
   $3X = 12$
   $3X/3 = 12/3$
   $X = 4$

5. $(X - 2)/3 = 7$
   $(X - 2)/3 \times 3 = 7 \times 3$
   $X - 2 = 21$
   $X - 2 + 2 = 21 + 2$
   $X = 23$

SELF-QUIZ #6: Solving Equations with a Single Unknown Variable

(Answers to this quiz can be found on page A-7.)

1. $7X = 42$
   $X =$
2. $87 - X + 16 = 57$
   $X =$
3. $X - 17 = -6$
   $X =$
4. $5X - 4 = 21$
   $X =$
5. $X - 10 = -4$
   $X =$
6. $X / 8 = 20$
   $X =$
7. $(X + 17)/3 = 10$
   $X =$
8. $2(X + 4) = 24$
   $X =$
9. $X(3 + 12) - 20 = 40$
   $X =$
10. $34 - X/6 = 27$
    $X =$

A.6 Answers to Diagnostic Test and Self-Quizzes

Answers to Diagnostic Test

Section 1

1. 28; 2. −24; 3. 29; 4. 48; 5. −4; 6. 19; 7. 4; 8. 18; 9. −14; 10. 7

Section 2

1. 0; 2. 28; 3. −1; 4. 16; 5. 20; 6. 16; 7. −4; 8. 2; 9. 7; 10. 16

Section 3

1. $^{42}/_{100}$ or $^{21}/_{50}$; 2. 0.6; 3. 80%; 4. $^{10}/_{13}$; 5. 0.336; 6. 48; 7. $^{24}/_{35}$; 8. 32; 9. $^{3}/_{4}$; 10. $^{20}/_{63}$

Section 4

1. 4; 2. 5; 3. 24; 4. 6; 5. 6; 6. 12; 7. 7; 8. 12; 9. 6; 10. 2

Answers for Self-Quiz #1: Symbols and Notation

1. 28; 2. 18; 3. −9; 4. −9; 5. 36; 6. 7; 7. −48; 8. 8; 9. 15; 10. 28

Answers for Self-Quiz #2: Order of Operations

1. 19; 2. 2; 3. −4; 4. 4; 5. 3; 6. −5; 7. 20; 8. 30; 9. 12; 10. 213

Answers for Self-Quiz #3: Fractions

1. $^{3}/_{5}$; 2. $^{8}/_{35}$; 3. $^{1}/_{3}$ or $^{5}/_{15}$ or $^{25}/_{75}$; 4. $^{24}/_{30}$ or $^{4}/_{5}$; 5. $^{5}/_{8}$; 6. $^{5}/_{32}$; 7. $^{5}/_{9}$; 8. $^{13}/_{21}$; 9. $^{12}/_{75}$ or $^{4}/_{25}$; 10. $^{3}/_{28}$

Answers for Self-Quiz #4: Decimals

1. 1.244; 2. 6.837; 3. 1.274; 4. $^{35}/_{70}$ or $^{1}/_{2}$ or 0.5; 5. 0.41028; 6. $^{82}/_{174}$ or $^{41}/_{87}$ or 0.47; 7. $^{125}/_{625}$ or $^{1}/_{5}$ or 0.2; 8. 0.07172; 9. 2.039; 10. 8.828

Answers for Self-Quiz #5: Percentages

1. 45; 2. 17.6; 3. 31.5; 4. 18.4; 5. 34.8; 6. 5.12; 7. 91.25; 8. 18; 9. 154.56; 10. 18.87

Answers for Self-Quiz #6: Solving Equations with a Single Unknown Variable

1. 6; 2. 46; 3. 11; 4. 5; 5. 6; 6. 160; 7. 13; 8. 8; 9. 4; 10. 42

Statistical Tables

TABLE B.1 THE *z* DISTRIBUTION

Normal curve columns represent percentages between the mean and the z scores and percentages beyond the z scores in the tail.

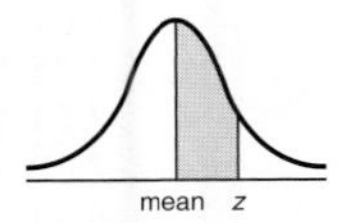

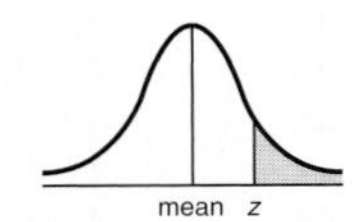

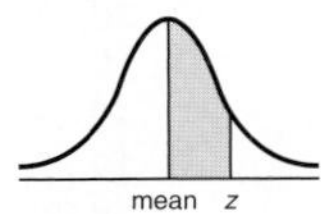

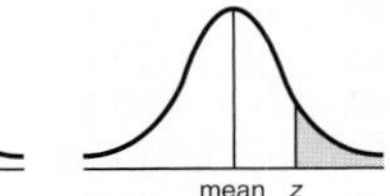

z	% MEAN TO *z*	% IN TAIL	*z*	% MEAN TO *z*	% IN TAIL
.00	0.00	50.00	.34	13.31	36.69
.01	0.40	49.60	.35	13.68	36.32
.02	0.80	49.20	.36	14.06	35.94
.03	1.20	48.80	.37	14.43	35.57
.04	1.60	48.40	.38	14.80	35.20
.05	1.99	48.01	.39	15.17	34.83
.06	2.39	47.61	.40	15.54	34.46
.07	2.79	47.21	.41	15.91	34.09
.08	3.19	46.81	.42	16.28	33.72
.09	3.59	46.41	.43	16.64	33.36
.10	3.98	46.02	.44	17.00	33.00
.11	4.38	45.62	.45	17.36	32.64
.12	4.78	45.22	.46	17.72	32.28
.13	5.17	44.83	.47	18.08	31.92
.14	5.57	44.43	.48	18.44	31.56
.15	5.96	44.04	.49	18.79	31.21
.16	6.36	43.64	.50	19.15	30.85
.17	6.75	43.25	.51	19.50	30.50
.18	7.14	42.86	.52	19.85	30.15
.19	7.53	42.47	.53	20.19	29.81
.20	7.93	42.07	.54	20.54	29.46
.21	8.32	41.68	.55	20.88	29.12
.22	8.71	41.29	.56	21.23	28.77
.23	9.10	40.90	.57	21.57	28.43
.24	9.48	40.52	.58	21.90	28.10
.25	9.87	40.13	.59	22.24	27.76
.26	10.26	39.74	.60	22.57	27.43
.27	10.64	39.36	.61	22.91	27.09
.28	11.03	38.97	.62	23.24	26.76
.29	11.41	38.59	.63	23.57	26.43
.30	11.79	38.21	.64	23.89	26.11
.31	12.17	37.83	.65	24.22	25.78
.32	12.55	37.45	.66	24.54	25.46
.33	12.93	37.07	.67	24.86	25.14

TABLE B.1 continued

z	% MEAN TO z	% IN TAIL	z	% MEAN TO z	% IN TAIL
.68	25.17	24.83	1.28	39.97	10.03
.69	25.49	24.51	1.29	40.15	9.85
.70	25.80	24.20	1.30	40.32	9.68
.71	26.11	23.89	1.31	40.49	9.51
.72	26.42	23.58	1.32	40.66	9.34
.73	26.73	23.27	1.33	40.82	9.18
.74	27.04	22.96	1.34	40.99	9.01
.75	27.34	22.66	1.35	41.15	8.85
.76	27.64	22.36	1.36	41.31	8.69
.77	27.94	22.06	1.37	41.47	8.53
.78	28.23	21.77	1.38	41.62	8.38
.79	28.52	21.48	1.39	41.77	8.23
.80	28.81	21.19	1.40	41.92	8.08
.81	29.10	20.90	1.41	42.07	7.93
.82	29.39	20.61	1.42	42.22	7.78
.83	29.67	20.33	1.43	42.36	7.64
.84	29.95	20.05	1.44	42.51	7.49
.85	30.23	19.77	1.45	42.65	7.35
.86	30.51	19.49	1.46	42.79	7.21
.87	30.78	19.22	1.47	42.92	7.08
.88	31.06	18.94	1.48	43.06	6.94
.89	31.33	18.67	1.49	43.19	6.81
.90	31.59	18.41	1.50	43.32	6.68
.91	31.86	18.14	1.51	43.45	6.55
.92	32.12	17.88	1.52	43.57	6.43
.93	32.38	17.62	1.53	43.70	6.30
.94	32.64	17.36	1.54	43.82	6.18
.95	32.89	17.11	1.55	43.94	6.06
.96	33.15	16.85	1.56	44.06	5.94
.97	33.40	16.60	1.57	44.18	5.82
.98	33.65	16.35	1.58	44.29	5.71
.99	33.89	16.11	1.59	44.41	5.59
1.00	34.13	15.87	1.60	44.52	5.48
1.01	34.38	15.62	1.61	44.63	5.37
1.02	34.61	15.39	1.62	44.74	5.26
1.03	34.85	15.15	1.63	44.84	5.16
1.04	35.08	14.92	1.64	44.95	5.05
1.05	35.31	14.69	1.65	45.05	4.95
1.06	35.54	14.46	1.66	45.15	4.85
1.07	35.77	14.23	1.67	45.25	4.75
1.08	35.99	14.01	1.68	45.35	4.65
1.09	36.21	13.79	1.69	45.45	4.55
1.10	36.43	13.57	1.70	45.54	4.46
1.11	36.65	13.35	1.71	45.64	4.36
1.12	36.86	13.14	1.72	45.73	4.27
1.13	37.08	12.92	1.73	45.82	4.18
1.14	37.29	12.71	1.74	45.91	4.09
1.15	37.49	12.51	1.75	45.99	4.01
1.16	37.70	12.30	1.76	46.08	3.92
1.17	37.90	12.10	1.77	46.16	3.84
1.18	38.10	11.90	1.78	46.25	3.75
1.19	38.30	11.70	1.79	46.33	3.67
1.20	38.49	11.51	1.80	46.41	3.59
1.21	38.69	11.31	1.81	46.49	3.51
1.22	38.88	11.12	1.82	46.56	3.44
1.23	39.07	10.93	1.83	46.64	3.36
1.24	39.25	10.75	1.84	46.71	3.29
1.25	39.44	10.56	1.85	46.78	3.22
1.26	39.62	10.38	1.86	46.86	3.14
1.27	39.80	10.20	1.87	46.93	3.07

TABLE B.1 continued

z	% MEAN TO z	% IN TAIL	z	% MEAN TO z	% IN TAIL
1.88	46.99	3.01	2.46	49.31	.69
1.89	47.06	2.94	2.47	49.32	.68
1.90	47.13	2.87	2.48	49.34	.66
1.91	47.19	2.81	2.49	49.36	.64
1.92	47.26	2.74	2.50	49.38	.62
1.93	47.32	2.68	2.51	49.40	.60
1.94	47.38	2.62	2.52	49.41	.59
1.95	47.44	2.56	2.53	49.43	.57
1.96	47.50	2.50	2.54	49.45	.55
1.97	47.56	2.44	2.55	49.46	.54
1.98	47.61	2.39	2.56	49.48	.52
1.99	47.67	2.33	2.57	49.49	.51
2.00	47.72	2.28	2.58	49.51	.49
2.01	47.78	2.22	2.59	49.52	.48
2.02	47.83	2.17	2.60	49.53	.47
2.03	47.88	2.12	2.61	49.55	.45
2.04	47.93	2.07	2.62	49.56	.44
2.05	47.98	2.02	2.63	49.57	.43
2.06	48.03	1.97	2.64	49.59	.41
2.07	48.08	1.92	2.65	49.60	.40
2.08	48.12	1.88	2.66	49.61	.39
2.09	48.17	1.83	2.67	49.62	.38
2.10	48.21	1.79	2.68	49.63	.37
2.11	48.26	1.74	2.69	49.64	.36
2.12	48.30	1.70	2.70	49.65	.35
2.13	48.34	1.66	2.71	49.66	.34
2.14	48.38	1.62	2.72	49.67	.33
2.15	48.42	1.58	2.73	49.68	.32
2.16	48.46	1.54	2.74	49.69	.31
2.17	48.50	1.50	2.75	49.70	.30
2.18	48.54	1.46	2.76	49.71	.29
2.19	48.57	1.43	2.77	49.72	.28
2.20	48.61	1.39	2.78	49.73	.27
2.21	48.64	1.36	2.79	49.74	.26
2.22	48.68	1.32	2.80	49.74	.26
2.23	48.71	1.29	2.81	49.75	.25
2.24	48.75	1.25	2.82	49.76	.24
2.25	48.78	1.22	2.83	49.77	.23
2.26	48.81	1.19	2.84	49.77	.23
2.27	48.84	1.16	2.85	49.78	.22
2.28	48.87	1.13	2.86	49.79	.21
2.29	48.90	1.10	2.87	49.79	.21
2.30	48.93	1.07	2.88	49.80	.20
2.31	48.96	1.04	2.89	49.81	.19
2.32	48.98	1.02	2.90	49.81	.19
2.33	49.01	.99	2.91	49.82	.18
2.34	49.04	.96	2.92	49.82	.18
2.35	49.06	.94	2.93	49.83	.17
2.36	49.09	.91	2.94	49.84	.16
2.37	49.11	.89	2.95	49.84	.16
2.38	49.13	.87	2.96	49.85	.15
2.39	49.16	.84	2.97	49.85	.15
2.40	49.18	.82	2.98	49.86	.14
2.41	49.20	.80	2.99	49.86	.14
2.42	49.22	.78	3.00	49.87	.13
2.43	49.25	.75	3.50	49.98	.02
2.44	49.27	.73	4.00	50.00	.00
2.45	49.29	.71	4.50	50.00	.00

TABLE B.2 THE *t* DISTRIBUTIONS

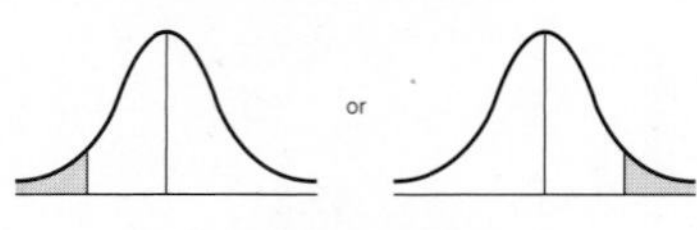

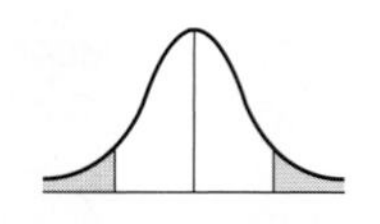

	One-Tailed Tests *p* level			Two-Tailed Tests *p* level		
df	.10	.05	.01	.10	.05	.01
1	3.078	6.314	31.821	6.314	12.706	63.657
2	1.886	2.920	6.965	2.920	4.303	9.925
3	1.638	2.353	4.541	2.353	3.182	5.841
4	1.533	2.132	3.747	2.132	2.776	4.604
5	1.476	2.015	3.365	2.015	2.571	4.032
6	1.440	1.943	3.143	1.943	2.447	3.708
7	1.415	1.895	2.998	1.895	2.365	3.500
8	1.397	1.860	2.897	1.860	2.306	3.356
9	1.383	1.833	2.822	1.833	2.262	3.250
10	1.372	1.813	2.764	1.813	2.228	3.170
11	1.364	1.796	2.718	1.796	2.201	3.106
12	1.356	1.783	2.681	1.783	2.179	3.055
13	1.350	1.771	2.651	1.771	2.161	3.013
14	1.345	1.762	2.625	1.762	2.145	2.977
15	1.341	1.753	2.603	1.753	2.132	2.947
16	1.337	1.746	2.584	1.746	2.120	2.921
17	1.334	1.740	2.567	1.740	2.110	2.898
18	1.331	1.734	2.553	1.734	2.101	2.879
19	1.328	1.729	2.540	1.729	2.093	2.861
20	1.326	1.725	2.528	1.725	2.086	2.846
21	1.323	1.721	2.518	1.721	2.080	2.832
22	1.321	1.717	2.509	1.717	2.074	2.819
23	1.320	1.714	2.500	1.714	2.069	2.808
24	1.318	1.711	2.492	1.711	2.064	2.797
25	1.317	1.708	2.485	1.708	2.060	2.788
26	1.315	1.706	2.479	1.706	2.056	2.779
27	1.314	1.704	2.473	1.704	2.052	2.771
28	1.313	1.701	2.467	1.701	2.049	2.764
29	1.312	1.699	2.462	1.699	2.045	2.757
30	1.311	1.698	2.458	1.698	2.043	2.750
35	1.306	1.690	2.438	1.690	2.030	2.724
40	1.303	1.684	2.424	1.684	2.021	2.705
60	1.296	1.671	2.390	1.671	2.001	2.661
80	1.292	1.664	2.374	1.664	1.990	2.639
100	1.290	1.660	2.364	1.660	1.984	2.626
120	1.289	1.658	2.358	1.658	1.980	2.617
∞	1.282	1.645	2.327	1.645	1.960	2.576

TABLE B.3 THE *F* DISTRIBUTIONS

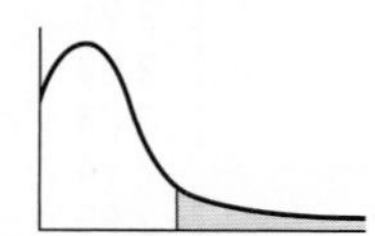

WITHIN-GROUPS *df*	SIGNIFICANCE (*p*) LEVEL	BETWEEN-GROUPS DEGREES OF FREEDOM 1	2	3	4	5	6
1	**.01**	**4,052**	**5,000**	**5,404**	**5,625**	**5,764**	**5,859**
	.05	162	200	216	225	230	234
	.10	*39.9*	*49.5*	*53.6*	*55.8*	*57.2*	*58.2*
2	**.01**	**98.50**	**99.00**	**99.17**	**99.25**	**99.30**	**99.33**
	.05	18.51	19.00	19.17	19.25	19.30	19.33
	.10	*8.53*	*9.00*	*9.16*	*9.24*	*9.29*	*9.33*

TABLE B.3 continued

WITHIN-GROUPS *df*	SIGNIFICANCE (*p*) LEVEL	BETWEEN-GROUPS DEGREES OF FREEDOM 1	2	3	4	5	6
3	**.01**	**34.12**	**30.82**	**29.46**	**28.71**	**28.24**	**27.91**
	.05	10.13	9.55	9.28	9.12	9.01	8.94
	.10	*5.54*	*5.46*	*5.39*	*5.34*	*5.31*	*5.28*
4	**.01**	**21.20**	**18.00**	**16.70**	**15.98**	**15.52**	**15.21**
	.05	7.71	6.95	6.59	6.39	6.26	6.16
	.10	*4.55*	*4.33*	*4.19*	*4.11*	*4.05*	*4.01*
5	**.01**	**16.26**	**13.27**	**12.06**	**11.39**	**10.97**	**10.67**
	.05	6.61	5.79	5.41	5.19	5.05	4.95
	.10	*4.06*	*3.78*	*3.62*	*3.52*	*3.45*	*3.41*
6	**.01**	**13.75**	**10.93**	**9.78**	**9.15**	**8.75**	**8.47**
	.05	5.99	5.14	4.76	4.53	4.39	4.28
	.10	*3.78*	*3.46*	*3.29*	*3.18*	*3.11*	*3.06*
7	**.01**	**12.25**	**9.55**	**8.45**	**7.85**	**7.46**	**7.19**
	.05	5.59	4.74	4.35	4.12	3.97	3.87
	.10	*3.59*	*3.26*	*3.08*	*2.96*	*2.88*	*2.83*
8	**.01**	**11.26**	**8.65**	**7.59**	**7.01**	**6.63**	**6.37**
	.05	5.32	4.46	4.07	3.84	3.69	3.58
	.10	*3.46*	*3.11*	*2.92*	*2.81*	*2.73*	*2.67*
9	**.01**	**10.56**	**8.02**	**6.99**	**6.42**	**6.06**	**5.80**
	.05	5.12	4.26	3.86	3.63	3.48	3.37
	.10	*3.36*	*3.01*	*2.81*	*2.69*	*2.61*	*2.55*
10	**.01**	**10.05**	**7.56**	**6.55**	**6.00**	**5.64**	**5.39**
	.05	4.97	4.10	3.71	3.48	3.33	3.22
	.10	*3.29*	*2.93*	*2.73*	*2.61*	*2.52*	*2.46*
11	**.01**	**9.65**	**7.21**	**6.22**	**5.67**	**5.32**	**5.07**
	.05	4.85	3.98	3.59	3.36	3.20	3.10
	.10	*3.23*	*2.86*	*2.66*	*2.54*	*2.45*	*2.39*
12	**.01**	**9.33**	**6.93**	**5.95**	**5.41**	**5.07**	**4.82**
	.05	4.75	3.89	3.49	3.26	3.11	3.00
	.10	*3.18*	*2.81*	*2.61*	*2.48*	*2.40*	*2.33*
13	**.01**	**9.07**	**6.70**	**5.74**	**5.21**	**4.86**	**4.62**
	.05	4.67	3.81	3.41	3.18	3.03	2.92
	.10	*3.14*	*2.76*	*2.56*	*2.43*	*2.35*	*2.28*
14	**.01**	**8.86**	**6.52**	**5.56**	**5.04**	**4.70**	**4.46**
	.05	4.60	3.74	3.34	3.11	2.96	2.85
	.10	*3.10*	*2.73*	*2.52*	*2.40*	*2.31*	*2.24*
15	**.01**	**8.68**	**6.36**	**5.42**	**4.89**	**4.56**	**4.32**
	.05	4.54	3.68	3.29	3.06	2.90	2.79
	.10	*3.07*	*2.70*	*2.49*	*2.36*	*2.27*	*2.21*
16	**.01**	**8.53**	**6.23**	**5.29**	**4.77**	**4.44**	**4.20**
	.05	4.49	3.63	3.24	3.01	2.85	2.74
	.10	*3.05*	*2.67*	*2.46*	*2.33*	*2.24*	*2.18*
17	**.01**	**8.40**	**6.11**	**5.19**	**4.67**	**4.34**	**4.10**
	.05	4.45	3.59	3.20	2.97	2.81	2.70
	.10	*3.03*	*2.65*	*2.44*	*2.31*	*2.22*	*2.15*
18	**.01**	**8.29**	**6.01**	**5.09**	**4.58**	**4.25**	**4.02**
	.05	4.41	3.56	3.16	2.93	2.77	2.66
	.10	*3.01*	*2.62*	*2.42*	*2.29*	*2.20*	*2.13*
19	**.01**	**8.19**	**5.93**	**5.01**	**4.50**	**4.17**	**3.94**
	.05	4.38	3.52	3.13	2.90	2.74	2.63
	.10	*2.99*	*2.61*	*2.40*	*2.27*	*2.18*	*2.11*

TABLE B.3 continued

WITHIN-GROUPS *df*	SIGNIFICANCE (*p*) LEVEL	BETWEEN-GROUPS DEGREES OF FREEDOM					
		1	2	3	4	5	6
20	**.01**	**8.10**	**5.85**	**4.94**	**4.43**	**4.10**	**3.87**
	.05	4.35	3.49	3.10	2.87	2.71	2.60
	.10	*2.98*	*2.59*	*2.38*	*2.25*	*2.16*	*2.09*
21	**.01**	**8.02**	**5.78**	**4.88**	**4.37**	**4.04**	**3.81**
	.05	4.33	3.47	3.07	2.84	2.69	2.57
	.10	*2.96*	*2.58*	*2.37*	*2.23*	*2.14*	*2.08*
22	**.01**	**7.95**	**5.72**	**4.82**	**4.31**	**3.99**	**3.76**
	.05	4.30	3.44	3.05	2.82	2.66	2.55
	.10	*2.95*	*2.56*	*2.35*	*2.22*	*2.13*	*2.06*
23	**.01**	**7.88**	**5.66**	**4.77**	**4.26**	**3.94**	**3.71**
	.05	4.28	3.42	3.03	2.80	2.64	2.53
	.10	*2.94*	*2.55*	*2.34*	*2.21*	*2.12*	*2.05*
24	**.01**	**7.82**	**5.61**	**4.72**	**4.22**	**3.90**	**3.67**
	.05	4.26	3.40	3.01	2.78	2.62	2.51
	.10	*2.93*	*2.54*	*2.33*	*2.20*	*2.10*	*2.04*
25	**.01**	**7.77**	**5.57**	**4.68**	**4.18**	**3.86**	**3.63**
	.05	4.24	3.39	2.99	2.76	2.60	2.49
	.10	*2.92*	*2.53*	*2.32*	*2.19*	*2.09*	*2.03*
26	**.01**	**7.72**	**5.53**	**4.64**	**4.14**	**3.82**	**3.59**
	.05	4.23	3.37	2.98	2.74	2.59	2.48
	.10	*2.91*	*2.52*	*2.31*	*2.18*	*2.08*	*2.01*
27	**.01**	**7.68**	**5.49**	**4.60**	**4.11**	**3.79**	**3.56**
	.05	4.21	3.36	2.96	2.73	2.57	2.46
	.10	*2.90*	*2.51*	*2.30*	*2.17*	*2.07*	*2.01*
28	**.01**	**7.64**	**5.45**	**4.57**	**4.08**	**3.75**	**3.53**
	.05	4.20	3.34	2.95	2.72	2.56	2.45
	.10	*2.89*	*2.50*	*2.29*	*2.16*	*2.07*	*2.00*
29	**.01**	**7.60**	**5.42**	**4.54**	**4.05**	**3.73**	**3.50**
	.05	4.18	3.33	2.94	2.70	2.55	2.43
	.10	*2.89*	*2.50*	*2.28*	*2.15*	*2.06*	*1.99*
30	**.01**	**7.56**	**5.39**	**4.51**	**4.02**	**3.70**	**3.47**
	.05	4.17	3.32	2.92	2.69	2.53	2.42
	.10	*2.88*	*2.49*	*2.28*	*2.14*	*2.05*	*1.98*
35	**.01**	**7.42**	**5.27**	**4.40**	**3.91**	**3.59**	**3.37**
	.05	4.12	3.27	2.88	2.64	2.49	2.37
	.10	*2.86*	*2.46*	*2.25*	*2.11*	*2.02*	*1.95*
40	**.01**	**7.32**	**5.18**	**4.31**	**3.83**	**3.51**	**3.29**
	.05	4.09	3.23	2.84	2.61	2.45	2.34
	.10	*2.84*	*2.44*	*2.23*	*2.09*	*2.00*	*1.93*
45	**.01**	**7.23**	**5.11**	**4.25**	**3.77**	**3.46**	**3.23**
	.05	4.06	3.21	2.81	2.58	2.42	2.31
	.10	*2.82*	*2.43*	*2.21*	*2.08*	*1.98*	*1.91*
50	**.01**	**7.17**	**5.06**	**4.20**	**3.72**	**3.41**	**3.19**
	.05	4.04	3.18	2.79	2.56	2.40	2.29
	.10	*2.81*	*2.41*	*2.20*	*2.06*	*1.97*	*1.90*
55	**.01**	**7.12**	**5.01**	**4.16**	**3.68**	**3.37**	**3.15**
	.05	4.02	3.17	2.77	2.54	2.38	2.27
	.10	*2.80*	*2.40*	*2.19*	*2.05*	*1.96*	*1.89*
60	**.01**	**7.08**	**4.98**	**4.13**	**3.65**	**3.34**	**3.12**
	.05	4.00	3.15	2.76	2.53	2.37	2.26
	.10	*2.79*	*2.39*	*2.18*	*2.04*	*1.95*	*1.88*

TABLE B.3 continued

WITHIN-GROUPS *df*	SIGNIFICANCE (*p*) LEVEL	BETWEEN-GROUPS DEGREES OF FREEDOM 1	2	3	4	5	6
65	**.01**	**7.04**	**4.95**	**4.10**	**3.62**	**3.31**	**3.09**
	.05	3.99	3.14	2.75	2.51	2.36	2.24
	.10	*2.79*	*2.39*	*2.17*	*2.03*	*1.94*	*1.87*
70	**.01**	**7.01**	**4.92**	**4.08**	**3.60**	**3.29**	**3.07**
	.05	3.98	3.13	2.74	2.50	2.35	2.23
	.10	*2.78*	*2.38*	*2.16*	*2.03*	*1.93*	*1.86*
75	**.01**	**6.99**	**4.90**	**4.06**	**3.58**	**3.27**	**3.05**
	.05	3.97	3.12	2.73	2.49	2.34	2.22
	.10	*2.77*	*2.38*	*2.16*	*2.02*	*1.93*	*1.86*
80	**.01**	**6.96**	**4.88**	**4.04**	**3.56**	**3.26**	**3.04**
	.05	3.96	3.11	2.72	2.49	2.33	2.22
	.10	*2.77*	*2.37*	*2.15*	*2.02*	*1.92*	*1.85*
85	**.01**	**6.94**	**4.86**	**4.02**	**3.55**	**3.24**	**3.02**
	.05	3.95	3.10	2.71	2.48	2.32	2.21
	.10	*2.77*	*2.37*	*2.15*	*2.01*	*1.92*	*1.85*
90	**.01**	**6.93**	**4.85**	**4.01**	**3.54**	**3.23**	**3.01**
	.05	3.95	3.10	2.71	2.47	2.32	2.20
	.10	*2.76*	*2.36*	*2.15*	*2.01*	*1.91*	*1.84*
95	**.01**	**6.91**	**4.84**	**4.00**	**3.52**	**3.22**	**3.00**
	.05	3.94	3.09	2.70	2.47	2.31	2.20
	.10	*2.76*	*2.36*	*2.14*	*2.01*	*1.91*	*1.84*
100	**.01**	**6.90**	**4.82**	**3.98**	**3.51**	**3.21**	**2.99**
	.05	3.94	3.09	2.70	2.46	2.31	2.19
	.10	*2.76*	*2.36*	*2.14*	*2.00*	*1.91*	*1.83*
200	**.01**	**6.76**	**4.71**	**3.88**	**3.41**	**3.11**	**2.89**
	.05	3.89	3.04	2.65	2.42	2.26	2.14
	.10	*273*	*2.33*	*2.11*	*1.97*	*1.88*	*1.80*
1000	**.01**	**6.66**	**4.63**	**3.80**	**3.34**	**3.04**	**2.82**
	.05	3.85	3.00	2.61	2.38	2.22	2.11
	.10	*2.71*	*2.31*	*2.09*	*1.95*	*1.85*	*1.78*
∞	**.01**	**6.64**	**4.61**	**3.78**	**3.32**	**3.02**	**2.80**
	.05	3.84	3.00	2.61	2.37	2.22	2.10
	.10	*2.71*	*2.30*	*2.08*	*1.95*	*1.85*	*1.78*

TABLE B.4 THE CHI-SQUARE DISTRIBUTIONS

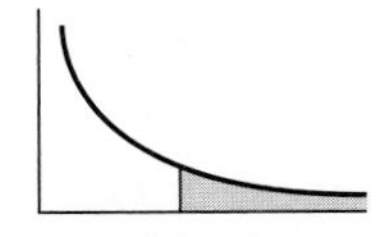

df	SIGNIFICANCE (*p*) LEVEL .10	.05	.01
1	2.706	3.841	6.635
2	4.605	5.992	9.211
3	6.252	7.815	11.345
4	7.780	9.488	13.277
5	9.237	11.071	15.087
6	10.645	12.592	16.812
7	12.017	14.067	18.475
8	13.362	15.507	20.090
9	14.684	16.919	21.666
10	15.987	18.307	23.209

TABLE B.5 THE q STATISTIC (TUKEY *HSD* TEST)

WITHIN-GROUPS *df*	SIGNIFICANCE (*p*) LEVEL	*k* = NUMBER OF TREATMENTS (LEVELS)										
		2	3	4	5	6	7	8	9	10	11	12
5	.05	3.64	4.60	5.22	5.67	6.03	6.33	6.58	6.80	6.99	7.17	7.32
	.01	**5.70**	**6.98**	**7.80**	**8.42**	**8.91**	**9.32**	**9.67**	**9.97**	**10.24**	**10.48**	**10.70**
6	.05	3.46	4.34	4.90	5.30	5.63	5.90	6.12	6.32	6.49	6.65	6.79
	.01	**5.24**	**6.33**	**7.03**	**7.56**	**7.97**	**8.32**	**8.61**	**8.87**	**9.10**	**9.30**	**9.48**
7	.05	3.34	4.16	4.68	5.06	5.36	5.61	5.82	6.00	6.16	6.30	6.43
	.01	**4.95**	**5.92**	**6.54**	**7.01**	**7.37**	**7.68**	**7.94**	**8.17**	**8.37**	**8.55**	**8.71**
8	.05	3.26	4.04	4.53	4.89	5.17	5.40	5.60	5.77	5.92	6.05	6.18
	.01	**4.75**	**5.64**	**6.20**	**6.62**	**6.96**	**7.24**	**7.47**	**7.68**	**7.86**	**8.03**	**8.18**
9	.05	3.20	3.95	4.41	4.76	5.02	5.24	5.43	5.59	5.74	5.87	5.98
	.01	**4.60**	**5.43**	**5.96**	**6.35**	**6.66**	**6.91**	**7.13**	**7.33**	**7.49**	**7.65**	**7.78**
10	.05	3.15	3.88	4.33	4.65	4.91	5.12	5.30	5.46	5.60	5.72	5.83
	.01	**4.48**	**5.27**	**5.77**	**6.14**	**6.43**	**6.67**	**6.87**	**7.05**	**7.21**	**7.36**	**7.49**
11	.05	3.11	3.82	4.26	4.57	4.82	5.03	5.20	5.35	5.49	5.61	5.71
	.01	**4.39**	**5.15**	**5.62**	**5.97**	**6.25**	**6.48**	**6.67**	**6.84**	**6.99**	**7.13**	**7.25**
12	.05	3.08	3.77	4.20	4.51	4.75	4.95	5.12	5.27	5.39	5.51	5.61
	.01	**4.32**	**5.05**	**5.50**	**5.84**	**6.10**	**6.32**	**6.51**	**6.67**	**6.81**	**6.94**	**7.06**
13	.05	3.06	3.73	4.15	4.45	4.69	4.88	5.05	5.19	5.32	5.43	5.53
	.01	**4.26**	**4.96**	**5.40**	**5.73**	**5.98**	**6.19**	**6.37**	**6.53**	**6.67**	**6.79**	**6.90**
14	.05	3.03	3.70	4.11	4.41	4.64	4.83	4.99	5.13	5.25	5.36	5.46
	.01	**4.21**	**4.89**	**5.32**	**5.63**	**5.88**	**6.08**	**6.26**	**6.41**	**6.54**	**6.66**	**6.77**
15	.05	3.01	3.67	4.08	4.37	4.59	4.78	4.94	5.08	5.20	5.31	5.40
	.01	**4.17**	**4.84**	**5.25**	**5.56**	**5.80**	**5.99**	**6.16**	**6.31**	**6.44**	**6.55**	**6.66**
16	.05	3.00	3.65	4.05	4.33	4.56	4.74	4.90	5.03	5.15	5.26	5.35
	.01	**4.13**	**4.79**	**5.19**	**5.49**	**5.72**	**5.92**	**6.08**	**6.22**	**6.35**	**6.46**	**6.56**
17	.05	2.98	3.63	4.02	4.30	4.52	4.70	4.86	4.99	5.11	5.21	5.31
	.01	**4.10**	**4.74**	**5.14**	**5.43**	**5.66**	**5.85**	**6.01**	**6.15**	**6.27**	**6.38**	**6.48**
18	.05	2.97	3.61	4.00	4.28	4.49	4.67	4.82	4.96	5.07	5.17	5.27
	.01	**4.07**	**4.70**	**5.09**	**5.38**	**5.60**	**5.79**	**5.94**	**6.08**	**6.20**	**6.31**	**6.41**
19	.05	2.96	3.59	3.98	4.25	4.47	4.65	4.79	4.92	5.04	5.14	5.23
	.01	**4.05**	**4.67**	**5.05**	**5.33**	**5.55**	**5.73**	**5.89**	**6.02**	**6.14**	**6.25**	**634**
20	.05	2.95	3.58	3.96	4.23	4.45	4.62	4.77	4.90	5.01	5.11	5.20
	.01	**4.02**	**4.64**	**5.02**	**5.29**	**5.51**	**5.69**	**5.84**	**5.97**	**6.09**	**6.19**	**6.28**
24	.05	2.92	3.53	3.90	4.17	4.37	4.54	4.68	4.81	4.92	5.01	5.10
	.01	**3.96**	**4.55**	**4.91**	**5.17**	**5.37**	**5.54**	**5.69**	**5.81**	**5.92**	**6.02**	**6.11**
30	.05	2.89	3.49	3.85	4.10	4.30	4.46	4.60	4.72	4.82	4.92	5.00
	.01	**3.89**	**4.45**	**4.80**	**5.05**	**5.24**	**5.40**	**5.54**	**5.65**	**5.76**	**5.85**	**5.93**

WITHIN-GROUPS *df*	SIGNIFICANCE (*p*) LEVEL	*k* = NUMBER OF TREATMENTS (LEVELS)										
		2	3	4	5	6	7	8	9	10	11	12
40	.05	2.86	3.44	3.79	4.04	4.23	4.39	4.52	4.63	4.73	4.82	4.90
	.01	**3.82**	**4.37**	**4.70**	**4.93**	**5.11**	**5.26**	**5.39**	**5.50**	**5.60**	**5.69**	**5.76**
60	.05	2.83	3.40	3.74	3.98	4.16	4.31	4.44	4.55	4.65	4.73	4.81
	.01	**3.76**	**4.28**	**4.59**	**4.82**	**4.99**	**5.13**	**5.25**	**5.36**	**5.45**	**5.53**	**5.60**
120	.05	2.80	3.36	3.68	3.92	4.10	4.24	4.36	4.47	4.56	4.64	4.71
	.01	**3.70**	**4.20**	**4.50**	**4.71**	**4.87**	**5.01**	**5.12**	**5.21**	**5.30**	**5.37**	**5.44**
∞	.05	2.77	3.31	3.63	3.86	4.03	4.17	4.28	4.39	4.47	4.55	4.62
	.01	**3.64**	**4.12**	**4.40**	**4.60**	**4.76**	**4.88**	**4.99**	**5.08**	**5.16**	**5.23**	**5.29**

TABLE B.6 THE PEARSON CORRELATION COEFFICIENT

To be significant, the sample correlation coefficient, *r*, must be greater than or equal to the critical value in the table.

	LEVEL OF SIGNIFICANCE FOR ONE-TAILED TEST *p* level			LEVEL OF SIGNIFICANCE FOR TWO-TAILED TEST *p* level	
df = *N* – 2	.005	.01	*df* = *N* – 2	.05	.01
1	.988	.9995	1	.997	.9999
2	.900	.980	2	.950	.990
3	.805	.934	3	.878	.959
4	.729	.882	4	.811	.917
5	.669	.833	5	.754	.874
6	.622	.789	6	.707	.834
7	.582	.750	7	.666	.798
8	.549	.716	8	.632	.765
9	.521	.685	9	.602	.735
10	.497	.658	10	.576	.708
11	.476	.634	11	.553	.684
12	.458	.612	12	.532	.661
13	.441	.592	13	.514	.641
14	.426	.574	14	.497	.623
15	.412	.558	15	.482	.606
16	.400	.542	16	.468	.590
17	.389	.528	17	.456	.575
18	.378	.516	18	.444	.561
19	.369	.503	19	.433	.549
20	.360	.492	20	.423	.537
21	.352	.482	21	.413	.526
22	.344	.472	22	.404	.515
23	.337	.462	23	.396	.505
24	.330	.453	24	.388	.496
25	.323	.445	25	.381	.487
26	.317	.437	26	.374	.479
27	.311	.430	27	.367	.471
28	.306	.423	28	.361	.463
29	.301	.416	29	.355	.456
30	.296	.409	30	.349	.449
35	.275	.381	35	.325	.418
40	.257	.358	40	.304	.393
45	.243	.338	45	.288	.372
50	.231	.322	50	.273	.354
60	.211	.295	60	.250	.325

TABLE B.6 continued

	LEVEL OF SIGNIFICANCE FOR ONE-TAILED TEST *p* level			LEVEL OF SIGNIFICANCE FOR TWO-TAILED TEST *p* level	
df = *N* – 2	.05	.01	*df* = *N* – 2	.05	.01
70	.195	.274	70	.232	.302
80	.183	.256	80	.217	.283
90	.173	.242	90	.205	.267
100	.164	.230	100	.195	.254

TABLE B.7 THE SPEARMAN CORRELATION COEFFICIENT

To be significant, the sample correlation coefficient, r_s, must be greater than or equal to the critical value in the table.

	LEVEL OF SIGNIFICANCE FOR ONE-TAILED TEST *p* level			LEVEL OF SIGNIFICANCE FOR TWO-TAILED TEST *p* level	
N	.05	.01	*N*	.05	.01
4	1.000	—	4	—	—
5	0.900	1.000	5	1.000	—
6	0.829	0.943	6	0.886	1.000
7	0.714	0.893	7	0.786	0.929
8	0.643	0.833	8	0.738	0.881
9	0.600	0.783	9	0.700	0.833
10	0.564	0.745	10	0.648	0.794
11	0.536	0.709	11	0.618	0.755
12	0.503	0.671	12	0.587	0.727
13	0.484	0.648	13	0.560	0.703
14	0.464	0.622	14	0.538	0.675
15	0.443	0.604	15	0.521	0.654
16	0.429	0.582	16	0.503	0.635
17	0.414	0.566	17	0.485	0.615
18	0.401	0.550	18	0.472	0.600
19	0.391	0.535	19	0.460	0.584
20	0.380	0.520	20	0.447	0.570
21	0.370	0.508	21	0.435	0.556
22	0.361	0.496	22	0.425	0.544
23	0.353	0.486	23	0.415	0.532
24	0.344	0.476	24	0.406	0.521
25	0.337	0.466	25	0.398	0.511
26	0.331	0.457	26	0.390	0.501
27	0.324	0.448	27	0.382	0.491
28	0.317	0.440	28	0.375	0.483
29	0.312	0.433	29	0.368	0.475
30	0.306	0.425	30	0.362	0.467
35	0.283	0.394	35	0.335	0.433
40	0.264	0.368	40	0.313	0.405
45	0.248	0.347	45	0.294	0.382
50	0.235	0.329	50	0.279	0.363
60	0.214	0.300	60	0.255	0.331
70	0.190	0.278	70	0.235	0.307
80	0.185	0.260	80	0.220	0.287
90	0.174	0.245	90	0.207	0.271
100	0.165	0.233	100	0.197	0.257

TABLE B.8A MANN-WHITNEY *U* FOR A *p* LEVEL OF .05 FOR A ONE-TAILED TEST

To be statistically significant, the smaller *U* must be equal to or less than the value in the table.

N_A/N_B	1	2	3	4	5	6	7	8	9	10	11	12	13	14	15	16	17	18	19	20
1	—	—	—	—	—	—	—	—	—	—	—	—	—	—	—	—	—	—	0	0
2	—	—	—	—	0	0	0	1	1	1	1	2	2	2	3	3	3	4	4	4
3	—	—	0	0	1	2	2	3	3	4	5	5	6	7	7	8	9	9	10	11
4	—	—	0	1	2	3	4	5	6	7	8	9	10	11	12	14	15	16	17	18
5	—	0	1	2	4	5	6	8	9	11	12	13	15	16	18	19	20	22	23	25
6	—	0	2	3	5	7	8	10	12	14	16	17	19	21	23	25	26	28	30	32
7	—	0	2	4	6	8	11	13	15	17	19	21	24	26	28	30	33	35	37	39
8	—	1	3	5	8	10	13	15	18	20	23	26	28	31	33	36	39	41	44	47
9	—	1	3	6	9	12	15	18	21	24	27	30	33	36	39	42	45	48	51	54
10	—	1	4	7	11	14	17	20	24	27	31	34	37	41	44	48	51	55	58	62
11	—	1	5	8	12	16	19	23	27	31	34	38	42	46	50	54	57	61	65	69
12	—	2	5	9	13	17	21	26	30	34	38	42	47	51	55	60	64	68	72	77
13	—	2	6	10	15	19	24	28	33	37	42	47	51	56	61	65	79	75	80	84
14	—	2	7	11	16	21	26	31	36	41	46	51	56	61	66	71	77	82	87	92
15	—	3	7	12	18	23	28	33	39	44	50	55	61	66	72	77	83	88	94	100
16	—	3	8	14	19	25	30	36	42	48	54	60	65	71	77	83	89	95	101	107
17	—	3	9	15	20	26	33	39	45	51	57	64	70	77	83	89	96	102	109	115
18	—	4	9	16	22	28	35	41	48	55	61	68	75	82	88	95	102	109	116	123
19	0	4	10	17	23	30	37	44	51	58	65	72	80	87	94	101	109	116	123	130
20	0	4	11	18	25	32	39	47	54	62	69	77	84	92	100	107	115	123	130	138

TABLE B.8B MANN-WHITNEY ***U*** FOR A *p* LEVEL OF .05 FOR A TWO-TAILED TEST

To be statistically significant, the smaller *U* must be equal to or less than the value in the table.

N_A/N_B	1	2	3	4	5	6	7	8	9	10	11	12	13	14	15	16	17	18	19	20
1	—	—	—	—	—	—	—	—	—	—	—	—	—	—	—	—	—	—	—	—
2	—	—	—	—	—	—	—	0	0	0	0	1	1	1	1	1	2	2	2	2
3	—	—	—	—	0	1	1	2	2	3	3	4	4	5	5	6	6	7	7	8
4	—	—	—	0	1	2	3	4	4	5	6	7	8	9	10	11	11	12	13	13
5	—	—	0	1	2	3	5	6	7	8	9	11	12	13	14	15	17	18	19	20
6	—	—	1	2	3	5	6	8	10	11	13	14	16	17	19	21	22	24	25	27
7	—	—	1	3	5	6	8	10	12	14	16	18	20	22	24	26	28	30	32	34
8	—	0	2	4	6	8	10	13	15	17	19	22	24	26	29	31	34	36	38	41
9	—	0	2	4	7	10	12	15	17	20	23	26	28	31	34	37	39	42	45	48
10	—	0	3	5	8	11	14	17	20	23	26	29	33	36	39	42	45	48	52	55
11	—	0	3	6	9	13	16	19	23	26	30	33	37	40	44	47	51	55	58	62
12	—	1	4	7	11	14	18	22	26	29	33	37	41	45	49	53	57	61	65	69
13	—	1	4	8	12	16	20	24	28	33	37	41	45	50	54	59	63	67	72	76
14	—	1	5	9	13	17	22	26	31	36	40	45	50	55	59	64	67	74	78	83
15	—	1	5	10	14	19	24	29	34	39	44	49	54	59	64	70	75	80	85	90
16	—	1	6	11	15	21	26	31	37	42	47	53	59	64	70	75	81	86	92	98
17	—	2	6	11	17	22	28	34	39	45	51	57	63	67	75	81	87	93	99	105
18	—	2	7	12	18	24	30	36	42	48	55	61	67	74	80	86	93	99	106	112
19	—	2	7	13	19	25	32	38	45	52	58	65	72	78	85	92	99	106	113	119
20	—	2	8	13	20	27	34	41	48	55	62	69	76	83	90	98	105	112	119	127

TABLE B.9 WILCOXON SIGNED-RANKS TEST FOR MATCHED PAIRS (*T*)

	LEVEL OF SIGNIFICANCE (*p* LEVEL) FOR ONE-TAILED TEST			LEVEL OF SIGNIFICANCE (*p* LEVEL) FOR TWO-TAILED TEST	
N	.05	.01	*N*	.05	.01
5	0	—	5	—	—
6	2	—	6	0	—
7	3	0	7	2	—
8	5	1	8	3	0
9	8	3	9	5	1
10	10	5	10	8	3
11	13	7	11	10	5
12	17	9	12	13	7
13	21	12	13	17	9
14	25	15	14	21	12
15	30	19	15	25	15
16	35	23	16	29	19
17	41	27	17	34	23
18	47	32	18	40	27
19	53	37	19	46	32
20	60	43	20	52	37
21	67	49	21	58	42
22	75	55	22	65	48
23	83	62	23	73	54
24	91	69	24	81	61
25	100	76	25	89	68
26	110	84	26	98	75
27	119	92	27	107	83
28	130	101	28	116	91
29	140	110	29	126	100
30	151	120	30	137	109
31	163	130	31	147	118
32	175	140	32	159	128
33	187	151	33	170	138
34	200	162	34	182	148
35	213	173	35	195	159
36	227	185	36	208	171
37	241	198	37	221	182
38	256	211	38	235	194
39	271	224	39	249	207
40	286	238	40	264	220
41	302	252	41	279	233
42	319	266	42	294	247
43	336	281	43	310	261
44	353	296	44	327	276
45	371	312	45	343	291
46	389	328	46	361	307
47	407	345	47	378	322
48	426	362	48	396	339
49	446	379	49	415	355
50	466	397	50	434	373

TABLE B.10 RANDOM DIGITS

19223	95034	05756	28713	96409	12531	42544	82853
73676	47150	99400	01927	27754	42648	82425	36290
45467	71709	77558	00095	32863	29485	82226	90056
52711	38889	93074	60227	40011	85848	48767	52573
95592	94007	69971	91481	60779	53791	17297	59335
68417	35013	15529	72765	85089	57067	50211	47487
82739	57890	20807	47511	81676	55300	94383	14893
60940	72024	17868	24943	61790	90656	87964	18883
36009	19365	15412	39638	85453	46816	83485	41979
38448	48789	18338	24697	39364	42006	76688	08708
81486	69487	60513	09297	00412	71238	27649	39950
59636	88804	04634	71197	19352	73089	84898	45785
62568	70206	40325	03699	71080	22553	11486	11776
45149	32992	75730	66280	03819	56202	02938	70915
61041	77684	94322	24709	73698	14526	31893	32592
14459	26056	31424	80371	65103	62253	50490	61181
38167	98532	62183	70632	23417	26185	41448	75532
73190	32533	04470	29669	84407	90785	65956	86382
95857	07118	87664	92099	58806	66979	98624	84826
35476	55972	39421	65850	04266	35435	43742	11937
71487	09984	29077	14863	61683	47052	62224	51025
13873	81598	95052	90908	73592	75186	87136	95761
54580	81507	27102	56027	55892	33063	41842	81868
71035	09001	43367	49497	72719	96758	27611	91596
96746	12149	37823	71868	18442	35119	62103	39244
96927	19931	36809	74192	77567	88741	48409	41903
43909	99477	25330	64359	40085	16925	85117	36071
15689	14227	06565	14374	13352	49367	81982	87209
36759	58984	68288	22913	18638	54303	00795	08727
69051	64817	87174	09517	84534	06489	87201	97245
05007	16632	81194	14873	04197	85576	45195	96565
68732	55259	84292	08796	43165	93739	31685	97150
45740	41807	65561	33302	07051	93623	18132	09547
27816	78416	18329	21337	35213	37741	04312	68508
66925	55658	39100	78458	11206	19876	87151	31260
08421	44753	77377	28744	75592	08563	79140	92454
53645	66812	61421	47836	12609	15373	98481	14592
66831	68908	40772	21558	47781	33586	79177	06928
55588	99404	70708	41098	43563	56934	48394	51719
12975	13258	13048	45144	72321	81940	00360	02428
96767	35964	23822	96012	94591	65194	50842	53372
72829	50232	97892	63408	77919	44575	24870	04178
88565	42628	17797	49376	61762	16953	88604	12724
62964	88145	83083	69453	46109	59505	69680	00900
19687	12633	57857	95806	09931	02150	43163	58636
37609	59057	66967	83401	60705	02384	90597	93600
54973	86278	88737	74351	47500	84552	19909	67181
00694	05977	19664	65441	20903	62371	22725	53340
71546	05233	53946	68743	72460	27601	45403	88692
07511	88915	41267	16853	84569	79367	32337	03316

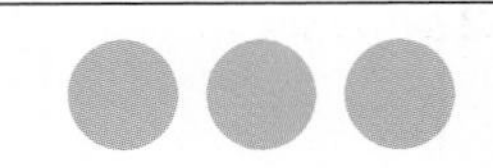

Solutions to Odd-Numbered End-of-Chapter Problems

CHAPTER 1

1.1 Descriptive statistics organize, summarize, and communicate a group of numerical observations. Inferential statistics use sample data to make general estimates about the larger population.

1.3 The four types of variables are nominal, ordinal, interval, and ratio. A nominal variable is used for observations that have categories, or names, as their values. An ordinal variable is used for observations that have rankings (i.e., 1st, 2nd, 3rd . . .) as their values. An interval variable has numbers as its values; the distance (or interval) between pairs of consecutive numbers is assumed to be equal. Finally, a ratio variable meets the criteria for interval variables but also has a meaningful zero point. Interval and ratio variables are both often referred to as scale variables.

1.5 Discrete variables can only be represented by specific numbers, usually whole numbers; continuous variables can take on any values, including those with great decimal precision (e.g., 1.597).

1.7 A confounding variable (also called a confound) is any variable that systematically varies with the independent variable so that we cannot logically determine which variable affects the dependent variable. Researchers attempt to control confounding variables in experiments by randomly assigning participants to conditions. The hope with random assignment is that the confounding variable will be spread equally across the different conditions of the study, thus neutralizing its effects.

1.9 An operational definition specifies the operations or procedures used to measure or manipulate an independent or dependent variable.

1.11 When conducting experiments, the researcher randomly assigns participants to conditions or levels of the independent variable. When random assignment is not possible, such as when studying something like gender or marital status, correlational research is used. Correlational research allows us to examine how variables are related to each other; experimental research allows us to make assertions about how an independent variable causes an effect in a dependent variable.

1.13
- **a.** "This was an experiment" (not "This was a correlational study")
- **b.** ". . . the independent variable of caffeine . . ." (not ". . . the dependent variable of caffeine. . . .")
- **c.** "A university assessed the validity . . ." (not "A university assessed the reliability . . .")
- **d.** "In a between-groups experiment . . ." (not "In a within-groups experiment . . .")

1.15 The sample is the 100 customers who completed the survey. The population is all of the customers at the grocery store.

1.17
- **a.** 73 people
- **b.** All people who shop in grocery stores similar to the one where data were collected
- **c.** Inferential statistic
- **d.** Answers may vary, but people could be labeled as having a "healthy diet" or an "unhealthy diet."
- **e.** Answers may vary, but there could be groupings such as "no items," "a minimal number of items," "some items," and "many items."
- **f.** Answers may vary, but the number of items could be counted or weighed.

1.19
- **a.** The independent variables are physical distance and emotional distance. The dependent variable is accuracy of memory.
- **b.** There are two levels of physical distance (within 100 miles and 100 miles or farther) and three levels of emotional distance (knowing no one who was affected, knowing people who were affected but lived, and knowing someone who died).
- **c.** Answers may vary, but accuracy of memory could be operationalized as the number of facts correctly recalled.

1.21
- **a.** The average weight for a 10-year-old girl was 77.4 pounds in 1963 and nearly 88 pounds in 2002.
- **b.** No; the CDC would not be able to weigh every single girl in the United States because it would be too expensive and time consuming.
- **c.** It is a descriptive statistic because it is a numerical summary of a sample. It is an inferential statistic because the researchers drew conclusions about the population's average weight based on this information from a sample.

1.23
- **a.** Ordinal
- **b.** Scale
- **c.** Nominal

1.25
- **a.** Discrete
- **b.** Continuous

c. Discrete

d. Discrete

e. Continuous

1.27 a. The independent variables are temperature and rainfall. Both are continuous, scale variables.

b. The dependent variable is experts' ratings. This is a discrete, scale variable.

c. The researchers wanted to know if the wine experts are consistent in their ratings—that is, if they're reliable.

d. This observation would suggest that Robert Parker's judgments are valid. His ratings seem to be measuring what they intend to measure—wine quality.

1.29 a. Age: teenagers and adults in their 30s; video game performance: final score on a video game or average reaction time on a video game task

b. Spanking: spanking and not spanking; violent behavior: parental measure of child aggression or number of aggressive acts observed in an hour of play

c. Meetings: go to meetings and participate online; weight loss: measured in pounds or kilograms, or by change in waist size

d. Studying: with others and alone; statistics performance: average test score for the semester or overall grade for the semester

e. Beverage: caffeinated and decaffeinated; time to fall asleep: minutes to fall asleep from when the participant goes to bed, or the actual time at which the participant falls asleep

1.31 a. An experiment requires random assignment to conditions. It would not be ethical to randomly assign some people to smoke and some people not to smoke, so this research had to be correlational.

b. Other unhealthy behaviors have been associated with smoking, such as poor diet and infrequent exercise. These other unhealthy behaviors might be confounded with smoking.

c. The tobacco industry could claim it was not the smoking that was harming people, but rather the other activities in which smokers tend to engage or fail to engage.

d. You could randomly assign people to either a smoking group or a nonsmoking group, and assess their health over time.

1.33 a. This is experimental because students are randomly assigned to one of the incentive conditions for recycling.

b. Answers may vary, but one hypothesis could be "Students fined for not recycling will report lower concerns for the environment, on average, than those rewarded for recycling."

1.35 a. Researchers could have randomly assigned some people who are HIV-positive to take the oral vaccine and other people who are HIV-positive not to take the oral vaccine. The second group would likely take a placebo.

b. This would have been a between-groups experiment because the people who are HIV-positive would have been in only one group: either vaccine or no vaccine.

c. This limits the researchers' ability to draw causal conclusions because the participants who received the vaccine may have been different in some way from those who did not receive the vaccine. There may have been a confounding variable that led to these findings. For example, those who received the vaccine might have had better access to health care and better sanitary conditions to begin with, making them less likely to contract cholera regardless of the vaccine's effectiveness.

d. The researchers might not have used random assignment because it would have meant recruiting participants, likely immunizing half, then following up with all of them. The researchers likely did not want to deny the vaccine to people who were HIV-positive because they might have contracted cholera and died without it.

e. We could have recruited a sample of people who were HIV-positive. Half would have been randomly assigned to take the oral vaccine; half would have been randomly assigned to take something that appeared to be an oral vaccine but did not have the active ingredient. They would have been followed to determine whether they developed cholera.

CHAPTER 2

2.1 Raw scores are the original data, to which nothing has been done.

2.3 A frequency table is a visual depiction of data that shows how often each value occurred; that is, it shows how many scores are at each value. Values are listed in one column, and the numbers of individuals with scores at that value are listed in the second column. A grouped frequency table is a visual depiction of data that reports the frequency within each given interval, rather than the frequency for each specific value.

2.5 Bar graphs typically provide scores for nominal data, whereas histograms typically provide frequencies for scale data. Also, the categories in bar graphs do not need to be arranged in a particular order and the bars should not touch, whereas the intervals in histograms are arranged in a meaningful order (lowest to highest) and the bars should touch each other.

2.7 A histogram looks like a bar graph but is usually used to depict scale data, with the values (or midpoints of intervals) of the variable on the x-axis and the frequencies on the y-axis. A frequency polygon is a line graph, with the x-axis representing values (or midpoints of intervals) and the y-axis representing frequencies; a dot is placed at the frequency for each value (or midpoint), and the points are connected.

2.9 In everyday conversation, you might use the word *distribution* in a number of different contexts, from the distribution of food to a marketing distribution. A statistician would use *distribution* only to describe the way that a set of scores, such as a set of grades, is distributed. A statistician is looking at the overall pattern of the data—what the shape is, where the data tend to cluster, and how they trail off.

2.11 With positively skewed data, the distribution's tail extends to the right, in a positive direction, and with negatively skewed data, the distribution's tail extends to the left, in a negative direction.

2.13 A ceiling effect occurs when there are no scores above a certain value; a ceiling effect leads to a negatively skewed distribution because the upper part of the distribution is constrained.

2.15 17.95% and 40.67%

2.17 0.10% and 96.77%

2.19 0.04, 198.22, and 17.89

2.21 The full range of data is 68 minus 2, plus 1, or 67. The range (67) divided by the desired seven intervals gives us an interval size of 9.57, or 10 when rounded. The seven intervals are: 0–9, 10–19, 20–29, 30–39, 40–49, 50–59, and 60–69.

2.23 25 shows

2.25 Serial killers would create positive skew, adding high numbers of murders to the data that are clustered around 1.

2.27 **a.** For the college population, the range of ages extends farther to the right (with a larger number of years) than to the left, creating positive skew.

b. The fact that youthful prodigies have limited access to college creates a sort of floor effect that makes low scores less possible.

2.29 **a.**

PERCENTAGE	FREQUENCY	PERCENTAGE
10	1	5.26
9	0	0.00
8	0	0.00
7	0	0.00
6	0	0.00
5	2	10.53
4	2	10.53
3	4	21.05
2	4	21.05
1	5	26.32
0	1	5.26

b. 10.53% of these schools had exactly 4% of their students report that they wrote between 5 and 10 twenty-page papers that year.

c. This is not a random sample. It includes schools that chose to participate in this survey and opted to have their results made public.

d.

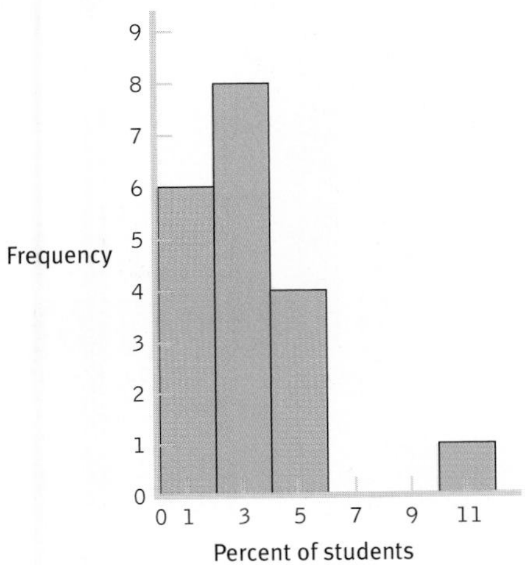

e. One

f. The data are clustered around 1% to 4%, with a high outlier, 10%.

2.31 **a.** The variable of alumni giving was operationalized by the percentage of alumni who donated to a given school. There are several other ways it could be operationalized. For example, the data might consist of the total dollar amount or the mean dollar amount that each school received.

b.

INTERVAL	FREQUENCY
60–69	1
50–59	0
40–49	6
30–39	15
20–29	21
10–19	24
0–9	3

c. There are many possible answers to this question. For example, we might ask whether sports team success predicts alumni giving or whether the prestige of the institution is a factor (the higher the ranking, the more alumni who donate).

d.

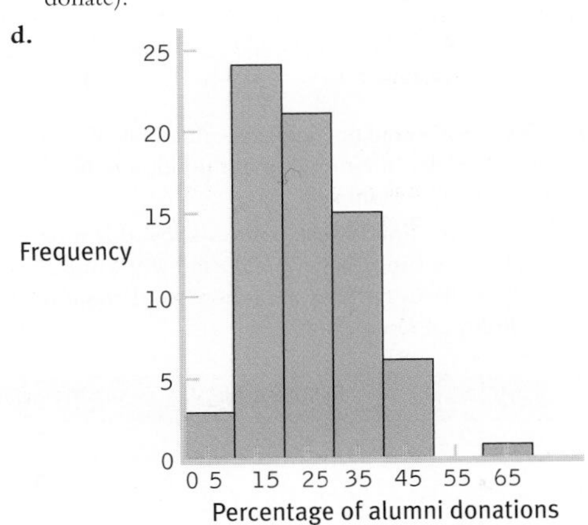

e.

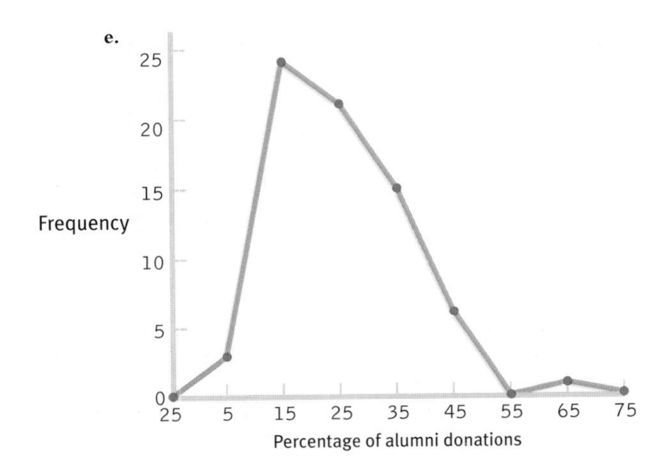

f. There is one unusual score—61. The distribution appears to be positively skewed. The center of the distribution seems to be in the 10–29 range.

2.33 a. Extroversion scores are most likely to have a normal distribution. Most people would fall toward the middle, with some people having higher levels and some having lower levels.

b. The distribution of finishing times for a marathon is likely to be positively skewed. The floor is the fastest possible time, a little over 2 hours; however, some runners take as long as 6 hours or more. Unfortunately for the very, very slow but unbelievably dedicated runners, many marathons shut down the finish line 6 hours after the start of the race.

c. The distribution of numbers of meals eaten in a dining hall in a semester on a three-meal-a-day plan is likely to be negatively skewed. The ceiling is three times per day, multiplied by the number of days; most people who choose to pay for the full plan would eat many of these meals. A few would hardly ever eat in the dining hall, pulling the tail in a negative direction.

2.35 a.

INTERVAL	FREQUENCY
300–339	4
260–299	7
220–259	9
180–219	3

b. This is not a random sample because only résumés from those applying for a receptionist position in his office were included in the sample.

c. This information lets the trainees know that most of these résumés contained between 220 and 299 words. This analysis tells us nothing about how word count might relate to quality of résumé.

2.37 a.

MONTHS	FREQUENCY	PERCENTAGE
12	1	5
11	0	0
10	1	5
9	1	5
8	0	0
7	1	5
6	1	5
5	0	0
4	1	5
3	4	20
2	2	10
1	3	15
0	5	25

b.

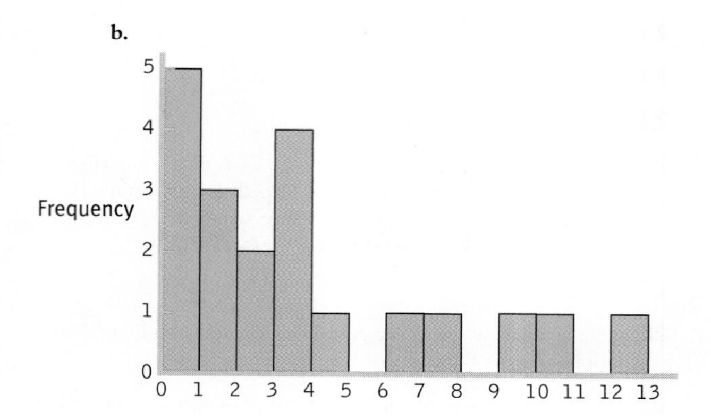

c.

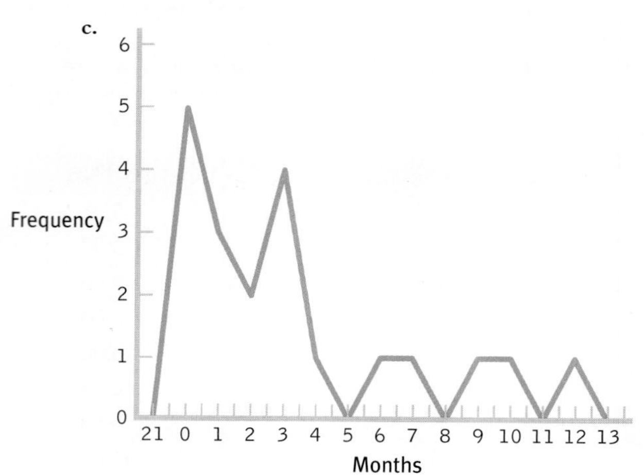

d.

INTERVAL	FREQUENCY
10–14 months	2
5–9 months	3
0–4 months	15

e.

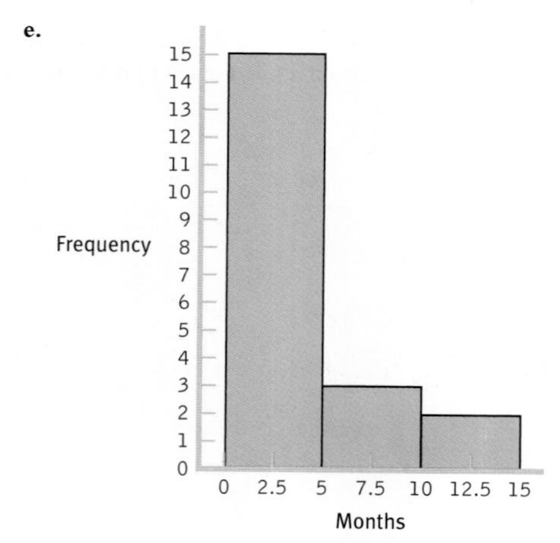

f.

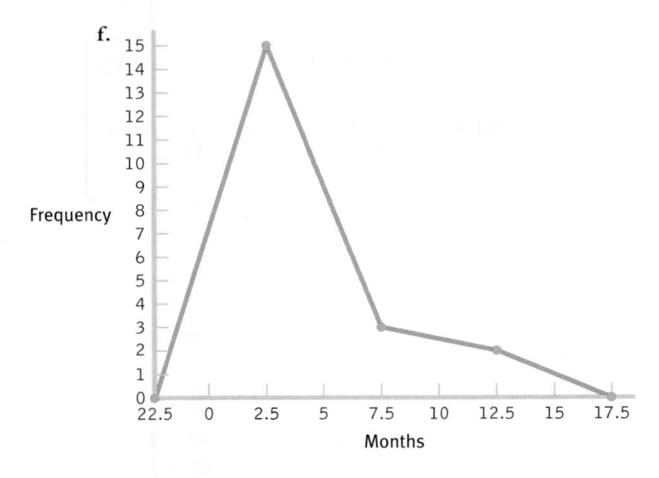

g. These data are centered around the 3-month period, with positive skew extending the data out to the 12-month period.

h. The bulk of the data would need to be shifted from the 3-month period to approximately 12 months, so that group of women might be the focus of attention. Perhaps early contact at the hospital and at follow-up visits after birth would help encourage mothers to breast-feed, and to breast-feed longer. One could also consider studying the women who create the positive skew to learn what unique characteristics or knowledge they have that influenced their behavior.

2.39

FORMER STUDENTS NOW IN TOP JOBS	FREQUENCY	PERCENTAGE
13	1	1.85
12	0	0.00
11	0	0.00
10	0	0.00
9	1	1.85
8	3	5.56
7	4	7.41
6	5	9.26
5	9	16.67
4	8	14.81
3	23	42.59

a.

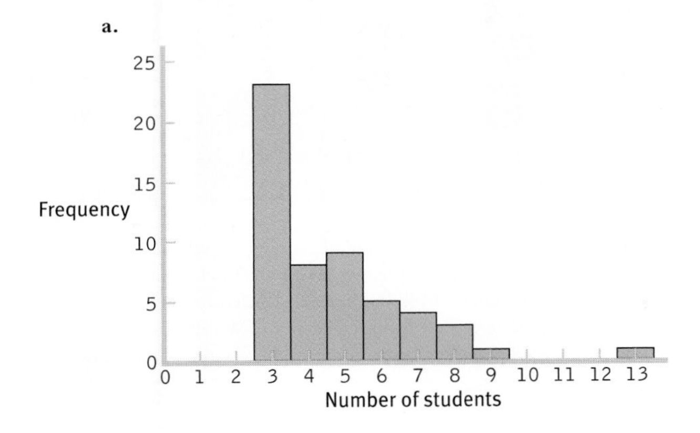

b.

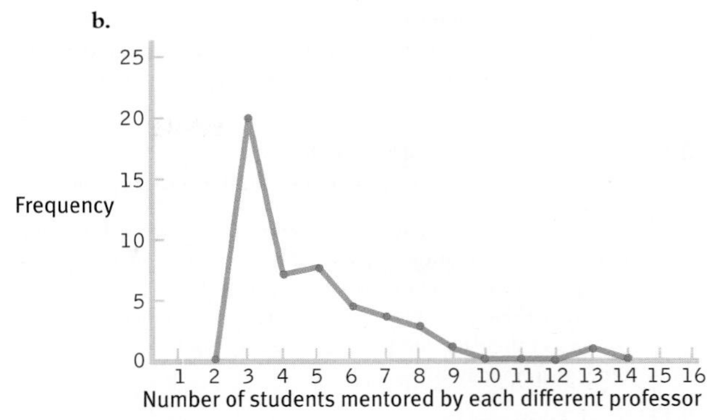

c. This distribution is positively skewed.

d. The researchers operationalized the variable of mentoring success as numbers of students placed into top professorial positions. There are many other ways this variable could have been operationalized. For example, the researchers might have counted numbers of student publications while in graduate school or might have asked graduates to rate their satisfaction with their graduate mentoring experiences.

e. The students might have attained their professor positions because of the prestige of their advisor, not because of his mentoring.

f. There are many possible answers to this question. For example, the attainment of a top professor position might be predicted by the prestige of the institution, the number of publications while in graduate school, or the graduate student's academic ability.

CHAPTER 3

3.1 The biased scale lie, the sneaky sample lie, the interpolation lie, the extrapolation lie, and the inaccurate values lie.

3.3 To convert a scatterplot to a range-frame, simply erase the axes below the minimum score and above the maximum score.

3.5 With scale data, a scatterplot allows for a helpful visual analysis of the relation between two variables. If the data points appear to fall approximately along a straight line, this indicates a linear relation. If the data form a line that changes direction along its path, a nonlinear relation may be present. If the data points show no particular relation, it is possible that the two variables are not related.

3.7 A bar graph is a visual depiction of data in which the independent variable is nominal or ordinal and the dependent variable is scale. Each bar typically represents the mean value of the dependent variable for each category. A Pareto chart is a specific type of bar graph in which the categories along the x-axis are ordered from highest bar on the left to lowest bar on the right.

3.9 A pictorial graph is a visual depiction of data typically used for a nominal independent variable with very few levels (categories) and a scale dependent variable. Each level uses a picture or symbol to represent its value on the scale dependent variable. A pie chart is a graph in the shape of a circle, with a slice for every level. The size of each slice represents the proportion (or percentage) of each category. In most cases, a bar graph is preferable to a pictorial graph or a pie chart.

3.11 The independent variable typically goes on the horizontal *x*-axis and the dependent variable goes on the vertical *y*-axis.

3.13 Moiré vibrations are any visual patterns that create a distracting impression of vibration and movement. A grid is a background pattern, almost like graph paper, on which the data representations, such as bars, are superimposed. Ducks are features of the data that have been dressed up to be something other than merely data.

3.15 Total dollars donated per year is scale data. A time plot would nicely show how donations varied across years.

3.17 **a.** The independent variable is gender and the dependent variable is video game score.

b. Nominal

c. Scale

d. The best graph for these data would be a bar graph because there is a nominal independent variable and a scale dependent variable.

3.19 Linear, because the data could be fit with a line drawn from the upper-left to the lower-right corner of the graph.

3.21 **a.** Bar graph

b. Line graph; more specifically, a time plot

c. The *y*-axis should go down to 0.

d. The lines in the background are grids, and the three-dimensional effect is a type of duck.

e. 3.20%, 3.22%, 2.80%

f. If the *y*-axis started at 0, all of the bars would appear to be about the same height. The differences would be minimized.

3.23 The minimum value is 0.04 and the maximum is 0.36, so the axis could be labeled from 0.00 to 0.40. We might choose to mark every 0.05 value:

0.00, 0.05, 0.10, 0.15, 0.20, 0.25, 0.30, 0.35, and 0.40

3.25 **a.** The independent variable is height and the dependent variable is attractiveness. Both are scale variables.

b. The best graph for these data would be a scatterplot (which also might include a line of best fit if the relation is linear) because there are two scale variables.

c. It would not be practical to start the axis at 0. With the data clustered from 58 to 71 inches, a 0 start to the axis would mean that a large portion of the graph would be empty. We would use cut marks to indicate that the axis did not include all values from 0 to 58. (However, we would include the full range of data—0 to 71—if omitting some of these numbers would be misleading.)

3.27 **a.** The independent variable is country and the dependent variable is male suicide rate.

b. Country is a nominal variable and suicide rate is a scale variable.

c. The best graph for these data would be a Pareto chart. Because there are 20 categories along the *x*-axis, it is best to arrange them in order from highest to lowest.

d. A time series plot could show year on the *x*-axis and suicide rate on the *y*-axis. Each country would be represented by a different color line.

3.29 **a.**

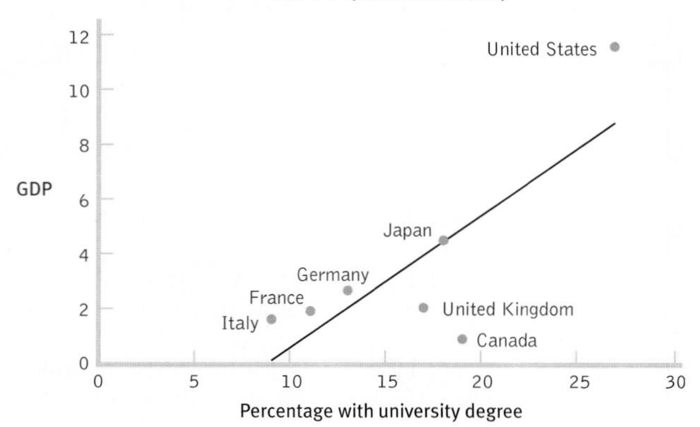

b. The percentage of residents with a university degree appears to be related to GDP. As the percentage with a university degree increases, so does GDP.

c. It is possible that an educated populace has the skills to make that country productive and profitable. Conversely, it is possible that a productive and profitable country has the money needed for the populace to be educated.

3.31 **a.** The independent variable is type of academic institution. It is nominal; the levels are private national, public national, and liberal arts.

b. The dependent variable is alumni donation rate. It is a scale variable; the units are percentages, and the range of values is from 9 to 66.

c. The defaults will differ, depending on which software is used.

Here is one example.

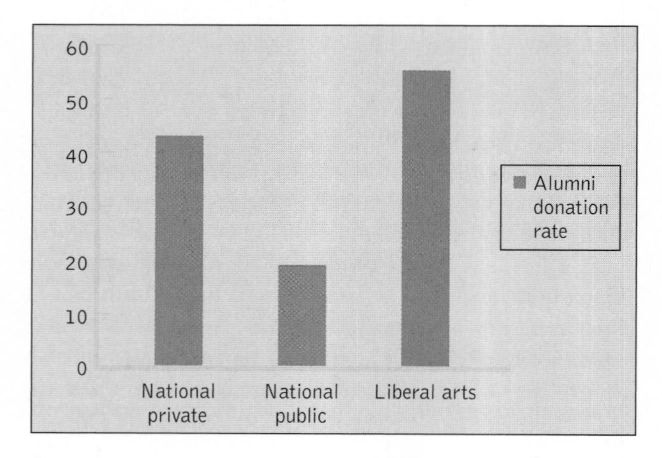

d. The redesigns will differ, depending on which software is used. In this example, we have added a clear title, labeled the *x*-axis, omitted the key, and labeled the *y*-axis (being sure that it reads from left to right). We also toned down the unnecessary color in the background and cut some of the extra numbers from the y-axis. Finally, we removed the black box from around the graph.

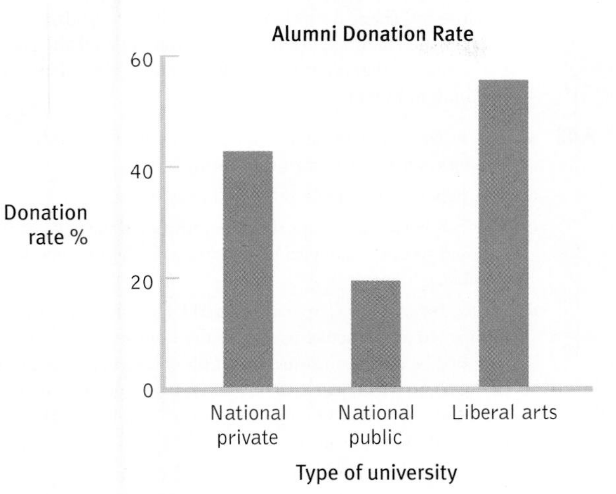

e. These data suggest that a higher percentage of alumni of liberal arts colleges than of national private or national public universities donate to their institutions. Moreover, a higher percentage of alumni of national private universities than of national public universities donate.

f. There are many possible answers to this question. One might want to identify characteristics of alumni who donate, methods of soliciting donations that result in the best outcomes, or characteristics of universities within a given category (e.g., liberal arts) that have the highest rates.

g. Pictures could be used instead of bars. For example, dollar signs might be used to represent the three quantities.

h. If the dollar signs become wider as they get taller, as often happens with pictorial graphs, the overall size would be proportionally larger than the increase in donation rate it is meant to represent. A bar graph is not subject to this problem because graphmakers are not likely to make bars wider as they get taller.

3.33 a. One independent variable is time frame; it has two levels: 1945–1950 and 1996–1998. The other independent variable is type of graduate program; it also has two levels: clinical psychology and experimental psychology.

b. The dependent variable is percentage of graduates who had a mentor while in graduate school.

c.

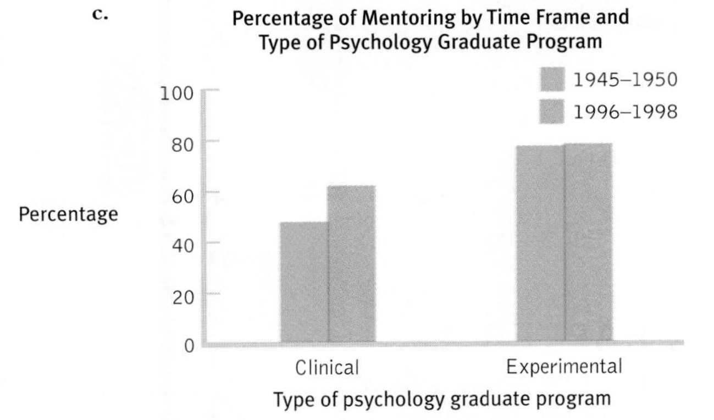

d. These data suggest that clinical psychology graduate students were more likely to have been mentored if they were in school in the 1996–1998 time frame than if they were in school during the 1945–1950 time frame. There does not appear to be such a difference among experimental psychology students.

e. This was not a true experiment. Students were not randomly assigned to time period or type of graduate program.

f. A time series plot would be inappropriate with so few data points. It would suggest that we could interpolate between these data points. It would suggest a continual increase in the likelihood of being mentored among clinical psychology students, as well as a stable trend, albeit at a high level, among experimental psychology students.

g. The story based on two time points might be falsely interpreted as a continual increase of mentoring rates for the clinical psychology students and a plateau for the experimental psychology students. The expanded data set suggests that the rates of mentoring have fluctuated over the years. Without the four time points, we might be seduced by interpolation into thinking that the two scores represent the end points of a linear trend. We cannot draw conclusions about time points for which we have no data—especially when we have only two points, but even when we have more points.

3.35 a. The details will differ, depending on the software used. Here is one example.

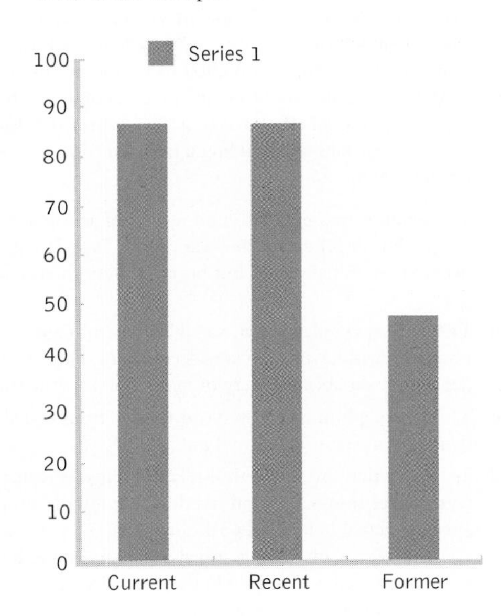

b. The default options that students choose to override will differ. For the bar graph here, we (1) added a title, (2) labeled the x-axis, (3) labeled the y-axis, (4) once we created the label on the y-axis, we rotated it so that it reads from left to right, (5) eliminated the box around the whole graph, and (6) eliminated the unnecessary key.

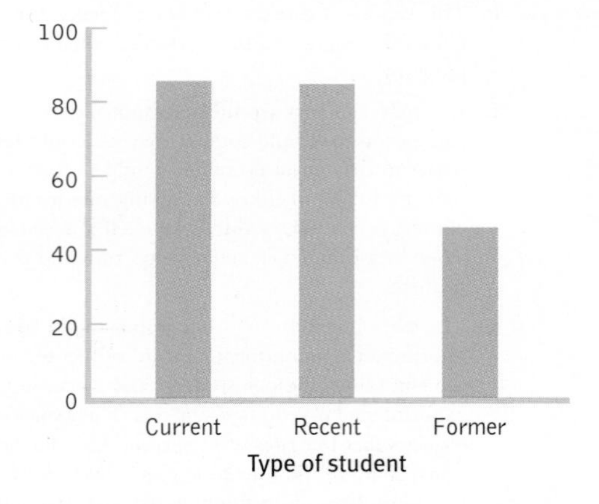

3.37 Each student's advice will differ. The following are examples of advice.

- **a.** The shrinking doctor: Replace the pictures with bars. Space the 3 years out in relation to their actual values (in the art shown, 1964 and 1975 are a good deal farther apart than are 1975 and 1990). Make the main title more descriptive.
- **b.** Workforce participation: Eliminate all the pictures. A falling line in the art shown indicates an *increase* in percentage; notice that 40% is at the top and 80% is at the bottom. Make the *y*-axis go from highest to lowest, starting from 0. Make the lines easier to compare by eliminating the three-dimensional effect. Make it clear where the data point for each year falls by including a tick mark for each number on the *x*-axis.

3.39
- **a.** The graph proposes that Type I regrets of action are initially intense but decline over the years, while Type II regrets of inaction are initially mild but become more intense over the years.
- **b.** There are two independent variables: type of regret (a nominal variable) and age (a scale variable). There is one dependent variable: intensity of regrets (also a scale variable).
- **c.** This is a graph of a theory. No data have been collected, so there are no statistics of any kind.
- **d.** The story that this theoretical relation suggests is that regrets over things a person has done are intense shortly after the actual behavior but decline over the years. In contrast, regrets over things a person has not done but wishes they had are initially low in intensity but become more intense as the years go by.

3.41
- **a.** When first starting therapy, the client showed a decline, as measured by the Mental Health Index (MHI). After 8 weeks of therapy, this trajectory reversed and there was a week-to-week improvement in the client's MHI.
- **b.** There are many possible answers. For example, the initial decline in the client's MHI may have been due to difficulties in adapting to therapy that were overcome as the client and therapist worked together. Alternatively, it may be that the client initially entered therapy due to difficult life circumstances that continued through the first weeks of therapy but resolved after several weeks.
- **c.** Because the client is not beneath the failure boundary, and because the client experienced improvement over the last few weeks of therapy, it may be beneficial for the client to continue in therapy.

3.43
- **a.** The independent variable is song type, with two levels: romantic song and nonromantic song.
- **b.** The dependent variable is dating behavior.
- **c.** This is a between-groups study because each participant is exposed to only one level or condition of the independent variable.
- **d.** Dating behavior was operationalized by giving one's phone number to an attractive person of the opposite sex. This may not be a valid measure of dating behavior, as we do not know if the participant actually intended to go on a date with the researcher. Giving one's phone number might not necessarily indicate an intention to date.
- **e.** We would use a bar graph because there is one nominal independent variable and one scale dependent variable.
- **f.** The default graph will differ, depending on which software is used. Here is one example:

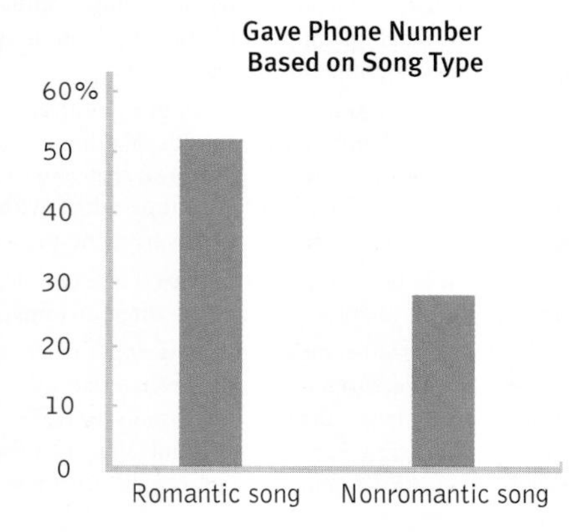

- **g.**

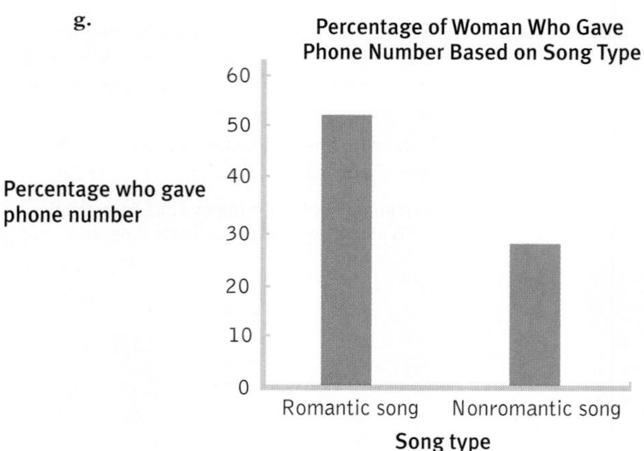

CHAPTER 4

4.1 The mean is the arithmetic average of a group of scores; it is calculated by summing all the scores and dividing by the total number of scores. The median is the middle score of all the scores when a group of scores is arranged in ascending order. If there is no single middle score, the median is the mean of the two middle scores. The mode is the most common score of all the scores in a group of scores.

4.3 The mean takes into account the actual numeric value of each score. The mean is the mathematic center of the data. It is the center balance point in the data, such that the sum of the deviations (rather than the number of deviations) below the mean equals the sum of deviations above the mean.

4.5 The mean might not be useful in a bimodal or multimodal distribution because in a bimodal or multimodal distribution the mathematical center of the distribution is not the number that describes what is typical or most representative of that distribution.

4.7 The mean is affected by outliers because the numeric value of the outlier is used in the computation of the mean. The median typically is not affected by outliers because its computation is based on the data in the middle of the distribution, and outliers lie at the extremes of the distribution.

4.9 The standard deviation is *the typical amount each score in a distribution varies from the mean of the distribution.*

4.11 The standard deviation is a measure of variability in terms of the values of the measure used to assess the variable, whereas the variance is squared values. Squared values simply don't make intuitive sense to us, so we take the square root of the variance and report this value, the standard deviation.

4.13 **a.** The mean is calculated as

$$M = \frac{\Sigma X}{N} = (15 + 34 + 32 + 46 + 22 + 36 + 34 + 28 + 52 + 28)/10 = 327/10 = 32.7$$

The median is found by arranging the scores in numeric order—15, 22, 28, 28, 32, 34, 34, 36, 46, 52—then dividing the number of scores, 10, by 2 and adding ½ to get 5.5. The mean of the 5th and 6th score in our ordered list of scores is our median—(32 + 34)/2 = 33—so 33 is the median.

The mode is the most common score. In these data, two scores appear twice, so we have two modes, 28 and 34.

b. Adding the value of 112 to the data from Exercise 4.15 changes the calculation of the mean in the following way:

$$(15 + 34 + 32 + 46 + 22 + 36 + 34 + 28 + 52 + 28 + 112)/11 = 439/11 = 39.91$$

The mean gets larger with this outlier.

There are now 11 data points, so the median is the 6th value in the ordered list, which is 34.

The modes are unchanged at 28 and 34.

This outlier increases the mean by approximately 7 values; it increases the median by 1; and it does not affect the mode at all.

c. The range is: $X_{highest} - X_{lowest} = 52 - 15 = 37$

The variance is: $SD^2 = \frac{\Sigma(X - M)^2}{N}$

We start by calculating the mean, which is 32.7. We then calculate the deviation of each score from the mean and the square of that deviation.

X	*X* − *M*	(*X* − *M*)²
15	−17.7	313.29
34	1.3	1.69
32	−0.7	0.49
46	13.3	176.89
22	−10.7	114.49
36	3.3	10.89
34	1.3	1.69
28	−4.7	22.09
52	19.3	372.49
28	−4.7	22.09

$$SD^2 = \frac{\Sigma(X - M)^2}{N} = \frac{1036.1}{10} = 103.61$$

The standard deviation is: $SD = \sqrt{SD^2}$ or

$$SD = \sqrt{\frac{\Sigma(X - M)^2}{N}} = \sqrt{103.61} = 10.18$$

4.15 **a.** The mean is calculated as

$$M = \frac{\Sigma X}{N} = [-3.7 + (-1.7) + 5.9 + 16.4 + 29.5 \ldots + 1.7]/12 = 244.2/12 = 20.35°F$$

The median is found by arranging the temperatures in numeric order:

−3.7, −1.7, 1.7, 5.9, 13.6, 16.4, 24, 29.5, 34.6, 38.5, 42.1, 43.3

There are 12 data points, so the mean of the 6th and 7th data points gives us the median: (16.4 + 24)/2 = 20.2°F.

b. The mean is calculated as

$$M = \frac{\Sigma X}{N} = [-47 + (-46) + (-38) + (-20) \ldots + -46]/12 = -163/12 = -13.58°F$$

The median is found by arranging the temperatures in numeric order:

−47, −46, −46, −38, −20, −20, −5, −2, 8, 9, 20, 24

There are 12 data points, so the mean of the 6th and 7th data points gives us the median: [−20 + −5]/2 = −25/2 = −12.5°F.

There are two modes: both −46 and −20 were recorded twice.

c. The mean is calculated as

$$M = \frac{\Sigma X}{N} = (173 + 166 + 180 \ldots + 178)/12 = 2022/12 = 168.5 \text{ mph}$$

The median is found by arranging the wind gusts in numeric order:

136, 142, 154, 161, 163, 164, 166, 173, 174, 178, 180, 231

There are 12 data points, so the mean of the 6th and 7th data points gives us the median: (164 + 166)/2 = 165 mph.

There is no mode among these wind gusts.

d. For the wind gust data, we could create 10-mph intervals and calculate the mode as the interval that occurs most often. There are four recorded gusts in the 160–169 mph interval, three in the 170–179 interval, and only one in the other intervals. So, the 160–169 mph interval could be presented as the mode.

e. The range is: $X_{highest} - X_{lowest} = 43.3 - (-3.7) = 47°F$

The variance is: $SD^2 = \frac{\Sigma(X-M)^2}{N}$

We start by calculating the mean, which is 20.35°F. We then calculate the deviation of each score from the mean and the square of that deviation.

X	X – M	(X – M)²
−3.7	−24.05	578.403
−1.7	−22.05	486.203
5.9	−14.45	208.803
16.4	−3.95	15.603
29.5	9.15	83.723
38.5	18.15	329.423
43.3	22.95	526.703
42.1	21.75	473.063
34.6	14.25	203.063
24	3.65	13.323
13.6	−6.75	45.563
1.7	−18.65	347.823

The variance is: $SD^2 = \frac{\Sigma(X-M)^2}{N} = \frac{3311.696}{12} = 275.975$

The standard deviation is: $SD = \sqrt{SD^2}$ or

$SD = \sqrt{\frac{\Sigma(X-M)^2}{N}} = \sqrt{275.975} = 16.61°F$

f. The range is $X_{highest} - X_{lowest} = 24 - (-47) = 71°F$

The variance is $SD^2 = \frac{\Sigma(X-M)^2}{N}$

We already calculated the mean, −13.583°F. We now calculate the deviation of each score from the mean and the square of that deviation.

X	X – M	(X – M)²
−47	33.417	1116.696
−46	−32.417	1050.862
−38	−24.417	596.190
−20	−6.417	41.178
−2	11.583	134.166
8	21.583	465.826
24	37.583	1412.482
20	33.583	1127.818
9	22.583	509.992
−5	8.583	73.668
−20	−6.417	41.178
−46	−32.417	1050.862

The variance is $SD^2 = \frac{\Sigma(X-M)^2}{N} = \frac{7620.018}{12}$

$= 635.077$

The standard deviation is $SD = \sqrt{SD^2}$ or

$SD = \sqrt{\frac{\Sigma(X-M)^2}{N}} = \sqrt{635.077} = 25.20°F$

g. For the peak wind gust data, the range is: $X_{highest} - X_{lowest} = 231 - 136 = 95$ mph

The variance is: $SD^2 = \frac{\Sigma(X-M)^2}{N}$

We start by calculating the mean, which is 168.5 mph. We then calculate the deviation of each score from the mean and the square of that deviation.

X	X – M	(X – M)²
173	4.5	20.25
166	−2.5	6.25
180	11.5	132.25
231	62.5	3906.25
164	−4.5	20.25
136	−32.5	1056.25
154	−14.5	210.25
142	−26.5	702.25
174	5.5	30.25
161	−7.5	56.25
163	−5.5	30.25
178	9.5	90.25

$SD^2 = \frac{\Sigma(X-M)^2}{N} = \frac{6261}{12} = 521.75$ is our variance.

The standard deviation is: $SD = \sqrt{SD^2}$ or

$SD = \sqrt{\frac{\Sigma(X-M)^2}{N}} = \sqrt{521.75} = 22.84$ mph

4.17 The mean for salary is often greater than the median for salary because the high salaries of top management inflate the mean but not the median. If we are trying to attract people to our company, we may want to present the typical salary as whichever value is higher—in most cases, the mean. However, if we are going to offer someone a low salary, presenting the median might make them feel better about that amount!

4.19 There are few participants in this study (only seven) so a single extreme score would influence the mean more than it would influence the median. The median is a more trustworthy indicator than the mean when there is only a handful of scores.

4.21 In April 1934, a wind gust of 231 mph was recorded. This data point is rather far from the next closest record of 180 mph. If this extreme score were excluded from analyses of central tendency, the mean would be lower, the median would change only slightly, and the mode would be unaffected.

4.23 There are many possible answers to this question. All answers will include a distribution that is skewed, perhaps one that has outliers. A skewed distribution would affect the mean but not the median. One example would be the variable of number of foreign countries visited; the few jet-setters who have been to many countries would pull the mean higher. The median is more representative of the typical score.

4.25 **a.** These ads are likely presenting outlier data.

b. To capture the experience of the typical individual who uses the product, the ad could include the mean result and the standard deviation. If the distribution of outcomes is skewed, it would be best to present the median result.

4.27 **a.** $M = \frac{\Sigma X}{N} = (0 + 5 + 3 + 3 + 1 \ldots + 3 + 5)/19$

$= 53/19 = 2.789$

b. The formula for variance is $SD^2 = \frac{\Sigma(X - M)^2}{N}$

We start by creating three columns: one for the scores, one for the deviations of the scores from the mean, and one for the squares of the deviations.

X	X − M	(X − M)²
0	−2.789	7.779
5	2.211	4.889
3	0.211	0.045
3	0.211	0.045
1	−1.789	3.201
10	7.211	51.999
2	−0.789	0.623
2	−0.789	0.623
3	0.211	0.045
1	−1.789	3.201
2	−0.789	0.623
4	1.211	1.467
2	−0.789	0.623
1	−1.789	3.201
1	−1.789	3.201
1	−1.789	3.201
4	1.211	1.467
3	0.211	0.045
5	2.211	4.889

We can now calculate variance: $SD^2 = \frac{\Sigma(X - M)^2}{N} =$ $(7.779 + 4.889 + 0.045 + 0.045 \ldots + 0.045 + 4.889)/19$ $= 91.167/19 = 4.798$

c. Standard deviation is calculated just like we calculated variance, but we then take the square root:

$$SD = \sqrt{\frac{\Sigma(X - M)^2}{N}} = \sqrt{4.798} = 2.19$$

d. The typical score is around 2.79, and the typical deviation from 2.79 is around 2.19.

4.29 There are many possible answers to these questions. The following are only examples.

a. 70, 70. There is no skew; the mean is not pulled away from the median.

b. 80, 70. There is positive skew; the mean is pulled up, but the median is unaffected.

c. 60, 70. There is negative skew; the mean is pulled down, but the median is unaffected.

4.31 It would probably be appropriate to use the mean because the data are scale; we would assume we have a large number of data points available to us; and the mean is the most commonly used measure of central tendency. Because of the large amount of data available, the effect of outliers is minimized. All of these factors would support the use of the mean for presenting information about the heights or weights of large numbers of people.

4.33 We cannot directly compare the mean ages reported by Denmark with the median ages reported by the United States because it is likely that there were some older outliers in both Denmark and the United States, and these outliers would affect the means reported by Denmark more than the medians reported by the United States.

4.35 **a.**

INTERVAL	FREQUENCY
60–69	1
50–59	7
40–49	10
30–39	7
20–29	2
10–19	3

b.

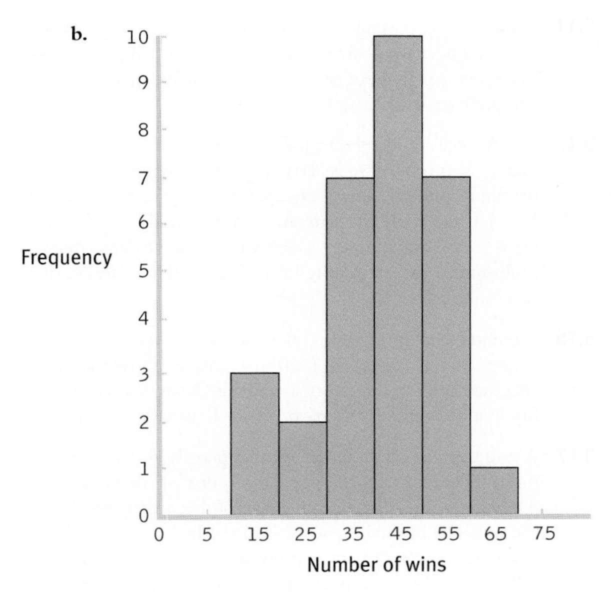

c. $M = \frac{\Sigma X}{N} = (45 + 43 + 42 + 33 \ldots + 45 + 18)/30$

$= 1230/30 = 41$

With 30 scores, the median would be between the 15th and 16th scores: $(30/2) + 0.5 + 15.5$. The 15th and 16th scores are 43 and 44, respectively, and so the median is 43.5. The mode is 45; there are three scores of 45.

d. Software reports that the range is 49 and the standard deviation is 12.69.

e. The summary will differ for each student but should include the following information: The data appear to be roughly symmetric and unimodal, maybe a bit negatively skewed. There are no glaring outliers.

f. Answers will vary. One example is whether number of wins is related to the average age of a team's players.

CHAPTER 5

5.1 It is rare to have access to an entire population. That is why we study samples and use inferential statistics to estimate what is happening in the population.

5.3 Generalizability refers to the ability of researchers to apply findings from one sample or in one context to other samples or contexts.

5.5 Random sampling means that every member of a population has an equal chance of being selected to participate in a study. Random assignment means that each selected participant has an equal chance of being in any of the experimental conditions.

5.7 Random assignment is a process in which every participant (regardless of how he or she was selected) has an equal chance of being in any of the experimental conditions. This avoids bias across experimental conditions.

5.9 An illusory correlation is a belief that two events are associated when in fact they are not.

5.11 Students' answers will vary. Personal probability is a person's belief about the probability of an event occurring; for example, someone's belief about the likelihood that she or he will complete a particular task.

5.13 In reference to probability, the term *trial* refers to each occasion that a given procedure is carried out. For example, each time we flip a coin, it is a trial. *Outcome* refers to the result of a trial. For coin-flip trials, the outcome is either heads or tails. *Success* refers to the outcome for which we're trying to determine the probability. If we are testing for the probability of heads, then success is heads.

5.15 The independent variable is the variable the researcher manipulates. Independent trials or events are those that do not affect each other; the flip of a coin is independent of another flip of a coin because the two events do not affect each other.

5.17 A null hypothesis is a statement that postulates that there is no mean difference between populations or that the mean difference is in a direction opposite of that anticipated by the researcher. A research hypothesis, also called an alternative hypothesis, is a statement that postulates that there is a mean difference between populations or sometimes, more specifically, that there is a mean difference in a certain direction, positive or negative.

5.19 We commit a Type I error when we reject the null hypothesis, but the null hypothesis is true. We commit a Type II error when we fail to reject the null hypothesis, but the null hypothesis is false.

5.21 In each of the six groups of 10 passengers that go through the checkpoint, we would check the 9th, 9th, 10th, 1st, 10th, and 8th passengers, respectively.

5.23 Only recording the numbers 1 to 5, the sequence appears as 5, 3, 5, 5, 2, 2, and 2. So, the first person is assigned to the fifth condition, the second person to the third condition, and so on.

5.25 Illusory correlation is particularly dangerous because people might perceive there to be an association between two variables that does not in fact exist. Because we often make decisions based on associations, it is important that those associations be real and be based on objective evidence. For example, a parent might perceive an illusory correlation between body piercings and trustworthiness, believing that a person with a large number of body piercings is untrustworthy. This illusory correlation might lead the parent to unfairly eliminate anyone with a body piercing from consideration when choosing babysitters.

5.27 The probability of winning is estimated as the number of people who have already won out of the total number of contestants, or $8/266 = 0.03$.

5.29 **a.** 0.627

b. 0.003

c. 0.042

5.31 **a.** Expected relative-frequency probability

b. Personal probability

c. Personal probability

d. Expected relative-frequency probability

5.33 Given that the population is high school students in Marseille and Lyon, it is possible that the researcher can compile a list of all members of the population, allowing her to use random selection. She could not, however, use random assignment because she could not assign the students to have lived in Marseille or Lyon.

5.35 **a.** The independent variable is type of news information, with two levels: information about an improving job market and information about a declining job market.

b. The dependent variable is psychologists' attitudes toward their careers.

c. The null hypothesis would be that, on average, the psychologists who received the positive article about the job market have the same attitude toward their career as those who read a negative article about the job market. The research hypothesis would be that a difference, on average, exists between the two groups.

5.37 Although we all believe we can think randomly if we want to, we do not, in fact, generate numbers independently of the ones that came before. We tend to glance at the preceding numbers in order to make the next ones "random." Yet once we do this, the numbers are not independent and therefore are not random. Moreover, even if we can keep ourselves from looking at the previous numbers, the numbers we generate are not likely to be random. For example, if we were born on the 6th of the month, then we may be more likely to choose 6's than other digits. Humans just don't think randomly.

5.39 **a.** The typical study volunteer is likely someone who cares deeply about U.S. college football. Moreover, it is particularly the fans of the top ACC teams, who themselves are likely extremely biased, who are most likely to vote.

b. External validity refers to the ability to generalize beyond the current sample. In this case, it is likely that fans of the top ACC teams are voting and that the poll results do not reflect the opinions of U.S. college football fans at large.

c. There are several possible answers to this question. As one example, only eight options were provided. Even though one of these options was "other," this limited the range of possible answers that respondents would be likely to provide. The sample is also biased in favor of those who know about and would spend time at the *USA Today* Web site in the first place.

5.41 **a.** These numbers are likely not representative. This is a volunteer sample.

b. Those most likely to volunteer are those who have stumbled across, or searched for, this Web site: a site that advocates for self-government. Those who respond are more likely to tend toward supporting self-government than are those who do not respond (or even find this Web site).

c. This description of libertarians suggests they would advocate for self-government, part of the name of the group that hosts this quiz, a likely explanation for the predominance of libertarians who responded to this survey. The repeated use of the word "Libertarian" (in the heading and in the icon) likely help preselect who would come to this Web site in the first place.

d. It doesn't matter how large a sample is if it's not representative. With respect to external validity, it would be far preferable to have a smaller but representative sample than a very large but unrepresentative sample.

5.43 Your friend's bias is an illusory correlation—he perceives a relation between gender and driving performance, when in fact there is none.

5.45 If a depressed person has negative thoughts about himself or herself and about the world, confirmation bias may make it difficult to change those thoughts because confirmation bias would lead this person to pay more attention to and better remember negative events than positive events. For example, he or she might remember the one friend who slighted him or her at a party but not the many friends who were excited to see him or her.

5.47 **a.** *Probability* refers to the proportion of aces that we expect to see in the long run. In the long run, given 4 aces out of 52 cards, we would expect the proportion of aces to be $4/52 = 0.077$, rounded to 0.08.

b. *Proportion* refers to the observed fraction of cards that are aces—the number of successes divided by the number of trials. In this case, the proportion of aces is $5/15 = 0.333$, rounded to 0.33.

c. *Percentage* refers to the proportion multiplied by 100: $0.333(100) = 33.3$. Thus, 33.3% of the cards drawn were aces.

d. Although 0.333 is far from 0.077, we would expect a great deal of fluctuation in the short run. These data are not sufficient to determine whether the deck is stacked.

5.49 **a.** The null hypothesis is that the average tendency to develop false memories is either unchanged or is lowered by the repetition of false information. The research hypothesis is that false memories are higher, on average, when false information is repeated than when it is not.

b. The null hypothesis is that the average outcome is the same or worse whether or not structured assessments are used. The research hypothesis is that the average outcome is better when structured assessments are used than when they are not used.

c. The null hypothesis is that average employee morale is the same whether employees work in enclosed offices or in cubicles. The research hypothesis is that average employee morale is different when employees work in enclosed offices versus in cubicles.

d. The null hypothesis is that ability to speak one's native language is the same, on average, whether or not a second language is taught from birth. The research hypothesis is that the ability to speak one's native language is different, on average, when a second language is taught from birth than when no second language is taught.

5.51 **a.** If this conclusion is incorrect, the researcher has made a Type I error. The researcher rejected the null hypothesis when the null hypothesis is really true. (Of course, he or she never knows whether there has been an error! She or he just has to acknowledge the possibility.)

b. If this conclusion is incorrect, the researcher has made a Type I error. She has rejected the null hypothesis when the null hypothesis is really true.

c. If this conclusion is incorrect, the researcher has made a Type II error. He has failed to reject the null hypothesis when the null hypothesis is not true.

d. If this conclusion is incorrect, the researcher has made a Type II error. She has failed to reject the null hypothesis when the null hypothesis is not true.

5.53 **a.** Confirmation bias has guided his logic in that he looked for specific events that occurred during the day to fit the horoscope but ignored the events that did not fit the prediction.

b. If this conclusion is incorrect, they have made a Type I error. Dean and Kelly would have failed to reject the null hypothesis when the null hypothesis was not true.

c. If an event occurs regularly or a research finding is replicated many times and by other researchers and in a range of contexts, then it is likely the event or finding is not occurring in error or by chance alone.

5.55 **a.** The population of interest is male students with alcohol problems. The sample is the 64 students who were ordered to meet with a school counselor.

b. Random selection was not used. The sample was comprised of 64 male students who had been ordered to meet with a school counselor; they were not chosen out of all male students with alcohol problems.

c. Random assignment was used. Each participant had an equal chance of being assigned to either of the two conditions.

d. The independent variable is type of counseling. It has two levels: BMI and AE. The dependent variable is number of alcohol-related problems at follow-up.

e. The null hypothesis is that the mean number of alcohol-related problems at follow-up is the same, regardless of type of counseling (BMI or AE). The research hypothesis is that students who undergo a BMI have different mean numbers of alcohol-related problems at follow-up than do students who participate in AE.

f. The researchers rejected the null hypothesis.

g. If the researchers were incorrect in their decision, then they made a Type I error, rejecting the null hypothesis when the null hypothesis was true. The consequences of this type of error are that a new treatment that is no better, on average, than the standard treatment would be implemented. This might lead to unnecessary costs to train counselors to implement the new treatment.

CHAPTER 6

6.1 In everyday conversation, the word *normal* is used to refer to events or objects that are common or that typically occur. Statisticians use the word to refer to distributions that conform to a specific bell-shaped curve, with a peak in the middle where most of the observations lie, and symmetric areas underneath the curve on either side of the midpoint. This normal curve represents the pattern of occurrence of many different kinds of events.

6.3 The distribution of sample scores approaches normal as the sample size increases, assuming the population is normally distributed.

6.5 A z score is a way to standardize data; it expresses how far a data point is from the mean of its distribution in terms of standard deviations.

6.7 The mean is 0 and the standard deviation is 1.0.

6.9 The symbol μ_M stands for the mean of the distribution of means. The μ indicates that it is the mean of a *population,* and the subscript M indicates that the population is composed of *sample means*—the means of all possible samples of a given size from a particular population of individual scores.

6.11 Standard deviation is the measure of spread for a distribution of scores in a single sample or in a population of scores. Standard error is the standard deviation (or measure of spread) in a distribution of means of all possible samples of a given size from a particular population of individual scores.

6.13 The z statistic tells us how many standard errors a sample mean is from the population mean.

6.15 **a.**

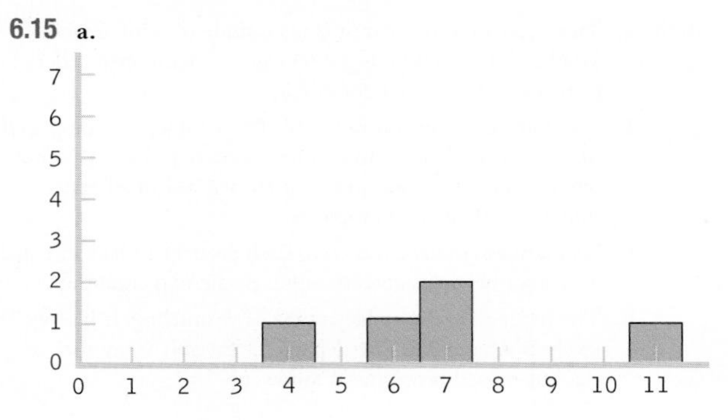

b.

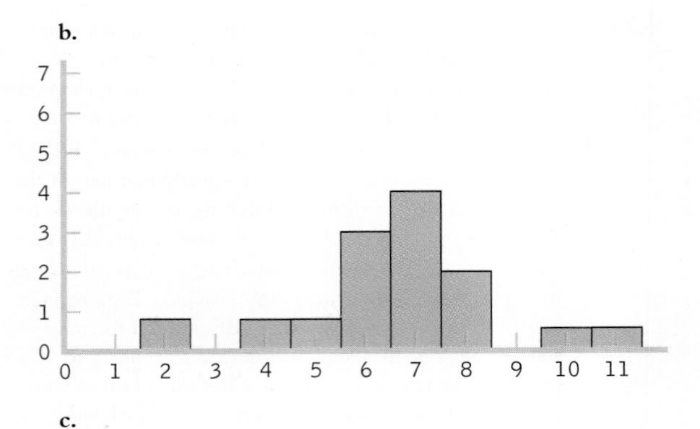

c.

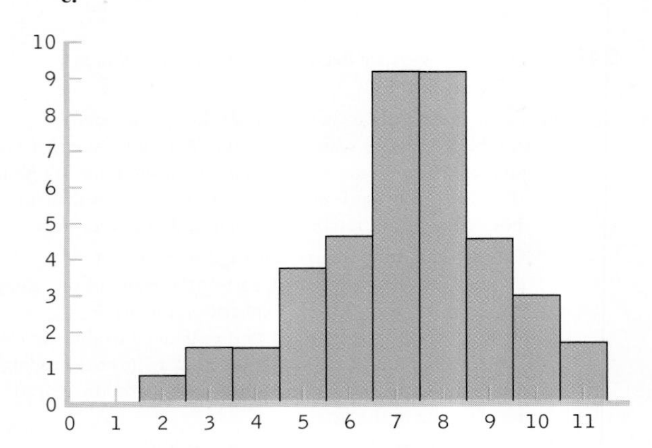

d. As the sample size increases, the distribution approaches the shape of the normal curve.

6.17 **a.** $z = \frac{1000 - 1179}{164} = -1.09$

b. $z = \frac{721 - 1179}{164} = -2.79$

c. $z = \frac{1531 - 1179}{164} = 2.15$

d. $z = \frac{1184 - 1179}{164} = 0.03$

6.19 $z = \frac{203 - 250}{47} = -1.0$

$z = \frac{297 - 250}{47} = 1.0$

Each of these scores is 47 points away from the mean, which is the value of the standard deviation. The z scores of -1.0 and 1.0 express that the first score, 203, is 1 standard deviation below the mean, whereas the other score, 297, is 1 standard deviation above the mean.

6.21 **a.** $X = z(\sigma) + \mu = -0.23(164) + 1179 = 1141.28$

b. $X = 1.41(164) + 1179 = 1410.24$

c. $X = 2.06(164) + 1179 = 1516.84$

d. $X = 0.03(164) + 1179 = 1183.92$

6.23 **a.** $X = z(\sigma) + \mu = 1.5(100) + 500 = 650$

b. $X = z(\sigma) + \mu = -0.5(100) + 500 = 450$

c. $X = z(\sigma) + \mu = -2.0(100) + 500 = 300$

6.25 **a.** $z = \dfrac{45 - 51}{4} = -1.5$

$z = \dfrac{732 - 765}{23} = -1.43$

b. Both of these scores fall below the means of their distributions, resulting in negative z scores. One score (45) is a little farther below its mean than the other (732).

6.27 **a.** 50%

b. 82% (34 + 34 + 14)

c. 4% (2 + 2)

d. 48% (34 + 14)

e. 100% or nearly 100%

6.29 **a.** $\mu_M = \mu = 55$, and $\sigma_M = \dfrac{8}{\sqrt{30}} = 1.46$

b. $\mu_M = \mu = 55$, and $\sigma_M = \dfrac{8}{\sqrt{300}} = 0.46$

c. $\mu_M = \mu = 55$, and $\sigma_M = \dfrac{8}{\sqrt{3000}} = 0.15$

6.31 **a.** $z = \dfrac{(M - \mu_M)}{\dfrac{\sigma}{\sqrt{N}}} = \dfrac{85 - 80}{\dfrac{20}{\sqrt{100}}} = \dfrac{5}{\dfrac{20}{10}} = \dfrac{5}{2} = 2.50$

$z = \dfrac{(M - \mu_M)}{\dfrac{\sigma}{\sqrt{N}}} = \dfrac{17 - 15}{\dfrac{5}{\sqrt{100}}} = \dfrac{2}{\dfrac{5}{10}} = \dfrac{2}{0.50} = 4.00$

b. The first sample had a mean that was 2.50 standard deviations above the population mean, whereas the second sample had a mean that was 4 standard deviations above the mean. Compared to the population mean (as measured by this scale), both samples are extreme scores; however, a z score of 4.0 is even more extreme than a z score of 2.5.

6.33 **a.** Histogram for the 10 scores:

b. Histogram for all 40 scores:

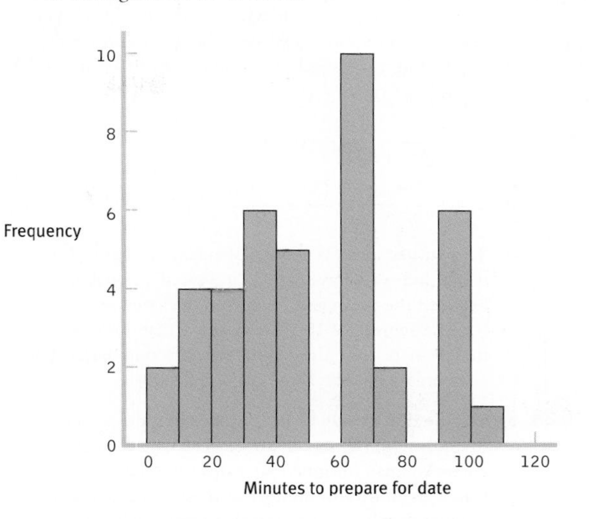

c. The shape of the distribution became more normal as the number of scores increased. If we added more scores, the distribution would become more and more normal. This occurs because many physical, psychological, and behavioral variables are normally distributed. With smaller samples, this might not be clear. But as the sample size approaches the size of the population, the shape of the sample distribution approaches that of the population.

d. These are distributions of scores, because each individual score is represented in the histograms on its own, not as part of a mean

e. There are several possible answers to this question. For example, instead of using retrospective self-reports, we could have had students call a number or send an e-mail as they began to get ready; they would then have called the same number or sent another e-mail when they were ready. This would have led to scores that would be closer to the actual time it took the students to get ready.

f. There are several possible answers to this question. For example, we could examine whether there was a mean gender difference in time spent getting ready for a date.

6.35 **a.** The mean of the z distribution is always 0.

b. $z = \dfrac{(X - \mu)}{\sigma} = \dfrac{(6.65 - 6.65)}{1.24} = 0$

c. The standard deviation of the z distribution is always 1.

d. A student 1 standard deviation above the mean would have a score of 6.65 + 1.24 = 7.89. This person's z score would be: $z = \dfrac{(X - \mu)}{\sigma} = \dfrac{(7.89 - 6.65)}{1.24} = 1$

e. The answer will differ for each student but will involve substituting one's own score for X in this equation:
$z = \dfrac{(X - 6.65)}{1.24}$

6.37 **a.** It would not make sense to compare the mean of this sample to the distribution of individual scores because, in a sample of means, the occasional extreme individual score is balanced by less extreme scores that are also part of the sample.

b. The null hypothesis would state that the population from which our sample was drawn has a mean of 3.51. The research hypothesis would state that the mean for the population from which our sample was drawn is not 3.51.

c. $\mu_M = \mu = 3.51$

$$\sigma_M = \frac{\sigma}{\sqrt{N}} = \frac{0.61}{\sqrt{40}} = 0.096$$

d. $z = \frac{(M - \mu_M)}{\sigma_M} = \frac{(3.62 - 3.51)}{0.096} = 1.15$

e. This sample mean is about 1 standard deviation above the mean, and we know that about 34% of a distribution falls between the mean and 1 standard deviation above the mean (i.e., a z statistic of 1). We also know that 50% fall below the mean, because the z distribution is symmetric. The percentile would be about 50 + 34 = 84%.

6.39 **a.** Yes, the distribution of the number of movies college students watch in a year would likely approximate a normal curve. You can imagine that a small number of students watch an enormous number of movies and that a small number watch very few but that most watch a moderate number of movies between these two extremes.

b. Yes, the number of full-page advertisements in magazines is likely to approximate a normal curve. We could find magazines that have no or just one or two full-page advertisements and some that are chock full of them, but most magazines have some intermediate number of full-page advertisements.

c. Yes, human birth weights in Canada could be expected to approximate a normal curve. Few infants would weigh in at the extremes of very light or very heavy, and the weight of most infants would cluster around some intermediate value.

6.41 **a.** $z = \frac{(X - \mu)}{\sigma} = \frac{(95 - 81.71)}{13.07} = 1.02$

b. $z = \frac{(X - \mu)}{\sigma} = \frac{(10 - 8.13)}{3.70} = 0.51$

c. According to these data, the Red Sox had a better regular season (they had a higher z score) than did the Patriots.

d. The Patriots would have had to have won 12 regular season games to have a slightly higher z score than the Red Sox:

$$z = \frac{(X - \mu)}{\sigma} = \frac{(12 - 8.13)}{3.70} = 1.05$$

e. There are several possible answers to this question. For example, we could have summed the teams' scores for every game (as compared to other teams' scores within their leagues).

6.43 **a.** $X = z(\sigma) + \mu = -0.18(10.83) + 81.00 = 79$ games (rounded to a whole number)

b. $X = z(\sigma) + \mu = -1.475(3.39) + 8.0 = 3$ games (rounded to a whole number)

c. Fifty percent of scores fall below the mean, so 34% (84 − 50 = 34) fall between the mean and the Steelers' score. We know that 34% of scores fall between the mean and a z score of 1.0, so the Steelers have a z score of 1.0. $X = z(\sigma) + \mu = 1(3.39) + 8.0 = 11$ games (rounded to a whole number).

d. We can examine our answers to be sure that negative z scores match up with answers that are below the mean and positive z scores match up with answers that are above the mean.

6.45 **a.** $\mu = 50; \sigma = 10$

b. $\mu_M = \mu = 50; \sigma_M = \frac{\sigma}{\sqrt{N}} = \frac{10}{\sqrt{95}} = 1.03$

c. When we calculate the mean of the scores for 95 individuals, the most extreme MMPI-2 depression scores will likely be balanced by scores toward the middle. It would be rare to have an extreme mean of the scores for 95 individuals. Thus, the spread is smaller than is the spread for all of the individual MMPI-2 depression scores.

6.47 **a.** These are the data for a distribution of scores rather than means because they have been obtained by entering each individual score into the analysis.

b. Comparing the sizes of the mean and the standard deviation suggests that there is positive skew. A person can't have fewer than zero friends, so the distribution would have to extend in a positive direction to have a standard deviation larger than the mean.

c. Because the mean is larger than either the median or the mode, it suggests that the distribution is positively skewed. There are extreme scores in the positive end of the distribution that are causing the mean to be more extreme than the median or mode.

d. You would compare this person to the distribution of scores. When making a comparison of an individual score, we must use the distribution of scores.

e. You would compare this sample to a distribution of means. When making a comparison involving a sample mean, we must use a distribution of means because it has a different pattern of variability from a distribution of scores (it has less variability).

f. $\mu_M = \mu = 7.44$. The number of individuals in the sample is 80. Substituting 80 in the standard error equation yields

$$\sigma_M = \frac{\sigma}{\sqrt{N}} = \frac{10.98}{\sqrt{80}} = 1.23.$$

g. The distribution of means is likely to be a normal curve. Because the sample of 80 is well above the 30 recommended to see the central limit theorem at work, we expect that the distribution of the sample means will approximate a normal distribution.

6.49 **a.** You would compare this sample mean to a distribution of means. When we are making a comparison involving a sample mean, we need to use the distribution of means because it is this distribution that indicates the variability we are likely to see in sample means.

b. $z = \frac{(M - \mu_M)}{\sigma_M} = \frac{(8.7 - 7.44)}{1.228} = 1.026$

This z statistic of 1.03 is approximately 1 standard deviation above the mean. Because 50% of the sample are below the mean and 34% are between the mean and 1 standard deviation above it, this sample would be at approximately the 84th percentile.

c. It does make sense to calculate a percentile for this sample. Given the central limit theorem and the size of the sample used to calculate the mean (80), we would expect the distribution of the sample means to be approximately normal.

6.51 **a.** The population is all patients treated for blocked coronary arteries in the United States. The sample is Medicare patients in Eylria, Ohio, who received angioplasty.

b. Medicare and the commercial insurer compared the angioplasty rate in Elyria to that in other towns. Given that the rate was so far above that of other towns, they decided that such a high angioplasty rate was unlikely to happen just by chance. Thus, they used probability to make a decision to investigate.

c. Medicare and the commercial insurer could look at the z distribution of angioplasty rates in cities from all over the country. Locating the rate of Elyria within that distribution would indicate exactly how extreme or unlikely its angioplasty rates are.

d. The error made would be a Type I error, as they would be rejecting the null hypothesis that there is no difference among the various towns in rates of angioplasty, and concluding that there is a difference, when there really is no difference.

e. Elyria's extremely high rates do not necessarily mean the doctors are committing fraud. One could imagine that an area with a population composed mostly of retirees (that is, more elderly people) would have a higher rate of angioplasty. Conversely, perhaps Elyria has a talented set of surgeons who are renowned for their angioplasty skills and people from all over the country come there to have angioplasty.

6.53 **a.** Cheating is operationally defined as the change in standardized test score for a given classroom. This variable is a scale variable.

b. Researchers could establish a cut-off z statistic at which those who had a mean change greater than that z statistic would be considered "suspicious." For example, a classroom with a z statistic of 2 or more may have cheated on this year's test.

c. A histogram or frequency polygon would provide an easy visual to see where a given classroom falls on the distribution. A researcher could even draw lines indicating the cutoffs and see which classrooms fall beyond them.

d. They would be committing a Type I error, because they would be rejecting the null hypothesis that there is no difference in a classroom's test scores from one year to the next when there really is no difference and they should have failed to reject the null hypothesis.

CHAPTER 7

7.1 A percentile tells you the percentage of scores that fall below a certain point on a distribution.

7.3 We add the percentage between the mean and the positive z score to 50%, which is the percentage of scores below the mean (50% of scores are on each side of the mean).

7.5 In statistics, *assumptions* are the characteristics we ideally require the population from which we are sampling to have so that we can make accurate inferences.

7.7 *Parametric tests* are statistical analyses based on a set of assumptions about the population. By contrast, *nonparametric tests* are statistical analyses that are not based on assumptions about the population.

7.9 *Critical values,* often simply called *cutoffs,* are the test statistic values beyond which we reject the null hypothesis. The *critical region* refers to the area in the tails of the distribution in which the null hypothesis will be rejected if the test statistic falls there.

7.11 A *statistically significant* finding is one in which we have rejected the null hypothesis because the pattern in the data differed from what we would expect by chance. The word *significant* has a particular meaning in statistics. "Statistical significance" does *not* mean that the finding is necessarily important or meaningful. Statistical significance only means that we are justified in believing that the pattern in the data is likely to reoccur, that is, the pattern is likely genuine.

7.13 *Critical region* may have been chosen because values of a test statistic describe the area beneath the normal curve that represents a statistically significant result.

7.15 For a one-tailed test, the critical region (usually 5%, or a p level of 0.05) is placed in only one tail of the distribution; for a two-tailed test, the critical region must be split in half and shared between both tails (usually 2.5%, or 0.025, in each tail).

7.17 **a.** If 22.96% are beyond this z score (in the tail), then 77.04% are below it (100% − 22.96%).

b. If 22.96% are beyond this z score, then 27.04% are between it and the mean (50% − 22.96%).

c. Because the curve is symmetric, the area beyond a z score of −0.74 is the same as that beyond 0.74. Expressed as a proportion, 22.96% appears as 0.2296.

7.19 **a.** The percentage above is the percentage in the tail, 4.36%.

b. The percentage below is calculated by adding the area below the mean, 50%, and the area between the mean and this z score, 45.64%, to get 95.64%.

c. The percentage at least as extreme is computed by doubling the amount beyond the z score, 4.36%, to get 8.72%.

7.21 **a.** 19%

b. 4%

c. 92%

7.23 **a.** 2.5% in each tail

b. 5% in each tail

c. 0.5% in each tail

7.25 $\mu_M = \mu = 500$

$$\sigma_M = \frac{\sigma}{\sqrt{N}} = \frac{100}{\sqrt{50}} = 14.14$$

7.27 **a.** Fail to reject the null hypothesis because 1.06 does not exceed the cutoff of 1.96.

b. Reject the null hypothesis because −2.06 is more extreme than −1.9.

c. Fail to reject the null hypothesis because a z statistic with 7% of the data in the tail occurs between ±1.48 and ±1.47, which do not exceed ±1.96.

7.29 **a.** Fail to reject the null hypothesis because 0.95 does not exceed 1.65.

b. Reject the null hypothesis because −1.77 exceeds −1.65.

c. Reject the null hypothesis because the critical value resulting in 2% in the tail falls within the 5% cutoff region.

7.31 **a.** $z = \frac{(X-\mu)}{\sigma} = \frac{5.4-7}{1.85} = \frac{-1.6}{1.85} = -0.86$

The percentage below is 19.49%.

b. $z = \frac{(X-\mu)}{\sigma} = \frac{8.5-7}{1.85} = \frac{1.5}{1.85} = 0.81;$

The percentage below is 50% + 29.10% = 79.10%.

c. $z = \frac{(X-\mu)}{\sigma} = \frac{8.9-7}{1.85} = \frac{1.9}{1.85} = 1.03$

The percentage below is 50% + 34.85% = 84.85%.

d. $z = \frac{(X-\mu)}{\sigma} = \frac{6.5-7}{1.85} = \frac{-0.5}{1.85} = -0.27$

The percentage below is 39.36%.

7.33 **a.** $z = \frac{(X-\mu)}{\sigma} = \frac{72-67}{3.19} = 1.57$

b. 44.18% of scores are between this z score and the mean. We need to add this to the area below the mean, 50%, to get the percentile score of 94.18%.

c. 94.18% of boys are shorter than Kona at this age.

d. If 94.18% of boys are shorter than Kona, that leaves 5.82% in the tail. To compute how many scores are at least as extreme, we double this to get 11.64%.

e. We look at the z table to find a critical value that puts 30% of scores in the tail, or as close as we can get to 30%. A z score of −0.52 puts 30.15% in the tail. We can use that z score to compute the raw score for height:

$$X = -0.52(3.19) + 67 = 65.34 \text{ inches}$$

At 72 inches tall, Kona is 6.66 inches taller than Ian.

7.35 **a.** $z = \frac{(M-\mu_M)}{\sigma_M} = \frac{69.5-67}{\frac{3.19}{\sqrt{13}}} = 2.83$

b. The z statistic indicates that this sample mean is 2.83 standard deviations above the expected mean for samples of size 13. In other words, this sample of boys is, on average, exceptionally tall.

c. The percentile rank is 99.77%, meaning that 99.77% of sample means would be of lesser value than the one obtained for this sample.

7.37 **a.** $\mu_M = \mu = 63.8$

$$\sigma_M = \frac{\sigma}{\sqrt{N}} = \frac{2.66}{\sqrt{14}} = 0.711$$

b. $z = \frac{(M-\mu_M)}{\sigma_M} = \frac{62.4-63.8}{0.711} = -1.97$

c. 2.44% of sample means would be shorter than this mean.

d. We double 2.44% to account for both tails, so we get 4.88% of the time.

e. The average height of this group of 15-year-old females is rare, or statistically significant.

7.39 **a.** This is a nondirectional hypothesis because the researcher is predicting that it will alter skin moisture, not just decrease it or increase it.

b. This is a directional hypothesis because better grades are expected.

c. This hypothesis is nondirectional because any change is of interest, not just a decrease or an increase in closeness of relationships.

7.41 **a.**

X	$(X-\mu)$	$(X-\mu)^2$
4.41	0.257	0.066
8.24	4.087	16.704
4.69	0.537	0.288
3.31	−0.843	0.711
4.07	−0.083	0.007
2.52	−1.633	2.667
10.65	6.497	42.211
3.77	−0.383	0.147
4.07	−0.083	0.007
0.04	−4.113	16.917
0.75	−3.403	11.580
3.32	−0.833	0.694

$$\mu = 4.153; SS = \Sigma(X-\mu)^2 = 91.999;$$

$$\sigma^2 = \frac{\Sigma(X-\mu)^2}{N} = (91.999)/12 = 7.667;$$

$$\sigma = \sqrt{\sigma^2} = \sqrt{7.667} = 2.769$$

August: $X = 3.77$

$$z = \frac{(X-\mu)}{\sigma} = \frac{(3.77-4.153)}{2.769} = -0.138$$

b. The table tells us that 44.43% of scores fall in the tail beyond a z score of −0.14. So, the percentile for August is 44.43%. This is surprising because it is below the mean, and it was the month in which a devastating hurricane hit New Orleans. (*Note:* It is helpful to draw a picture of the curve when calculating this answer.)

c. Paragraphs will be different for each student but will include the fact that a monthly total based on missing data is inaccurate. The mean and the standard deviation based on this population, therefore, are inaccurate. Moreover, even if we had these data points, they would likely be large and would increase the total precipitation for August; August would likely be an outlier, skewing the overall mean. The median would be a more accurate measure of central tendency than the mean under these circumstances.

d. We would look up the z score that has 10% in the tail. The closest z score is 1.28, so the cutoffs are −1.28 and 1.28. (*Note:* It is helpful to draw a picture of the curve that includes these z scores.) We can then convert these z scores to raw scores. $X = z(\sigma) + \mu = -1.28(2.769) + 4.153 = 0.609$; $X = z(\sigma) + \mu = 1.28(2.769) + 4.153 = 7.697$. Only October (0.04) is below 0.609. Only February (8.24) and July (10.65) are above 7.697. These data are likely inaccurate, however, because the mean and the standard deviation of the population are based on an inaccurate mean from August. Moreover, it is quite likely that August would have been in the most extreme upper 10% if there were complete data for this month.

7.43 **a.** Population 1 is adult psychiatric inpatients. Population 2 is adult nonpatients.

b. The comparison distribution would be a distribution of means. Boone would compare his sample of 150 psychiatric inpatients to a distribution of the means of all possible samples of 150 individuals.

c. We would use a z test because there is one sample and we're comparing it to a population for which we know the mean and the standard deviation.

d. (1) From the description, the dependent variable, intrasubtest scatter, seems to be a scale variable. (2) The sample includes 150 adult psychiatric inpatients. It is unlikely that they were randomly selected from all adult psychiatric inpatients; thus, we must be cautious about generalizing from this sample. (3) We do not know if the population distribution is normal, but we have more than 30 participants (to be specific, we have 150), so the sampling distribution is likely to be normal.

e. Boone uses the word *significantly* as an indication that he rejected the null hypothesis.

7.45 **a.** The independent variable is the division. Teams were drawn from either Division I-A or Division I-AA. The dependent variable is the spread.

b. Random selection was not used. Random selection would entail having some process for randomly selecting Division I-AA games for inclusion in the sample. We did not describe such a process and, in fact, took all the Division I-AA teams from one league within that division.

c. The populations of interest are football games between teams in the upper divisions of the NCAA (Division I-A and Division I-AA).

d. The comparison distribution would be the distribution of sample means.

e. The first assumption—that the dependent variable is a scale variable—is met in this example. The dependent variable is point spread, which is a scale measure. The second assumption—that participants are randomly selected—is not met. As described in part (b), the teams for inclusion in the sample were not randomly selected. The third assumption—that the distribution of scores in the population of interest must be normal—is not likely to have been met. The standard deviation is almost as large as the mean, an indication that one or more outliers is creating positive skew. Moreover, we only have a sample size of 4, not the 30 we would need to have a normal distribution of means.

7.47 **a.** *Step 3:* $\mu_M = \mu = 16.189; \sigma_M = \frac{\sigma}{\sqrt{N}} = \frac{12.128}{\sqrt{4}} = 6.064$

Step 4: When we adopt 0.05 as the p level for significance and have a two-tailed hypothesis, we need to divide the 0.05 by 2 to obtain the z score cutoff for each end of the distribution (high and low). Dividing 0.05 by 2 yields 0.025. The z score corresponding to a probability of 0.025 is 1.96. Therefore, the cutoffs are -1.96 and 1.96.

Step 5: We first must obtain the mean spread in the sample. The games and their spreads are listed here:

GAME	SPREAD
Holy Cross, 27/Bucknell, 10	7
Lehigh, 23/Colgate, 15	8
Lafayette, 31/Fordham, 24	7
Georgetown, 24/Marist, 21	3

Mean spread = 8.75

$$z = \frac{(M - \mu_M)}{\sigma_M} = \frac{(8.75 - 16.189)}{6.064} = -1.23$$

Step 6: Given that the z statistic of -1.23 is not beyond the cutoff of -1.96, we would fail to reject the null hypothesis. We can conclude only that we do not have sufficient evidence that the point spread of Division I-AA teams is different, on average, from that of Division I-A teams.

b. It would be unwise to generalize these findings beyond the sample. The sample of games was not randomly selected from all Division I-AA team games that week. It is possible that this particular league differs from other leagues and therefore is not representative of Division I-AA as a whole.

7.49 **a.** The independent variable is whether a patient received the DVD with information about orthodontics. One group received the DVD; the other group did not. The dependent variable is the number of hours per day patients wore their appliances.

b. The researcher did not use random selection when choosing his sample. He selected the next 15 patients to come into his clinic.

c. *Step 1:* Population 1 is patients who did not receive the DVD. Population 2 is patients who received the DVD. The comparison distribution will be a distribution of means. The hypothesis test will be a z test because we have only one sample and we know the population mean and the standard deviation. This study meets the assumption that the dependent variable is a scale measure. We might expect the distribution of number of hours per day people wear their appliances to be normally distributed, but from the information provided it is not possible to tell for sure. Additionally, the sample includes fewer than 30, so the central limit theorem may not apply here. The distribution of sample means may not approach normality. Finally, the participants were not randomly selected. Therefore, we may not want to generalize the results beyond this sample.

Step 2: Null hypothesis: Patients who received the DVD do not wear their appliances a different mean number of hours per day than patients who did not receive the DVD; $H_0: \mu_1 = \mu_2$.

Research hypothesis: Patients who received the DVD wear their appliances a different mean number of hours per day than patients who did not receive the DVD; $H_1: \mu_1 \neq \mu_2$.

Step 3: $\mu_M = \mu = 14.78; \sigma_M = \frac{\sigma}{\sqrt{N}} = \frac{5.31}{\sqrt{15}} = 1.371$

Step 4: The cutoff z statistics, based on a p level of 0.05 and a two-tailed test, are -1.96 and 1.96. (*Note:* It is helpful to draw a picture of the normal curve and include these z statistics on it.)

Step 5: $z = \frac{(M - \mu_M)}{\sigma_M} = \frac{(17 - 14.78)}{1.371} = 1.62$

(*Note:* It is helpful to add this z statistic to your drawing of the normal curve that includes the cutoff z statistics.)

Step 6: Fail to reject the null hypothesis. We cannot conclude that receiving the DVD improves average patient compliance.

d. The researcher would have made a Type II error. He would have failed to reject the null hypothesis when a mean difference actually existed between the two populations.

CHAPTER 8

8.1 There may be a statistically significant difference between group means, but the difference might not be meaningful or have a real-life application.

8.3 Confidence intervals add details to the hypothesis test. Specifically, they tell us a range within which the population mean would fall 95% of the time if we were to conduct repeated hypothesis tests using samples of the same size from the same population.

8.5 In everyday language, we use the word *effect* to refer to the outcome of some event. Statisticians use the word in a similar way when they look at effect sizes. They want to assess a given outcome. For statisticians, the outcome is any change in a dependent variable, and the event creating the outcome is an independent variable. When statisticians calculate an effect size, they are calculating the size of an outcome.

8.7 In many hypothesis tests, we compare whether two distributions are truly different. If the two distributions overlap a lot, then we would probably find a small effect size and not be willing to conclude that the distributions are necessarily different. If the distributions do not overlap much, this would be evidence for a larger effect or a real difference between them.

8.9 According to Cohen's guidelines for interpreting the d statistic, a small effect is around 0.2, a medium effect is around 0.5, and a large effect is around 0.8.

8.11 In everyday language, we use the word *power* to mean either an ability to get something done or an ability to make others do things. Statisticians use the word *power* to refer to the ability to detect an effect, given that one exists.

8.13 80%

8.15 A researcher could increase statistical power by (1) increasing the alpha level; (2) performing a one-tailed test instead of a two-tailed test; (3) increasing the sample size; (4) maximizing the difference in the levels of the independent variable (e.g., giving a larger dose of a medication); (5) decreasing variability in the distributions by using, for example, reliable measures and homogeneous samples. You want statistical power in your studies, and each of these techniques will increase the probability of discovering an effect that genuinely exists. In many instances, the most practical way to increase statistical power is (3) to increase the sample size.

8.17 The goal of a meta-analysis is to find the mean of the effect sizes from many different studies that all manipulated the same independent variable and measured the same dependent variable.

8.19 (i) σ_M is incorrect. (ii) The correct symbol is σ. (iii) Because we are calculating Cohen's *d*, a measure of effect size, we divide by the standard deviation, σ, not the standard error of the mean. We use standard deviation rather than standard error because effect size is independent of sample size.

8.21 18.5% to 25.5% of respondents were suspicious of steroid use among swimmers.

8.23 **a.** 20%

b. 15%

c. 1%

8.25 **a.** A z of 0.85 leaves 19.77% in the tail.

b. A z of 1.04 leaves 14.92% in the tail.

c. A z of 2.33 leaves 0.99% in the tail.

8.27 We know that the cutoffs for the 95% confidence interval are $z = \pm 1.96$. The standard error is calculated as:

$$\sigma_M = \frac{\sigma}{\sqrt{N}} = \frac{1.3}{\sqrt{78}} = 0.147$$

Now we can calculate the lower and upper bounds of the confidence interval.

$$M_{lower} = -z(\sigma_M) + M_{sample} = -1.96(0.147) + 4.1 = 3.812 \text{ hours}$$

$$M_{upper} = z(\sigma_M) + M_{sample} = 1.96(0.147) + 4.1 = 4.388 \text{ hours}$$

The 95% confidence interval can be expressed as [3.81, 4.39].

8.29 z values of ± 2.58 put 0.49% in each tail, without going over, so we will use those as the critical values for the 99% confidence interval. The standard error is calculated as:

$$\sigma_M = \frac{\sigma}{\sqrt{N}} = \frac{1.3}{\sqrt{78}} = 0.147$$

Now we can calculate the lower and upper bounds of the confidence interval.

$$M_{lower} = -z(\sigma_M) + M_{sample} = -2.58(0.147) + 4.1 = 3.721 \text{ hours}$$

$$M_{upper} = z(\sigma_M) + M_{sample} = 2.58(0.147) + 4.1 = 4.479 \text{ hours}$$

The 99% confidence interval can be expressed as [3.72, 4.48].

8.31 **a.** $\sigma_M = \frac{136}{\sqrt{12}} = 39.261$

$$z = \frac{(M - \mu_M)}{\sigma_M} = \frac{1057 - 1014}{39.261} = 1.10$$

b. $\sigma_M = \frac{136}{\sqrt{39}} = 21.777$

$$z = \frac{1057 - 1014}{21.777} = 1.97$$

c. $\sigma_M = \frac{136}{\sqrt{188}} = 9.919$

$$z = \frac{1057 - 1014}{9.919} = 4.34$$

8.33 **a.** Cohen's $d = \frac{(M - \mu)}{\sigma} = \frac{(480 - 500)}{100} = -0.20$

b. Cohen's $d = \frac{(M - \mu)}{\sigma} = \frac{(520 - 500)}{100} = 0.20$

c. Cohen's $d = \frac{(M - \mu)}{\sigma} = \frac{(610 - 500)}{100} = 1.10$

8.35 **a.** Large

b. Medium

c. Small

d. No effect (very close to zero)

8.37 **a.** The percentage beyond the z statistic of 2.23 is 1.29%. Doubled to take into account both tails, this is 2.58%. Converted to a proportion by dividing by 100, we get a p value of 0.0258, or 0.03.

b. For −1.82, the percentage in the tail is 3.44%. Doubled, it is 6.88%. As a proportion, it is 0.0688, or 0.07.

c. For 0.33, the percentage in the tail is 37.07%. Doubled, it is 74.14%. As a proportion, it is 0.7414, or 0.74.

8.39 We would fail to reject the null hypothesis because the confidence interval around the mean effect size includes 0.

8.41 Your friend is not considering the fact that the two distributions, that of IQ scores of Burakumin and that of IQ scores of other Japanese, will have a great deal of overlap. The fact that one mean is higher than another does not imply that all members of one group have higher IQ scores than all members of another group. Any individual member of either group, such as your friend's former student, might fall well above the mean for his or her group (and the other group) or well below the mean for his or her group (and the other group). Research reports that do not give an indication of the overlap between two distributions risk misleading their audience.

8.43 **a.** *Step 3:*

$$\mu_M = \mu = 20.4;\ \sigma_M = \frac{\sigma}{\sqrt{N}} = \frac{3.2}{\sqrt{3}} = 1.848$$

Step 4: The cutoff z statistics are −1.96 and 1.96.

Step 5:

$$z = \frac{(M - \mu_M)}{\sigma_M} = \frac{(17.5 - 20.4)}{1.848} = -1.57$$

Step 6: Fail to reject the null hypothesis; we can conclude only that there is not sufficient evidence that Canadian adults have different average GNT scores from English adults. The conclusion has changed, but the actual difference between groups has not. The smaller sample size led to a larger standard error and a smaller test statistic. This makes sense because an extreme mean based on just a few participants is more likely to have occurred by chance than is an extreme mean based on many participants.

b. *Step 3:*

$$\mu_M = \mu = 20.4;\ \sigma_M = \frac{\sigma}{\sqrt{N}} = \frac{3.2}{\sqrt{100}} = 0.32$$

Step 5:

$$z = \frac{(M - \mu_M)}{\sigma_M} = \frac{(17.5 - 20.4)}{0.32} = -9.06$$

Step 6: Reject the null hypothesis. It appears that Canadian adults have lower average GNT scores than English adults. The test statistic has increased along with the increase in sample size.

c. *Step 3:*

$$\mu_M = \mu = 20.4;\ \sigma_M = \frac{\sigma}{\sqrt{N}} = \frac{3.2}{\sqrt{20{,}000}} = 0.023$$

Step 5:

$$z = \frac{(M - \mu_M)}{\sigma_M} = \frac{(17.5 - 20.4)}{0.023} = -126.09$$

The test statistic is now even larger, as the sample size has grown even larger. Step 6 is the same as in part (b).

d. As sample size increases, the test statistic increases. A mean difference based on a very small sample could have occurred just by chance. Based on a very large sample, that same mean difference is less likely to have occurred just by chance.

e. The underlying difference between groups has not changed. This might pose a problem for hypothesis testing because the same mean difference is statistically significant under some circumstances but not others. A very large test statistic might not indicate a very large difference between means; therefore, a statistically significant difference might not be an important difference.

8.45 **a.** No, we cannot tell which student will do better on the LSAT. It is likely that the distributions of LSAT scores for the two groups (humanities majors and social science majors) have a great deal of overlap. Just because one group, on average, does better than another group does not mean that every student in one group does better than every student in another group.

b. Answers to this will vary, but the two distributions should overlap and the mean of the distribution for the social sciences majors should be farther to the right (i.e., higher) than the mean of the distribution for the humanities majors.

8.47 **a.** Given $\mu = 16.189$ and $\sigma = 12.128$, we calculate $\sigma_M = \frac{\sigma}{\sqrt{N}} = \frac{12.128}{\sqrt{4}} = 6.064$. To calculate the 95% confidence interval, we find the z values that mark off the most extreme 0.025 in each tail, which are −1.96 and 1.96. We calculate the lower end of the interval as $M_{lower} = -z(\sigma_M) + M_{sample} = -1.96(6.064) + 8.75 = -3.14$ and the upper end of the interval as $M_{upper} = z(\sigma_M) + M_{sample} = 1.96(6.064) + 8.75 = 20.64$. The confidence interval around the mean of 8.75 is [−3.14, 20.64].

b. Because 16.189, the null-hypothesized value of the population mean, falls within this confidence interval, it is plausible that the point spreads of Division I-AA schools are the same, on average, as the point spreads of Division I-A schools. It is plausible that they come from the same population of point spreads.

c. Because the confidence interval includes 16.189, we know that we would fail to reject the null hypothesis if we conducted a hypothesis test. It is plausible that the sample came from a population with $\mu = 16.189$ by chance. We do not have sufficient evidence to conclude that the point spreads of Division I-AA schools are from a different population than the point spreads of Division I-A schools.

d. In addition to letting us know that it is plausible that the Division I-AA point spreads are from the same population as those for the Division I-A schools, the confidence interval tells us a range of plausible values for the mean point spread.

8.49 **a.** The appropriate measure of effect size for a z statistic is Cohen's d, which is calculated as

$$d = \frac{M - \mu}{\sigma} = \frac{8.75 - 16.189}{12.128} = -0.61$$

b. Based on Cohen's conventions, this is a medium-to-large effect size.

c. The hypothesis test tells us only whether a sample mean is likely to have been obtained by chance, whereas the effect size gives us the additional information of how much overlap there is between the distributions. Cohen's *d*, in particular, tells us how far apart two means are in terms of standard deviation. Because it's based on standard deviation, not standard error, Cohen's *d* is independent of sample size and therefore has the added benefit of allowing us to compare across studies. In summary, effect size tells us the magnitude of the effect, giving us a sense of how important or practical this finding is, and allows us to standardize the results of the study. Here, we know that there's a medium-to-large effect.

8.51 **a.** We know that the cutoffs for the 95% confidence interval are $z = \pm 1.96$. Standard error is calculated as:

$$\sigma_M = \frac{\sigma}{\sqrt{N}} = \frac{16}{\sqrt{18}} = 3.771$$

Now we can calculate the lower and upper bounds of the confidence interval.

$$M_{lower} = -z(\sigma_M) + M_{sample} = -1.96(3.771) + 38 = \$30.61$$

$$M_{upper} = z(\sigma_M) + M_{sample} = 1.96(3.771) + 38 = \$45.39$$

The 95% confidence interval can be expressed as [$30.61, $45.39].

b. Standard error is now calculated as:

$$\sigma_M = \frac{\sigma}{\sqrt{N}} = \frac{16}{\sqrt{180}} = 1.193$$

Now we can calculate the lower and upper bounds of the confidence interval.

$$M_{lower} = -z(\sigma_M) + M_{sample} = -1.96(1.193) + 38 = \$35.66$$

$$M_{upper} = z(\sigma_M) + M_{sample} = 1.96(1.193) + 38 = \$40.34$$

The 95% confidence interval can be expressed as [$35.66, $40.34].

c. The null-hypothesized mean of $45 falls in the 95% confidence interval when *N* is 18. Because of this, we cannot claim that things are lower in 2009 than what we would normally expect. When *N* is increased to 180, the confidence interval becomes narrower because standard error is reduced. As a result, the mean of $45 no longer falls within the interval, and we can now conclude that Valentine's Day spending is different in 2009 from what was expected based on previous population data.

d. Cohen's $d = \frac{M - \mu}{\sigma} = \frac{(38 - 45)}{16} = -0.44$, just around a medium effect size.

8.53 **a.** Standard error is calculated as:

$$\sigma_M = \frac{\sigma}{\sqrt{N}} = \frac{12}{\sqrt{26}} = 2.353$$

Now we can calculate the lower and upper bounds of the confidence interval.

$$M_{lower} = -z(\sigma_M) + M_{sample} = -1.96(2.353) + 123 = 118.39 \text{ mph}$$

$$M_{upper} = z(\sigma_M) + M_{sample} = 1.96(2.353) + 123 = 127.61 \text{ mph}$$

The 95% confidence interval can be expressed as [118.39, 127.61].

Because the population mean of 118 mph does not fall within the confidence interval around the new mean, we can conclude that the program had an impact. In fact, we can conclude that the program increased the average speed of women's serves.

b. Cohen's $d = \frac{(M - \mu)}{\sigma} = \frac{(123 - 118)}{12} = 0.42$, a medium effect.

c. Because standard error, which utilizes sample size in its calculation, is part of the calculations for confidence interval, the interval becomes narrower as the sample size increases; however, because sample size is eliminated from the calculation of effect size, the effect size does not change.

8.55 **a.** The topic is the effectiveness of culturally adapted therapies.

b. The researchers used Cohen's *d* as a measure of effect size for each study in the analysis.

c. The mean effect size they found was 0.45. According to Cohen's conventions, this is a medium effect.

d. The researchers could use the group means and standard deviations to calculate a measure of effect size.

8.57 **a.** A statistically significant difference just indicates that the difference between the means is unlikely to be due to chance. It does not tell us that there is *no* overlap in the distributions of the two populations we are considering. It is likely that there is overlap between the distributions and that some players with three children actually perform better than some players with two or fewer children. The drawings of distributions will vary; the two curves will overlap, but the mean of the distribution representing two or fewer children should be farther to the right than the mean of the distribution representing three or more children.

b. A difference can be statistically significant even if it is very small. In fact, if there are enough observations in a sample, even a tiny difference will approach statistical significance. Statistical significance does not indicate the importance or size of an effect—we need measures of effect size, which are not influenced by sample size, to understand the importance of an effect. These measures of effect size allow us to compare different predictors of performance. For example, in this case, it is likely that other aspects of a player's stats are more strongly associated with his performance and therefore would have a larger effect size. We could make the decision about whom to include in the fantasy team on the basis of the largest predictors of performance.

c. Even if the association is true, we cannot conclude that having a third child causes a decline in baseball performance. There are a number of possible causal explanations for this relation. It could be the reverse; perhaps those players who are not performing as well in their careers end up devoting more time to family, so not playing well could lead to having more children. Alternatively, a third variable could explain both (a) having three children, and (b) poorer baseball performance. For example, perhaps less competitive or more laid-back players have more children and also perform more poorly.

d. The sample size for this analysis is likely small, so the statistical power to detect an effect is likely small as well.

CHAPTER 9

9.1 The t distributions are used when we do not know the population standard deviation and are comparing two groups.

9.3 For both tests, standard error is calculated as the standard deviation divided by the square root of N. For the z test, the population standard deviation is calculated with N in the denominator. For the t test, the standard deviation for the population is estimated by dividing the sum of squared deviations by $N - 1$.

9.5 t stands for the t statistic, M is the sample mean, μ_M is the mean of the distribution of means, and s_M is the standard error as estimated from a sample.

9.7 *Free to vary* refers to the number of scores that can take on different values if we know a given parameter.

9.9 As the sample size increases, we can feel more confident in the estimate of the variability in the population. Remember, this estimate of variability (s) is calculated with $N - 1$ in the denominator in order to inflate the estimate somewhat. As the sample increases from 10 to 100, for example, and then up to 1000, subtracting 1 from N has less of an on the overall calculation. As this happens, the t distributions approach the z distribution, where we in fact knew the population standard deviation and did not need to estimate it.

9.11 A distribution of mean differences is constructed by calculating the difference scores for a sample of individuals and then averaging those differences. This process is performed repeatedly, using the same population and samples of the same size. Once a collection of mean differences is gathered, they can be displayed on a graph (in most cases, they form a bell-shaped curve).

9.13 The term *paired=samples* is used to describe a test that compares an individual's scores in two conditions; it is also called a *paired-samples t test*. The term *independent=samples* refers to groups that do not overlap in any way, including membership; the observations made in one group in no way relate to or depend on the observations made in another group.

9.15 Unlike a single-sample t test, in the paired-samples t test we have two scores for every participant; we take the difference between these scores before calculating the sample mean difference that will be used in the t test.

9.17 If the confidence interval around the mean difference score includes the value of 0, then 0 is a plausible mean difference. If we conduct a hypothesis test for these data, we would fail to reject the null hypothesis.

9.19 As with other hypothesis tests, the conclusions from both the single-sample t test or paired-samples t test and the confidence interval are the same, but the confidence interval gives us more information—an interval estimate, not just a point estimate.

9.21 **a.** First we need to calculate the mean:

$$M = \frac{\Sigma X}{N} = \frac{93 + 97 + 91 + 88 + 103 + 94 + 97}{7} = \frac{663}{7} = 94.714$$

We then calculate the deviation of each score from the mean and the square of that deviation.

X	X − M	(X − M)^2
93	−1.714	2.938
97	2.286	5.226
91	−3.714	13.794
88	−6.714	45.078
103	8.286	68.658
94	−0.714	0.510
97	2.286	5.226

The standard deviation is:

$$SD = \sqrt{\frac{\Sigma(X - M)^2}{N}} = \sqrt{\frac{141.430}{7}} = \sqrt{20.204} = 4.49$$

b. When estimating the population variability, we calculate s:

$$s = \sqrt{\frac{\Sigma(X - M)^2}{N - 1}} = \sqrt{\frac{141.430}{7 - 1}} = \sqrt{23.572} = 4.855$$

c. $s_M = \frac{s}{\sqrt{N}} = \frac{4.855}{\sqrt{7}} = 1.835$

d. $t = \frac{(M - \mu_M)}{s_M} = \frac{(94.714 - 96)}{1.835} = -0.70$

9.23 **a.** Because 73 *df* is not on the table, we go to 60 *df* (we do not go to the closest value, which would be 80, because we want to be conservative and go to the next-lowest value for *df*) to find the critical value of 1.296 in the upper tail. If we are looking in the lower tail, the critical value is −1.296.

b. ±1.984

c. Either −2.438 or +2.438

9.25 **a.** This is a two-tailed test with $df = 25$, so the critical t values are ±2.060.

b. $df = 17$, so the critical t value is +2.567, assuming you're anticipating an increase in marital satisfaction.

c. $df = 17$, so the critical t value is +2.567, assuming you're anticipating an increase in marital satisfaction.

d. $df = 63$, so the critical t values are ±2.001

e. $df = 33$, so the critical t values are ±2.043

9.27 **a.** $t = \frac{(M - \mu_M)}{s_M} = \frac{(8.5 - 7)}{\left(\frac{2.1}{\sqrt{41}}\right)} = 4.57$

b. $M_{lower} = -t(s_M) + M_{sample} = -2.705(0.328) + 8.5 = 7.61$

$M_{upper} = t(s_M) + M_{sample} = 2.705(0.328) + 8.5 = 9.39$

c. $d = \frac{(M - \mu)}{s} = \frac{(8.5 - 7)}{2.1} = 0.71$

9.29 **a.**

EXAM I	EXAM II	DIFFERENCE	X − M	(X − M)²
92	84	−8	−9.25	85.563
67	75	8	6.75	45.563
95	97	2	0.75	0.563
82	87	5	3.75	14.063
73	68	−5	−6.25	39.063
59	63	4	2.75	7.563
90	88	−2	−3.25	10.563
72	78	6	4.75	22.563

$M_{difference} = 1.25$

$SS = \Sigma(X - M)^2 = 225.504$

$$s = \sqrt{\frac{SS}{N-1}} = \sqrt{\frac{225.504}{7}} = 5.676$$

$$s_M = \frac{s}{\sqrt{N}} = \frac{5.676}{\sqrt{8}} = 2.007$$

$$t = \frac{(M - \mu)}{s_M} = \frac{(1.25 - 0)}{2.007} = 0.62$$

b. With $df = 7$, the critical t values are ± 2.365. The calculated t statistic of 0.62 does not exceed the critical value. Therefore, we fail to reject the null hypothesis.

c. When increasing N to 1000, we need to recalculate s_M and the t test.

$$s_M = \frac{s}{\sqrt{N}} = \frac{5.676}{\sqrt{1000}} = 0.179$$

$$t = \frac{(1.25 - 0)}{0.179} = 6.98$$

Our critical values with $df = 999$ are $t = \pm 1.98$. Because the calculated t exceeds one of the t critical values, we reject the null hypothesis.

d. Increasing the sample size increased the value of the t statistic and decreased the critical t values, making it easier for us to reject the null hypothesis.

9.31 **a.**

SCORE 1	SCORE 2	DIFFERENCE	X − M	(X − M)²
45	62	17	5.429	29.474
34	56	22	10.429	108.764
22	40	18	6.429	41.332
45	48	3	−8.571	73.462
15	26	11	−0.571	0.326
51	56	5	−6.571	43.178
28	33	5	−6.571	43.178

$M_{difference} = 11.571$

$SS = \Sigma(X - M)^2 = 339.714$

$$s = \sqrt{\frac{SS}{N-1}} = \sqrt{\frac{339.714}{6}} = 7.525$$

$$s_M = \frac{s}{\sqrt{N}} = \frac{7.525}{\sqrt{7}} = 2.844$$

$$t = \frac{(M - \mu)}{s_M} = \frac{(11.571 - 0)}{2.844} = 4.07$$

b. With $N = 7$, $df = 6$, $t = \pm 2.447$:

$M_{lower} = -t(s_M) + M_{sample} = -2.447(2.844) + 11.571 = 4.61$

$M_{upper} = t(s_M) + M_{sample} = 2.447(2.844) + 11.571 = 18.53$

c. $$d = \frac{(M - \mu)}{s} = \frac{(11.571 - 0)}{7.525} = 1.54$$

9.33 **a.** $$s_M = \frac{s}{\sqrt{N}} = \frac{1.42}{\sqrt{13}} = 0.394$$

$$t = \frac{(-0.77 - 0)}{0.394} = -1.95$$

b. $M_{lower} = -t(s_M) + M_{sample} = -2.179(0.394) + (-0.77) = -1.63$

$M_{upper} = t(s_M) + M_{sample} = 2.179(0.394) + (-0.77) = 0.09$

c. $$d = \frac{(M - \mu)}{s} = \frac{(-0.77 - 0)}{1.42} = -0.54$$

9.35 **a.** ± 1.96

b. Either -2.33 or $+2.33$, depending on the tail of interest

c. ± 1.96

d. The critical z values are lower than the critical t values, making it easier to reject the null hypothesis when conducting a z test. Decisions using the t distributions are more conservative because of the chance we may have poorly estimated the population standard deviation.

9.37 **a.** *Step 1:* Population 1 is male U.S. Marines following a month-long training exercise. Population 2 is college men. The comparison distribution will be a distribution of means. The hypothesis test will be a single-sample t test because we have only one sample and we know the population mean but not the standard deviation. This study meets one of the three assumptions and may meet another. The dependent variable, anger, appears to be scale. The data were not likely randomly selected, so we must be cautious with respect to generalizing to all Marines who complete this training. We do not know whether the population is normally distributed, and there are not at least 30 participants. However, the data from our sample do not suggest a skewed distribution.

Step 2: Null hypothesis: Male U.S. Marines after a month-long training exercise have the same average anger levels as college men—H_0: $\mu_1 = \mu_2$.

Research hypothesis: Male U.S. Marines after a month-long training exercise have different average anger levels than college men—H_1: $\mu_1 \neq \mu_2$.

Step 3: $\mu_M = \mu = 8.90$; $s_M = 0.494$

X	$X - M$	$(X - M)^2$
14	0.667	0.445
12	−1.333	1.777
13	−0.333	0.111
12	−1.333	1.777
14	0.667	0.445
15	1.667	2.779

$M = 13.333$

$SS = \Sigma(X - M)^2 = \Sigma(0.445 + 1.777 + 0.111 + 1.777 + 0.445 + 2.779) = 7.334$

$$s = \sqrt{\frac{\Sigma(X-M)^2}{N-1}} = \sqrt{\frac{SS}{(N-1)}} = \sqrt{\frac{7.334}{6-1}}$$
$$= \sqrt{1.467} = 1.211$$

$$s_M = \frac{s}{\sqrt{N}} = \frac{1.211}{\sqrt{6}} = 0.494$$

Step 4: $df = N - 1 = 6 - 1 = 5$; the critical values, based on 5 degrees of freedom, a p level of 0.05, and a two-tailed test, are −2.571 and 2.571. (*Note:* It is helpful to draw a curve that includes these cutoffs.)

Step 5: $t = \frac{(M - \mu_M)}{s_M} = \frac{(13.333 - 8.90)}{0.494} = 8.97$

(*Note:* It is helpful to add this t statistic to the curve that you drew in step 4.)

Step 6: Reject the null hypothesis. It appears that male U.S. Marines just after a month-long training exercise have higher average anger levels than college men; $t(5) = 8.96$, $p < 0.05$.

b. $t = \frac{(M - \mu_M)}{s_M} = \frac{(13.333 - 9.20)}{0.494} = 8.37$; reject the null hypothesis; it appears that male U.S. Marines just after a month-long training exercise have higher average anger levels than adult men; $t(5) = 8.35$, $p < 0.05$.

c. $t = \frac{(M - \mu_M)}{s_M} = \frac{(13.333 - 13.5)}{0.494} = -0.34$ fail to reject the null hypothesis; we conclude that there is no evidence from this study to support the research hypothesis; $t(5) = -0.34$, $p > 0.05$.

d. We can conclude that Marines' anger scores just after high-altitude, cold-weather training are, on average, higher than those of college men and adult men. We cannot conclude, however, that they are different, on average, than those of male psychiatric outpatients. With respect to the latter difference, we can only conclude that there is no evidence to support that there is a difference between Marines' mean anger scores and those of male psychiatric outpatients.

9.39 **a.** *Step 1:* Population 1 is the Devils players in the 2007–2008 season. Population 2 is the Devils players in the 2008–2009 season. The comparison distribution is a distribution of mean differences. We meet one assumption: The dependent variable, goals, is scale. We do not, however, meet the assumption that our participants are randomly selected from the population. We may also not meet the assumption that the population distribution of scores is normally distributed (the scores do not appear normally distributed and we do not have an N of at least 30).

Step 2: Null hypothesis: The team performed no differently, on average, in the 2007–2008 and 2008–2009 seasons—H_0: $\mu_1 = \mu_2$.

Research hypothesis: The team scored a different number of goals, on average, in the 2007–2008 and 2008–2009 seasons—H_1: $\mu_1 \neq \mu_2$.

Step 3: $\mu = 0$ and $s_M = 3.682$

2007–2008	2008–2009	DIFFERENCE	$X - M$	$(X - M)^2$
20	31	11	4.833	23.358
14	20	6	−0.167	0.028
12	5	−7	−13.167	173.370
13	29	16	9.833	96.688
22	20	−2	−8.167	66.670
32	45	13	6.833	46.690

$M_{difference} = 6.167$

$SS = \Sigma(X - M)^2 = 406.804$

$$s = \sqrt{\frac{SS}{N-1}} = \sqrt{\frac{406.804}{5}} = 9.020$$

$$s_M = \frac{s}{\sqrt{N}} = \frac{9.020}{\sqrt{6}} = 3.682$$

Step 4: The critical t values with a two-tailed test, p level of 0.05, and $df = 5$ are ±2.571.

Step 5: $t = \frac{(M - \mu)}{s_M} = \frac{(6.167 - 0)}{3.682} = 1.67$

Step 6: Fail to reject the null hypothesis because the calculated t statistic of 1.67 does not exceed the critical t value.

b. $t(5) = 1.67, p > 0.05$ (*Note:* If we had used software, we would provide the actual p value.)

c. $M_{lower} = -t(s_M) + M_{sample} = -2.571(3.682) + 6.167 = -3.30$

$M_{upper} = t(s_M) + M_{sample} = 2.571(3.682) + 6.167 = 15.63$

Because the confidence interval includes 0, we fail to reject the null hypothesis. This is consistent with the results of the hypothesis test conducted in part (a).

d. $d = \frac{(M - \mu)}{s} = \frac{(6.167 - 0)}{9.020} = 0.68$

9.41 **a.** The professor would use a paired-samples t-test.

b. No. A change or a difference in mean score might not be statistically significant, particularly with a small sample.

c. It would be easier to reject the null hypothesis for a given mean difference with the class with 700 students than with the class with 7 students because the t value would be higher with the larger sample.

9.43 **a.** The appropriate mean: $\mu_M = \mu = 11.72$

The calculations for the appropriate standard deviation (in this case, standard error, s_M):

$$M = \frac{\Sigma X}{N} = \frac{(25.62 + 13.09 + 8.74 + 17.63 + 2.80 + 4.42)}{6}$$
$$= 12.05$$

X	X − M	(X − M)²
25.62	13.57	184.145
13.09	1.04	1.082
8.74	−3.31	10.956
17.63	5.58	31.136
2.80	−9.25	85.563
4.42	−7.63	58.217

Numerator: $\Sigma(X - M)^2 = \Sigma(184.145 + 1.082 + 10.956 + 31.136 + 85.563 + 58.217) = 371.099$

$$s = \sqrt{\frac{\Sigma(X-M)^2}{(N-1)}} = \sqrt{\frac{371.099}{(6-1)}} = \sqrt{74.220} = 8.615$$

$$s_M = \frac{s}{\sqrt{N}} - \frac{8.615}{\sqrt{6}} = 3.517$$

b. $t = \frac{(M - \mu_M)}{s_M} = \frac{(12.05 - 11.72)}{3.517} = 0.09$

c. There are several possible answers to this question. Among the hypotheses that could be examined are whether the length of stay on death row depends on gender, race, or age. Specifically, given prior evidence of a racial bias in the implementation of the death penalty, we might hypothesize that black and Hispanic prisoners have shorter times to execution than do prisoners overall.

d. We would need to know the population standard deviation. If we were really interested in this, we could calculate the standard deviation from the entire online execution list.

e. The null hypothesis states that the average time spent on death row in recent years is equal to what it has been historically (no change)—$H_0: \mu_1 = \mu_2$. The research hypothesis is that there has been a change in the average time spent on death row—$H_1: \mu_1 \neq \mu_2$.

f. The t statistic we calculated was 0.09. The critical t values for a two-tailed test, alpha or p of 0.05 and df of 5, are ±2.571. We fail to reject the null hypothesis and conclude that we do not have sufficient evidence to indicate a change in time spent on death row.

g. $M_{lower} = -t(s_M) + M_{sample} = -2.571(3.517) + 12.05 = 3.01$ years

$M_{upper} = t(s_M) + M_{sample} = 2.571(3.517) + 12.05 = 21.09$ years

h. Because the population mean of 11.72 years is within the very large range of the confidence interval, we fail to reject the null hypothesis. This confidence interval is so large that it is not useful. The large size of the confidence interval is due to the large variability in the sample (s_M) and the small sample size (resulting in a large critical t value).

i. $d = \frac{(M - \mu)}{s} = \frac{(12.05 - 11.72)}{8.615} = 0.12$

j. This is a small effect.

9.45 **a.** The independent variable is presence of posthypnotic suggestion with two levels: suggestion or no suggestion. The dependent variable is Stroop reaction time in seconds.

b. *Step 1:* Population 1 is highly hypnotizable individuals who receive a posthypnotic suggestion. Population 2 is highly hypnotizable individuals who do not receive a posthypnotic suggestion. The comparison distribution will be a distribution of mean differences. The hypothesis test will be a paired-samples t test because we have two samples and all participants are in both samples. This study meets one of the three assumptions and may meet another. The dependent variable, reaction time in seconds, is scale. The data were not likely randomly selected, so we should be cautious when generalizing beyond the sample. We do not know whether the population is normally distributed and there are not at least 30 participants, but the sample data do not suggest skew.

Step 2: Null hypothesis: Highly hypnotizable individuals who receive a posthypnotic suggestion will have the same average Stroop reaction times as highly hypnotizable individuals who receive no posthypnotic suggestion—$H_0: \mu_1 = \mu_2$.

Research hypothesis: Highly hypnotizable individuals who receive a posthypnotic suggestion will have different average Stroop reaction times than will highly hypnotizable individuals who receive no posthypnotic suggestion—$H_1: \mu_1 \neq \mu_2$.

Step 3: $\mu_M = \mu = 0; s_M = 0.420$

(*Note:* Remember to cross out the original scores once you have created the difference scores so you won't be tempted to use them in your calculations.)

X	Y	DIFFERENCE	X − M	(X − M)²
12.6	8.5	−4.1	−1.05	1.103
13.8	9.6	−4.2	−1.15	1.323
11.6	10.0	−1.6	1.45	2.103
12.2	9.2	−3.0	0.05	0.003
12.1	8.9	−3.2	−0.15	0.023
13.0	10.8	−2.2	0.85	0.723

$M_{difference} = -3.05$

$SS = \Sigma(X - M)^2 = \Sigma(1.103 + 1.323 + 2.103 + 0.003 + 0.023 + 0.723) = 5.278$

$$s = \sqrt{\frac{\Sigma(X-M)^2}{(N-1)}} = \sqrt{\frac{SS}{(N-1)}} = \sqrt{\frac{5.278}{(6-1)}}$$

$$= \sqrt{1.056} = 1.028$$

$$s_M = \frac{s}{\sqrt{N}} = \frac{1.028}{\sqrt{6}} = 0.420$$

Step 4: $df = N - 1 = 6 - 1 = 5$; the critical values, based on 5 degrees of freedom, a p level of 0.05, and a two-tailed test, are −2.571 and 2.571.

Step 5: $t = \frac{(M_{difference} - \mu_{difference})}{s_M} = \frac{(-3.05 - 0)}{0.420} = -7.26$

Step 6: Reject the null hypothesis; it appears that highly hypnotizable people have faster Stroop reaction times when they receive a posthypnotic suggestion than when they do not.

c. $t(5) = -7.26, p < 0.05$

d. *Step 2:* Null hypothesis: The average Stroop reaction time of highly hypnotizable individuals who receive a posthypnotic suggestion is greater than or equal to that of highly hypnotizable individuals who receive no posthypnotic suggestion—$H_0: \mu_1 \geq \mu_2$.

Research hypothesis: Highly hypnotizable individuals who receive a posthypnotic suggestion will have faster (i.e., lower number) average Stroop reaction times than highly hypnotizable individuals who receive no posthypnotic suggestion—$H_1: \mu_1 < \mu_2$.

Step 4: $df = N - 1 = 6 - 1 = 5$; the critical value, based on 5 degrees of freedom, a p level of 0.05, and a one-tailed test, is -2.015. (*Note:* It is helpful to draw a curve that includes this cutoff.)

Step 6: Reject the null hypothesis; it appears that highly hypnotizable people have faster Stroop reaction times when they receive a posthypnotic suggestion than when they do not.

It is easier to reject the null hypothesis with a one-tailed test. Although we rejected the null hypothesis under both conditions, the critical t value is less extreme with a one-tailed test because the entire 0.05 (5%) critical region is in one tail instead of divided between two.

The difference between the means of the samples is identical, as is the test statistic. The only aspect that is affected is the critical value.

e. *Step 4:* $df = N - 1 = 6 - 1 = 5$; the critical values, based on 5 degrees of freedom, a p level of 0.01, and a two-tailed test, are -4.032 and 4.032. (*Note:* It is helpful to draw a curve that includes these cutoffs.)

Step 6: Reject the null hypothesis; it appears that highly hypnotizable people have faster Stroop reaction times when they receive a posthypnotic suggestion than when they do not.

A p level of 0.01 leads to more extreme critical values than a p level of 0.05. When the tails are limited to 1% versus 5%, the tails beyond the cutoffs are smaller and the cutoffs are more extreme. So it is easier to reject the null hypothesis with a p level of .05 than the null hypothesis with a p level of .01.

The difference between the means of the samples is identical, as is the test statistic. The only aspect that is affected is the critical value.

f. *Step 3:* $\mu_M = \mu = 0$; $s_M = 0.850$

(*Note:* Remember to cross out the original scores once you have created the difference scores so you won't be tempted to use them in your calculations.)

X	Y	DIFFERENCE	X − M	(X − M)²
12.6	8.5	−4.1	−0.8	0.64
13.8	9.6	−4.2	−0.9	0.81
11.6	10.0	−1.6	1.7	2.89

$M_{difference} = -3.3$

$SS = \Sigma(X - M)^2 = \Sigma(0.64 + 0.81 + 2.89) = 4.34$

$$s = \sqrt{\frac{\Sigma(X - M)^2}{(N - 1)}} = \sqrt{\frac{SS}{(N - 1)}} = \sqrt{\frac{4.34}{(3 - 1)}}$$

$$= \sqrt{2.17} = 1.473$$

$$s_M = \frac{s}{\sqrt{N}} = \frac{1.473}{\sqrt{3}} = 0.850$$

Step 4: $df = N - 1 = 3 - 1 = 2$; the critical values, based on 2 degrees of freedom, a p level of 0.05, and a two-tailed test, are -4.303 and 4.303. (*Note:* It is helpful to draw a curve that includes these cutoffs.)

$$\textit{Step 5: } t = \frac{(M_{difference} - \mu_{difference})}{s_M} = \frac{(-3.3 - 0)}{0.850} = -3.88$$

(*Note:* It is helpful to add this t statistic to the curve that you drew in step 4.)

This test statistic is no longer beyond the critical value. Reducing the sample size makes it more difficult to reject the null hypothesis because it results in a larger standard error and therefore a smaller test statistic. It also results in more extreme critical values.

CHAPTER 10

10.1 An independent-samples t test is used when we do not know the population parameters and are comparing two groups that are composed of nonoverlapping, unrelated participants or observations.

10.3 Independent events are things that do not affect each other. For example, the lunch you buy today does not impact the hours of sleep the authors of this book will get tonight.

10.5 The comparison distribution for the paired-samples t test is made up of *mean differences*—the average of many difference scores. The comparison distribution for the independent-samples t test is made up of *differences between means,* or the differences we can expect to see between group means if the null hypothesis is true.

10.7 Both of these represent corrected variance within a group (s^2), but one is for the X variable and the other is for the Y variable. Because these are corrected measures of variance, $N - 1$ is in the denominator of the equations.

10.9 We assume that larger samples do a better job of estimating the population than smaller samples do, so we would want the variability measure based on the larger sample to count more.

10.11 We can take the confidence interval's upper bound and lower bound, compare those to the point estimate in the numerator, and get the margin of error. So, if we predict a score of 7 with a confidence interval of [4.3, 9.7], we can also express this as a margin of error of 2.7 points (7 ± 2.7). Confidence interval and margin of error are simply two ways to say the same thing.

10.13 Larger ranges mean less precision in making predictions, just as widening the goal posts in rugby or in American football mean that you can be less precise when trying to kick the ball between the posts. Smaller ranges indicate we are doing a better job of predicting the phenomenon within the population. For example, a 95% confidence interval that spans a range from 2 to 12 is larger than a 95% confidence interval from 5 to 6. Although the percentage range has stayed the same, the width of the distribution has changed.

10.15 We would take several steps back from the final calculation of standard error to the step in which we calculated pooled variance. Pooled variance is the variance version, or squared version, of standard deviation. To convert pooled variance to the pooled standard deviation, we take its square root.

10.17 Guidelines for interpreting the size of an effect based on Cohen's d were presented in Chapter 8. Those guidelines state that 0.2 is a small effect, 0.5 is a medium effect, and 0.8 is a large effect.

10.19 **a.** Group 1 is treated as the X variable; $M_X = 95.8$.

X	X − M	(X − M)²
97	1.2	1.44
83	−12.8	163.84
105	9.2	84.64
102	6.2	38.44
92	−3.8	14.44

$$s_X^2 = \frac{\Sigma(X-M)^2}{N-1} = \frac{(1.44+163.84+84.64+28.44+14.44)}{5-1} = 75.7$$

Group 2 is treated as the Y variable; $M_Y = 104$.

Y	Y − M	(Y − M)²
111	7	49
103	−1	1
96	−8	64
106	2	4

$$s_Y^2 = \frac{\Sigma(Y-M)^2}{N-1} = \frac{(49+1+64+4)}{4-1} = 39.333$$

b. Treating group 1 as X and group 2 as Y, $df_X = N - 1 = 5 - 1 = 4$, $df_Y = 4 - 1 = 3$, and $df_{total} = df_X + df_Y = 4 + 3 = 7$.

c. **−2.365**, 2,365

d. $$s_{pooled}^2 = \left(\frac{df_X}{df_{total}}\right)s_X^2 + \left(\frac{df_Y}{df_{total}}\right)s_Y^2 = \left(\frac{4}{7}\right)75.7 + \left(\frac{3}{7}\right)39.333 = 43.257 + 16.857 = 60.114$$

e. For group 1: $$s_{M_X}^2 = \frac{s_{pooled}^2}{N_Y} = \frac{60.114}{5} = 12.023$$

For group 2: $$s_{M_Y}^2 = \frac{s_{pooled}^2}{N_Y} = \frac{60.114}{4} = 15.029$$

f. $$s_{difference}^2 = s_{M_X}^2 + s_{M_Y}^2 = 12.023 + 15.029 = 27.052$$

The standard deviation of the distribution of differences between means is:

$$s_{difference} = \sqrt{s_{difference}^2} = \sqrt{27.052} = 5.201$$

g. $$t = \frac{(M_X - M_Y) - (\mu_X - \mu_Y)}{s_{difference}} = \frac{(95.8 - 104) - (0)}{5.201} = -1.58$$

h. The critical t values for the 95% confidence interval for a df of 7 are −2.365 and 2.365.

$(M_X - M_Y)_{lower} = -t(s_{difference}) + (M_X - M_Y)_{sample}$
$(M_X - M_Y)_{lower} = -2.365(5.201) + (-8.2) = -20.50$
$(M_X - M_Y)_{upper} = t(s_{difference}) + (M_X - M_Y)_{sample}$
$(M_X - M_Y)_{upper} = 2.365(5.201) + (-8.2) = 4.10$

The confidence interval is [−20.50, 4.10].

i. To calculate Cohen's d, we need to calculate the pooled standard deviation for the data:

$$s_{pooled} = \sqrt{s_{pooled}^2} = \sqrt{60.114} = 7.753$$

$$\text{Cohen's } d = \frac{(M_X - M_Y) - (\mu_X - \mu_Y)}{s_{pooled}} = \frac{(95.8 - 104) - (0)}{7.753} = -1.06$$

10.21 **a.** df_{total} is 35, and the cutoffs are −2.030 and 2.030.

b. df_{total} is 26, and the cutoffs are −2.779 and 2.779.

c. −1.740 and 1.740

10.23 **a.** *Step 1:* Population 1 is highly hypnotizable people who receive a posthypnotic suggestion. Population 2 is highly hypnotizable people who do not receive a posthypnotic suggestion. The comparison distribution will be a distribution of differences between means. The hypothesis test will be an independent-samples t test because we have two samples and every participant is in only one sample. This study meets one of the three assumptions and may meet another. The dependent variable, reaction time in seconds, is scale. The data were not likely randomly selected, so we should be cautious when generalizing beyond the sample. We do not know whether the population is normally distributed, and there are fewer than 30 participants, but the sample data do not suggest skew.

Step 2: Null hypothesis: Highly hypnotizable individuals who receive a posthypnotic suggestion have the same average Stroop reaction times as highly hypnotizable individuals who receive no posthypnotic suggestion—$H_0: \mu_1 = \mu_2$.

Research hypothesis: Highly hypnotizable individuals who receive a posthypnotic suggestion have different average Stroop reaction times than highly hypnotizable individuals who receive no posthypnotic suggestion—$H_1: \mu_1 \neq \mu_2$.

Step 3: $(\mu_1 - \mu_2) = 0$; $s_{difference} = 0.463$

Calculations:

$M_X = 12.55$

X	X − M	(X − M)²
12.6	0.05	0.003
13.8	1.25	1.563
11.6	−0.95	0.903
12.2	−0.35	0.123
12.1	−0.45	0.203
13.0	0.45	0.203

$$s_Y^2 = \frac{\Sigma(X-M)^2}{N-1} = \frac{(0.003+1.563+0.903+0.123+0.203+0.203)}{6-1} = 0.600$$

$M_Y = 9.5$

Y	Y − M	(Y − M)²
8.5	−1.0	1.000
9.6	0.1	0.010
10.0	0.5	0.250
9.2	−0.3	0.090
8.9	−0.6	0.360
10.8	1.3	1.690

$$s_X^2 = \frac{\Sigma(Y - M)^2}{N - 1}$$
$$= \frac{(1.0 + 0.01 + 0.25 + 0.09 + 0.36 + 1.69)}{6 - 1}$$
$$= 0.680$$

$$df_X = N - 1 = 6 - 1 = 5$$
$$df_Y = N - 1 = 6 - 1 = 5$$
$$df_{total} = df_X + df_Y = 5 + 5 = 10$$

$$s_{pooled}^2 = \left(\frac{df_X}{df_{total}}\right)s_X^2 + \left(\frac{df_Y}{df_{total}}\right)s_Y^2$$
$$= \left(\frac{5}{10}\right)0.600 + \left(\frac{5}{10}\right)0.680$$
$$= 0.300 + 0.340 = 0.640$$

$$s_{M_X}^2 = \frac{s_{pooled}^2}{N} = \frac{0.640}{6} = 0.107$$
$$s_{M_Y}^2 = \frac{s_{pooled}^2}{N} = \frac{0.640}{6} = 0.107$$
$$s_{difference}^2 = s_{M_X}^2 + s_{M_X}^2 = 0.107 + 0.107 = 0.214$$
$$s_{difference} = \sqrt{s_{difference}^2} = \sqrt{0.214} = 0.463$$

b. $t(10) = 6.59, p < 0.05$ (*Note:* If we used software to conduct the t test, we would report the actual p value associated with this test statistic.)

c. When there are two separate samples, the t statistic becomes smaller. Thus, it becomes more difficult to reject the null hypothesis with a between-groups design than with a within-groups design.

d. In the within-groups design and the calculation of the paired-samples t test, we create a set of difference scores and conduct a t test on that set of difference scores. This means that any overall differences that participants have on the dependent variable are subtracted out and do not go into the measure of overall variability that is in the denominator of the t statistic.

e. To calculate the 95% confidence interval, first calculate

$$s_{difference} = \sqrt{s_{difference}^2} = \sqrt{s_{M_X}^2 + s_{M_Y}^2}$$
$$= \sqrt{\frac{s_{pooled}^2}{N_X} + \frac{s_{pooled}^2}{N_Y}} = \sqrt{\frac{0.640}{6} + \frac{0.640}{6}}$$
$$= \sqrt{0.214} = 0.463$$

The critical t statistics for a distribution with $df = 10$ that correspond to a p level of 0.05—that is, the values that mark off the most extreme 0.025 in each tail—are -2.228 and 2.228. Then calculate:

$$(M_X - M_Y)_{lower} = -t(s_{difference}) + (M_X - M_Y)_{sample}$$
$$= -2.228(0.463) + (12.55 - 9.5)$$
$$= -1.032 + 3.05 = 2.02$$

$$(M_X - M_Y)_{upper} = t(s_{difference}) + (M_X - M_Y)_{sample}$$
$$= 2.228(0.463) + (12.55 - 9.5)$$
$$= 1.032 + 3.05 = 4.08$$

The 95% confidence interval around the difference between means of 3.05 is [2.02, 4.08].

f. Were we to draw repeated samples (of the same sizes) from these two populations, 95% of the time the confidence interval would contain the true population parameter.

g. Because the confidence interval does not include 0, it is not plausible that there is no difference between means. Were we to conduct a hypothesis test, we would be able to reject the null hypothesis and could conclude that the means of the two samples are different.

h. In addition to determining statistical significance, the confidence interval allows us to determine a range of plausible differences between means. An interval estimate gives us a better sense than does a point estimate of how precisely we can estimate from this study.

i. The appropriate measure of effect size for a t statistic is Cohen's d, which is calculated as

$$d = \frac{(M_X - M_Y) - (\mu_X - \mu_Y)}{s_{pooled}}$$
$$= \frac{(12.55 - 9.5) - (0)}{\sqrt{0.640}} = 3.81$$

j. Based on Cohen's conventions, this is a large effect size.

k. It is useful to have effect-size information because the hypothesis test tells us only whether we were likely to have obtained our sample mean by chance. The effect size tells us the magnitude of the effect, giving us a sense of how important or practical this finding is, and allows us to standardize the results of the study so that we can compare across studies. Here, we know that there's a large effect.

10.25 **a.** *Step 1:* Population 1 consists of men. Population 2 consists of women. The comparison distribution is a distribution of differences between means. We will use an independent-samples t test because men and women cannot be in both conditions, and we have two groups. Of the three assumptions, we meet one because the dependent variable, number of words uttered, is a scale variable. We do not know whether the data were randomly selected or whether the population is normally distributed, and we have a small N, so we should be cautious in drawing conclusions.

Step 2: Null hypothesis: There is no mean difference in the number of words uttered by men and women—$H_0: \mu_1 = \mu_2$.

Research hypothesis: Men and women utter a different number of words, on average—$H_1: \mu_1 \neq \mu_2$.

Step 3: $(\mu_1 = \mu_2) = 0$; $s_{difference} = 684.869$

Calculations (treating women as X and men as Y):

$$M_Y = 16{,}091.600$$

X	X − M	(X − M)²
17,345	1253.400	1,571,011.560
15,593	−498.600	248,601.960
16,624	532.400	283,449.760
16,696	604.400	365,299.360
14,200	−1891.600	3,578,150.560

$$s_X^2 = \frac{\Sigma(X - M)^2}{N - 1} = \frac{6{,}046{,}513.200}{5 - 1} = 1{,}511{,}628.300$$

$M_X = 16{,}160.600$

Y	Y − M	(Y − M)²
16,345	184.400	34,003.360
17,222	1061.400	1,126,569.960
15,646	−514.600	264,813.160
14,889	−1271.600	1,616,966.560
16,701	540.400	292,032.160

$$s_Y^2 = \frac{\Sigma(Y-M)^2}{N-1} = \frac{3{,}334{,}385.200}{5-1} = 833{,}596.300$$

$$df_X = N - 1 = 5 - 1 = 4$$
$$df_Y = N - 1 = 5 - 1 = 4$$
$$df_{total} = df_X + df_Y = 8$$

$$s_{pooled}^2 = \left(\frac{df_X}{df_{total}}\right)s_X^2 + \left(\frac{df_Y}{df_{total}}\right)s_Y^2$$
$$= \left(\frac{4}{8}\right)1{,}511{,}628.300 + \left(\frac{4}{8}\right)833{,}596.300$$
$$= 1{,}172{,}612.300$$

$$s_{M_X}^2 = \frac{s_{pooled}^2}{N_X} = \frac{1{,}172{,}612.300}{5} = 234{,}522.460$$

$$s_{M_Y}^2 = \frac{s_{pooled}^2}{N_Y} = \frac{1{,}172{,}612.300}{5} = 234{,}522.460$$

$$s_{difference}^2 = s_{M_X}^2 + s_{M_Y}^2$$
$$= 234{,}522.460 + 234{,}522.460 = 469{,}044.920$$

$$s_{difference} = \sqrt{s_{difference}^2} = \sqrt{469{,}044.920} = 684.869$$

Step 4: The critical values, based on a two-tailed test, a p level of 0.05, and a df_{total} of 8, are −2.306 and 2.306.

Step 5: $t = \frac{(16{,}091.600 - 16{,}160.600) - 0}{684.869} = -0.101$

Step 6: We fail to reject the null hypothesis. The calculated t statistic of −0.101 is not more extreme than the critical t values.

b. $t(8) = -0.10, p > 0.05$ (*Note:* If we used software to conduct the t test, we would report the actual p value associated with this test statistic.)

c. $(M_X - M_Y)_{lower} = -t(s_{difference}) + (M_X - M_Y)_{sample}$
$= -2.306(684.869) + (-69.000)$
$= -1648.308$

$(M_X - M_Y)_{upper} = t(s_{difference}) + (M_X - M_Y)_{sample}$
$= 2.306(684.869) + (-69.000)$
$= 1510.308$

d. The 95% confidence interval around the observed mean difference of −69.00 is [−1648.31, 1510.31].

e. This confidence interval indicates that if we were to repeatedly sample differences between the means, 95% of the time our mean would fall between −1648.308 and 1510.308.

f. First, we need the appropriate measure of variability. In this case, we calculate pooled standard deviation by taking the square root of the pooled variance:

$$s_{pooled} = \sqrt{s_{pooled}^2} = \sqrt{1{,}172{,}612.300} = 1082.872$$

Now we can calculate Cohen's d:

$$d = \frac{M_X - M_Y}{s} = \frac{16{,}091.600 - 16{,}160.600}{1082.872} = -0.06$$

g. This is a small effect.

h. Effect size tells us how big the difference we observed between means was, uninfluenced by sample size. Often, this measure will help us understand whether we want to continue along our current research lines; that is, if a strong effect is indicated but we fail to reject the null hypothesis, we might want to continue collecting data to increase the statistical power. In this case, however, the failure to reject the null hypothesis is accompanied by a small effect.

10.27 **a.** *Step 1:* Population 1 consists of mothers, and population 2 is nonmothers. The comparison distribution will be a distribution of differences between means. We will use an independent-samples t test because someone is either identified as being a mother or not being a mother; both conditions, in this case, cannot be true. Of the three assumptions, we meet one because the dependent variable, decibel level, is a scale variable. We do not know whether the data were randomly selected and whether the population is normally distributed, and we have a small N, so we will be cautious in drawing conclusions.

Step 2: Null hypothesis: There is no mean difference in sound sensitivity, as reflected in the minimum level of detection, between mothers and nonmothers—$H_0: \mu_1 = \mu_2$.

Research hypothesis: There is a mean difference in sensitivity between the two groups—$H_1: \mu_1 \neq \mu_2$.

Step 3: $(\mu_1 = \mu_2) = 0$; $s_{difference} = 9.581$

Calculations:

$M_X = 47$

X	X − M	(X − M)²
33	−14	196
55	8	64
39	−8	64
41	−6	36
67	20	400

$$s_X^2 = \frac{\Sigma(X-M)^2}{N-1} = \frac{(196+64+64+36+400)}{5-1} = 190$$

$M_Y = 58.333$

Y	Y − M	(Y − M)²
56	−2.333	5.443
48	−10.333	106.771
71	12.667	160.453

$$s_Y^2 = \frac{\Sigma(Y-M)^2}{N-1} = \frac{(5.443+106.771+160.453)}{3-1} = 136.334$$

$$df_X = N - 1 = 5 - 1 = 4$$
$$df_Y = N - 1 = 3 - 1 = 2$$
$$df_{total} = df_X + df_Y = 4 + 2 = 6$$

$$s^2_{pooled} = \left(\frac{df_X}{df_{total}}\right)s^2_X + \left(\frac{df_Y}{df_{total}}\right)s^2_Y = \left(\frac{4}{6}\right)190 + \left(\frac{2}{6}\right)136.334$$
$$= 126.667 + 45.445 = 172.112$$

$$s^2_{M_X} = \frac{s^2_{pooled}}{N_X} = \frac{172.112}{5} = 34.422$$

$$s^2_{M_Y} = \frac{s^2_{pooled}}{N_Y} = \frac{172.112}{3} = 57.371$$

$$s^2_{difference} = s^2_{M_X} + s^2_{M_Y} = 34.422 + 57.371 = 91.793$$
$$s_{difference} = \sqrt{s^2_{difference}} = \sqrt{91.793} = 9.581$$

Step 4: The critical values, based on a two-tailed test, a p level of 0.05, and a df_{total} of 6, are -2.447 and 2.447.

Step 5: $t = \frac{(M_X - M_Y) - (\mu_X - \mu_Y)}{s_{difference}}$
$$= \frac{(47 - 58.333) - (0)}{9.581} = -1.183$$

Step 6: Fail to reject the null hypothesis. We do not have enough evidence, based on these data, to conclude that mothers have more sensitive hearing, on average, when compared to nonmothers.

b. $t(6) = -1.183, p > 0.05$ (*Note:* If we used software to conduct the t test, we would report the actual p value associated with this test statistic.)

c. $(M_X - M_Y)_{lower} = -t(s_{difference}) + (M_X - M_Y)_{sample}$
$(M_X - M_Y)_{lower} = -2.447(9.581) + (47 - 58.333)$
$= -34.778$

$(M_X - M_Y)_{upper} = t(s_{difference}) + (M_X - M_Y)_{sample}$
$(M_X - M_Y)_{upper} = 2.447(9.581) + (47 - 58.333) = 12.112$

The 95% confidence interval around the difference between means of -11.333 is $[-34.778, 12.112]$.

d. What we learn from this confidence interval is that there is great variability in the plausible difference between means for these data, reflected in the wide range. We also notice that 0 is within the confidence interval, so we cannot assume a difference between these groups.

e. Whereas point estimates result in one value (-11.333 in this case) in which we have no estimate of confidence, the interval estimate gives us a range of scores about which we have known confidence.

f. $s_{pooled} = \sqrt{s^2_{pooled}} = \sqrt{172.112} = 13.119$

Cohen's $d = \frac{(M_X - M_Y) - (\mu_X - \mu_Y)}{s_{pooled}}$
$$= \frac{(47 - 58.333) - (0)}{13.119}$$
$$= -0.864$$

g. This is a large effect.

h. Effect size tells us how big the difference we observed between means was, without the influence of sample size. Often, this measure helps us decide whether we want to continue along our current research lines. In this case, the large effect would encourage us to collect more data to increase statistical power.

10.29 **a.** We would use a single-sample t test because we have one sample of figure skaters and are comparing that sample to a population (women with eating disorders) for which we know the mean.

b. We would use an independent-samples t test because we have two samples, and no participant can be in both samples. One cannot have both a high level and a low level of knowledge about a topic.

c. We would use a paired-samples t test because we have two samples, but every student is assigned to both samples—one night of sleep loss and one night of no sleep loss.

10.31 **a.** We would use an independent-samples t test because there are two samples, and no participant can be in both samples. One cannot be a pedestrian en route to work and a tourist pedestrian at the same time.

b. Null hypothesis: People en route to work tend to walk at the same pace, on average, as people who are tourists—$H_0\colon \mu_1 = \mu_2$. Research hypothesis: People en route to work tend to walk at a different pace, on average, than do those who are tourists—$H_1\colon \mu_1 \neq \mu_2$.

10.33 **a.** Waters is predicting lower levels of obesity among children who are in the Edible Schoolyard program than among children who are not in the program. Waters and others who believe in her program are likely to notice successes and overlook failures. Solid research is necessary before instituting such a program nationally, even though it sounds extremely promising.

b. Students could be randomly assigned to participate in the Edible Schoolyard program or to continue with their usual lunch plan. The independent variable is the program, with two levels (Edible Schoolyard, control), and the dependent variable could be weight. Weight is easily operationalized by weighing children, perhaps after one year in the program.

c. We would use an independent-samples t test because there are two samples and no student is in both samples.

d. *Step 1:* Population 1 is all students who participated in the Edible Schoolyard program. Population 2 is all students who did not participate in the Edible Schoolyard program. The comparison distribution will be a distribution of differences between means. The hypothesis test will be an independent-samples t test. This study meets all three assumptions. The dependent variable, weight, is scale. The data would be collected using a form of random selection. In addition, there would be more than 30 participants in the sample, indicating that the comparison distribution would likely be normal.

e. *Step 2:* Null hypothesis: Students who participate in the Edible Schoolyard program weigh the same, on average, as students who do not participate—$H_0\colon \mu_1 = \mu_2$.

Research hypothesis: Students who participate in the Edible Schoolyard program have different weights, on average, than students who do not participate—$H_1\colon \mu_1 \neq \mu_2$.

f. The dependent variable could be nutrition knowledge, as assessed by a test, or body mass index (BMI).

g. There are many possible confounds when we do not conduct a controlled experiment. For example, the Berkeley school might be different to begin with. After all, the school allowed Waters to begin the program, and perhaps it had already emphasized nutrition. Random selection allows us to have faith in the ability to generalize beyond our sample. Random assignment allows us to eliminate confounds, other variables that may explain any differences between groups.

CHAPTER 11

11.1 An ANOVA is a hypothesis test with at least one nominal independent variable (with at least three total groups) and a scale dependent variable.

11.3 Between-groups variance is an estimate of the population variance based on the differences among the means; within-groups variance is an estimate of the population variance based on the differences within each of the three (or more) sample distributions.

11.5 The three assumptions are that the participants were randomly selected, the underlying populations are normally distributed, and the underlying variances of the different conditions are similar, or *homoscedastic*.

11.7 The F statistic is calculated as the ratio of two variances. Variability, and the variance measure of it, is always positive—it always exists. Variance is calculated as the sum of squared deviations, and squaring both positive and negative values makes them positive.

11.9 With sums of squares, we add up all the squared values. Deviations from the mean always sum to 0. By squaring these deviations, we can sum them and they will not sum to 0. Sums of squares are measures of variability of scores from the mean.

11.11 The grand mean is the mean of every score in a study, regardless of which sample the score came from.

11.13 Cohen's d; R^2

11.15 *Post hoc* means "after this." These tests are needed when an ANOVA is significant and we want to discover where the significant differences exist between the groups.

11.17 **a.** *Standard error* is wrong. The professor is reporting the spread for a distribution of scores, the *standard deviation*.

b. *t statistic* is wrong. We do not use the population standard deviation to calculate a t statistic. The sentence should say *z statistic* instead.

c. *Parameters* is wrong. Parameters are numbers that describe populations, not samples. The researcher calculated the *statistics*.

d. *z statistic* is wrong. Evelyn is comparing two means; thus, she would have calculated a *t statistic*.

11.19 The four assumptions are that (1) the data are randomly selected; (2) the underlying population distributions are normal; (3) the variability is similar across groups, or homoscedasticity; and (4) there are no order effects.

11.21 The "subjects" variability is noise in the data caused by each participant's personal variability compared with the other participants. It is calculated by comparing each person's mean response across all levels of the independent variable with the grand mean, the overall mean response across all levels of the independent variable.

11.23 Counterbalancing involves exposing participants to the different levels of the independent variable in different orders.

11.25 To calculate the sum of squares for subjects, we first calculate an average of each participant's scores across the levels of the independent variable. Then we subtract the grand mean from each participant's mean. We repeat this subtraction for each score the participant has—that is, for as many times as there are levels of the independent variable. Once we have the deviation scores, we square each of them and then sum the squared deviations to get the sum of squares for participants.

11.27 If we have a between-groups study in which different people are participating in the different conditions, then we can turn it into a within-groups study by having all the people in the sample participate in all the conditions.

11.29 The calculations for R^2 for a one-way within-groups ANOVA and a one-way between-groups ANOVA are similar. In both one-way ANOVAs, the numerator is a measure of the variability that takes into account just the differences among means, $SS_{between}$. The denominator, however, is different for the within-groups ANOVA, as it takes into account the total variability, SS_{total}, similarly to the between-groups ANOVA, but removes the variability due to differences among participants, $SS_{subjects}$, as well as the variability due to differences within each group, SS_{within}. This enables us to determine the variability explained only by between-groups differences.

11.31 **a.** $df_{between} = N_{groups} - 1 = 3 - 1 = 2$

b. $df_{within} = df_1 + df_2 + \ldots + df_{last} = (4 - 1) + (3 - 1) + (5 - 1) = 3 + 2 + 4 = 9$

c. $df_{total} = df_{between} + df_{within} = 2 + 9 = 11$

d. The critical value for a between-groups degrees of freedom of 2 and a within-groups degrees of freedom of 9 at a p level of 0.05 is 4.26.

e. $M_{1970} = \frac{\Sigma(X)}{N} = \frac{45 + 211 + 158 + 74}{4} = 122$

$M_{1980} = \frac{\Sigma(X)}{N} = \frac{92 + 128 + 382}{3} = 200.667$

$M_{1990} = \frac{\Sigma(X)}{N} = \frac{273 + 396 + 178 + 248 + 374}{5} = 293.80$

$$GM = \frac{\Sigma(X)}{N_{total}} = \frac{\begin{pmatrix}45 + 211 + 158 + 74 + 92 + 128 + \\ 382 + 273 + 396 + 178 + 248 + 374\end{pmatrix}}{12} = 213.25$$

f. (*Note:* The total sum of squares will not exactly equal the sum of the between-groups and within-groups sums of squares because of rounding decisions.)

The total sum of squares is calculated here as $SS_{total} = \Sigma(X - GM)^2$:

SAMPLE	X	$(X - GM)$	$(X - GM)^2$
1970	45	−168.25	28,308.063
M_{1970} = 122	211	−2.25	5.063
	158	−55.25	3,052.563
	74	−139.25	19,390.563
1980	92	−121.25	14,701.563
M_{1980} = 200.667	128	−85.25	7,267.563
	382	168.75	28,476.563
1990	273	59.75	3,570.063
M_{1990} = 293.8	396	182.75	33,397.563
	178	−35.25	1,242.563
	248	34.75	1,207.563
	374	160.75	25,840.563
	GM = 213.25	SS_{total} =	**166,460.256**

g. The within-groups sum of squares is calculated here as $SS_{within} = \Sigma(X - M)^2$:

SAMPLE	X	(X – M)	(X – M)²
1970	45	−77	5,929.00
M_{1970} = 122	211	89	7,921.00
	158	36	1,296.00
	74	−48	2,304.00
1980	92	−108.667	11,808.517
M_{1980} = 200.667	128	−72.667	5,280.493
	382	181.333	32,881.657
1990	273	−20.8	432.64
M_{1990} = 293.8	396	102.2	10,444.84
	178	−115.8	13,409.64
	248	−45.8	2,097.64
	374	80.2	6,432.04
	GM = 213.25		SS_{within} = **100,237.467**

h. The between-groups sum of squares is calculated here as $SS_{between} = \Sigma(M - GM)^2$:

SAMPLE	X	(M – GM)	(M – GM)²
1970	45	−91.25	8326.563
M_{1970} = 122	211	−91.25	8326.563
	158	−91.25	8326.563
	74	−91.25	8326.563
1980	92	−12.583	158.332
M_{1980} = 200.667	128	−12.583	158.332
	382	−12.583	158.332
1990	273	80.55	6488.303
M_{1990} = 293.8	396	80.55	6488.303
	178	80.55	6488.303
	248	80.55	6488.303
	374	80.55	6488.303
	GM = 213.25		$SS_{between}$ = **66,222.763**

i. $MS_{between} = \frac{SS_{between}}{df_{between}} = \frac{66{,}222.763}{2} = 33{,}111.382$

$MS_{within} = \frac{SS_{within}}{df_{within}} = \frac{100{,}237.467}{9} = 11{,}137.607$

$F = \frac{MS_{between}}{MS_{within}} = \frac{33{,}111.382}{11{,}137.607} = 2.97$

SOURCE	SS	df	MS	F
Between	66,222.763	2	33,111.382	**2.97**
Within	100,238.467	9	11,137.496	
Total	166,460.256	11		

$R^2 = \frac{SS_{between}}{SS_{total}} =$

j. Effect size is calculated as

$\frac{66{,}222.763}{166{,}460.256} = 0.40$. According to Cohen's conventions for R^2, this is a very large effect.

11.33 **a.** $F = \frac{\text{between-groups variance}}{\text{within-groups variance}} = \frac{321.83}{177.24} = 1.82$

b. $F = \frac{2.79}{2.20} = 1.27$

c. $F = \frac{34.45}{41.60} = 0.83$

11.35

SOURCE	SS	df	MS	F
Between	43	2	21.500	**2.66**
Within	89	11	8.091	
Total	132	13		

11.37 With four groups, there would be a total of six different comparisons.

11.39 **a.** $df_{between} = N_{groups} - 1 = 3 - 1 = 2$

b. $df_{subjects} = n - 1 = 4 - 1 = 3$

c. $df_{within} = (df_{between})(df_{subjects}) = (2)(3) = 6$

d. $df_{total} = df_{between} + df_{subjects} + df_{within} = 2 + 3 + 6 = 11$, or we can calculate it as $df_{total} = N_{total} - 1 = 12 - 1 = 11$

e. $SS_{total} = \Sigma(X - GM)^2 = 754$

LEVEL	RATING (X)	(X – GM)	(X – GM)²
1	7	−10	100
1	16	−1	1
1	3	−14	196
1	9	−8	64
2	15	−2	4
2	18	1	1
2	18	1	1
2	13	−4	16
3	22	5	25
3	28	11	121
3	26	9	81
3	29	12	144
	GM = 17		Σ(X – GM)² = 754

f. $SS_{between} = \Sigma(M - GM)^2 = 618.504$

LEVEL	RATING (X)	GROUP MEAN	(M − GM)	(M − GM)²
1	7	8.75	−8.25	68.063
1	16	8.75	−8.25	68.063
1	3	8.75	−8.25	68.063
1	9	8.75	−8.25	68.063
2	15	16	−1	1.000
2	18	16	−1	1.000
2	18	16	−1	1.000
2	13	16	−1	1.000
3	22	26.25	9.25	85.563
3	28	26.25	9.25	85.563
3	26	26.25	9.25	85.563
3	29	26.25	9.25	85.563
	GM = 17		Σ(M − GM)² = 618.504	

g. $SS_{subjects} = \Sigma(M_{participant} - GM)^2 = 62.001$

PARTICIPANT	LEVEL	RATING (X)	PARTICIPANT MEAN	(M_PARTICIPANT − GM)	(M_PARTICIPANT − GM)²
1	1	7	14.667	−2.333	5.443
2	1	16	20.667	3.667	13.447
3	1	3	15.667	−1.333	1.777
4	1	9	17	0	0
1	2	15	14.667	−2.333	5.443
2	2	18	20.667	3.667	13.447
3	2	18	15.667	−1.333	1.777
4	2	13	17	0	0
1	3	22	14.667	−2.333	5.443
2	3	28	20.667	3.667	13.447
3	3	26	15.667	−1.333	1.777
4	3	29	17	0	0
	GM = 17			Σ(M_participant − GM)² = 62.001	

h. $SS_{within} = SS_{total} - SS_{between} - SS_{subjects} = 754 - 618.504 - 62.001 = 73.495$

i. $MS_{between} = \frac{SS_{between}}{df_{between}} = \frac{618.504}{2} = 309.252$

$$MS_{subjects} = \frac{SS_{subjects}}{df_{subjects}} = \frac{62.001}{3} = 20.667$$

$$MS_{within} = \frac{SS_{within}}{df_{within}} = \frac{73.495}{6} = 12.249$$

$$F_{between} = \frac{MS_{between}}{MS_{within}} = \frac{309.252}{12.249} = 25.247$$

$$F_{subjects} = \frac{MS_{subjects}}{MS_{within}} = \frac{20.667}{12.249} = 1.687$$

SOURCE	SS	df	MS	F
Between-groups	618.504	2	309.252	25.25
Subjects	62.001	3	20.667	1.69
Within-groups	73.495	6	12.249	
Total	754	11		

j. $R^2 = \frac{SS_{between}}{(SS_{total} - SS_{subjects})} = \frac{618.504}{(754 - 62.001)} = 0.89$

k. $s_M = \sqrt{\frac{MS_{within}}{N}} = \sqrt{\frac{12.249}{4}} = 1.750$

The Tukey *HSD* statistic comparing level 1 and level 3 would be:

$$HSD = \frac{M_{level\ 1} - M_{level\ 3}}{s_M} = \frac{8.75 - 26.25}{1.750} = -10$$

11.41 **a.**

SOURCE	SS	df	MS	F
Between	941.102	2	470.551	10.16
Subjects	3807.322	10	380.732	8.22
Within	926.078	20	46.304	
Total	5674.502	32		

b. $R^2 = \frac{SS_{between}}{(SS_{total} - SS_{subjects})} = \frac{941.102}{(5674.502 - 3807.322)} = 0.50$

11.43 **a.** The independent variable is type of program. The levels are *The Daily Show* and network news. The dependent variable is the amount of substantive video and audio reporting per second.

b. The hypothesis test that Fox would use is an independent-samples t test.

c. The independent variable is still type of program, but now the levels are *The Daily Show,* network news, and cable news. The hypothesis test would be a one-way between-groups ANOVA.

11.45 **a.** A t distribution; we are comparing the mean IQ of a sample of 10 to the population mean of 100; this student knows only the population mean—not the population standard deviation.

b. An F distribution; we are comparing the mean ratings of four samples—families with no books visible, with only children's books visible, with only adult books visible, and with both types of books visible.

c. A t distribution; we are comparing the average vocabulary of two groups.

11.47 **a.** The independent variable in this case is the type of program in which students were enrolled; the levels were arts and sciences, education, law, and business. Because every student is enrolled in only one program, the researcher would use a one-way between-groups ANOVA.

b. Now the independent variable is year, with levels of first, second, or third. Because the same participants are repeatedly measured, the researcher would use a one-way within-groups ANOVA.

c. The independent variable in this case is type of degree, and its levels are master's, doctoral, and professional. Because every student is in only one type of degree program, the researcher would use a one-way between-groups ANOVA.

d. The independent variable in this case is stage of training, and its levels are master's, doctoral, and postdoctoral. Because the same students are repeatedly measured, the researcher would use a one-way within-groups ANOVA.

11.49 **a.** The independent variable is political viewpoint, with the levels Republican, Democrat, and neither.

b. The dependent variable is religiosity.

c. The populations are all Republicans, all Democrats, and all who categorize themselves as neither. The samples are the Republicans, Democrats, and people who say they are neither among the 180 students.

d. Because every student identified only one type of political viewpoint, the researcher would use a one-way between-groups ANOVA. No participant could be in more than one level of the independent variable.

e. First, we would calculate the between-groups variance. This involves calculating a measure of variability among the three sample means—the religiosity scores of the Republicans, Democrats, and others. Then we would calculate the within-groups variance; this is essentially an average of the variability within each of the three samples. Finally, we would divide the between-groups variance by the within-groups variance. If the variability among the means is much larger than the variability within each sample, this provides evidence that the means are different from one another.

11.51 **a.** $F = \frac{MS_{between}}{MS_{within}} = \frac{4.623}{0.522} = 8.856$

b. $t = \sqrt{F} = \sqrt{8.856} = 2.98$

c. The "Sig." for t is the same as that for the ANOVA, 0.005, because the F distribution reduces to the t distribution when we are dealing with two groups.

11.53 **a.** Null hypothesis: People experience the same mean amount of fear across all three levels of dog size—H_0: $\mu_1 = \mu_2 = \mu_3$. Research hypothesis: People do not experience the same mean amount of fear across all three levels of dog size.

b. We do not know how the participants were selected, so the first assumption of random selection might not be met. We do not know how the dogs were presented to the participants, so we cannot assess whether order effects are present.

c. The effect size was 0.89, which is a large effect. This indicates that the effect might be important, meaning the size of a dog might have a large impact on the amount of fear experienced by people.

d. The Tukey *HSD* test statistic was −10. According to the q statistic table, the critical value for the Tukey *HSD* when there are six within-groups degrees of freedom and three treatment levels is 4.34. We can conclude that the mean difference in fear when a small versus large dog is presented is statistically significant, with the large dog evoking greater fear.

11.55 **a.** *Step 5:* We must first calculate df and SS to fill in the source table.

$$df_{between} = N_{groups} - 1 = 2$$

$$df_{subjects} = n - 1 = 4$$

$$df_{within} = (df_{between})(df_{subjects}) = 8$$

$$df_{total} = N_{total} - 1 = 14$$

For sums of squares total: $SS_{total} = \Sigma(X - GM)^2 = 73.6$

TIME	X	X − GM	(X − GM)²
Past	18	−1.6	2.56
Past	17.5	−2.1	4.41
Past	19	−0.6	0.36
Past	16	−3.6	12.96
Past	20	0.4	0.16
Present	18.5	−1.1	1.21
Present	19.5	−0.1	0.01
Present	20	0.4	0.16
Present	17	−2.6	6.76
Present	18	−1.6	2.56
Future	22	2.4	5.76
Future	24	4.4	19.36
Future	20	0.4	0.16
Future	23.5	3.9	15.21
Future	21	1.4	1.96
	GM = 19.6		SS_{total} = 73.6

For sum of squares between: $SS_{between} = \Sigma(M - GM)^2 = 47.5$

TIME	X	GROUP MEAN	M − GM	(M − GM)^2
Past	18	18.1	−1.5	2.25
Past	17.5	18.1	−1.5	2.25
Past	19	18.1	−1.5	2.25
Past	16	18.1	−1.5	2.25
Past	20	18.1	−1.5	2.25
Present	18.5	18.6	−1	1
Present	19.5	18.6	−1	1
Present	20	18.6	−1	1
Present	17	18.6	−1	1
Present	18	18.6	−1	1
Future	22	22.1	2.5	6.25
Future	24	22.1	2.5	6.25
Future	20	22.1	2.5	6.25
Future	23.5	22.1	2.5	6.25
Future	21	22.1	2.5	6.25
	GM = 19.6		$SS_{between}$ = 47.5	

For sum of squares subjects: $SS_{subjects} = \Sigma(M_{participant} - GM)^2 = 44.278$

PARTICIPANT	TIME	X	PARTICIPANT MEAN	$M_{PARTICIPANT}$ − GM	($M_{PARTICIPANT}$ − GM)^2
1	Past	18	19.500	−0.100	0.010
2	Past	17.5	20.333	0.733	0.538
3	Past	19	19.667	0.067	0.004
4	Past	16	18.833	−0.767	0.588
5	Past	20	19.667	0.067	0.004
1	Present	18.5	19.500	−0.100	0.010
2	Present	19.5	20.333	0.733	0.538
3	Present	20	19.667	0.067	0.004
4	Present	17	18.833	−0.767	0.588
5	Present	18	19.667	0.067	0.004
1	Future	22	19.500	−0.100	0.010
2	Future	24	20.333	0.733	0.538
3	Future	20	19.667	0.067	0.004
4	Future	23.5	18.833	−0.767	0.588
5	Future	21	19.667	0.067	0.004
		GM = 19.6		$SS_{subjects}$ = 3.432	

$SS_{within} = SS_{total} - SS_{between} - SS_{subjects} = 22.668$

SOURCE	SS	df	MS	F
Between	47.5	2	23.750	8.38
Subjects	3.432	4	0.858	0.30
Within	22.667	8	2.834	
Total	73.6	14		

Step 6: The F statistic, 8.28, is beyond 4.46, the critical F value at a p level of 0.05. We would reject the null hypothesis. There is a difference, on average, among the past, present, and future self-reported life satisfaction of pessimists.

b. First, we calculate s_M: $s_M = \sqrt{\frac{MS_{within}}{N}} = \sqrt{\frac{2.834}{5}} = 0.753$

Next, we calculate HSD for each pair of means.

For past versus present:

$$HSD = \frac{(18.1 - 18.6)}{0.753} = -0.66$$

For past versus future:

$$HSD = \frac{(18.1 - 22.1)}{0.753} = -5.31$$

For present versus future:

$$HSD = \frac{(18.6 - 22.1)}{0.753} = -4.65$$

The critical value of q at a p level of 0.05 is 4.04. Thus, we reject the null hypothesis for the past versus future comparison and for the present versus future comparison, but not for the past versus present comparison. These results indicate that the mean self-reported life satisfaction of pessimists is not significantly different for their past and present, but they expect to have greater life satisfaction in the future, on average.

c. $R^2 = \frac{SS_{between}}{(SS_{total} - SS_{subjects})} = \frac{47.5}{(73.6 - 3.432)} = 0.68$

11.57 We must first calculate df and SS to fill in the source table.

$$df_{between} = N_{groups} - 1 = 2$$
$$df_{subjects} = n - 1 = 4$$
$$df_{within} = (df_{between})(df_{subjects}) = 8$$
$$df_{total} = N_{total} - 1 = 14$$

For sums of squares total: $SS_{total} = \Sigma(X - GM)^2 = 4207.333$

STIMULUS	X	X − GM	(X − GM)²
Owner	69	20.667	427.125
Owner	72	23.667	560.127
Owner	65	16.667	277.789
Owner	75	26.667	711.129
Owner	70	21.667	469.459
Cat	28	−20.333	413.431
Cat	32	−16.333	266.767
Cat	30	−18.333	336.099
Cat	29	−19.333	373.765
Cat	31	−17.333	300.433
Dog	45	−3.333	11.109
Dog	43	−5.333	28.441
Dog	47	−1.333	1.777
Dog	45	−3.333	11.109
Dog	44	−4.333	18.775
	GM = 48.333		SS_{total} = 4207.333

For sum of squares between: $SS_{between} = \Sigma(M - GM)^2 = 4133.733$

STIMULUS	X	GROUP MEAN	M − GM	(M − GM)²
Owner	69	70.2	21.867	478.166
Owner	72	70.2	21.867	478.166
Owner	65	70.2	21.867	478.166
Owner	75	70.2	21.867	478.166
Owner	70	70.2	21.867	478.166
Cat	28	30	−18.333	336.099
Cat	32	30	−18.333	336.099
Cat	30	30	−18.333	336.099
Cat	29	30	−18.333	336.099
Cat	31	30	−18.333	336.099
Dog	45	44.8	−3.533	12.482
Dog	43	44.8	−3.533	12.482
Dog	47	44.8	−3.533	12.482
Dog	45	44.8	−3.533	12.482
Dog	44	44.8	−3.533	12.482
	GM = 48.333		$SS_{between}$ = 4133.733	

For sum of squares subjects: $SS_{subjects} = \Sigma(M_{participant} - GM)^2 = 12.667$

STIMULUS	X	PARTICIPANT MEAN	$M_{PARTICIPANT}$ − GM	($M_{PARTICIPANT}$ − GM)²
Owner	69	47.333	−1.000	0.999
Owner	72	49.000	0.667	0.445
Owner	65	47.333	−1.000	0.999
Owner	75	49.667	1.334	1.779
Owner	70	48.333	0.000	0.000
Cat	28	47.333	−1.000	0.999
Cat	32	49.000	0.667	0.445
Cat	30	47.333	−1.000	0.999
Cat	29	49.667	1.334	1.779
Cat	31	48.333	0.000	0.000
Dog	45	47.333	−1.000	0.999
Dog	43	49.000	0.667	0.445
Dog	47	47.333	−1.000	0.999
Dog	45	49.667	1.334	1.779
Dog	44	48.333	0.000	0.000
		GM = 48.333		$SS_{subjects}$ = 12.667

$SS_{within} = SS_{total} - SS_{between} - SS_{subjects} = 60.933$

SOURCE	*SS*	*df*	*MS*	*F*
Between	4133.733	2	2066.867	271.35
Subjects	12.667	4	3.167	0.42
Within	60.933	8	7.617	
Total	4207.333	14		

11.59 At a *p* level of 0.05, the critical *F* value is 4.46. Because the calculated *F* statistic does not exceed the critical *F* value, we would fail to reject the null hypothesis. Because we failed to reject the null hypothesis, it would not be appropriate to perform post-hoc comparisons.

11.61 **a.** Level of trust in the leader is the independent variable. It has three levels: low, moderate, and high.

b. The dependent variable is level of agreement with a policy supported by the leader or supervisor.

c. *Step 1:* Population 1 is employees with low trust in their leader. Population 2 is employees with moderate trust in their leader. Population 3 is employees with high trust in their leader. The comparison distribution will be an *F* distribution. The hypothesis test will be a one-way between-groups ANOVA. We do not know if employees were randomly selected. We also do not know if the underlying distributions are normal, and the sample sizes are small so we must proceed with caution. To check the final assumption, that we have homoscedastic variances, we will calculate variance for each group.

SAMPLE	LOW TRUST	MODERATE TRUST	HIGH TRUST
Squared deviations	16	100	3.063
	1	121	18.063
	4	1	60.063
	25		27.563
Sum of squares	46	222	108.752
$N - 1$	3	2	3
Variance	**15.33**	**111**	**36.25**

Because the largest variance, 111, is much more than twice as large as the smallest variance, we can conclude we have heteroscedastic variances. Violation of this third assumption of homoscedastic samples means we should proceed with caution. Because these data are intended to give you practice calculating statistics, proceed with your analyses. When conducting real research, we would want to have much larger sample sizes and to more carefully consider meeting the assumptions.

Step 2: Null hypothesis: There are no mean differences between these three groups: the mean level of agreement with a policy does not vary across the three trust levels—H_0: $\mu_1 = \mu_2 = \mu_3$.

Research hypothesis: There are mean differences between some or all of these groups: the mean level of agreement depends on trust.

Step 3: $df_{between} = N_{groups} - 1 = 3 - 1 = 2$

$$df_{within} = df_1 + df_2 + \ldots + df_{last}$$
$$= (4 - 1) + (3 - 1) + (4 - 1) = 3 + 2 + 3 = 8$$
$$df_{total} = df_{between} + df_{within} = 2 + 8 = 10$$

The comparison distribution will be an F distribution with 2 and 8 degrees of freedom.

Step 4: The critical value for the F statistic based on a p level of 0.05 is 4.46.

Step 5: $GM = 21.727$

Total sum of squares is calculated here as $SS_{total} = \Sigma(X - GM)^2$:

SAMPLE	X	$(X - GM)$	$(X - GM)^2$
Low trust	9	−12.727	161.977
$M_{low} = 13$	14	−7.727	59.707
	11	−10.727	115.069
	18	−3.727	13.891
Moderate trust	14	−7.727	59.707
$M_{mod} = 24$	35	13.273	176.173
	23	1.273	1.621
High trust	27	5.273	27.805
$M_{high} = 28.75$	33	11.273	127.081
	21	−0.727	0.529
	34	12.273	150.627
	$GM = 21.727$		$SS_{total} =$ **894.187**

Within-groups sum of squares is calculated here as $SS_{within} = \Sigma(X - M)^2$:

SAMPLE	X	$(X - M)$	$(X - M)^2$
Low trust	9	−4	16.00
$M_{low} = 13$	14	1	1.00
	11	−2	4.00
	18	5	25.00
Moderate trust	14	−10	100.00
$M_{mod} = 24$	35	11	121.00
	23	−1	1.00
High trust	27	−1.75	3.063
$M_{high} = 28.75$	33	4.25	18.063
	21	−7.75	60.063
	34	5.25	27.563
	$GM = 21.727$		$SS_{within} =$ **376.752**

Between-groups sum of squares is calculated here as $SS_{between} = \Sigma(M - GM)^2$:

SAMPLE	X	$(M - GM)$	$(M - GM)^2$
Low trust	9	−8.727	76.161
$M_{low} = 13$	14	−8.727	76.161
	11	−8.727	76.161
	18	−8.727	76.161
Moderate trust	14	2.273	5.167
$M_{mod} = 24$	35	2.273	5.167
	23	2.273	5.167
High trust	27	7.023	49.323
$M_{high} = 28.75$	33	7.023	49.323
	21	7.023	49.323
	34	7.023	49.323
	$GM = 21.727$		$SS_{between} =$ **517.437**

$$MS_{between} = \frac{SS_{between}}{df_{between}} = \frac{517.437}{2} = 258.719$$

$$MS_{within} = \frac{SS_{within}}{df_{within}} = \frac{376.752}{8} = 47.094$$

$$F = \frac{MS_{between}}{MS_{within}} = \frac{258.715}{47.094} = 5.49$$

SOURCE	SS	df	MS	F
Between	517.437	2	258.719	5.49
Within	376.752	8	47.094	
Total	894.187	10		

Step 6: The F statistic, 5.49, is beyond the cutoff of 4.46, so we can reject the null hypothesis. The mean level of agreement with a policy supported by a supervisor varies

across level of trust in that supervisor. Remember, the research design and data did not meet the three assumptions of this statistical test, so we should be careful in interpreting this finding.

d. $F(2,8) = 5.49, p < 0.05$; Note: We would include the actual p value if we used software to conduct this analysis.

e. Because we have unequal sample sizes, we must calculate a weighted sample size.

$$N' = \frac{N_{groups}}{\Sigma\left(\frac{1}{N}\right)} = \frac{3}{\left(\frac{1}{4}+\frac{1}{3}+\frac{1}{4}\right)} = \frac{3}{0.25+0.333+0.25}$$
$$= \frac{3}{0.833} = 3.601$$

$$s_M = \sqrt{\frac{MS_{within}}{N'}} = \sqrt{\frac{47.094}{3.601}} = 3.616$$

Now we can compare the three groups.

Low trust ($M = 13$) versus moderate trust ($M = 24$):

$$HSD = \frac{13-24}{3.616} = -3.04$$

Low trust ($M = 13$) versus high trust ($M = 28.75$):

$$HSD = \frac{13-28.75}{3.616} = -4.36$$

Moderate trust ($M = 24$) versus high trust ($M = 28.75$):

$$HSD = \frac{24-28.75}{3.616} = -1.31$$

According to the q table, the critical value is 4.04 for a p level of 0.05 when we are comparing three groups and have within-groups degrees of freedom of 8. We obtained one q value (-4.36) that exceeds this cutoff. Based on the calculations, there is a statistically significant difference between the mean level of agreement by employees with low trust in their supervisors compared to those with high trust. Because the sample sizes here were so small and we did not meet the three assumptions of ANOVA, we should be careful in making strong statements about this finding. In fact, these preliminary findings would encourage additional research.

f. It is not possible to do a t test in this situation because there are more than two groups or levels of the independent variable.

g. It is not possible to conduct this study with a within-groups design because participants cannot be in more than one of the groups or levels of the independent variable. In other words, employees only have one level of trust in their supervisor.

11.63 a. The independent variable is the type of substance placed beneath the eyes, and its levels are black grease, black antiglare stickers, and petroleum jelly.

b. The dependent variable is eye glare.

c. This is a one-way within-groups ANOVA.

d. The first assumption of ANOVA is that the samples are randomly selected from their populations. It is unlikely that the researchers met this assumption. The study description indicates that the researchers were from Yale University and does not mention any techniques the researchers might have used to obtain participants from across the nation. So it is likely that the Yale researchers used a sample of participants from their local area.

e. The second assumption is that the population distribution is normal. Although we do not know the exact distribution of the population of scores, there are more than 30 participants in the study. When there are at least 30 participants in a sample, the distribution of sample means will be approximately normal even if the underlying distribution of scores is not. So it is likely that the distribution of sample means is normal and that this assumption was met.

f. The third assumption is homoscedasticity—that the samples come from populations with equal variances. It is not possible to tell whether this assumption was met on the basis of the study description. The researchers could assess whether this assumption was met by comparing the variance of each of the three treatment groups to ensure that the largest variance is no larger than two times the smallest variance.

g. The fourth assumption that is specific to the within-groups ANOVA is that there are no order effects. To protect against order effects, the researcher would want to have counterbalanced the order in which the participants experienced the treatment conditions.

h. *Step 5:* We must first calculate df and SS to fill in the source table.

$$df_{between} = N_{groups} - 1 = 2;$$
$$df_{subjects} = n - 1 = 3;$$
$$df_{within} = (df_{between})(df_{subjects}) = 6$$
$$df_{total} = N_{total} - 1 = 11$$

For sums of squares total: $SS_{total} = \Sigma(X - GM)^2 = 16.523$

CONDITION	X	X − GM	(X − GM)²
Black grease	19.8	2.175	4.731
Black grease	18.2	0.575	0.331
Black grease	19.2	1.575	2.481
Black grease	18.7	1.075	1.156
Antiglare stickers	17.1	−0.525	0.276
Antiglare stickers	17.2	−0.425	0.181
Antiglare stickers	18	0.375	0.141
Antiglare stickers	17.9	0.275	0.076
Petroleum jelly	15.9	−1.725	2.976
Petroleum jelly	16.3	−1.325	1.756
Petroleum jelly	16.2	−1.425	2.031
Petroleum jelly	17	−0.625	0.391
	GM = 17.625		SS_{total} = 16.523

For sum of squares between: $SS_{between} = \Sigma(M - GM)^2 = 13.815$

CONDITION	X	GROUP MEAN	M − GM	(M − GM)²
Black grease	19.8	18.975	1.35	1.823
Black grease	18.2	18.975	1.35	1.823
Black grease	19.2	18.975	1.35	1.823
Black grease	18.7	18.975	1.35	1.823
Antiglare stickers	17.1	17.55	−0.075	0.006
Antiglare stickers	17.2	17.55	−0.075	0.006
Antiglare stickers	18	17.55	−0.075	0.006
Antiglare stickers	17.9	17.55	−0.075	0.006
Petroleum jelly	15.9	16.35	−1.275	1.626
Petroleum jelly	16.3	16.35	−1.275	1.626
Petroleum jelly	16.2	16.35	−1.275	1.626
Petroleum jelly	17	16.35	−1.275	1.626
	GM = 17.625		$SS_{between}$ = 13.815	

For sum of squares subjects: $SS_{subjects} = \Sigma(M_{participant} - GM)^2 = 0.729$

PARTICIPANT	CONDITION	X	PARTICIPANT MEAN	$M_{PARTICIPANT} - GM$	$(M_{PARTICIPANT} - GM)^2$
1	Black grease	19.8	17.600	−0.025	0.001
2	Black grease	18.2	17.233	−0.392	0.153
3	Black grease	19.2	17.800	0.175	0.031
4	Black grease	18.7	17.867	0.242	0.058
1	Antiglare stickers	17.1	17.600	−0.025	0.001
2	Antiglare stickers	17.2	17.233	−0.392	0.153
3	Antiglare stickers	18	17.800	0.175	0.031
4	Antiglare stickers	17.9	17.867	0.242	0.058
1	Petroleum jelly	15.9	17.600	−0.025	0.001
2	Petroleum jelly	16.3	17.233	−0.392	0.153
3	Petroleum jelly	16.2	17.800	0.175	0.031
4	Petroleum jelly	17	17.867	0.242	0.058
		GM = 17.625		$SS_{subjects}$ = 0.729	

$SS_{within} = SS_{total} - SS_{between} - SS_{subjects} = 1.979$

SOURCE	SS	df	MS	F
Between	13.815	2	6.908	20.93
Subjects	0.729	3	0.243	0.74
Within	1.979	6	0.330	
Total	16.523	11		

Step 6: The F statistic, 20.93, is beyond 5.14, the critical F value at a p level of 0.05. We would reject the null hypothesis. There is a difference, on average, in the visual acuity of participants while wearing different substances beneath their eyes.

i. First, we calculate s_M: $s_M = \sqrt{\frac{MS_{within}}{N}} = \sqrt{\frac{0.330}{4}} = 0.287$

Next, we calculate HSD for each pair of means.

For grease versus stickers:

$$HSD = \frac{(18.975 - 17.550)}{0.287} = 4.97$$

For grease versus jelly:

$$HSD = \frac{(18.975 - 16.35)}{0.287} = 9.15$$

For stickers versus jelly:

$$HSD = \frac{(17.55 - 16.35)}{0.287} = 4.18$$

The critical value of q at a p level of 0.05 is 4.34. Thus, we reject the null hypothesis for the grease versus stickers comparison and for the grease versus jelly comparison, but not for the stickers versus jelly comparison. These results indicate that black grease beneath the eyes leads to better visual acuity, on average, than either antiglare stickers or petroleum jelly.

j. $R^2 = \frac{SS_{between}}{(SS_{total} - SS_{subjects})} = \frac{13.815}{(16.523 - 0.729)} = 0.87$

k. This study could be conducted using a between-groups design if football players were assigned to only one of the three conditions; thus, they would be exposed to the black grease, the antiglare stickers, or the petroleum jelly, rather than all three.

CHAPTER 12

12.1 A two-way ANOVA is a hypothesis test that includes two nominal (or sometimes ordinal) independent variables and a scale dependent variable.

12.3 In everyday conversation, the word *cell* conjures up images of a prison or a small room in which someone is forced to stay, or of one of the building blocks of a plant or animal. In statistics, the word *cell* refers to a single condition in a factorial ANOVA that is characterized by its values on each of the independent variables.

12.5 A two-way ANOVA has two independent variables. When we express that as a 2 × 3 ANOVA, we get added detail; the first number tells us that the first independent variable has two levels, and the second number tells us that the other independent variable has three levels.

12.7 A marginal mean is the mean of a row or a column in a table that shows the cells of a study with a two-way ANOVA design.

12.9 Bar graphs allow us to visually depict the relative changes across the different levels of each independent variable. By adding lines that connect the bars within each series, we can assess whether the lines appear parallel, significantly different from parallel, or intersecting. Intersecting and significantly nonparallel lines are indications of interactions.

12.11 First, we may be able to reject the null hypothesis for the interaction. (If the interaction is statistically significant, then it might not matter whether the main effects are significant; if they are also significant, then those findings are usually qualified by the interaction and they are not described separately. The overall pattern of cell means can tell the whole story.) Second, if we are not able to reject the null hypothesis for the interaction, then we focus on any significant main effects, drawing a specific directional conclusion for each. Third, if we do not reject the null hypothesis for either main effect or the interaction, then we can only conclude that there is insufficient evidence from this study to support the research hypotheses.

12.13 This is the formula for the between-groups sum of squares for the interaction; we can calculate this by subtracting the other between-groups sums of squares (those for the two main effects) and the within-groups sum of squares from the total sum of squares. (The between-groups sum of squares for the interaction is essentially what is left over when the main effects are accounted for.)

12.15 We can use R^2 to calculate effect size similarly to how we did for a one-way ANOVA according to Cohen's conventions. An effect size can be calculated for each main effect and for the interaction.

12.17 **a.** There are two independent variables or factors: gender and sporting event. Gender has two levels, male and female, and sporting event has two levels, Sport 1 and Sport 2.

b. Type of campus is one factor that has two levels: dry and wet. The second factor is type of college, which has three levels, including state, private, and religious.

c. Age group is the first factor, with three levels: 12–13, 14–15, and 16–17. Gender is a second factor, with two levels: female and male. Family composition is the last factor, with three levels: two parents, single parent, no identified authority figure.

12.19 **a.**

	SPORT 1	SPORT 2	
MEN	M = (19 + 17 + 18 + 17)/4 = 17.75	M = (6 + 4 + 8 + 3)/4 = 5.25	(17.75 + 5.25)/2 = 11.50
WOMEN	M = (13 + 14 + 18 + 8)/4 = 13.25	M = (11 + 7 + 4 + 14)/4 = 9	(13.25 + 9)/2 = 11.125
	(17.75 + 13.25)/2 = 15.5	(5.25 + 9)/2 = 7.125	

b.

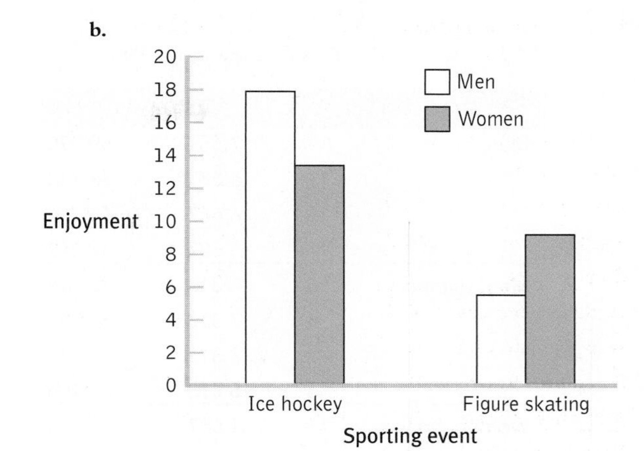

c. $df_{rows(gender)} = N_{rows} - 1 = 2 - 1 = 1$

$df_{columns(sport)} = N_{columns} - 1 = 2 - 1 = 1$

$df_{interaction} = (df_{rows})(df_{columns}) = (1)(1) = 1$

$df_{within} = df_{M,H} + df_{M,S} + df_{W,H} + df_{W,S} = 3 + 3 + 3 + 3 = 12$

$df_{total} = N_{total} - 1 = 16 - 1 = 15$

We can also check that this answer is correct by adding all of the other degrees of freedom together:

$$1 + 1 + 1 + 12 = 15$$

The critical value for an F distribution with 1 and 12 degrees of freedom, at a p level of 0.01, is 9.33.

d. $GM = 11.313.$
$SS_{total} = \Sigma(X - GM)^2$ for each score $= 475.438$

	X	(X − GM)	(X − GM)²
Men, hockey	19	7.687	59.090
	17	5.687	32.342
	18	6.687	44.716
	17	5.687	32.342
Men, skating	6	−5.313	28.228
	4	−7.313	53.480
	8	−3.313	10.976
	3	−8.313	69.106
Women, hockey	13	1.687	2.846
	14	2.687	7.220
	18	6.687	44.716
	8	−3.313	10.976
Women, skating	11	−0.313	0.098
	7	−4.313	18.602
	4	−7.313	53.480
	14	2.687	7.220
			Σ = 475.438

e. Sum of squares for gender: $SS_{between(rows)} = \Sigma(M_{row} - GM)^2$ for each score $= 0.560$

	X	(M_row − GM)	(M_row − GM)²
Men, hockey	19	0.187	0.035
	17	0.187	0.035
	18	0.187	0.035
	17	0.187	0.035
Men, skating	6	0.187	0.035
	4	0.187	0.035
	8	0.187	0.035
	3	0.187	0.035
Women, hockey	13	−0.187	0.035
	14	−0.187	0.035
	18	−0.187	0.035
	8	−0.187	0.035
Women, skating	11	−0.187	0.035
	7	−0.187	0.035
	4	−0.187	0.035
	14	−0.187	0.035
			Σ = 0.560

f. Sum of squares for sporting event: $SS_{between(columns)} = \Sigma(M_{column} - GM)^2$ for each score $= 280.560$

	X	(M_column − GM)	(M_column − GM)²
Men, hockey	19	4.187	17.531
	17	4.187	17.531
	18	4.187	17.531
	17	4.187	17.531
Men, skating	6	−4.188	17.539
	4	−4.188	17.539
	8	−4.188	17.539
	3	−4.188	17.539
Women, hockey	13	4.187	17.531
	14	4.187	17.531
	18	4.187	17.531
	8	4.187	17.531
Women, skating	11	−4.188	17.539
	7	−4.188	17.539
	4	−4.188	17.539
	14	−4.188	17.539
			Σ = 280.560

g. $SS_{within} = \Sigma(X - M_{cell})^2$ for each score $= 126.256$

	X	(X − M_cell)	(X − M_cell)²
Men, hockey	19	1.25	1.563
	17	−0.75	0.563
	18	0.25	0.063
	17	−0.75	0.563
Men, skating	6	0.75	0.563
	4	−1.25	1.563
	8	2.75	7.563
	3	−2.25	5.063
Women, hockey	13	−0.25	0.063
	14	0.75	0.563
	18	4.75	22.563
	8	−5.25	27.563
Women, skating	11	2	4.000
	7	−2	4.000
	4	−5	25.000
	14	5	25.000
			Σ = 126.256

h. We use subtraction to find the sum of squares for the interaction. We subtract all other sources from the total sum of squares, and the remaining amount is the sum of squares for the interaction.

$$SS_{gender \times sport} = SS_{total} - (SS_{gender} + SS_{sport} + SS_{within})$$

$$SS_{gender \times sport} = 475.438 - (0.560 + 280.560 + 126.256) = 68.062$$

i.

SOURCE	*SS*	*df*	*MS*	*F*
Gender	0.560	1	0.560	0.05
Sporting event	280.560	1	280.560	26.67
Gender × sport	68.062	1	68.062	6.47
Within	126.256	12	10.521	
Total	475.438	15		

12.21

SOURCE	*SS*	*df*	*MS*	*F*
Gender	248.25	1	248.25	8.07
Parenting style	84.34	3	28.113	0.91
Gender × style	33.60	3	11.20	0.36
Within	1107.2	36	30.756	
Total	1473.39	43		

12.23 For the main effect A:

$$R^2_{rows} = \frac{SS_{rows}}{(SS_{total} - SS_{columns} - SS_{interaction})} = \frac{30.006}{(652.291 - 33.482 - 1.720)} = 0.049$$

According to Cohen's conventions, this is approaching a medium effect size.

For the main effect B:

$$R^2_{columns} = \frac{SS_{columns}}{(SS_{total} - SS_{rows} - SS_{interaction})} = \frac{33.482}{(652.291 - 30.006 - 1.720)} = 0.054$$

According to Cohen's conventions, this is approaching a medium effect size.

For the interaction:

$$R^2_{interaction} = \frac{SS_{interaction}}{(SS_{total} - SS_{rows} - SS_{columns})} = \frac{1.720}{(652.291 - 30.006 - 33.482)} = 0.003$$

According to Cohen's conventions, this is smaller than a small effect size.

12.25 **a.** This study would be analyzed with a between-groups ANOVA because different groups of participants were assigned to the different treatment conditions.

b. This study could be redesigned to use a within-groups ANOVA by testing the same group of participants on some myths repeated once and some repeated three times both when they are young and then again when they are old.

12.27 **a.** There are two independent variables. The first is gender, and its levels are male and female. The second is the gender of the person being sought, and its levels are same-sex and opposite-sex.

b. The dependent variable is the preferred maximum age difference.

c. He would use a two-way between-groups ANOVA.

d. He would use a 2 × 2 between-groups ANOVA.

e. The ANOVA would have four cells. This number is obtained by multiplying the number of levels of each independent variable (2 × 2).

f.

	MALE	FEMALE
SAME-SEX	Same-sex; male	Same-sex; female
OPPOSITE-SEX	Opposite-sex; male	Opposite-sex; female

12.29 **a.** The first independent variable is the gender said to be most affected by the illness, and its levels are men and women. The second independent variable is the gender of the participant, and its levels are male and female. The dependent variable is level of comfort, on a scale of 1–7.

b. The researchers conducted a two-way between-groups ANOVA.

c. The reported statistics do indicate that there is a significant interaction because the probability associated with the *F* statistic for the interaction is less than 0.05.

d.

	FEMALE PARTICIPANTS	MALE PARTICIPANTS
ILLNESS AFFECTS WOMEN	4.88	3.29
ILLNESS AFFECTS MEN	3.56	4.67

e. Bar graph for the interaction:

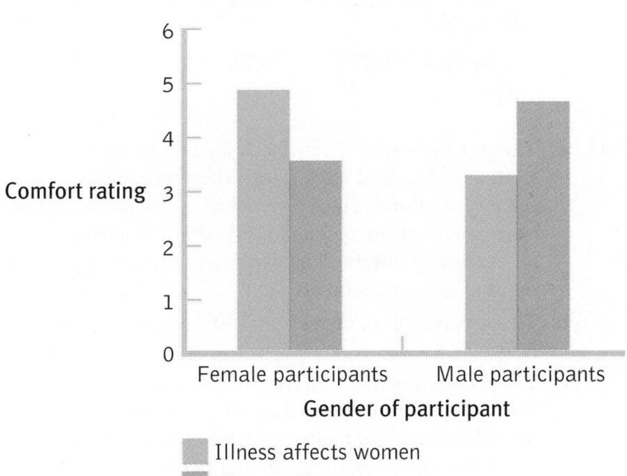

f. This is a qualitative interaction. Female participants indicated greater average comfort for attending a meeting regarding an illness that affects women than for attending a meeting regarding an illness that affects men. Male participants had the opposite pattern of results; male participants indicated greater average comfort for attending a meeting regarding an illness that affects men as opposed to one that affects women.

g.

	FEMALE PARTICIPANTS	MALE PARTICIPANTS
ILLNESS AFFECTS WOMEN	4.88	**4.80**
ILLNESS AFFECTS MEN	3.56	4.67

Note: There are several cell means that would work.

h. Bar graph for the new means:

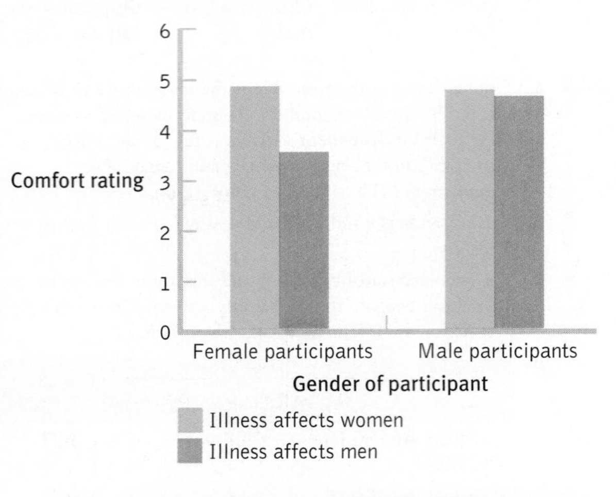

i.

	FEMALE PARTICIPANTS	MALE PARTICIPANTS
ILLNESS AFFECTS WOMEN	4.88	**5.99**
ILLNESS AFFECTS MEN	3.56	4.67

12.31 **a.** The first independent variable is the race of the face, and its levels are white and black. The second independent variable is the type of instruction given to the participants, and its levels are no instruction and instruction to attend to distinguishing features. The dependent variable is the measure of recognition accuracy.

b. The researchers conducted a two-way between-groups ANOVA.

c. The reported statistics indicate that there is a significant main effect of race. On average, the white participants who saw white faces had higher recognition scores than did white participants who saw black faces.

d. The main effect is misleading because those who received instructions to attend to distinguishing features actually had lower mean recognition scores for the white faces than did those who received no instruction, whereas those who received instructions to attend to distinguishing features had higher mean recognition scores for the black faces than did those who received no instruction.

e. The reported statistics do indicate that there is a significant interaction because the probability associated with the *F* statistics for the interaction is less than 0.05.

f.

	BLACK FACE	WHITE FACE
NO INSTRUCTION	1.04	1.46
DISTINGUISHING FEATURES INSTRUCTION	1.23	1.38

g. Bar graph of findings:

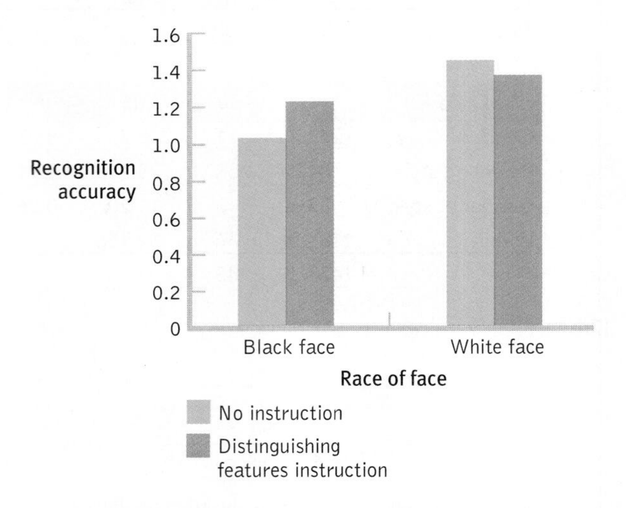

h. When given instructions to pay attention to distinguishing features of the faces, participants' average recognition of the black faces was higher than when given no instructions, whereas their average recognition of the white faces was worse than when given no instruction. This is a qualitative interaction because the direction of the effect changes between black and white.

12.33 **a.** The first independent variable is gender of the seeker, and its levels are men and women. The second independent variable is gender of the person being sought, and its levels are men and women. The dependent variable is the oldest acceptable age of the person being sought.

b.

	WOMEN SEEKERS	MEN SEEKERS
MEN SOUGHT	34.80	35.40
WOMEN SOUGHT	36.00	27.20

c. *Step 1:* Population 1 (women, men) is women seeking men. Population 2 (men, women) is men seeking women. Population 3 (women, women) is women seeking women. Population 4 (men, men) is men seeking men. The comparison distributions will be *F* distributions. The hypothesis test will be a two-way between-groups ANOVA. Assumptions: The data are not from random samples, so we must generalize with caution. The assumption of homogeneity of variance is violated because the largest variance (29.998) is much larger than the smallest variance (1.188). For the purposes of this exercise, however, we will conduct this ANOVA.

Step 2: Main effect of first independent variable—gender of seeker:

Null hypothesis: On average, men and women report the same oldest acceptable ages for a partner—$\mu_M = \mu_W$.

Research hypothesis: On average, men and women report different oldest acceptable ages for a partner—$\mu_M \neq \mu_W$.

Main effect of second independent variable—gender of person sought:

Null hypothesis: On average, those seeking men and those seeking women report the same oldest acceptable ages for a partner—$\mu_M = \mu_W$.

Research hypothesis: On average, those seeking men and those seeking women report different oldest acceptable ages for a partner—$\mu_M \neq \mu_W$.

Interaction: seeker × sought:

Null hypothesis: The effect of the gender of the seeker does not depend on the gender of the person sought.

Research hypothesis: The effect of the gender of the seeker does depend on the gender of the person sought.

Step 3: $df_{columns(seeker)} = 2 - 1 = 1$

$df_{rows(sought)} = 2 - 1 = 1$

$df_{interaction} = (1)(1) = 1$

$df_{within} = df_{W,M} + df_{M,W} + df_{W,W} + df_{M,M} = 4 + 4 + 4 + 4 = 16$

Main effect of gender of seeker: F distribution with 1 and 16 degrees of freedom

Main effect of gender of sought: F distribution with 1 and 16 degrees of freedom

Interaction of seeker and sought: F distribution with 1 and 16 degrees of freedom

Step 4: Cutoff F for main effect of seeker: 4.49

Cutoff F for main effect of sought: 4.49

Cutoff F for interaction of seeker and sought: 4.49

Step 5: $SS_{total} = \Sigma(X - GM)^2 = 454.550$

$SS_{column(seeker)} = \Sigma(M_{column(seeker)} - GM)^2 = 84.050$

$SS_{row(sought)} = \Sigma(M_{row(sought)} - GM)^2 = 61.250$

$SS_{within} = \Sigma(X - M_{cell})^2 = 198.800$

$SS_{interaction} = SS_{total} - (SS_{row} + SS_{column} + SS_{within}) = 110.450$

SOURCE	*SS*	*df*	*MS*	*F*
Seeker gender	84.050	1	84.050	6.77
Sought gender	61.250	1	61.250	4.93
Seeker × sought	110.450	1	110.450	8.89
Within	198.800	16	12.425	
Total	454.550	19		

Step 6: There is a significant main effect of gender of the seeker; it appears that women are willing to accept older dating partners, on average, than are men. There is also a significant main effect of gender of the person being sought; it appears that those seeking men are willing to accept older dating partners, on average, than are those seeking women. Additionally, there is a significant interaction between the gender of the seeker and the gender of the person being sought. Because there is a significant interaction, we ignore the main effects and report only the interaction. This is a quantitative interaction because there is a difference for male seekers, but not for female seekers. We are not seeing a reversal of direction, necessary for a qualitative interaction.

d.

e. For the main effect of seeker gender:

$$R^2_{seeker} = \frac{SS_{seeker}}{(SS_{total} - SS_{sought} - SS_{interaction})} = \frac{84.05}{(454.55 - 61.25 - 110.45)} = 0.30$$

According to Cohen's conventions, this is a large effect size.

For the main effect of sought gender:

$$R^2_{sought} = \frac{SS_{sought}}{(SS_{total} - SS_{seeker} - SS_{interaction})} = \frac{61.25}{(454.55 - 84.05 - 110.45)} = 0.24$$

According to Cohen's conventions, this is a large effect size.

For the interaction:

$$R^2_{interaction} = \frac{SS_{interaction}}{(SS_{total} - SS_{seeker} - SS_{sought})} = \frac{110.45}{(454.55 - 84.05 - 61.25)} = 0.36$$

According to Cohen's conventions, this is a large effect size.

12.35 **a.** The independent variables are type of payment, still with two levels, and level of payment, now with three levels (low, moderate, and high). The dependent variable is still willingness to help, as assessed with the 11-point scale.

b.

	LOW AMOUNT	MODERATE AMOUNT	HIGH AMOUNT	
CASH PAYMENT	4.75	7.50	8.00	6.75
CANDY PAYMENT	6.25	6.00	6.50	6.25
	5.50	6.75	7.25	

c.

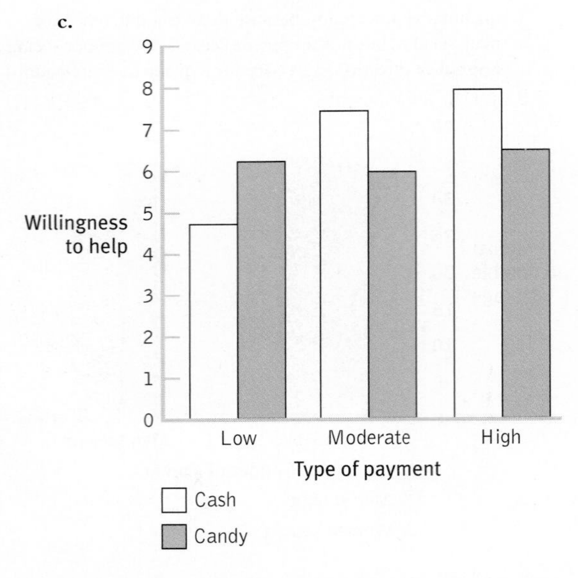

d. There does still seem to be the same qualitative interaction, such that the effect of the level of payment depends on the type of payment. When candy payments are used, the level seems to have no mean impact. However, when cash payments are used, a low level leads to a lower willingness to help, on average, than when candy is used, and a moderate or high level leads to a higher willingness to help, on average, than when candy is used.

e. Post hoc tests would be needed. Specifically, we would need to compare the three levels of payment to see where specific significant differences exist. Based on the graph we created, it appears as if willingness to help in the low payment condition is significantly lower, on average, than in the moderate and high conditions for payments.

CHAPTER 13

13.1 A correlation coefficient is a statistic that quantifies the relation between two variables.

13.3 A perfect relation occurs when the data points fall exactly on the line we fit through the data. A perfect relation results in a correlation coefficient of −1.0 or 1.0.

13.5 According to Cohen (1988), a correlation coefficient of 0.50 is a large correlation, and 0.30 is a medium one. However, it is unusual in social science research to have a correlation as high as 0.50. The decision of whether a correlation is worth talking about is sometimes based on whether it is statistically significant, as well as what practical effect a correlation of a certain size indicates.

13.7 When used to capture the relation between two variables, the correlation coefficient is a descriptive statistic. When used to draw conclusions about the greater population, such as with hypothesis testing, the coefficient serves as an inferential statistic.

13.9 Positive products of deviations, indicating a positive correlation, occur when both members of a pair of scores tend to result in a positive deviation or when both members tend to result in a negative deviation. Negative products of deviations, indicating a negative correlation, occur when members of a pair of scores tend to result in opposite-valued deviations (one negative and the other positive).

13.11 (1) We calculate the deviation of each score from its mean, multiply the two deviations for each participant, and sum the products of the deviations. (2) We calculate a sum of squares for each variable, multiply the two sums of squares, and take the square root of the product of the sums of squares. (3) We divide the sum from step 1 by the square root in step 2.

13.13 Test–retest reliability involves giving the same group of people the exact same test with some amount of time (perhaps a week) between the two administrations of the test. Test–retest reliability is then calculated as the correlation between their scores on the two administrations of the test. Calculation of coefficient alpha does not require giving the same test two times. Rather, coefficient alpha is based on correlations between different halves of the test items from a single administration of the test.

13.15 **a.** These data appear to be negatively correlated.

b. These data appear to be positively correlated.

c. Neither; these data appear to have a very small correlation, if any.

13.17 **a.** −0.28 is a medium correlation.

b. 0.79 is a large correlation.

c. 1.0 is a perfect correlation.

d. −0.015 is almost no correlation.

13.19 **a.**

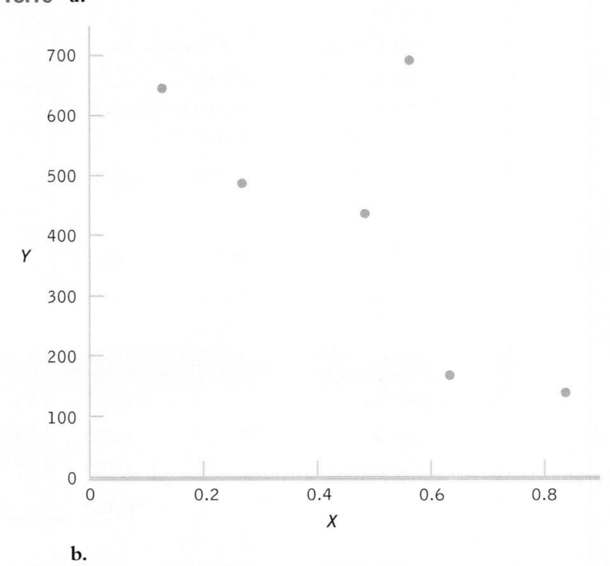

b.

X	$(X - M_X)$	Y	$(Y - M_Y)$	$(X - M_X)(Y - M_Y)$
0.13	−0.36	645	218.50	−78.660
0.27	−0.22	486	59.50	−13.090
0.49	0.00	435	8.50	0.000
0.57	0.08	689	262.50	21.000
0.84	0.35	137	−289.50	−101.325
0.64	0.15	167	−259.50	−38.925
$M_X = 0.49$		$M_Y = 426.5$		$\Sigma[(X - M_X)(Y - M_Y)] = -211.0$

c.

X	$(X - M_X)$	$(X - M_X)^2$	Y	$(Y - M_Y)$	$(Y - M_Y)^2$
0.13	−0.36	0.130	645	218.50	47,742.25
0.27	−0.22	0.048	486	59.50	3,540.25
0.49	0.00	0.000	435	8.50	72.25
0.57	0.08	0.006	689	262.50	68,906.25
0.84	0.35	0.123	137	−289.50	83,810.25
0.64	0.15	0.023	167	−259.50	67,340.25
		$\Sigma(X - M_X)^2 =$ 0.330			$\Sigma(Y - M_Y)^2 =$ 271,411.50

$$\sqrt{(SS_X)(SS_Y)} = \sqrt{(0.330)(271{,}411.50)} = \sqrt{89{,}565.795} = 299.275$$

d. $$r = \frac{\Sigma[(X - M_X)(Y - M_Y)]}{\sqrt{(SS_X)(SS_Y)}} = \frac{-211}{299.275} = -0.71$$

e. $df_r = N - 2 = 6 - 2 = 4$

f. −0.811 and 0.811

13.21 a.

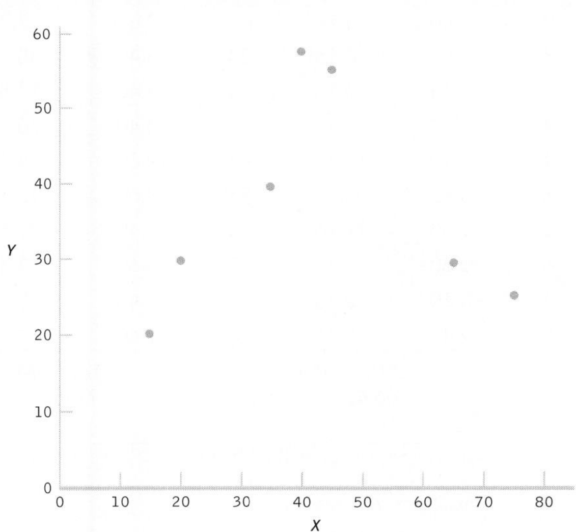

b.

X	$(X - M_X)$	Y	$(Y - M_Y)$	$(X - M_X)(Y - M_Y)$
40	−2.143	60	22.857	−48.983
45	2.857	55	17.857	51.017
20	−22.143	30	−7.143	158.167
75	32.857	25	−12.143	−398.983
15	−27.143	20	−17.143	465.312
35	−7.143	40	2.857	−20.408
65	22.857	30	−7.143	−163.268
$M_X =$ 42.143		$M_Y =$ 37.143		$\Sigma[(X - M_X)(Y - M_Y)] =$ 42.854

c.

X	$(X - M_X)$	$(X - M_X)^2$	Y	$(Y - M_Y)$	$(Y - M_Y)^2$
40	−2.143	4.592	60	22.857	522.442
45	2.857	8.162	55	17.857	318.872
20	−22.143	490.312	30	−7.143	51.022
75	32.857	1079.582	25	−12.143	147.452
15	−27.143	736.742	20	−17.143	293.882
35	−7.143	51.022	40	2.857	8.162
65	22.857	522.442	30	−7.143	51.022
		$\Sigma(X - M_X)^2 =$ 2892.854			$\Sigma(Y - M_Y)^2 =$ 1392.854

$$\sqrt{(SS_X)(SS_Y)} = \sqrt{(2892.854)(1392.854)} = \sqrt{4{,}029{,}323.265} = 2007.317$$

d. $$r = \frac{\Sigma[(X - M_X)(Y - M_Y)]}{\sqrt{(SS_X)(SS_Y)}} = \frac{42.854}{2007.317} = 0.021$$

e. $df_r = N - 2 = 7 - 2 = 5$

f. −0.754 and 0.754

13.23 When using a measure to diagnose individuals, having a reliability of at least 0.90 is important—and the more reliable the test, the better. So, based on reliability information alone, we would recommend she use the test with 0.95 reliability.

13.25 **a.** Newman's data do not suggest a correlation between Mercury's phases and breakdowns. There was no consistency in the report of breakdowns during one of the phases.

b. Massey may believe there is a correlation because she already believes that there is a relation between astrological events and human events. As you learned in Chapter 5, the confirmation bias refers to the tendency to pay attention to those events that confirm our prior beliefs. The confirmation bias may lead Massey to observe an illusory correlation (i.e., she perceives a correlation that does not actually exist) because she attends only to those events that confirm her prior belief that the phase of Mercury is related to breakdowns.

c. Given that there are two phases of Mercury (and assuming they're equal in length), half of the breakdowns that occur would be expected to occur during the retrograde phase and the other half during the nonretrograde phase, just by chance. Expected relative-frequency probability refers to the expected frequency of events. So in this example we would expect 50% of breakdowns to occur during the retrograde phase and 50% during the nonretrograde phase. If we base our conclusions on only a small number of observations of breakdowns, the observed relative-frequency probability is more likely to differ from the expected relative-frequency probability because we are less likely to have a representative sample of breakdowns.

d. This correlation would not be useful in predicting events in your own life because no relation would be observed in this limited time span.

e. Available data do not support the idea that a correlation exists between Mercury's phases and breakdowns.

13.27 **a.** The accompanying scatterplot depicts the relation between hours of exercise and number of friends. Note that you could have chosen to put exercise along the y-axis and friends along the x-axis.

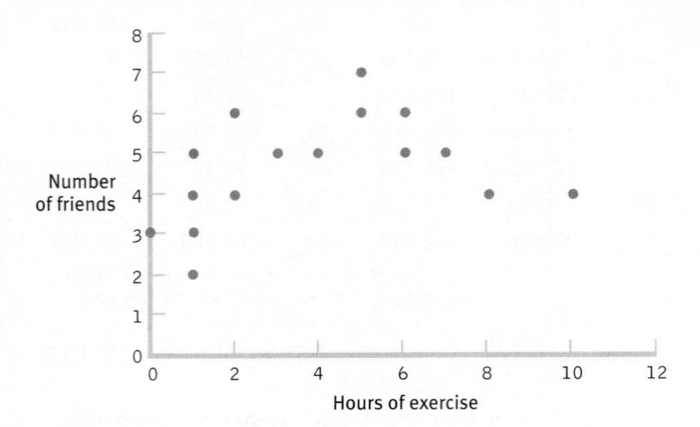

b. The scatterplot suggests that as the number of hours of exercise each week increases from 0 to 5, there is an increase in the number of friends, but as the hours of exercise continues to increase past 5, there is a decrease in the number of friends.

c. It would not be appropriate to calculate a Pearson correlation coefficient with this set of data. The scatterplot suggests a nonlinear relation between exercise and number of friends, and the Pearson correlation coefficient measures only the extent of linear relation between two variables.

13.29 **a.** *Step 1:* Population 1: Adolescents like those we studied. Population 2: Adolescents for whom there is no relation between externalizing behavior and anxiety. The comparison distribution is made up of correlation coefficients based on many, many samples of our size, 10 people, randomly selected from the population.

We do not know if the data were randomly selected (first assumption), so we must be cautious when generalizing our findings. We also do not know if the underlying population distribution for externalizing behaviors and anxiety in adolescents is normally distributed (second assumption). The sample size is too small to make any conclusions about this assumption, so we should proceed with caution. The third assumption, unique to correlation, is that the variability of one variable is equal across the levels of the other variable. Because we have such a small data set, it is difficult to evaluate this. However, we can see from the scatterplot that the data are somewhat consistently variable.

Step 2: Null hypothesis: There is no correlation between externalizing behavior and anxiety among adolescents—H_0: $\rho = 0$.

Research hypothesis: There is a correlation between externalizing behavior and anxiety among adolescents—H_1: $\rho \neq 0$.

Step 3: The comparison distribution is a distribution of Pearson correlations, r, with the following degrees of freedom: $df_r = N - 2 = 10 - 2 = 8$.

Step 4: The critical values for an r distribution with 8 degrees of freedom for a two-tailed test with a p level of 0.05 are -0.632 and 0.632.

Step 5: The Pearson correlation coefficient is calculated in three steps. First, we calculate the numerator:

X	$(X - M_X)$	Y	$(Y - M_Y)$	$(X - M_X)(Y - M_Y)$
9	2.40	37	7.60	18.24
7	0.40	23	−6.40	−2.56
7	0.40	26	−3.40	−1.36
3	−3.60	21	−8.40	30.24
11	4.40	42	12.60	55.44
6	−0.60	33	3.60	−2.16
2	−4.60	26	−3.40	15.64
6	−0.60	35	5.60	−3.36
6	−0.60	23	−6.40	3.84
9	2.40	28	−1.40	−3.36
$M_X = 6.60$		$M_Y = 29.40$		$\Sigma[(X - M_X)(Y - M_Y)] = 110.60$

Second, we calculate the denominator:

X	$(X - M_X)$	$(X - M_X)^2$	Y	$(Y - M_Y)$	$(Y - M_Y)^2$
9	2.40	5.76	37	7.60	57.76
7	0.40	0.16	23	−6.40	40.96
7	0.40	0.16	26	−3.40	11.56
3	−3.60	12.96	21	−8.40	70.56
11	4.40	19.36	42	12.60	158.76
6	−0.60	0.36	33	3.60	12.96
2	−4.60	21.16	26	−3.40	11.56
6	−0.60	0.36	35	5.60	31.36
6	−0.60	0.36	23	−6.40	40.96
9	2.40	5.76	28	−1.40	1.96
		$\Sigma(X - M_X)^2 =$ 66.40			$\Sigma(Y - M_Y)^2 =$ 438.40

$$\sqrt{(SS_X)(SS_Y)} = \sqrt{(66.40)(438.40)} = \sqrt{29{,}109.76} = 170.616$$

Finally, we compute r:

$$r = \frac{\Sigma[(X - M_X)(Y - M_Y)]}{\sqrt{(SS_X)(SS_Y)}} = \frac{110.60}{170.616} = 0.65$$

f. The test statistic, $r = 0.65$, is larger in magnitude than the critical value of 0.632. We can reject the null hypothesis and conclude that there is a strong positive correlation between the number of externalizing behaviors performed by adolescents and their level of anxiety.

13.31 **a.** You would expect a person who owns a lot of cats to tend to have many mental health problems. Because the two variables are positively correlated, as cat ownership increases, the number of mental health problems tends to increase.

b. You would expect a person who owns no cats or just one cat to tend to have few mental health problems. Because the variables are positively correlated, people who have a low score on one variable are also likely to have a low score on the other variable.

c. You would expect a person who owns a lot of cats to tend to have few mental health problems. Because the two variables are negatively related, as one variable increases, the other variable tends to decrease. This means a person owning lots of cats would likely have a low score on the mental health variable.

d. You would expect a person who owns no cats or just one cat to tend to have many mental health problems. Because the two variables are negatively related, as one variable decreases, the other variable tends to increase, which means that a person with fewer cats would likely have more mental health problems.

13.33 **a.** The accompanying scatterplot depicts a negative linear relation between perceived femininity and perceived trauma. Because the relation appears linear, it is appropriate to calculate the Pearson correlation coefficient for these data. (*Note:* The number (2) indicates that two participants share that pair of scores.)

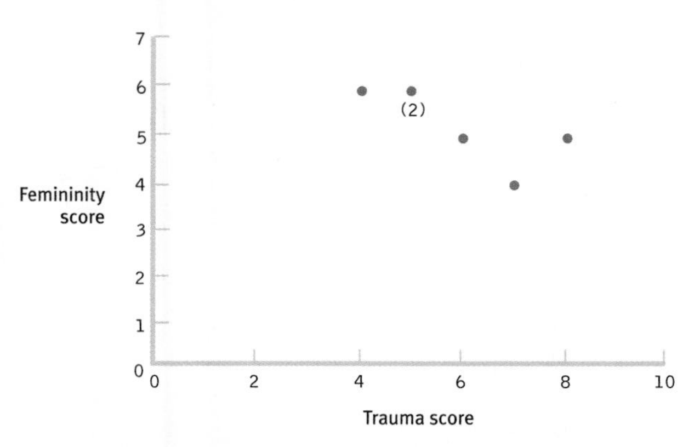

b. The Pearson correlation coefficient is calculated in three steps. Step 1 is calculating the numerator:

X	$(X - M_X)$	Y	$(Y - M_Y)$	$(X - M_X)(Y - M_Y)$
5	−0.833	6	0.667	−0.556
6	0.167	5	−0.333	−0.056
4	−1.833	6	0.667	−1.223
5	−0.833	6	0.667	−0.556
7	1.167	4	−1.333	−1.556
8	2.167	5	−0.333	−0.722
$M_X =$ 5.833		$M_Y =$ 5.333		$\Sigma[(X - M_X)(Y - M_Y)] =$ −4.669

Step 2 is calculating the denominator:

X	$(X - M_X)$	$(X - M_X)^2$	Y	$(Y - M_Y)$	$(Y - M_Y)^2$
5	−0.833	0.694	6	0.667	0.445
6	0.167	0.028	5	−0.333	0.111
4	−1.833	3.360	6	0.667	0.445
5	−0.833	0.694	6	0.667	0.445
7	1.167	1.362	4	−1.333	1.777
8	2.167	4.696	5	−0.333	0.111
		$\Sigma(X - M_X)^2 =$ 10.834			$\Sigma(Y - M_Y)^2 =$ 3.334

$$\sqrt{(SS_X)(SS_Y)} = \sqrt{(10.834)(3.334)} = \sqrt{36.121} = 6.010$$

Step 3 is computing *r*:

$$r = \frac{\Sigma[(X - M_X)(Y - M_Y)]}{\sqrt{(SS_X)(SS_Y)}} = \frac{-4.669}{6.010} = -0.78$$

c. The correlation coefficient reveals a strong negative relation between perceived femininity and perceived trauma; as trauma increases, perceived femininity tends to decrease.

d. Those participants who had positive deviation scores on trauma tended to have negative deviation scores on femininity (and vice versa), meaning that when a person's score on one variable was above the mean for that variable (positive deviation), his or her score on the second variable was typically below the mean for that variable (negative deviation). So, having a high score on one variable was associated with having a low score on the other, which is a negative correlation.

13.35 **a.** The accompanying scatterplot depicts a positive linear relation between perceived trauma and perceived masculinity. The data appear to be linearly related; therefore, it is appropriate to calculate a Pearson correlation coefficient.

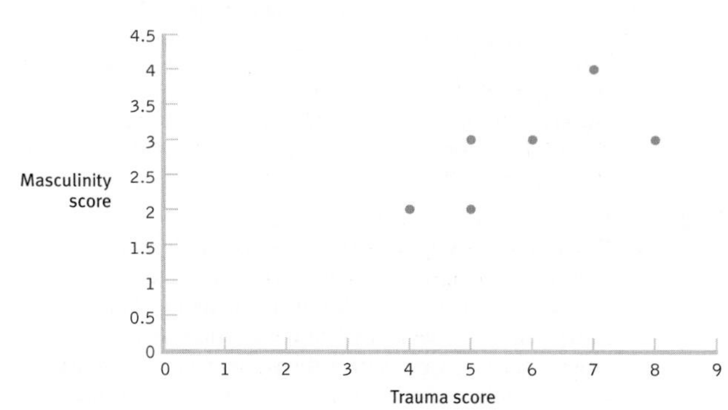

b. The Pearson correlation coefficient is calculated in three steps. Step 1 is calculating the numerator:

X	$(X - M_X)$	Y	$(Y - M_Y)$	$(X - M_X)(Y - M_Y)$
5	−0.833	3	0.167	−0.139
6	0.167	3	0.167	0.028
4	−1.833	2	−0.833	1.527
5	−0.833	2	−0.833	0.694
7	1.167	4	1.167	1.362
8	2.167	3	0.167	0.362
$M_X = 5.833$		$M_Y = 2.833$		$\Sigma[(X - M_X)(Y - M_Y)] = 3.834$

Step 2 is calculating the denominator:

X	$(X - M_X)$	$(X - M_X)^2$	Y	$(Y - M_Y)$	$(Y - M_Y)^2$
5	−0.833	0.694	3	0.167	0.028
6	0.167	0.028	3	0.167	0.028
4	−1.833	3.360	2	−0.833	0.694
5	−0.833	0.694	2	−0.833	0.694
7	1.167	1.362	4	1.167	1.362
8	2.167	4.696	3	0.167	0.028
		$\Sigma(X - M_X)^2 =$ 10.834			$\Sigma(Y - M_Y)^2 =$ 2.834

$$\sqrt{(SS_X)(SS_Y)} = \sqrt{(10.834)(2.834)} = \sqrt{30.704} = 5.541$$

Step 3 is computing *r*:

$$r = \frac{\Sigma[(X - M_X)(Y - M_Y)]}{\sqrt{(SS_X)(SS_Y)}} = \frac{3.834}{5.541} = 0.69$$

c. The correlation indicates that a large positive relation exists between perceived trauma and perceived masculinity.

d. For most of the participants, the sign of the deviation for the traumatic variable is the same as that for the masculinity variable, which indicates that those participants scoring above the mean on one variable also tended to score above the mean on the second variable (and likewise with the lowest scores). Because the scores for each participant tend to fall on the same side of the mean, this is a positive relation.

e. When the person was a woman, the perception of the situation as traumatic was strongly negatively correlated with the perception of the woman as feminine. This relation is opposite that observed when the person was a man. When the person was a man, the perception of the situation as traumatic was strongly positively correlated with the perception of the man as feminine. Regardless of whether the person was a man or a woman, there was a positive correlation between the perception of the situation as traumatic and perception of masculinity, but the observed correlation was stronger for the perceptions of women than for the perceptions of men.

13.37 **a.** Because your friend is running late, she is likely more concerned about traffic than she otherwise would be. Thus, she may take note of traffic only when she is running late, leading her to believe that the amount of traffic correlates with how late she is. Furthermore, having this belief, in the future she may think only of cases that confirm her belief that a relation exists between how late she is and traffic conditions, reflecting a confirmation bias. Alternatively, traffic conditions might be worse when your friend is running late, but that could be a coincidence. A more systematic study of the relation between your friend's behavior and traffic conditions would be required before she could conclude that a relation exists.

b. There are a number of possible answers to this question. For example, we could operationalize the degree to which she is late as the number of minutes past her intended departure time that she gets in the car. We could operationalize the amount of traffic as the number of minutes the car is being driven at less than the speed limit (given that your friend would normally drive right at the speed limit).

13.39 **a.** The reporter suggests that convertibles are not generally less safe than other cars.

b. Convertibles may be driven less often than other cars, as they may be considered primarily a recreational vehicle. If they are driven less, owners have fewer chances to get into accidents while driving them.

c. A more appropriate comparison may be to determine the number of fatalities that occur per every 100 hours driven in various kinds of cars.

13.41 **a.** It appears that the data are somewhat positively correlated.

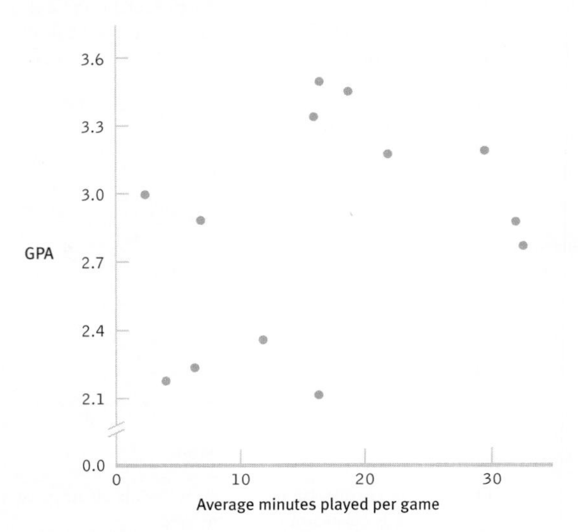

b. The Pearson correlation coefficient is calculated in three steps. Step 1 is calculating the numerator:

X	$(X - M_X)$	Y	$(Y - M_Y)$	$(X - M_X)(Y - M_Y)$
29.70	13.159	3.20	0.343	4.514
32.14	15.599	2.88	0.023	0.359
32.72	16.179	2.78	−0.077	−1.246
21.76	5.219	3.18	0.323	1.686
18.56	2.019	3.46	0.603	1.217
16.23	−0.311	2.12	−0.737	0.229
11.80	−4.741	2.36	−0.497	2.356
6.88	−9.661	2.89	0.033	−0.319
6.38	−10.161	2.24	−0.617	6.269
15.83	−0.711	3.35	0.493	−0.351
2.50	−14.041	3.00	0.143	−2.008
4.17	−12.371	2.18	−0.677	8.375
16.36	−0.181	3.50	0.643	−0.116
$M_X =$ 16.541		$M_Y =$ 2.857		$\Sigma[(X - M_X)(Y - M_Y)] =$ 20.965

Step 2 is calculating the denominator:

X	$(X - M_X)$	$(X - M_X)^2$	Y	$(Y - M_Y)$	$(Y - M_Y)^2$
29.70	13.159	173.159	3.20	0.343	0.118
32.14	15.599	243.329	2.88	0.023	0.001
32.72	16.179	261.760	2.78	−0.077	0.006
21.76	5.219	27.238	3.18	0.323	0.104
18.56	2.019	4.076	3.46	0.603	0.364
16.23	−0.311	0.097	2.12	−0.737	0.543
11.80	−4.741	22.477	2.36	−0.497	0.247
6.88	−9.661	93.335	2.89	0.033	0.001
6.38	−10.161	103.246	2.24	−0.617	0.381
15.83	−0.711	0.506	3.35	0.493	0.243
2.50	−14.041	197.150	3.00	0.143	0.020
4.17	−12.371	153.042	2.18	−0.677	0.458
16.36	−0.181	0.033	3.50	0.643	0.413
		$\Sigma(X - M_X)^2 =$ 1279.448			$\Sigma(Y - M_Y)^2 =$ 2.899

$$\sqrt{(SS_X)(SS_Y)} = \sqrt{(1279.448)(2.899)} = \sqrt{3709.120} = 60.903$$

Step 3 is computing *r*:

$$r = \frac{\Sigma[(X - M_X)(Y - M_Y)]}{\sqrt{(SS_X)(SS_Y)}} = \frac{20.964}{60.903} = 0.34$$

c. We computed the correlation coefficient for these data to explore whether there was a relation between GPA and playing time for the members of this team. If we were interested in making a statement about athletes in general, an inferential analysis, we would want to collect more data from a random or representative sample and conduct a hypothesis test.

d. *Step 1:* Population 1: Athletes like those we studied. Population 2: Athletes for whom there is no relation between minutes played and GPA. The comparison distribution is made up of many, many correlation coefficients based on samples of our size, 13 people, randomly selected from the population.

We know that these data were not randomly selected (first assumption), so we must be cautious when generalizing our findings. We also do not know if the underlying population distributions are normally distributed (second assumption). The sample size is too small to make any conclusions about this assumption, so we should proceed with caution. The third assumption, unique to correlation, is that the variability of one variable is equal across the levels of the other variable. Because we have such a small data set, it is difficult to evaluate this.

Step 2: Null hypothesis: There is no correlation between participation in athletics, as measured by minutes played on average, and GPA—H_0: $\rho = 0$.

Research hypothesis: There is a correlation between participation in athletics and GPA—H_1: $\rho \neq 0$.

Step 3: The comparison distribution is a distribution of Pearson correlation coefficients, *r*, with the following degrees of freedom: $df_r = N - 2 = 13 - 2 = 11$.

Step 4: The critical values for an *r* distribution with 11 degrees of freedom for a two-tailed test with a *p* level of 0.05 are −0.553 and 0.553.

Step 5: $r = 0.34$, as calculated in b.

Step 6: The test statistic, $r = 0.34$, is not larger in magnitude than the critical value of 0.553, so we fail to reject the null hypothesis. We cannot conclude that a relation exists between these two variables. Because the sample size is rather small and we calculated a medium correlation with this small sample, we would be encouraged to collect more data to increase statistical power so that we may more fully explore this relation.

e. Because the results are not statistically significant, we cannot draw any conclusion, except that we do not have enough information.

f. We could have collected these data randomly, rather than looking at just one team. We also could have collected a larger sample size. In order to say something about causation, we could manipulate average minutes played to see whether that manipulation results in a change in GPA. Because very few coaches would be willing to let us do that, we would have a difficult time conducting such an experiment.

13.43 **a.** If students were marked down for talking about the rooster rather than the cow, the reading test would not meet the established criteria. The question asked on the test is ambiguous because the information regarding what caused the cow's behavior to change is not explicitly stated in the story. Furthermore, the correct answer to the question provided on the Web site is not actually an answer to the question itself. The question states, "What caused Brownie's behavior to change?" The answer that the cow started out kind and ended up mean is a description of *how* her behavior changed, not what caused her behavior to change. This question does not appear to be a valid question because it does not appear to provide an accurate assessment of students' *writing* ability.

b. One possible third variable that could lead to better performance in some schools over others is the average socioeconomic status of the families whose children attend the school. Schools in wealthier areas or counties would have students of higher socioeconomic status, who might be expected to perform better on a test of writing skill. A second possible third variable that could lead to better performance in some schools over others is the type of reading and writing curriculum implemented in the school. Different ways of teaching the material may be more effective than others, regardless of the effectiveness of the teachers who are actually presenting the material.

13.45 **a.** The participants in the study are the various countries on which rates were obtained.

b. The two variables are health care spending and health, as assessed by life expectancy. Health care spending was operationalized as the amount spent per capita on health care, whereas life expectancy is the average age of death. Another way to operationalize health could be rates of various diseases, such as heart disease, or obesity via body mass index (BMI).

c. The study finding was that there is a negative correlation between health care spending and life expectancy, in which countries, such as the United States, that have higher rates of spending on health care per capita have lower life expectancies. One would suspect the opposite to be true, that the more a country spends on health care, the healthier the population would be, thus resulting in higher life expectancy.

d. Other possible third variables could be the typical body weight in a country, the typical exercise levels in a country, accident rates, access to health care, access or knowledge of preventative health measures, stereotypes, or a country's typical diet.

e. This study is a correlational study, not a true experiment, because countries were not assigned to certain levels of health care spending, and then assessed for life expectancy. The data were obtained from naturally occurring events.

f. It would not be possible to conduct a true experiment on this topic as this would require a manipulation in the health care spending for various countries for the entire population for a long period of time, which would not be realistic, practical, or ethical to implement.

CHAPTER 14

14.1 Regression allows us to make predictions based on the relation established in the correlation. Regression also allows us to consider the contributions of several variables.

14.3 There is no difference between these two terms. They are two ways to express the same thing.

14.5 a is the *intercept,* the predicted value for Y when X is equal to 0, which is the point at which the line crosses, or intercepts, the y-axis. b is the *slope,* the amount that Y is predicted to increase for an increase of 1 in X.

14.7 The intercept is not meaningful or useful when it is impossible to observe a value of 0 for X. If height is being used to predict weight, it would not make sense to talk about the weight of someone with no height.

14.9 The line of best fit in regression means that we couldn't make the line a little steeper, or raise or lower it, in any way that would allow it to represent those dots any better than it already does. This is why we can look at the scatterplot around this line and observe that the line goes precisely through the middle of the dots. Statistically, this is the line that leads to the least amount of error in prediction.

14.11 Data points clustered closely around the line of best fit are described by a small standard error of the estimate, and we enjoy a high level of confidence in the predictive ability of the independent variable as a result. Data points clustered far away from the line of best fit are described by a large standard error of the estimate, and as a result we have a low level of confidence in the predictive ability of our independent variable.

14.13 If regression to the mean did not occur, every distribution would look bimodal, like a valley. Instead, the end result of the phenomenon of regression to the mean is that things look unimodal, like a hill or what we call the normal, bell-shaped curve. Remember that the center of the bell-shaped curve is the mean, and this is where the bulk of data cluster, thanks to regression to the mean.

14.15 The sum of squares total, SS_{total}, represents the worst-case scenario, the total error we would have in the predictions if there was no regression equation and we had to predict the mean for everybody.

14.17 (1) Determine the error associated with using the mean as the predictor. (2) Determine the error associated with using the regression equation as the predictor. (3) Subtract the error associated with the regression equation from the error associated with the mean. (4) Divide the difference (calculated in step 3) by the error associated with using the mean.

14.19 An orthogonal variable is an independent variable that makes a separate and distinct contribution in the prediction of a dependent variable, as compared with another independent variable.

14.21 **a.** $z_x = \frac{X - M_x}{SD_x} = \frac{2.9 - 1.9}{0.6} = 1.667$

b. $z_{\hat{Y}} = (r_{XY})(z_X) = (0.31)(1.667) = 0.517$

c. $\hat{Y} = z_{\hat{Y}}(SD_Y) + M_Y = (0.517)(3.2) + 10 = 11.65$

14.23 **a.** $z_X = \frac{X - M_X}{SD_X} = \frac{76 - 55}{12} = 1.75$

b. $z_{\hat{Y}} = (r_{XY})(z_X) = (-0.19)(1.75) = -0.333$

c. $\hat{Y} = z_{\hat{Y}}(SD_Y) + M_Y = (-0.333)(95) + 1000 = 968.37$

d. The y intercept occurs when X is equal to 0. We start by finding a z score:

$$z_X = \frac{X - M_X}{SD_X} = \frac{0 - 55}{12} = -4.583$$

This is the z score for an X of 0. Now we need to figure out the predicted z score on Y for this X value:

$$z_{\hat{Y}} = (r_{XY})(z_X) = (-0.19)(-4.583) = 0.871$$

The final step is to convert the predicted z score on this predicted Y to a raw score:

$$\hat{Y} = z_{\hat{Y}}(SD_Y) + M_Y = (0.871)(95) + 1000 = 1082.745$$

This is the y intercept.

e. The slope can be found by comparing the predicted Y value for an X value of 0 (the intercept) and an X value of 1. Using the same steps as in part (a), we can compute the predicted Y score for an X value of 1.

$$z_X = \frac{X - M_X}{SD_X} = \frac{1 - 55}{12} = -4.5$$

This is the z score for an X of 1. Now we need to figure out the predicted z score on Y for this X value:

$$z_{\hat{Y}} = (r_{XY})(z_X) = (-0.19)(-4.5) = 0.855$$

The final step is to convert the predicted z score on this predicted Y to a raw score:

$$\hat{Y} = z_{\hat{Y}}(SD_Y) + M_Y = (0.855)(95) + 1000 = 1081.225$$

We compute the slope by measuring the change in Y with this 1-unit increase in X:

$$1081.225 - 1082.745 = -1.52$$

This is the slope.

f. $\hat{Y} = 1082.745 - 1.52(X)$

g. In order to draw the line, we have one more $\hat{Y}$ value to compute. This time we can use the regression equation to make the prediction:

$$\hat{Y} = 1082.745 - 1.52(48) = 1009.785$$

Now we can draw the regression line.

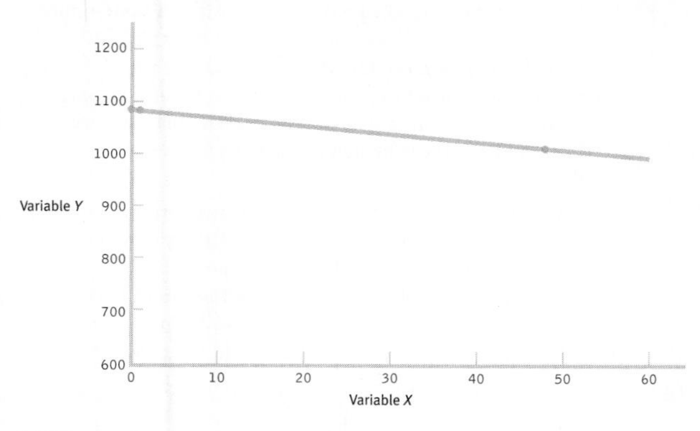

14.25 **a.** $\hat{Y} = 49 + (-0.18)(X) = 49 + (-0.18)(-31) = 54.58$

b. $\hat{Y} = 49 + (-0.18)(65) = 37.3$

c. $\hat{Y} = 49 + (-0.18)(14) = 46.48$

14.27 **a.** The sum of squared error for the mean, SS_{total}:

X	Y	M_Y	ERROR	SQUARED ERROR
4	6	6.75	−0.75	0.563
6	3	6.75	−3.75	14.063
7	7	6.75	0.25	0.063
8	5	6.75	−1.75	3.063
9	4	6.75	−2.75	7.563
10	12	6.75	5.25	27.563
12	9	6.75	2.25	5.063
14	8	6.75	1.25	1.563

$SS_{total} = \Sigma(Y - M_Y)^2 = 59.504$

b. The sum of squared error for the regression equation, SS_{error}:

X	Y	REGRESSION EQUATION	$\hat{Y}$	ERROR $(Y - \hat{Y})$	SQUARED ERROR
4	6	$\hat{Y} = 2.643 + 0.469(4)$	= 4.519	1.481	2.193
6	3	$\hat{Y} = 2.643 + 0.469(6)$	= 5.457	−2.457	6.037
7	7	$\hat{Y} = 2.643 + 0.469(7)$	= 5.926	1.074	1.153
8	5	$\hat{Y} = 2.643 + 0.469(8)$	= 6.395	−1.395	1.946
9	4	$\hat{Y} = 2.643 + 0.469(9)$	= 6.864	−2.864	8.202
10	12	$\hat{Y} = 2.643 + 0.469(10)$	= 7.333	4.667	21.781
12	9	$\hat{Y} = 2.643 + 0.469(12)$	= 8.271	0.729	0.531
14	8	$\hat{Y} = 2.643 + 0.469(14)$	= 9.209	−1.209	1.462

$SS_{error} = \Sigma(Y - \hat{Y})^2 = 43.306$

c. The proportionate reduction in error for these data:

$$r^2 = \frac{(SS_{total} - SS_{error})}{SS_{total}} = \frac{(59.504 - 43.306)}{59.504} = 0.272$$

d. This calculation of r^2, 0.272, equals the square of the correlation coefficient, $r^2 = (0.52)(0.52) = 0.270$. These numbers are slightly different due to rounding decisions.

e. The standardized regression coefficient is equal to the correlation coefficient for simple linear regression, 0.52. We can also check that this is correct by computing β:

X	$(X - M_X)$	$(X - M_X)^2$	Y	$(Y - M_Y)$	$(Y - M_Y)^2$
4	−4.75	22.563	6	−0.75	0.563
6	−2.75	7.563	3	−3.75	14.063
7	−1.75	3.063	7	0.25	0.063
8	−0.75	0.563	5	−1.75	3.063
9	0.25	0.063	4	−2.75	7.563
10	1.25	1.563	12	5.25	27.563
12	3.25	10.563	9	2.25	5.063
14	5.25	27.563	8	1.25	1.563
		$\Sigma(X - M_X)^2 =$ 73.504			$\Sigma(Y - M_Y)^2 =$ 59.504

$$\beta = (b)\frac{\sqrt{SS_X}}{\sqrt{SS_Y}} = 0.469\frac{\sqrt{73.504}}{\sqrt{59.504}} = 0.469(8.573/7.714) = 0.521$$

14.29 **a.** $\hat{Y} = 1.675 + (0.001)(X_{SAT}) + (-0.008)(X_{rank})$; or $\hat{Y} = 1.675 + 0.001\ (X_{SAT}) - 0.008(X_{rank})$

b. $\hat{Y} = 1.675 + (0.001)(1030) - 0.008(41) = 1.675 + 1.03 - 0.328 = 2.377$

c. $\hat{Y} = 1.675 + (0.001)(860) - 0.008(22) = 1.675 + 0.86 - 0.176 = 2.359$

d. $\hat{Y} = 1.675 + (0.001)(1060) - 0.008(8) = 1.675 + 1.06 - 0.064 = 2.671$

14.31 **a.** Outdoor temperature is the independent variable.

b. Number of hot chocolates sold is the dependent variable.

c. As the outdoor temperature increases, we would expect the sale of hot chocolate to decrease.

d. There are several possible answers to this question. For example, the number of fans in attendance may positively predict the number of hot chocolates sold. The number of children in attendance may also positively predict the number of hot chocolates sold. The number of alternative hot beverage choices may negatively predict the number of hot chocolates sold.

14.33 **a.** $X = z(\sigma) + \mu = -1.2(0.61) + 3.51 = 2.778$. This answer makes sense because the raw score of 2.778 is a bit more than 1 standard deviation below the mean of 3.51.

b. $X = z(\sigma) + \mu = 0.66(0.61) + 3.51 = 3.913$. This answer makes sense because the raw score of 3.913 is slightly more than 0.5 standard deviation above the mean of 3.51.

14.35 **a.** 3.12

b. 3.16

c. 3.18

d. The accompanying graph depicts the regression line for GPA and hours studied.

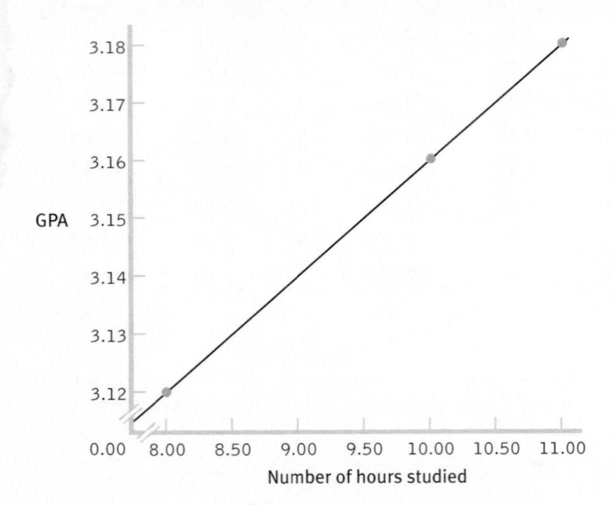

e. We can calculate the number of hours one would need to study in order to earn a 4.0 by substituting 4.0 for $\hat{Y}$ in the regression equation and solving for X: $4.0 = 2.96 + 0.02(X)$. To isolate the X, we subtract 2.96 from the left side of the equation and divide by 0.02: $X = (4.0 - 2.96)/0.02 = 52$. This regression equation predicts that we would have to study 52 hours a week in order to earn a 4.0. It is misleading to make predictions about what will happen when a person studies this many hours because the regression equation for prediction is based on a sample that studied far fewer hours. Even though the relation between hours studied and GPA was linear within the range of studied scores, outside of that range it may have a different slope or no longer be linear, or the relation may not even exist.

14.37 **a.** We cannot conclude that cola consumption causes a decrease in bone mineral density because there are a number of different kinds of causal relations that could lead to the predictive relation observed by Tucker and colleagues (2006). There may be some characteristic about these older women that both causes them to drink cola and leads to a decrease in bone mineral density. For example, perhaps overall poorer health habits lead to an increased consumption of cola and a decrease in bone mineral density.

b. Multiple regression allows us to assess the contributions of more than one independent variable to the outcome, the dependent variable. Performing this multiple regression allowed the researchers to explore the unique contributions of a third variable, such as physical activity, in addition to bone density.

c. Physical activity might produce an increase in bone mineral density, as exercise is known to increase bone density. Conversely, it is possible that physical activity might produce a decrease in cola consumption because people who exercise more might drink beverages that are more likely to keep them hydrated (such as water or sports drinks).

d. Calcium intake should produce an increase in bone mineral density, thereby producing a positive relation between calcium intake and bone density. It is possible that consumption of cola means less consumption of beverages with calcium in them, such as milk, producing a negative relation between cola consumption and bone density.

14.39 **a.** $\hat{Y} = 24.698 + 0.161(X)$ or predicted year 3 anxiety = 24.698 + 0.161 (year 1 depression)

b. As depression at year 1 increases by 1 point, predicted anxiety at year 3 increases, on average, by the slope of the regression equation, which is 0.161.

c. We would predict that her year 3 anxiety score would be 26.31.

d. We would predict that his year 3 anxiety score would be 25.02.

14.41 **a.** The independent variable in this study was marital status, and the dependent variable was chance of breaking up.

b. It appears the researchers initially conducted a simple linear regression and then conducted a multiple regression analysis to account for the other variables (e.g., age, financial status) that may have been confounded with marital status in predicting the dependent variable.

c. Answers will differ, but the focus should be on the statistically significant contribution these other variables had in predicting the dependent variable, which appear to be more important than, and perhaps explain, the relation between marital status and the break-up of the relationship.

d. Another "third variable" in this study could have been length of relationship before child was born. Married couples may have been together longer than cohabitating couples, and it may be that those who were together longer before the birth of the child, regardless of their marital status, are more likely to stay together than those who had only been together for a short period of time prior to the birth.

14.43 **a.** To predict the number of hours he studies per week, we use the formula $z_{\hat{Y}} = (r_{XY})(z_X)$ to find the predicted z score for the number of hours he studies; then we can transform the predicted z score into his raw score. First, translate his predicted raw score for age into a z score for age: $z_X = \frac{(X - M_X)}{SD_X} = \frac{(24 - 21)}{1.789} = 1.677$. Then calculate his predicted z score for number of hours studied: $z_{\hat{Y}} = (r_{XY})(z_X) = (0.49)(1.677) = 0.82$. Finally, translate the z score for hours studied into the raw score for hours studied: $\hat{Y} = 0.82(5.582) + 14.2 = 18.777$.

b. First, translate age raw score into an age z score: $z_X = \frac{(X - M_X)}{SD_X} = \frac{(19 - 21)}{1.789} = -1.118$. Then calculate the predicted z score for hours studied: $z_{\hat{Y}} = (r_{XY})(z_X) = (0.49)(-1.118) = -0.548$. Finally, translate the z score for hours studied into the raw score for hours studied: $\hat{Y} = -0.548(5.582) + 14.2 = 11.141$.

c. Seung's age is well above the mean age of the students sampled. The relation that exists for traditional-aged students may not exist for students who are much older. Extrapolating beyond the range of the observed data may lead to erroneous conclusions.

d. From a mathematical perspective, the word *regression* refers to a tendency for extreme scores to drift toward the mean. In the calculation of regression, the predicted score is closer to its mean (i.e., less extreme) than the score used for prediction. For example, in part (a) the z score used for predicting was 1.677 and the predicted z score was 0.82, a less extreme score. Similarly, in part (b) the z score used for predicting was −1.118 and the predicted z score was −0.548—again, a less extreme score.

e. First, we calculate what we would predict for Y when X equals 0; that number, −17.908, is the intercept.

$$z_X = \frac{(X - M_X)}{SD_X} = \frac{(0 - 21)}{1.789} = -11.738$$

$$z_{\hat{Y}} = (r_{XY})(z_X) = (0.49)(-11.738) = -5.752$$

$$\hat{Y} = z_{\hat{Y}}(SD_Y) + M_Y = -5.752(5.582) + 14.2 = -17.908$$

Note that the reason this prediction is negative (it doesn't make sense to have a negative number of hours) is that the number for age, 0, is not a number that would actually be used in this situation—it's another example of the dangers of extrapolation, but it still is necessary to determine the regression equation.

Then we calculate what we would predict for Y when X equals 1: the amount that that number, −16.378, differs from the prediction when X equals 0 is the slope.

$$z_X = \frac{(X - M_X)}{SD_X} = \frac{(1 - 21)}{1.789} = -11.179$$

$$z_{\hat{Y}} = (r_{XY})(z_X) = (0.49)(-11.179) = -5.478$$

$$\hat{Y} = z_{\hat{Y}}(SD_Y) + M_Y = -5.478(5.582) + 14.2 = -16.378$$

When X equals 0, −17.908 is the prediction for Y. When X equals 1, −16.378 is the prediction for Y. The latter number is 1.530 higher [−16.378 − (−17.908) = 1.530]—that is, more positive—than the former. Remember when you're calculating the difference to consider whether the prediction for Y was more positive or more negative when X increased from 0 to 1.

Thus, the regression equation is: $\hat{Y} = -17.91 + 1.53(X)$.

f. Substituting 17 for X in the regression equation for part (a) yields 8.1. Substituting 22 for X in the regression equation yields 15.75. We would predict that a 17-year-old would study 8.1 hours and a 22-year-old would study 15.75 hours.

g. The accompanying graph depicts the regression line for predicting hours studied per week from a person's age.

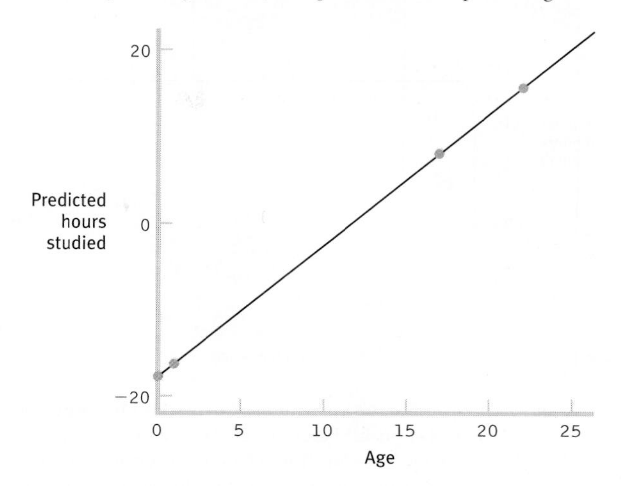

h. It is misleading to include young ages such as 0 and 5 on the graph because people of that age would never be college students.

i. The accompanying graph shows the scatterplot and regression line relating age and number of hours studied. Vertical lines from each observed data point are drawn to the regression line to represent the error prediction from the regression equation.

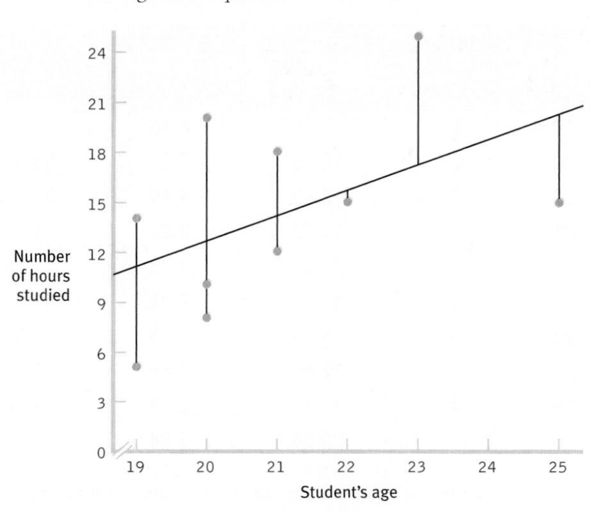

j. The accompanying scatterplot relating age and number of hours studied includes a horizontal line at the mean number of hours studied. Vertical lines between the observed data points and the mean represent the amount of error in predicting from the mean.

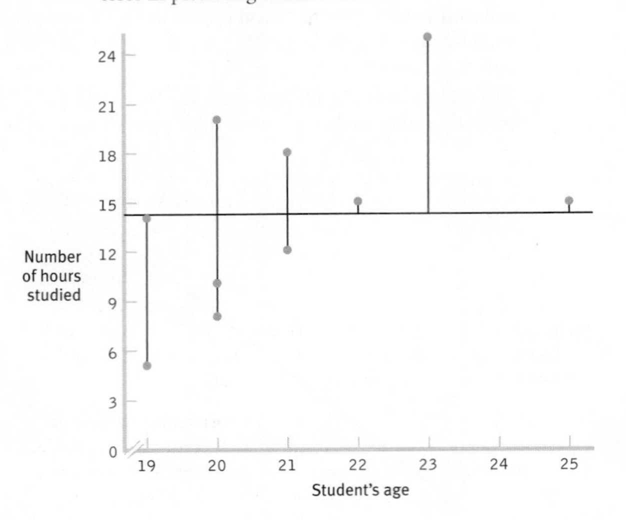

k. There appears to be less error in part (a), where the regression line is used to predict hours studied. This occurs because the regression line is the line that minimizes the distance between the observed scores and the line drawn through them. That is, the regression line is the *one* line that can be drawn through the data that produces the minimum error.

l. To calculate the proportionate reduction in error the long way, we first calculate the predicted Y scores (3rd column) for each of the observed X scores in the data set and determine how much those predicated Y scores differ from the observed Y scores (4th column), and then we square them (5th column).

AGE	OBSERVED HOURS STUDIED	PREDICTED HOURS STUDIED	OBSERVED – PREDICTED	SQUARE OF OBSERVED – PREDICTED	OBSERVED – MEAN	SQUARE OF OBSERVED – MEAN
19	5	11.16	−6.16	37.946	−9.2	84.64
20	20	12.69	7.31	53.436	5.8	33.64
20	8	12.69	−4.69	21.996	−6.2	38.44
21	12	14.22	−2.22	4.928	−2.2	4.84
21	18	14.22	3.78	14.288	3.8	14.44
23	25	17.28	7.72	59.598	10.8	116.64
22	15	15.75	−0.75	0.563	0.8	0.64
20	10	12.69	−2.69	7.236	−4.2	17.64
19	14	11.16	2.84	8.066	−0.2	0.04
25	15	20.34	−5.34	28.516	0.8	0.64

We then calculate SS_{error}, which is the sum of the squared error when using the regression equation as the basis of prediction. This sum, calculated by adding the numbers in column 5, is 236.573. We then subtract the mean from each score (column 6), and square these differences (column 7). Next, we calculate SS_{total}, which is the sum of the squared error when using the mean as the basis of prediction. This sum is 311.6. Finally, we calculate the proportionate reduction in error as $r^2 = \frac{(SS_{total} - SS_{error})}{SS_{total}} = 0.24$.

m. The r^2 calculated in part (a) indicates that 24% of the variability in hours studied is accounted for by a student's age. By using the regression equation, we have reduced the error of our prediction by 24% as compared with using the mean.

n. To calculate the proportionate reduction in error the short way, we would square the correlation coefficient. The correlation between age and hours studied is 0.49. Squaring 0.49 yields 0.24. It makes sense that the correlation coefficient could be used to determine how useful the regression equation will be because the correlation coefficient is a measure of the strength of association between two variables. If two variables are strongly related, we are better able to use one of the variables to predict the values of the other.

o. Here are the computations needed to compute β:

X	$(X - M_X)$	$(X - M_X)^2$	Y	$(Y - M_Y)$	$(Y - M_Y)^2$
19	−2	4	5	−9.2	84.64
20	−1	1	20	5.8	33.64
20	−1	1	8	−6.2	38.44
21	0	0	12	−2.2	4.84
21	0	0	18	3.8	14.44
23	2	4	25	10.8	116.64
22	1	1	15	0.8	0.64
20	−1	1	10	−4.2	17.64
19	−2	4	14	−0.2	0.04
25	4	16	15	0.8	0.64
		$\Sigma(X - M_X)^2 = 32$			$\Sigma(Y - M_Y)^2 = 311.6$

$$\beta = (b)\frac{\sqrt{SS_X}}{\sqrt{SS_Y}} = 1.53\frac{\sqrt{32}}{\sqrt{311.6}} = 0.490$$

p. The standardized regression coefficient is equal to the correlation coefficient, 0.49, for simple linear regression.

q. The hypothesis test for regression is the same as that for correlation. The critical values for r with 8 degrees of freedom at a p level of 0.05 are −0.632 and 0.632. With a correlation of 0.49, we fail to exceed the cutoff and therefore fail to reject the null hypothesis. The same is true then for the regression equation. We do not have a statistically significant regression and should be careful not to claim that the slope is different from 0.

CHAPTER 15

15.1 Nominal data are those that are categorical in nature; they cannot be ordered in any meaningful way, and they are often thought of as simply named. Ordinal data can be ordered, but we cannot assume even distances between points of equal separation. For example, the difference between the second and third scores may not be the same as the difference between the seventh and the eighth. Scale data are measured on either the interval or ratio level; we can assume equal intervals between points along these measures.

15.3 The chi-square test for goodness of fit is a nonparametric hypothesis test used with one nominal variable. The chi-square test for independence is a nonparametric test used with two nominal variables.

15.5 Throughout the book, we have referred to independent variables, those variables that we hypothesize to have an effect on the dependent variable. We also described how statisticians refer to observations that are independent of one another, such as a between-groups research design requiring that observations be taken from independent samples. Here, with regard to chi square, *independence* takes on a similar meaning. We are testing whether the effect of one variable is independent of the other—that the proportion of cases across the levels of one variable does not depend on the levels of the other variable.

15.7 In most previous hypothesis tests, the degrees of freedom have been based on sample size. For the chi-square hypothesis tests, however, the degrees of freedom are based on the numbers of categories, or cells, in which participants can be counted. For example, the degrees of freedom for the chi-square test for goodness of fit is the number of categories minus 1: $df_{\chi^2} = k - 1$. Here, k is the symbol for the number of categories.

15.9 The contingency table presents the observed frequencies for each cell in the study.

15.11 This is the formula to calculate the chi-square statistic, which is the sum, for each cell, of the squared difference between each observed frequency and its matching expected frequency, divided by the expected value for its cell.

15.13 Relative likelihood indicates the relative chance of an outcome (i.e., how many times more likely the outcome is, given the group membership of an observation). For example, we might determine the relative likelihood that a person would be a victim of bullying, given that the person is a boy versus a girl.

15.15 Relative likelihood and relative risk are exactly the same measure, but relative likelihood is typically called *relative risk* when it comes to health and medical situations because it describes a person's risk for a disease or health outcome.

15.17 When we are concerned about meeting the assumptions of a parametric test, we can convert scale data to ordinal data and use a nonparametric test.

15.19 When transforming scale data to ordinal data, the scale data are rank ordered. This means that even a very extreme scale score will have a rank that makes it continuous with the rest of the data when rank ordered.

15.21 In all correlations, we assess the relative position of a score on one variable with its position on the other variable. In the case of the Spearman rank-order correlation, we examine how ranks on one variable relate to ranks on the other variable. For example, with a positive correlation, scores that rank low on one variable tend to rank low on the other, and scores that rank high on one variable tend to rank high on the other. For a negative correlation, low ranks on one variable tend to be associated with high ranks on the other.

15.23 Values for the Spearman rank-order correlation coefficient range from −1.00 to +1.00, just like its parametric equivalent, the Pearson correlation coefficient. Similarly, the conventions for interpreting the magnitude of the Pearson correlation coefficient can also be applied to the Spearman correlation coefficient (small is roughly 0.10, medium 0.30, and large 0.50).

15.25 The Wilcoxon signed-rank test is appropriate to use when one is comparing two sets of dependent observations (scores from the same participants) and the dependent variable is either ordinal or does not meet the assumptions required by the paired-samples t test.

15.27 The assumptions are that (1) the data are ordinal, (2) random selection was used, and (3) no ranks are tied.

15.29 The Kruskal–Wallis H test is appropriate to use when one is comparing three or more groups of independent observations (i.e., the independent variable has three or more levels) and the dependent variable is ordinal or does not meet the assumptions of the parametric test.

15.31 **a.** The independent variable is gender, which is nominal (men or women). The dependent variable is number of loads of laundry, which is scale.

b. The independent variable is need for approval, which is ordinal (rank). The dependent variable is miles on a car, which is scale.

c. The independent variable is place of residence, which is nominal (on or off campus). The dependent variable is whether the student is an active member of a club, which is also nominal (active or not active).

15.33 **a.** $df_{\chi^2} = k - 1 = 4 - 1 = 3$

b.

CATEGORY	OBSERVED (O)	EXPECTED (E)	O − E	$(O - E)^2$	$\frac{(O - E)^2}{E}$
1	750	625	750 − 625 = 125	15,625	25
2	650	625	650 − 625 = 25	625	1
3	600	625	600 − 625 = −25	625	1
4	500	625	500 − 625 = −125	15,625	25

c. $\chi^2 = \Sigma\left[\frac{(O-E)^2}{E}\right] = 25 + 1 + 1 + 25 = 52$

15.35 The conditional probability of being a smoker, given that a person is female is $\frac{13}{199} = 0.065$, and the conditional probability of being a smoker, given that a person is male is $\frac{723}{905} = 0.799$. The relative likelihood of being a smoker given that one is female rather than male is $\frac{0.065}{0.799} = 0.08$. These Turkish women with lung cancer were less than one-tenth as likely to be smokers as were the male lung cancer patients.

15.37 a.

COUNT	VARIABLE X	RANK X	VARIABLE Y	RANK Y
1	134.5	3	64.00	7
2	186	10	60.00	1
3	157	9	61.50	2
4	129	1	66.25	10
5	147	7	65.50	8.5
6	133	2	62.00	3.5
7	141	5	62.50	5
8	147	7	62.00	3.5
9	136	4	63.00	6
10	147	7	65.50	8.5

b.

COUNT	RANK X	RANK Y	DIFFERENCE	SQUARED DIFFERENCE
1	3	7	−4	16
2	10	1	9	81
3	9	2	7	49
4	1	10	−9	81
5	7	8.5	−1.5	2.25
6	2	3.5	−1.5	2.25
7	5	5	0	0
8	7	3.5	3.5	12.25
9	4	6	−2	4
10	7	8.5	−1.5	2.25

$$r_S = 1 - \frac{6(\Sigma D^2)}{N(N^2-1)} = 1 - \frac{6(250)}{10(100-1)} = 1 - \frac{1500}{990}$$
$$= 1 - 1.515 = -0.515$$

15.39 a. When calculating the Spearman correlation coefficient, we must first transform the variable "hours trained" into a rank-ordered variable. We then take the difference between the two ranks and square those differences:

RACE RANK	HOURS TRAINED	HOURS RANK	DIFFERENCE	SQUARED DIFFERENCE
1	25	1.5	−0.5	0.25
2	25	1.5	0.5	0.25
3	22	3	0	0
4	18	5.5	−1.5	2.25
5	19	4	1	1
6	18	5.5	0.5	0.25
7	12	10	−3	9
8	17	7	1	1
9	15	9	0	0
10	16	8	2	4
				$\Sigma D^2 = 18$

We calculate the Spearman correlation coefficient as:

$$r_S = 1 - \frac{6(\Sigma D^2)}{N(N^2-1)} = 1 - \frac{6(18)}{10(10^2-1)} = 1 - \frac{108}{990} = 1 - 0.109$$
$$= 0.891$$

b. The critical r_S with an N of 10, a p level of 0.05, and a two-tailed test is 0.648. The calculated r_S is 0.89, which exceeds the critical value. So we reject the null hypothesis. Finishing place was positively associated with the number of hours spent training.

15.41 $\Sigma R_{group1} = 1 + 2.5 + 8 + 4 + 6 + 10 = 31.5$

$\Sigma R_{group2} = 11 + 9 + 2.5 + 5 + 7 + 12 = 46.5$

15.43 a. To conduct the Mann–Whitney U test, we first obtain the rank of every person in the data set. We then separately sum the ranks of the two groups, men and women:

STUDENT	GENDER	CLASS STANDING	RANK	MALE RANKS	FEMALE RANKS
1	Male	98	11	11	
2	Female	72	9		9
3	Male	15	3	3	
4	Female	3	1		1
5	Female	102	12		12
6	Female	8	2		2
7	Male	43	7	7	
8	Male	33	6	6	
9	Female	17	4		4
10	Female	82	10		10
11	Male	63	8	8	
12	Male	25	5	5	

We sum the ranks for the men: $\Sigma R_m = 11 + 3 + 7 + 6 + 8 + 5 = 40$.

We sum the ranks for the women: $\Sigma R_w = 9 + 1 + 12 + 2 + 4 + 10 = 38$.

We calculate U for the men: $U_m = (n_m)(n_w) + \frac{n_m(n_m+1)}{2} - \Sigma R_m = (6)(6) + \left(\frac{6(6+1)}{2}\right) - 40 = 17.$

We calculate U for the women: $U_w = (n_m)(n_w) + \frac{n_w(n_w + 1)}{2} - \Sigma R_m = (6)(6) + \left(\frac{6(6+1)}{2}\right) - 38 = 19.$

b. The critical value for the Mann–Whitney U test with two samples of size 6, a p level of 0.05, and a two-tailed test is 5. We compare the smaller of the two U values to the critical value and reject the null hypothesis if it is smaller than the critical value. Because the smaller U of 17 is not less than 5, we fail to reject the null hypothesis. There is no evidence for a difference in the class standing of men and women.

15.45 **a.** A nonparametric test would be appropriate because both of the variables are nominal: gender and major.

b. A nonparametric test is more appropriate because the sample size is small and the data are unlikely to be normal; the "top boss" is likely to have a much higher income than the other employees. This outlier would lead to a nonnormal distribution.

c. A parametric test would be appropriate because the independent variable (type of student: athlete versus nonathlete) is nominal and the dependent variable (grade point average) is scale.

d. A nonparametric test would be appropriate because the independent variable (athlete versus nonathlete) is nominal and the dependent variable (class rank) is ordinal.

e. A nonparametric test would be appropriate because the research question is about the relation between two nominal variables: seat-belt wearing and degree of injuries.

f. A parametric test would be appropriate because the independent variable (seat-belt use: no seat belt versus seat belt) is nominal and the dependent variable (speed) is scale.

15.47 **a.** (i) Year. (ii) Grades received. (iii) This is a category III research design because the independent variable, year, is nominal and the dependent variable, grade (A or not), could also be considered nominal.

b. (i) Type of school. (ii) Average GPA of graduating students. (iii) This is a category II research design because the independent variable, type of school, is nominal and the dependent variable, GPA, is scale.

c. (i) SAT scores of incoming students. (ii) College GPA. (iii) This is a category I research design because both the independent variable and the dependent variable are scale.

15.49 **a.**

	MEXICAN	WHITE	BLACK
MARRIED			
SINGLE			

b.

	MARRIED HEAD OF HOUSEHOLD	
	IMMIGRANT NEIGHBORHOOD	NONIMMIGRANT NEIGHBORHOOD
COMMITTED CRIME		
NO CRIME		

	UNMARRIED HEAD OF HOUSEHOLD	
	IMMIGRANT NEIGHBORHOOD	NONIMMIGRANT NEIGHBORHOOD
COMMITTED CRIME		
NO CRIME		

c.

	FIRST GENERATION	SECOND GENERATION	THIRD GENERATION
COMMITTED CRIME			
NO CRIME			

15.51 **a.** There is one variable, the gender of the op-ed writers. Its levels are men and women.

b. A chi-square test for goodness of fit would be used because we have data on a single nominal variable from one sample.

c. *Step 1:* Population 1 is op-ed contributors, in proportions of males and females that are like those in our sample. Population 2 is op-ed contributors, in proportions of males and females that are like those in the general population. The comparison distribution is a chi-square distribution. The hypothesis test will be a chi-square test for goodness-of-fit because we have only one nominal variable. This study meets three of the four assumptions. (1) The variable under study is nominal. (2) Each observation is independent of all the others. (3) There are more than five times as many participants as there are cells (there are 124 op-ed articles and only 2 cells). (4) This is not, however, a randomly selected sample of op-eds, so we must generalize with caution; specifically, we should not generalize beyond the *New York Times*.

Step 2: Null hypothesis: The proportions of male and female op-ed contributors are the same as those in the population as whole.

Research hypothesis: The proportions of male and female op-ed contributors are different from those in the population as a whole.

Step 3: The comparison distribution is a chi-square distribution with 1 degree of freedom: $df_{\chi^2} = 2 - 1 = 1$.

Step 4: The critical χ^2, based on a p level of 0.05 and 1 degree of freedom, is 3.841.

Step 5:

OBSERVED (PROPORTIONS OF MEN AND WOMEN)	
MEN	WOMEN
103	21

EXPECTED (FROM THE GENERAL POPULATION)	
MEN	WOMEN
62	62

CATEGORY	OBSERVED (O)	EXPECTED (E)	O − E	(O − E)^2	$\frac{(O-E)^2}{E}$
Men	103	62	41	1681	27.113
Women	21	62	−41	1681	27.113

$$\chi^2 = \Sigma\left[\frac{(O-E)^2}{E}\right] = 27.113 + 27.113 = 54.226$$

Step 6: Reject the null hypothesis. The calculated chi-square statistic exceeds the critical value. It appears that the proportion of op-eds written by women versus men is not the same as the proportion of men and women in the population. Specifically, there are fewer women than in the general population.

d. $\chi^2(1, N = 124) = 54.23, p < 0.05$

15.53 **a.** The accompanying table shows the conditional proportions.

	EXCITING	ROUTINE	DULL	
SAME CITY	0.424	0.521	0.055	1.00
SAME STATE/ DIFFERENT CITY	0.468	0.485	0.047	1.00
DIFFERENT STATE	0.502	0.451	0.047	1.00

b. The accompanying graph shows these conditional proportions.

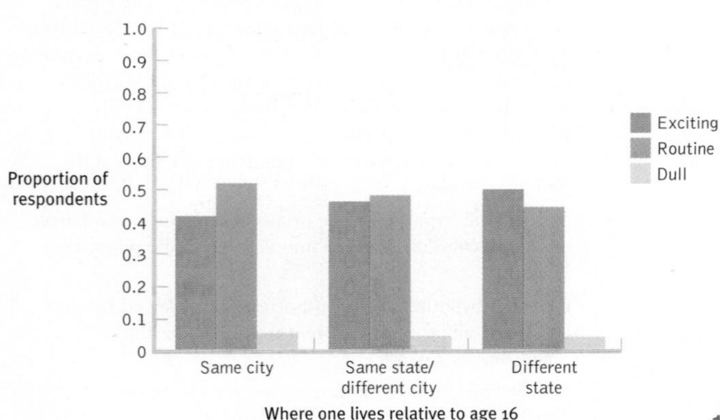

c. The relative likelihood of finding life exciting if one lives in a different state as opposed to the same city is $\frac{0.502}{0.424} = 1.18$.

15.55 **a.** The Mann–Whitney *U* test would be most appropriate because it is a nonparametric equivalent to the independent-samples *t* test. It is used when we have a nominal independent variable with two levels (here, they are north and south of the equator), a between-groups research design, and an ordinal dependent variable (here, it is the ranking of the city).

b. The Wilcoxon signed-rank test would be most appropriate because we have a nominal independent variable with two levels (the time of the previous study versus 2005), a within-groups research design, and an ordinal dependent variable (ranking).

c. The Spearman rank-order correlation would be most appropriate because we are asking a question about the relation between two ordinal variables.

15.57 **a.** The first variable of interest is test grade, which is a scale variable. The second variable of interest is the order in which students completed the test, which is an ordinal variable.

b. The accompanying table shows test grade converted to ranks, difference scores, and squared differences.

GRADE PERCENTAGE	GRADE SPEED	RANK	D	D^2
98	1	1	0	0
93	6	2	4	16
92	4	3	1	1
88	5	4	1	1
87	3	5	−2	4
74	2	6	−4	16
67	8	7	1	1
62	7	8	−1	1

We calculate the Spearman correlation coefficient as:

$$r_S = 1 - \frac{6(\Sigma D)^2}{N(N^2-1)} = 1 - \frac{6(40)}{8(64-1)} = 1 - \frac{240}{504} = 0.524$$

c. The coefficient tells us that there is a rather large positive relation between the two variables. Students who completed the test more quickly also tended to score higher.

d. We could not have calculated a Pearson correlation coefficient because one of our variables, order in which students turned in the test, is ordinal.

e. This correlation does not indicate that students should attempt to take their tests as quickly as possible. Correlation does not provide evidence for a particular causal relation. A number of underlying causal relations could produce this observed correlation.

f. A third variable that might cause both speedy test-taking and a good test grade is knowledge of the material. Students with better knowledge of and more practice with the material would be able to get through the test more quickly and get a better grade.

15.59 **a.** We would use a Kruskal–Wallis *H* test because we have a nominal independent variable (type of funding) with three levels (primarily fellowships, primarily teaching assistantships, don't have full funding for most students) and an ordinal dependent variable (ranking).

b. We would use the Spearman rank-order correlation coefficient because we want to know the relation between one ordinal and one scale variable. Even though GRE scores are scale, to be analyzed with school rankings, they will need to be transformed to ordinal.

c. We would use the Mann–Whitney *U* test because we have an independent variable (distance), which has two independent conditions (schools within a 3-hour drive and schools beyond a 3-hour drive), and an ordinal dependent variable (ranking).

15.61 **a.** The independent variable is type of state, and its levels are red and blue. The dependent variable is the percentage of registered voters who voted.

b. This is a between-groups design because each state is either a red state or a blue state but cannot be both.

c. *Step 1:* We need to convert the data to an ordinal measure. The states were randomly selected, so we can assume that they are representative of their populations. Finally, there are no tied ranks.

Step 2: Null hypothesis: There is no difference between the voter turnout in red and blue states.

Research hypothesis: There is a difference between the voter turnout in red and blue states.

Step 3: There are eight red and eight blue states.

Step 4: The critical value for a Mann–Whitney U test with two groups of eight, a p level of 0.05, and a two-tailed test is 13. The smaller calculated statistic needs to be less than or equal to this critical value to be considered statistically significant.

Step 5:

STATE	TURNOUT	TURNOUT RANK	STATE TYPE	RED RANK	BLUE RANK
Wisconsin	76.73	1	Blue		1
Maine	73.4	2	Blue		2
Oregon	70.5	3	Blue		3
Washington	67.42	4	Blue		4
Missouri	66.89	5	Red	5	
Vermont	66.19	6	Blue		6
Idaho	64.89	7	Red	7	
New Jersey	64.54	8	Blue		8
Montana	64.36	9	Red	9	
Virginia	61.5	10	Red	10	
Louisiana	60.78	11	Red	11	
Illinois	60.73	12	Blue		12
California	60.01	13	Blue		13
Georgia	57.38	14	Red	14	
Indiana	55.69	15	Red	15	
Texas	53.35	16	Red	16	

$$\Sigma R_{red} = 5 + 7 + 9 + 10 + 11 + 14 + 15 + 16 = 87$$

$$\Sigma R_{blue} = 1 + 2 + 3 + 4 + 6 + 8 + 12 + 13 = 49$$

$$U_{red} = (8)(8) + \frac{8(8+1)}{2} - 87 = 13$$

$$U_{blue} = (8)(8) + \frac{8(8+1)}{2} - 49 = 51$$

Step 6: The smaller calculated U, 13, is equal to the critical value of 13. In order to reject the null hypothesis for the Mann–Whitney U tests, the calculated value must be less than or equal to the critical value. So we reject the null hypothesis. There is a statistically significant difference between voter turnout in red and blue states. Voter turnout tends to be higher in blue states than in red states.

d. $U = 13, p = 0.05$

15.63 **a.** Hours studied per week appears to be roughly normal, with observations across the range of values—from 0 through 20. Monthly cell phone bill appears to be positively skewed, with one observation much higher than all the others.

b. The histogram confirms the impression that the monthly cell phone bill is positively skewed. It appears that there is an outlier in the distribution.

c. Parametric tests assume that the underlying population data are normally distributed or that there is a large enough sample size that the sampling distribution will be normal anyway. These data seem to indicate that the underlying distribution is not normally distributed; moreover, there is a fairly small sample size ($N = 29$). We would not want to use a parametric test.

15.65 **a.** There are two variables in this study. The independent variable is the referred child's gender (boy, girl) and the dependent variable is the diagnosis (problem, no problem but below norms, no problem and normal height).

b. A chi-square test for independence would be used because we have data on two nominal variables.

c. *Step 1:* Population 1 is referred children like those in this sample. Population 2 is referred children from a population in which growth problems do not depend on the child's gender. The comparison distribution is a chi-square distribution. The hypothesis test will be a chi-square test for independence because we have two nominal variables. This study meets three of the four assumptions. (1) The two variables are nominal. (2) Every participant is in only one cell. (3) There are more than five times as many participants as there are cells (there are 278 participants and 6 cells). (4) The sample, however, was not randomly selected, so we must use caution when generalizing.

Step 2: Null hypothesis: The proportion of boys in each diagnostic category is the same as the proportion of girls in each category.

Research hypothesis: The proportion of boys in each diagnostic category is different from the proportion of girls in each category.

Step 3: The comparison distribution is a chi-square distribution that has 2 degrees of freedom:

$$df_{\chi^2} = (k_{row} - 1)(k_{column} - 1) = (2 - 1)(3 - 1) = 2.$$

Step 4: The critical χ^2, based on a p level of 0.05 and 2 degrees of freedom, is 5.99.

Step 5:

	OBSERVED			
	MEDICAL PROBLEM	NO PROBLEM/ BELOW NORM	NO PROBLEM/ NORMAL HEIGHT	
BOYS	27	86	69	182
GIRLS	39	38	19	96
	66	124	88	278

$$\frac{Total_{column}}{N}(Total_{row}) = \frac{66}{278}(182) = 43.209$$

$$\frac{Total_{column}}{N}(Total_{row}) = \frac{66}{278}(96) = 22.791$$

$$\frac{Total_{column}}{N}(Total_{row}) = \frac{124}{278}(182) = 81.180$$

$$\frac{Total_{column}}{N}(Total_{row}) = \frac{124}{278}(96) = 42.820$$

$$\frac{Total_{column}}{N}(Total_{row}) = \frac{88}{278}(182) = 57.612$$

$$\frac{Total_{column}}{N}(Total_{row}) = \frac{88}{278}(96) = 30.388$$

	EXPECTED MEDICAL PROBLEM	NO PROBLEM/ BELOW NORM	NO PROBLEM/ NORMAL HEIGHT	
BOYS	43.209	81.180	57.612	182
GIRLS	22.791	42.820	30.388	96
	66	124	88	278

CATEGORY	OBSERVED (O)	EXPECTED (E)	O − E	(O − E)^2	$\frac{(O-E)^2}{E}$
Boy; med prob	27	43.209	−16.209	262.732	6.080
Boy; no prob/below	86	81.180	4.82	23.232	0.286
Boy; no prob/norm	69	57.612	11.388	129.687	2.251
Girl; med prob	39	22.791	16.209	262.732	11.528
Girl; no prob/below	38	42.820	−4.82	23.232	0.543
Girl; no prob/norm	19	30.388	−11.388	129.687	4.268

$$\chi^2 = \Sigma\left[\frac{(O-E)^2}{E}\right] = 6.08 + 0.286 + 2.251 + 11.528$$
$$+ 0.543 + 4.268 = 24.956$$

Step 6: Reject the null hypothesis. The calculated chi-square value exceeds the critical value. It appears that the proportion of boys in each diagnostic category is not the same as the proportion of girls in each category.

d. Cramer's $V = \sqrt{\frac{\chi^2}{(N)(df_{row/column})}} = \sqrt{\frac{24.956}{(278)(1)}} = 0.300$

According to Cohen's conventions, this is a small-to-medium effect size.

e. $\chi^2(2, N = 278) = 24.96, p < 0.05$, Cramer's $V = 0.30$

f. The accompanying table shows the conditional proportions.

	MEDICAL PROBLEM	NO PROBLEM/ BELOW NORM	NO PROBLEM/ NORMAL HEIGHT	
BOYS	0.148	0.473	0.379	1.00
GIRLS	0.406	0.396	0.198	1.00

g. The accompanying graph shows all six conditions.

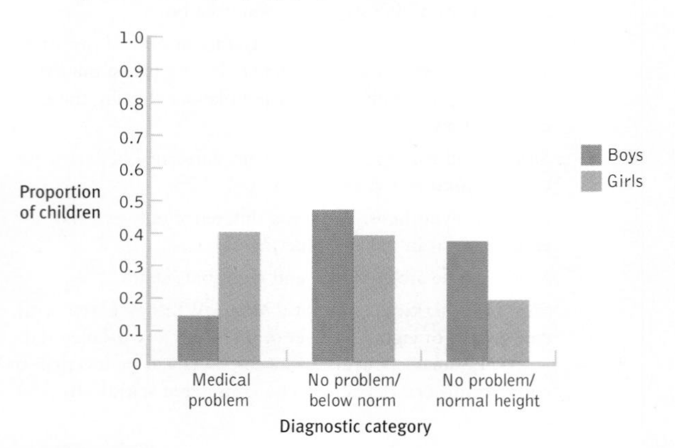

h. Of the 113 boys below normal height, 27 were diagnosed with a medical problem. Of the 77 girls below normal height, 39 were diagnosed with a medical problem. The conditional proportion for boys is 0.239 and for girls is 0.506. This makes the relative risk for having a medical condition, given that one is a boy as opposed to a girl $\frac{0.239}{0.506} = 0.472$.

i. Boys below normal height are about half as likely to have a medical condition as are girls below normal height.

j. The relative risk for having a medical condition, given that one is a girl is $\frac{0.506}{0.239} = 2.117$.

k. Girls below normal height are about twice as likely to have a medical condition as are boys below normal height.

l. The two relative risks give us complementary information. Saying that boys are half as likely to have a medical condition implies that girls are twice as likely to have a medical condition.

APPENDIX D

Solutions to Check Your Learning Problems

CHAPTER 1

1-1 Data from samples are used in inferential statistics (to make an inference about the larger population).

1-2 **a.** The average grade for your statistics class would be a descriptive statistic because it is being used only to describe the tendency of people in your class with respect to a statistics grade.

b. In this case, the average grade would be an inferential statistic because it is being used to estimate the results of a population of students taking statistics.

1-3 **a.** 100 selected students

b. 12,500 students at the university

c. The 100 students in the sample have an average score of 18, a moderately high stress level.

d. The entire population of students at this university has a moderately high stress level, on average. The sample mean, 18, is an estimate of the unknown population mean.

1-4 Discrete observations can take on only specific values, usually whole numbers; continuous observations can take on a full range of values.

1-5 **a.** These data are continuous because they can take on a full range of values.

b. The variable is a ratio observation because there is a true zero point.

c. On an ordinal scale, Lorna's score would be 2 (or 2nd).

1-6 **a.** The nominal variable is gender. The levels of gender, male and female, have no numerical meaning even if they are arbitrarily labeled 1 and 2.

b. The ordinal variable is hair length. The three levels of hair length (short, mid-length, and very long) are arranged in order, but we do not know the magnitude of the differences in length.

c. The scale variable is the probability that the student will be harassed. It is a scale variable because the distances between probability scores are assumed to be equal.

1-7 Independent; dependent

1-8 **a.** There are two independent variables: beverage and subject to be remembered. The dependent variable is memory.

b. Beverage has two levels: caffeine and no caffeine. The subject to be remembered has three levels: numbers, word lists, and aspects of a story.

1-9 **a.** Whether or not a student declared a major

b. Declared a major; did not declare a major

c. Anxiety score

d. The scores would be consistent over time unless a student's anxiety level changed.

e. The anxiety scale was actually measuring anxiety.

1-10 Experimental research involves random assignment to conditions; correlational research examines associations where random assignment is not possible and variables are not manipulated.

1-11 Random assignment helps to distribute confounding variables evenly across all conditions so that the levels of the independent variable are what truly vary across groups or conditions.

1-12 Rank in high school class and high school grade point average (GPA) are good examples.

1-13 **a.** Researchers could randomly assign a certain number of women to be told about a gender difference on the test and randomly assign a certain number of other women to be told that no gender difference existed on this test.

b. If researchers did not use random assignment, any gender differences might be due to confounding variables. The women in the two groups might be different in some way (e.g., in math ability or belief in stereotypes) to begin with.

c. There are many possible confounds. Women who already believed the stereotype might do so because they had always performed poorly in mathematics, whereas those who did not believe the stereotype might be those who always did particularly well in math. Women who believed the stereotype might be those who were discouraged from studying math because "girls can't do math," whereas those who did not believe the stereotype might be those who were encouraged to study math because "girls are just as good as boys in math."

d. Math performance is operationalized as scores on a math test.

e. Researchers could have two math tests that are similar in difficulty. All women would take the first test after being told that women tend not to do as well as men on this test. Once they had taken that test, they would be given the second test after being told that women tend to do as well as men on this test.

CHAPTER 2

2-1 Frequency tables, grouped frequency tables, histograms, and frequency polygons

2-2 A frequency is a count of how many times a score appears. A grouped frequency is a count for a defined interval, or group, of scores.

2-3 **a.**

INTERVAL	FREQUENCY
50–59	2
40–49	1
30–39	1
20–29	2
10–19	4
0–9	7

b.

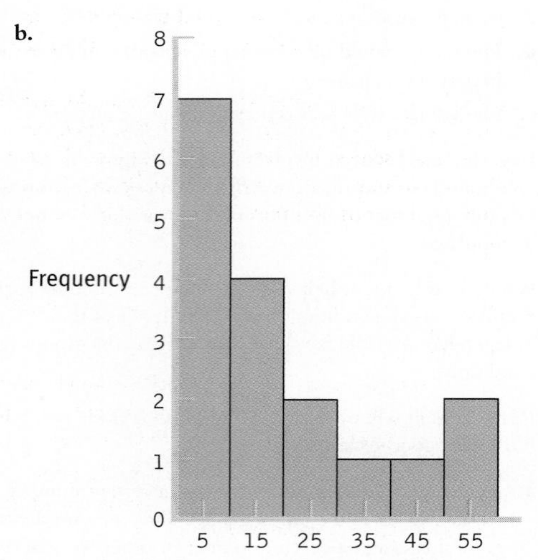

c.

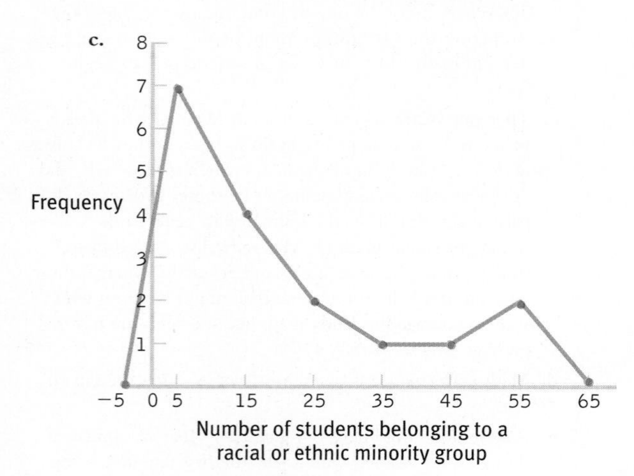

2-4 **a.** We can now get a sense of the overall pattern of the data.

b. Percentages might be more useful because they allow us to compare programs. Two programs might have the same number of minority students, but if one program has far more students overall, it is less diverse than one with fewer students overall.

c. It is possible that schools did not provide data if they had no or few minority students. Thus, this data set might be composed of the schools with more diverse student bodies. This is a volunteer sample; schools are not obligated to report these data.

2-5 A normal distribution is a specific distribution that is symmetric around a center high point: It looks like a bell. A skewed distribution is asymmetric or lopsided to the left or to the right, with a long tail of data to one side.

2-6 Negative; positive

2-7 Positive skew

2-8 **a.** Early-onset Alzheimer's disease would create negative skew in the distribution for age of onset.

b. Because all humans eventually die, there is a sort of ceiling effect.

2-9 Being aware of these exceptional early-onset cases allows medical practitioners to be open to such surprising diagnoses. In addition, exceptional cases like these often give us great insight into the underlying mechanisms of disease.

CHAPTER 3

3-1 The purpose of a graph is to reveal and clarify relations between variables.

3-2 Five miles per gallon change (from 22 to 27) and $\frac{5}{22}(100) =$ 22.73% change

3-3 The graph on the left is misleading. It shows a sharp decline in annual traffic deaths in Connecticut from 1955 to 1956, but we cannot draw valid conclusions from just two data points. The graph on the right is a more accurate and complete depiction of the data. It includes nine, rather than two, data points and suggests that the sharp 1-year decline was the beginning of a clear downward trend in traffic fatalities that extended through 1959. It also shows that there had been previous 1-year declines of similar magnitude—from 1951 to 1952 and from 1953 to 1954. Also, the *y*-axis does not go down to zero, which exaggerates any differences.

3-4 Scatterplots and line graphs both depict the relation between two scale variables.

3-5 The data can almost always be presented more clearly in a table or in a bar graph.

3-6 The line graph known as a time plot or time series plot allows us to do so.

3-7 **a.** A scatterplot is the best graph choice to depict the relation between two scale variables such as depression and stress.

b. A time plot, or time series plot, is the best graph choice to depict the change in a scale variable, such as number of facilities, over time.

c. For one scale variable, such as number of siblings, the best graph choice would be either a frequency histogram or frequency polygon.

d. In this case, there is a nominal variable (region of the United States) and a scale variable (years of education). The best choice would be a bar graph, with one bar depicting the mean years of education for each region. A Pareto chart would arrange the bars from highest to lowest, allowing for easier comparisons.

e. Calories and hours are both scale variables, and the question is about prediction rather than relation. In this case, we'd calculate and graph a line of best fit.

3-8 Chartjunk is any unnecessary information or feature in a graph that detracts from the viewer's understanding.

3-9 **a.** Scatterplot or line graph

b. Bar graph

c. Scatterplot or line graph

3-10

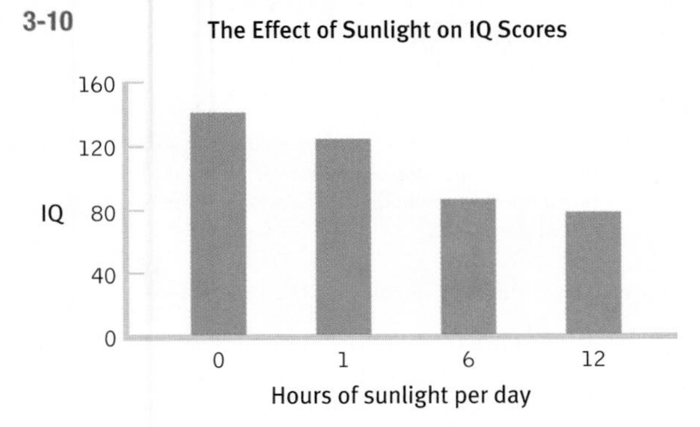

The accompanying graph improves on the chartjunk graph in several ways. First, it has a clear, specific caption. Second, all axes are labeled left to right. Third, there are no abbreviations. The units of measurement, IQ and hours of sunlight per day, are included. The y-axis has 0 as its minimum, the colors are simple and muted, and all chartjunk has been eliminated. This graph wasn't as much fun to create, but it offers a far clearer presentation of the data! (*Note:* We are treating hours as an ordinal variable.)

CHAPTER 4

4-1 Statistics are calculated for samples; they are usually symbolized by Latin letters (e.g., M). Parameters are calculated for populations; they are usually symbolized by Greek letters (e.g., μ).

4-2 Mean, because the calculation of the mean takes into account the numeric value of each data point, including that outlier.

4-3 **a.** The mean, 10.18, is calculated using the following formula:

$M = \frac{\Sigma X}{N} = (10 + 8 + 22 + 5 + 6 + 1 + 19 + 8 + 13 + 12 + 8)/11 = 112/11 = 10.18$

The median is found by arranging the scores in numeric order—1, 5, 6, 8, 8, 8, 10, 12, 13, 19, 22—then dividing the total number of scores, 11, by 2 and adding 0.5 to get 6. The 6th score in our ordered list of scores is the median, and in this case the 6th score is the number 8. The mode is the most common score. In these data, the score 8 occurs most often (three times), so 8 is the mode.

b. The mean, 122.75, is calculated using the following formula:

$M = \frac{\Sigma X}{N} = (122.5 + 123.8 + 121.2 + 125.8 + 120.2 + 123.8 + 120.5 + 119.8 + 126.3 + 123.6)/10 = 1227.5/10 = 122.75$

The data ordered are: 119.8, 120.2, 120.5, 121.2, 122.5, 123.6, 123.8, 123.8, 125.8, 126.3. Again, we find the median by ordering the data, and then dividing the number of scores (here there are 10 scores) by 2 and adding ½. In this case, we get 5.5, so the mean of the 5th and 6th data points is the median. The median is (122.5 + 123.6)/2 = 123.05. The mode is 123.8, which occurs twice in these data.

c. The mean, 0.429, is calculated using the following formula:

$M = \frac{\Sigma X}{N} = (0.100 + 0.866 + 0.781 + 0.555 + 0.222 + 0.245 + 0.234)/7 = 3.003/7 = 0.429$

Note that three decimal places are included here (rather than the standard two places used throughout this book) because the data are carried out to three decimal places.

The median is found by first ordering the data: 0.100, 0.222, 0.234, 0.245, 0.555, 0.781, 0.866. Then the total number of scores, 7, is divided by 2 to get 3.5, to which ½ is added to get 4. So, the 4th score, 0.245, is our median. There is no mode in these data. All scores occur once.

4-4 **a.** $M = \frac{\Sigma X}{N} = (1 + 0 + 1 + 2 + 5 + \ldots 4 + 6)/20 = 50/20 = 2.50$

b. In this case, the scores would comprise a sample taken from the whole population, and this mean would be a statistic. The symbol, therefore, would be either M or $\overline{X}$.

c. In this case, the scores would constitute the entire population of interest, and the mean would be a parameter. Thus, the symbol would be μ.

d. To find the median, we would arrange the scores in order: 0, 0, 1, 1, 1, 1, 2, 2, 2, 2, 2, 2, 3, 3, 3, 3, 4, 5, 6, 7. We would then divide the total number of scores, 20, by 2 and add ½, which is 10.5. The median, therefore, is the mean of the 10th and 11th scores. Both of these scores are 2; therefore, the median is 2.

e. The mode is the most common score—in this case, there are six 2's, so the mode is 2.

f. The mean is a little higher than the median. This indicates that there are potential outliers pulling the mean higher; outliers would not affect the median.

4-5 Variability is the concept of variety in data, often measured as deviation around some center.

4-6 The range tells us the span of the data, from highest to lowest score. It is based on just two scores. The standard deviation tells us how far the typical score falls from the mean. The standard deviation takes every score into account.

4-7 **a.** The range is: $X_{highest} - X_{lowest} = 22 - 1 = 21$

The variance is: $SD^2 = \frac{\Sigma(X - M)^2}{N}$

We start by calculating the mean, which is 10.18. We then calculate the deviation of each score from the mean and the square of that deviation.

X	X − M	(X − M)²
10	−0.182	0.033
8	−2.182	4.761
22	11.818	139.665
5	−5.182	26.853
6	−4.182	17.489
1	−9.182	84.309
19	8.818	77.757
8	−2.182	4.761
13	2.818	7.941
12	1.818	3.305
8	−2.182	4.761

$$SD^2 = \frac{\Sigma(X-M)^2}{N} = \frac{371.61}{11} = 33.785$$

The standard deviation is: $SD = \sqrt{SD^2}$ or

$$SD = \sqrt{\frac{\Sigma(X-M)^2}{N}} = \sqrt{33.785} = 5.81$$

b. The range is: $X_{highest} - X_{lowest} = 126.3 - 119.8 = 6.5$

The variance is: $SD^2 = \frac{\Sigma(X-M)^2}{N}$

We start by calculating the mean, which is 122.75. We then calculate the deviation of each score from the mean and the square of that deviation.

X	X − M	(X − M)²
122.500	−0.250	0.063
123.800	1.050	1.102
121.200	−1.550	2.402
125.800	3.050	9.302
120.200	−2.550	6.502
123.800	1.050	1.102
120.500	−2.250	5.063
119.800	−2.950	8.703
126.300	3.550	12.603
123.600	0.850	0.722

$$SD^2 = \frac{\Sigma(X-M)^2}{N} = \frac{47.564}{10} = 4.756$$

The standard deviation is: $SD = \sqrt{SD^2}$ or

$$SD = \sqrt{\frac{\Sigma(X-M)^2}{N}} = \sqrt{4.756} = 2.18$$

c. The range is: $X_{highest} - X_{lowest} = 0.866 - 0.100 = 0.766$

The variance is: $SD = \frac{\Sigma(X-M)^2}{N}$

We start by calculating the mean, which is 0.429. We then calculate the deviation of each score from the mean and the square of that deviation.

X	X − M	(X − M)²
0.100	−0.329	0.108
0.866	0.437	0.191
0.781	0.352	0.124
0.555	0.126	0.016
0.222	−0.207	0.043
0.245	−0.184	0.034
0.234	−0.195	0.038

$$SD^2 = \frac{\Sigma(X-M)^2}{N} = \frac{0.554}{7} = 0.079$$

The standard deviation is: $SD = \sqrt{SD^2}$ or

$$SD = \sqrt{\frac{\Sigma(X-M)^2}{N}} = \sqrt{0.079} = 0.28$$

4-8 **a.** Range $= X_{highest} - X_{lowest} = 1460 - 450 = 1010$

b. We do not know whether scores cluster at some point in the distribution—for example, near one end of the distribution—or whether the scores are more evenly spread out.

c. The formula for variance is $SD^2 = \frac{\Sigma(X-M)^2}{N}$.

The first step is to calculate the mean, which is 927.50. We then create three columns: one for the scores, one for the deviations of the scores from the mean, and one for the squares of the deviations.

X	X − M	(X − M)²
450	−477.5	228,006.25
670	−257.5	66,306.25
1130	202.5	41,006.25
1460	532.5	283,556.25

We can now calculate variance: $SD^2 = \frac{\Sigma(X-M)^2}{N} =$ (228,006.25 + 66,306.25 + 41,006.25 + 283,556.25)/4 = 618,875/4 = 154,718.75.

d. We calculate standard deviation just like we calculated variance, but we then take the square root.

$$SD = \sqrt{\frac{\Sigma(X-M)^2}{N}} = \sqrt{154{,}718.75} = 393.34$$

e. If the researcher was interested only in these four students, these scores would represent the entire population of interest, and the variance and standard deviation would be parameters. Therefore, the symbols would be σ^2 and σ, respectively.

f. If the researcher hoped to generalize from these four students to all students at the university, these scores would represent a sample, and the variance and standard deviation would be statistics. Therefore, the symbols would be SD^2, $s2$, or MS for variance and SD or s for standard deviation.

CHAPTER 5

5-1 The risks of sampling are that we might not have a representative sample, and sometimes it is difficult to know whether a sample is representative. If a sample is not representative, we might draw conclusions about the population that are inaccurate.

5-2 The numbers in the fourth row, reading across, are 59808 08391 45427 26842 83609 49700 46058. Each person is assigned a number from 01 to 80. We then read the numbers from the table as two-digit numbers: 59, 80, 80, 83, 91, 45, 42, 72, 68, and so on. We ignore repeat numbers (e.g., 80) and numbers that exceed the sample of 80. So, the six people chosen would have the assigned numbers 59, 80, 45, 42, 72, and 68.

5-3 Reading down from the first column, then the second, and so on, noting only the appearance of 0's and 1's, we see the numbers 0, 0, 0, 0, 1, and 0 (ending in the 6th column). Using these numbers, we could assign the first through the fourth people and the sixth person to the group designated as 0, and the fifth person to the group designated as 1. If we want an equal number of people in each of the two groups, we would assign the first three people to the 0 group and the last three to the 1 group because we pulled three 0's first.

5-4 **a.** The likely population is all patients who will undergo surgery; the researcher would not be able to access this population, and therefore random selection could not be used. Random assignment, however, could be used. The psychologist could randomly assign half of the patients to counseling and half to a control group.

b. The population is all children in this school system; the psychologist could identify all of these children and thus could use random selection. The psychologist could also use random assignment. She could randomly assign half the children to the online textbook and half to the printed textbook.

c. The population is patients in therapy; because the whole population could not be identified, random selection could not be used. Moreover, random assignment could not be used. It is not possible to assign people to either have or not have a diagnosed personality disorder.

5-5 We regularly make personal assessments about how probable we think an event is, but we base these evaluations on our opinions about things rather than on systematically collected data. Statisticians are interested in objective probabilities, based on unbiased research.

5-6 **a.** Probability = successes/trials = 5/100 = 0.05

b. 8/50 = 0.16

c. 130/1044 = 0.12

5-7 **a.** In the short run, we might see a wide range of numbers of successes. It would not be surprising to have several in a row or none in a row. In the short run, our observations seem almost like chaos.

b. Given the assumptions listed for this problem, in the long run, we'd expect 0.50, or 50%, to be women, although there would likely be strings of men and of women along the way.

5-8 When we reject the null hypothesis, we are saying we reject the idea that there is no mean difference in the dependent variable across the levels of our independent variable. Rejecting the null hypothesis means we can support our research hypothesis that there is a mean difference.

5-9 The null hypothesis assumes no mean difference would be observed, so the mean difference in grades would be zero.

5-10 **a.** The null hypothesis is that a decrease in temperature does not affect mean academic performance (or does not decrease mean academic performance).

b. The research hypothesis is that a decrease in temperature does affect mean academic performance (or decreases mean academic performance).

c. The researchers would reject the null hypothesis.

d. The researchers would fail to reject the null hypothesis.

5-11 We make a Type I error when we reject the null hypothesis, but the null hypothesis is correct. We make a Type II error when we fail to reject the null hypothesis, but the null hypothesis is false.

5-12 In this scenario, a Type I error would be imprisoning a person who is really innocent, and 7 convictions out of 280 innocent people calculates to be 0.025, or 2.5%.

5-13 In this scenario, a Type II error would be failing to convict a guilty person, and 11 acquittals for every 35 guilty people calculates to be 0.314, or 31.4%.

5-14 **a.** If the virtual-reality glasses really don't have any effect, the researchers have made a Type I error: They rejected the null hypothesis, but the null hypothesis is really true.

b. If the virtual-reality glasses really do have an effect, the researchers have made a Type II error: They failed to reject the null hypothesis, but the null hypothesis is not true.

CHAPTER 6

6-1 Unimodal means there is one mode or high point to the curve. Symmetric means the left and right sides of the curve have the same shape and are mirror images of each other.

6-2 **a.**

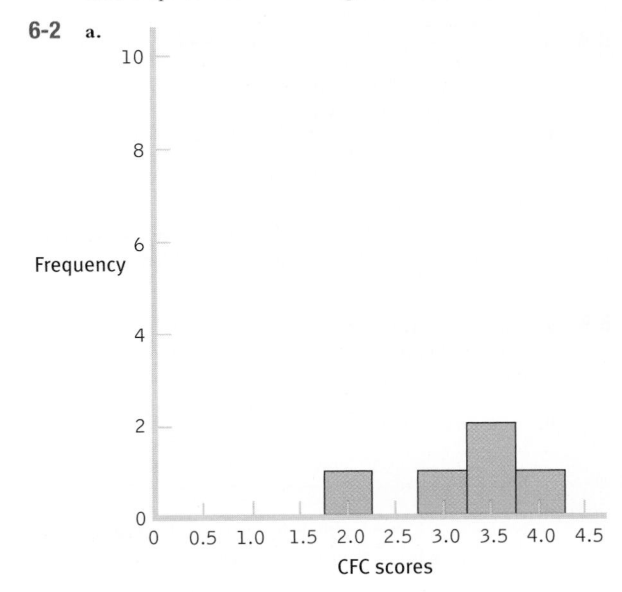

b.

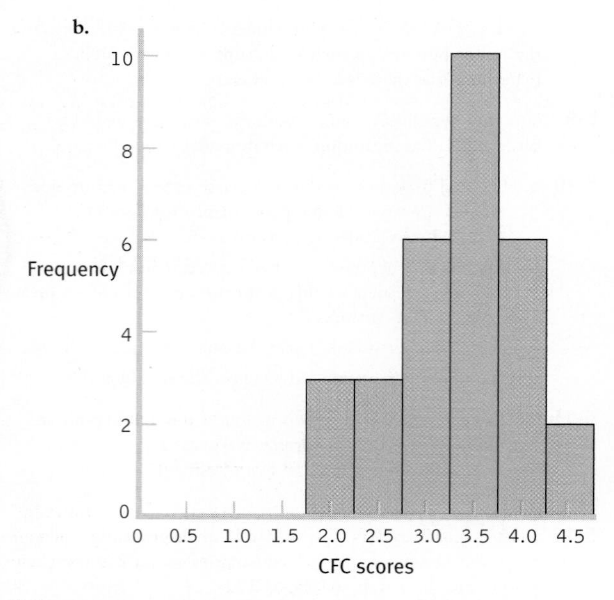

6-3 The shape of the distribution becomes more normal as the size of the sample increases (although the larger sample appears to be somewhat negatively skewed).

6-4 In standardization, we convert individual scores to standardized scores for which we know the percentiles.

6-5 The numeric value tells us how many standard deviations a score is from the mean of the distribution. The sign tells us whether the score is above or below the mean.

6-6 **a.** $z = \frac{(X - \mu)}{\sigma} = \frac{11.5 - 14}{2.5} = -1.0$

b. $z = \frac{(X - \mu)}{\sigma} = \frac{18 - 14}{2.5} = 1.6$

6-7 **a.** $X = z(\sigma) + \mu = 2(2.5) + 14 = 19$

b. $X = z(\sigma) + \mu = -1.4(2.5) + 14 = 10.5$

6-8 **a.** $z = \frac{(X - \mu)}{\sigma} = \frac{2.3 - 3.51}{0.61} = -1.98$; approximately 2% of students have a CFC score of 2.3 or less.

b. $z = \frac{(X - \mu)}{\sigma} = \frac{4.7 - 3.5}{0.61} = 1.97$; this score is at approximately the 98th percentile.

c. This student has a z score of 1.

d. $X = z(\sigma) + \mu = 1(0.61) + 3.51 = 4.12$; this answer makes sense because 4.12 is above the mean of 3.51, as a z score of 1 would indicate.

6-9 **a.** Nicole is in better health because her score is above the mean for her measure, whereas Samantha's score is below the mean.

b. Samantha's z score is $z = \frac{(X - \mu)}{\sigma} = \frac{84 - 93}{4.5} = -2.0$

Nicole's z score is $z = \frac{(X - \mu)}{\sigma} = \frac{332 - 312}{20} = 1.0$

Nicole is in better health, being 1 standard deviation above the mean, whereas Samantha is 2 standard deviations below the mean.

c. We can conclude that approximately 98% of the population is in better health than Samantha, who is 2 standard deviations below the mean. We can conclude that approximately 16% of the population is in better health than Nicole, who is 1 standard deviation above the mean.

6-10 The central limit theorem asserts that a distribution of sample means approaches the shape of the normal curve as sample size increases. It also asserts that the spread of the distribution of sample means gets smaller as the sample size gets larger.

6-11 A distribution of means is composed of many means that are calculated from all possible samples of a particular size from the same population.

6-12 $\sigma_M = \frac{\sigma}{\sqrt{N}} = \frac{11}{\sqrt{35}} = 1.86$

6-13 **a.** The scores range from 2.0 to 4.5, which gives us a range of $4.5 - 2.0 = 2.5$.

b. The means are 3.4 for the first row, 3.4 for the second row, and 3.15 for the third row [e.g., for the first row, $M = (3.5 + 3.5 + 3.0 + 4.0 + 2.0 + 4.0 + 2.0 + 4.0 + 3.5 + 4.5)/10 = 3.4$]. These three means range from 3.15 to 3.40, which gives us a range of $3.40 - 3.15 = 0.25$.

c. The range is smaller for the means of samples of 10 scores than for the individual scores because the more extreme scores are balanced by lower scores when samples of 10 are taken. Individual scores are not attenuated in that way.

d. The mean of the distribution of means is the same as the mean of the individual scores: $\mu_M = \mu = 3.32$. The standard error is smaller than the standard deviation; we must divide by the square root of the sample size of 10:

$$\sigma_M = \frac{\sigma}{\sqrt{N}} = \frac{0.69}{\sqrt{10}} = 0.22$$

CHAPTER 7

7-1 We need to know the mean, μ, and standard deviation, σ, of the population in order to use the z table.

7-2 Raw scores are used to compute z scores, and z scores are used to determine what percentage of scores fall below and above that particular position on the distribution. A z score can also be used to compute a raw score.

7-3 Because the curve is symmetric, the same percentage of scores (41.47%) lies between the mean and a z score of -1.37 as between the mean and a z score of 1.37.

7-4 Fifty percent of scores fall below the mean, and 12.93% fall between the mean and a z score of 0.33.

$$50\% + 12.93\% = 62.93\%$$

7-5 **a.** $\mu_M = \mu = 156.8$

$$\sigma_M = \frac{\sigma}{\sqrt{N}} = \frac{14.6}{\sqrt{36}} = 2.433$$

$$z = \frac{(M - \mu_M)}{\sigma_M} = \frac{(164.6 - 156.8)}{2.433} = 3.21$$

50% below the mean; 49.9% above the mean; $50 + 49.9 = 99.9$th percentile

b. 100 − 99.9 = 0.1% of samples of this size scored higher than the students at Baylor.

c. At the 99.9th percentile, these 36 students from Baylor are truly outstanding. If these students are representative of their majors, clearly these results reflect positively on Baylor's Psychology and Neuroscience Department.

7-6 For most parametric hypothesis tests, we assume that (1) the dependent variable is assessed on a scale measure—that is, equal changes are reflected by equal distances on the measure; (2) the participants are randomly selected, meaning everyone has the same chance of being selected; and (3) the distribution of the population of interest is approximately normal.

7-7 We compare the test statistic to the critical values. If a test statistic is more extreme than the critical value, then we reject the null hypothesis. If a test statistic is less extreme than the critical value, then we fail to reject the null hypothesis.

7-8 If the null hypothesis is true, he will reject it 8% of the time.

7-9 **a.** 0.15; **b.** 0.03; **c.** 0.055

7-10 **a.** (1) The dependent variable—diagnosis (correct versus incorrect)—is nominal, not scale, so this assumption is not met. Based only on this, we should not proceed with a hypothesis test based on a z distribution. (2) The samples include only outpatients seen over 2 specific months and only those at one community mental health center. The sample is not randomly selected, so we must be cautious about generalizing from it. (3) The populations are not normally distributed because the dependent variable is nominal.

b. (1) The dependent variable, health score, is likely scale. (2) The participants were randomly selected; all wild cats in zoos in North America had an equal chance of being selected for this study. (3) The data are not normally distributed; we are told that a few animals had very high scores, so the data are likely positively skewed. Moreover, there are fewer than 30 participants in this study. It is probably not a good idea to proceed with a hypothesis test based on a z distribution.

7-11 A directional test indicates that either a mean increase or a mean decrease in the dependent variable is hypothesized, but not both. A nondirectional test does not indicate a direction of mean difference for the research hypothesis, it just indicates that there is a mean difference.

7-12 $\mu_M = \mu = 1090$

$$\sigma_M = \frac{\sigma}{\sqrt{N}} = \frac{87}{\sqrt{53}} = 11.95$$

7-13 $$z = \frac{(M - \mu_M)}{\sigma_M} = \frac{(1094 - 1090)}{11.95} = 0.33$$

7-14 *Step 1:* Population 1 is coffee drinkers who spend the day in coffee shops/cybercafés. Population 2 is all coffee drinkers in the United States. The comparison distribution will be a distribution of means. The hypothesis test will be a z test because we have only one sample and we know the population mean and standard deviation. This study meets two of the three assumptions and may meet the third. The dependent variable, the number of cups coffee drinkers drank, is scale. In addition, there are more than 30 participants in the sample, indicating that the comparison distribution is normal. The data were not randomly selected, however, so we must be cautious when generalizing.

Step 2: The null hypothesis is that people who spend the day working in the coffee shop/cybercafé drink the same amount of coffee, on average, as those in the general U.S. population ($H_0: \mu_1 = \mu_2$).

The research hypothesis is that people who spend the day in coffee shops/cybercafés drink a different amount of coffee, on average, than those in the general U.S. population ($H_1: \mu_1 \neq \mu_2$).

Step 3: $\mu_M = \mu = 3.10$

$$\sigma_M = \frac{\sigma}{\sqrt{N}} = \frac{0.9}{\sqrt{34}} = 0.154$$

Step 4: The cutoff z statistics are −1.96 and 1.96.

Step 5: $$z = \frac{(M - \mu_M)}{\sigma_M} = \frac{(3.17 - 3.10)}{0.154} = 0.45$$

Step 6: Because the z statistic does not exceed the cutoffs, we fail to reject the null hypothesis. We did not find any evidence that the sample was different from what was normally expected according to the null hypothesis.

CHAPTER 8

8-1 Interval estimates provide a range of scores in which we have some confidence the population statistic will fall, whereas point estimates use just a single value to describe the population.

8-2 The interval estimate is 17% to 25% (21% − 4% = 17% and 21% + 4% = 25%), whereas the point estimate is 21%.

8-3 **a.** First, we draw a normal curve with the sample mean, 3.7, in the center. Then we put the bounds of the 95% confidence interval on either end, writing the appropriate percentages under the segments of the curve: 2.5% beyond the cutoffs on either end and 47.5% between the mean and each cutoff. Now we look up the z statistics for these cutoffs; the z statistic associated with 47.5%, the percentage between the mean and the z statistic, is 1.96. Thus, the cutoffs are −1.96 and 1.96. Next, we calculate standard error so that we can convert these z statistics to raw means:

$$\sigma_M = \frac{\sigma}{\sqrt{N}} = \frac{0.61}{\sqrt{45}} = 0.091$$

$$M_{lower} = -z(\sigma_M) + M_{sample} = -1.96(0.091) + 3.7 = 3.52$$

$$M_{upper} = z(\sigma_M) + M_{sample} = 1.96(0.091) + 3.7 = 3.88$$

Finally, we check to be sure the answer makes sense by demonstrating that each end of the confidence interval is the same distance from the mean: 3.52 − 3.7 = −0.18 and 3.88 − 3.7 = 0.18. The confidence interval is [3.52, 3.88].

b. If we were to conduct this study over and over, with the same sample size, we would expect the population mean to fall in that interval 95% of the time. Thus, it provides a range of plausible values for the population mean. Because the null-hypothesized population mean of 3.51 is not a plausible value, we can conclude that those who attended the discussion group have higher CFC scores than those who did not. This conclusion matches that of the hypothesis test, in which we rejected the null hypothesis.

c. The confidence interval is superior to the hypothesis test because not only does it lead to the same conclusion but it also gives us an interval estimate, rather than a point estimate, of the population mean.

8-4 Statistical significance means that the observation met the standard for special events, typically something that occurs less than 5% of the time. Practical importance means that the outcome really matters.

8-5 Effect size is a standardized value that indicates the size of a difference with respect to a measure of spread but is not affected by sample size.

8-6 $\text{Cohen's } d = \frac{(M - \mu)}{\sigma} = \frac{(105 - 100)}{15} = 0.33$

8-7 **a.** We calculate Cohen's *d*, the effect size appropriate for data analyzed with a *z* test. We use standard deviation in the denominator, rather than standard error, because effect sizes are for distributions of scores rather than distributions of means.

$$\text{Cohen's } d = \frac{(M - \mu)}{\sigma} = \frac{(3.7 - 3.51)}{0.61} = 0.31$$

b. Cohen's conventions indicate that 0.2 is a small effect and 0.5 is a medium effect. This effect size, therefore, would be considered a small-to-medium effect.

c. If the career discussion group is easily implemented in terms of time and money, the small-to-medium effect might be worth the effort. For university students, a higher level of Consideration of Future Consequences might translate into a higher level of readiness for life after graduation, a premise that we could study.

8-8 Three ways to increase power are to increase alpha, to conduct a one-tailed test rather than a two-tailed test, and to increase *N*. All three of these techniques serve to increase the chance of rejecting the null hypothesis. (We could also increase the difference between means, or decrease variability, but these are more difficult.)

8-9 *Step 1:* We know the following about population 1: $\mu = 3.51$, $\sigma = 0.61$. We assume the following about population 2 based on the information from the sample: $N = 45$, $M = 3.7$. We need to calculate standard error based on the standard deviation for population 1 and the size of the sample:

$$\sigma_M = \frac{\sigma}{\sqrt{N}} = \frac{0.61}{\sqrt{45}} = 0.091$$

Step 2: Because the sample mean is higher than the population mean, we will conduct this one-tailed test by examining only the high end of the distribution. We need to find the cutoff that marks where 5% of the data fall in the tail. We know that the *z* cutoff for a one-tailed test is 1.64. Using that *z* statistic, we can calculate a raw score.

$$M = z(\sigma_M) + \mu_M = 1.64(0.091) + 3.51 = 3.659$$

This mean of 3.659 marks the point beyond which 5% of all means based on samples of 45 observations will fall.

Step 3: For the second distribution, centered around 3.7, we need to calculate how often means of 3.659 (the cutoff) and greater occur. We do this by calculating the *z* statistic for the raw mean of 3.659 with respect to the sample mean of 3.7.

$$z = \frac{3.659 - 3.7}{0.091} = -0.451$$

We now look up this *z* statistic on the table and find that 32.64% falls toward the tail and 17.36% falls between this *z* statistic and the mean. We calculate power as the proportion of observations between this *z* statistic and the tail of interest, which is at the high end. So we would add 17.36% and 50% to get statistical power of 67.36%.

8-10 **a.** The statistical power calculation means that, if the second population really does exist, we have a 67.36% chance of observing a sample mean, based on 45 observations, that will allow us to reject the null hypothesis. We fall somewhat short of the desired 80% statistical power.

b. We can increase statistical power by increasing the sample size; extending or enhancing the career discussion group such that we create a bigger effect; or by changing alpha.

CHAPTER 9

9-1 The *t* statistic indicates the distance of a sample mean from a population mean in terms of the estimated standard error.

9-2 First we need to calculate the mean:

$$M = \frac{\Sigma X}{N} = (6 + 3 + 7 + 6 + 4 + 5)/6 = 31/6 = 5.167$$

We then calculate the deviation of each score from the mean and the square of that deviation.

X	X − M	(X − M)²
6	0.833	0.694
3	−2.167	4.696
7	1.833	3.360
6	0.833	0.694
4	−1.167	1.362
5	−0.167	0.028

Numerator: $\Sigma(X - M)^2 = (0.694 + 4.696 + 3.360 + 0.694 + 1.362 + 0.028) = 10.834$

The standard deviation is:

$$SD = \sqrt{\frac{\Sigma(X - M)^2}{N}} = \sqrt{\frac{10.834}{6}} = \sqrt{1.806} = 1.344$$

When estimating the population variability, we calculate *s*:

$$s = \sqrt{\frac{\Sigma(X - M)^2}{N - 1}} = \sqrt{\frac{10.834}{6 - 1}} = \sqrt{2.167} = 1.472$$

9-3 $s_M = \frac{s}{\sqrt{N}} = \frac{1.472}{\sqrt{6}} = 0.601$

9-4 **a.** We will use a distribution of means, specifically a *t* distribution. It is a distribution of means because we have a sample consisting of more than one individual. It is a *t* distribution because we are comparing one sample to a population, but we know only the population mean, not its standard deviation.

b. The appropriate mean: $\mu_M = \mu = 25$

The calculations for the appropriate standard deviation (in this case, standard error, s_M) follow:

$$M = \frac{\Sigma X}{N} = (20 + 19 + 27 + 24 + 18)/5 = 21.6$$

X	X − M	(X − M)^2
20	−1.6	2.56
19	−2.6	6.76
27	5.4	29.16
24	2.4	5.76
18	−3.6	12.96

Numerator: $\Sigma(X - M)^2 = (2.56 + 6.76 + 29.16 + 5.76 + 12.96) = 57.2$

$$s = \sqrt{\frac{\Sigma(X-M)^2}{(N-1)}} = \sqrt{\frac{57.2}{5-1}} = \sqrt{14.3} = 3.782$$

$$s_M = \frac{s}{\sqrt{N}} = \frac{3.782}{\sqrt{5}} = 1.691$$

c. $$t = \frac{(M - \mu_M)}{s_M} = \frac{(21.6 - 25)}{1.691} = -2.01$$

9-5 *Degrees of freedom* is the number of scores that are free to vary, or take on any value, when estimating a population parameter from a sample.

9-6 A single-sample t test has more uses than a z test because we only need to know the population mean (not the population standard deviation) for the single-sample t test.

9-7 **a.** $df = N - 1 = 35 - 1 = 34$
b. $df = N - 1 = 14 - 1 = 13$

9-8 **a.** ± 2.201
b. Either -2.584 or $+2.584$, depending on the tail of interest

9-9 *Step 1:* Population 1 is the sample of six students. Population 2 is all university students. The distribution will be a distribution of means, and we will use a single-sample t test. We meet the assumption that the dependent variable is scale. We do not know if the sample was randomly selected, and we do not know if the population variable is normally distributed. Some caution should be exercised when drawing conclusions from these data.

Step 2: The null hypothesis is $H_0: \mu_1 = \mu_2$; that is, students we're working with miss the same number of classes, on average, as the population. The research hypothesis is $H_1: \mu_1 \neq \mu_2$; that is, students we're working with miss a different number of classes, on average, than the population.

Step 3: $\mu_M = \mu = 3.7$

$$M = \frac{\Sigma X}{N} = (6 + 3 + 7 + 6 + 4 + 5)/6 = 31/6 = 5.167$$

$$s = \sqrt{\frac{\Sigma(X-M)^2}{N-1}} = \sqrt{\frac{10.834}{6-1}} = \sqrt{2.167} = 1.472$$

$$s_M = \frac{s}{\sqrt{N}} = \frac{1.472}{\sqrt{5}} = 0.601$$

Step 4: $df = N - 1 = 6 - 1 = 5$

For a two-tailed test with a p level of 0.05 and 5 degrees of freedom, the cutoffs are -2.571 and 2.571.

Step 5: $$t = \frac{(M - \mu_M)}{s_M} = \frac{(5.167 - 3.7)}{0.601} = 2.441$$

Step 6: Because the calculated t value falls short of the critical values, we fail to reject the null hypothesis.

9-10 For a paired-samples t test, we calculate a difference score for every individual. We then compare the average difference observed to the average difference we would expect based on the null hypothesis. If there is no difference, then all difference scores should average to 0.

9-11 An individual difference score is a calculation of change or difference for each participant. For example, we might subtract weight before the holiday break from weight after the break to evaluate how many pounds an individual lost or gained.

9-12 The null hypothesis for the paired-samples t test is that the mean difference score is 0; that is, $\mu_M = 0$. Therefore, if the confidence interval around the mean difference does not include 0, we know that the sample mean is unlikely to have come from a distribution with a mean of 0 and we can reject the null hypothesis.

9-13 We calculate Cohen's d by subtracting 0 (the population mean based on the null hypothesis) from the sample mean and dividing by the standard deviation of the difference scores.

9-14 We want to subtract the before-lunch energy level from the after-lunch energy level to get values that reflect loss of energy as a negative value and an increase of energy with food as a positive value.

The mean of these differences is -1.4.

BEFORE LUNCH	AFTER LUNCH	AFTER − BEFORE
6	3	3 − 6 = −3
5	2	2 − 5 = −3
4	6	6 − 4 = +2
5	4	4 − 5 = −1
7	5	5 − 7 = −2

9-15 **a.** We first find the t values associated with a two-tailed hypothesis test and alpha of 0.05. These are ± 2.776. We then calculate s_M by dividing s by the square root of the sample size, which results in $s_M = 0.548$.

$$M_{lower} = -t(s_M) + M_{sample} = -2.776(0.548) + 1.0 = -0.52$$

$$M_{upper} = t(s_M) + M_{sample} = 2.776(0.548) + 1.0 = 2.52$$

The confidence interval can be written as [−0.52, 2.52]. Because this confidence interval includes 0, we would fail to reject the null hypothesis. Zero is one of the plausible mean differences we would get when repeatedly sampling from a population with a mean difference score of 1.

b. We calculate Cohen's d as:

$$d = \frac{(M - \mu)}{s} = \frac{(1 - 0)}{1.225} = 0.82$$

This is a large effect size.

9-16 **a.** *Step 1:* Population 1 is students for whom we're measuring energy levels before lunch. Population 2 is students for whom we're measuring energy levels after lunch. The comparison distribution is a distribution of mean difference scores. We use the paired-samples t test because each participant contributes a score to each of the two samples we are comparing. We meet the assumption that the dependent variable is a scale measurement. However, we do not know if our participants were randomly selected or if the population is normally distributed, and the sample is less than 30.

Step 2: The null hypothesis is that there is no difference in mean energy levels before and after lunch—$H_0: \mu_1 = \mu_2$. The research hypothesis is that there is a mean difference in energy levels—$H_1: \mu_1 \neq \mu_2$.

Step 3:

DIFFERENCE SCORES	DIFFERENCE – MEAN DIFFERENCE	SQUARED DEVIATION
−3	−1.6	2.56
−3	−1.6	2.56
+2	3.4	11.56
−1	0.4	0.16
−2	−0.6	0.36

$M_{difference} = -1.4$

$$s = \sqrt{\frac{\Sigma(X-M)^2}{(N-1)}} = \sqrt{\frac{17.2}{(5-1)}} = \sqrt{4.3} = 2.074$$

$$s_M = \frac{s}{\sqrt{N}} = \frac{2.074}{\sqrt{5}} = 0.928$$

$\mu_M = 0, s_M = 0.928$

Step 4: The degrees of freedom are $5 - 1 = 4$, and the cutoffs, based on a two-tailed test and a p level of 0.05, are ± 2.776.

Step 5: $t = \frac{(-1.4-0)}{0.928} = -1.51$

Step 6: Because the test statistic, −1.51, failed to exceed the critical value of −2.776, we fail to reject the null hypothesis.

9-17 **a.** $M_{lower} = -t(s_M) + M_{sample} = -2.776(0.928) + (-1.4) = -3.98$

$M_{upper} = t(s_M) + M_{sample} = 2.776(0.928) + (-1.4) = 1.18$

The confidence interval can be written as [−3.98, 1.18]. Notice that the confidence interval spans 0, the null-hypothesized difference between mean energy levels before and after lunch. Because the null value is within the confidence interval, we fail to reject the null hypothesis.

b. $d = \frac{(M-\mu)}{s} = \frac{(-1.4-0)}{2.074} = -0.68$

This is a medium-to-large effect size, according to Cohen's guidelines.

CHAPTER 10

10-1 When the data we are comparing were collected using the same participants in both conditions, a paired-samples t test is used; each participant contributes two values to the analysis. When we are comparing two independent groups and no participant is in more than one condition, we use an independent-samples t test.

10-2 Pooled variance is a weighted combination of the variability in both groups in an independent-samples t test.

10-3 **a.** Group 1 is treated as the X variable; its mean is 3.0.

X	X − M	(X − M)²
3	0	0
2	−1	1
4	1	1
6	3	9
1	−2	4
2	−1	1

$$s_X^2 = \frac{\Sigma(X-M)^2}{N-1} = \frac{(0+1+1+9+4+1)}{6-1} = 3.2$$

Group 2 is treated as the Y variable; its mean is 4.6.

Y	Y − M	(Y − M)²
5	0.4	0.16
4	−0.6	0.36
6	1.4	1.96
2	−2.6	6.76
6	1.4	1.96

$$s_Y^2 = \frac{\Sigma(Y-M)^2}{N-1} = \frac{(0.16+0.36+1.96+6.76+1.96)}{5-1} = 2.8$$

b. $df_X = N - 1 = 6 - 1 = 5$

$df_Y = N - 1 = 5 - 1 = 4$

$df_{total} = df_X + df_Y = 5 + 4 = 9$

$$s_{pooled}^2 = \left(\frac{df_X}{df_{total}}\right)s_X^2 + \left(\frac{df_Y}{df_{total}}\right)s_Y^2 = \left(\frac{5}{9}\right)3.2 + \left(\frac{4}{9}\right)2.8 = 1.778 + 1.244 = 3.022$$

c. The variance version of standard error is calculated for each sample as:

$$s_{M_X}^2 = \frac{s_{pooled}^2}{N_X} = \frac{3.022}{6} = 0.504$$

$$s_{M_Y}^2 = \frac{s_{pooled}^2}{N_Y} = \frac{3.022}{5} = 0.604$$

d. The variance of the distribution of differences between means is:

$$s_{difference}^2 = s_{M_X}^2 + s_{M_Y}^2 = 0.504 + 0.604 = 1.108$$

This can be converted to standard deviation units by taking the square root:

$$s_{difference} = \sqrt{s_{difference}^2} = \sqrt{1.108} = 1.053$$

e. $t = \frac{(M_X - M_Y) - (\mu_X - \mu_Y)}{s_{difference}} = \frac{(3 - 4.6) - (0)}{1.053} = -1.519$

10-4 a. The null hypothesis asserts that there are no average between-group differences; employees with low trust in their leader show the same mean level of agreement with decisions as those with high trust in their leader. Symbolically, this would be written $H_0: \mu_1 = \mu_2$. The research hypothesis asserts that the mean level of agreement is different between the two groups—$H_1: \mu_1 \neq \mu_2$.

b. The critical values, based on a two-tailed test, a p level of 0.05, and df_{total} of 9, are −2.262 and 2.262. The t value we calculated, −1.519, does not exceed the cutoff of −2.262, so we fail to reject the null hypothesis.

c. Based on these results, we did not find evidence that the mean level of agreement with a decision is different across the two levels of trust: $t(9) = -1.519, p > 0.05$.

d. Despite having similar means for the two groups, we failed to reject the null hypothesis, whereas the original researchers rejected the null hypothesis. Our failure to reject the null hypothesis is likely due to the low statistical power from the small samples we used.

10-5 We calculate confidence intervals to determine a range of plausible values for the population parameter, based on the data.

10-6 Effect size tells us how large or small the difference we observed is, regardless of sample size. Even when a result is statistically significant, it might not be important. Effect size helps us evaluate practical significance.

10-7 a. The upper and lower bounds of the confidence interval are calculated as:

$(M_X - M_Y)_{lower} = -t(s_{difference}) + (M_X - M_Y)_{sample} = -2.262(1.053) + (-1.6) = -3.98$

$(M_X - M_Y)_{upper} = t(s_{difference}) + (M_X - M_Y)_{sample} = 2.262(1.053) + (-1.6) = 0.78$

The confidence interval is [−3.98, 0.78].

b. To calculate Cohen's d, we need to calculate the pooled standard deviation for the data:

$$s_{pooled} = \sqrt{s^2_{pooled}} = \sqrt{3.022} = 1.738$$

$$\text{Cohen's } d = \frac{(M_X - M_Y) - (\mu_X - \mu_Y)}{s_{pooled}} = \frac{(3 - 4.6) - (0)}{1.738} = -0.92$$

10-8 The confidence interval tells us a range of differences between means in which we could expect the population mean difference to fall 95% of the time, based on samples of this size. Whereas the hypothesis test evaluates the point estimate of the difference between means—which is (3 − 4.6), or −1.6, in this case—the confidence interval gives us a range, or interval estimate, of [−3.98, 0.78].

10-9 The effect size we calculated, Cohen's d of −0.92, is a large effect, according to Cohen's guidelines. Beyond the hypothesis test and confidence interval, which both lead us to fail to reject the null hypothesis, the size of the effect indicates that we might be on to a real effect here. We might want to increase statistical power by collecting more data in an effort to better test this hypothesis.

CHAPTER 11

11-1 The F statistic is a ratio of between-groups variance and within-groups variance.

11-2 The two types of research design are within-groups design and between-groups design.

11-3 a. $F = \frac{\text{between-groups variance}}{\text{within-groups variance}} = \frac{8.6}{3.7} = 2.324$

b. $F = \frac{102.4}{123.77} = 0.827$

c. $F = \frac{45.2}{32.1} = 1.408$

11-4 a. We would use an F distribution because there are more than two groups.

b. We would determine the variance among the three sample means—the means for those in the control group, for those in the 2-hour communication ban, and for those in the 4-hour communication ban.

c. We would determine the variance within each of the three samples, and we would take a weighted average of the three variances.

11-5 If the F statistic is beyond the cutoff, then we can reject the null hypothesis—meaning that there is a significant mean difference (or differences) somewhere in the data, but we do not know where the difference lies.

11-6 When calculating $SS_{between}$, we subtract the grand mean (GM) from the mean of each group (M). We do this for every score.

11-7 a. $df_{between} = N_{groups} - 1 = 3 - 1 = 2$

b. $df_{within} = df_1 + df_2 + \ldots + df_{last} = (4 - 1) + (4 - 1) + (3 - 1) = 3 + 3 + 2 = 8$

c. $df_{total} = df_{between} + df_{within} = 2 + 8 = 10$

11-8 $GM = \frac{\Sigma(X)}{N_{total}} = \frac{(37 + 30 + 22 + 29 + 49 + 52 + 41 + 39 + 36 + 49 + 42)}{11} = 38.727$

11-9 a. The total sum of squares is calculated here as $SS_{total} = \Sigma(X - GM)^2$:

SAMPLE	X	$(X - GM)$	$(X - GM)^2$
Group 1	37	−1.727	2.983
$M_1 = 29.5$	30	−8.727	76.161
	22	−16.727	279.793
	29	−9.727	94.615
Group 2	49	10.273	105.535
$M_2 = 45.25$	52	13.273	176.173
	41	2.273	5.167
	39	0.273	0.075
Group 3	36	−2.727	7.437
$M_3 = 42.333$	49	10.273	105.535
	42	3.273	10.713
	$GM = 38.727$		$SS_{total} =$ **864.187**

b. Within-groups sum of squares is calculated here as $SS_{within} = \Sigma(X - M)^2$:

SAMPLE	X	(X − M)	(X − M)²
Group 1	37	7.500	56.250
M_1 = 29.5	30	0.500	0.250
	22	−7.500	56.250
	29	−0.500	0.250
Group 2	49	3.750	14.063
M_2 = 45.25	52	6.750	45.563
	41	−4.250	18.063
	39	−6.250	39.063
Group 3	36	−6.333	40.107
M_3 = 42.333	49	6.667	44.449
	42	−0.333	0.111
	GM = 38.727	SS_{within} =	**314.419**

c. Between-groups sum of squares is calculated here as $SS_{between} = \Sigma(M - GM)^2$:

SAMPLE	X	(M − GM)	(M − GM)²
Group 1	37	−9.227	85.138
M_1 = 29.5	30	−9.227	85.138
	22	−9.227	85.138
	29	−9.227	85.138
Group 2	49	6.523	42.550
M_2 = 45.25	52	6.523	42.550
	41	6.523	42.550
	39	6.523	42.550
Group 3	36	3.606	13.003
M_3 = 42.333	49	3.606	13.003
	42	3.606	13.003
	GM = 38.727	$SS_{between}$ =	**549.761**

11-10 $MS_{between} = \frac{SS_{between}}{df_{between}} = \frac{549.761}{2} = 274.881$

$$MS_{within} = \frac{SS_{within}}{df_{within}} = \frac{314.419}{8} = 39.302$$

$$F = \frac{MS_{between}}{MS_{within}} = \frac{274.881}{39.302} = 6.99$$

SOURCE	SS	df	MS	F
Between	549.761	2	**274.881**	**6.99**
Within	314.419	8	**39.302**	
Total	864.187	10		

11-11 **a.** According to the null hypothesis, there are no mean differences in efficacy among these three treatment conditions—H_0: $\mu_1 = \mu_2 = \mu_3$; they would all come from one underlying distribution. The research hypothesis states that there are mean differences in efficacy across some or all of these treatment conditions.

b. There are three assumptions: that the participants were selected randomly, that the underlying populations are normally distributed, and that the underlying populations have similar variances. Although we can't say much about the first two assumptions, we can assess the last one using the sample data.

SAMPLE	GROUP 1	GROUP 2	GROUP 3
Squared deviations	56.25	14.063	40.107
of scores from	0.25	45.563	44.449
sample means:	56.25	18.063	0.111
	0.25	39.063	
Sum of squares:	113	116.752	84.667
N − 1:	3	3	2
Variance:	**37.67**	**38.92**	**42.33**

Because these variances are all close together, with the biggest being no more than twice as large as the smallest, we can conclude that we met the third assumption of homoscedastic samples.

c. The critical value for F with a p value of 0.05, two between-groups degrees of freedom, and 8 within-groups degrees of freedom, is 4.46. The F statistic exceeds this cutoff, so we can reject the null hypothesis. There are mean differences between these three groups, but we do not know where.

11-12 If we are able to reject the null hypothesis when conducting an ANOVA, then we must also conduct a post hoc test, such as a Tukey HSD test, to determine which pairs of means are significantly different from one another.

11-13 R^2 tells us the proportion of variance in the dependent variable that is accounted for by the independent variable.

11-14 **a.** $R^2 = \frac{SS_{between}}{SS_{total}} = \frac{336.360}{522.782} = 0.64$

b. Children's grade level accounted for 64% of the variability in reaction time. This is a very large effect.

11-15 The number of levels of the independent variable is 3 and df_{within} is 32. At a p level of 0.05, the critical values of the q statistic would be −3.49 and 3.49.

11-16 Because we have unequal sample sizes, we must calculate a weighted sample size.

$$N' = \frac{N_{groups}}{\Sigma\left(\frac{1}{N}\right)} = \frac{3}{\left(\frac{1}{4} + \frac{1}{4} + \frac{1}{3}\right)} = \frac{3}{0.25 + 0.25 + 0.333} = \frac{3}{0.833} = 3.601$$

$$s_M = \sqrt{\frac{MS_{within}}{N'}}, \text{ then equals } \sqrt{\frac{39.302}{3.601}} = 3.304$$

Now we can compare the three treatment groups.

Psychodynamic therapy (M = 29.50) versus interpersonal therapy (M = 45.25):

$$HSD = \frac{29.50 - 45.25}{3.304} = -4.77$$

Psychodynamic therapy (M = 29.50) versus cognitive-behavioral therapy (M = 42.333):

$$HSD = \frac{29.50 - 42.333}{3.304} = -3.88$$

Interpersonal therapy (M = 45.25) versus cognitive-behavioral therapy (M = 42.333):

$$HSD = \frac{45.25 - 42.333}{3.304} = 0.88$$

We look up the critical value for this post-hoc test on the q table. We look in the row for 8 within-groups degrees of freedom, and then in the column for 3 treatment groups. At a p level of 0.05, the value in the q table is 4.04, so the cutoffs are −4.04 and 4.04.

We have just one significant difference between psychodynamic therapy and interpersonal therapy: Tukey $HSD = -4.77$. Specifically, clients responded at statistically significantly higher rates to interpersonal therapy than to psychodynamic therapy, with an average difference of 15.75 points on this scale.

11-17 Effect size is calculated as $R^2 = \frac{SS_{between}}{SS_{total}} = \frac{549.761}{864.187} = 0.64$. According to Cohen's conventions for R^2, this is a very large effect.

11-18 For the one-way within-groups ANOVA, we calculate two types of variability that occur within groups: subjects variability and within-groups variability. Subjects variability assesses how much each person's mean differs from the others', assessed by comparing each person's mean score to the grand mean. We then compute within-groups variability as the remainder once between-groups and subjects variability are subtracted from the total sum of squares.

11-19 **a.** $df_{between} = N_{groups} - 1 = 3 - 1 = 2$

b. $df_{subjects} = n - 1 = 3 - 1 = 2$

c. $df_{within} = (df_{between})(df_{subjects}) = (2)(2) = 4$

d. $df_{total} = df_{between} + df_{subjects} + df_{within} = 2 + 2 + 4 = 8$; or we can calculate it as $df_{total} = N_{total} - 1 = 9 - 1 = 8$

11-20 **a.** $SS_{total} = \Sigma(X - GM)^2 = 24.886$

GROUP	RATING (X)	(X − GM)	(X − GM)²
1	7	0.111	0.012
1	9	2.111	4.456
1	8	1.111	1.234
2	5	−1.889	3.568
2	8	1.111	1.234
2	9	2.111	4.456
3	6	−0.889	0.790
3	4	−2.889	8.346
3	6	−0.889	0.790
	GM = 6.889		Σ(X − GM)²= 24.886

b. $SS_{between} = \Sigma(M - GM)^2 = 11.556$

GROUP	RATING (X)	GROUP MEAN	(M − GM)	(M − GM)²
1	7	8	1.111	1.234
1	9	8	1.111	1.234
1	8	8	1.111	1.234
2	5	7.333	0.444	0.197
2	8	7.333	0.444	0.197
2	9	7.333	0.444	0.197
3	6	5.333	−1.556	2.421
3	4	5.333	−1.556	2.421
3	6	5.333	−1.556	2.421
	GM = 6.889			Σ(M − GM)² = 11.556

c. $SS_{subject} = \Sigma(M_{participant} - GM)^2 = 4.221$

PARTICIPANT	GROUP	RATING (X)	PARTICIPANT MEAN	(M_PARTICIPANT − GM)	(M_PARTICIPANT − GM)²
1	1	7	6	−0.889	0.790
2	1	9	7	0.111	0.012
3	1	8	7.667	0.778	0.605
1	2	5	6	−0.889	0.790
2	2	8	7	0.111	0.012
3	2	9	7.667	0.778	0.605
1	3	6	6	−0.889	0.790
2	3	4	7	0.111	0.012
3	3	6	7.667	0.778	0.605
		GM = 6.889			Σ(M_participant − GM)² = 4.221

d. $SS_{within} = SS_{total} - SS_{between} - SS_{subjects} = 24.886 - 11.556 - 4.221 = 9.109$

11-21 $MS_{between} = \frac{SS_{between}}{df_{between}} = \frac{11.556}{2} = 5.778$

$$MS_{subjects} = \frac{SS_{subjects}}{df_{subjects}} = \frac{4.221}{2} = 2.111$$

$$MS_{within} = \frac{SS_{within}}{df_{within}} = \frac{9.109}{4} = 2.277$$

$$F_{between} = \frac{MS_{between}}{MS_{within}} = \frac{5.778}{2.277} = 2.538$$

$$F_{subjects} = \frac{MS_{subjects}}{MS_{within}} = \frac{2.111}{2.277} = 0.927$$

SOURCE	SS	df	MS	F
Between-groups	11.556	2	5.778	2.54
Subjects	4.221	2	2.111	0.93
Within-groups	9.109	4	2.277	
Total	24.886	8		

11-22 **a.** Null hypothesis: People rate the driving experience of these three cars the same, on average—$H_0: \mu_1 = \mu_2 = \mu_3$. Research hypothesis: People do not rate the driving experience of these three cars the same, on average.

b. Order effects are addressed by counterbalancing. We could create a list of random orders of the three cars to be driven. Then, as a new customer arrives, we would assign him or her the next random order on the list. With a large enough sample size (much larger than the three participants we used in this example), we could feel confident that this assumption would be met with this approach.

c. The critical value for the F statistic for a p level of 0.05 and 2 and 4 degrees of freedom is 6.95. The between-groups F statistic of 2.54 does not exceed this critical value. We cannot reject the null hypothesis, so we cannot conclude that there are differences among mean ratings of cars.

11-23 In both cases, the numerator in the ratio is $SS_{between}$, but the denominators differ in the two cases. For the between-groups ANOVA, the denominator of the R^2 calculation is SS_{total}. For the within-groups ANOVA, the denominator of the R^2 calculation is $SS_{total} - SS_{subjects}$, which takes into account the fact that we are subtracting out variability due to subjects from the measure of error.

11-24 There are no differences in the way that the Tukey HSD is calculated. The formula for the calculation of the Tukey HSD is exactly the same for both the between-groups ANOVA and the within-groups ANOVA.

11-25 **a.** First, we calculate s_M:

$$s_M = \sqrt{\frac{MS_{within}}{N}} = \sqrt{\frac{771.256}{6}} = 11.338$$

Next, we calculate HSD for each pair of means.

For time 1 versus time 2:

$$HSD = \frac{(155.833 - 206.833)}{11.338} = -4.498$$

For time 1 versus time 3:

$$HSD = \frac{(155.833 - 251.667)}{11.338} = 8.452$$

For time 2 versus time 3:

$$HSD = \frac{(206.833 - 251.667)}{11.338} = -3.954$$

b. We have an independent variable with three levels and $df_{within} = 10$, so the q cutoff value at a p level of 0.05 is 3.88. Because we are performing a two-tailed test, the cutoff values are 3.88 and −3.88.

c. We reject the null hypothesis for all three of the mean comparisons because all of the HSD calculations exceed the critical value of −3.88. This tells us that all three of the group means are statistically significantly different from one another.

11-26 $R^2 = \frac{SS_{between}}{(SS_{total} - SS_{subjects})} = \frac{27{,}590.486}{(52{,}115.111 - 16{,}812.189)} = 0.78$

11-27 **a.** $R^2 = \frac{SS_{between}}{(SS_{total} - SS_{subjects})} = \frac{11.556}{(24.886 - 4.221)} = 0.56$. This is a large effect size.

b. Because the F statistic did not exceed the critical value, we failed to reject the null hypothesis. As a result, Tukey HSD tests are not necessary.

CHAPTER 12

12-1 A factorial ANOVA is a statistical analysis used with one scale dependent variable and at least two nominal (or sometimes ordinal) independent variables (also called factors).

12-2 A statistical interaction occurs in a factorial design when the two independent variables have an effect in combination that we do not see when we examine each independent variable on its own.

12-3 **a.** There are two factors: diet programs and exercise programs.

b. There are three factors: diet programs, exercise programs, and metabolism type.

c. There is one factor: gift certificate value.

d. There are two factors: gift certificate value and store quality.

12-4 **a.** The participants are the stocks themselves.

b. One independent variable is the type of ticker-code name, with two levels: pronounceable and unpronounceable. The second independent variable is time lapsed since the stock was initially offered, with four levels: 1 day, 1 week, 6 months, and 1 year.

c. The dependent variable is the selling price of the stocks.

d. This would be a two-way mixed-design ANOVA.

e. This would be a 2 × 4 mixed-design ANOVA.

f. This study would have eight cells: 2 × 4 = 8. We multiplied the numbers of levels of each of the two independent variables.

12-5 A quantitative interaction is an interaction in which one independent variable exhibits a strengthening or weakening of its effect at one or more levels of the other independent variable, but the direction of the initial effect does not change. More specifically, the effect of one independent variable is modified in the presence of another independent variable. A qualitative interaction is a particular type of quantitative interaction of two (or more) independent variables in which one independent variable reverses its effect depending on the level of the other independent variable. In a qualitative interaction, the effect of one variable doesn't just become stronger or weaker; it actually reverses direction in the presence of another variable.

12-6 An interaction indicates that the effect of one independent variable depends on the level of the other independent variable(s). The main effect alone cannot be interpreted because the effect of that one variable depends on another.

12-7 **a.** There are four cells.

		IV 2	
		LEVEL A	LEVEL B
IV 1	LEVEL A		
	LEVEL B		

b.

		IV 2	
		LEVEL A	LEVEL B
IV 1	LEVEL A	$M = (2 + 1 + 1 + 3)/4$ $= 1.75$	$M = (2 + 3 + 3 + 3)/4$ $= 2.75$
	LEVEL B	$M = (5 + 4 + 3 + 4)/4$ $= 4$	$M = (3 + 2 + 2 + 3)/4$ $= 2.5$

c. Because the sample size is the same for each cell, we can compute marginal means as simply the average between cell means.

		IV 2		MARGINAL
		LEVEL A	LEVEL B	MEANS
IV 1	LEVEL A	1.75	2.75	2.25
	LEVEL B	4	2.5	3.25
	Marginal Means	2.875	2.625	

d.

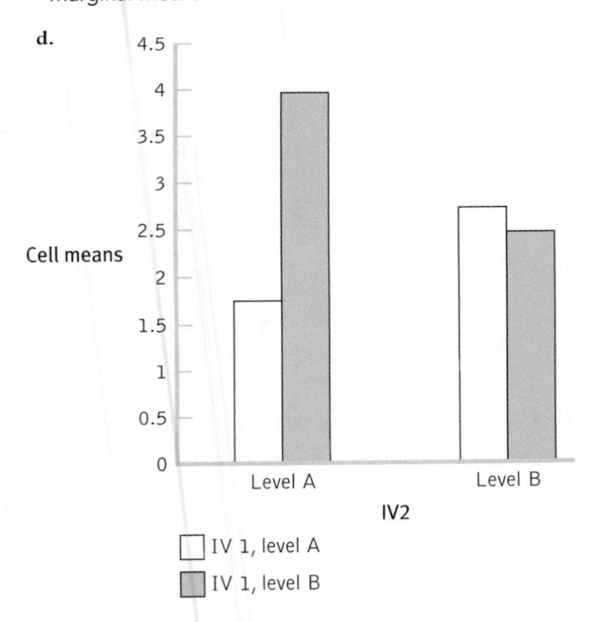

12-8 **a. i.** Independent variables: student (Caroline, Mira); class (philosophy, psychology)

ii. Dependent variable: performance in class

iii.

	CAROLINE	MIRA
PHILOSOPHY CLASS		
PSYCHOLOGY CLASS		

iv. This describes a qualitative interaction because the direction of the effect reverses. Caroline does worse in philosophy class than in psychology class, whereas Mira does better.

b. i. Independent variables: game location (home, away); team (own conference, other conference)

ii. Dependent variable: number of runs

iii.

	HOME	AWAY
OWN CONFERENCE		
OTHER CONFERENCE		

iv. This describes a qualitative interaction because the direction of the effect reverses. The team does worse at home against teams in the other conference but does well against those teams while away; the team does better at home against teams in its own conference but performs poorly against teams in its own conference when away.

c. i. Independent variables: amount of caffeine (caffeine, none); exercise (worked out, did not work out)

ii. Dependent variable: amount of sleep

iii.

	CAFFEINE	NO CAFFEINE
WORKING OUT		
NOT WORKING OUT		

iv. This describes a quantitative interaction because the effect of working out is particularly strong in the presence of caffeine versus no caffeine (and the presence of caffeine is particularly strong in the presence of working out versus not). The direction of the effect of either independent variable, however, does not change depending on the level of the other independent variable.

12-9 Because we have the possibility of two main effects and an interaction, each step from a one-way between-groups ANOVA is broken down into three parts for a two-way between-groups ANOVA; we have three sets of hypotheses, three comparison distributions, three critical F values, three F statistics, and three conclusions.

12-10 Variability is associated with the two main effects, the interaction, and the within-groups component.

12-11 $df_{IV\,1} = df_{rows} = N_{rows} - 1 = 2 - 1 = 1$

$df_{IV\,2} = df_{columns} = N_{columns} - 1 = 2 - 1 = 1$

$df_{interaction} = (df_{rows})(df_{columns}) = (1)(1) = 1$

$df_{within} = df_{1A,2A} + df_{1A,2B} + df_{1B,2A} + df_{1B,2B}$
$= 3 + 3 + 3 + 3 = 12$

$df_{total} = N_{total} - 1 = 16 - 1 = 15$

We can also check that this calculation is correct by adding all of the other degrees of freedom together: $1 + 1 + 1 + 12 = 15$.

12-12 The critical value for the main effect of the first independent variable, based on a between-groups degrees of freedom of 1 and a within-groups degrees of freedom of 12, is 4.75. The critical value for the main effect of the second independent variable, based on 1 and 12 degrees of freedom, is 4.75. The critical value for the interaction, based on 1 and 12 degrees of freedom, is 4.75.

12-13 **a.** Population 1 is students who received an initial grade of C who received e-mail messages aimed at bolstering self-esteem. Population 2 is students who received an initial grade of C who received e-mail messages aimed at bolstering their sense of control over their grades. Population 3 is students who received an initial grade of C who received e-mails that just included review questions. Population 4 is students who received an initial grade of D or F who received e-mail messages aimed at bolstering self-esteem. Population 5 is students who received an initial grade of D or F who received e-mail messages aimed at bolstering their sense of control over their grades. Population 6 is students who received an initial grade of D or F who received emails with just review questions.

b. *Step 2:* For the main effect of the first independent variable, initial grade, the null hypothesis is: The mean final exam grade of students with an initial grade of C is the same as that of students with an initial grade of D or F; $H_1: \mu_C = \mu_{D/F}$. The research hypothesis is: The mean final exam grade of students with an initial grade of C is not the same as that of students with an initial grade of D or F; $H_0: \mu_C \neq \mu_{D/F}$.

For the main effect of the second independent variable, type of email, the null hypothesis is: On average, the mean exam grades among those receiving different types of e-mails are the same—$H_0: \mu_{SE} = \mu_{CG} = \mu_{TR}$. The research hypothesis is: On average, the mean exam grades among those receiving different types of e-mails are not the same.

The interaction is: Initial grade × type of e-mail. The null hypothesis is: The effect of type of e-mail is not dependent on the levels of initial grade. The research hypothesis is: The effect of type of e-mail depends on the levels of initial grade.

c. *Step 3:* $df_{between/grade} + N_{groups} - 1 = 2 - 1 = 1$

$df_{between/e\text{-}mail} = N_{groups} - 1 = 3 - 1 = 2$

$df_{interaction} = (df_{between/grade})(df_{between/e\text{-}mail}) = (1)(2) = 2$

$df_{C,SE} = N - 1 = 14 - 1 = 13$

$df_{C,C} = N - 1 = 14 - 1 = 13$

$df_{C,TR} = N - 1 = 14 - 1 = 13$

$df_{D/F,SE} = N - 1 = 14 - 1 = 13$

$df_{D/F,C} = N - 1 = 14 - 1 = 13$

$df_{D/F,TR} = N - 1 = 14 - 1 = 13$

$df_{within} = df_{C,SE} + df_{C,C} + df_{C,TR} + df_{D/F,SE} + df_{D/F,C} + df_{D/F,TR} = 13 + 13 + 13 + 13 + 13 + 13 = 78$

Main effect of initial grade: F distribution with 1 and 78 degrees of freedom

Main effect of type of e-mail: F distribution with 2 and 78 degrees of freedom

Main effect of interaction of initial grade and type of e-mail: F distribution with 2 and 78 degrees of freedom

d. *Step 4:* Note that when the specific degrees of freedom is not in the F table, you should choose the more conservative—that is, larger—cutoff. In this case, use the cutoffs for a within-groups degrees of freedom of 75 rather than 80. The three cutoffs are:

Main effect of initial grade: 3.97

Main effect of type of e-mail: 3.12

Interaction of initial grade and type of e-mail: 3.12

e. *Step 6:* There is a significant main effect of initial grade because the F statistic, 20.84, is larger than the critical value of 3.97. The marginal means, seen in the accompanying table, tell us that students who earned a C on the initial exam have higher scores on the final exam, on average, than do students who earned a D or an F on the initial exam. There is no statistically significant main effect of type of e-mail, however. The F statistic of 1.69 is not larger than the critical value of 3.12. Had this main effect been significant, we would have conducted a post hoc test to determine where the differences were. There also is not a significant interaction. The F statistic of 3.02 is not larger than the critical value of 3.12. (Had we used a cutoff based on a p level of 0.10, we would have rejected the null hypothesis for the interaction. The cutoff for a p level of 0.10 is 2.77.) If we had rejected the null hypothesis for the interaction, we would have examined the cell means in tabular and graph form.

	SELF-ESTEEM	TAKE RESPONSIBILITY	CONTROL GROUP	MARGINAL MEANS
C	67.31	69.83	71.12	69.42
D/F	47.83	60.98	62.13	56.98
Marginal means	57.57	65.41	66.63	

CHAPTER 13

13-1 (1) The correlation coefficient can be either positive or negative. (2) The correlation coefficient always falls between −1.00 and 1.00. (3) It is the strength, also called the *magnitude*, of the coefficient, not its sign, that indicates how large it is.

13-2 When two variables are correlated, there can be multiple explanations for that association. The first variable can cause the second variable; the second variable can cause the first variable; or a third variable can cause both the first and second variables. In fact, there may be more than one "third" variable causing both the first and second variables.

13-3
a. According to Cohen, this is a large (strong) correlation. Note that the sign (negative in this case) is not related to the assessment of strength.
b. This is just above a medium correlation.
c. This is lower than the guideline for a small correlation, 0.10.

13-4 Students will draw a variety of different scatterplots. The important thing to note is the closeness of data points to an imaginary line drawn through the data.

a. A scatterplot for a correlation coefficient of −0.60 might look like this:

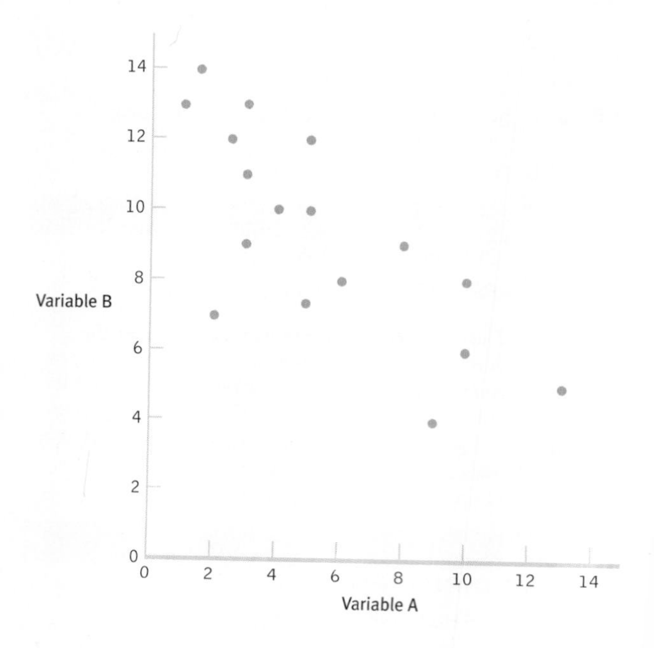

b. A scatterplot for a correlation coefficient of 0.35 might look like this:

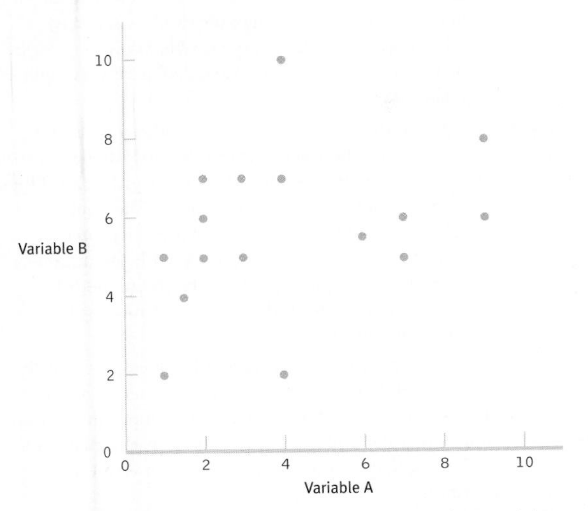

c. A scatterplot for a correlation coefficient of 0.04 might look like this:

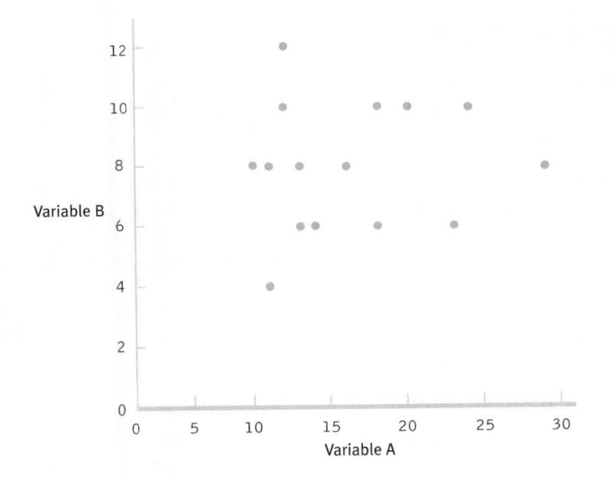

13-5 It is possible that training while listening to music (A) causes an increase in a country's average finishing time (B), perhaps because music decreases one's focus on running. It is also possible that high average finishing times (B) cause an increase in the percentage of marathon runners in a country who train while listening to music (A), perhaps because slow runners tend to get bored and need music to get through their runs. It also is possible that a third variable, such as a country's level of wealth (C), causes a higher percentage of runners who train while listening to music (because of the higher presence of technology in wealthy countries) (A) and also causes higher (slower) finishing times (perhaps because long-distance running is a less popular sport in wealthy countries with access to so many sport and entertainment options) (B). Without a true experiment, we cannot know the direction of causality.

13-6 The Pearson correlation coefficient is a statistic that quantifies a linear relation between two scale variables. Specifically, it describes the direction and strength of the relation between the variables.

13-7 The two issues are variability and sample size.

13-8

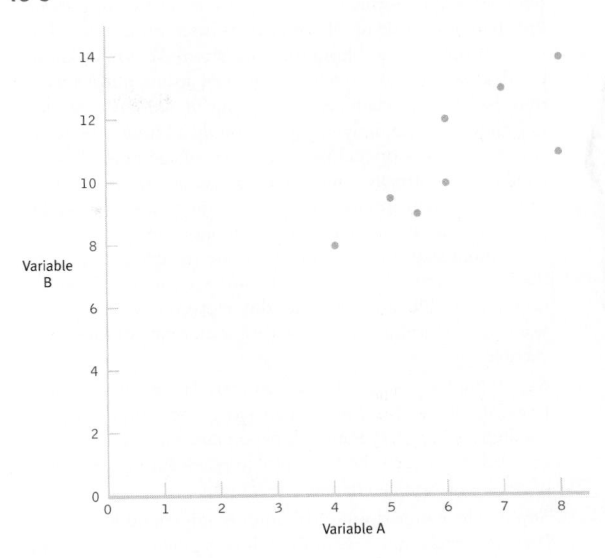

13-9

VARIABLE A (X)	$(X - M_X)$	VARIABLE B (Y)	$(Y - M_Y)$	$(X - M_X)(Y - M_Y)$
8	1.812	14	3.187	5.775
7	0.812	13	2.187	1.776
6	−0.188	10	−0.813	0.153
5	−1.188	9.5	−1.313	1.560
4	−2.188	8	−2.813	6.155
5.5	−0.688	9	−1.813	1.247
6	−0.188	12	1.187	−0.223
8	1.812	11	0.187	0.339
$M_X =$ 6.118		$M_Y =$ 10.813		$\Sigma[(X - M_X)(Y - M_Y)] = 16.780$

VARIABLE A (X)	$(X - M_X)$	$(X - M_X)^2$	VARIABLE B (Y)	$(Y - M_Y)$	$(Y - M_Y)^2$
8	1.812	3.283	14	3.187	10.157
7	0.812	0.659	13	2.187	4.783
6	−0.188	0.035	10	−0.813	0.661
5	−1.188	1.411	9.5	−1.313	1.724
4	−2.188	4.787	8	−2.813	7.913
5.5	−0.688	0.473	9	−1.813	3.287
6	−0.188	0.035	12	1.187	1.409
8	1.812	3.283	11	0.187	0.035
	$\Sigma(X - M_X)^2 = 13.966$			$\Sigma(Y - M_Y)^2 = 29.969$	

$$\sqrt{(SS_X)(SS_Y)} = \sqrt{(13.966)(29.969)} = \sqrt{418.547} = 20.458$$

$$r = \frac{\Sigma[(X - M_X)(Y - M_Y)]}{\sqrt{(SS_X)(SS_Y)}} = \frac{16.780}{20.458} = 0.82$$

13-10 *Step 1:* Population 1: Children like those we studied. Population 2: Children for whom there is no relation between observed and performed acts of aggression. The comparison distribution is made up of correlations based on samples of this size, 8 people, selected from the population. We do not know if the data were randomly selected, the first assumption, so we must be cautious when generalizing our findings. We also do not know if the underlying population distributions for witnessed aggression and performed acts of aggression by children are normally distributed. The sample size is too small to make any conclusions about this assumption, so we should proceed with caution. The third assumption, unique to correlations, is that the variability of one variable is equal across the levels of the other variable. Because we have such a small data set, it is difficult to evaluate this. However, we can see from the scatterplot that the data are somewhat consistently variable.

Step 2: Null hypothesis: There is no correlation between the levels of witnessed and performed acts of aggression among children—$H_0: \rho = 0$. Research hypothesis: There is a correlation between the levels of witnessed and performed acts of aggression among children—$H_1: \rho \neq 0$.

Step 3: The comparison distribution is a distribution of Pearson correlations, *r*, with the following degrees of freedom: $df_r = N - 2 = 8 - 2 = 6$.

Step 4: The critical values for an *r* distribution with 6 degrees of freedom for a two-tailed test with a *p* level of 0.05 are −0.707 and 0.707.

Step 6: The test statistic, $r = 0.82$, is larger in magnitude than the critical value of 0.707. We can reject the null hypothesis and conclude that a strong positive correlation exists between the number of witnessed acts of aggression and the number of acts of aggression performed by children.

13-11 Psychometricians calculate correlations when assessing the reliability and validity of measures and tests.

13-12 Coefficient alpha is a measure of reliability. To calculate coefficient alpha, we take, in essence, the average of all split-half correlations. That is, the items on a test are split in half and the correlation of those two halves is calculated. This process is done repeatedly for all possible "halves" of the test, and then the average of those correlations is obtained.

13-13 **a.** The test does not have sufficient reliability to be used as a diagnostic tool. As stated in the chapter, when using tests to diagnose or make statements about the ability of individuals, a reliability of at least 0.90 is necessary.

b. We do not have enough information to determine whether the test is valid. The coefficient alpha tells us only about the reliability of the test.

c. To assess the validity of the test, we would need the correlation between this measure and other measures of reading ability or between this measure and students' grades in the nonremedial class (i.e., do students who perform very poorly in the nonremedial reading class also perform poorly on this measure?).

13-14 **a.** The psychometrician could assess test–retest reliability by administering the quiz to 100 heterosexual female readers and then one week later readministering the test to the same 100 female readers. If their scores at the two times are highly correlated, the test would have high test–retest reliability. She also could calculate a coefficient alpha using computer software. The computer would essentially calculate correlations for every possible two groups of five items and then calculate the average of all of these split-half correlations.

b. The psychometrician could assess validity by choosing criteria that she believed assessed the underlying construct of interest, a boyfriend's devotion to his girlfriend. There are many possible criteria. For example, she could correlate the amount of money each participant's boyfriend spent on her last birthday with the number of minutes the participant spent on the phone with her boyfriend today or with the number of months the relationship ends up lasting.

c. Of course, we assume that these other measures actually assess the underlying construct of a boyfriend's devotion, which may or may not be true! For example, the amount of money that the boyfriend spent on the participant's last birthday might be a measure of his income, not his devotion.

CHAPTER 14

14-1 Simple linear regression is a statistical tool that lets statisticians predict the score on a scale dependent variable from the score on a scale independent variable.

14-2 The regression line allows us to make predictions about one variable based on what we know about another variable. It gives us a visual representation of what we believe is the underlying relation between the variables, based on the data we have available to us.

14-3 **a.** $z_X = \frac{X - M_X}{SD_X} = \frac{67 - 64}{2} = 1.5$

b. $z_{\hat{Y}} = (r_{XY})(z_X) = (0.28)(1.5) = 0.42$

c. $\hat{Y} = z_{\hat{Y}}(SD_Y) + M_Y = (0.42)(15) + 155$
$= 161.3$ pounds

14-4 **a.** $\hat{Y} = 12 + 0.67(X) = 12 + 0.67(78) = 64.26$

b. $\hat{Y} = 12 + 0.67(-14) = 2.62$

c. $\hat{Y} = 12 + 0.67(52) = 46.84$

14-5 **a.** The *y* intercept, 2.586, is the GPA we might expect if someone played no minutes. The slope of 0.016 is the increase in GPA that we would expect for each 1-minute increase in playing time. Because the correlation is positive, it makes sense that the slope is also positive.

b. The standardized regression coefficient is equal to the correlation coefficient for simple linear regression, 0.344. We can also check that this is correct by computing β:

X	(X − M_X)	(X − M_X)²	Y	(Y − M_Y)	(Y − M_Y)²
29.70	13.159	173.159	3.20	0.343	0.118
32.14	15.599	243.329	2.88	0.023	0.001
32.72	16.179	261.760	2.78	−0.077	0.006
21.76	5.219	27.238	3.18	0.323	0.104
18.56	2.019	4.076	3.46	0.603	0.364
16.23	−0.311	0.097	2.12	−0.737	0.543
11.80	−4.741	22.477	2.36	−0.497	0.247
6.88	−9.661	93.335	2.89	0.033	0.001
6.38	−10.161	103.246	2.24	−0.617	0.381
15.83	−0.711	0.506	3.35	0.493	0.243
2.50	−14.041	197.150	3.00	0.143	0.020
4.17	−12.371	153.042	2.18	−0.677	0.458
16.36	−0.181	0.033	3.50	0.643	0.413
		Σ(X − M_X)² = 1279.448			Σ(Y − M_Y)² = 2.899

$$\beta = (b)\frac{\sqrt{SS_X}}{\sqrt{SS_Y}} = (0.016)\frac{\sqrt{1279.448}}{\sqrt{2.899}} = 0.336$$

There is some difference due to rounding decisions, but in both cases, these numbers would be expressed as 0.34.

c. Strong correlations indicate strong linear relations between variables. Because of this, when we have a strong correlation, we have a more useful regression line, resulting in more accurate predictions.

d. According to the hypothesis test for the correlation, this r value, 0.34, fails to reach statistical significance. The critical values for r with 11 degrees of freedom at a p level of 0.05 are −0.553 and 0.553. In this chapter, we learned that the outcome of hypothesis testing is the same for simple linear regression as it is for correlation, so we know we also do not have a statistically significant regression.

14-6 The standard error of the estimate is a statistic that indicates the typical distance between a regression line and the actual data points. When we do not have enough information to compute a regression equation, we often use the mean as our "best guess." The error of prediction when the mean is used is typically greater than the standard error of the estimate.

14-7 Strong correlations mean highly accurate predictions with regression. This translates into a large proportionate reduction in error.

14-8 **a.**

X	Y	(M_Y) MEAN FOR Y	(Y − M_Y) ERROR	SQUARED ERROR
5	6	5.333	0.667	0.445
6	5	5.333	−0.333	0.111
4	6	5.333	0.667	0.445
5	6	5.333	0.667	0.445
7	4	5.333	−1.333	1.777
8	5	5.333	−0.333	0.111

$SS_{total} = \Sigma(Y - M_Y)^2 = 3.334$

b.

X	Y	REGRESSION EQUATION	$\hat{Y}$	ERROR ($Y - \hat{Y}$)	SQUARED ERROR
5	6	$\hat{Y}$ = 7.846 − 0.431 (5) =	5.691	0.309	0.095
6	5	$\hat{Y}$ = 7.846 − 0.431 (6) =	5.260	−0.260	0.068
4	6	$\hat{Y}$ = 7.846 − 0.431 (4) =	6.122	−0.122	0.015
5	6	$\hat{Y}$ = 7.846 − 0.431 (5) =	5.691	0.309	0.095
7	4	$\hat{Y}$ = 7.846 − 0.431 (7) =	4.829	−0.829	0.687
8	5	$\hat{Y}$ = 7.846 − 0.431 (8) =	4.398	0.602	0.362

$SS_{error} = \Sigma(Y - \hat{Y})^2 = 1.322$

c. We have reduced error from 3.334 to 1.322, which is a reduction of 2.012. Now we calculate this reduction as a proportion of the total error:

$$\frac{2.012}{3.334} = 0.603$$

This can also be written as:

$$r^2 = \frac{(SS_{total} - SS_{error})}{SS_{total}} = \frac{(3.334 - 1.322)}{3.334} = 0.603$$

We have reduced 0.603, or 60.3%, of error using the regression equation as an improvement over the use of the mean as our predictor.

d. $r^2 = (-0.77)(-0.77) = 0.593$, which closely matches our calculation of r^2 above, 0.603. These numbers are slightly different due to rounding decisions.

14-9 Tell Coach Parcells that prediction suffers from the same limitations as correlation. First, just because two variables are associated doesn't mean one causes the other. This is not a true experiment, and if we didn't randomly assign athletes to appear on a *Sports Illustrated* cover or not, then we cannot determine if a cover appearance causes sporting failure. Moreover, we have a limited range; by definition, those lauded on the cover are the best in sports. Would the association be different among those with a wider range of athletic ability? Finally, and most important, there is the very strong possibility of regression to the mean. Those chosen for a cover appearance are at the very top of their game. There is nowhere to go but down, so it is not surprising that those who merit a cover appearance would soon thereafter experience a decline. There's likely no need to avoid that cover, Coach.

14-10 Multiple regression is a statistical tool that predicts a dependent variable by using two or more independent variables as predictors. It is an improvement over simple linear regression, which only allows one independent variable to inform predictions.

14-11 $\hat{Y} = 5.251 + 0.06(X_1) + 1.105(X_2)$

14-12 **a.** $\hat{Y} = 5.251 + 0.06(40) + 1.105(14) = 23.12$

b. $\hat{Y} = 5.251 + 0.06(101) + 1.105(39) = 54.41$

c. $\hat{Y} = 5.251 + 0.06(76) + 1.105(20) = 31.91$

14-13 **a.** $\hat{Y} = 2.695 + 0.069(X_1) + 0.015(X_2) - 0.072(X_3)$

b. $\hat{Y} = 2.695 + 0.069(6) + 0.015(20) - 0.072(4) = 3.12$

c. The negative sign in the slope (−0.072) tells us that those with higher levels of admiration for Pamela Anderson tend to have lower GPAs, and those with lower levels of admiration for Pamela Anderson tend to have higher GPAs.

CHAPTER 15

15-1 A nonparametric test is a statistical analysis that is not based on a set of assumptions about the population, whereas parametric tests are based on assumptions about the population.

15-2 We use nonparametric tests when the data violate the assumptions about the population that parametric tests make. The three most common situations that call for nonparametric tests are: (1) having a nominal dependent variable, (2) having an ordinal dependent variable, and (3) having a small sample size with possible skew.

15-3 a. The independent variable is city, a nominal variable. The dependent variable is whether a woman is pretty or not so pretty, an ordinal variable.

b. The independent variable is city, a nominal variable. The dependent variable is beauty, assessed on a scale of 1–10. This is a scale variable.

c. The independent variable is intelligence, likely a scale variable. The dependent variable is beauty, assessed on a scale of 1–10. This is a scale variable.

d. The independent variable is ranking on intelligence, an ordinal variable. The dependent variable is ranking on beauty, also an ordinal variable.

15-4 a. We'd choose a hypothesis test from category III. We'd use a nonparametric test because the dependent variable is not scale and would not meet the primary assumption of a normally distributed dependent variable, even with a large sample.

b. We'd choose a test from category II because the independent variable is nominal and the dependent variable is scale. (In fact, we'd use a one-way between-groups ANOVA because there is only one independent variable and it has more than two levels.)

c. We'd choose a hypothesis test from category I because we have a scale independent variable and a scale dependent variable. (If we were assessing the relation between these variables, we'd use the Pearson correlation coefficient. If we wondered whether intelligence predicted beauty, we'd use simple linear regression.)

d. We'd choose a hypothesis from category III because both the independent and dependent variables are ordinal. We would not meet the assumption of having a normal distribution of the dependent variable, even if we had a large sample.

15-5 We use chi-square tests when all variables are nominal.

15-6 Observed frequencies indicate how often something happens in a given category with the data we collected. Expected frequencies indicate how often something happens in a given category based on what we know about the population or based on the null hypothesis.

15-7 The measure of effect size for chi square is Cramer's *V*. It is calculated by first multiplying the total *N* by the *df* for either the rows or columns (whichever is smaller) and then dividing the calculated chi-square value by this number. Finally, we take the square root—and that is Cramer's *V*.

15-8 a. $df_{\chi^2} = k - 1 = 2 - 1 = 1$

b. Observed:

CLEAR BLUE SKIES	UNCLEAR SKIES
59 days	19 days

Expected:

CLEAR BLUE SKIES	UNCLEAR SKIES
(78)(0.80) = 62.4	(78)(0.20) = 15.6 days

c.

CATEGORY	OBSERVED (O)	EXPECTED (E)	$O - E$	$(O - E)^2$	$\frac{(O - E)^2}{E}$
Clear blue skies	59	62.4	−3.4	11.56	0.185
Unclear skies	19	15.6	3.4	11.56	0.741

$$\chi^2 = \Sigma\left[\frac{(O-E)^2}{E}\right] = 0.185 + 0.741 = 0.93$$

15-9 To calculate the relative likelihood, we first need to calculate two conditional probabilities: the conditional probability of being a Republican given that a person is a business major, which is $\frac{54}{92} = 0.587$, and the conditional probability of being a Republican given that a person is a psychology major, which is $\frac{36}{67} = 0.537$. Now we divide the conditional probability of being a Republican given that a person is a business major by the conditional probability of being a Republican given that a person is a psychology major: $\frac{0.587}{0.537} = 1.09$. The relative likelihood of being a Republican given that a person is a business major as opposed to a psychology major is 1.09.

15-10 a. The participants are the lineups. The independent variable is type of lineup (simultaneous, sequential), and the dependent variable is outcome of the lineup (suspect identification, other identification, no identification).

b. *Step 1:* Population 1 is police lineups like those we observed. Population 2 is police lineups for which type of lineup and outcome are independent. The comparison distribution is a chi-square distribution. The hypothesis test is a chi-square test for independence because we have two nominal variables. This study meets three of the four assumptions. The two variables are nominal; every participant (lineup) is in only one cell; and there are more than five times as many participants as cells (548 participants and 6 cells).

Step 2: Null hypothesis: Lineup outcome is independent of type of lineup. Research hypothesis: Lineup outcome depends on type of lineup.

Step 3: The comparison distribution is a chi-square distribution with 2 degrees of freedom:

$$df_{\chi^2} = (k_{row} - 1)(k_{column} - 1) = (2 - 1)(3 - 1) = (1)(2) = 2$$

Step 4: The cutoff chi-square statistic, based on a *p* level of 0.05 and 2 degrees of freedom, is 5.992. (*Note:* It is helpful to include a drawing of the chi-square distribution with the cutoff.)

Step 5:

	OBSERVED SUSPECT ID	OTHER ID	NO ID	
SIMULTANEOUS	191	8	120	319
SEQUENTIAL	102	20	107	229
	293	28	227	548

We can calculate the expected frequencies in one of two ways. First, we can think about it. Out of the total of 548 lineups, 293 led to identification of the suspect, an identification rate of 293/548 = 0.535, or 53.5%. If identification was independent of type of lineup, we would expect the same rate for each type of lineup. For example, for the 319 simultaneous lineups, we would expect: (0.535)(319) = 170.665. For the 229 sequential lineups, we would expect: (0.535)(229) = 122.515. Or we can use the formula. For these same two cells (the column labeled "Suspect ID"), we calculate:

$$\frac{Total_{column}}{N}(Total_{row}) = \frac{293}{548}(319) = (0.535)(319) = 170.665$$

$$\frac{Total_{column}}{N}(Total_{row}) = \frac{293}{548}(229) = (0.535)(229) = 122.515$$

For the column labeled "Other ID":

$$\frac{Total_{column}}{N}(Total_{row}) = \frac{28}{548}(319) = (0.051)(319) = 16.269$$

$$\frac{Total_{column}}{N}(Total_{row}) = \frac{28}{548}(229) = (0.051)(229) = 11.679$$

For the column labeled "No ID":

$$\frac{Total_{column}}{N}(Total_{row}) = \frac{227}{548}(319) = (0.414)(319) = 132.066$$

$$\frac{Total_{column}}{N}(Total_{row}) = \frac{227}{548}(229) = (0.414)(229) = 94.806$$

	EXPECTED SUSPECT ID	OTHER ID	NO ID	
SIMULTANEOUS	170.665	16.269	132.066	319
SEQUENTIAL	122.515	11.679	94.806	229
	293	28	227	548

CATEGORY	OBSERVED (O)	EXPECTED (E)	O − E	(O − E)²	(O − E)²/E
Sim; suspect	191	170.665	20.335	413.512	2.423
Sim; other	8	16.269	−8.269	68.376	4.203
Sim; no	120	132.066	−12.066	145.588	1.102
Seq; suspect	102	122.515	−20.515	420.865	3.435
Seq; other	20	11.679	8.321	69.239	5.929
Seq; no	107	94.806	12.194	148.694	1.568

$$\chi^2 = \Sigma\left(\frac{(O-E)^2}{E}\right) = (2.423 + 4.203 + 1.102 + 3.435 + 5.929 + 1.568) = 18.660$$

Step 6: Reject the null hypothesis. It appears that the outcome of a lineup depends on the type of lineup. In general, simultaneous lineups tend to lead to a higher rate than expected of suspect identification, lower rates than expected of identification of other members of the lineup, and lower rates than expected of no identification at all. (*Note:* It is helpful to add the test statistic to the drawing that included the cutoff).

c. $\chi^2(1, N = 548) = 18.66, p < 0.05$

d. The findings of this study were opposite to what had been expected by the investigators; the report of results noted that, prior to this study, police departments believed that the sequential lineup led to more accurate identification of suspects. This situation occurs frequently in behavioral research, a reminder of the importance of conducting two-tailed hypothesis tests. (Of course, the fact that this study produced different results doesn't end the debate. Future researchers should explore why there are different findings in different contexts in an effort to target the best lineup procedures based on specific situations.)

e. Cramer's $V = \sqrt{\frac{\chi^2}{(N)(df_{row/column})}} = \sqrt{\frac{18.660}{(548)(1)}} = \sqrt{0.034} =$ 0.184. This is a small-to-medium effect.

f. To create a graph, we must first calculate the conditional proportions by dividing the observed frequency in each cell by the row total. These conditional proportions appear in the table below and are graphed in the figure.

	ID SUSPECT	ID OTHER	NO ID
SIMULTANEOUS	0.599	0.025	0.376
SEQUENTIAL	0.445	0.087	0.467

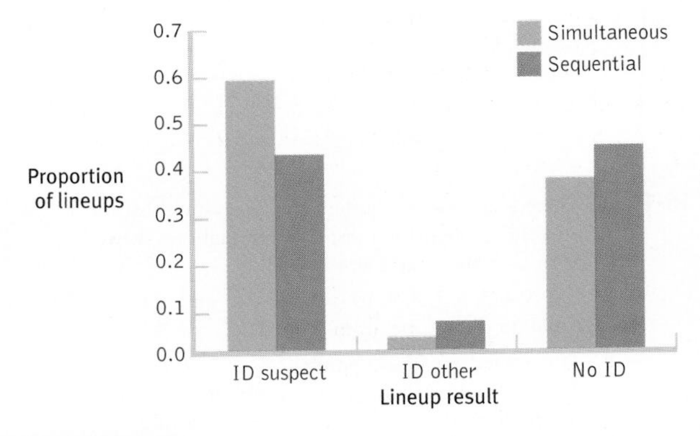

g. We must first calculate two conditional probabilities: the conditional probability of obtaining a suspect identification in the simultaneous lineups, which is $\frac{191}{319} = 0.599$, and the conditional probability of obtaining a suspect identification in the sequential lineups, which is $\frac{102}{229} = 0.445$. We then divide 0.599 by 0.445 to obtain the relative likelihood of 1.35. Suspects are 1.35 times more likely to be identified in simultaneous as opposed to sequential lineups.

15-11 We use such tests when we have an ordinal dependent variable.

15-12 Nonparametric tests are performed on ordinal data, so any data that are scale must be converted to ordinal before we compute the nonparametric test.

15-13

OBSERVATION	VARIABLE 1 SCORE	VARIABLE 1 RANK	VARIABLE 2 SCORE	VARIABLE 2 RANK
1	1.30	3	54.39	5
2	1.80	4.5	50.11	3
3	1.20	2	53.39	4
4	1.06	1	44.89	1
5	1.80	4.5	48.50	2

15-14

OBSERVATION	VARIABLE 1 RANK	VARIABLE 2 RANK	DIFFERENCE	SQUARED DIFFERENCE
1	3	5	−2	4
2	4.5	3	1.5	2.25
3	2	4	−2	4
4	1	1	0	0
5	4.5	2	2.5	6.25

15-15 **a.** There is an extreme outlier, 139, suggesting that the underlying population distribution might be skewed. Moreover, the sample size is small.

b. 1, 2, 3, 4, 5, 6, 7, 8, 9, 10 (we chose to rank this way, but you could do the reverse, from 10 to 1).

c. The outlier was 25 IQ points (139 − 114 = 25) behind the next-highest score of 114. It now is ranked 10, compared to the next-highest score's rank of 9.

15-16 *Step 1:* We convert the data from scale to ordinal. The researchers did not indicate whether they used random selection to choose the countries in the sample, so we must be cautious when generalizing from these results. There are some ties, but we will assume that there are not so many as to render the results of the test invalid.

Step 2: Null hypothesis: Countries in which English is a primary language and countries in which English is not a primary language do not tend to differ in accomplishment-related national pride. Research hypothesis: Countries in which English is a primary language and countries in which English is not a primary language tend to differ in accomplishment-related national pride.

Step 3: There are seven countries in the English-speaking group and seven countries in the non-English-speaking group.

Step 4: The cutoff, or critical value, for a Mann–Whitney U test with two groups of seven participants (countries), a p level of 0.05, and a two-tailed test is 8.

Step 5: (*Note:* E stands for English-speaking, and NE stands for non-English-speaking.)

COUNTRY	PRIDE SCORE	PRIDE RANK	ENGLISH LANGUAGE	OTHER RANKS	RANKS
United States	4.00	1	E	1	
Australia	2.90	2.5	E	2.5	
Ireland	2.90	2.5	E	2.5	
South Africa	2.70	4	E	4	
New Zealand	2.60	5	E	5	
Canada	2.40	6	E	6	
Chile	2.30	7	NE		7
Great Britain	2.20	8	E	8	
Japan	1.80	9	NE		9
France	1.50	10	NE		10
Czech Republic	1.30	11.5	NE		11.5
Norway	1.30	11.5	NE		11.5
Slovenia	1.10	13	NE		13
South Korea	1.00	14	NE		14

$$\Sigma R_E = 1 + 2.5 + 2.5 + 4 + 5 + 6 + 8 = 29$$

$$\Sigma R_{NE} = 7 + 9 + 10 + 11.5 + 11.5 + 13 + 14 = 76$$

$$U_E = (n_E)(n_{NE}) + \frac{n_E(n_E + 1)}{2} - \Sigma R_E$$

$$= (7)(7) + \frac{7(7+1)}{2} - 29 = 49 + 28 - 29 = 48$$

$$U_{NE} = (n_E)(n_{NE}) + \frac{n_{NE}(n_{NE} + 1)}{2} - \Sigma R_{NE}$$

$$= (7)(7) + \frac{7(7+1)}{2} - 76 = 49 + 28 - 76 = 1$$

Step 6: The smaller test statistic, 1, is smaller than the critical value, 8. We can reject the null hypothesis; it appears that English-speaking countries tend to have higher accomplishment-related national pride than non-English-speaking countries.

Choosing the Appropriate Statistical Test

We have learned four categories of statistical tests:

1. Statistical tests in which all variables are scale
2. Statistical tests in which the independent variable (or variables) is nominal, but the dependent variable is scale
3. Statistical tests in which all variables are nominal
4. Statistical tests in which any variable is ordinal

Category 1: Two Scale Variables

If we have a research design with only scale variables, we have two choices about how to analyze the data. The only question we have to ask ourselves is whether the research question pertains to an association (or relation) between two variables or to the degree to which one variable predicts the other. If the research question is about association, then we choose the Pearson correlation coefficient. If it is about prediction, then we choose regression. The decisions for category 1 are represented in Table E-1. (Note that the relation must be linear.)

TABLE E-1 Category 1 Statistics

When both the independent variable and the dependent variable are scale, we calculate either a Pearson correlation coefficient or a regression equation.

Research Question: Association (Relation)	Research Question: Prediction
Pearson correlation coefficient	Regression equation

Category 2: Nominal Independent Variable(s) and Scale Dependent Variable

If the research design includes one or more nominal independent variables and a scale dependent variable, then we have several choices. The next question pertains to the number of independent variables.

1. If there is *just one independent variable,* then we ask ourselves how many levels it has.
   a. If there are *two levels, but just one sample*—that is, one level is represented by the sample and one level by the population—then we use either a z test or a single-sample t test. It is unusual to know enough about a population that we only need to collect data from a single sample. If this is the case, however, and we know *both* the population mean and the population standard deviation, then we can use a z test. If this is the case and we know *only* the population mean (but not the population standard deviation), then we use the single-sample t test.
   b. If there are *two levels, each represented by a sample* (either a single sample in which everyone participates in both levels or two different samples, one for each level), then we use either a paired-samples t test (if all participants are in both levels of the independent variable) or an independent-samples t test (if participants are in only one level of the independent variable).
   c. If there are three or more levels, then we use a form of a one-way ANOVA. We examine the research design to determine if it is a between-groups ANOVA (participants in just one level of the independent variable) or a within-groups ANOVA (participants in all levels of the independent variable).
2. If there are *at least two independent variables,* we must use a form of ANOVA. Remember, we name ANOVAs according to the number of independent variables (one-way, two-way, three-way) and the research design (between-groups, within-groups).

The decisions about data that fall into category 2 and have one independent variable are summarized in Table E-2. For those with two or more independent variables, see Table 12-1.

Category 3: One or Two Nominal Variables

If we have a design with only nominal variables—that is, counts, not means, in the cells—then we have two choices; both nonparametric tests. The only question we have to ask ourselves is whether there are one or two nominal variables. If there is one nominal variable, then we choose the chi-square test for goodness of fit. If there are two nominal variables, then we choose the chi-square test for independence. The decision for category 3 is represented in Table E-3.

TABLE E-2 Category 2 Statistics

When there are one or more nominal independent variables and a scale dependent variable, we have several choices. Start by selecting the appropriate number of independent variables. For *one independent variable*, use the accompanying chart. To use this chart, look at the first two columns, those that identify the number of levels of the independent variable and the number of samples. For two levels but one sample, it's a choice between the *z* test and single-sample *t* test; for two levels and two samples, it's a choice between the paired-samples *t* test and the independent-samples *t* test. For three or more levels (and the matching number of samples), we use either a one-way within-groups ANOVA or a one-way between-groups ANOVA. For *two independent variables* or *three independent variables*, we'll use a form of ANOVA and refer to Table 12-1 on naming ANOVAs.

NUMBER OF LEVELS OF INDEPENDENT VARIABLE	NUMBER OF SAMPLES	INFORMATION ABOUT POPULATION	HYPOTHESIS TEST
Two	One (compared with the population)	Mean and standard deviation	*z* test
Two	One (compared with the population)	Mean only	Single-sample *t* test

NUMBER OF LEVELS OF INDEPENDENT VARIABLE	NUMBER OF SAMPLES	RESEARCH DESIGN	HYPOTHESIS TEST
Two	Two	Within-groups	Paired-samples *t* test
Two	Two	Between-groups	Independent-samples *t* test
Three (or more)	Three (or more)	Between-groups	One-way between-groups ANOVA
Three (or more)	Three (or more)	Within-groups	One-way within-groups ANOVA

TABLE E-3 Category 3 Statistics

When we have only nominal variables, then we choose one of the two chi-square tests.

ONE NOMINAL VARIABLE	TWO NOMINAL VARIABLES
Chi-square test for goodness of fit	Chi-square test for independence

Category 4: At Least One Ordinal Variable

If we have a design with even one ordinal variable or a design in which it makes sense to convert the data from scale to ordinal, then we have several choices, as seen in Table E-4. All these choices have parallel parametric hypothesis tests, as seen in Table E-5. For situations in which we want to investigate the relation between two ordinal variables, we use the Spearman rank-order correlation coefficient. For situations in which we have a within-groups research design and two groups, we use the Wilcoxon signed-rank test. When we have a between-groups design with two groups, we use a Mann–Whitney *U* test. And when we have a between-groups design with more than two groups, we use a Kruskal–Wallis *H* test.

The decisions we've outlined are summarized in Figure E-1.

TABLE E-4 Category 4 Statistics

When at least one variable is ordinal, we have several choices. If both variables are ordinal, or can be converted to ordinal, and we are interested in quantifying the relation between them, we use the Spearman rank-order correlation coefficient. If the independent variable is nominal and the dependent variable is ordinal, we choose the correct nonparametric test based on the research design and the number of levels of the independent variable.

TYPE OF INDEPENDENT VARIABLE (AND NUMBER OF LEVELS, IF APPLICABLE)	RESEARCH DESIGN	QUESTION TO BE ANSWERED	HYPOTHESIS TEST
Ordinal	Not applicable	Are two variables related?	Spearman rank-order correlation coefficient
Nominal (two levels)	Within-groups	Are two groups different?	Wilcoxon signed-rank test
Nominal (two levels)	Between-groups	Are two groups different?	Mann–Whitney *U* test
Nominal (three or more levels)	Between-groups	Are three or more groups different?	Kruskal–Wallis *H* test

TABLE E-5 Nonparametric Statistics and Their Parametric Counterparts

Every parametric hypothesis test has at least one nonparametric counterpart. If the data are far from meeting the assumptions for a parametric test or at least one variable is ordinal, we should use the appropriate nonparametric test instead of a parametric test.

DESIGN	PARAMETRIC TEST	NONPARAMETRIC TEST
Association between two variables	Pearson correlation coefficient	Spearman rank-order correlation coefficient
Two groups; within-groups design	Paired-samples *t* test	Wilcoxon signed-rank test
Two groups; between-groups design	Independent-samples *t* test	Mann–Whitney *U* test
More than two groups; between-groups design	One-way between-groups ANOVA	Kruskal–Wallis *H* test

FIGURE E-1.
Choosing the Appropriate Hypothesis Test

By asking the right questions about our variables and research design, we can choose the appropriate hypothesis test for our research.

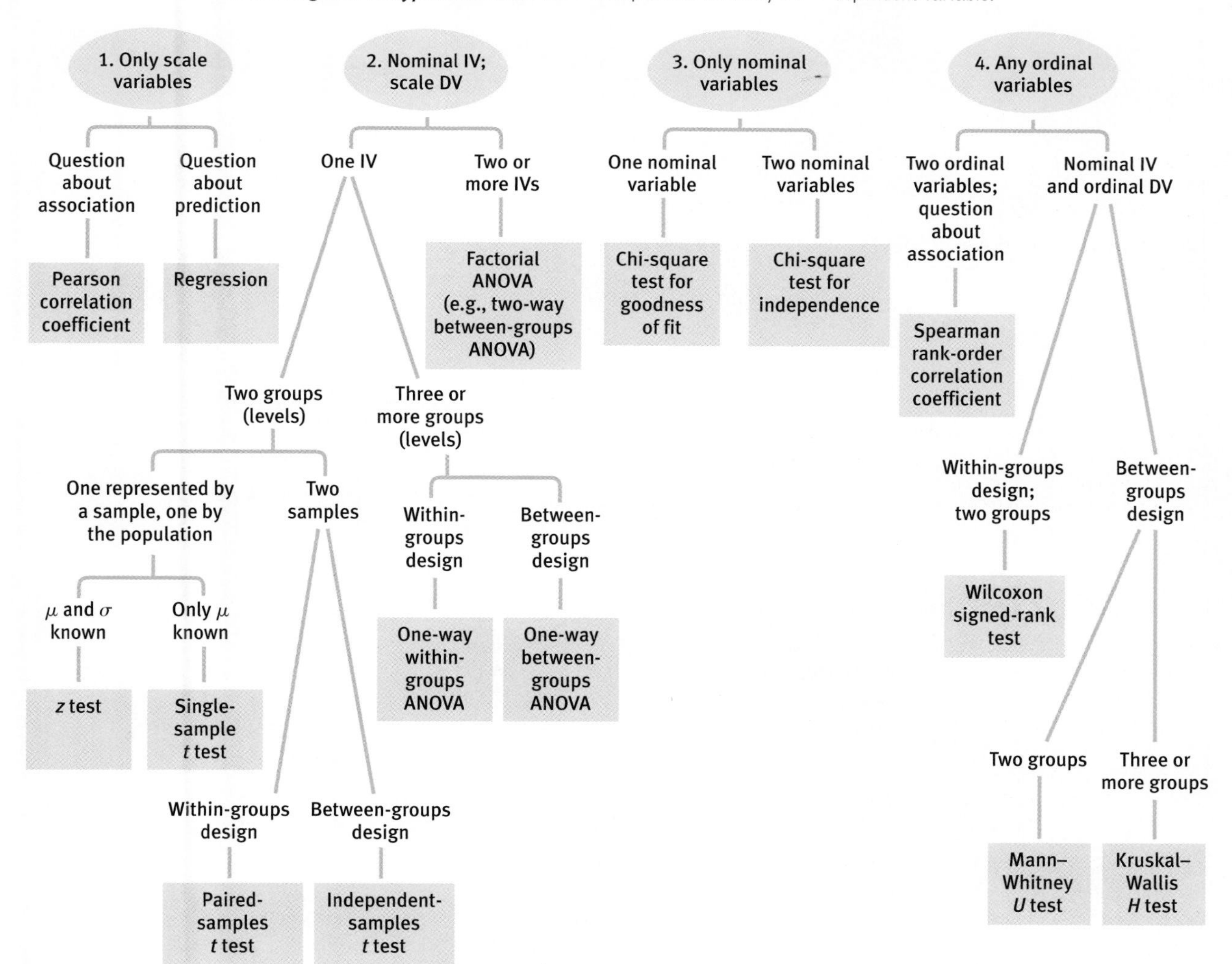

Reporting Statistics

Overview of Reporting Statistics

In Chapter 10, How It Works 10.1, we described a study about gender differences in humor (Azim, Mobbs, Jo, Menon, & Reiss, 2005). Let's recap the results of the analyses and then use this information to report statistics in the Methods and Results sections of a paper written in the style of the American Psychological Association (APA). In the analyses of the humor data, we used fictional data that had the same means as the actual study by Azim and colleagues. We used the following raw data:

Percentage of cartoons labeled as "funny"

Women: 84, 97, 58, 90

Men: 88, 90, 52, 97, 86

We conducted an independent-samples *t* test on these data and found a test statistic, *t*, of −0.03. This test statistic was not beyond the cutoff, so we failed to reject the null hypothesis. We *cannot* conclude that men and women, on average, find different percentages of the cartoons to be funny; we can only conclude that this study did not provide evidence that women and men are different on this variable.

The statistics for this test, as reported in a journal article, would include the symbol for the statistic, the degrees of freedom, the value of the test statistic, and, for statistics calculated by hand, whether the *p* value associated with the test statistic was less than or greater than the cutoff *p* level of 0.05. In the humor example, the statistics would read:

$$t(7) = -0.03, p > 0.05$$

(Note that when we conducted this hypothesis test using SPSS, we got an exact *p* value of 0.977, so we would say $p = 0.98$ instead of $p > 0.05$ if we had used software.) In addition to the statistics, we also would report the means and standard deviations for the two samples:

Women: $M = 82.25$, $SD = 17.02$; Men: $M = 82.60$, $SD = 18.13$

We can also calculate that the percentage difference between women and men is $82.25 - 82.60 = -0.35$. There is just a 0.35% difference between women and men.

In How It Works 10.2, we calculated a confidence interval for these data. The 95% confidence interval, centered around the difference between means of $82.25 - 82.60 = -0.35$, is [−27.88, 27.18].

In How It Works 10.3, we calculated the effect size for this study, a Cohen's *d* of −0.02. We now have sufficient information to write up these findings.

There are three topics to consider when reporting statistics, all covered in various sections of the *Publication Manual of the American Psychological Association* (APA, 2010):

1. In the Methods section, we justify our study by including information about the statistical power, reliability, and validity of any measures we used.
2. In the Results section, we report the traditional statistics, which include any relevant descriptive statistics and often the results of hypothesis testing.
3. In the Results section, we also report the newer statistics that are now required by the APA, including effect sizes and confidence intervals (APA, 2010).

Justify the Study

Researchers should first report the results of the statistical power analyses that were conducted before the data were collected. Then researchers should report any information related to the reliability and validity of the measured variables. This information usually goes in the Methods section of the paper.

To summarize this aspect of the findings, we include in the Methods section:

- Statistical power analyses
- Psychometric data for each scale used (reliability and validity information)

Report Traditional Statistics

The Results section should include any relevant summary statistics. For analyses with a scale dependent variable, include means, standard deviations, and sample sizes for each cell in the research design. For analyses with a nominal dependent variable (chi-square analyses), include the frequencies (counts) for each cell; there won't be means or standard deviations because there are no scores on a scale measure.

Summary statistics are sometimes presented first in a Results section but are more typically presented after a description of each hypothesis test. If there are only two or three cells, then the summary statistics are typically presented in the text; if there are more cells, then these statistics should be displayed in a table or figure.

Reports of hypothesis tests typically begin by reiterating the hypothesis to be tested and then describing the test that was conducted, including the independent and dependent variables. The results of the hypothesis test are then presented, usually including the symbol for the statistic, the degrees of freedom, the actual value of the statistic, and, if using software, the actual *p* value associated with that statistic. The format for reporting this information has been presented after each hypothesis test in this text and is presented again in Table F-1.

After the statistics are presented, a brief statement summarizes the results, indicating the direction of any effects. This brief statement does not draw conclusions beyond the actual finding. In the Results section, re-

TABLE F-1. The Format for the Results of a Hypothesis Test

There is a general format for reporting the results of hypothesis tests. The symbol for the statistic is followed by the degrees of freedom in parentheses, then the value of the test statistic, and finally the exact *p* value associated with that test statistic. This table presents the way that format would be implemented for several of the test statistics discussed in this text. (Note that you will only have the exact *p* value if you use software. If you conduct a test by hand, you may report whether the *p* value is greater than or less than 0.05.)

Symbol	Degrees of Freedom	Value of Test Statistic	Information About the Cutoff	Effective Size	Example
z	(df)	= XX,	p = 0.XX	d = XX	$z(54) = 0.60, p = 0.45, d = 0.08$
t	(df)	= XX,	p = 0.XX	d = XX	$t(146) = 2.29, p = 0.024, d = 0.50$
F	($df_{between}$, df_{within})	= XX,	p = 0.XX	R^2 = XX	$F(2, 142) = 6.63, p = 0.002, R^2 = 0.09$
χ^2	(df, N = XX)	= XX,	p = 0.XX	V = XX	$\chi^2 (1, N = 147) = 0.58, p = 0.447, V = 0.06$
T	None	= XX,	p = 0.XX	None	$T = 7, p = 0.04$
U	None	= XX,	p = 0.XX	None	$U = 19, p = 0.14$

searchers do not discuss the finding in terms of the general theories in the field or in terms of its potential implications or applications (which go, appropriately enough, in the Discussion section). Researchers should present the results of all hypothesis tests that they conducted, even those in which they failed to reject the null hypothesis.

To summarize this aspect of Results sections:

- Include summary statistics: means, standard deviations, and sample sizes for each cell when the dependent variable is scale, and frequencies (counts) for each cell when the dependent variable is nominal. These are often included after each hypothesis test.
- For each hypothesis test conducted:
    - Include a brief summary of the hypotheses and hypothesis test.
    - Report the results of hypothesis testing: the symbol for the statistic used, degrees of freedom, the actual value of the statistic, and the *p* value associated with this statistic (if using software).
    - Provide a statement that summarizes the results of hypothesis testing.
    - Use tables and figures to clarify patterns in the findings.
- Include all results, even for findings that are not statistically significant.

The statistics for the study from How It Works in Chapter 10 that compared the mean percentages of cartoons that women and men found funny might be reported as follows:

> To examine the hypothesis that women and men, on average, find different percentages of cartoons funny, we conducted an independent-samples *t* test. The independent variable was gender, with two levels: female and male. The dependent variable was the percentage of cartoons deemed funny. There was not a statistically significant effect of gender, $t(7) = -0.03, p = 0.98$; this study does not provide evidence that women (M = 82.25, SD = 17.02) and men (M = 82.60, SD = 18.13) deem, on average, different percentages of cartoons to be funny. The difference between the mean percentages for women and men is just 0.35%.

Reporting Newer Statistics

It is no longer enough to simply present the descriptive statistics and the results of the hypothesis test. As of the 2010 edition of its *Publication Manual,* the APA requires that effect sizes and confidence intervals be included when relevant. The effect-size estimate is often included as part of the report of the statistics, just after the *p* value. There is often a statement that indicates the size of the effect in words, not just in numbers. The confidence interval can be presented after the effect size, abbreviated as "95% CI" with the actual interval in brackets. Note that nonparametric tests often do not have associated measures of effect size or confidence intervals. In these cases, researchers should provide enough descriptive information for readers to interpret the findings.

To summarize this aspect of the Results sections, we include:

- Effect size, along with a statement about the size of the effect.
- Confidence intervals when possible, along with a statement interpreting the confidence interval in the context of the study.

For the study on humor, we might report the effect size as part of the traditional statistics that we described above: There was not a statistically significant effect of gender, $t(7) = -0.03, p = 0.98, d = -0.02$, 95% CI [−28.37, 27.67]; this was a small, almost nonexistent, effect. In fact, there is only a 0.35% difference between the mean percentages for women and men. This study does not provide evidence that men and women, on average, rate different percentages of cartoons as funny.

For the humor study, we can now pull the parts together. Here is how the results would be reported:

> To examine the hypothesis that women and men, on average, find different percentages of cartoons funny, we conducted an independent-samples *t* test. The independent variable was gender, with two levels: female and male. The dependent variable was the percentage of cartoons deemed funny. There was not a statistically significant effect of gender, $t(7) = -0.03, p = 0.98, d = -0.02$, 95% CI [−28.37, 27.67]; this was a small, almost nonexistent, effect.

Based on the hypothesis test and the confidence interval, this study does not provide evidence that women (M = 82.25, SD = 17.02) and men (M = 82.60, SD = 18.13) deem, on average, different percentages of cartoons to be funny. In fact, there is only a very small difference between the mean percentages for women and men, just 0.35%.

APPENDIX G

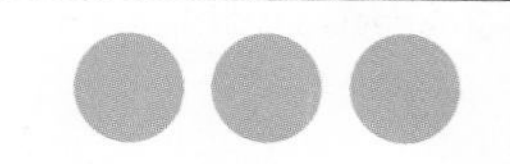

Building Better Graphs Using Excel

In terms of default settings, there is no graphing software that adheres specifically to all of the APA guidelines; however, you can learn to use software to adapt the default graph into a clear and useful visualization of data. For example, Microsoft Excel is a program that is used in many settings to create graphs. Like other graphing software, Excel software changes with every application revision, so you might encounter slight variations from the instructions that follow.

This appendix demonstrates how to create a simple, clean-looking graph, using the snowfall data from the opening example in Chapter 12. You may recall that ski resorts reported greater snowfall amounts than weather stations did, an effect that was even more pronounced on weekends. This effect is likely due to the benefit ski resorts receive when people think there is more snow.

Remember, the purpose of a graph is to *let the data speak*. You can introduce insight (rather than confusion) into an Excel graph by following the specific dos and don'ts in Chapter 3. In APA style, various kinds of visual displays of data are all referred to as "Figures." For example, Figure G-1 displays two nominal independent variables and one scale dependent variable in a spreadsheet that Excel refers to as a "workbook."

FIGURE G-1
Summary Data in an Excel Spreadsheet

To transform these summary data in an Excel spreadsheet into an APA-style figure, first highlight all nine cells (A1 through C3) and then select the kind of chart you wish to create.

	A	B	C	D	E	F
1		Ski Resort	Weather Station			
2	Weekdays	1.29	1.05			
3	Weekends	1.59	1.15			
4						
5						
6						
7						
8						

Step 1. Identify the independent and dependent variables, and determine which type of observation each is. In the snowfall example, the dependent variable is a scale variable, and the independent variables are nominal variables. Specifically, snowfall in inches is the scale dependent variable; the source of the snow report (levels: ski resort, weather station) is a nominal independent variable; and time of the week (levels: weekdays, weekends) is a second independent variable. (Think of these particular independent variables as predictor variables because they cannot be manipulated by the experimenter.)

Step 2. Enter the summary data (the averages of the dependent variable within each cell) in the Excel spreadsheet, as shown in Figure G-2. Then highlight the data and labels, and select the type of figure you want to create. To do this, click "Insert," "Column," and under "2-D Column," select the left-most graph icon, called "Clustered Column." Do not select a 3-D graph because it will introduce chartjunk. Figure G-2 shows what the graph will look like at this point.

FIGURE G-2
The Default Graph Created by Excel

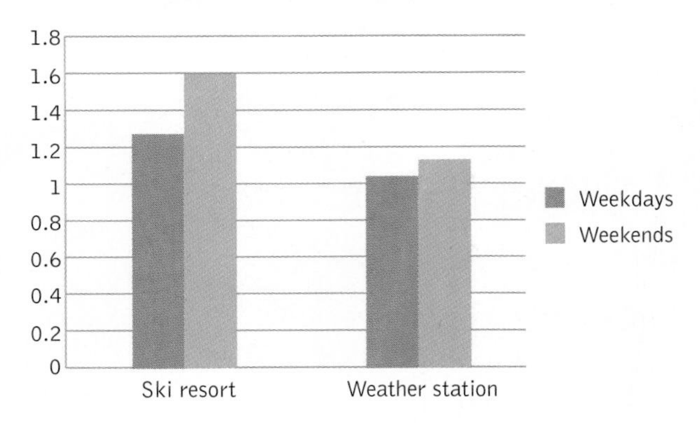

Step 3. Change the defaults so that the graph adheres to the graphing checklist from Chapter 3. If you click on the graph and then look at "Chart Tools" along the top of the Excel toolbar, you'll see a number of options for changing the graph. Under "Chart Layouts," we modified the chart by using these tools and by entering information directly into the text boxes that are part of the chart. For instance, in Step 2, we had selected a chart that included a title; we then put the cursor in the title box and retitled the chart to what you can see in Figure G-3 is "Biased reporting: Snowfall reports based on time of week and source of report." We then used the "Axis Titles" menu to change the orientation of the *y*-axis label so that it is horizontal. To make the data values clear, we used the "Data Labels" menu and selected "Outside End" to display the values for each bar.

FIGURE G-3

A Graph Created in Excel That Has Been Altered to Adhere to the Graphing Guidelines

Notice that the bars are two-dimensional and that the label for the *y*-axis is turned horizontally so that it is easy to read.

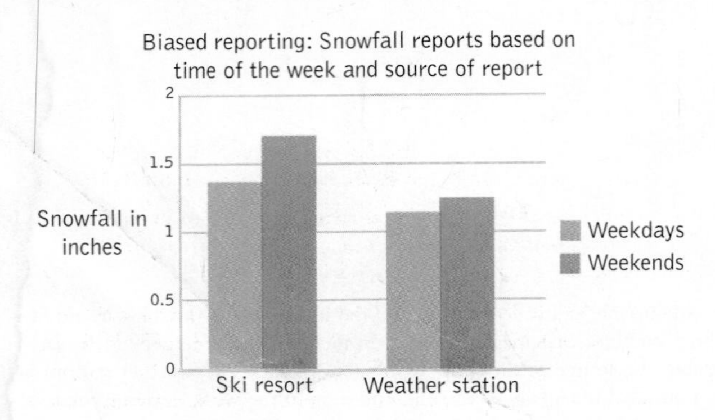

We right-clicked on each of the data bars to change its color using the "Shape Fill" tool so that we could set the colors of the bars as variations on a single color. We wanted to reduce the chartjunk, so we first made the grid lines lighter by using the "Line Color" tool under the "Gridlines" menu to increase the solid line transparency; then we used the same "Gridlines" menu to display only major gridlines for both the horizontal and the vertical axis. Figure G-3 shows our choices.

Know your audience. A figure designed for submission to a scholarly journal will look a little different than a PowerPoint presentation to a class, just as a presentation at a sales meeting will look different than a research report submitted to your boss. Focus on clarity rather than gimmicks. Remember: *Let the data speak.*

GLOSSARY

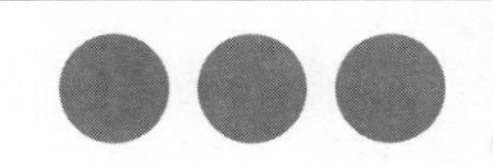

A

alpha The chance of making a Type I error and another name for the *p* level; symbolized as α.

analysis of variance (ANOVA) A hypothesis test typically used with one or more nominal (and sometimes ordinal) independent variables (with at least three groups overall) and a scale dependent variable.

assumption A requirement that the population from which we are sampling has specific characteristics that will allow us to make accurate inferences.

B

bar graph A visual depiction of data when the independent variable is nominal or ordinal and the dependent variable is scale. Each bar typically represents the average value of the dependent variable for each category.

between-groups ANOVA A hypothesis test in which there are more than two samples, and each sample is composed of different participants.

between-groups research design An experimental design in which participants experience one, and only one, level of the independent variable.

between-groups variance An estimate of the population variance based on the differences among the means.

bimodal A distribution that has two modes, or most common scores.

C

ceiling effect A situation in which a constraint prevents a variable from taking on values above a given number.

cell A box that depicts one unique combination of levels of the independent variables in a factorial design.

central limit theorem The idea that a distribution of sample means is a more normal distribution than a distribution of scores, even when the population distribution is not normal.

central tendency A descriptive statistic that best represents the center of a data set, the particular value that all the other data seem to be gathering around.

chartjunk Any unnecessary information or feature in a graph that distracts from a viewer's ability to understand the data.

chi-square test for goodness-of-fit A nonparametric hypothesis test used with one nominal variable.

chi-square test for independence A nonparametric hypothesis test used with two nominal variables.

coefficient alpha A commonly used estimate of a test's or measure's reliability that is calculated by taking the average of all possible split-half correlations and symbolized as α; sometimes called *Cronbach's alpha*.

Cohen's *d* A measure of effect size that assesses the difference between two means in terms of standard deviation, not standard error.

confidence interval An interval estimate, based on the sample statistic, that includes the population mean a certain percentage of the time, if we sample from the same population repeatedly.

confirmation bias Our usually unintentional tendency to pay attention to evidence that confirms what we already believe and to ignore evidence that would disconfirm our beliefs.

confounding variable A variable that systematically varies with the independent variable so that we cannot logically determine which variable is at work; also called a *confound*.

continuous observation Observed data point that can take on a full range of values (e.g., numbers out to many decimal points); an infinite number of potential values exists.

control group A level of the independent variable that is designed to match the experimental group in all ways but the experimental manipulation itself.

convenience sample A subset of a population whose members are chosen strictly because they are readily available, as opposed to randomly selecting participants from the entire population of interest.

correlation An association between two or more variables.

correlation coefficient A statistic that quantifies a relation between two variables.

Cramer's *V* The standard effect size used with the chi-square test for independence; also called *Cramer's phi*, symbolized as Φ.

critical region The area in the tails of the comparison distribution in which the null hypothesis can be rejected.

critical value Test statistic value beyond which we will reject the null hypothesis; often called *cutoff*.

D

defaults The options that a software designer has preselected. These are the built-in decisions that the software will implement if we do not instruct it otherwise.

degrees of freedom The number of scores that are free to vary when estimating a population parameter from a sample.

dependent variable The outcome variable that we hypothesize to be related to, or caused by, changes in the independent variable.

descriptive statistic Statistical technique used to organize, summarize, and communicate a group of numerical observations.

deviation from the mean The amount that a score in a sample differs from the mean of the sample; also called *deviation*.

discrete observation Observed data point that can take on only specific values (e.g., whole numbers); no other values can exist between these numbers.

distribution of means A distribution composed of many means that are calculated from all possible samples of a given size, all taken from the same population.

duck A form of chartjunk where a feature of the data has been dressed up in a graph to be something other than merely data.

E

effect size A standardized value that indicates the size of a difference but is not affected by sample size.

expected relative-frequency probability The likelihood of an event occurring based on the actual outcome of many, many trials.

experiment A study in which participants are randomly assigned to a condition or level of one or more independent variables.

experimental group A level of the independent variable that receives the treatment or intervention of interest in an experiment.

F

***F* statistic** A ratio of two measures of variance: (1) between-groups variance, which indicates differences among sample means, and (2) within-groups variance, which is essentially an average of the sample variances.

factor A term used to describe an independent variable in a study with more than one independent variable.

factorial ANOVA A statistical analysis used with one scale dependent variable and at least two nominal independent variables (also called *factors*); also called a *multifactorial ANOVA*.

floor effect A situation in which a constraint prevents a variable from taking values below a certain point.

frequency distribution A distribution that describes the pattern of a set of numbers by displaying a count or proportion for each possible value of a variable.

frequency polygon A line graph with the *x*-axis representing values (or midpoints of intervals) and the *y*-axis representing frequencies. A dot is placed at the frequency for each value (or midpoint), and the dots are connected.

frequency table A visual depiction of data that shows how often each value occurred; that is, how many scores were at each value. Values are listed in one column, and the numbers of individuals with scores at that value are listed in the second column.

G

generalizability Researchers' ability to apply findings from one sample or in one context to other samples or contexts; also called *external validity.*

grand mean The mean of every score in a study, regardless of which sample the score came from.

grid A form of chartjunk that takes the form of a background pattern, almost like graph paper, on which the data representations, such as bars, are superimposed on a graph.

grouped frequency table A visual depiction of data that reports the frequencies within a given interval rather than the frequencies for a specific value.

H

heteroscedastic A term given to populations that have different variances.

histogram A graph similar to a bar graph typically used to depict scale data with the values of the variable on the *x*-axis and the frequencies on the *y*-axis.

homoscedastic A term given to populations that have the same variance; also called *homogeneity of variance.*

hypothesis testing The process of drawing conclusions about whether a particular relation between variables is supported by the evidence.

I

illusory correlation The phenomenon of believing that one sees an association between variables when no such association exists.

independent variable A variable that we either manipulate or observe to determine its effects on the dependent variable.

independent-samples *t* test A hypothesis test used to compare two means for a between-groups design, a situation in which each participant is assigned to only one condition.

inferential statistic Statistical technique that uses sample data to make general estimates about the larger population.

interaction The statistical result achieved in a factorial design when two or more independent variables have an effect in combination that we do not see when we examine each independent variable on its own.

intercept The predicted value for *Y* when *X* is equal to 0, or the point at which the line crosses, or intercepts, the *y*-axis.

interval estimate An estimate based on a sample statistic, providing a range of plausible values for the population parameter.

interval variable A variable used for observations that have numbers as their values; the distance (or interval) between pairs of consecutive numbers is assumed to be equal.

K

Kruskal–Wallis *H* test A nonparametric hypothesis test used when there are more than two groups, a between-groups design, and an ordinal dependent variable.

L

level A discrete value or condition that a variable can take on.

line graph A graph used to illustrate the relation between two scale variables.

linear relation A relation between two variables best described by a straight line.

M

main effect A result occurring in a factorial design when one of the independent variables has an influence on the dependent variable.

Mann–Whitney *U* test A nonparametric hypothesis test used when there are two groups, a between-groups design, and an ordinal dependent variable.

marginal mean The mean of a row or a column in a table that shows the cells of a study with a two-way ANOVA design.

mean The arithmetic average of a group of scores. It is calculated by summing all the scores and dividing by the total number of scores.

median The middle score of all the scores in a sample when the scores are arranged in ascending order. If there is no single middle score, the median is the mean of the two middle scores.

meta-analysis A type of statistical analysis that simultaneously examines as many studies as possible for a given research topic, and involves the calculation of a mean effect size from the individual effect sizes of these studies.

mode The most common score of all the scores in a sample.

moiré vibration Any visual pattern that creates a distracting impression of vibration and movement.

multimodal A distribution that has more than two modes, or most common scores.

multiple regression A statistical technique that includes two or more predictor variables in a prediction equation.

N

negative correlation An association between two variables in which participants with high scores on one variable tend to have low scores on the other variable.

negatively skewed data An asymmetric distribution whose tail extends to the left, in a negative direction.

nominal variable A variable used for observations that have categories, or names, as their values.

nonlinear relation A relation between variables best described by a line that breaks or curves in some way.

nonparametric test Inferential statistical analysis that is not based on a set of assumptions about the population.

normal curve A specific bell-shaped curve that is unimodal, symmetric, and defined mathematically.

normal distribution A specific frequency distribution in the shape of a bell-shaped, symmetric, unimodal curve.

null hypothesis A statement that postulates that there is no difference between populations or that the difference is in a direction opposite from that anticipated by the researcher.

O

one-tailed test A hypothesis test in which the research hypothesis is directional, positing either a mean decrease or a mean increase in the dependent variable, but not both, as a result of the independent variable.

one-way ANOVA A hypothesis test that includes both a single nominal independent variable with more than two levels and a single scale dependent variable.

operational definition The operations or procedures used to measure or manipulate a variable.

ordinal variable A variable used for observations that have rankings (i.e., 1st, 2nd, 3rd, . . .) as their values.

orthogonal variable An independent variable that makes a separate and distinct contribution to the prediction of a dependent variable, as compared with another variable.

outcome In reference to probability, the result of a trial.

P

***p* level** The probability used to determine the critical values, or cutoffs, in hypothesis testing; often called *alpha.*

paired-samples *t* test A test used to compare two means for a within-groups design, a situation in which every participant is in both samples; also called a *dependent-samples* t *test.*

parameter A number based on the whole population; it is usually symbolized by a Greek letter.

parametric test Inferential statistical analysis that is based on a set of assumptions about the population.

Pareto chart A type of bar graph in which the categories along the *x*-axis are ordered from highest bar on the left to lowest bar on the right.

Pearson correlation coefficient A statistic that quantifies a linear relation between two scale variables.

personal probability The likelihood of an event occurring based on an individual's opinion or judgment; also called *subjective probability.*

pictorial graph A visual depiction of data typically used for an independent variable with very few levels (categories) and a scale dependent variable. Each level uses a picture or symbol to represent its value on the scale dependent variable.

pie chart A graph in the shape of a circle with a slice for every level (category). The size of each slice represents the proportion (or percentage) of each level.

point estimate A summary statistic from a sample that is just one number used as an estimate of the population parameter.

pooled variance A weighted average of the two estimates of variance—one from each sample—that are calculated when conducting an independent-samples *t* test.

population All of the possible observations about which we'd like to know something.

positive correlation An association between two variables such that participants with high scores on one variable tend to have high scores on the other variable as well, and those with low scores on one variable tend to have low scores on the other variable as well.

positively skewed data An asymmetric distribution whose tail extends to the right, in a positive direction.

post-hoc test A statistical procedure frequently carried out after we reject the null hypothesis in an ANOVA; it allows us to make multiple comparisons among several means; often referred to as a *follow-up test.*

probability The likelihood that a certain outcome—out of all possible outcomes—will occur.

proportionate reduction in error A statistic that quantifies how much more accurate predictions are when we use the regression line instead of the mean as a prediction tool; also called *coefficient of determination,* symbolized as R^2.

psychometricians The statisticians and psychologists who develop tests and measures.

psychometrics The branch of statistics used in the development of tests and measures.

Q

qualitative interaction A particular type of quantitative interaction of two (or more) independent variables in which one independent variable reverses its effect depending on the level of the other independent variable.

quantitative interaction An interaction in which one independent variable exhibits a strengthening or weakening of its effect at one or more levels of the other independent variable, but the direction of the initial effect does not change.

R

R^2 The proportion of variance in the dependent variable that is accounted for by the independent variable.

random assignment The protocol established for an experiment whereby every participant in a study has an equal chance of being assigned to any of the groups, or experimental conditions, in the study.

random sample A subset of a population selected using a method that ensures that every member of the population has an equal chance of being selected into the study.

range A measure of variability calculated by subtracting the lowest score (the minimum) from the highest score (the maximum).

range-frame A scatterplot or related graph that indicates only the range of the data on each axis; the lines extend only from the minimum to the maximum scores.

ratio variable A variable that meets the criteria for an interval variable but also has a meaningful zero point.

raw score A data point that has not yet been transformed or analyzed.

regression to the mean The tendency of scores that are particularly high or low to drift toward the mean over time.

relative risk A measure created by making a ratio of two conditional proportions; also called *relative likelihood* or *relative chance.*

reliability The consistency of a measure.

replication The duplication of scientific results, ideally in a different context or with a sample that has different characteristics.

research hypothesis A statement that postulates that there is a difference between populations or sometimes, more specifically, that there is a difference in a certain direction, positive or negative; also called the *alternative hypothesis.*

robust A term given to a hypothesis test that produces fairly accurate results even when the data suggest that the population might not meet some of the assumptions.

S

sample A set of observations drawn from the population of interest.

scale variable A variable that meets the criteria for an interval variable or a ratio variable.

scatterplot A graph that depicts the relation between two scale variables. The values of each variable are marked along the two axes, and a mark is made to indicate the intersection of the two scores for each participant.

simple linear regression A statistical tool that lets us predict a person's score on a dependent variable from his or her score on one independent variable.

single-sample *t* test A hypothesis test in which we compare data from one sample to a population for which we know the mean but not the standard deviation.

skewed distribution A distribution in which one of the tails of the distribution is pulled away from the center.

slope The amount that Y is predicted to increase for an increase of 1 in X.

source table A table that presents the important calculations and final results of an ANOVA in a consistent and easy-to-read format.

Spearman rank-order correlation coefficient A nonparametric statistic that quantifies the association between two ordinal variables.

standard deviation The typical amount that each score in a sample varies, or deviates, from the mean; it is the square root of the average of the squared deviations from the mean.

standard error The name for the standard deviation of a distribution of means.

standard error of the estimate A statistic indicating the typical distance between a regression line and the actual data points.

standard normal distribution A normal distribution of z scores.

standardization A process that converts individual scores from different normal distributions to a shared normal distribution with a known mean, standard deviation, and percentiles.

standardized regression coefficient A standardized version of the slope in a regression equation, it is the predicted change in the dependent variable in terms of standard deviations for an increase of 1 standard deviation in the independent variable; often called the *beta weight*.

statistic A number based on a sample taken from a population; it is usually symbolized by a Latin letter.

statistically significant A name given to a finding in which the data differ from what we would expect by chance if there were, in fact, no actual difference.

statistical power A measure of our ability to reject the null hypothesis given that the null hypothesis is false.

success In reference to probability, the outcome for which we are trying to determine the probability.

sum of squares The sum of each score's squared deviation from the mean; symbolized as SS.

T

***t* statistic** A statistic that indicates the distance of a sample mean from a population mean in terms of the standard error.

test–retest reliability A method that determines whether the scale being used provides consistent information every time the test is taken.

time plot or **time series plot** A graph that plots a scale variable on the y-axis as it changes over an increment of time (e.g., second, day, century) labeled on the x-axis.

trial In reference to probability, each occasion that a given procedure is carried out.

Tukey *HSD* test A widely used post-hoc test that determines the differences between means in terms of standard error; the *HSD* is compared to a critical value; sometimes called the *q test*.

two-tailed test A hypothesis test in which the research hypothesis does not indicate a direction of mean difference or change in the dependent variable, but merely indicates that there will be a mean difference.

two-way ANOVA A hypothesis test that includes two nominal independent variables, regardless of their numbers of levels, and a scale dependent variable.

Type I error The result when we reject the null hypothesis, but the null hypothesis is correct.

Type II error The result when we fail to reject the null hypothesis, but the null hypothesis is false.

U

unimodal A distribution that has one mode, or most common score.

V

validity The extent to which a test actually measures what it was intended to measure.

variability A numerical way of describing how much spread there is in a distribution.

variable Any observation of a physical, attitudinal, or behavioral characteristic that can take on different values.

variance The average of the squared deviations from the mean.

volunteer sample A special kind of convenience sample in which participants actively choose to participate in a study; also called a *self-selected sample.*

W

Wilcoxon signed-rank test A nonparametric hypothesis test used when there are two groups, a within-groups design, and an ordinal dependent variable.

within-groups ANOVA A hypothesis test in which there are more than two samples, and each sample is composed of the same participants; also called a *repeated-measures ANOVA*.

within-groups research design An experimental design in which all participants in the study experience the different levels of the independent variable; also called a *repeated-measures design.*

within-groups variance An estimate of the population variance based on the differences within each of the three (or more) sample distributions.

Z

***z* distribution** A normal distribution of standardized scores.

***z* score** The number of standard deviations a particular score is from the mean.

REFERENCES

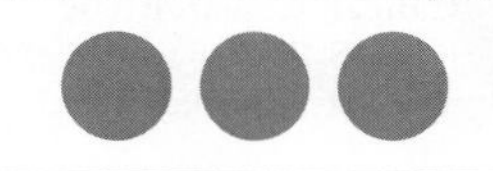

Abumrad, J., & Krulwich, R. (Hosts). (2009, September 11). Stochasticity [Radio series episode]. In Wheeler, S., & Abumrad, J. (Producers), *Radiolab.* New York: WNYC.

Alter, A., & Oppenheimer, D. (2006). Predicting short-term stock fluctuations by using processing fluency. *Proceedings of the National Academy of Sciences of the United States of America, 103,* 9369–9372.

American Academy of Physician Assistants. (2005). Income reported by PAs who graduated in 2004. *American Academy of Physician Assistants.* Retrieved November 20, 2006, from http://www.aapa.org/research/05newgrad-income.pdf

American Psychological Association. (2010). *Publication manual of the American Psychological Association* (6th ed.). Washington, DC: American Psychological Association.

Archibold, R. C. (1998, February 18). Just because the grades are up, are Princeton students smarter? *New York Times.* Retrieved November 14, 2006, from http://www.nytimes.com

Arcidiacono, P., Bayer, P., & Hizmo, A. (2008). *Beyond signaling and human capital* (NBER Working Paper No. 13591). Cambridge, MA: National Bureau of Economic Research. Retrieved March 4, 2013, from http://www.nber.org/papers/w13951

Aron, A., & Aron, E. N. (2002). *Statistics for psychology* (3rd ed.). Upper Saddle River, NJ: Pearson Education.

Aron, A., Fisher, H., Mashek, D. J., Strong, G., Li, H., & Brown, L. L. (2005). Reward, motivation, and emotion systems associated with early-stage intense romantic love. *Journal of Neurophysiology, 94,* 327–337.

Azim, E., Mobbs, D., Jo, B., Menon, V., & Reiss, A. L. (2005). Sex differences in brain activation elicited by humor. *Proceedings of the National Academy of Sciences, 102,* 16496–16501. Retrieved February 10, 2006, from http://www.pnas.org/cgi/doi/10.1073/pnas.0408456102

Bailey, D. G., & Dresser, G. K. (2004). Natural products and adverse drug interactions. *Canadian Medical Association Journal, 170,* 1531–1532.

Baird, M. (2010, July 2). Woman wins millions from Texas lottery for 4th time. ABC News. Retrieved July 12, 2010, from http://www.abc15.com/dpp/news/national/Copy_of_Woman-wins-millions-from-Texas-lottery-for-4th-time_35170322-ews-SHNS-wmar1278098941198

Banks, J., Marmot, M., Oldfield, Z., & Smith, J. P. (2006). Disease and disadvantage in the United States and England. *Journal of the American Medical Association, 295,* 2037–2045.

Bardwell, W. A., Ensign, W. Y., & Mills, P. J. (2005). Negative mood endures after completion of high-altitude military training. *Annals of Behavioral Medicine, 29,* 64–69.

Bartlett, C. P., Harris, R. J., & Bruey, C. (2008). The effect of the amount of blood in a violent video game on aggression, hostility, and arousal. *Journal of Experimental Social Psychology, 44,* 539–546.

Behenam, M., & Pooya, O. (2007). Factors affecting patients cooperation during orthodontic treatment. *The Orthodontic CYBER journal.* Retrieved October 18, 2012, from http://orthocj.com/2007/01/patient-cooperation-orthodontic-treatment/

Bellosta, S., Paoletti, R., & Corsini, A. (2004). Safety of statins: Focus on clinical pharmacokinetics and drug interactions. *Circulation, 109,* III-50–III-57.

Benbow, C. P., & Stanley, J. C. (1980). Sex differences in math ability: Fact or artifact? *Science, 210,* 1262–1264.

Berger, J., & Fitzsimons, G. (2008). Dogs on the street, Pumas on your feet: How cues in the environment influence product evaluation and choice. *Journal of Marketing Research, XLV,* 1–14.

Bernstein, P. L. (1996). *Against the gods: The remarkable story of risk.* New York: Wiley.

Bollinger, B., Leslie, P., & Sorenson, A. (2010). *Calorie posting in chain restaurants* (NBER Working Paper No. 15648). Cambridge: MA: National Bureau of Economic Research. Retrieved May 31, 2010, from http://www.gsb.stanford.edu/news/StarbucksCaloriePosting Study.pdf

Boone, D. E. (1992). WAIS-R scatter with psychiatric inpatients: I. Intrasubtest scatter. *Psychological Reports, 71,* 483–487.

Borsari, B., & Carey, K. B. (2005). Two brief alcohol interventions for mandated college students. *Psychology of Addictive Behaviors, 19,* 296–302.

Bowen, W. G., & Bok, D. (2000). *The shape of the river: Long-term consequences of considering race in college and university admissions.* Princeton, NJ: Princeton University Press.

Box, J. (1978). *R. A. Fisher: The life of a scientist.* New York: Wiley.

Brinn, D. (2006, June 25). Israeli "clown therapy" boosts fertility treatment birthrate. *Health.* Retrieved April 6, 2007, from http://israel21c.org/

Buekens, P., Xiong, X., & Harville, E. (2006). Hurricanes and pregnancy. *Birth, 33,* 91–93.

Busseri, M. A., Choma, B. L., & Sadava, S. W. (2009). "As good as it gets" or "the best is yet to come"? How optimists and pessimists view their past, present, and anticipated future life satisfaction. *Personality and Individual Differences, 47,* 352–356. doi: 10.1016/j.paid.209.04.002

Canadian Institute for Health Information (CIHI). (2005). More patients receiving transplants than 10 years ago, despite stagnant organ donation rate. Retrieved July 15, 2005, from http://secure. cihi.ca/cihiweb/dispPage.jsp?cw_page=media_13apr2005_e

Cancer Research UK. (2003). *Cancer deaths in the UK.* Retrieved June 15, 2005, from http://info.cancerresearchuk.org/cancerstats/mortality/cancerdeaths/

Centers for Disease Control (CDC). (2004). *Americans slightly taller, much heavier than four decades ago.* Retrieved May 26, 2005, from http://www.cdc.gov/nchs/pressroom/04news/americans.htm.

Centers for Disease Control National Center for Health Statistics. (2000). *National health and nutrition examination survey, CDC growth charts: United States.* Retrieved January 6, 2006, from http://www.cdc.gov/growthcharts/

Chang, A., Sandhofer, C. M., & Brown, C. S. (2011). Gender biases in early number exposure to preschool-aged children. *Journal of Language and Social Psychology, 30,* 440–450. doi: 10.1177/0261927X11416207

CNN. (2005, October 5). Vancouver is 'best city to live.' Retrieved March 4, 2013, from http://www.cnn.com/2005/WORLD/europe/10/04/eui.survey/

Cohen, J. (1988). *Statistical power analysis for the behavioral sciences* (2nd ed.). Hillsdale, NJ: Erlbaum.

Cohen, J. (1990). Things I have learned (so far). *American Psychologist, 45,* 1304–1312.

Cohen, J. (1992). A power primer. *Psychological Bulletin, 112,* 155–159.

Cooper, M. J., Gulen, H., & Ovtchinnikov, A. V. (2007). *Corporate political contributions and stock returns.* Available at SSRN: http://ssrn.com/abstract-940790

Corsi, A., & Ashenfelter, O. (2001, April). *Wine quality: Experts' ratings and weather determinants* [Electronic version]. Poster session presented at the annual meeting of the European Association of Agricultural Economists, Zaragoza, Spain.

Cortina, J. M. (1993). What is coefficient alpha? An examination of theory and applications. *Journal of Applied Psychology, 78,* 98–104.

Coulson, M., Healey, M., Fidler, F., & Cumming, G. (2010). Confidence intervals permit, but do not guarantee, better inference than statistical significance testing. *Frontiers in Psychology: Quantitative Psychology and Measurement, 1,* 1–9. doi: 10.3389/fpsyg.2010.00026

Cox, R. H., Thomas. T. R., Hinton, P. S., & Donahue, W. M. (2006). Effects of acute bouts of aerobic exercise of varied intensity on subjective mood experiences in women of different age groups across time. *Journal of Sport Behavior, 29,* 40–59.

Cunliffe, S. (1976). Interaction. *Journal of the Royal Statistical Society, A, 139,* 1–19.

Czerwinski, M., Smith, G., Regan, T., Meyers, B., Robertson, G., & Starkweather, G. (2003). Toward characterizing the productivity benefits of very large displays. In M. Rauterberg et al. (Eds.), *Human-computer interaction-INTERACT '03* (pp. 9–16). Amsterdam, Netherlands: IOS Press.

Darlin, D. (2006, July 1). Air fare made easy (or easier). *New York Times.* Retrieved July 1, 2006, from http://www.nytimes.com

Dean, G., & Kelly, I. W. (2003). Is astrology relevant to consciousness and PSI? *Journal of Consciousness Studies, 10,* 175–198.

DeBroff, B. M., & Pahk, P. J. (2003). The ability of periorbitally applied antiglare products to improve contrast sensitivity in conditions of sunlight exposure. *Archives of Ophthalmology, 121,* 997–1001.

Delucchi, K. L. (1983). The use and misuse of chi square: Lewis and Burke revisited. *Psychological Bulletin, 94,* 166–176.

DeVellis, R. F. (1991). *Scale development: Theory and applications.* Newbury Park, CA: Sage.

Diamond, M. & Stone, M. (1981). Nightingale on Quetelet. I: The passionate statistician, II: The marginalia, III. Essay in memoriam. *Journal of the Royal Statistical Society, A, 144,* 66–79, 176–213, 332–351.

Dijksterhuis, A., Bos, M. W., Nordgren, L. F., & van Baaren, R. B. (2006). On making the right choice: The deliberationwithout-attention effect. *Science, 311,* 1005–1007.

Ditto, P. H., & Lopez, D. L. (1992). Motivated skepticism: Use of differential decision criteria for preferred and nonpreferred conclusions. *Journal of Personality and Social Psychology, 63,* 568–584.

Dogster. (2009). What dog breed are you? Retrieved March 4, 2013, from http://www.dogster.com/quizzes/what_dog_breed_are_you/)

Dubner, S. J., & Levitt, S. D. (2006a, May 7). A star is made: The birth-month soccer anomaly. *New York Times.* Retrieved May 7, 2006, from http://www.nytimes.com

Dubner, S. J., & Levitt, S. D. (2006b, November 5). The way we live now: Freakonomics; The price of climate change. *New York Times,* Retrieved March 7, 2007, from http://www. nytimes.com

Ellison, N. B., Steinfeld, C., & Lampe, C. (2007). The benefits of Facebook "friends": Social capital and college students' use of online social network sites. *Journal of Computer-Mediated Communication, 12,* at http://jcmc.indiana.edu/vol12/issue4/ellison.html

Engle-Friedman, M., Riela, S., Golan, R., Ventuneac, A. M., Davis, C. M., Jefferson, A. D., & Major, D. (2003). The effect of sleep loss on next day effort. *Journal of Sleep Research, 12,* 113–124.

Erdfelder, E., Faul, F., & Buchner, A. (1996). GPOWER: A general power analysis program. *Behavior Research Methods, Instruments, and Computers, 28,* 1–11.

Eurofound. (2009). Mean age at first marriage. *EurLIFE Interactive Database.* Retrieved July 21, 2010, from http://www.eurofound.europa.eu/areas/qualityoflife/eurlife/index.php?template=3&radioindic=186&idDomain=5

Fackler, M. (2012, January 21). Japanese struggle to protect their food supply. *New York Times.* Retrieved February 9, 2012, from http://www.nytimes.com/2012/01/22/world/asia/wary-japanese-take-food-safety-into-their-own-hands.html?_r=2&sq=radiation%20levels%20in%20japan&st=cse&scp=1&pagewanted=all

Fallows, J. (1999, August 31). Booze you can use: Getting the best beer for your money. *Slate.* Retrieved March 4, 2013, from http://www.slate.com/articles/life/shopping/1999/08/booze_you_can_use.html

Feng, J., Spence, I., & Pratt, J. (2007). Playing an action video game reduces gender differences in spatial cognition. *Psychological Science, 18,* 850–855.

Fisher, R. A. (1935/1971). *The design of experiments* (9th ed.) New York: Macmillan.

Forsyth, D. R., & Kerr, N. A. (1999, August). *Are adaptive illusions adaptive?* Poster presented at the annual meeting of the American Psychological Association, Boston, MA.

Forsyth, D. R., Lawrence, N. K., Burnette, J. L., & Baumeister, R. F. (2007). *Attempting to improve the academic performance of struggling college students by bolstering their self-esteem: An intervention that backfired.* Unpublished manuscript.

Friendly, M. (2005). Gallery of data visualization. Retrieved July 21, 2005, from http://www.math.yorku.ca/SCS/Gallery/

Gallagher, R. P. (2009). National survey of counseling center directors. *Monographs of the International Association of Counseling Services, Inc.* (Monograph Series No. 8R). Alexandria, VA: International Association of Counseling Services.

Geier, A. B., Rozin, P., & Doros, G. (2006). Unit bias: A new heuristic that helps explain the effect of portion size on food intake. *Psychological Science, 17,* 521–525.

Georgiou, C. C., Betts, N. M., Hoerr, S. L., Keim, K., Peters, P. K., Stewart, B., & Voichick, J. (1997). Among young adults, college students and graduates practiced more healthful habits and made more healthful food choices than did nonstudents. *Journal of the American Dietetic Association, 97,* 754–759.

Gill, G. (2005). *Nightingales: The extraordinary upbringing and curious life of Miss Florence Nightingale.* New York: Random House.

Gilovich, T. (1991). *How we know what isn't so: The fallibility of human reason in everyday life.* New York: Free Press.

Gilovich, T., & Medvec, V. H. (1995). The experience of regret: What, when, and why. *Psychological Review, 102,* 379–395.

Gilovich, T., Vallone, R., & Tversky, A. (1985). The hot hand in basketball: On the misperception of random sequences. *Cognitive Psychology, 17,* 295–314.

Golder, S. A., & Macy, M. W. (2011). Diurnal and seasonal mood vary with work, sleep, and daylength across diverse cultures. *Science, 333,* 1878–1881. doi:10.1126/science.1202775

Goodman, A., & Greaves, E. (2010). *Cohabitation, marriage and relationship stability* (IFS Briefing Note BN107). London: Institute for Fiscal Studies.

Gossett, W. S. (1908). The probable error of a mean. *Biometrics, 6,* 1–24.

Gossett, W. S. (1942). *"Student's" collected papers* (E. S. Pearson & J. Wishart, Eds). Cambridge, UK: Cambridge University Press.

Grimberg, A., Kutikov, J. K., & Cucchiara, A. J. (2005). Sex differences in patients referred for evaluation of poor growth. *Journal of Pediatrics, 146,* 212–216.

Griner, D., & Smith, T. B. (2006). Culturally adapted mental health intervention: A meta-analytic review. *Psychotherapy: Research, Practice, Training, 43,* 531–548.

Guéguen, N., Jacob, C., & Lamy, L. (2010). 'Love is in the air': Effects of songs with romantic lyrics on compliance with a courtship request. *Psychology of Music, 38,* 303–307.

Harding, K. (2010, March 19). Dispelling Sandra Bullock's Oscar curse. *Salon.* Retrieved July 10, 2010, from http://www.salon.com/life/broadsheet/2010/03/19/best_actress_curse

Hatchett, G. T. (2003). Does psychopathology predict counseling duration? *Psychological Reports, 93,* 175–185.

Hatfield, E., & Sprecher, S. (1986). Measuring passionate love in intimate relationships. *Journal of Adolescence, 9,* 383–410.

Hays, W. L. (1994). *Statistics* (5th ed.). Fort Worth, TX: Harcourt Brace College Publishers.

Headquarters Counseling Center. (2005). Myths and facts about suicide. Retrieved February 12, 2007, from http://www. hqcc .lawrence.ks.us/suicide_prevention/myths_facts.html

Healey, J. R. (2006, October 13). Driving the hard (top, that is) way. *USA Today,* p. 1B.

Healy, J. (1990). *Endangered minds: Why our children don't think.* New York: Simon & Schuster.

Henrich, J., Ensminger, J., McElreath, R., Barr, A., Barrett, C., Bolyanatz, A., et al. (2010). Markets, religion, community size, and the evolution of fairness and punishment. *Science, 327,* 1480–1484 and supporting online material retrieved from http://www.sciencemag.org/content/full/327/5972/1480/DC1

Herszenhorn, D. M. (2006, May 5). As test-taking grows, testmakers grow rarer. *New York Times.* Retrieved May 5, 2006, from http://www.nytimes.com

Heyman, J., & Ariely, D. (2004). Effort for payment. *Psychological Science, 15,* 787–793.

Hockenbury, D. H., & Hockenbury, S. E. (2003). *Psychology* (3rd ed.). New York: Worth.

Holiday, A. (2007). *Perceptions of depression based on etiology and gender.* Unpublished manuscript.

Hollon, S. D., Thase, M. E., & Markowitz, J. C. (2002). Treatment and prevention of depression. *Psychological Science in the Public Interest, 3,* 39–77.

Holm-Denoma, J. M., Joiner, T. E., Vohs, K. D., & Heatherton, T. F. (2008). The "freshman fifteen" (the "freshman five" actually): Predictions and possible explanations. *Health Psychology, 27,* s3–s9.

Howard, K. I., Moras, K., Brill, P. L., Martinovich, Z., & Lutz, W. (1996). Efficacy, effectiveness, and patient progress. *American Psychologist, 51,* 1059–1064.

Hugenberg, K., Miller, J., & Claypool, H. (2007). Categorization and individuation in the cross-race recognition deficit: Toward a solution to an insidious problem. *Journal of Experimental Social Psychology, 43,* 334–340.

Hull, H. R., Radley, D., Dinger, M. K., & Fields, D. A. (2006). The effect of the Thanksgiving holiday on weight gain. *Nutrition Journal, 5,* 29.

Hyde, J. S. (2005). The gender similarities hypothesis. *American Psychologist, 60,* 581–592.

Hyde, J. S., Fennema, E., & Lamon, S. J. (1990). Gender differences in mathematics performance: A meta-analysis. *Psychological Bulletin, 107,* 139–155.

IMD International (2001). Competitiveness rankings as of April 2001. Retrieved June 29, 2006, from http://www.photius.com/wfb1999/rankings/competitiveness.html

Indiana University Media Relations. (2006). It's no joke: IU study finds *The Daily Show* with Jon Stewart to be as substantive as network news. Retrieved December 10, 2006, from http://newsinfo .iu.edu/news/page/normal/4159.html

Irwin, M. L., Tworoger, S. S., Yasui, Y., Rajan, B., McVarish, L., LaCroix, K., et al. (2004). Influence of demographic, physiologic, and psychosocial variables on adherence to a year-long moderate-intensity exercise trial in postmenopausal women. *Preventive Medicine, 39,* 1080–1086.

Jacob, C., Guéguen, N., & Boulbry, G. (2010). Effects of songs with prosocial lyrics on tipping behavior in a restaurant. *International Journal of Hospitality Management, 29,* 761–763.

Jacob, J. E., & Eccles, J. (1982). Science and the media: Benbow and Stanley revisited. Report funded by the National Institute of Education, Washington, DC. ERIC # ED235925.

Jacob, J. E., & Eccles, J. (1986). Social forces shape math attitudes and performance. *Signs, 11,* 367–380.

Jacobs, T. (2010). Ink on skin doesn't necessarily indicate sin. *MillerMcCune News Blog.* Retrieved January 4, 2010, from http://www.miller-mccune.com/news/ink-on-skin-doesn't-necessarily-indicate-sin-1712

Johnson, W. B., Koch, C., Fallow, G. O., & Huwe, J. M. (2000). Prevalence of mentoring in clinical versus experimental doctoral programs: Survey findings, implications, and recommendations. *Psychotherapy: Theory, Research, Practice, Training, 37,* 325–334.

Kida, T. (2006). *Don't believe everything you think.* Amherst, NY: Prometheus Books.

Koch, J. R., Roberts, A. E., Armstrong, M. L., & Owens, D. C. (2010). Body art, deviance, and American college students. *The Social Science Journal, 47,* 151–161.

Koehler, J. J., & Conley, C. A. (2003). The "hot hand" myth in professional basketball. *Journal of Sport & Exercise Psychology, 25,* 253–259.

Konner, J., Risser, F., & Wattenberg, B. (2001). Television's performance on election night 2000: A report for CNN. Retrieved November 25, 2008, from http://www.cnn.com/2001/ALLPOLITICS/stories/02/02/cnn.report/cnn.pdf

Krugman, P. (2006, May 5). Our sick society. *New York Times.* Retrieved May 5, 2006, from http://www.nytimes.com

Kuck, V. J., Buckner, J. P., Marzabadi, C. H., & Nolan, S. A. (2007). A review and study on graduate training and academic hiring of chemists. *Journal of Chemical Education, 84,* 277–284.

Kuper, S., & Szymanski, S. (2009). *Soccernomics: Why England loses, why Germany and Brazil win, and why the U.S., Japan, Australia, Turkey—and even Iraq—are destined to become the kings of the world's most popular sport.* New York: Nation Books.

Lam, R. W., & Kennedy, S. H. (2005). Using meta-analysis to evaluate evidence: Practical tips and traps. *Canadian Journal of Psychiatry, 50,* 167–174.

Landrum, E. (2005). Core terms in undergraduate statistics. *Teaching of Psychology, 32,* 249–251.

Leung, D. P. K., Ng, A. K. Y., & Fong, K. N. K. (2009). Effect of small group treatment of the modified constraint induced movement therapy for clients with chronic stroke in a community setting. *Human Movement Science, 28,* 798–808.

Levitt, S. D., & Dubner, S. J. (2005). *Freakonomics: A rogue economist explores the hidden side of everything.* New York: Morrow.

Lifescript. (2009). Which dog is right for you? Lifescript.com. Retrieved March 4, 2013, from http://www.lifescript.com/Quizzes/Pets/Which_Dog_Is_Right_For_You.aspx

Lloyd, C. (2006, December 14). Saved, or sacrificed? *Salon.com.* Retrieved February 26, 2007, from http://www.salon.com/mwt/broadsheet/2006/12/14/selection/index.html

Lucas, M. E. S., Deen, J. L., von Seidlein, L., Wang, X., Ampuero, J., Puri, M., et al. (2005). Effectiveness of mass oral cholera vaccination in Beira, Mozambique. *New England Journal of Medicine, 352,* 757–767.

Luo, L., Hendriks, T., & Craik, F. (2007). Age differences in recollection: Three patterns of enhanced encoding. *Psychology and Aging, 22,* 269–280.

Maner, J. K., Miller, S. L., Rouby, D. A., & Gailliot, M. T. (2009). Intrasex vigilance: The implicit cognition of romantic rivalry. *Journal of Personality and Social Psychology, 97,* 74–87.

Marist Poll. (2009, October 7). "Whatever . . ." takes top honors as most annoying. Retrieved March 4, 2013, from http://maristpoll.marist.edu/107-whatever-takes-top-honors-as-most-annoying/

Mark, G., Gonzalez, V. M., & Harris, J. (2005, April). No task left behind? Examining the nature of fragmented work. *Proceedings of the Association for Computing Machinery Conference on Human Factors in Computing Systems* (ACM CHI 2005), Portland, OR, 321–330. New York: ACM Press.

Markoff, J. (2005, July 18). Marrying maps to data for a new web service. *New York Times.* Retrieved July 18, 2005, from http://www.nytimes.com

Matlin, M. W., & Kalat, J. W. (2001). Demystifying the GRE psychology test: A brief guide for students. *Eye on Psi Chi, 5,* 22–25.

McCollum, J. F., & Bryant, J. (2003). Pacing in children's television programming. *Mass Communication and Society, 6,* 115–136.

Mecklenburg, S. H., Malpass, R. S., & Ebbesen, E. (2006, March 17). Report to the legislature of the state of Illinois: the Illinois pilot program on sequential double-blind identification procedures. Retrieved April 19, 2006, from http://eyewitness.utep.edu

Mehl, M. R., Vazire, S., Ramirez-Esparza, N., Slatcher, R. B., & Pennebaker, J. W. (2007). Are women really more talkative than men? *Science, 317,* 82. doi 10.1126/science.1139940

Micceri, T. (1989). The unicorn, the normal curve, and other improbable creatures. *Psychological Bulletin, 105,* 156–166.

Mitchell, P. (1999). Grapefruit juice found to cause havoc with drug uptake. *Lancet, 353,* 1355.

Möller, I., & Krahé, B. (2009). Exposure to violent video games and aggression in German adolescents: A longitudinal analysis. *Aggressive Behavior, 35,* 75–89.

Munro, G., & Munro, J. (2000). Using daily horoscopes to demonstrate expectancy confirmation. *Teaching of Psychology, 27,* 114–116. doi:10.1207/S15328023TOP2702_08

Murphy, K. R., & Myors, B. (2004). *Statistical power analysis: A simple and general model for traditional and modern hypothesis tests.* Mahwah, NJ: Erlbaum.

Myers, D. G., & Diener, E. (1995). Who is happy? *Psychological Science, 6,* 10–19.

Nail, P. R., Harton, H. C., & Decker, B. P. (2003). Political orientation and modern versus aversive racism: Tests of Dovidio and Gaertner's (1998) integrated model. *Journal of Personality and Social Psychology, 84,* 754–770.

Neighbors, L., & Sobal, J. (2008). Weight and weddings: Women's weight ideals and weight management behaviors for their wedding day. *Appetite, 50,* 550–554.

Newman, A. (2006, November 11). Missed the train? Lost a wallet? Maybe it was all Mercury's fault. *New York Times,* p. B3.

Newton, R. R., & Rudestam, K. E. (1999). *Your statistical consultant: Answers to your data analysis questions.* Thousand Oaks, CA: Sage.

New York City Department of City Planning. (2006). Pedestrian level of service study: Phase I. New York: NYC DCP.

Nicol, A. A. M., & Pexman, P. M. (2010). *Displaying your findings: A practical guide for creating figures, posters, and presentations.* Washington DC: American Psychological Association.

Noel, J., Forsyth, D. R., & Kelley, K. (1987). Improving the performance of failing students by overcoming their self-serving attributional biases. *Basic and Applied Social Psychology, 8,* 151–162.

Nolan, S. A., Flynn, C., & Garber, J. (2003). Prospective relations between rejection and depression in young adolescents. *Journal of Personality and Social Psychology, 85,* 745–755.

Nunnally, J. C., & Bernstein, I. H. (1994). *Psychometric theory* (3rd ed.). New York: McGraw-Hill.

NYTimes.com (2004, September 9). A look at 1000 who died. *New York Times.* Retrieved March 4, 2013, from http://www.nytimes.com/packages/html/national/20040909_THOUSAND_GRAPHIC/index_GRAPHIC.html.

Oberst, U., Charles, C., & Chamarro, A. (2005). Influence of gender and age in aggressive dream content of Spanish children and adolescents. *Dreaming, 15,* 170–177.

Ogbu, J. U. (1986). The consequences of the American caste system. In U. Neisser (Ed.), *The school achievement of minority children: New perspectives* (pp. 19–56). Hillsdale, NJ: Erlbaum.

Palazzo, D. J., Lee, Y-J, Warnakulasooriya, R., & Pritchard, D. E. (2010). Patterns, correlates, and reduction of homework copying. *Physical Review Special Topics—Physics Education Research, 6.* Retrieved June 4, 2012, from http://prst-per.aps.org/pdf/PRSTPER/v6/i1/e010104. doi: 10.1103/PhysRevSTPER.6.010104

Parker-Pope, T. (2005, December 13). A weight guessing game: Holiday gains fall short of estimates, but pounds hang on. *Wall Street Journal,* p. 31.

Petrocelli, J. V. (2003). Factor validation of the Consideration of Future Consequences Scale: Evidence for a shorter version. *Journal of Social Psychology, 143,* 405–413.

Plassman, H., O'Doherty, J., Shiv, B., & Rangel, A. (2008). Marketing actions can modulate neural representations of experienced pleasantness. *Proceedings of the National Academy of Sciences, 105,* 1050–1054. doi 10.1073/pnas.0706929105

Popkin, S. J. & Woodley, W. (2002). *Hope VI Panel Study.* Urban Institute: Washington, D.C.

Postman, N. (1985). *Amusing ourselves to death.* New York: Penguin Books.

Press, E. (2006, December 3). Do immigrants make us safer? *New York Times Magazine,* pp. 20–24.

Public vs. private schools [Editorial]. (2006, July 19). *New York Times.* Retrieved July 19, 2006, from http://www. nytimes.com

Quaranta, A., Siniscalchi, M., & Vallortigara, G. (2007). Asymmetric tail-wagging responses by dogs to different emotive stimuli. *Current Biology, 17,* 199–201.

Rajecki, D. W., Lauer, J. B., & Metzner, B. S. (1998). Early graduate school plans: Uninformed expectations. *Journal of College Student Development, 39,* 629–632.

Rampell, C. (2010, July 26). The two-track lawyer market. [Economix blog post.] *New York Times.* Retrieved July 26, 2010, from http://economix.blogs.nytimes.com/2010/07/26/the-two-track-lawyer-market/?scp=8&sq=median&st=cse

Ratner, R. K., & Miller, D. T. (2001). The norm of self-interest and its effects on social action. *Journal of Personality and Social Psychology, 81,* 5–16.

Raz, A., Fan, J., & Posner, M. I. (2005). Hypnotic suggestion reduces conflict in the human brain. *Proceedings of the National Academy of Sciences, 102,* 9978–9983.

Richards, S. E. (2006, March 22). Women silent on abortion on NYT op-ed page. *Salon.com.* Retrieved March 22, 2006, from http://www.salon.com

Rockwell, P. (2006, June 23). Send in the clowns: No joke: "Medical clowning" seems to help women conceive. *Salon.com.* Retrieved June 25, 2006, from http://www.salon.com

Roberts, P. M. (2003). Performance of Canadian adults on the Graded Naming Test. *Aphasiology, 17,* 933–946.

Roberts, S. B., & Mayer, J. (2000). Holiday weight gain: Fact or fiction? *Nutrition Review, 58,* 378–379.

Rosenthal, R. (1995). Writing meta-analytic reviews. *Psychological Bulletin, 118,* 183–192.

Rosser, J. C., Lynch, P. J., Cuddihy, L., Gentile, D. A., Klonsky, J., & Merrell, R. (2007). The impact of video games on training surgeons in the 21st century. *Archives of Surgery, 142,* 181–186.

Ruby, C. (2006). *Coming to America: An examination of the factors that influence international students' graduate school choices.* Draft of dissertation.

Ruhm, C. J. (2000). Are recessions good for your health? *Quarterly Journal of Economics, 115,* 617–650.

Ruhm, C. J. (2006). *Healthy living in hard times* (NBER Working Paper No. 9468). Cambridge, MA: National Bureau of Economic Research. Retrieved May 30, 2006, from http://www.nber.org/papers/w9468

Ryan, C. (2006, June 21). "Therapeutic clowning" boosts IVF. *BBC News.* Retrieved June 25, 2006, from http://news.bbc.co.uk

Sandberg, D. E., Bukowski, W. M., Fung, C. M., & Noll, R. B. (2004). Height and social adjustment: Are extremes a cause for concern and action? *Pediatrics, 114,* 744–750.

Schackman, B. R., Gebo, K. A., Walensky, R. P., Losina, E., Muccio, T., Sax, P. E., et al. (2006). The lifetime cost of current human immunodeficiency virus care in the United States. *Medical Care, 44,* 990–997.

Schmidt, M. E., & Vandewater, E. A. (2008). Media and attention, cognition, and school achievement. *Future of Children, 18,* 39–61.

Seymour, C. (2006). Listen while you run. *Runner's World.* Retrieved May 24, 2006, from http://msn.runnersworld.com

Sherman, J. D., Honegger, S. D., & McGivern, J. L. (2003). *Comparative indicators of education in the United States and other G-8 countries: 2002,* NCES 2003-026. Washington, D.C.: U.S. Department of Education, National Center for Health Statistics, http://scsvt.org/resource/global_ed_compare2002.pdf

Simmons, T., & Dye, J. L. (2004). What has happened to median age at first marriage data? U.S. Census Bureau, Population Division. *Annual Meeting of the American Sociological Association.* Retrieved July 21, 2010, from http://www.census.gov/acs/www/Downloads/library/2004/2004_Simmons_01.pdf

Skurnik, I., Yoon, C., Park, D. C., & Schwarz, N. (2005). How warnings about false claims become recommendations. *Journal of Consumer Research, 31,* 713–724.

Smith, T. W., & Kim, S. (2006). National pride in cross-national and temporal perspective. *International Journal of Public Opinion Research, 18,* 127–136.

Spencer, S. J., Steele, C. M., & Quinn, D. M. (1999). Stereotype threat and women's math performance. *Journal of Experimental Social Psychology, 35,* 1–28.

Stampone, E. (1993). *Effects of gender of rater and a woman's hair length on the perceived likelihood of being sexually harassed.* Paper presented at the 46th Annual Undergraduate Psychology Conference, Mount Holyoke College, South Hadley, MA.

Steele, J. P., & Pinto, J. N. (2006). Influences of leader trust on policy agreement. *Psi Chi Journal of Undergraduate Research, 11,* 21–26.

Steinman, G. (2006). Mechanisms of twinning: VII. Effect of diet and heredity on human twinning rate. *Journal of Reproductive Medicine, 51,* 405–410.

Stigler, S. M. (1999). *Statistics on the table: The history of statistical concepts and methods.* Cambridge, MA: Harvard University Press.

Suicide Prevention Action Network. (2004). National Strategy for Suicide Prevention Benchmark Survey. Retrieved July 7, 2005, from http://www.spanusa.org.pdf/NSSP_Benchmark_Survey_Results.pdf

Talarico, J. M., & Rubin, D. C. (2003). Confidence, not consistency, characterizes flashbulb memories. *Psychological Science, 14,* 455–461.

Taylor, G. M. & Ste-Marie, D. M. (2001). Eating disorders symptoms in Canadian female pair and dance figure skaters. *International Journal of Sports Psychology, 32,* 21–28.

Tierney, J. (2008a). Health halo can hide the calories. *New York Times.* Retrieved December 7, 2008, from http://www.nytimes.com/2008/12/02/science/02tier.html

Tierney, J. (2008b). The perils of "healthy" food. *New York Times.* Retrieved December 7, 2008, from http://tierneylab.blogs.nytimes.com

Tucker, K. L., Morita, K., Qiao, N., Hannan, M. T., Cupples, A., & Kiel, D. P. (2006). Colas, but not other carbonated beverages, are associated with low bone mineral density in older women: The Framingham Osteoporosis Study. *American Journal of Clinical Nutrition, 84,* 936–942.

Tufte, E. R. (2001). *The visual display of quantitative information.* Cheshire, CT: Graphics Press.

Tufte, E. R. (2005). *Visual explanations* (2nd ed.) Cheshire, CT: Graphics Press. (Original work published 1997)

Tufte, E. R. (2006a). *Beautiful evidence.* Cheshire, CT: Graphics Press.

Tufte, E. R. (2006b). *The visual display of quantitative information* (2nd ed.) Cheshire, CT: Graphics Press.

Twenge, J., & Campbell, W. K. (2001). Age and birth cohort differences in self-esteem: A cross-temporal meta-analysis. *Personality and Social Psychology Review, 5,* 321–344.

Upton, P, & Eiser, C. (2006). School experiences after treatment for a brain tumour. *Child: Care, Health and Development, 32,* 9–17.

U.S. News & World Report. (2003, September 1). America's best colleges 2004. *U.S. News & World Report.*

Vergin, R. C. (2000). Winning streaks in sports and the misperception of momentum. *Journal of Sports Behavior, 23,* 181–197.

Vinten-Johansen P., Brody H., Paneth N., Rachman, S., & Rip, M. (2003). *Cholera, chloroform, and the science of medicine: A life of John Snow.* Oxford University Press New York.

Vogel, C. (2011, December 13). Art world star doesn't change his spots. *New York Times.* Retrieved December 20, 2011, from http://www.nytimes.com/2011/12/14/arts/design/damien-hirsts-spot-paintings-will-fill-all-11-gagosians.html?pagewanted=all

Walker, S. (2006). *Fantasyland: A season on baseball's lunatic fringe.* New York: Penguin.

Wansink, B., & van Ittersum, K. (2003). Bottoms up! The influence of elongation and pouring on consumption volume. *Journal of Consumer Research, 30,* 455–463.

Waters, A. (2006, February 24). Eating for credit. *New York Times.* Retrieved February 24, 2006, from http://www.nytimes.com.

The Week Staff. (2010). Can Facebook predict your breakup? *The Week.* Retrieved January 14, 2013, from http://theweek.com/article/index/203122/can-facebook-predict-your-breakup

Weinberg, B. A., Fleisher, B. M., & Hashimoto, M. (2007). *Evaluating methods for evaluating instruction: The case of higher education* (NBER Working Paper No. 12844). Cambridge, MA: National Bureau of Economic Research.

White, B., Driver, S., & Warren, A. (2010). Resilience and indicators of adjustment during rehabilitation from a spinal cord injury. *Rehabilitation Psychology, 55,* 23–32. doi: 10.1037/a0018451

Wiley, J. (2005). A fair and balanced look at the news: What affects memory for controversial arguments. *Journal of Memory and Language, 53,* 95–109.

Wolff, A. (2002). Is the SI jinx for real? *Sports Illustrated,* January 26.

Wood, J. M., Nezworski, M. T., Lilienfeld, S. O., & Garb, H. N. (Eds.). (2003). *What's wrong with the Rorschach? Science confronts the controversial inkblot test.* San Francisco: Jossey-Bass.

World Health Organization. (2007). Myths and realities in disaster situations. Retrieved February 12, 2007, from http://www. who.int/hac/techguidance/ems/myths/en/index.html

Yanovski, J. A., Yanovski, S. Z., Sovik, K. N., Nouven, T. T., O'Neil, P. M., & Sebring, N. G. A. (2000). A prospective study of holiday weight gain. *New England Journal of Medicine, 23,* 861–867.

Yilmaz, A., Baran, R., Bayramgurler, B., Karahalli, E., Unutmaz, S., & Uskul, T. B. (2000). Lung cancer in non-smokers. *Turkish Respiratory Journal, 2,* 13–15.

Zarate, C. A. (2006). A randomized trial of an N-methyl-D-aspartate antagonist in treatment-resistant major depression. *Archives of General Psychiatry, 63,* 856–864.

Zinman, J. & Zitzewitz, E. (2009, June 29). Snowed: Deceptive advertising by ski resorts. Social Science Resource Network. Retrieved March 14, 2013, from http://ssrn.com/abstract=1427490

INDEX

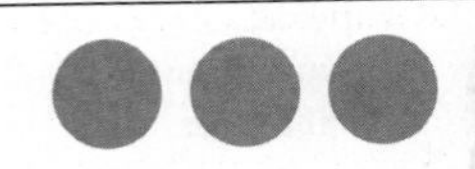

Note: Page numbers followed by f indicate figures; those followed by t indicate tables.

FORMULAS

CHAPTER 11 (Chapter 11 formulas continued from inside front cover)

Tukey *HSD* post-hoc test

$$HSD = \frac{(M_1 - M_2)}{s_M}$$, for any two sample means

$$s_M = \sqrt{\frac{MS_{within}}{N}}$$, if equal sample sizes

$$N' = \frac{N_{groups}}{\Sigma(1/N)}$$

$$s_M = \sqrt{\frac{MS_{within}}{N'}}$$, if unequal sample sizes

One-Way Within-Groups ANOVA

$$df_{subjects} = n - 1$$

$$df_{within} = (df_{between})(df_{subjects})$$

$$df_{total} = df_{between} + df_{subjects} + df_{within}$$

$$SS_{subjects} = \Sigma(M_{participant} - GM)^2$$ for each score

$$SS_{within} = SS_{total} - SS_{between} - SS_{subjects}$$

$$MS_{subjects} = \frac{SS_{subjects}}{df_{subjects}}$$

$$F_{subjects} = \frac{MS_{subjects}}{MS_{within}}$$

$$R^2 = \frac{SS_{between}}{(SS_{total} - SS_{subjects})}$$

CHAPTER 12

Two-Way Between-Groups ANOVA

$$df_{rows} = N_{rows} - 1$$

$$df_{columns} = N_{columns} - 1$$

$$df_{interaction} = (df_{rows})(df_{columns})$$

$$SS_{total} = \Sigma(X - GM)^2$$ for each score

$$SS_{between(rows)} = \Sigma(M_{row} - GM)^2$$ for each score

$$SS_{between(columns)} = \Sigma(M_{column} - GM)^2$$ for each score

$$SS_{within} = \Sigma(X - M_{cell})^2$$ for each score

$$SS_{between(interaction)} = SS_{total} - (SS_{between(rows)} + SS_{between(columns)} + SS_{within})$$

$$R^2_{rows} = \frac{SS_{rows}}{(SS_{total} - SS_{columns} - SS_{interaction})}$$

$$R^2_{columns} = \frac{SS_{columns}}{(SS_{total} - SS_{rows} - SS_{interaction})}$$

$$R^2_{interaction} = \frac{SS_{interaction}}{(SS_{total} - SS_{rows} - SS_{columns})}$$